THE ILLUSTRATED
PETROLEUM REFERENCE DICTIONARY

4th EDITION

THE ILLUSTRATED
PETROLEUM
REFERENCE
DICTIONARY

4th EDITION

EDITED BY

ROBERT D. LANGENKAMP

PennWell Books

PENNWELL PUBLISHING COMPANY
TULSA, OKLAHOMA

Copyright © 1980, 1982, 1985, 1994 by
PennWell Publishing Company
1421 South Sheridan Road/P.O. Box 1260
Tulsa, Oklahoma 74101

Library of Congress Cataloging-in-Publication Data

The illustrated petroleum reference dictionary / edited by Robert D.
Langenkamp. — 4th ed.
     p.   cm.
   ISBN 0-87814-423-4
   1. Petroleum—Dictionaries. 2. Petroleum engineering—
Dictionaries. I. Langenkamp, R. D. II. Title: Petroleum
reference dictionary.
TN865.I43 1994   94-34489
665.5′03—dc20   CIP

Printed in the United States of America

CONTENTS

FOREWORD

Welcome to the revised and enlarged edition of the Illustrated Petroleum Reference Dictionary, Fourth Edition. Included in the new book are 1,280 new entries, a third more than the previous edition, along with fitting illustrations.

Additional subjects include current technology in high-angle and horizontal drilling; well control and completion; sand control; coiled tubing drilling; multilaterals; underbalanced drilling; selective perforating; the latest in cementing materials and techniques; closed-circuit disposal systems at drilling sites, on and offshore; partnerships and partnering (there is a difference); new regulations concerning oxygenated and synthetic fuels; advances in three-dimensional seismology; supervisory control and data acquisition, SCADA systems, for remote, unmanned operations, and considerably more—all made readily understandable for the non-technical reader.

To have the new edition in hand is a great satisfaction to the author. It is hoped that the reader will find it enjoyable as well as revealing and, in specific instances, useful. It's the result of years of *rodent work*, digging out significant new material from the wealth of regulatory, scientific, operational and contractual information generated by the dynamism of oil, an industry that in each new decade has become more determinedly global in reach and celebrated for its achievements. Its hard-won successes in exploration, production, and transportation, often in hostile areas of the world, are notable. Today's oil men and women know few barriers as they explore the new frontiers, each in its way as demanding of foresight, energy, and innovation as any encountered in oil's often turbulent but unfailingly colorful, 125-year history.

Also included as important adjuncts to the Fourth Edition are Steven Gerolde's *Universal Conversion Factors* and the Desk and Derrick Club's *Abbreviator*.

This new issue is a piece of work that all involved hope will meet with ready acceptance, not as a coffee-table decoration, but as a useful, working book, a well-thumbed, often-used reference dictionary, a publication that will, per chance, add a bit of luster to the whole enterprise.

—R. D. Langenkamp

A.O.D.C.
American Association of Oilwell Drilling Contractors.

A.A.P.G.
American Association of Petroleum Geologists.

ABANDONED OIL
Oil permitted to escape from storage tanks or pipeline by an operator. If the operator makes no effort to recover the oil, the landowner on whose property the oil has run may trap the oil for his own use.

ABANDONED WELL
A well no longer in use; a dry hole that, in most states, must be properly plugged.

ABSOLUTE ALCOHOL
One hundred percent ethyl alcohol.

ABSOLUTE PERMEABILITY
The ability of a rock or a formation to conduct a fluid (oil, gas, or water) when 100 percent saturated, i.e., at 100 percent saturation.

ABSORPTION
The taking in or assimilation of a gas by a liquid; soaking up of a substance by another. *See* Absorption Plant.

ABSORPTION OIL
An oil used to remove heavier hydrocarbons from natural gas in an absorption tower.

ABSORPTION PLANT
An oilfield facility that removes liquid hydrocarbons from natural gas, especially casinghead gas. The gas is run through oil of a proper character that absorbs the liquid components of the gas. The liquids are then recovered from the oil by distillation.

Absorption tower

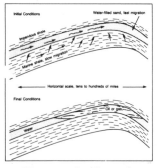

Oil or gas migration and
accumulation

Accumulator system
(Courtesy Hydril)

ABSORPTION TOWER
A tower or column in which contact is made between a rising gas and a falling liquid so that part of the gas is taken up or absorbed by the liquid.

ABSTRACT OF TITLE
A collection of all recorded instruments affecting the title to a tract of land. Some abstracts contain complete copies of instruments on record, but others are summaries of the effect of the various instruments. In most states, title examination is made using an abstract of title.

ACCELERATED AGING TEST
A procedure whereby an oil product may be subjected to intensified but controlled conditions of heat, pressure, radiation, or other variables to produce, in a short time, the effects of long-time storage or use under normal conditions.

ACCOMMODATION MODULE
Offshore crews' quarters: dormitories, dining, and recreation facilities.

ACCOMMODATION RIG
See Rig, Accommodation.

ACCUMULATION OF OIL AND GAS
Hydrocarbons accumulate in porous and permeable formations and stratify or form in layers; gas at the highest level, oil in the second level beneath the gas, and water (if there is any) on the bottom level. Oil and gas accumulate in the highest parts of a reservoir, which makes the top and upper flanks of an anticline a good place to drill for oil. Petroleum accumulations require a great deal of time (a million years or so) to form as the oil and gas percolate upward from their source beds through more-or-less permeable rock to the reservoir rock where, with luck, it is discovered by a wildcatter.

ACCUMULATOR
A small tank or vessel to hold air or liquid under pressure for use in a hydraulic or air-actuated system. Accumulators, in effect, store a source of pressure for use at a regulated rate in mechanisms or equipment in a plant or in drilling or production operations.

ACCUMULATOR SYSTEM
A hydraulic system designed to provide power to all closure elements of the rig's blowout-preventer stack. Hydraulic oil is forced into one or more vessels by a high-pressure, small-volume pump and its charge of inert gas, usually nitrogen. The gas is compressed and stores potential energy. When the system is actuated, the oil under high pressure is released and opens or closes the valves on the B.O.P. stack.

ACETONE
A flammable, liquid compound used widely in industry as a solvent for many organic substances.

ACETYLENE
A colorless, highly flammable gas with a sweetish odor used with oxygen in oxyacetylene welding. It is produced synthetically by incomplete combustion of coal gas and also by the action of water on calcium carbide (C_aC_2). Also can be made from natural gas.

ACID-BOTTLE INCLINOMETER
A device used in a well to determine the degree of deviation from the vertical of the well bore. The acid is used to etch a horizontal line on the container. From the angle the line makes with the wall of the container, the angle of the well's course can be determined. *See* Inclinometer.

ACID GAS
Sour gas; hydrogen sulfide (H_2S); a gas with a strong rotten-egg odor, sometimes produced with natural gas. Even in small amounts, sour gas can be lethal.

ACIDIZING A WELL
A technique for increasing the flow of oil from a well. Hydrochloric acid is pumped into the well under high pressure to reopen and enlarge the pores in the oil-bearing limestone formations.

ACID OIL
Sour oil, i.e., oil with a high concentration of hydrogen sulfide (H_2S). Acid gas is sour gas with a high percentage of H_2S.

ACID-RECOVERY PLANT
An auxiliary facility at some refineries where acid sludge is separated into acid oil, tar, and weak sulfuric acid. The sulfuric acid is then reconcentrated.

ACID SLUDGE
The residue left after treating petroleum oil with sulfuric acid for the removal of impurities. The sludge is a black, viscous substance containing the spent acid and the impurities that the acid has removed from the oil.

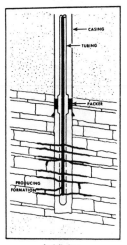

Acidizing

ACID TREATMENT
A refining process in which unfinished petroleum products such as gasoline, kerosene, diesel fuels, and lubricating stocks are treated with sulfuric acid to improve color, odor, and other properties.

ACOUSTIC LOG
A generic term for a well log that displays any of several measurements of acoustic waves in rocks exposed in a borehole, e.g., compressional-

wave transit time over an interval (sonic log) or relative amplitude (cement bond log).

ACOUSTIC PLENUM
A soundproof room; an office or "sanctuary" aboard an offshore drilling platform protected from the noise of drilling engines and pipe handling.

ACOUSTIC REENTRY
A method used in deepwater operations offshore to reposition a drill-ship over a borehole previously drilled and cased. The technique employs acoustic signals to locate the pipe and guide the ship into position.

ACOUSTIC WAVE
A sound wave; sonic wave.

ACQUIRED RIGHTS CLAUSE, AFTER
A clause in a joint venture, farmout, or other agreement designed to afford parties to the agreement the right to share in specified future acquisitions by another party to the agreement.

ACREAGE CONTRIBUTION AGREEMENT
In this type of support agreement, the contributing party agrees to contribute a lease or interest in leases in the immediate vicinity of the well being drilled. Part of the agreement requires that the test well be drilled to a certain depth and that certain information be made available. *See* Bottom-Hole Letter.

ACRE-FOOT
A unit of measurement applied to petroleum reserves; an acre of producing formation 1 foot thick.

A.C.S.
American Chemical Society.

ACT OF GOD CLAUSE
See Force Majeure Clause.

A.C.T SYSTEM
Automatic Custody Transfer System. *See* Lease Automatic Custody Transfer.

ACTUATOR
See Operator.

A.C.V.
Air-cushion vehicle. *See* Air-Cushion Transport.

Actuator *(Courtesy Fisher Controls Co.)*

ADAMANTINE LUSTER

A brilliant mineral luster characteristic of minerals with a high index of refraction (deflects a ray of light with little change in the light ray's velocity). Diamonds have such a luster, as does cerussite.

A.D.A. MUD

A material that may be added to drilling mud to condition it in order to obtain satisfactory core samples.

ADAPTER

A device to provide a connection between two dissimilar parts or between similar parts of different sizes. *See* Swage.

ADDITIVE

A chemical added to oil, gasoline, or other products to enhance certain characteristics or to give them other desirable properties.

ADOLESCENT ROCK

See Immature Rock.

ADSORPTION

The attraction exhibited by the surface of a solid for a liquid or a gas when they are in contact. *Compare with* Absorption.

ADVANCE PAYMENT AGREEMENT

A transaction in which one operator advances a sum of money or credit to another operator to assist in developing an oil or gas field. The agreement provides an option to the "lender" to buy a portion or all of the production resulting from the development work.

ADVANCE PAYMENT FINANCING

See Production Payment.

A.E.C.

Atomic Energy Commission.

AEOLIAN

See Eolian.

AERATED DRILLING FLUIDS

Aerated or foamed drilling fluids are created by introducing air or nitrogen or other inert gases into the circulating system and pumping the mixture downhole. To maintain an underbalanced system, foamed water-base or oil-base drilling fluid is used. This is particularly useful in drilling with coiled tubing. *See* Overbalanced Drilling.

AERIFY

To change into a gaseous form; to infuse with or force air into; gasify.

A-frame

Air-balanced pumping
unit

Underwater seismic
explosion, now replaced
by air bursts

A.F.E

Authority For Expenditure. In the oil patch, there are several A.F.E. levels: the field office, district office, division office, and headquarters, with expenditure authorizations ranging from a few thousand dollars at the field level to perhaps a million at headquarters. This would, of course, depend on the size and financial condition or stability of the company.

A.F.R.A.

Average Freight Rate Assessment (for tankers).

A-FRAME

A two-legged metal or wooden support in the form of the letter A for hoisting or exerting a vertical pull with block and tackle or winch line attached to the apex of the A-frame.

AFTER MARKET FACILITY

A large under-roof area for the repair and resale of offshore drilling and production components.

A.G.A.

American Gas Association.

AGENCY CONTRACT

A type of agreement that in many cases has replaced the concession as the form of petroleum development agreement in the Middle East and with O.P.E.C. countries elsewhere. Under an agency contract, title to oil installations and oil produced is held by the host government, but the government bears none of the costs of initial exploration. Also, the foreign company does not have a long-term, exclusive right to exploit the minerals as is the case under a concession agreement.

A.I.Ch.E.

American Institute of Chemical Engineers.

A.I.M.M.E.

American Institute of Mining and Metallurgical Engineers.

AIR-BALANCED PUMPING UNIT

See Pumping Unit, Air-Balanced.

AIR BOTTLE

A cylinder of oxygen for oxyacetylene welding; an air chamber.

AIR BURSTS

A geophysical technique used in marine seismic work in which bursts of compressed air from an air gun towed by the seismographic vessel are used to produce sound waves. Air bursts do not destroy marine life as did explosive charges.

AIR CHAMBER
A small tank or "bottle" connected to a reciprocating pumps discharge chamber or line to absorb and dampen the surges in pressure from the rhythmic pumping action. Air chambers are charged with sufficient air pressure to provide an air cushion that minimizes the pounding and vibration associated with the pumping of fluids with plunger pumps.

AIR-COOLED ENGINE
An engine in which heat from the combustion chamber and friction are dissipated to the atmosphere through metal fins integral to the engine's cylinder head and block assemblies. The heat generated flows through the engine head and cylinder walls and into the fins by conductance and is given off by the fins acting as radiators. A small, two-cycle engine without water jacketing, water pump, or conventional radiator.

AIR-CUSHION TRANSPORT
A vehicle employing the hovercraft principle of downthrusting airstream support, developed to transport equipment and supplies in the Arctic regions. The air cushion protects the tundra from being cut by the wheels or treads of conventional vehicles.

Air-cushion transport
(Courtesy Bell
Aerospace Co.)

AIR CUT
The accidental or inadvertent incursion of air into a liquid system. *See* Aired Up.

AIR DRILLING
See Drilling, Air.

AIRED UP
Refers to a condition in a plunger pump when the suction chamber is full of air or gas blocking the intake of oil into the chamber. Before the pump will operate efficiently, the air must be bled off—vented to the atmosphere—through a bleeder line or by loosening the suction valve covers to permit the escape of the air.

Air drilling rig

AIR-FILLED BOREHOLE
An empty borehole, no water, no drilling fluids, just air.

AIR GUN
A device used in geophysical or seismic surveys in a water environment that creates seismic signals (sound waves) with bursts of compressed

air. Air bursts from air guns trailed behind a geophysical ship are as effective as explosive detonations but do not damage marine life. *See* Seismic Sea Streamer.

AIR HOIST

A hoist; a mechanism for lifting operated by a compressed air motor; pneumatic hoist.

AIR-INJECTION METHOD

A type of secondary recovery to increase production by forcing the oil from the reservoir into the well bore. Because of the dangers inherent in the use of air, this method is not a common practice except in areas where there is insufficient gas for repressuring.

AIR LIFT

See Gas Lift.

AIR WEIGHT OF CASING

The weight of a string of casing without the buoyant effect of the drilling fluid is the air weight. If the maximum hook load of a derrick is 1,900,000 lbs. and the air weight of the casing string is 2,100,000 (200,000 lbs. more), with the borehole full of drilling mud, it is possible to handle safely the 2,100,000 lb. string.

AIR WRENCH

See Impact Wrench.

ALGAL LIMESTONE

(1) A limestone made up largely of the remains of calcium carbonate-producing algae. (2) A limestone in which algae bind together the fragments of other calcium carbonate-producing organisms.

ALGAL REEF

An organic reef in which algae were the principal organisms, secreting calcium carbonate to build the reef.

ALIPHATICS

One of the two classes of organic petrochemicals; the other is the aromatics. The most important aliphatics are the gases ethylene, butylene, acetylene, and propylene.

ALKYLATION

A refining process that, simply stated, is the reverse of cracking. The alkylation process starts with small molecules and ends up with larger ones. To a refining engineer, alkylation is the reaction of butylene or

Air hoist *(Courtesy Ingersoll-Rand)*

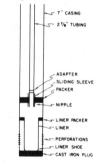

Air-injection well *(After Franco)*

Alkylation section of a refinery *(Courtesy Econo-Therm Corp.)*

8

propylene with isobutane to form an isoparaffin, alkylate—a superior gasoline blending component.

ALLOWABLE
The amount of oil or gas a well or a leasehold is permitted to produce under proration orders of a state regulatory body.

ALL-THREAD NIPPLE
A short piece of small-diameter pipe with threads over its entire length; a close nipple.

ALLUVIAL FAN
(1) A fan-shaped area of soil and small rock sediment deposited by mountain or highland streams as their flow meets the relatively flat desert floor. (2) The silt, clay, sand, and other sediment deposited by a stream or river as it spreads out on a plain or continental shelf. Alluvial fans are usually cut by numerous distributary channels that divide the main stream to form the common fan shapes similar to those occurring in deltas. Large alluvial fans are a feature of the southwest United States, where fast-flowing mountain streams meet the flat land, slow down to a crawl, and drop their suspended bed load, load of sediment.

ALLUVIAL TALUS
An accumulation of pebbles and rock fragments deposited by rainwash after a storm or by melting snow.

ALTERNATE FUELS
Fuels—gas, gasoline, heating oil—made from coal, oil shales, or tar sands by various methods. Alternate fuels may also include steam from geothermal wells where superheated water deep in the earth is used to generate steam for electric power generation.

ALUMINUM CHLORIDE
A chemical used as a catalytic agent in oil refining and for the removal of odor and color from cracked gasoline.

AMERIPOL
The trade name for products made from a type of synthetic rubber.

AMINE
Organic base used in refining operations to absorb acidic gases (H_2S, COS, CO_2) occurring in process streams. Two common amines are monoethanolamine (MEA) and diethanolamine (DEA).

Refinery coolers for a stream of DEA

AMINE UNIT

A natural gas treatment unit for removing contaminants—H_2S, COS, CO_2—by the use of amines. Amine units are often skid-mounted so they can be moved to the site of new gas production. Gas containing H_2S and other impurities must be cleaned up before it is acceptable to gas transmission pipelines.

AMMONIA (FERTILIZER)

An extremely pungent, colorless, gaseous alkaline compound of nitrogen and hydrogen (NH_3) that is soluble in water. The gas can be condensed to a liquid by severe cooling and pressure. Ammonia is one of the valuable products made principally from natural gas (CH_4).

AMMONIUM SULFATE

A salt having commercial value, which is obtained in the distillation of shale oils.

AMORPHOUS

A mineral or other substance that lacks a crystalline structure or whose internal arrangement is so irregular that there is no characteristic external form. A term once used to describe a mass of rock with no apparent divisions.

AMPHIBOLE

A group of dark, ferromagnesian silicate minerals widely distributed in igneous and metamorphic rocks. Hornblende is a member of this group.

AMPLITUDE

The extent of a vibratory movementor of an oscillation. The maximum numerical value of a periodically varying quantity. In seismic application, the reflection coefficient.

AMYL HYDRIDE

This fraction in the distillation of petroleum was used as an anesthetic by J. Bigelow and B. Richardson in 1865.

ANADARKO BASIN

A deep geological basin in western Oklahoma which has substantial oil and gas reserves. Some of the deepest gas wells in the U.S. (27,000 ft.+) are located in this basin.

ANCHOR, PIPELINE

See Pipeline Anchor.

ANCHOR BOLT

A stud bolt; a large bolt for securing an engine or other item of equipment to its foundation.

Installing pipeline anchors

ANCHOR STRING
A short string of casing run in the hole in offshore wells that serves as an anchor or base for the installation of wellhead equipment. On land, an anchor string is called *surface pipe,* which may be from 200 to 2,000 feet long. It also serves as the foundation or anchor for all subsequent drilling activity. The anchor string is cemented securely before the borehole is taken down to guard against a blowout should high downhole pressure be encountered. A blowout around the anchor string is a near disaster because there is no way, short of heroic measures, to control the escaping pressure. *See* Killer Well.

Large-diameter well casing; surface pipe

ANEMOMETER
An instrument for measuring and indicating the force or speed of the wind.

ANGLE BUILDING
The technique of drilling slanted or directional boreholes. This is accomplished by special bottom-hole assemblies, i.e., drilling, stabilizing, and reaming tools attached to the drillstring in a certain sequence. This permits the hole to be drilled at a predetermined angle from the vertical. *See* Angle-Building Assemblies.

ANGLE-BUILDING ASSEMBLIES
Special bottom-hole assemblies used in the field for directional or slant-hole drilling and for drilling near-horizontal drain holes. Three assemblies in general use are the turbo drill or positive displacement mud motor with a bent sub; a drill bit, a near-bit reamer or stabilizer, and a drill collar of reduced diameter; and a bit, a reamer, and a knuckle-joint assembly.

A.N.G.T.S.
Alaska Natural Gas Transmission System. At this writing, 10 companies are involved in the A.N.G.T.S. system, which is to move natural gas from the Prudhoe Bay area and across the southwestern corner of Canada to the Lower 48.

ANGULAR
Having sharp angles or edges. Refers to sedimentary particles showing little or no evidence of abrasion, their corners and edges still sharp.

ANGULAR DISCORDANCE
See Nonconformity.

ANGULAR UNCONFORMITY
See Unconformity, Angular.

ANHYDRITE
A mineral (C_aSO_4) closely related to gypsum that occurs in thick layers comparable to beds of limestone. Geologists assume that anhydrite was crystallized from solution when a shallow sea or arm of the sea evaporated during ages past.

ANHYDROUS
Refers to a mineral that is without water. Anhydrous minerals contain no water in their chemical makeup.

ANNULAR BLOWOUT PREVENTER
See Spherical Blowout Preventer.

ANNULAR CHANNELING
Fluid breakthrough in the cement between the casing and the wall of the borehole as the result of the extreme pressures that develop during the fracing operation. *See* Cement Squeeze.

ANNULAR GAS-FLOW PROBLEM
At various times one or a combination of the following treatments are employed to prevent gas flow, gas migration, or gas leakage, all one and the same. Gas is escaping upward through the cement pumped into the annulus between the casing and the wall of the borehole. Possible treatment: Improve fluid-loss control; increase fluid density; shorten cement column (stage cementing); use special thixotropic or compressible cement slurries.

ANNULAR GAS LEAKAGE
Gas leakage or gas flow between the casing and the wall of the borehole. Sometimes this type of leakage is difficult to stop or shut off. Being a distinct danger to personnel and equipment, there are a number of remedial techniques employed to seal off the annular space to block the percolating gas. Before the casing is run, the wall of the borehole is scraped and washed down; to center the casing in the hole, spacers are run. Then a high-density cement is pumped downhole; thixotropic cement or compressible cement are also used, and two-stage cementing.

ANNULAR SPACE
The space between the well's casing and the wall of the borehole.

ANNULUS OF A WELL
The space between the surface casing and the producing or well bore casing.

ANNUNCIATOR
An electronically controlled device that signals or sounds an alarm when conditions deviate from normal or from predetermined levels of pressure, heat, or speed in a process or in operating equipment.

Blowout preventer stack with spherical BOP at the top

ANODE

A block of nonferrous metal buried near a pipeline, storage tank, or other facility and connected to the structure to be protected. The anode sets up a weak electric current that flows to the structure, thus reversing the flow of current that is associated with the corrosion of iron and steel. *See* Rectifier Bed.

ANODE, BUOYANT

A source of electric current (D.C.) for protecting offshore platforms and other steel structures resting on the sea floor against corrosion. The anode is anchored to the seafloor a few hundred feet away from a structure but is held off bottom by its buoyancy. The anode is connected to a source of D.C. current on the platform by an insulated cable. The weak current is supplied by a transformer-rectifier, the negative terminal of which is grounded to the steel structure. Thus the completion of the circuit from rectifier to anode to structure is through the seawater. The weak current moving from anode to the structure reverses the flow of current associated with the corrosion of metal. *See* Rectifier Bed.

An anode attached to an offshore platform

ANODE, SACRIFICIAL

An anode made of material that is expendable and is sacrificed to the good of the installation: tank, building, or pipeline. The anode, wired to the structure being protected, is gradually corroded away by the weak chemoelectric current that causes certain types of corrosion.

ANOMALY

Something that is different from the normal or the expected; a geological feature, especially in the subsurface, that is identified by geological, geochemical, or geophysical methods to be different from the general surroundings. This quite often indicates the presence of a salt pillar, salt dome, anticline, or other type of stratigraphic trap, which could mean an accumulation of oil and gas.

ANOMALY, NEGATIVE-GRAVITY

With the use of a gravity meter (gravimeter), the differences in the Earth's gravity can be measured over areas of the surface. When there is a significant difference in the gravitational pull (as over a salt dome, for example) compared to the surrounding area, the lower reading identifies the area over the salt dome as a negative-gravity anomaly.

ANOMOLY, RADIOACTIVE

A deviation from expected results when making a radioactivity survey. Such anomalies are important signs or markers in mineral exploration.

A.N.S.I.

American National Standards Institute.

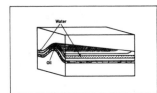

Anticline

ANWR

API logo

ANTICLINAL FOLD
A subsurface formation resembling an anticline.

ANTICLINAL THEORY
The theory first set forth by I.C. White in 1885 that oil and gas tend to accumulate in anticlines or anticlinal structures.

ANTICLINE
A subsurface geological structure in the form of a sine curve or an elongated dome. The formation is favorable to the accumulation of oil and/or gas.

ANTICLINE, BALD-HEADED
An anticline whose crest has been eroded before the deposit of sedimentary layers above it, which results in an overlying unconformity. *See* Unconformity.

ANTICLINE, BREACHED
An anticline whose top or crust has been so deeply eroded that all that remains of the structure are the inward-leaning flanks or sides.

ANTIKNOCK COMPOUNDS
Certain chemicals that are added to automotive gasolines to improve their performance—to reduce "ping" or knock—in high-compression internal-combustion engines. Tetraethyl lead is one well-known antiknock compound.

ANTI-TWO-BLOCK WARNING SYSTEM
An electronic device that sounds a warning if two blocks in a block and tackle or other hoisting rig-up are in danger of coming together, or if a block is about to pull up to the end of a boom. This could cause loss of the load and other serious damage as the cable breaks. There is no simple name for this system.

A.N.W.R.
An area of controversy between conservationists and the U.S. Interior Department. The Department has proposed leasing the Coastal Plain area of the wildlife refuge to oil companies for exploration, but so far Congress has not given its permission. National planners know it is inevitable because the United States badly needs another large oil discovery if the nation is to avoid becoming critically dependent on imported oil from unstable areas.

A.P.&A. WELL
A plugged and abandoned well.

A.P.I.
(1) The American Petroleum Institute; (2) the proper way to do a job; "strictly A.P.I."

A.P.I. BID SHEET AND WELL SPECIFICATIONS

A form many operators use in soliciting bids on a well to be drilled and completed. The form is submitted to the drilling contractors in the area of the proposed well. The operator asking for bids fills out the part of the form giving name and location of the proposed well, commencement date, depth or formation to be drilled to, and other information.

When the drilling contractor submits a bid, he lists the rig and equipment to be furnished by him: drawworks, mud pumps, derrick or mast size, make and capacity, drillpipe, tool joints, etc. The bid sheet brings operator and contractor together, as it were; they then arrive at rates and other matters.

A.P.I. GRAVITY

Gravity (weight per unit of volume) of crude oil or other liquid hydrocarbon as measured by a system recommended by the A.P.I. A.P.I. gravity bears a relationship to true specific gravity but is more convenient to work with than the decimal fractions that would result if petroleum were expressed in specific gravity.

A.P.I. NEUTRON UNIT

A working unit set up by the American Petroleum Institute for the calibration of neutron well logs.

APPALACHIAN BASIN

A sedimentary basin with thick deposits in the interior, becoming thinner as they approach the edges, extending from New York to Alabama. Topographically, it forms the Appalachian Mountains, and westward the Allegheny Plateau.

APPRAISAL DRILLING

Wells drilled in the vicinity of a discovery or wildcat well in order to evaluate the extent and the importance of the find.

APRON RING

The bottommost ring of steel plates in the wall of an upright cylindrical tank.

AQUAGEL

A specifically prepared bentonite (clay) widely used as a conditioning material in drilling mud.

AQUATORY

Underseas territory; offshore and coastline parcels of land offered for lease by a foreign government.

AQUEOUS FRACTURING

The use of a water-base fracturing fluid, which may be successfully done if there are no fresh-water sensitive sections. When shale sections with interruptions or banding of clay are encountered, special stabilizing agents can be added to control the osmosic action, the absorption of the water in the fracture fluid.

AQUIFER
Water-bearing rock strata. In a water-drive oil field, the aquifer is the water zone of the reservoir.

AQUIFER, CONFINED
An aquifer bounded above and below by impervious beds; also one containing trapped ground water.

ARABIAN LIGHT
A marker crude oil produced in Saudi Arabia that is high quality and against which other crudes, particularly those in the Middle East, are measured for quality and price.

ARBITRAGE, PRODUCT
The buying, selling, or trading of petroleum or products in various markets to make a profit from short-term differences in prices in one market as compared to those in another. A sophisticated method of trading in world petroleum markets.

ARC WELDER
(1) An electric welding unit consisting of an engine and D.C. generator, usually skid-mounted. (2) A person who uses such a machine in making welds.

Welder adjusting automatic arc welding machines

AREAL GEOLOGY
The branch of geology that pertains to the distribution, position, and form of the areas of the Earth's surface occupied by different types of rocks or geologic formations; also, the making of maps of such areas.

AREAL MAP
See Map, Areal.

AREA OF INTEREST
The area immediately surrounding a successful well in which the investors (in the good well) have an implied right to participate in any future wells drilled by the same operator.

AREA OF MUTUAL INTEREST (A.M.I.)
See Area of Interest.

AREOMETER
An instrument for measuring the specific gravity of liquids; a hydrometer.

ARGILLACEOUS
Clayey or clay bearing; shaly and having little if any permeability.

ARGILLACEOUS LIMESTONE
A limestone containing a significant amount of clay (but less than 50 percent); cement rock.

ARGILLACEOUS SANDSTONE
An impure sandstone containing varying amounts of silt and clay; a weak, friable sandstone; a clayey sandstone.

ARGON
An inert, colorless, odorless gaseous element sometimes, and in some locations, produced with natural gas.

ARKANSAS STONE
A variety of novaculite found in the Ouachita Mountains of western Arkansas. A whet stone made of this material.

ARKOSE
A coarse-grained, pinkish sandstone rich in feldspar that resembles granite. Arkose is composed principally of quartz.

AROMATICS
A group of hydrocarbon fractions that forms the basis of most organic chemicals so far synthesized. The name aromatics is derived from their rather pleasant odor. The unique ring structure of their carbon atoms makes it possible to transform aromatics into an almost endless number of chemicals. Benzene, toluene, and xylene are the principal aromatics and are commonly referred to as the B.T.X. group.

ARTIFICIAL DRIVES
Methods of producing oil from a reservoir when natural drives—gas-cap, solution-gas, water, etc.—are not present or have been depleted. Waterflood, repressuring or recycling, and in situ combustion are examples of artificial drives.

ARTIFICIAL LIFT
Pumping an oil well with a rod, tubing, or bottom-hole centrifugal pump may be termed artificially lifting crude oil to the surface or doing so by mechanical means. For other means of producing wells when natural drives have been depleted, see Artificial Drives.

A.S.K. SYSTEM
Automatic station-keeping system; the name applied to a sophisticated drillship-positioning technique consisting of subsurface acoustical equipment linked to shipboard computers that control ship's thrusters. The thrusters fore and aft reposition the ship, compensating for drift, wind drag, current, and wave action. *See* Dynamic Stationing.

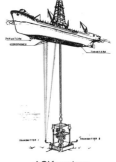
ASK system

A.S.M.E.
American Society of Mechanical Engineers.

ASPHALT

A solid hydrocarbon found as a natural deposit. Crude oil of high asphaltic content, when subjected to distillation to remove the lighter fractions such as naphtha and kerosene, leaves asphalt as a residue. Asphalt is dark brown or black in color and at normal temperatures is a solid. *See* Brea.

ASPHALT-BASE CRUDE

Crude oil containing very little paraffin wax and a residue primarily asphaltic. Sulfur, oxygen, and nitrogen are often relatively high. This type of crude is particularly suitable for making high-quality gasoline, lubricating oil, and asphalt. *See* Paraffin-Base Crude.

ASPHALTENES

At the very bottom of the crude-oil barrel are the asphaltenes, composed of complex molecules. Asphaltenes are polyaromatic compounds with high carbon-hydrogen ratios in their molecules from which asphalt is made.

ASPHALTIC PETROLEUM

Petroleum that contains sufficient amounts of asphalt in solution to make recovery commercially practical by merely distilling off the solvent oils.

ASPHALTIC SAND

Natural mixture of asphalts with varying proportions of loose sand. The quantity of bituminous cementing material extracted from the sand may run as high as 12 percent. This bitumen is composed of soft asphalt.

ASSEMBLY

A term to describe a number of special pieces of equipment fitted together to perform a particular function; e.g., a drill assembly may include other pieces of downhole equipment besides the drill bit, such as drill collars, damping subs, stabilizers, etc.

ASSET, WASTING

See Wasting Asset.

ASSIGNEE

A recipient of an interest in property or a contract; in oil and gas usage, the recipient of an oil or gas lease; a transferee.

ASSIGNMENT

In oil and gas usage, assignment is a transfer of a property or an interest in an oil or gas property; most commonly, the transfer of an oil or gas lease. The assignor does the transferring and the assignee receives the interest or property.

ASSOCIATED GAS
Gas that occurs with oil, either as free gas or in solution. Gas occurring alone in a reservoir is unassociated gas.

A.S.T.M.
American Society for Testing Materials.

A.S.T.M. DISTILLATION
A test of an oil's distillation properties standardized by the American Society for Testing Materials. A sample of oil is heated in a flask; the vapors pass through a tube where they are cooled and condensed; the liquid is collected in a graduated cylinder. When the first drop of distillate is obtained, the temperature at which this occurs is the initial boiling point of the oil. The test is continued until all distillable fractions have distilled over and have been measured and their properties examined.

ASTROBLEME
An unusual structural feature discovered near Ames, Oklahoma, was identified as an impact crater, an astrobleme, created by asteroid impact. The crater, known as the Ames Hole, is 8 to 10 miles in diameter and is 455 million years old. Numerous tests were conducted in the Arbuckle dolomite before the actual origin of the feature was recognized. At this writing (1994), thirty-eight wells have been completed in the Ames structure. The crater-floor wells—from brecciated (broken into sharp fragments) granite, granite wash, and dolomite—are highly productive. This is the largest known productive astrobleme.

ATMOSPHERE, ONE
The pressure of the ambient air at sea level; 14.69 pounds per square inch, 29.92 inches of mercury, or 33.90 feet of water.

ATMOSPHERIC STILL
See Still, Atmospheric.

ATOMIZER, FUEL-OIL
A nozzle or spraying device used to break up fuel oil into a fine spray so the oil may be brought into more intimate contact with the air in the combustion chamber. *See* Ultrasonic Atomizer.

ATTIC OIL
An unscientific but descriptive term for the oil above the borehole in horizontal wells; oil in the top few feet of a productive interval which will gravitate or be pressured into the horizontal drain hole.

AUSTRALIAN OFFSET
A humorous reference to a well drilled miles away from proven production.

AUTHEGENIC CARBONATE ROCK
A precipitate from the bacterial oxidation of oil and gas found at oil seeps in the Gulf of Mexico.

AUTOFRETTAGE
Prestressing equipment, e.g., pump barrels, liners, valve pots, by hydrostatic pressure to condition the equipment for extremely high-pressure service.

AUTOMATED WELLHEADS
The remote control of pumping oil wells and flowing gas wells is a computer age development employed in large fields with widely dispersed producing wells. With such a system, it is possible to receive at a central point, a district production office, measurements of gas production, real-time information on oil production and produced water. Wells can be programmed to a pump-rest or on-off regimen. At the wellhead, automated systems include a remote terminal unit (R.T.U.), an adjustable choke with electric actuator, a grouping of batteries, a solar panel to keep them charged, an antenna, transmitters to measure temperature and wellhead static and differential pressures, and a radio to relay data to a computer in the district production office. The operations of large, middle-aged fields with hundreds of wells have been adapted successfully to automation. Offsetting the initial cost of such a system are the savings in man-hours per well, in custodial travel, in record keeping, and in net production of oil and gas.

AUTOMATIC CUSTODY TRANSFER
See Lease Automatic Custody Transfer.

AUTOMATIC MUD VALVE
See Valve, Lower Kelly.

AUTOMATIC TANK BATTERY
A lease tank battery (two or more tanks) equipped with automatic measuring, switching (full tank to empty and full tank into the pipeline), and recording devices. *See* Lease Automatic Custody Transfer.

AUTOMATIC WELDING MACHINE
See Welding, Automatic.

AVIATION FUEL
See J–4 Fuel.

AXIAL COMPRESSOR
See Compressor, Axial.

AXLE GREASE
A cold-setting grease made of rosin oil, hydrated lime, and petroleum oils. *See* Grease.

Automatic welding
machine

BABBITT
A soft, silver-colored metal alloy of relatively low melting point used for engine and pump bearings; an alloy containing tin, copper, and antimony invented by Isaac Babbitt in 1862.

BACKFILL
To replace the earth dug from a ditch or trench; also, the earth removed from an excavation.

BACKFLOW GATE
See Gate, Backflow.

BACKHOE
A self-propelled ditching machine with a hydraulically operated arm equipped with a toothed shovel that scoops earth as the shovel is pulled back toward the machine.

BACK-IN AFTER PAYMENT
See Back-In Provision.

BACK-IN FARMOUT
A farmout agreement in which a retained nonoperating interest of the lessor may be converted, at a later date, into a specified individual working interest.

Backhoe *(Courtesy Robinson-Gerrard Inc.)*

BACK-IN PROVISION
A term used to describe a provision in a farmout agreement whereby the person granting the farmout (the farmor) has the option to exchange a retained override for a share of the working interest.

BACK OFF
To raise the drill bit off the bottom of the hole; to slack off on a cable or winch line; to unscrew.

BACK-OFF JOINT
A section of pipe with left-hand thread on one end and right-hand or conventional thread on the other. A back-off joint is used in setting a liner. When a liner is lowered in and landed, the drill column can be disengaged from the liner by rotating the drill pipe to the right. This mo-

tion unscrews the left-hand threaded back-off joint from the liner, and it keeps all threaded joints above the back-off joint tight.

BACK-OFF WHEEL
See Stripper Wheel.

BACK OUT
A term meaning to replace or be equivalent to. For example: "Windfarm (a group of power-generating windmills) to back out 600,000 barrels a year of oil equivalent." The windfarm will generate the same number of kilowatts of electric power as an oil-fired plant using 600,000 bbl/year.

BACK PRESSURE
The pressure against the face of the reservoir rock caused by the control valves at the wellhead, hydrostatic head of the fluid in the hole, chokes, and piping. Maintenance of back pressure reduces the pressure differential between the formation and the borehole so that oil moves into the well with a smaller pressure loss. This results in the expenditure of smaller volumes of gas from the reservoir, improves the gas-oil ratio, and ensures the recovery of more oil.

BACK-PRESSURE VALVE
A check valve. *See* Valve, Check.

BACKSIDE PUMPING
See Pumping, Backside.

BACKUP MAN
The person who holds (with a wrench) one length of pipe while another length is being screwed into or out of it.

BACKWASHING
Reversing the fluid flow through a filter to clean out sediment that has clogged the filter or reduced its efficiency. Backwashing is done on closed-system filters and on open-bed gravity filters.

BAD OIL
See Cut oil.

BAFFLES
Plates or obstructions built into a tank or other vessel that change the direction of the flow of fluids or gases.

BAG HOUSE
A construction housing receptacles (bags) that capture and hold dry chemicals, dust, and other particulate matter removed from refinery stack gas by a cleaning or scrubbing facility.

BAIL
The heavy metal arms or links that connect the swivel to the hook of the traveling block. The bail bears the weight of the drill string as does the swivel.

BAIL DOWN
To reduce the level of liquid in a well bore by bailing.

BAILER
A cylindrical, bucket-like piece of equipment used in cable-tool drilling to remove mud and rock cuttings from the borehole.

BAILER DART
The protruding "tongue" of the valve on the bottom of a bailer. When the dart reaches the bottom of the hole, it is thrust upward, opening the valve to admit the mud-water slurry.

BAIT BOX
A pipeliner's lunch pail.

BALANCING
See Makeup Gas.

BALD-HEADED ANTICLINE
See Anticline, Bald-Headed.

BALL-AND-SEAT VALVE
See Valve, Ball-and-Seat.

BALL BEARING
See Bearing, Ball.

BALLING OF THE BIT
The fouling of a rotary drilling bit in sticky, gumbo-like shale that causes a serious drag on the bit and loss of circulation.

BALL JOINT
A connector in a subsea marine-riser assembly whose ball-and-socket design permits an angular deflection of the riser pipe caused by horizontal movement of the drillship or floating platform of 10° or so in all directions.

BALL STOPS
See Plug Valve.

BALL VALVE
See Valve, Ball.

Ball bearings

Ball joint

Ball valve

BANDING, REGIONAL
Bedding, usually thin, produced by the deposition of different minerals or materials in alternating layers so as to appear laminated when viewed in cross section; the existence of layering in an outcrop, the laying down of successive types and colors of sediment.

BANDWHEEL
In a cable-tool rig, the large vertical wheel that transmits power from the drilling engine to the crank and pitman assembly that actuates the walking beam. Used in former years in drilling with cable tools. Old pumping wells still use a bandwheel.

BAREFOOT CHARTER
A contract or charter agreement between the owner of a drilling rig, semisubmersible, or drillship and a second party in which the owner rents or leases his equipment (usually short-term) barefoot, i.e., without the owner or his representative taking any part in the operation or maintenance of the equipment. The lessee agrees to man or staff the equipment and operate it without assistance from or responsibility by the owner. Also bareboat charter for boats or ships.

BAREFOOT COMPLETION
Wells completed in firm sandstone or limestone that show no indication of caving or disintegration may be finished "barefoot," i.e., without casing through the producing interval.

Reel barge

BARGE, REEL
See Reel Barge.

BARGE RIG
See Rig, Barge.

BARITE
A mineral used as weighting material in drilling mud; a material to increase the density or weight per gallon or cubic foot of mud.

Submersible drilling barge

BARITE DOLLAR
A Texas and Oklahoma term for a small disk-shaped piece of barite found in sandstone or shale. Barite is a white or yellowish mineral occurring in tabular crystals or as a compact mass resembling marble.

BARKER
A whistle-like device attached to the exhaust pipe of a one-cylinder oilfield engine so the lease pumper can tell from a distance whether the engine is running. The noise the device makes resembles the bark of a hoarse fox.

The floating drilling vessel *Discoverer*

Marine pipeline construction *(Courtesy Reading & Bates Construction Co.)*

BARNSDALL, WILLIAM

William Barnsdall and W.H. Abbott built the first refinery in Pennsylvania in 1860, shortly after Colonel Edwin Drake discovered oil near Titusville in 1859. By the end of the Civil War, there were more than 100 plants using 6,000 barrels a day. Kerosene was the main product.

BAROID

A specially processed barite (barium sulfate) to which Aquagel has been added, used as a conditioning material in drilling mud in order to obtain satisfactory cores and formation samples.

BARREL

(1) Petroleum barrel; a unit of measure for crude oil and oil products equal to 42 U.S. gallons. (2) Pump barrel; cylindrical body of an oil-well pump.

BARREL HOUSE

A building on the refinery grounds where barrels are filled with various grades of lubricating and other oils, sealed, and made ready for shipment; oil house. *See* Drum.

BARREL-MILE

The cost to move a barrel of oil or an equivalent amount of product one mile.

BASALT

A general term for a dark-colored mafic (ferromagnesian) igneous rock that may be extrusive, erupted onto the surface of the Earth by volcanic action, or intrusive, as in dikes or sills where the igneous rock in a molten state was forced upward between the planes or surfaces of other rock formations.

BASE LINE

(1) A carefully surveyed line that serves as a reference to which land surveys are coordinated and correlated. (2) One of a pair of coordinate axes (the other is the principal meridian) used in the U.S. Public Land Survey system.

BASE MAP

See Map, Base.

BASEMENT ROCK

Igneous or metamorphic rock lying below the sedimentary formations in the earth's crust. Basement rock does not contain petroleum deposits.

BASE STATION

An observation point used in geophysical surveys as a reference to which measurements at other points can be compared.

BASIC SEDIMENT

Impurities and foreign matter contained in oil produced from a well. *See* B.S.&W.

BASIC WASH

A term for material eroded from outcrops of igneous rock and re-deposited to form rock of about the same mineral makeup as the original rock; granite wash.

BASIN

A synclinal structure in the subsurface, once the bed of a prehistoric sea. Basins, composed of sedimentary rock, are regarded as good prospects for oil exploration.

BASKET PRICE

The blanket or average price of crude oil on the world market. For example the basket price of $18.00/bbl. could mean average price of average gravity. Lower-gravity crude with high-transit cost would bring less than $18.00, and conversely, higher gravity crude with low sulfur and close to market would bring a premium—a basket of crude oils of differing gravities, sulfur content, sweet and sour.

BASKET SUB

A fishing tool run just above the drill or milling tool to recover small, non-drillable pieces of junk metal that have been dropped in the borehole or are parts of broken equipment. As the drilling fluid is circulated, the small metal pieces are washed into the basket; a junk basket. *See* Junk Basket

BASTARD

(1) Any nonstandard piece of equipment. (2) A kind of file. (3) A word used in grudging admiration or as a term of opprobrium.

BATCH

A measured amount of oil or refined product in a pipeline or a tank; a shipment of oil or product by pipeline.

BATCHING SPHERE

An inflated, hard-rubber sphere used in product pipelines to separate "incompatible" batches of product being pumped one behind the other. Fungible products are not physically separated, but gasoline is separated from diesel fuel and heating oils by batching spheres.

BATCH INTERFACE

See Interface.

BATHOLITH

A great mass of intruded igneous or metamorphosed rock found at or near the surface of the Earth. The presence of a batholith, often referred to as a shield, usually precludes drilling for oil or gas as there are no sedimentary formations above it. The largest batholith in the United States is in Idaho, underlying nearly two-thirds of the state.

BATHYMETRY
The measurement of the depths of bodies of water; also information gathered for such measurements.

BATTERY
Two or more tanks connected together to receive oil production on a lease; tank battery.

BATTERY, TRICKLE-CHARGED
See Trickle-Charged Battery.

BAUME, ANTOINE
The French chemist who devised a simple method to measure the relative weights of liquids using the hydrometer.

BAUXITE
A mineral, off-white, brown, yellow, or reddish-brown in color, composed of a mixture of amorphous or crystalline hydrous aluminum oxides along with silica and clay minerals. It is a common residual of clay deposits found in tropical or semitropical areas. Bauxite occurs in various forms: concretionary, oolitic, compact, or earthy. Bauxite is the main source of aluminum.

B.C.D.
Barrels per calendar day (bcd). *See* Stream Day.

BEAD
A course of molten metal laid down by a welder (electric or oxyacetylene) in joining two pieces of metal. *See* Welding, Pipeline.

BEAKER SAMPLER
A metal or glass container with a small opening fitted with a stopper that is lowered into a tank of oil to obtain a sample.

BEAM-BALANCED PUMPING UNIT
See Pumping Unit, Beam-Balanced.

BEAM STEERING
A patented seismic program of overcoming attenuation and band limiting, impediments to high resolution.

BEAM WELL
A well whose fluid is being lifted by rods and pump actuated by a walking beam.

Tank battery *(Courtesy Cities Service)*

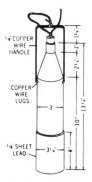

Beaker sampler

Beam-balanced pumping unit

BEAN
A choke used to regulate the flow of fluid from a well. *See* Flow Bean.

BEAN, FLOW
See Flow Bean.

BEAN JOINT
In early pipeline parlance, the joint of line pipe laid just before the break for lunch. When the bean joint was bucked in, the pipeliners grabbed lunch buckets from the gang truck and found a comfortable place to eat.

BEARING, BALL
A type of revolving bearing. The other type is the roller bearing.

Ball bearings

BEARING, INSERT
Thin, bimetal, half-round bearings that fit in the journal box around a shaft to provide a smooth, hard surface. One-half of the insert (in cross section, a semicircle) fits into the journal box, the other half into the journal box cap. Insert bearings are designated bimetal because, although the bearing surface is made of babbitt, it is backed with a layer of bronze, brass, or steel. There are also trimetal insert bearings. They are made with steel backing, a soft alloy middle layer, and a babbitt outer layer.

BEARING, OUTBOARD
A shaft-supporting bearing outside the body or frame of a pump's gear box or engine's crankcase; a bearing on a pump's pinion shaft outside the gear box; a line-shaft bearing.

BEARING, RADIAL
A roller bearing in a circular or cylindrical configuration. The roller bearings are held in a track or cage that may have either a cylindrical or flat circular shape.

BEARING, ROLLER
Cylindrical, pin-like steel bearings that are held in a circular or cylindrical configuration by a metal track or cage so they can be inserted in a journal box or slipped on an engine or pump shaft.

Roller bearing

BEARING, SADDLE
A broad, heavy bearing located on top of the Samson post to support the walking beam on a cable-tool drilling rig or an oil-well pumping jack.

BEARING, STIRRUP
A bearing and its frame in the shape of a saddle stirrup; e.g., the bearing connecting the pitman and the walking beam on an early cable-tool drilling or pumping well.

BEARING, THRUST
A bearing to support the endwise or downward thrust or weight of a machinery part against another. Thrust bearings can be constructed of ball bearings or cylindrical roller bearings held in a circular frame or housing that fits over a shaft.

BEDDED
Formed or deposited in layers or beds; refers especially to sedimentary rocks or strata deposited in recognizable layers.

BEDDING
The stratification or layering of sediment or deposits that is typical of sedimentary rock formations. Bedding can be continuous or there can be variations in the thickness of layers of the same type of rock, sometimes the result of weathering, which can be seen in outcroppings.

BEDDING PLANE
The plane of a bedded formation that visibly separates each successive layer of stratified rock from the one above and below. A plane of deposition that shows a change in the manner or rate of deposits. The bedding plane may show a difference in color and it need not be horizontal; it may be bent or folded and still be recognized as a bedding plane.

BEDDING SURFACE
An easily recognized surface within a mass of stratified rock representing a line of original deposition; the interface between two beds of sedimentary rock. If the surface is fairly regular and is a plane, it may be referred to as a bedding plane.

BED LOAD
Sediment suspended in streams that eventually drops to the river bottom as the stream spreads out and/or slows down. *See* Alluvial Fan.

BEHIND THE PIPE
Refers to oil and gas reservoirs penetrated or passed through by wells but never tapped or produced. Behind the pipe usually refers to tight formations of low permeability that, although recognized, were passed through because they were uneconomical to produce at the time. Today, however, with the growing scarcity of oil and high prices, many of these passed-through formations are getting a second look by producers.

BELL-AND-SPIGOT JOINT
A threaded pipe joint; the spigot or male end is threaded and screwed into the bell or female coupling. The female end of a coupling has threads on the inside circumference. Line pipe screwed together one joint at a time forms a bell-and-spigot connection. *See* Box-and-Pin Joint.

BELL HOLE

An excavation dug beneath a pipeline to provide room for the use of tools by workers; a hole larger in diameter at the bottom than at the top.

BELL-HOLE WELDER

A welder who can do oxyacetylene or electric welding in a bell hole. This requires a great deal of skill as the molten metal from the welding rod is being laid on upside down and tends to fall away from the weld; a skilled welder.

BELL NIPPLE

A large swage nipple for attaching casinghead fittings to the well's casing above the ground or at the surface. The bell nipple is threaded on the casing end and has a plain or weld end to take the casinghead valves.

BELLOWS-SEALED VALVE

See Valve, Bellows-Sealed.

BELT GUARD

A housing or cage made of sheet metal or heavy wire mesh built over or around the sheaves and the flat or V-belts of an engine and pump or engine and other driven equipment. The belt guard prevents contact with the moving sheaves and belts and also protects workmen should a belt break while running.

Packless or bellows
valve

BELT HALL

A wooden shed built to protect the wide belt that runs from the engine to the bandwheel on a cable-tool rig or an old beam pumping well. The belt hall extends from the engine house to the derrick.

BENCH-SCALE TEST

Testing of methods or materials on so small a scale that it can be carried out on a laboratory table or specially constructed bench.

BENTONITE

A soft, porous, plastic, light-colored rock composed mainly of clay minerals and silica. The rock is greasy to the touch and has the ability to absorb quantities of water, which increases its volume about eight times. This property makes it ideal for thickening or adding body to drilling mud. It was named bentonite for the Benton formation in the Rock Creek district in eastern Wyoming where it was first identified and named by Knight in 1898.

BENT SUB

A short, heavy tubular section or connector made with a bend of a few degrees in its long axis. Bent subs are used to connect a mud motor and drill bit to the drillpipe. This permits the bit to drill at an angle from the vertical or from the former direction of the borehole. Bent subs and mud motors are used in sidetracking and directional drilling.

BENZ, KARL
A German inventor who developed a gasoline-burning, internal-combustion engine in 1885 that propelled a carriage-like vehicle that carried passengers. *See* Diamler, Gottlieb.

BENZENE
A solvent and/or motor fuel.

BENZINE
An old term for light petroleum distillates in the gasoline and naphtha range.

BENZOL
The general term that refers to commercial benzene that may contain other aromatic hydrocarbons.

BEVELING MACHINE
An oxyacetylene pipe-cutting machine. A device that holds an acetylene cutting torch so that the ends of joints of pipe may be trimmed off at an angle to the pipe's long axis. Line pipe is beveled in preparation for welding joints together.

B.H.P.
Brake horsepower.

B.H.T.
Bottom-hole temperature. In deep wells, 15,000 feet and deeper, bottom-hole temperatures are above the boiling point of water, ranging up to 400°F. At these depths and temperatures, water-base drilling muds can not be used, only oil-based. *See* Temperature Gradient.

BID SHEET, A.P.I.
See A.P.I. Bid Sheet.

BIG-INCH PIPELINE
A 24-inch pipeline from Longview, Texas, to Norris City, Illinois, built during World War II to meet the problem caused by tanker losses at sea as a result of submarine attacks. Later during the war the pipeline was extended to Pennsylvania. Following the war the line was sold to a private company and converted to a gas line.

BIG SPROCKET, ON THE
See On the Big Sprocket.

Benzene, C_6H_6
(an aromatic)

Beveling machine

33

BINDER

The material that produces consolidation in sediments that are loosely aggregated or held together; a mineral cement that is precipitated in the spaces between the grains of sediment and cements them together into a coherent mass. A binder may be a clay that fills the pore spaces between the grains of sediment; a soil binder.

BIOCHEMICAL CONVERSION

The use of bacteria to separate kerogen from oil shale. Certain bacteria will biodegrade the minerals in oil shale, releasing the kerogen from the shale in liquid or semiliquid form. (From studies made by Dr. Ten Fu Yen and Dr. Milo D. Appleman, University of Southern California, Professors of Bacteriology.)

BIOCLASTIC ROCK

See Fragmental Rock.

BIODEGRADABLE

Capable of being decomposed by microorganisms; more loosely, subject to decomposition by ultraviolet rays in sunlight or natural chemical action as well as by bacteria and like creatures.

BIOGENESIS

Formed by the presence or the actions of living organisms, for example, coral reefs and atolls. Biogenesis is also the theory that all life is derived from previously living organisms.

BIOGENETIC ROCK

Rock formed or produced by the activities (living and dying) of organisms, both plant and animal, for example, coral reefs, certain limestones, coal, and peat. An organic rock.

BIOGEOLOGY

The biological phases of geology; paleontology, for example, is the study of sedimentation produced by organisms, plants, and animals.

BIOGRADATION

The breaking up and removal of a crude-oil spill on water by the introduction of oil-eating bacteria to the spill area.

BIOHERM

A mound or reef-like mass of rock built up by sedimentary, marine creatures, such as coral, algae, mollusks and composed almost entirely of their calcareous skeletal remains; an organic reef or a nonreef, just a limestone mound.

BIOMASS

Wood and other plant material used to make methanol as a supplement to petroleum.

BIOREMEDIATION
The use of microbes and other microorganisms in cleaning up oil spills. The tiny organisms break up the oil by eating it and changing oil's character and make up so it is no longer detrimental to man nor lethal to plants and animals.

BIOTECHNOLOGY
See Ergonomics.

BIRD CAGE
(1) To flatten and spread the strands of a cable or wire rope. (2) The slatted or mesh-enclosed cage used to hoist workmen from crew boats to offshore platforms.

BIRDCAGED WIRE
Wire rope used for hoisting heavy loads that has had its steel strands distorted into the shape of a bird cage by a sudden release of the load, as when the rope parts or slips.

BIRD DOG
To pay close attention to a job or to follow a person closely with the intent to learn or to help; to follow up on a job until it is finished.

BIRDNESTING
A problem associated with borehole washover milling operations. Metal cuttings are so large and stringy that they are not drilling-mud suspendable. When the large, elongated cuttings from the milled or shaved pipe fall to the bottom of the borehole they birdnest, pack and intertwine and cannot be flushed up the hole to the surface by the circulating mud. New milling tools have been developed by Tri-State Oil Tool Industries, Bossier City, La. that are able to cut small, mud-suspendable cuttings which are readily circulated out of the hole.

BIT
The cutting or pulverizing tool or head attached to the drillpipe in boring a hole in underground formations.

BIT, BUTTON
An insert bit; a drill bit with tungsten carbide or other superhard metal inserts or buttons pressed into the face of the bit's cutting cones.

BIT, CORE
A special drill bit for cutting and removing a plug-shaped rock sample from the bottom of the well bore.

BIT, DIAMOND
A drill bit with many small industrial (man-made) diamonds set in the nose or cutting surface of the bit. Diamonds are many times harder than

Bird cage

A drill bit *(Courtesy Reed Tool Co.)*

Core bit

Polycrystalline diamond bit *(Courtesy Strata Bit Co.)*

Insert drill bit

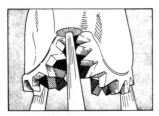

Mill-tooth bit

Polycrystalline bit *(Courtesy Hughes Tool Co.)*

the hardest steel, so a diamond bit makes possible longer bit runs before a round trip is necessary to change bits.

BIT, DIAMOND SHEAR
A recently developed (in the past 10 years) rotary drill bit that is different from the conventional roller cone and bull-nose diamond bit. The diamond shear bit does not gouge or pulverize but makes the rock fail in shear; that is, the rock is shaved or broken across the face of the rock as a chef slices a carrot or makes shaved ice. The bit body has sintered diamonds (synthetic, man-made) set in a tungsten carbide body that has 5–7 screw-in nozzles for the stream of drilling mud that jets out and carries the cuttings to the surface. This bit is very expensive but justifies its cost by outlasting other bits when used in medium-hard formations. The diamond shear bit is also known as a polycrystalline diamond compact bit or a Stratapax bit, a trade name.

BIT, DRAG
A type of old-style drilling tool in which the cutting tooth or teeth were the shape of a fishtail. Drilling was accomplished by the tearing and gouging action of the bit and was efficient in soft formations; the forerunner of the modern three-cone roller bit. A fishtail bit; finger bit.

BIT, FINGER
See Bit, Drag.

BIT, FISHTAIL
A drag bit. *See* Bit, Drag.

BIT, INSERT
A bit with superhard metal lugs or cutting points inserted in the bit's cutting cones; a rock bit with cutting elements added that are harder and more durable than the teeth of a mill-tooth bit.

BIT, MILL-TOOTH
A bit with cutting teeth integral to the metal of the cones of the bit; a noninsert bit. Mill-tooth bits are used in relatively soft formations found at shallow depths.

BIT, POLYCRYSTALLINE DIAMOND COMPACT
See Bit, Diamond Shear.

BIT, ROLLER
The rock-cutting tool on the bottom of the drillstring made with three or four shanks welded together to form a tapered body. Each shank supports a cone-like wheel with case-hardened teeth that rotate on steel bearings.

BIT, ROTARY

The tool attached to the lower end of the drillpipe; a heavy steel head equipped with various types of cutting or grinding teeth. Some are fixed; some turn on bearings. A hole in the bottom of the drill permits the flow of drilling mud being pumped down through the drillpipe to wash the cuttings to the surface and also cool and lubricate the bit.

BIT, ROTARY-PERCUSSION

A drill bit that rotates in a conventional manner but at the same time acts as a high-frequency pneumatic hammer, producing both a boring and a fracturing action simultaneously. The hammer-like mechanism is located just above the bit and is actuated by air, liquid, or high-frequency sound waves.

BIT, SPUDDING

A bit used to start the borehole; a bit that is some variation of the fishtail or drag bit, one used in soft, unconsolidated, near-surface material.

BIT BREAKER

A heavy metal plate that fits into the rotary table and holds the bit while it is being made up or broken out of the drillstring.

BIT RECORD

A detailed, written record, kept by the driller, of the drill bits used on a well: type of bit, feet drilled, formation drilled, condition of bit when removed, condition of "dulls," and notations of special problems.

BITUMEN

Bitumen is defined as "any of various mixtures of hydrocarbons together with their nonmetallic derivatives": asphalts and tars.

BITUMEN, OIL-SANDS

See Oil-Sands Bitumen.

BITUMINOUS SAND

Tar sand; a mixture of asphalt and loose sand that, when processed, may yield as much as 12 percent asphalt.

BLACK LIGHT

An oil prospector's term for ultraviolet light used to detect fluorescence in a mineral sample. Also the instrument, usually portable, that produces ultraviolet light for use at the rig site.

BLACK OIL

(1) A term denoting residual oil; oil used in ships' boilers or in large heating or generating plants; bunker oil. (2) Black-colored oil used for

lubricating heavy, slow-moving machinery where the use of higher-grade lubes would be impractical. (3) Asphalt-base crudes.

BLACK OILS MARKET
See Resid Market.

BLANK CASING
Well casing that has not been perforated; casing above and below the pay zone.

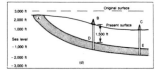

Widespread (blanket) sand

BLANKET DEPOSIT
A sedimentary deposit extending over a wide area and usually of fairly uniform thickness; a blanket sand or sandstone; blanket limestone; sheet sand.

BLASTING
Refers to the effect of fine particles wearing away an exposed surface as they are blown by the wind or carried by a fast-moving stream; sand-blasting.

BLAST JOINT
The bottom joint in the well's tubing string that, in a flowing well, is subjected to the abrasive action of oil and gas forcibly entering the well bore from a high-pressure formation. Blast joints are made of special steels that resist the severe conditions in this situation.

BLEED
To draw off a liquid or gas slowly. To reduce pressure by allowing fluid or gas to escape slowly; to vent the air from a pump.

BLEEDER VALVE
See Valve, Bleeder.

BLEEDING
The tendency of a liquid component to separate from a lubricant, as oil from a grease; to seep out.

BLEEDING CORE
A core sample of rock highly saturated and of such good permeability that oil drips from the core.

BLEED LINE
A line on the wellhead or blowout-preventer stack through which gas pressure can be bled to prevent a threatened blowout.

BLENDING
The process of mixing two or more oils having different properties to obtain a lubricating oil of intermediate or desired properties. Certain

classes of lube oils are blended to a specified viscosity. Other products, notably gasolines, are also blended to obtain desired properties.

BLENDING STOCK
A quantity of lubricating oil, gasoline, or other liquid product that is used to mix or blend with other batches of the same product. Motor gasolines are a blend of several different gasolines, each having certain desirable properties.

BLIND FLANGE
A companion flange with a disc bolted to one end to seal off a section of pipe.

BLIND POOL
Money put into a drilling fund that is held by the fund managers until likely prospects for drilling are found or come along. The rationale for the blind fund is that with ready money, the fund managers can act quickly when good opportunities for investment arise. Blind fund money usually is kept in an interest-bearing account while waiting for a hot prospect.

BLIND RAM
See Ram, Blind.

B.L.M.
Bureau of Land Management.

BLOCK
(1) A pulley or sheave in a rigid frame. (2) To prevent the flow of liquid or gas in a line. (3) A chock. (4) A large angular chunk of rock, showing little erosion or other changes in its original shape as it broke away from the original rock mass. A huge, newly broken chunk of rock. (5) An area of land made up of a number of contiguous leaseholds (leases) large enough to drill an exploratory well. Before drilling a deep, expensive exploratory well, an operator normally will acquire a large block of leases surrounding the proposed wellsite. This maneuver is to protect himself from loss of local drainage from adjacent areas should other operators decide to lease and drill next to his well.

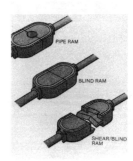

Ram elements, including blind ram

BLOCK-AND-BLEED VALVE
See Valve, Block-and-Bleed.

BLOCK AND TACKLE
An arrangement of ropes and blocks (pulleys) used to hoist or pull.

BLOCK FAULTING
A type of normal faulting in which the crust is divided into structural blocks or fault blocks of different elevations and positions.

BLOCK GREASE

A grease of high melting point that can be handled in block or stick form. Block grease is used on large, slow-moving machinery, axles, and crude bearings. In contact with a hot journal bearing, the grease melts, slowly lubricating the bearing.

BLOCKING

Pumping crude oil or refined products in batches or blocks through a pipeline.

BLOCK LEASE

See Lease, Block.

BLOCK TREE

A type of well-completion Christmas tree in which a number of control and production valves are made as a unit in one block of steel. Valve pockets for the special valve assembly are bored in the steel forging, which makes the valve assembly a strong, rigid unit integral to the forging. Block trees are often used on multiwell offshore platforms to conserve space.

BLOCK VALVE

See Valve, Block.

Block valve

BLOOIE PIPE

A horizontal vent pipe extending from the wellhead a couple of hundred feet from the rig floor to a burn pit. The blooie pipe, named for the noise it makes, vents the returns during air or gas drilling. In air drilling, no mud is used; the pulverized rock from the action of the bit is brought up from the bottom of the hole by compressed air and blown through the blooie pipe into the burn pit. Should gas or oil be encountered, it too is vented to the burn pit. If the well needs to be controlled because of oil or gas in quantity and under high pressure, the well must be mudded up. Drilling mud is pumped into the hole and circulated as in conventional rotary drilling.

Blooie pipe

BLOOM

The iridescent cast of color in lubricating oil.

BLOWBY

The escape of combustion or unburned fuel past the engine's piston and piston rings into the crankcase. Blowby occurs during the power stroke, but unburned fuel can also escape during the compression stroke on spark-ignition engines.

BLOWDOWN

The venting of pressure in a vessel or pipeline; the emptying of a refinery vessel by relieving pressure at a discharge valve to direct the contents into another vessel or to the atmosphere.

BLOWDOWN, N₂

Evacuating the well's borehole or casing by injecting gaseous nitrogen under pressure into the hole. Ridding the borehole or casing of mud and water by the injection of gaseous nitrogen under pressure into the hole to effect a blowdown.

BLOWDOWN STACK

A vent or stack into which the contents of a processing unit are emptied when an emergency arises. Steam is injected into the tank to prevent ignition of volatile material or a water quench is sometimes used.

Blowout

BLOWING A WELL

Opening a well to let it blow for a short period to free the well tubing or casing of accumulations of water, sand, or other deposits.

BLOWING DOWN THE GAS CAP

After the oil in a gas-cap reservoir nears depletion, the wells in the field begin producing gas, thus blowing down the gas cap, which signals the time for secondary recovery, waterflood or a two-phase miscible flood. *See* Miscible Flood.

BLOWING THE DRIP

To open the valve on a drip to drain off the "drip gasoline" and to allow the natural gas to "blow" for a moment to clear the line and drip of all liquid.

BLOWING-WELL CONTROL

See Momentum Kill of Well Blowout.

Blowout preventer

BLOWOUT

Out-of-control gas and/or oil pressure erupting from a well being drilled; a dangerous, uncontrolled eruption of gas and oil from a well; a wild well.

BLOWOUT-BACK PRESSURE GAUGE

See Pressure Gauge, Blowout-Back.

BLOWOUT PREVENTER

A stack or an assembly of heavy-duty valves attached to the top of the casing to control well pressure; a Christmas tree.

BLOWOUT PREVENTER, SPHERICAL

See Spherical Blowout Preventer.

BLOWPIPE

See Welding Torch.

BLUE 8, AUTOMATE

A dye concentrate that, under an E.P.A. directive, must be added to diesel fuel to mark higher-sulfur, off-highway use, exclusively. The dye

Blowout preventer stack with spherical BOP at the top

marker is a patented product of Morton International, Inc., Chicago. Use of the dye to mark off-highway use is nonoptional . . . a euphemism for mandatory. The fine for noncompliance is draconian: after a first warning, $25,000/ violation day—serious business.

BLUE SKY LAW
A statute that regulates the issuance and sale of securities. The term usually is restricted to state statutes. The corresponding federal statutes and regulations are the Federal Securities Act and the Securities and Exchange (S.E.C.) regulations. States differ in subjecting the sale of property interests in oil and gas to Blue Sky regulations.

B.N.O.C.
See British National Oil Corporation.

BOBTAIL
A short-bodied truck.

BOBTAIL ABSTRACT
A summary of instruments of record in the chain of title rather than copies of the instruments themselves.

BOBTAIL PLANT
A gas plant that extracts liquid hydrocarbons from natural gas but does not break down the liquid product into its separate components.

BODY
Colloquial term for the viscosity of an oil.

B.O.E.
Barrels of oil equivalent. Cubic feet of natural gas or barrels of gas liquids.

BOGIES
Colloquial term for small transport dollies. A low, sturdy frame or small platform with multiple wheels (four to eight) for moving heavy objects short distances.

BOILERHOUSE
(1) A lightly constructed building to house steam boilers; (2) to make a report without doing the work; to fake a report.

BOILING POINT
See Initial Boiling Point; *also* End Point.

BOILOFF
The vaporization or gasification of liquefied natural gas (L.N.G.) or other gases liquefied by applying high pressure and severe cooling. Boiloff occurs when the holding vessel's insulation fails to maintain the

low temperature required to keep the gas in liquid form. Boiloff is a problem for shippers of L.N.G. in the specially built ocean carriers.

BOLL WEEVIL
An inexperienced worker or "green hand" on a drilling crew.

BOLSTER
A support on a truck bed used for hauling pipe. The heavy wooden or metal beam rests on a pin that allows the forward end of the load to pivot as the truck turns a corner.

BONNET
The upper part of a gate or globe valve that encloses the packing gland and supports the valve stem.

BONUS
Usually, the bonus is the money paid by the lessee for the execution of an oil and gas lease by the landowner. Another form is called an oil or royalty bonus. This may be in the form of an overriding royalty reserved to the landowner in addition to the usual one-eighth royalty.

BONUS, PENALTY
See Nonconsent Penalty.

BONUS BIDDING
Competitive bidding for oil and gas leases in which the lease providing for a fixed royalty is offered to the prospective lessee offering to pay the largest bonus to the lessor. *See* Royalty Bidding.

BOOM
A beam extending out from a fixed foundation or structure for lifting or hoisting; a movable arm with a pulley and cable at the outer end for hoisting or exerting tension on an object. *See* Boom Cats.

Oil Spill Boom

BOOM CATS
Caterpillar tractors equipped with side booms and winches; used in pipeline construction to lift joints of pipe and to lower sections of the line into the ditch.

BOOMER
(1) A link-and-lever mechanism used to tighten a chain or cable holding a load of pipe or other material. (2) A worker who moves from one job to another. *See* Pipeline Cat.

A boom cat at work

BOOSTER STATION
A pipeline pumping station usually on a main line or trunk line; an intermediate station; a field station that pumps into a tank farm or main station.

BOOT

A tall section of 12- or 14-inch pipe used as a surge column at a lease tank battery, downstream from the oil/gas separator. The column,15–25 feet high, provides a means to separate oil from produced water in stripper wells or small producers.

B.O.P.D.

Barrels of oil per day; bo/d.

B.O.P. STACK

Blowout-preventer stack.

BORE AND STROKE

See Pump Specifications.

BOREHOLE

The hole in the earth made by the drill; the uncased drill hole from the surface to the bottom of the well.

BOP stack *(Courtesy Atlantic-Richfield Co.)*

BOREHOLE SWELLING

This condition is usually caused by the action of water-absorbing clay which has been drilled through. As it takes up water from the drilling fluid, it swells, closing in on the drillpipe. This action of the clay causes severe drag and balling of the bit, and it often results in stuck pipe. Any increase in the power from the drilling engines subjects the pipe to additional torque which can result in a twist off, a parting of the drill pipe. *See* Balling of the Bit.

BOREHOLE WASHOUT

The enlarging of the well's borehole by the velocity and abrasive effect of the returns (drilling mud laced with rockchips returning from downhole) on the uncased hole.

BOREHOLE WASHOUTS & FILL

A downhole condition often resulting when drilling through certain types of shale. The hydrophilic (water-loving or thirsty) shales absorb the fluid in the drilling mud and soften enough to slough and cave, filling the hole, sometimes sticking the drillpipe. If the shale section is 800 to 1200 feet thick, there can be enough severe caving to significantly enlarge this section of the hole, making it difficult or impossible to circulate the volume of cement for a successful cement job. To combat this condition, various additives have been developed to inhibit shale from swelling, softening, and sloughing.

Boring a road crossing

BORING MACHINE

A power-driven, large-diameter auger used to bore under roads, railroads, and canals for the purpose of installing casing or steel conduits to hold a pipeline.

BOTTLE, WELDING
See Welding Bottle.

BOTTLENECKING
The deformation of the ends of the casing or tubing in the hanger resulting from excessive weight of the string of pipe and the squeezing action of the slips.

BOTTOM FRACTION
The last cut; the bottom of the barrel in petroleum distillation.

BOTTOM-HOLE ASSEMBLY (B.H.A.)
A drilling string comprised of a drill bit and several drill collars is a simple bottom-hole assembly. Such an assembly may also include a bottom-hole reamer above the bit or above the first drill collar. When in addition to drill collars and reamers, there are two or three stabilizers in the string, it is referred to as a packed-hole assembly. The main purpose of a packed-hole assembly is to keep the bit drilling as straight down as possible.

BOTTOM-HOLE CHOKE
A device placed at the bottom of the tubing to restrict the flow of oil or to regulate the gas-oil ratio.

BOTTOM-HOLE HEATER
Equipment used in the bottom of the well bore to increase bottom-hole temperature in an effort to increase the recovery of low-gravity or heavy oil.

BOTTOM-HOLE LETTER
An agreement by which an operator, planning to drill a well on his own land, secures the promise from another operator to contribute to the cost of the well. In contrast to a dry-hole letter, the former requires payment upon completion of the well whether it produces or not. A bottom-hole letter is often used by the operator as security for obtaining a loan to finance the drilling of the well.

BOTTOM-HOLE PRESSURE
The reservoir or formation pressure at the bottom of the hole. If measured under flowing conditions, readings are usually taken at different rates of flow in order to arrive at a maximum productivity rate. A decline in pressure indicates some depletion of the reservoir.

BOTTOM-HOLE PUMP
See Pump, Bottom-Hole.

BOTTOM LEASE
A bottom lease covers rights to explore and produce oil and gas from all depths and formations below the deepest oil-and-gas-producing forma-

tions covered by existing leases, which have not been explored by the present lessee. A bottom lease might specify, "any and all producing formations" below 5,000 feet or 7,500 feet, for example, or below the Skinner sand or the Wilcox.

BOTTOM OUT
To reach total depth; to drill to a specified depth.

BOTTOMS CRACKING
A term that refers to the further refining, the cracking or breaking down of the larger more complicated molecules of the heavy residual oil, the "bottom of the barrel," into the smaller, simpler molecules of light oil and residual fuel.

BOTTOM WATER
Free water in a permeable reservoir rock beneath the space in the reservoir trap that contains oil and gas. If the water zone underlies the entire reservoir, it is called bottom water; if it occurs at the sides of the reservoir only, it is referred to as edge water.

BOURDON TUBE
A small, crescent-shaped tube closed at one end, connected to a source of gas pressure at the other, used in pressure-recording devices or in pilot-operated control mechanisms. With increases in gas pressure, the Bourdon tube flexes (attempts to straighten). This movement, through proper linkage, actuates recording instruments.

Bourdon tube

BOWL
A device that fits in the rotary table and holds the wedges or slips that support a string of tubing or casing.

BOW LINE
A knot used to form a loop in a rope that will neither slip nor jam.

BOW-TIE EFFECT
A slang term used in seismology for a false reading which appears to be a structural high or anticline but, in fact, is a seismic reflection of a lower point.

Bow-tie Effect

BOX-AND-PIN JOINT
A type of screw coupling used to connect sucker rods and drillpipe. The box is a thick-walled collar with threads on the inside; the pin is threaded on the outer circumference and is screwed into the box.

BOYCOTT EFFECT, THE
When drilling fluid or mud at rest in the borehole of a high-angle or horizontal well is not being circulated, the solids in the slurry drop out, and

by piling up result in discontinuities—density variations which are referred to as a "sag." Sags can result in stuck pipe; when the pumps are started again, and an attempt is made to resume circulating, the increase in pressure may harm the formation, or even cause fractures where they are not wanted. The dropping out of the solids is the boycott effect, named rather fancifully for Charles Boycott of 1879, a British land agent and rent collector who was sincerely disliked for not lowering rents and so was ostracized.

BOYLE'S LAW
"The volume of any weight of gas is inversely proportional to the absolute pressure, provided the temperature remains constant."

B.P.M.
Barrels per minute. The pumping rate of small rotary pumps.

BRADENHEAD GAS
Casinghead gas. Bradenhead was an early-day name for the wellhead or casinghead.

BRAKE HORSEPOWER (B.H.P)
The power developed by an engine as measured at the drive-shaft; the actual or delivered horsepower as contrasted to indicated horsepower.

BRASS POUNDER
A telegrapher, especially one who uses a telegraph key. Until the 1940s or so, much of the communication from oil patch to division and head offices was by telegraph.

BREA
A viscous, asphaltic material formed at oil seepages when the lighter fractions of the oil have evaporated, leaving the black, tar-like substance.

BREACHED ANTICLINE
See Anticline, Breached.

BREAK
Drilling) A definite and recognizable change in the geologic column, the rock formation being drilled through that alerts the geologist to the possibility that the drill is on top of the producing formation or interval. To the driller, a break is a change in the penetration rate of the drill bit. Usually this is an increase in the rate of penetration when drilling, which indicates that the bit has entered a softer formation, i.e., shale, sandstone, or carbonate rock. *Stratigraphic)* (1) An abrupt change at a specific or definite horizon in a sequence of sedimentary rocks, usually indicating a disconformity. (2) An interruption of a normal geologic sequence, especially the continuity of a stratigraphic section; stratigraphic break.

BREAK CIRCULATION

To resume the movement of drilling fluid down the drillpipe, through the "eyes" of the bit, and upward through the annulus to the surface.

BREAKDOWN PRESSURE

The hydrostatic pressure of a hole full of drilling fluid at the point it is greater than the reservoir pressure and breaks in, invading the formation. Good drilling practice avoids overbalancing the hole, which is permitting the hydrostatic pressure of the column of drilling mud to exceed the formation pressure (the pore pressure) which can damage the formation by plastering it over, plugging the pores.

BREAKER, DELAYED

In formation fracturing (breaking), proppant material (usually very small-diameter ceramic beads about the size of millet seeds) is suspended in the frac fluid. It is pumped under very high pressure into the minute cracks, crevices and other interstices in the formation. Patented additives for the frac fluid make possible higher concentrations of proppant without adversely affecting the viscosity. When the frac job is complete, the fluid becomes less viscous, drops its load of proppant material (the beads), becomes easily pumpable, and can be cleaned up readily. One of the developers of delayed breakers is Halliburton Services, Duncan, Ok.

BREAKING DOWN THE PIPE

Unscrewing stands of drillpipe in 1-joint lengths, usually in preparation for stacking and moving to another well location.

BREAK OUT

(1) To isolate pertinent figures from a mass of data; to retrieve relevant information from a comprehensive report. (2) To loosen a threaded pipe joint. (3) To be promoted, "He broke out as a driller at Midland"; to begin a new job after being promoted.

BREAK-OUT TANKAGE

Tankage at a take-off point or delivery point on a large crude-oil or products pipeline.

BREAK THRUST

An overthrust caused by the deformation of an anticline that happens when the folding becomes a fracture and the crustal units overthrust one another along the fault surface or fault plane.

BREAK TOUR

To begin operating 24 hours a day on three 8-hour shifts after rigging up on a new well. Until the derrick is in place and rigged up, mud pits dug, pipe racked, and other preparatory work done, the drill crew works a

Break-out tankage at a take-off point on a products line

regular 8-hour day. When drilling commences, the crews break tour and begin working the three 8-hour tours.

BREATHING
The movement of oil vapors and air in and out of a storage tank, owing to the alternate heating by day and cooling by night of the vapors above the oil in the tank.

BRECCIA
A coarse-grained clastic rock composed of angular rock fragments held together by a mineral cement or other fine-grained matrix. Breccia differs from conglomerate in that the fragments have sharp, unworn edges. Breccia may result from the accumulation of talus laid down or deposited in a sedimentary process.

BRIDGEOVER
The collapse of the walls of the borehole around the drill column.

BRIDGE PLUG
An expandable plug used in a well's casing to isolate producing zones or to plug back to produce from a shallower formation; also to isolate a section of the borehole to be filled with cement when a well is plugged.

BRIDLE
A sling made of steel cable fitted over the "horsehead" on a pumping jack and connected to the pump rod; the cable link between horsehead and pump rod on a pumping well.

Bridle

BRIGHT SPOTS
White areas on seismographic recording strips that may signal to the geologist or trained observer close up, the presence of hydrocarbons.

Bright spots *(Courtesy Shell Ecolibrium)*

BRIGHT STOCKS
High-viscosity, fully refined and dewaxed lubricating oils used for blending with lower-viscosity oils. The name originated from the clear, bright appearance of the dewaxed lubes.

BRIMSTONE
A common name for sulfur, especially native sulfur or that found free of other minerals.

BRINES, OILFIELD
Saltwater produced with oil and gas which is brinier than sea water. Oilfield brines are not connate water but bottom or edge water.

BRING BOTTOMS UP
To wash rock cuttings from the bottom of the hole to the surface by maintaining circulation after halting the drilling operation. This allows

time for the closer inspection of the cuttings, and for a decision as to how to proceed when encountering a certain formation.

BRISTLE PIG

Bristle or foam pig

A type of pipeline pig or scraper made of tough plastic covered with flame-hardened steel bristles. Bristle or foam pigs are easy to run, do not get hung up in the line, and are easy to "catch." They are usually run in newly constructed lines to remove rust and mill scale.

BRITISH NATIONAL OIL CORPORATION (B.N.O.C.)

The U.K. government agency that "participates" in drilling and production activities in the British sectors of the North Sea with U.S. oil companies and others; the "corporation" through which Britain assumes ownership of the United Kingdom's share of the North Sea oil.

BRITTLE

Refers to rock that breaks or fractures when bent or deformed even slightly, 3–5 percent deformation.

BROKEN SAND

A sandstone containing shaly layers or other sequences mixed with the layers of sandstone; an interrupted sandstone.

BRONC

A new driller promoted from helper; a new toolpusher up from driller; any newly promoted oilfield worker whose performance is still untried.

BROWNSVILLE LOOP

The name given to an arrangement made to qualify Mexican crude oil for the overland exemption to the system of import quotas for oil. Because there was no pipeline linking Mexico's producing areas and U.S. consuming areas, the only economically feasible means of transporting Mexican crude was by tanker from Mexico to East Coast refineries. Such tanker imports would not qualify for the overland exemption. To qualify, oil was shipped by ocean tanker to Brownsville, Texas (where it was treated as being landed in bond), loaded on trucks, hauled across the border back into Mexico, and immediately brought over the border into the United States, reloaded on coastal tankers and shipped to the East Coast. The second entry qualified for the overland exemption, whereas the first entry being under bond was not counted under the Mandatory Program. In 1971, the Brownsville Loop was converted into what amounted to a country-of-origin quota for Mexico. Synonyms: Brownsville U-Turn and El Loophole.

BRUCKER SURVIVAL CAPSULE

Brucker survival capsule

A patented, self-contained survival vessel that can be lowered from an offshore drilling platform or semisubmersible in the event of a fire or other emergency. The vessel, of spheroid shape, is self-propelled and is equipped with first-aid and life-support systems. Some models can accommodate 26 persons. *See* Whittaker System.

BRUSH HOG
A heavy-duty, power-driven brush cutter for clearing rights-of-way of bushes, small saplings, and brush.

B.S.&W.
Short for basic sediment and water often found in crude oil.

B.S.D.
Barrels per stream day. *See* Stream Day.

B.T.U.
British thermal unit (Btu); the amount of heat required to raise one pound of water one degree Fahrenheit.

B.T.U. TAX
A proposed tax based on the Btu or heat content of the particular fuel, oil, gas or coal; even a load of wood.

B.T.X.
Benzene-toluene-xylene; basic aromatics used in the manufacture of paints, synthetic rubber, agricultural chemicals, and chemical intermediates. The initials are used by refinery men in designating a unit of the refinery.

BUBBLE-CAP TRAYS
Shelves or horizontal baffles inside a fractionating tower or column that are perforated to allow the fluid charge to run down to the bottom of the column and the vapors to rise through the trays to the top where they are drawn off. The perforations in the trays are made with small, umbrella-like caps called bubble caps whose purpose is to force the rising vapors to bubble through the several inches of liquid standing on each tray before the vapors move upward to the next tray. The hot vapors bubbling through the liquid keep the liquid charge heated.

BUBBLE-POINT PRESSURE
The pressure at which gas, held in solution in crude oil, breaks out of solution as free gas; saturation pressure.

BUBBLE-POINT PUMP
See Pump, Bubble-Point.

BUBBLE TOWER
Any of the tall cylindrical towers at an oil refinery. *See* Fractionator.

Bubble tower

BUCKING THE TONGS
Working in a pipeline gang laying screw pipe; hitting the hooks.

BUCK UP
To tighten pipe joints with a wrench.

BUG, WELDING
See Welding Bug.

BUG BLOWERS
Large fans used on or near the floor of the drilling rig to keep mosquitoes and other flying insects off the rig crew.

BUGS
Bugs, like gremlins, are hard to pin down or define, but every engineer and plant operator knows there are such things waiting to foul up the best-laid plans. They cause plant startup mischief: furnaces flame out, air tanks spring leaks, switches will not open or close at the proper time, engines get out of balance and vibrate, fail-safe devices just fail, and bearings get hot. It takes a while to get the bugs out, and the gremlins run off.

BUILD RATES
The rate of deviation from the vertical of a high-angle or horizontal well bore. Building an angle of 90° from vertical begins at kick-off point and proceeds to build at varying rates depending on the radii, but a medium-radius profile build rate of 8° to 20° per 100 feet is within practical operating limits.

BUILDUP RATE
Refers to angle building, from vertical to high angle or horizontal; 40° in 100 feet is achievable in advanced, steerable, mud-motor drilling.

BULK PLANT
A distribution point for petroleum products. A bulk plant usually has tank-car unloading facilities and warehousing for products sold in packages or in barrels.

BULLDOGGED
Said of a fishing tool lowered into the well bore that has latched onto lost pipe or another object being fished out and, owing to a malfunction of the tool, won't unlatch or cannot be disengaged.

BULLET TANK
Colloquial term for horizontal pressure tanks made in the shape of a very fat bullet. Bullet tanks are for storing gasoline or butane under pressure. Other liquefied petroleum gases (L.P.G.) with higher vapor

Bullet tank

pressures are stored in Hortonspheres or spheroids that can withstand higher pressures per square inch.

BULL GANG
Common laborers who do the ditching and other heavy work on a pipeline construction job.

BULL GEAR
The large-diameter circular gear in the mud pump that meshes with and is driven by the pinion gear on the drive shaft; a ring gear.

BULLHEAD THE BUBBLE BACK, TO
A colloquial expression referring to the procedure of controlling a gas kick in which the gas "bubble" or other formation fluid threatens to blow out the well or produce a blowout. First, the kick is contained or controlled. Then by the weight of drilling mud pumped into the borehole, the bubble is forced back into the formation from where it came. This is a complicated procedure. It is mentioned here only to indicate that such feats are possible and that the technology of drilling and well control is as exacting and effective as the technology of space exploration.

BULL NOSE
A screw-end pipeline plug; a pipeline fitting, one end of which is closed and tapered to resemble a bull's nose; a nipple-like fitting, one end threaded, the other end closed.

BULL PLUG
A short, tapered pipefitting used to plug the open end of a pipe or throat of a valve.

BULL WAGON
A casing wagon.

BULL WHEELS
On a cable-tool rig, the large wheels and axle located on one side of the derrick floor used to hold the drilling line. *See also* Calf Wheel.

BUMPER SUB
(1) A slip joint that is part of the string of drillpipe used in drilling from a drillship to absorb the vertical motion of the ship caused by wave action. The slip joint is inserted above the heavy drill collars in order to maintain the weight of the collars on the drill bit as the drillpipe above the slip joint moves up and down with the motion of the ship. (2) A hydraulically actuated tool installed in the fishing string above the fishing tool to produce a jarring action. When the fishing tool has a firm hold on the lost drillpipe or tubing, which may also be stuck fast in the hole, the bumper sub imparts a jarring action to help free the "fish."

BUMP OFF A WELL
To disconnect a rod-line well from a central power unit.

BUNDLE
As used in the oil patch, a grouping of parallel lines, wires, or tubes reaching from the point of operation to the point of control. The reasons for bundles are convenience, safety, and efficiency. An example of a bundle is the hundreds of telephone wires wrapped together beneath the streets of larger cities. Another is the tube bundle in a heat exchanger or the control line bundle to subsea producing wells from the production platform.

BUNKER "C" FUEL OIL
A heavy, residual fuel oil used in ships' boilers and large heating and generating plants.

BUNKERING
To supply fuel to vessels for use in the ships' boilers; the loading of bunker fuel onboard ship for use by the ship's boilers.

BUNKHOUSE
Crew quarters; usually a portable building used on remote well locations to house the drilling crew and for supplies; quarters for single oilfield workers in the days when transportation to a nearby town was primitive or unavailable .

BURNER
A device for the efficient combustion of a mixture of fuel and air. *See* Ultrasonic Atomizer.

BURNING POINT
The lowest temperature at which a volatile oil in an open vessel will continue to burn when ignited by a flame. This temperature determines the degree of safety with which kerosene and other illuminants may be used.

BURN PIT
An excavation in which waste oil and other material are burned.

BURST TEST, LINE PIPE
Tests made on samples of pipe of different specifications: quality of steel, size, and wall thickness. The rupturing agent, under controlled hydraulic pressure, is water. The compressibility of water is negligible so when the test section of pipe gives way under the rupturing pressure, there is no great destructive, explosive effect on the test equipment and observers, as would be the case if air were used. *See* Propagating Fractures.

BURTON, WILLIAM M.
The petroleum chemist who developed the first profitable means of cracking low-value middle distillates into lighter fractions (gasolines) by the use of heat and pressure.

BUSHING
(1) A type of bearing in which a small shaft or spindle turns; a bronze bushing is an insert pressed into a mating piece making a bearing surface. (2) An insert to reduce the size of a hole or to carry or guide an item of equipment. *See* Kelly Bushing.

BUSHING, KELLY
See Kelly Bushing.

BUTANE
A hydrocarbon fraction; at ordinary atmospheric conditions, butane is a gas but it is easily liquefied; one of the most useful L.P.-gases; widely used household fuel.

BUTANE SPLITTER
A type of fractionator vessel at a gas reformer plant that produces commercial propane as well as normal and isobutanes. Splitters are fired with natural gas to provide heat for the distillation.

BUTTERFLY VALVE
See Valve, Butterfly.

BUTTON BIT
See Bit, Button.

BUTT-WELDED PIPE
Pipe made from a rectangular sheet of steel that is formed on mandrels. The two edges of the sheet are butted together and welded automatically.

BUY-BACK CRUDE
In foreign countries, buy-back oil is the host government's share of participation crude; it permits the company holding the concession (the producer) to buy back. This occurs when the host government has no market for its share of oil received under the joint-interest or participation agreement. *See* Phase-in Crude; *also* Participation Crude.

BUY DOWN, GAS CONTRACT
Payments made to the seller of gas to reduce the price to be paid for future production and future deliveries, to be taken by the original purchaser under an amended or successor contract. How's that again?

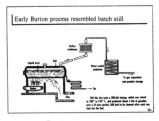

Burton process

Butterfly valve with manual actuator *(Courtesy Fisher)*

BUY OUTS, GAS CONTRACT
Payments made to the supplier to end the purchaser's obligation to take volumes of gas in the future.

BYPASSED OIL
See Oil Behind The Pipe.

BYPASS VALVE
See Valve, Bypass.

CABLE, DRILLING
See Drilling Cable.

CABLE FAIRING
A covering or surface added to moored cables or small piping which reduces the vibration and the strumming from flowing or turbulent water action. One type of cable fairing is a plastic fringe attached to the cable below water level that dampens the microvibrations that cause humming or thumming.

Drilling cable

CABLE-TOOL DRILLING
See Drilling, Cable-Tool.

CABLE TOOLS
The equipment necessary for cable-tool drilling of a well. A heavy metal bar, sharpened to a chisel-like point on the lower end, is attached to a drilling rope or wire line (cable) that is fastened to a walking beam above the rig floor that provides an up-and-down motion to the line and the metal drilling tool. The drilling tool, called a bit, comes in a variety of cutting-edge configurations.

CADASTRAL MAP
A large-scale map showing boundaries of subdivisions of land, drawn to describe and record land ownership. The French *cadastre* is an official register of the real property of a political subdivision, with details of the area, ownership, and value for tax purposes.

C.A.F.E.
The acronym for Corporation Average Fuel Economy. By law, new cars sold each year must meet fuel economy goals established by the E.P.A., Environmental Protection Agency. The average fuel economy is arrived at by offsetting the gas-guzzlers by the smaller, more efficient cars.

CALCAREOUS
Refers to a substance that contains calcium carbonate. When used in naming a rock, it means that as much as 50 percent of the rock is calcium carbonate. Some examples of calcareous rock are limestone, chalk, tufa, and shelly sandstone.

CALCINING

Calcining is the heating of a substance to drive off moisture and other gaseous impurities or to make it more friable or crushable. Petroleum coke is calcined, crushed, and heated to drive off any remaining liquid hydrocarbons and water.

CALCITE

A rock-forming mineral ($CACO_3$). It is usually white or colorless or may occur as pale gray, yellow, or blue. Calcite has rhombohedral cleavage and a hardness of three on the Mohs scale. It is the main constituent of limestone; it is also found crystalline in marble, unconsolidated in chalk, and is the principal mineral ingredient in stalactites found hanging from the roofs of caves.

CALCITIC DOLOMITE

A dolomite containing 10 to 50 percent calcite and 50 to 90 percent dolomite.

CALCULATED ABSOLUTE OPEN FLOW

The measurement of the potential of a flowing gas well. It would be dangerous to permit a very large, high-pressure gas well making 10 to 40 million cubic feet a day to blow uncontrolled. The flow is through a choke and from this figure the open flow is calculated.

CALF WHEELS

The spool or winch located across the derrick floor from the bull wheels on a cable-tool rig. The casing is usually run with the use of the calf wheels, which are powered by the bandwheel. A line from the calf wheels runs to the crown block and down to the rig floor.

CALICHE

A term used in the Southwest, New Mexico and Arizona particularly, for a brownish, buff, or white calcareous material found in layers at or near the surface in arid or semiarid localities. It is made up largely of crusts of soluble calcium salts along with gravel, sand, and clay. It may occur as a thin, friable layer within the soil, but usually it is 6 inches to 3 or 4 feet thick, impermeable, and quite hard. The cementing material i, for the most part calcium carbonate, but may include magnesium carbonate, silica, or gypsum. The term caliche is sometimes used for the calcium carbonate cementing material itself.

CALIPER

A thumb-and-finger adjustable measuring tool calibrated in thousandths (.000) of an inch or centimeter. The caliper jaws, or "fingertips," span the work to be measured: wire, sheet, filament, or an indentation. An essential tool of the engineer, designer, and machinist.

CALIPER (CORROSION) SURVEY

The use of super-sensitive calipers downhole to spot possible internal corrosion in production tubing. Multi-finger, multi-stylus caliper surveys provide perhaps the most effective means to determine the interior wall condition of production tubulars. Corrosion coupons or iron-loss methods are effective for limited areas only. Calipers on a wire line "feel" and record any corroded areas from perforations to wellhead.

CALIPER LOG

See Log, Caliper.

CALIPER PIG

An in-the-line pig that measures the inside diameter of the pipe; locates dents, flat places, and corrosion in the line. The pig, equipped with flexible arms and rollers which are in contact with the inner wall of the pipeline, records all imperfections and variations in diameter. The pig is inserted in the line at a "pig trap" and is pushed along by line pressure.

CALL ON PRODUCTION

Call on production is an option to purchase oil and gas from a proposed (not yet drilled) test well. The option or call is often retained by the farmor (sic) in a farmout agreement or other lease transaction. If the option is limited to crude oil, it is referred to as a call on oil; if gas only, it is a call on gas.

CALM SYSTEM

A catenary, anchor-leg mooring system. Such a system is employed in deep water (360 m) where other types of mooring are not feasible. The six anchor/chain-legs are heavy marine mooring chains or 5- to 6-inch wire rope attached to a tanker-loading buoy, like the legs of a giant arachnid spider. The buoy, supplied with crude oil from subsea wells, controlled through umbilicals, is equipped with a weather-vaning, loading swivel and an emergency disconnect joint. (Catenary describes the inevitable sag or bellying-down of a chain, cable, or rope stretched between two points.)

CAMP, COMPANY

A small community of oilfield workers; a settlement of oil-company employees living on a lease in company housing. In the early days, oil companies furnished housing, lights, gas, and water free or at a nominal charge to employees working on the lease and at nearby company installations—pumping stations, gasoline plants, tank farms, loading racks, etc. Camps were known by company lease or simply the lease name (e.g., Gulf Wolf Camp, Carter Camp, and Tom Butler).

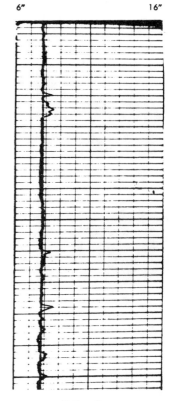

Caliper log

Canning line

CANNING LINE
A facility at a refinery where cans are filled with lubricating oil, sealed, and put in cases. Modern canning lines are fully automated.

C.A.O.F.
Calculated Absolute Open Flow.

CAP BEAD
The final bead or course of metal laid on a pipeline weld. The cap bead goes on top of the hot passes or filler beads to finish the weld.

CAPILLARITY
The action by which a fluid is drawn up into small tubes or interstices as a result of the fluid's surface tension. Capillary action.

CAPILLARY ATTRACTION
The attraction of the surface of a liquid to the surface of a solid. Capillary attraction or capillarity adversely affects the recovery of crude oil from a porous formation because a portion of the oil clings to the surface of each pore in the rock. Flooding the formation with certain chemicals reduces the capillary attraction, the surface tension, permitting the oil to drain out of the pores of the rock. *See* Tertiary Recovery.

CAPITAL ASSETS
Assets acquired for investment and not for sale, and requiring no personal services or management duties. In federal income tax law, oil and gas leases are, ordinarily, property used in the taxpayer's trade or business and are not capital assets. Royalty, if held for investment, is usually considered a capital asset.

CAPITAL EXPENDITURES
Nondeductible expenditures that must be recovered through depletion or depreciation. In the oil industry, these items illustrate expenditures that must be capitalized: geophysical and geologic costs, well equipment, and lease bonuses paid by lessee.

CAPITAL-GAP DILEMMA
The growing disproportion of capital investment to oil reserves discovered; the increasing need for investment capital coupled with diminishing results in terms of oil and gas discovered; spending more money to find less oil.

CAPITAL-INTENSIVE INDUSTRY
Said of the oil industry because of the great amounts of investment capital required to search for and establish petroleum reserves.

CAPITAL STRING
Another name for the production string.

CAPPING
Closing in a well to prevent the escape of gas or oil.

CAP ROCK
A hard, impervious formation forming a cap over permeable layers of sedimentary rock. This prevents the further upward migration of oil and gas and results in a stratigraphic trap.

CAP SCREW
A bolt made with an integral, hexagonal head; cap screws are commonly used to fasten water jackets and other auxiliary pieces to an engine or pump and have slightly pointed ends, below the threads, to aid in getting the "screw" into the tapped hole and started straight.

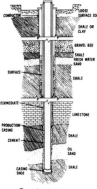

Capital string

CAPTURED BOLT
A bolt held in place by a fixed nut or threaded piece. The bolt can be tightened or loosened, but cannot be removed completely because of a shoulder at the end of the bolt. Captured bolts are in reality a part of an adjustable piece and are so made to preclude the chance of being removed and dropped, or because of limited space and accessibility in an item of equipment.

CARBONATE ROCK
Limestone, dolomite, and chalk are the principal rocks that are examples of limey deposits, rich in calcium carbonate.

CARBONATE ROCK, AUTHIGENIC
In deep-water, Gulf of Mexico macroseeps, bacterial oxidation of oil and gas produces carbon dioxide which precipitates as authigenic (formed on the spot, at that location) carbonate rocks.

CARBON BLACK
A fine, bulky carbon obtained as soot by burning natural gas in large horizontal "ovens" with insufficient air.

CARBON CLOCK
See Carbon Dating.

CARBON DATING or CARBON-14 DATING
A method of determining an age, in years, of organic material by measuring the concentration of carbon-14 remaining in the material—formerly living matter or tissue. Carbon dating, worked out in 1946–1951 by the U.S. Chemist W.F. Libby, is based on the assumption

that assimilation of carbon-14 ceased upon the death of the organism, and that it remained a closed system thereafter. *Also* Carbon Clock.

CARBON DIOXIDE (CO₂)

A heavy, colorless gas found in great quantities in certain wells in several states. It is used extensively in secondary oil-recovery projects and for repressuring a depleted gas-cap reservoir. A weak dibasic acid is formed by dissolving carbon dioxide in water (H_2CO_3).

CARBON PLANT

A plant for the production of carbon black by burning natural gas in the absence of sufficient air. Carbon plants are located close to a source of gas and in more-or-less isolated sections of the country because of the heavy emission of smoke.

CARRIED INTEREST

A fractional interest in an oil or gas property, most often a lease, the holder of which has no obligation for operating costs. These are paid by the owner or owners of the remaining fraction who reimburse themselves out of profits from production. The person paying the costs is the carrying party; the other person is the carried party.

CARRIED-INTEREST PAYOUT

See Payout, Carried-Interest.

CARRIED WORKING INTEREST

A fractional interest in an oil and gas property conveyed or assigned to another party by the operator or owner of the working interest. In its simplest form, a carried working interest is exempt from all costs of development and operation of the property. However, the carried interest may specify "to casing point," "to setting of tanks," or "through well completion." If the arrangement specifies through well completion, then the carried interest may assume the equivalent fractional interest of operating costs upon completion of the well. There are many different types of carried interests, the details varying considerably from arrangement to arrangement. One authority has observed, "The numerous forms this interest is given from time to time make it apparent the term 'carried interest' does not define any specific form of agreement but serves only as a guide in preparing and interpreting instruments."

CARRIER BAR

A yoke or clamp fastened to the pumping well's polished rod and to which the bridle of the pumping unit's horsehead is attached. *See* Bridle.

CARVED-OUT INTEREST

An interest; an oil payment or overriding royalty conveyed to another party by the owner of a larger interest, i.e., a working interest. The

Carrier bar fastened to bridle

owner of the working interest in a producing property may grant an oil payment to a bank to pay off a loan. For other considerations, the owner of the larger interest may convey an overriding royalty—one-sixteenth, for example —which he has "sliced off" or carved out of his interest.

CASH BALANCING (GAS CONTRACT)
In gas-contract terminology, the right of an underproduced party to be paid in cash in lieu of make-up gas, when depletion of the reservoir makes balancing by delivering make-up gas impossible. Cash balancing agreements usually set forth the guiding conditions.

CASH BONUS
See Bonus.

CASING
Steel pipe used in oil wells to seal off fluids from the borehole and to prevent the walls of the hole from sloughing off or caving. There may be several strings of casing in a well, one inside the other. *See* Production String; *also* Surface Casing.

CASING, COLLAPSED
This condition is defined as casing with enough deformation that it will not permit the passage of necessary tools up or down the hole. Casing in the hole need not be crushed flat to be considered collapsed; it need only be sufficiently out of round or dented to prevent the use of downhole tools. *See* Scab Liner.

CASING, CONDUCTOR
A well's surface pipe used to seal off near-surface water, to prevent the caving or sloughing of the walls of the hole, and to conduct drilling mud through loose, unconsolidated shallow layers of sand, clays, and shales. *See* Casing.

CASING, HIGH-COLLAPSE
Special well casing with wall thicknesses of 0.250 inch (1/4 in.) to 0.689 inch (almost 3/4 in.) for downhole pressures to 15,000 psi and temperatures to 500°F. It is also run or set in a well where there is the danger of unstable formations, such as shale that swells and "flows" into the hole, collapsing ordinary casing.

CASING, PUMP
See Pump, Casing.

CASING, SHALLOW-WELL
Small-diameter casing of lighter weight than conventional casing used in deep wells. The lighter-weight casing is less costly, easier to handle, and adequate for certain kinds of shallow, low-pressure wells.

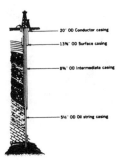

Casing

Casing spool with attached valve

CASING, SPIRAL-WALLED

Well casing made with spiral grooves on the outer circumference of the pipe. The purpose of the patented axial grooves is to aid in running casing or a liner in deviated or crooked holes. The pipe with its grooves, like screw threads, is said to be less susceptible to wall or differential sticking.

CASING & TUBING INTERNAL PATCH

A corrugated metal liner run inside a leaking joint or corroded tubular. After spotting (being put in place), the corrugated liner is subjected to pressure which causes it to assume a cylindrical shape, pressing against and covering the affected area in the casing or tubing.

CASING AND TUBING ROLLERS

See Rollers, Casing and Tubing.

CASING AND TUBING SPOOL

See Spool, Casing and Tubing.

CASING-COLLAR LOG

See Log, Casing-Collar.

Casinghead

CASINGHEAD

The top of the casing set in a well; the part of the casing that protrudes above the surface and to which the control valves and flow pipes are attached.

CASINGHEAD FITTING

A heavy, threaded casting screwed to the well's casing at ground level. The heavy-duty fitting supports control valves and Christmas tree connections. To the wellhead fitting are screwed blowout preventers, Christmas tree valves, and take off connections.

CASINGHEAD GAS

Gas produced with oil from an oil well as distinguished from gas from a gas well. The casinghead gas is taken off at the top of the well or at the separator.

CASINGHEAD GASOLINE

Liquid hydrocarbons separated from casinghead gas by the reduction of pressure at the wellhead or by a separator or absorption plant. Casinghead gasoline, or natural gasoline, is a highly volatile, water-white liquid.

CASING JACK

A long-stroke hydraulic jacking device for lowering in a part of a very heavy string of casing, e.g., a load too much for the rig's rated capacity of 900,000 lb. hook load. The casing jack, using hydraulic power, can lower in a joint at a time the last 50–70 joints of casing which, if handled by the traveling block and hook, would overload the derrick. Casing jacks are also used to pull casing out of the hole when plugging and abandoning a well. *See* Floating the Casing.

CASING PATCHES, INTERNAL

A patented corrugated steel liner run inside the casing to shut off leaks at pipe joints, spot corrosion holes, and to blank off previously made perforations. The small-interval corrugations, when in place, flatten out under pressure, making a close-fitting liner and a satisfactory seal.

CASING POINT

A term that designates a time when a decision must be made whether casing is to be run and set or the well abandoned and plugged. In a joint operating agreement, casing point refers to the time when a well has been drilled to objective depth, tests made, and the operator notifies the drilling parties of his recommendation with respect to setting casing and a production string and completing the well. On a marginal well, the decision to set pipe is often difficult. To case a well often costs as much as the drilling. On a very good well there is no hesitation; the operators are glad to run casing and complete the well.

CASING PRESSURE

Pressure between the casing and the well's tubing.

CASING PROGRAMS

Informational outlines as to the proposed depth of the well, the type, grade, size of casing strings from surface pipe to long string or production string. Sometimes, owing to unforeseen circumstances, a casing program must be altered. If the well and casing specs are set down, deviations in plans can be made in an orderly fashion. Exact measurements are essential because of the downhole tools that are needed in drilling and completing a well. A driller must know where all of his pipe is, tops and bottoms of each size and string. He enters this information in his log. *See* Log, Driller's.

CASING PULLER

See Casing Jack.

CASING SCAB

A length of casing, 50 to 100 feet, which is cemented in the well bore, the open hole, across a problem zone: water incursion, sloughing walls, or a desaturation section, a thief zone.

CASING SHOE

A reinforcing collar of steel screwed onto the bottom joint of casing to prevent abrasion or distortion of the casing as it forces its way past an

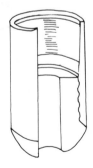

Casing shoe

obstruction on the wall of the borehole. Casing shoes are about an inch thick and 10 to 16 inches long and are an inch or so larger in diameter in order to clear a path for the casing.

CASING SPOOL
A heavy, flanged steel fitting attached to the first string of casing, the surface casing, or surface pipe. The casing spool or casinghead provides a housing for the slips and packing assemblies, also permits other strings of casing to be suspended in the hole. A casing spool also provides a means to seal off the annulus gas tight; a casinghead.

CASING WAGON
A small, low cart for moving casing from the pipe rack to the derrick floor. Two wagons are used. The forward wagon holds the pipe in a V-shaped cradle; the rear wagon is, in reality, a lever on wheels that raises the end of the casing so it is free to be pulled.

CAT
Short for Caterpillar tractor, a crawler-type tractor that moves on metal tracks made in segments and connected with pins to form an "endless" tread; a skilled and experienced pipeliner.

CATAGENESIS
The process by which the organic matter in inorganic sediments is altered or changed by increasing pressure and temperature; the thermal generation of hydrocarbons (oil and gas) from organic matter of sedimentary rock.

CATALYSIS
The increase or speeding up of a chemical reaction caused by a substance, a catalyst, that remains chemically unchanged at the end of the reaction. Any reaction brought about by a separate agent.

CATALYST
A substance that hastens or retards a chemical reaction without undergoing a chemical change itself during the process.

CATALYTIC AGENT
A catalyst.

Cat cracker *(Courtesy Phillips Petroleum Co.)*

CAT CRACKER
A large refinery vessel for processing reduced crude oil, naphthas, or other intermediates in the presence of a catalyst. *See* Fluid Catalytic Cracking Unit.

CATENARY
Describes the inevitable sag or bellying-down of a working chain, cable, or rope stretched more or less horizontally between two points.

CATHEAD
A spool-shaped hub on a winch shaft around which a rope is wound for pulling and hoisting; a power-take-off spool used by the driller as he operates the cat line.

Cathead

CATHEAD, BREAKOUT
The power-take-off spool on the catshaft of the draw works, used by the driller to breakout or break loose joints of drillpipe or tubing. The lead (leed) pipe tongs, suspended horizontally in the derrick or on the mast, are operated by a length of wire line connected to a length of hemp rope which is wound around the polished cathead. To operate the heavy tongs after they are latched onto the pipe, the driller tightens the rope on the cathead, which pulls the wrench, applying torque to the pipe joint.

CATHODIC PROTECTION
An anticorrosion technique for metal installations—pipelines, tanks, buildings—in which weak electric currents are set up to offset the current associated with metal corrosion. Carbon or nonferrous anodes buried near the pipeline are connected to the pipe. Current flowing from the corroding anode to the metal installation controls the corrosion of the installation .

CATION
A positively charged ion.

CAT LINE
A hoisting or pulling rope operated from a cathead. On a drilling rig, the rope used by the driller to exert a pull on pipe tongs in tightening (making up) or loosening (breaking out) joints of pipe.

CAT SHAFT
The shaft on the drawworks on which the catheads are mounted. One cathead is a drum, and by using a large rope wrapped around it a few turns, the drilling crew can do such jobs as makeup and breakout and light hoisting. The other end of the cat shaft has a manual or air-actuated, quick-release friction clutch and drum to which the tong jerk line or spinning chain is attached.

CATSTILL
A short, trenchant, vernacular word for a fluid catalytic cracking unit.

CATTLE GUARD
A ground-level, trestle-like crossing placed at an opening in a pasture fence to prevent cattle from getting out while permitting vehicles to cross over the metal or wooden open framework.

Cattle Guard

Catwalk

CATWALK

A raised, narrow walkway between tanks or other installations.

CATWORKS

The drawworks of a drilling or workover derrick; specifically the cat-shaft with its polished steel drum-like pulley, the power-take-off (the cathead) on one end and a quick-action clutch on the other. *See* Catshaft.

CAVEY FORMATION

A formation that tends to cave or slough into the well's borehole. In the parlance of cable-tool drillers, "the hole doesn't stand up."

CAVINGS

Dislodged rock fragments that fall into the well bore, thus contaminating the well cuttings or blocking the hole. Cavings can be serious enough to fill a section of the hole and stick the drillstring. *See* Bridgeover.

CAVITATION

The creation of a partial vacuum or a cavity by a high-speed impeller blade or boat propeller moving in or through a liquid. Cavitation is also caused by a suction pump drawing in liquid where there is an insufficient suction or hydrostatic head to keep the line supplied.

C.D.

Contract depth; the depth of a well called for or specified in the drilling contract.

C.D.R. FLOW IMPROVER

A patented chemical that reduces drag in pipelines pumping crude oil; it is used downhole in drilling wells to unstick pipe in doglegs and keyseats.

A wellhead cellar

CELLAR

An excavation dug at the drillsite before erecting the derrick to provide working space for the casinghead equipment beneath the derrick floor. Blowout-preventer valves (B.O.P. stack) are also located beneath the derrick floor in the cellar.

CELLAR DECK

Lower deck on a large, double-decked semisubmersible drilling platform.

CELSIUS, ANDERS

The Swiss astronomer who devised a system of temperature readings now used in the scientific community. Water freezes at 0°C.

CEMENT, TO

(1) To fix the casing firmly in the hole with cement, which is pumped through the drillpipe to the bottom of the casing and up into the annu-

lar space between the casing and the walls of the well bore. After the cement sets (hardens), it is drilled out of the casing. The casing is then perforated to allow oil and gas to enter the well. (2) *Sedimentary.* Mineral material, usually precipitated chemically, that fills the spaces between individual grains of a consolidated (hard) sedimentary rock; the binding material that holds the grains together. The most common binders are silica, carbonates, and certain iron oxides. Other cements are clay minerals, barite, gypsum, anhydrite, and pyrite.

Cementers and their equipment *(Courtesy Halliburton)*

CEMENT (CONCRETE)

In oilfield usage, one squeezes a well, sets surface pipe, and blocks gas migration with cement; one builds forms, manifold enclosures, and pump and engine foundations with concrete. The difference is more than semantic and quite important. Only a boll weevil would squeeze a well with concrete.

CEMENT, FOAMED

A patented, low-density cement made by introducing foamed nitrogen (N_2) into a conventional cement slurry resulting in a product as light as six pounds to the gallon. Halliburton Services, the developer, says it can be used for cementing across weak zones, zones full of capillary channels and crevices. In such weak or porous zones, the loss of drilling fluid can cause a loss of circulation.

CEMENT, NEAT

Just cement, no sand or gravel in the mix.

CEMENTATION

Sand grains and rock fragments have to be cemented together in order to form a stone. Cementation adheres individual grains to each other, and very often the cementing agent is quartz or calcite. Porosity is influenced by the degree of cementation—poorly cemented rock has a higher degree of porosity than one that is well cemented.

CEMENT FLOAT VALVE, BALL-TYPE

The ball-type cement float valve has a free ball as a closure element. It has a two or more piece body with a sealing surface in one member and a ball-retention arrangement in the other. The sealing configuration is generally a ball on a cone with an elastomer sealing device. The seal is provided by elastomer coating the ball, molded elastomer bonded to the body (seat), or by making the ball or body with elastomeric properties. The ball is held in the open position by gravity. Valve closure occurs when reverse flow is attempted and the ball is lifted into the valve seat shutting off return flow.

CEMENT FLOAT VALVE, FLAPPER-TYPE

The flapper (swing) check valve consists of a closure disc (flapper), with an integral hinge attached to the body/seat by a pivot. The attachment

Cement slurry sampling

Cement truck on a
squeeze job *(Courtesy
Dowell)*

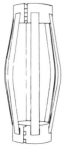

Centralizer

is such that the spring-loaded flapper is free to swing through an arc of 80° to 90°, making way for the flow of cement. The valve seat may have a rubber coating which makes for a leak-proof closure.

CEMENT FLOAT VALVE, PLUNGER-TYPE
The plunger-type float valve has a conical closure with a guide shaft in a multi-piece steel body. One body or frame piece contains the sealing surface, the seat; the other contains the plunger, which is "spring-biased closed," engineering jargon for spring-loaded. The valve is opened by the force of the down-flowing cement pumped from the surface through the casing. When pumping ceases, a compression spring closes the valve and prevents backflow. When the cement in the annulus sets, the small amount of cement remaining in the lower few feet of the casing, plus the valve itself, is drilled out.

CEMENTING, STAB-IN
See Stab-in Cementing.

CEMENTING BARGE
An unmanned, shallow draft barge for transporting dry cement, additives, gels, and weighting material with onboard facilities for mixing and blending cement slurries. The towed barge is used in cementing offshore wells.

CEMENT ROCK
See Argillaceous Limestone.

CEMENT SLURRY
A cement slurry is dry cement mixed with water to form a pumpable solution. However, a cement slurry is not a true solution, as is a cup of tea in which a spoonful of sugar is dissolved. The cement particles do not dissolve, but are suspended in the water. When the slurry is pumped downhole under great pressure, the liquid phase (the water) is forced into the rock matrix while the solids (the cement particles) bridge on the surface of the rock. This dehydration or loss of the liquid phase causes the slurry to become unpumpable, and is referred to as a flash set. *See* Flash Set.

CEMENT SQUEEZE
See Squeezing A Well.

CENTRALIZERS, CEMENTING
Cylindrical, cage-like devices fitted to a well's casing as it is run to keep the pipe centered in the borehole. Cementing centralizers are made with two bands that fit the pipe tightly with spring steel ribs that arch out to press against the wall of the borehole. By keeping the pipe centered, a more uniform cementing job is assured. Centralizers are especially useful in deep or deviated holes.

CENTRAL POWER
A well-pumping installation consisting of an engine powering a large-diameter, horizontal bandwheel with shackle-rod lines attached to its circumference. The bandwheel is an eccentric, and as it revolves on a vertical axle, a reciprocating motion is imparted to the shackle rods. A central power may pump from 10 to 25 wells on a lease.

CENTRIFUGAL PUMP
See Pump, Centrifugal.

CENTRIFUGE
A motor-driven machine in which samples of oil or other liquids are rotated at high speed, causing suspended material to be forced to the bottom of a graduated sample tube so that the percent of impurities or foreign matter may be observed. Some centrifuges are hand-operated. *See* Shake Out.

Single-stage centrifugal pump

CENTRIFUGE, DECANTING
A large centrifuge machine for separating or removing pulverized rock and fines from drilling mud returning from downhole. A decanting centrifuge located between the rig and mud pits removes the fine particles of rock from the mud by centrifugal action and discharges the clean mud to the working pits.

CERUSSITE
A derivative of galena (PbS), an important lead ore.

CESSATION OF PRODUCTION
The termination of production from a well. It may be owing to mechanical breakdown, reworking operations, governmental orders, or depletion of oil or gas. Temporary cessation usually does not affect the lease, but a permanent shutdown terminates the ordinary oil and gas lease.

Decanting centrifuge
(Courtesy Pioneer)

CETANE NUMBER
A measure of the ignition quality of diesel fuel. The cetane number of diesel fuel corresponds to the percentage of cetane (C16 H34) in a mixture of cetane and alpha-methyl naphthalene. When this mixture has the same ignition characteristic in a test engine as the diesel fuel, the diesel fuel has a cetane number equal to the percentage of cetane in the mixture. Regular diesel is 40-45 cetane; premium is 45-50.

C.F.M.
Cubic feet per minute.

CH₃ (CH₂)₅ CH₃
Heptane, a liquid hydrocarbon of the paraffin series.

$CH_3(CH_2)_4 CH_3$
Hexane, a hydrocarbon fraction of the paraffin series.

CHAIN TONGS
A pipe wrench with a flexible chain to hold the toothed wrench head in contact with the pipe. The jointed chain can be looped around pipes of different diameters and made fast in dogs on the wrench head.

Chain wheels

CHAIN WHEEL
Some gate valves are operated from a distance either for safety or convenience. Such valves have a gate wheel made to accept a chain in the wheel's outer circumference. The chain is reeved or passed over a drum or windlass that the operator turns to open or close the valve from a distance.

CHALCEDONY
A variety of quartz referred to as cryptocrystalline, i.e., with a crystalline structure too small to be seen with a microscope. It may be translucent or semitransparent and has a waxy luster. It occurs as black, brown, gray, or pale blue stone. Among gemologists, the name chalcedony refers to the common variety, which is blue-gray.

CHALK
A soft, fine-textured limestone of marine origin consisting almost entirely of calcite. Chalk is porous, friable, and barely coherent. Its colors are white (predominantly), light gray, or beige; chalk rock.

CHANNEL
A "vacation" or void in a cement squeeze job allowing salt water or other fluid into the production zone or another interval in the annular space. Also, in waterflooding, a natural void or "path" in a formation permitting the injection fluid to break through to a producing well from the injection well subverting the waterflooding project. *See* Squeezing a Well.

CHANNELING
A condition that arises in oil production when water bypasses the oil in the formation and enters the well bore through fissures or fractures. There are two general types of channeling: (1) coning off, in which a small amount of oil rides on top of the encroaching water; and (2) bypassing, where water breaks through to the well bore through fractures or more permeable streaks or sections of the formation, leaving the oil behind.

CHAR
A combustible residue remaining after the destructive distillation of coal; charcoal.

CHARCOAL TEST
A test to determine the gasoline content of natural gas.

CHARGE STOCK
Oil that is to be "charged" or treated in a particular refinery unit.

CHART, STRIP
See Strip Chart.

CHASE THREADS
To straighten and clean threads of any type.

CHATTER
A noisy indication that a mechanical part is behaving erratically and destructively.

CHEATER
A length of pipe used to increase the leverage of a wrench; anything used to lengthen a handle to increase the applied leverage.

CHECKERBOARD LEASING
The acquisition of mineral rights (oil and gas) in a checkerboard pattern. A company may be forced to lease land over a wide area before it has completed geological and geophysical studies. Leases then may be taken on one-quarter section (160 acres) in each section of land.

CHECK VALVE
See Valve, Check.

CHECK VALVE, PISTON
A check valve that does not have the usual clapper and seat arrangement; instead, the backflow in the line is controlled or stopped, by a sliding piston-and-sleeve valve mechanism. Piston-and-sleeve checks are found mainly on relatively small-diameter piping.

CHEESE BOX STILL
An early-day, square, box-like refining vessel; a still to heat crude oil for distilling the products in those days—kerosene, gas oil, and lubricating oil.

CHEMICAL FEEDER PUMP
See Pump, Chemical Feeder.

CHEMICAL INJECTION PUMP
See Pump, Chemical Injection.

CHEMICAL WEATHERING
See Weathering.

CHEMOSYNTHETIC ORGANISMS
Organisms that thrive in the deep cold water of the Gulf of Mexico and other areas where hydrocarbon-based bacterial activity creates a favorable environment.

A feeder pump (center foreground) feeding chemicals to heater treater

Manual production choke

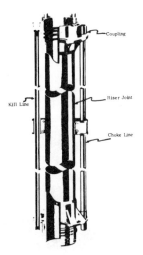

Kill-and-choke lines attached to marine riser

Christmas tree

CHERT
A hard, extremely dense microcrystalline sedimentary rock consisting mainly of interlocking crystals of quartz. Its fracture pattern is splintery to conchoidal, and it occurs in a variety of colors: white, gray, green, blue, pink, brown, and black. The dark variety is commonly referred to as flint.

CHILEAN NATIONAL OIL COMPANY
Empressa Nacional de Petroleo (E.N.A.P.).

CHILLERS
Refinery apparatus in which the temperature at which paraffin distillates is lowered preparatory to filtering out the solid wax.

CHOCK
A wedge or block to prevent a vehicle or other movable object from shifting position; a chunk.

CHOKE
A heavy steel nipple inserted into the production tubing that closes off the flow of oil except through an orifice in the nipple. Chokes are of various sizes. It is customary to refer to the production of a well as so many barrels through (or on) a (e.g.) 22/64th-inch choke. *See* Bottom-Hole Choke; *also* Production Choke; Wellhead Choke.

CHOKE-AND-KILL LINES
See Kill-and-Choke Lines.

CHRISTMAS TREE
(1) An assembly of valves mounted on the casinghead through which a well is produced. The Christmas tree also contains valves for testing the well and for shutting it in if necessary. (2) A subsea production system similar to a conventional land tree except it is assembled complete for remote installation on the seafloor with or without diver assistance. The marine tree is installed from the drilling platform; it is lowered into position on guide cables anchored to foundation legs implanted in the ocean floor. The tree is then latched mechanically or hydraulically to the casinghead by remote control.

CHURN DRILLING
Another name for cable-tool drilling because of the up-and-down, churning motion of the drill bit.

C.I.D.
Cubic inch displacement.

C.I. PLUG
A cast-iron plug; a flat plug used to close the end of a pipe or a valve.

CIRCLE JACK
A device used on the floor of a cable-tool rig to make up and break out (tighten and loosen) joints of drilling tools, casing, or tubing; a jacking device operated on a toothed or notched metal, circular track placed around the pipe joint protruding from the borehole above the floor. The jack is operated manually with a handle and is connected to a wrench that tightens the pipe joint as the jack is advanced, notch by notch.

CIRCULATE
To pump drilling fluid into the borehole through the drillpipe and back up the annulus between the pipe and the wall of the hole; to cease drilling but to maintain circulation for any reason. When closer inspection of the formation rock just encountered is desired, drilling is halted as circulation is continued to "bring bottoms up."

CIRCULATE & WEIGHT
A method of killing or controlling well pressure caused by a gas kick. Circulation of drilling fluids is started at once, or as soon as the condition is recognized, and mud weight is brought up to a point that the unwanted well pressure is brought under control. *See also* Wait & Weight.

CIRCULATED BOTTOMS UP
To pump cement or other fluid to the bottom of the borehole through the casing and up between the casing and the wall of the borehole, the annulus. The conventional pumping pattern in cementing casing or performing a squeeze job to shut off encroaching water or gas.

CIRCULATION
The round trip made by drilling mud: down through the drillpipe and up on the outside of the drillpipe, between the pipe and the walls of the borehole. If circulation is lost, the flow out of the well is less than the flow into the well, the mud may be escaping into some porous formation or a cavity downhole. *See* Lose Returns.

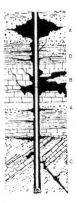

Lost circulation

C.I.S.
Commonwealth of Independent States; the states of the former Soviet Union.

CITY GATE
The measuring point at which a gas distributing utility receives gas from a gas transmission company.

CIVIL CROSSING
The crossing over or boring under streams, roads, cemeteries, and municipal water-supply lakes.

CLADDING

Coating one material with another; to cover one metal with another by bonding the two. In the oil patch, cladding or trimming steel pipe and valves with corrosion-resistant metal alloy is necessary when H_2S (hydrogen sulfide) and other corrosive gases must be handled. Cladding of pipe, valves, and fittings is less costly than making them of expensive, anticorrosion alloys. *See* Sour-Service Trim.

Pipeline lineup clamp

CLAMP, PIPELINE LINEUP

See Lineup Clamp.

CLAMP, RIVER

See River Clamp.

CLAMSHELL BUCKET

A hinged, jaw-like digging implement suspended at the end of a cable running down from the boom of an excavating machine. A drag-line bucket.

CLAPPER

The internal moving part, the "tongue," of a check valve that permits a liquid or gas to flow in one direction only in a pipeline. Like a trap door, the check-valve clapper works on a hinge attached to the body of the valve. When at rest, the clapper is a few degrees off the vertical or, as in certain valves, completely horizontal.

CLASTIC ROCK

Sandstone and conglomerate are two examples of clastic rocks, which require preexisting rocks for the source of the particles—grains, pebbles, minute fragments—of which they are composed. Clastic is from the Greek *klastos,* broken. So clastic rock is made from broken, eroded, or fragmented older rock cemented together to form younger rock. Siltstone and shale are also clastic rocks, but their deposition and formation is different from that of the prime examples, sandstone and conglomerate. *See also* Nonclastic Rock.

CLAUS PROCESS

A process for the conversion of hydrogen sulfide (H_2S) to plain sulfur developed in 1885 by the German chemist Claus.

CLAY

The filtering medium, especially Fuller's earth, used in refining; a substance that tends to adsorb the coloring materials present in oils that pass through it.

CLAY FORMATIONS, REACTIVE

Clay intervals encountered downhole that are hydrophilic, absorb water readily and swell, sticking the pipe and drill bit. When clay formations are encountered, the drilling fluid is changed from water-base drilling mud to oil-base. Also, there are additives for water-base mud that retard the reaction, the swelling, of the clay.

CLAY PERCOLATOR
Refinery filtering equipment employing a type of clay to remove impurities or to change the color of lubricating oils.

CLAYSTONE
A type of mudstone.

CLEAN CARGO
Refined products—distillates, kerosene, gasoline, jet fuel—carried by tankers, barges, and tank cars; all refined products except bunker fuels and residuals.

CLEAN CIRCULATION
The circulation of drilling mud free of rock cuttings from the bottom of the borehole. This condition may be caused by a worn bit, by circulating to clean the hole, or by a broken or parted drillstring.

CLEAN OIL
Crude oil containing less than one percent basic sediment and water (B.S.&W.); pipeline oil; oil clean enough to be accepted by a pipeline for transmission.

CLEAN-OUT BOX
A square or rectangular opening on the side of a tank or other vessel through which the sediment that has accumulated can be removed. The opening is closed with a sheet of metal (a door) bolted in place.

CLEAN-UP TRIP
Running the drillpipe into the hole for circulation of mud only; to clean the borehole of cuttings.

CLEAR OCTANE
Unleaded gasoline; gasoline without additives of any kind. It is manufactured by careful and precise blending of several components to achieve the right octane and other desirable properties.

CLEAVAGE
A very distinctive property of some minerals. Cleavage is the ability or propensity of a mineral to break or cleave along definite planes that parallel one another. An example of nearly perfect cleavage in one direction is mica. This mineral splits in successively thinner layers until very thin transparent sheets are left. In the early days, thin sheets of mica were used as small windows in the famous Franklin stove. Other

Clevis or shackle *(Courtesy Crosby)*

geometric forms of cleavage are cubic (halite or rock salt); prismatic (hornblende); and conchoidal (smooth, curved surface as in quartz and obsidian).

CLEVIS
A U-shaped metal link or shackle with the ends of the U drilled to hold a pin or bolt; used as a connecting link for a chain or cable.

CLINOFORM STRUCTURE
An underwater land form similar to the continental slope of the oceans or to the foreset beds of a delta.

CLINOFORM SURFACE
An offshore, sloping depositional surface commonly associated with strata prograding, the building outward by river-borne sediment and accumulation, into deeper and deeper water.

CLOGGING, FRACTURE
See Screen-Out, Fracturing.

CLONE
The entry section of a cyclone desander or desilter—the part of the cyclone cylinder that imparts a centrifugal or whirling motion to the fluid to be desanded, to be cleaned up.

CLOSED IN
Refers to a well, capable of producing, which is shut in.

CLOSE NIPPLE
A very short piece of pipe having threads over its entire length; an all-thread nipple.

CLOSE THE LOOP, TO
To contain and dispose of all liquids and solids generated at the drill site; to drill it out and put it back in the same hole closes the loop. The drilled solids and any liquid pollutants are pumped back into the well, into the annulus between casing and the wall of the borehole. A disposal method initially developed by Arco and Apallo Service, Lafayette, La.

CLOUD POINT
The temperature at which paraffin wax begins to crystallize or separate from the solution, imparting a cloudy appearance to the oil as it is chilled under prescribed conditions.

C.N.G.
Compressed Natural Gas—a fuel.

CO_2 INJECTION
A secondary-recovery technique in which carbon dioxide is injected into service wells in a field as part of a miscible recovery program. CO_2 is used in conjunction with waterflooding.

CO_2-SHIELDED WELDING
See Welding, CO_2-Shielded.

COALBED METHANE
A source of natural gas gathered from bituminous coalbeds. Some of the more productive coalbed regions are in the Uinta basin, Utah. Others are in Black Warrior basin, Alabama and in Piceance basin, Colorado.

CO_2-shielded welding

COALESCERS
In oil-water separation technology, a coalescer is equipment which provides a large flat surface area available for contact and wetting by the minute droplets dispersed in oilfield water. Upon contacting this surface, the droplets combine to form larger drops. The larger drops then disengage themselves from the surface; and being large enough to assert their buoyancy, rise to the surface where they can be skimmed or drawn off. Another more costly method for cleaning up oily water before it is discharged is the use of heat and/or chemicals to break down oil-water emulsions. The oil droplets' surface tension is reduced by the action of heat or chemicals or both, causing the droplets to release their hold on the droplet of water. In emulsions what appears as a droplet of oil is, in fact, an ultra thin film of oil surrounding a droplet of water.

COAL GAS
Also referred to as town gas. An artificial gas produced by pyrolysis (heating in the absence of air) of coal. Coal gas has a Btu content of 450 per cubic foot; natural gas, on an average, has 1,030 Btu per cubic foot, more than twice the thermal value of coal gas.

COAL-GAS D.M.E.
Dimethyl ether.

COAL GASIFICATION
A process for producing natural gas from coal. Coal is heated and brought in contact with steam. Hydrogen atoms in the vapor combine with coal's carbon atoms to produce a hydrocarbon product similar to natural gas.

COAL OIL
Kerosene made from distilling crude oil in early-day pot stills; illuminating and heating oil obtained from the destructive distillation of bituminous coal.

COAL-SEAM GAS

See Unconventional Gas.

COATING AND WRAPPING

A field operation in preparing a pipeline to be put in the ditch (lowered in). The line is coated with a tar-like substance and then spiral-wrapped with tough, chemically impregnated paper. Machines that ride the pipe coat and wrap in one continuous operation. Coating and wrapping protects the pipeline from corrosion. For large pipeline construction jobs, the pipe may be coated and wrapped at the mill or in yards set up at central points along the right of way.

Coating and wrapping a
section of pipeline

COAXIAL PIPE

A dual-wall pipe; a pipe within a pipe; a special-service pipeline consisting of a large pipe with a small one inside surrounded by a matrix or grout of cement or plastic resin. Coaxial pipelines are laid from a shore station to an offshore oil or gas well, and it is the pipeline engineer's and corrosion engineer's answer to pipeline integrity in critical locations such as the California coast adjacent to populated areas.

COGENERATION PLANT

A coal- or gas-fired plant that generates both process (commercial) steam and electricity for in-plant use or for sale.

COGEN UNIT

Short for cogeneration unit. *See* Cogeneration Plant.

COHERENT

Said of a deposit or a rock that is consolidated, i.e., firm and not easily broken. A rock that has it all together. An incoherent rock is unconsolidated, not very compact, and breaks apart easily.

COIL CAR (OR TRUCK)

A tank car or transport truck equipped with heating coils in order to handle viscous liquids that will not flow at ordinary temperatures.

COILED TUBING

A reel of continuous, various-sized steel tubing brought to the job site on an oilfield float or specialized trailer. Coiled tubing is used for various well work-over operations: to clean out, desand, acidize, frac, and in some instances as production tubing, the production string. The use of coiled tubing proliferated during the decades of the 1970s and 1980s. Advances in metallurgy have made possible the repeated use of a reel of tubing (unspooling and spooling) without excessive or prohibitive metal fatigue. A large reel of 1 1/2-inch tubing may contain 2 1/2 to 3 miles (13,000 to 15,000 feet) of the flexible pipe.

COILED TUBING, BOTTOM-HOLE ASSEMBLY
In drilling a slim hole, coiled tubing is used. A typical B.H.A. for 1 1/2-inch coiled tubing (bottom-hole assembly): first, a screw connector; a double-flapper check valve; a disconnect sub; 3 1/8-inch drill collar; 3 1/2-inch positive displacement mud motor connected to and driving a 3 7/8-inch P.D.C. bit.

COILED TUBING, WELLHEAD ASSEMBLY
A typical wellhead assembly for a slim-hole well being drilled with a tubing-conveyed positive displacement mud motor: (from the ground up) 4 1/8-inch tubing rams; 4 1/8-inch blind rams; spool; 7 1/8-inch annular B.O.P.; a union; coiled tubing 3 1/16-inch coiled tubing B.O.P.; the stripper; and on top, the injector head through which the B.H.A. enters to begin digging the well.

COKE, NEEDLE
A form of petroleum coke that gets its name from its microscopic, elongated, crystalline structure. Needle coke is of a higher quality than the more ordinary sponge coke. The manufacture of needle coke requires special feeds to the coker and more severe operating conditions. Severe conditions in refining parlance usually means higher temperatures and pressures in a process.

COKE, PETROLEUM
Solid or fixed carbon that remains in refining processes after distillation of all volatile hydrocarbons; the hard, black substance remaining after oils and tars have been driven off by distillation.

COKE, SPONGE
Petroleum coke that looks like a sponge, hence the name. Sponge coke is used for electrodes and anodes. The weak physical structure of sponge coke makes it unfit for use in blast furnaces and foundry work. *See* Coke, Needle.

COKE DRUMS
Large, vertical, cylindrical vessels that receive their charge of residue at very high temperatures (1,000°F). Any cracked lighter products rise to the top of the drum and are drawn off. The remaining heavier product remains and, because it is still very hot, cracks or is converted to petroleum coke, a solid coal-like substance. In a large refinery that makes a lot of coke, the drums are in batteries of four to eight drums.

Coking drums

COKING
(1) The process of distilling a petroleum product to dry residue. With the production of lighter, distillable hydrocarbons, an appreciable deposit of carbon or coke is formed and settles to the bottom of the still. (2) The undesirable building up of carbon deposits on refinery vessels.

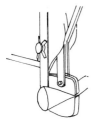

A cold-pinch tool

Drill collar with bit *(Courtesy Union Oil)*

Collet connector

COLD CUT

Severing a pipeline with mechanical pipe cutters. Cold cuts are made in a line when oil or gas is present, which could be ignited if a cutting torch were used. In repairing a section of pipeline that has been ruptured because of a washout or slide, the damaged section is usually cut out mechanically, with cutters or saw, and a sleeve installed to connect the two ends. After any spilled oil is cleaned up, the sleeve ends are welded to the pipeline.

COLD PINCH

To flatten the end of a pipe with a hydraulically powered set of pinchers. Pinching the pipe end is done to make a quick, temporary closure in the event a loaded pipeline is accidentally ruptured.

COLD STACK, TO

Refers to mobile, offshore drilling rigs that are idle but are not moved from location or mothballed in the usual sense. Cold stacking is simply abandoning the rig because of a slump in the drilling business, which happens periodically. The Coast Guard requires certain lighting and fog signals on all offshore rigs whether active or idle. Cold stacking is not recommended because the ocean environment quickly damages all unprotected equipment.

COLLAPSED CASING

See Casing, Collapsed.

COLLAR

A coupling for two lengths of pipe; a pipe fitting with threads on the inside for joining two pieces of threaded pipe of the same size.

COLLAR CLAMP

A device fitted with rubber gaskets bolted around a leaking pipe collar. The clamp is effective in stopping small leaks but is used only as a temporary measure until permanent repairs can be made.

COLLAR POUNDER OR PECKER

A pipeline worker who beats time with a hammer on the coupling into which a joint of pipe is being screwed by a tong gang. The purpose is twofold: to keep the tong men pulling in unison and to warm up the collar so that a tighter screw joint can be made.

COLLET CONNECTOR

A component of a subsea drilling system; a mechanically or hydraulically operated latching collar connecting the marine riser to the blowout-preventer stack.

COLUMN, PACKED

See Packed Column.

COMBINATION DRIVE
A condition in an oil reservoir where there are two or more natural drive mechanisms present to force the oil to the surface, e.g., water drive and gas-cap drive.

COMBINATION TRAP
A subsurface formation that exhibits characteristics of both a structural and a stratigraphic trap. For example, a monocline that loses porosity and permeability updip is a combination trap. The monocline is its structural character; the change in the reservoir rock gives it the characteristic of a stratigraphic trap.

COME-ALONG
A lever and short lengths of chain with hooks attached to the ends of the chains used for tightening or pulling a chain. The hooks are alternately moved forward on the chain being tightened.

COMMERCIAL WELL
A well of sufficient net production that it could be expected to pay out in a reasonable time and yield a profit for the operator. A shallow, 50-barrel-a-day well in a readily accessible location onshore could be a commercial well, whereas such a well in the North Sea or in the Arctic islands would not be considered commercial.

COMMINGLED WELL
A well producing crude oil from two or more formations or intervals through a common string of casing and tubing. The production from all formations, each tapped by perforating the casing at the proper level, mingles and is brought to the surface as oil from a single well and is treated so by the tax man.

COMMINGLING
(1) The intentional mixing of petroleum products having similar specifications. In some instances, products of like specification are commingled in a product pipeline for efficient and convenient handling. (2) Producing two pay zones in the same well bore.

COMMON CARRIER
A person or company having state or federal authority to perform public transportation for hire; an organization engaged in the movement of petroleum products—oil, gas, refined products—as a public utility and common carrier.

COMMON MID-POINT
A seismic shooting and analysis technique in which a single shot point is straddled by numerous geophones that record the same impulse.

COMMONWEALTH OF INDEPENDENT STATES
C.I.S. Formerly the U.S.S.R., The Soviet Union.

COMMUNITY LEASE
A single lease covering two or more separately owned parcels of land. A community lease may result from the execution of a single lease by the owners of separate tracts or by the execution of separate but identical leases by the owners of separate tracts when each lease purports to cover the entire consolidated acreage. Usually the result of the execution of a community lease is the apportionment of royalties in proportion to the interests owned in the entire leased acreage.

COMPANION FLANGE
A two-part connector or coupling, one part convex, the other concave. The two halves are held together by bolts and nuts. This type flange or "union" is used on small-diameter piping.

COMPANION PIECE OR UNIT
An item of equipment made to fit another so as to form a workable unit.

COMPANY CAMP
See Camp, Company.

COMPARATOR
A portable, easily handled instrument for measuring and comparing threads on tubular goods—casing, tubing, and drillpipe. The optical comparator has magnification of 30 to 40x, and it is as accurate as a micrometer. There are many thread sizes and designs so it is imperative when changing pipe sizes and types that the threads are compatible. A new batch of casing of the same size may have a different thread pattern so the first joint run on the original casing may not make up properly, bubble tight; the threads may be pulled.

COMPENSATOR
See Heave Compensator.

COMPENSATORY ROYALTY
See Royalty, Compensatory.

COMPETENT ROCK
Refers to a volume or mass of rock able to withstand tectonic forces, i.e., folding or compression, and still maintain its essential form. Such a volume of rock may be competent or incompetent a number of times during the long history of its deformation, depending upon environmental conditions, the degree and time of folding or fracturing. *See* Incompetent Rock.

COMPLETION

To finish a well so that it is ready to produce oil or gas. After reaching total depth (T.D.), casing is run and cemented; casing is perforated opposite the producing zone, tubing is run, and control and flow valves are installed at the wellhead. Well completions vary according to the kind of well, depth, and the formation from which the well is to produce.

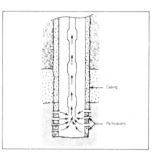

Single completion using
open-ended tubing

COMPLETION FUNDS

Completion funds are formed to invest in well completions, to finance the completing and equipping of a potentially productive well. After a well is drilled into a productive formation, there remain the costs of setting pipe (casing the well); perforating, testing, acidizing, or fracturing the formation; and running production tubing and installing pumping equipment, separators, stock tanks, etc. The operator who drills the well may not have the financial resources to complete the well, so he may sell part or all of his interest to a completion fund. Completion funds are not as risky an investment as drilling funds but are less certain than income funds and royalty funds.

COMPLETION TECHNIQUES, WELL

The type of well completion depends on the characteristic of the formation from which the well is to be produced. A limestone, dolomite, or sandstone interval, for example, may be produced open hole; that is, without casing to stabilize the borehole. If the producing formation is unconsolidated sand, the well completion may call for a gravel pack behind the casing to control the sand. Some wells are completed without tubing; the well either flows through the casing or is pumped with a casing pump. If two or more producing formations are to be produced separately (for separate royalty owners, say), the well is dually completed; there are two strings of tubing, each string servicing one formation. At the surface each stream goes to its own heater-treater and tank battery. As many as four separator strings of tubing have been run in a four-pay well. A tight fit to be sure, but it can be done with 2-inch or 2 1/2-inch tubing. One last type of completion is the so-called "miniature completion," which is done with production tubing smaller than 4 1/2 inches in diameter. This might be found in slim hole drilling.

COMPRESSION CUP

A grease cup; a container for grease made either with a screw cap or spring-loaded cap for forcing the grease onto a shaft bearing.

COMPRESSION-IGNITION ENGINE

A diesel engine; a four-cycle engine whose fuel charge is ignited by the heat of compression as the engine's piston comes up on the compression stroke. *See* Hot-Plug Engine.

Compressor *(Courtesy Union Pump Co.)*

Gas pipeline compressor station

Skid-mounted compressor *(Courtesy Fry Assoc. Inc.)*

COMPRESSION RATIO

The ratio of the volume of an engine's cylinder at the beginning of the compression stroke to the volume at the end or the top of the stroke. High-compression engines are generally more efficient in fuel utilization than those with lower compression ratios. A cylinder of 10 cubic inch volume at the beginning of the compression stroke and 1 cubic inch at the top of the stroke indicates a 10:1 compression ratio.

COMPRESSOR, AXIAL

A gas compressor that takes in gas at the inlet and moves the charge axially over the compressor's long axis to the discharge port. This is accomplished by the action of a central impeller shaft studded with hundreds of short, fixed blades. The impeller and its paddle-like blades rotate at speeds of 3,000–6,000 rpm. Large compressors move up to 300,000 cubic feet per minute.

COMPRESSOR, ROTARY-SCREW

A small-volume, rotary compressor with its principle working part as an "endless screw" rotating in a cylinder or barrel. On rotary-screw compressors, tolerances are very close.

COMPRESSOR, SKID-MOUNTED

A "portable" gas compressor and engine module for use in repressuring or injecting gas into a high-pressure gas trunk line.

COMPRESSOR PLANT

A pipeline installation to pump natural gas under pressure from one location to another through a pipeline.

COMPUTER VISUALIZATION

The visualization of vertical and horizontal seismic slices of the acoustically imaged subsurface permits geoscientists to "see" the internal structure and faulting in the Earth's crust. Advances in corporate and vendor software make possible even more animation, rotation, and shading of 3-D material whether it is seismic data, well bore information, or reservoir simulation models.

COMSAT

The copyrighted name for a satellite, marine communication system, "spanning the globe."

CONCENTRATED SUSPENSIONS

Fluids carrying high concentrations of foreign material; proppants, for example.

CONCESSION

An agreement (usually with a foreign government) to permit an oil company to prospect for and produce oil in the area covered by the agreement.

CONCESSION ACREAGE

The area of land measured in acres or hectares that makes up a concession; the land involved in a concession agreement.

CONCRETION

A hard, compact mass of mineral material formed by the precipitation from an aqueous solution upon or around a center such as a seashell, bone, or fossil in the pores of a sedimentary rock. The concretion usually is very different in composition from the rock in which it is found. Concretions are a concentration of some constituent of the enclosing rock or of a cementing material such as silica, calcite, dolomite, iron oxide, or gypsum. They range in size from pea size to almost 10 feet in diameter. *See* Geode.

CONDEMNATION

The taking of land by purchase, at fair market value, for public use and benefit by state or federal government as well as by certain other agencies and utility companies having power of eminent domain.

CONDENSATE

Liquid hydrocarbons produced with natural gas that are separated from the gas by cooling and various other means. Condensate generally has an A.P.I. gravity of 50° to 120° and is water-white, straw, or bluish in color.

CONDENSATE, LEASE

See Lease Condensate.

CONDENSATE, RETROGRADE GAS

See Retrograde Gas Condensate.

CONDENSATE GAS RESERVOIR

See Retrograde Gas Condensate.

CONDENSATE WATER

Water vapor in solution with natural gas in the formation. When the gas is produced, the water vapor condenses into liquid as both pressure and temperature are reduced. *See* Retrograde Gas Condensate.

CONDENSATION

The transformation of a vapor or gas to a liquid by cooling or an increase in pressure or both simultaneously.

CONDENSER

A water-cooled heat exchanger used for cooling and liquefying vapors.

CONDUCTOR CASING

See Casing, Conductor.

CONE BIT, METAL-SEALED

Metal-sealed, rolling-cone bits are made with two metal O-ring-sealed journal bearings supported by elastomer O-rings, called "energizers." The metal-sealed bits out last elastomer-sealed bits, particularly in high-speed drilling, and in deep, hot holes.

CONE ROOF

A type of tank roof built in the form of a flat, inverted cone; an old-style roof for large crude storage tanks but still employed on tanks storing less volatile products. *See* Floating Roof.

CONE VALVE

A type of valve (usually spring-loaded) in which the seat is in the shape of a truncated cone. An example of a cone valve is a float valve used in well cementing.

CONFIRMATION WELL

A well drilled to "prove" the formation or producing zone encountered by an exploratory or wildcat well. *See* Step-Out Well.

CONFORMITY

An uninterrupted sequence of sedimentary strata or rock, the younger (more recent) layers deposited on top of the older layers with no evidence of erosion, of tilting or folding of the older or lower strata, and no appreciable time (geologically speaking) between the deposition of older and younger units.

CONGLOMERATE

A type of sedimentary rock compounded of pebbles and rock fragments of various sizes held together by a cementing material, the same type of material that holds sandstone together. Conglomerates are a common form of reservoir rock.

Conglomerate *(Courtesy Core Lab)*

CONIC ROOF

See Cone Roof.

CONICAL-TOWER PLATFORM

A type of offshore drilling platform made of reinforced concrete for use in Arctic waters where pack ice prevents the use of conventional platform construction. The structure is a truncated cone supporting a platform from which the wells are drilled.

CONNATE WATER

The water present in a petroleum reservoir in the same zone occupied by oil and gas. Connate water is not to be confused with bottom or edge water. Connate water occurs as a film of water around each grain of sand in granular reservoir rock and is held in place by capillary attraction.

CONNECTION
See Tapped or Flanged Connection.

CONNECTION FOREMAN
The supervisor; the boss of a pipeline connection gang.

CONNECTION GANG
A pipeline crew that lays field gathering lines, connects stock tanks to gathering lines, and repairs pipelines and field pumping units in their district. Connection gangs also install manifolds and do pipe work in and around pumping stations. A typical gang of 8 or 10 men has a welder and a helper, a gang-truck driver and swamper (helper), 3 or 4 pipeliners, and a connection foreman.

CONSORTIUM
An international business association organized to pursue a common objective, e.g., to explore, drill, and produce oil.

CONSTANT CURVATURE (of the well-bore trajectory)
An advanced concept in horizontal drilling. Constant curvature drilling substitutes an even, consistent arc and a longer radius for the more severe, early-day drilling trajectory with its relatively short radius for angle building.

CONSTANT CURVATURE DRILLING
The constant curvature method of angle building from vertical to horizontal is based on the drilling tendencies of the bottom-hole assembly and the technical advances in directional drilling. Constant curvature yields a trajectory with less dogleg severity, less drag, and less torque on the drillstring compared to the conventional radius-of-curvature and constant-turn rate methods. Engineering reports show that the well-path trajectory of constant curvature is consistent with the directional performance of deflection tools, which increases fatigue life of the drillstring elements moving through the curved segments of the hole.

CONSUMER GAS
Gas sold by an interstate gas pipeline company to a utility company for resale to consumers.

CONTINENTAL SHELF
See Outer Continental Shelf (O.C.S.).

CONTOUR LINE
A line (as on a map) connecting points on a land surface that have the same elevation above or below sea level.

CONTOUR MAP
See Map, Contour.

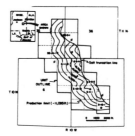

Structure contour map
(Courtesy Mace)

Control panel *(Courtesy Gulf Oil Canada Ltd.)*

Electrically operated control valve *(Courtesy Magnatrol)*

Cooling tower

Cord road

CONTRACT DEPTH
The depth of a well called for or specified in the drilling contract.

CONTRACTOR
See Drilling Contractor.

CONTRIBUTION AGREEMENTS
See Support Agreements.

CONTROL DRILLING
The practice of limiting the penetration rate of the drill to a value less than could be accomplished based on drillability of the formation or rock being drilled.

CONTROL PANEL
An assembly of indicators and recording instruments—pressure gauges, warning lamps, and other digital or audio signals—for monitoring and controlling a system or process.

CONTROL VALVE, ELECTRICALLY OPERATED
See Valve, Electrically Operated Control.

COOLING TOWER
A large louvered structure (usually made of wood) over which water flows to cool and aerate it. Although most cooling towers are square or rectangular in shape, some are cylindrical and open at the bottom and top, which produces strong air currents through the center of the structure for more rapid cooling.

COQUINA
A type of limestone made up of a matted or "felted" mass of seashells, loosely cemented together. Good examples of this light-colored, porous limestone are found in Florida where it is used for road building and in construction. *Coquina* is derived from the Spanish for cockle or shellfish.

CORD ROAD
A passable road made through a swampy, boggy area by laying logs or heavy timbers side by side to make a bumpy but firm surface; a log road.

CORE
(Earth) The Earth's core is estimated to be about 4,200 miles in diameter and consists of an outer core 1,300 miles in thickness that is liquid and a solid inner core approximately 1,500 miles in diameter. Enveloping the Earth's core is the mantle and surrounding the mantle are the lithosphere and the crust, the upper or outer layers of the lithosphere. *See also* Core Sample.

CORE BARREL
A device with which core samples of rock formations are cut and brought to the surface; a tube with cutting edges on the bottom circumference, lowered into the well bore on the drillpipe and rotated to cut the plug-like sample.

CORE BIT
See Bit, Core.

CORE BOAT
A seagoing vessel for drilling core holes in offshore areas.

CORE RECORD
A record showing the depth, character, and fluid content of cores taken from a well.

CORE SAMPLE
A solid column of rock, usually from 2 to 4 inches in diameter, taken from the bottom of a well bore as a sample of an underground formation. Cores are also taken in geological studies of an area to determine the oil and gas prospects.

CORPORATION OUTSOURCING
Relying on outside-the-corporation service companies for engineering, geologic, or counseling (lawyer) service. This practice is consistent with downsizing as practiced by many corporations, large and small, in the 1990s. The ostensible purpose is to reduce overhead, which includes cost of insurance, and to reduce the number of employees eligible in later years for pensions.

CORRECTED TO 60°F
Refers to A.P.I. gravity of crude oil which is arrived at by correcting the observed hydrometer reading of gravity and temperature. Crude oil is bought and sold at a standard 60°F or 15.57°C.

CORRELATIVE RIGHTS
The inherent right of an owner of oil or gas in a field to his share of the "reservoir energy" and his right to be protected from wasteful practices by others in the field.

CORROSION
The eating away of metal by chemical action or an electrochemical action. The rusting and pitting of pipelines, steel tanks, and other metal structures is caused by a complex electrochemical action. *See* Anode.

CORROSION, FORMS OF
(1) Uniform (general); (2) Pitting; (3) Galvanic corrosion; (4) Crevice corrosion; (5) Erosion corrosion; (6) Stress corrosion; (7) Inter-granular corrosion; (8) Selective leaching.

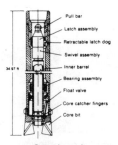

Core barrel

Core bit

Core samples

Testing a Core

Corroded pipe showing
deep pits

CORROSION COUPONS

Small rectangular pieces of representative metal placed in the same environment (acid, heat, pressure) to which tubulars and other downhole equipment are subjected to check on corrosion. After a test period, the coupons are removed and inspected for any signs of corrosion or other physical or chemical changes, and by inference, the condition of the downhole pipe and/or tools.

CO_2-SHIELDED WELDING

See Welding, CO_2-Shielded.

COST CRUDE OIL

Crude oil produced from an operator's own wells; oil produced at "cost" on a lease or concession acreage as compared to purchased crude.

COST OIL

The share of oil produced on a lease or from a well which is applied to the recovery of costs under a Production Sharing contract.

COST PROJECT

A cost project drills a well or wells to obtain information about soil conditions before an offshore lease transaction is entered into. The drilling is done for core samples of the subsoil and not to develop or take oil or gas.

COUNTRY ROCK

Rock enclosing a mineral deposit or intruded by and surrounding an igneous rock intrusion.

COUPLING

A collar; a short pipe fitting with both ends threaded on the inside circumference used for joining two lengths of line pipe, casing, or tubing.

Coupling

COUPLING POLE

The connecting member between the front and rear axles of a wagon or four-wheel trailer. To lengthen the frame of the vehicle, a pin in the pole can be removed and the rear-axle yoke (which is fastened to the pole by the pin) moved back to another hole. On pipe-carrying oilfield trailers, the coupling pole is a telescoping length of steel tubing. The trailer can be made as long as necessary for the load.

COVENANT, IMPLIED

An unwritten obligation or promise implied in an agreement; rights stemming from what is customary in a given situation.

CRACKING

The refining process of breaking down the larger, heavier, and more complex hydrocarbon molecules into simpler and lighter molecules.

Cracking is accomplished by the application of heat and pressure and, in certain advanced techniques, by the use of a catalytic agent. Cracking is an effective process for increasing the yield of gasoline from crude oil.

CRACKING A VALVE
To open a valve so slightly as to permit a small amount of fluid or gas to escape.

CRANE BARGE
A floating platform; a barge-like vessel equipped with one or more large cranes for offshore duty. Some large crane barges are self-propelled; others are true barges and are towed to location.

Crane barge

CRANK
An arm attached at right angles to the end of a shaft or axle for transmitting power to or from a connecting rod or pitman.

CRANK-BALANCED PUMPING UNIT
See Pumping Unit, Crank-Balanced.

CRANKCASE VENTILATION SYSTEM, POSITIVE
See Positive Crankcase Ventilation System.

CRATER
(1) A bowl-shaped depression around a blown-out well caused by the caving in and collapse of the surrounding earth structure. (2) To fail or fall apart (colloquial).

Crank-balanced pumping unit *(Courtesy Lufkin Industries Inc.)*

CREEKOLOGY
An ironic term for the unscientific methods of choosing drilling sites or prospective oil and gas acreage based on the appearance of outcrops or the general lay of the land. In the very early days in Pennsylvania and West Virginia, when Colonel Edwin Drake was drilling 69-foot wells with primitive cable tools, creekology was about all there was to spot a well. And the successes of the first intrepid wildcatters were impressive.

CRENOLOGY
An old term for the geology of surface features.

CRIBBING
Temporary support made of blocks and small, easily handled, timbers for sections of pipeline; a large valve, an engine or pump while work is being done or a foundation is made ready.

CRITICAL PRESSURE
The pressure required to condense a gas at the critical temperature, above which, regardless of pressure, the gas cannot be liquefied.

Top view of crown block

Side view of crown block

Crown platform

CROOKED-HOLE COUNTRY

Said of an area or a field in which there has been a high incidence of crooked holes drilled, boreholes that have deviated alarmingly from the vertical. Such a condition results when the underground formations, particularly hard-rock formations, are lying at angles from the horizontal, causing the drill bit to drift from the vertical. *See* Pendulum Drill Assembly; *also* Fanning the Bottom.

CROSSBEDDING

Although sedimentary rocks are most often laid down in layers parallel to one another, sometimes the layers are not horizontal or parallel. In some instances the layers may be at a broad angle to the lower strata. Crossbedding or cross stratification forms in sand dunes that were once a part of the surface and were blown into upsloping and downsloping configurations and remained so until consolidated.

CROSSHEAD

A sliding support for a pump or compressor's connecting rod. The rod is attached to a heavy "head" that moves to and fro on a lubricated slide in the pump's frame. Screwed into the other end of the crosshead is the pump's piston rod or plunger rod. A crosshead moves back and forth in a horizontal plane or up and down in a vertical plane, transmitting the power from the connecting rod to the pump's piston rod.

CROSSOVER

A stile; a step-and-platform unit to provide access to a work platform or an elevated crossing. *See* Stile.

CROSSOVER SUB

A short length of pipe used between two sizes or two different thread patterns in a drillstring assembly; a crossover or change over.

CROWBAR CONNECTION

A humorous reference to an assembly of pipe fittings so far out of alignment that a crowbar is required to force them to fit.

CROWN BLOCK

A stationary pulley system located at the top of the derrick used for raising and lowering the string of drilling tools; the sheaves and supporting members to which the lines of the traveling block and hook are attached.

CROWN PLATFORM

A platform at the very top of the derrick that permits access to the sheaves of the crown block and provides a safe area for work involving the gin pole.

CRUDE OIL
Oil as it comes from the well; unrefined petroleum.

CRUDE OIL, BUY-BACK
See Buy-Back Crude.

CRUDE OIL, REDUCED
See Reduced Crude Oil.

CRUDE (OIL) SLATE
The kind or type of crude oil a refinery is set up to handle: sweet, high-gravity; low-gravity, sour or some of both.

CRUDE OIL, VOLATILE-LADEN
A crude oil stream carrying condensate, natural gasoline, and butane. Sometimes it is convenient and economical to move certain natural gas liquids to refineries by injecting them into crude oil pipelines to be pumped with the crude.

CRUDE PRICING, ON ARRIVAL
The pricing and billing for crude oil upon arrival at its final destination into storage or at the refinery. Under customary pricing systems which can be termed F.O.B., the buyer bore the risk of price fluctuations while his purchased crude oil or products were en route. In arrival pricing or destination pricing, the seller—not the buyer—accepts the risk of price declines. And since in-transit time can be a month or more, arrival pricing has its advantages for the buyer. In times of oil surpluses, it is a buyers' market.

CRUDE STILL
A primary refinery unit; a large vessel in which crude oil is heated and various components taken off by distillation.

CRUMB BOSS
A person responsible for cleaning and keeping an oilfield bunkhouse supplied with towels, bed linen, and soap; a construction camp housekeeper.

CRUMB OUT
To shovel out the loose earth in the bottom of a ditch; also, to square up the floor and side of the ditch in preparation for laying pipe.

CRYOGENICS
A branch of physics that relates to the production and effects of very low temperatures. The process of reducing natural gas to a liquid by pressure and cooling to very low temperatures employs the principles of cryogenics.

CRYPTOCRYSTALLINE
Having a crystalline structure too small to be seen with a microscope.

CUBES
Short for cubic inch displacement; C.I.D.

CUBIC INCH DISPLACEMENT
The volume "swept out" or evacuated by the pistons of an engine in one working stroke; used to describe the size (and, by implication, the power) of an automobile engine.

CULTIVATOR WRENCH
Any square-jawed, adjustable wrench that is of poor quality or worn out. *See* Knuckle-Buster.

CUP
Disc with edges turned at right angles to the body used on plungers in certain kinds of pumps; disc of durable plastic or other tough, pliable material used on pipeline pigs or scrapers to sweep the line.

CUP GREASE
Originally, a grease used in compression cups; today the term refers to grease having a calcium fatty-acid soap base. *See* Grease.

CUSHION GAS
An amount of natural gas in an underground, salt cavern storage that is put in initially as an easy-to-calculate, so-called "cushion" at a few pounds per square inch. Gas to be stored in the same cavern is put in under high pressure. When the stored gas is removed, it is taken out until the pressure is brought down to the cushion pressure. Thus, all stored gas is accounted for.

CUSTODY TRANSFER
The transfer of oil or gas from seller to buyer or his representative. Crude oil or products are gauged in the tank and witnessed by both buyer and seller or their agents. The tank valves into the pipeline are opened. At this time, the oil leaves the custody of the seller or producer and enters the custody of the buyer or new owner. This comprises custody transfer. Gas is metered and in some cases, oil is also metered by positive displacement meter or volumetric meter.

CUT
(1) A petroleum fraction; a product such as gasoline or naphtha distilled from crude oil. (2) Crude oil contaminated with water so as to make an oil-water emulsion. (3) To dilute or dissolve.

CUT OIL
Crude oil partially emulsified with water; oil and water mixed in such a way as to produce an emulsion in which minute droplets of water are encased in a film of oil. In such case the water, although heavier, cannot

separate and settle to the bottom of a tank until the mixture is heated or treated with a chemical. *See* Roll a Tank.

CUT POINT
The point in a process at which the product reaches specifications; the temperature and/or the pressure in a process that reaches set limits.

CUTTING OILS
Special oils used to lubricate and cool metal-cutting tools.

CUTTINGS
Chips and small fragments of rock as the result of drilling that are brought to the surface by the flow of the drilling mud as it is circulated. Cuttings are important to the geologist, who examines them for information concerning the type of rock being drilled. *See* Sample.

CUTTINGS BED
Rock fragments and cuttings in a high-angle or horizontal well bore that, owing to a restriction in circulation rate or lack of circulation entirely, have yielded to gravity and settled to the lower side of the borehole, making a bed of fragments and pulverized rock, forming what is referred to as "sag."

CUTTINGS DISPOSAL, ON-SITE
A method of disposing of well cuttings, contaminated, oil-base drilling mud, by pumping them back into an acceptable downhole formation or into the annulus between casing and wall of the borehole. The cuttings, drilling slurries, and any oily residues are, in some instances, pumped into a predetermined formation at frac pressures—an environmentally acceptable method of disposing of contaminating material. *See also* Closing the Loop.

Cutting torches in a
holder for beveling pipe
ends

CUTTING TORCH
A piece of oxyacetylene welding and cutting equipment; a hand-held burner to which the oxygen and acetylene hoses are attached. The gases when ignited by the welder's lighter produce a small, intense flame that cuts metal by melting it. *See* Welding Torch.

CYCLING
Return to a gas-drive reservoir the gas remaining after extraction of liquid hydrocarbons for the purpose of maintaining pressure in the reservoir and thus increasing the ultimate recovery of liquids from the reservoir.

CYCLING PLANT
An oilfield installation that processes natural gas from a field, strips out the gas liquids, and returns the dry gas to the producing formation to maintain reservoir pressure.

CYLINDER OIL

Oils used to lubricate the cylinders and valves of steam engines.

CYLINDER STOCK

A class of highly viscous oils so called because originally their main use was in preparation of products to be used for steam cylinder lubrication.

D

DAMPING SUB
Essentially, a downhole shock absorber for a string of drilling tools; a 6-to 8-foot-long device, a part of the drill assembly, that acts to dampen bit vibration and impact loads during drilling operations. Damping subs are of the same diameter as the drillpipe into which they are screwed to form a part of the drillstring.

D.&P. PLATFORM
A drilling and production offshore platform.

D&P platform

DARCY
A unit of permeability of rock. A rock of 1 darcy permeability is one in which fluid of 1 centipoise viscosity will flow at a velocity of 1 centimeter per second under a pressure gradient of 1 atmosphere per centimeter. Since a darcy is too large a unit for most oil producing rocks, permeabilities used in the oil industry are expressed in units one-thousandth as large, i.e., millidarcies (0.001 darcy). Commercial oil and gas sands exhibit permeabilities ranging from a few millidarcies to several thousand.

D'ARCY'S LAW
During experimental studies on the flow of water through consolidated sand filter beds, Henry D'Arcy, in 1856, formulated a law that bears his name. D'Arcy's Law states that the velocity of a homogeneous fluid in a porous medium is proportional to the pressure gradient and inversely proportional to the fluid's viscosity. This law has been extended to describe, with certain limitations, the movement of other fluids, including miscible fluids in consolidated rocks and other porous substances.

DATA, SOFT AND HARD
Seismic data is referred to as "soft data," well data is "hard." The goal of the geologist, the reservoir engineer, is to merge statistically both sets to arrive at a credible set of guidelines.

DATA TRANSMISSION, DOWNHOLE
There are two primary means of transmitting data from a steered drill to the surface. One is pulse telemetry that encodes the data in a binary format and sends it to the surface station by positive or negative pressure

pulses generated in the drilling fluid. The other means of transmission is continuous-wave telemetry, a form of positive pulse, which employs a rotating device that generates a fixed-frequency signal, which also sends binary information encoded on a pressure wave to the surface through the mud column.

DAY RATE
Drilling and other related well work paid for by the day as opposed to a turnkey job in which a well is paid for when it is completed. If, however, under a turnkey contract the operator changes the target, the work automatically changes to a day rate until an agreed-upon point of the original well-plan is arrived at. Moral: Don't change plans in the middle of your well.

DAY-WORK BASIS
Refers to a drilling contract in which the work of drilling and completing a well is paid for by the days required for the job instead of by the feet drilled. *See* Turnkey Contract.

D.C.-D.C. RIG
See Rig, Electric.

D.E.A.-44
A horizontal-well technology exchange participated in by 200 representatives of the industry.

D.E.A.-67
An international exchange project on coiled tubing and slim-hole drilling.

DEAD LINE
The anchored end of the drilling line that comes down from the crown block through a fixed sheave at ground level, called a dead-line anchor, and onto a storage drum. When stringing up the drilling line, the big traveling block is set on the rig floor and the free end of the line is threaded over the crown block and through the traveling block a sufficient number of times to lift the anticipated load with a good margin of safety. The free end is then attached to the drawworks drum, which is rotated until one layer of the line is spooled on. The traveling block is then hoisted into the derrick. The other end of the threaded or reeved line is the dead line or, one might say, the anchored line. *See* Fast Line.

DEAD OIL
Crude oil containing no dissolved gas when it is produced.

DEAD WELL
A well that will not flow and in order to produce must be put on the pump.

Dead line and tie down

DEAD-LINE ANCHOR
See Dead Line.

DEADMAN
(1) A substantial timber or plug of concrete buried in the earth to which a guy wire or line is attached for bracing a mast or tower; a buried anchoring piece capable of withstanding a heavy pull. (2) A land-side mooring device used with lines and cables when docking a vessel.

DEADMAN CONTROL
A device for shutting down an operation should the attendant become incapacitated. The attendant using such a device must consciously exert pressure on a hold-down handle or lever to work the job. When pressure is relaxed owing to some emergency, the operation will automatically come to a halt.

DEADWOOD
Material inside a tank or other vessel such as pipes, supports, and construction members that reduces the true volume of the tank by displacing some of the liquid contents.

DE-BUBBLING
In marine seismic application, de-bubbling is eliminating the effect of bubble noise.

DECANTING CENTRIFUGE
See Centrifuge, Decanting.

Decanting centrifuge

DECK BLOCK
A pulley or sheave mounted in a steel frame that is securely fixed to the metal deck of a ship or barge. Deck blocks that lie horizontal to the vessel's deck are for horizontal pulls with hawser or cable.

DEEP GAS
Gas produced from 15,000 feet or below, which commands an incentive price because of the high cost of drilling 3 miles deep into hot, high-pressure formations. Such wells cost in the high-rent neighborhood of 4 to 8 million dollars, depending on the difficulties encountered.

DEEP RIG
See Rig, Deep.

Deep rig

DEEP RIGHTS
The rights to oil and gas below a certain depth retained by the lessor or landowner. A lease may drill his leasehold to 5,500 feet for one group of investors and drill down to 10,000 feet, for example, for another group of investors. If hydrocarbons are discovered at both depths, the well

would be a dual-completion well. The production from each depth would be piped to its own tankage.

DEEPSTAR (D.S.R.)
A consortium of 11 major oil companies and as many service and supply contractors to develop a production concept for the deepwater Gulf of Mexico. The name Deepstar stands for Deepwater Staged Recovery. The project's aim is to conquer water depths to 6,000 feet, a daunting challenge.

DEFICIENCY GAS
The difference between a quantity of gas a purchaser is obligated by contract either to take or pay for, if not taken and the amount actually taken.

DEFORMATION
A term for the geologic process of folding, faulting, shearing, or compression by various earth forces.

DEGASSER
A vessel into which drilling mud is pumped to remove gas entrained in the drilling mud. Modern, compact degassers use three commonly recognized methods of degassification: centrifugal force, turbulence, and vacuum. *See* Degassing Drilling Mud.

Degasser

DEGASSING DRILLING MUD
An important part of well-drilling operations is keeping the drilling mud free of entrained gas, or bubbles that enter the mud as it circulates downhole through gaseous formations. One of the three functions of mud is to provide sufficient hydrostatic head to control a kick when high-pressure oil or gas is encountered. When mud of a certain density is circulated, it can become infused with gas to an extent that, although the volume of mud may increase, the density is severely reduced. To guard against this dangerous situation, the mud is degassed at the surface. Several kinds of equipment get the gas out, but all have one aim in common: to make it possible for the gas bubbles to free themselves. One method flows the mud over wide sheets so the slurry is no more than one-eighth to three-eighths thick and the bubbles may come to the surface and escape. Another method sprays the mud against a baffle in a spray tank that squeezes out the gas. A third method directs the mud through a vacuum tank where, under reduced atmospheric pressure, the bubbles of gas expand and break out of the slurry.

DEHYDRATED AIR
Air that has all traces of moisture removed; instrument air.

DEHYDRATION REACTION

The reverse of hydration; dehydration is the transfer of H_2O from a mineral into the fluid phase or into free water. *See* Hydration.

DEHYDRATOR

A tank or tower through which gas is run to remove entrained water. A common method of gas dehydration is through the use of various glycols—diethylene, triethylene, and tetraethylene. Dehydration is accomplished by contact of the wet gas with a pure or "lean" glycol solution. Gas is fed in to the bottom of a trayed or packed column in the presence of the glycol solution. As the gas percolates upward through the solution, the lean glycol absorbs the entrained water and dry gas is taken off at the top of the tower. Gas must be extremely dry to meet pipeline specifications; it may not contain more than 7 pounds of water per million standard cubic feet.

Battery of dehydrators

DELAY RENTAL

Money payable to the lessor by the lessee for the privilege of deferring drilling operations or commencement of production during the primary term of the lease. Under an "unless" lease, the nonpayment of delay rental by the due date is the cause for automatic termination of the lease unless the lease is being held by drilling operations, by production, or by some special provision of the lease.

DELAY TIME (seis)

In seismic refraction work, the additional time required to traverse any ray path over that required to traverse the horizontal component at highest velocity encountered on the ray path. It refers either to the source or receiver end of the trajectory.

DELINEATION DRILLING

Wells drilled to establish the outer limits or boundaries of a pool, a field, or even a trend.

DELINEATION WELLS

Wells drilled outward from a successful wildcat well to determine the extent of the oil find, and the boundaries of the productive formation. *See* Development Wells.

DELTA FLOW

A balanced flow: the same volume of drilling mud returning from downhole as being pumped into the hole through the drillpipe. A delta flow indicates there is no loss of drilling fluid into caverns, cracks, or voids of any kind—an honest hole; a discreet hole.

DELTAIC BASIN

Sedimentary deposits laid down at the mouth of a stream or river, at a coastline, and in many cases extending miles into a gulf or ocean, e.g., the Mississippi River's deltaic basin.

DELTAICS

The formation, the laying down of sedimentary material to form a delta at the mouth of rivers, even middling streams. All streams carry some erosional material, and when the flow of water slackens, the fine material being carried settles out, and an alluvial fan or a delta is formed. *See* Alluvial Fan.

DEMULSIFIER

A chemical used to break down crude oil-water emulsions. The chemical reduces the surface tension of the film of oil surrounding the droplets of water. Thus freed, the water settles to the bottom of the tank.

DEMURRAGE

The charge incurred by the shipper for detaining a vessel, freight car, or truck. High loading rates for oil tankers are of utmost importance in order to speed turnaround and minimize demurrage charges.

DENSE

To a geologist, dense is one thing; to a teacher it is something else. Geologically, a dense rock is usually an igneous rock with grains that are so small and compact and a texture so fine that individual particles can not be distinguished with the unaided eye, the naked eye. Also, a dense rock has a relatively high specific gravity, and it sinks readily.

DENSITY LOG

See Log, Density.

DENSMORE, AMOS

The man who first devised a method of shipping crude oil by rail. In 1865 he mounted two iron-banded wooden tanks on a railway flatcar. The tanks or tubs held a total of 90 barrels. Densmore's innovation was the forerunner of the unit train for hauling oil and products and the latest development, Tank Train.

DEPLETION ALLOWANCE

See Percentage Depletion.

DEPOCENTER

The thickest part of a stratigraphic unit in a depositional basin or area. An area of maximum deposit.

DEPOSIT

An accumulation of oil or gas capable of being produced commercially.

DEPOSITION
The placing or laying down of any material in the accumulation of beds, veins, or masses of any kind of loose rock material by a natural process; the settling of sediment from suspension in water; the chemical precipitation of minerals by evaporation from solution; the accumulation of organic matter on the death of plants and animals. A deposit of material; sedimentation.

DEPOSITIONAL SYSTEM, BARRIER ISLAND
One of a number of such systems of interest to the exploration geologist and others. A barrier island system is sometimes described as a micro-tidal (small tidal inlet to a protected leaward lagoon) and wave dominated (sand-carrying) sandstone reservoir.

DEPROPANIZER
A unit of a processing plant where propane, a liquid hydrocarbon, is separated from natural gas.

DERRICK
A wooden or steel structure built over a wellsite to provide support for drilling equipment and a tall mast for raising and lowering drillpipe and casing; a drilling rig.

DERRICK, PUMPING
In the early days before the widespread use of portable units for pulling and reconditioning a well, the original derrick used for drilling was often replaced by a smaller, shorter derrick called a pumping derrick or pumping rig. Well workovers could be done with these rigs; the well also could be pumped by pumping jack or by a walking beam.

DERRICK BARGE
A type of work boat on which a large crane is mounted for use offshore or for other over-water work. The larger derrick or crane barges are self-propelled and are, in effect, a boat or ship with full-revolving crane, a helicopter pad, and tools and equipment for various tender work. A crane barge.

DERRICK BOARD
Sturdy board used in the rig as a platform for the derrick man; the monkey board.

DERRICK FLOOR
The platform (usually 10 feet or more above the ground) of a derrick on which drilling operations are carried on; rig floor.

DERRICKMAN
A member of the drilling crew who works in the derrick on the tubing board racking tubing or drillpipe as it is pulled from the well and unscrewed by other crew members on the derrick floor.

Old Wooden
Derrick

Derrick barge

Derrick floor

Desander

DESALTING PLANT

An installation that removes salt water and crystalline salt from crude oil streams. Some plants use electrostatic precipitation; others employ chemical processes to remove the salt.

DESANDER

See Desilter-Desander.

DESATURATION INTERVAL

An interval or a rock unit in a downhole formation that is hydrophilic, has a great affinity for water and aquaceous solutions. Water-base drilling fluids are vulnerable to such formations and lose significant amounts of their liquid unless the borehole is cased. *See* Casing Scab.

DESERT VARNISH

See Varnish.

DESICCANT DRYING

The use of a drying agent to remove moisture from a stream of air or gas. In certain product pipelines great effort is made to remove all water vapor before putting the line in service. To accomplish this, desiccant-dried air or an inert gas is pumped through the line to absorb the moisture that may be present even in the ambient air in the line.

DESIGNER FUELS

Reformulated or oxygenated fuels; fuels (gasolines) designed or formulated for different areas of the country; diesel fuel for off-road, agriculture use as opposed to diesel, over-the-road truck and bus fuel.

DESILTER-DESANDER

A filtering device on a drilling well's mud system that removes harmful abrasive material from the mud stream.

DESTRUCTIVE DISTILLATION

See Distillation, Destructive.

DESTRUCTIVE TESTING

Testing an item of equipment under normal operating parameters of pressure, heat, lubrication, and cycling until the equipment fails. Other tests to destruction submit the equipment to exaggerated conditions in an effort to determine changes to be made in design, materials, or use.

DETERGENT OILS

Lubricating oils containing additives that retard the formation of gums, varnishes, and other harmful engine deposits. The detergents act to keep all products of oxidation and other foreign matter in suspension, which permits it to be removed by the engine's filtering system.

DETRITAL
Pertains to rocks, minerals, and sediments formed from detritus, i.e., loose sand, silt, clay, etc. *See* Detritus.

DETRITUS
A term for mineral material or loose rock fragments resulting from abrasion or the breaking up of older rocks, moved from the place of origin by wind or water. Examples of detritus or detrital material are sand, silt, and clay.

DETROIT IRON
A humorous reference to a large, old car or truck.

DEVELOPMENT
The drilling and bringing into production of wells in addition to the discovery well on a lease. The drilling of development wells may be required by the express or implied covenants of a lease.

DEVELOPMENT CLAUSE
The drilling and delay rental clause of a lease; also, express clauses specifying the number of development wells to be drilled.

DEVELOPMENT WELLS
Wells drilled in an area already proved to be productive.

DEVIATED HOLE
A well bore that is off the vertical either by design or accident.

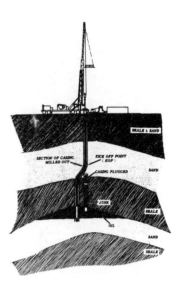

Deviated hole

DEWATERING
Removing produced water from produced oil. This is a necessary and time-consuming step in processing crude oil for the sales line or for temporary storage. Improvements in dewatering technologies have resulted in reducing significantly the oil/water separator-residence time. *See* Hydrocyclone.

DEW POINT
The temperature at which water vapor condenses out of a gas at 14.7 psia (pounds per square inch absolute) or at sea level.

DIAGENESIS
The transformation of sediment, after it is deposited, into rock. It involves processes that change a mass of sediment into coherent rock—rock that holds together firmly as a unit. Processes, both physical and chemical, operating during diagenesis can increase or decrease volume, enlarge or reduce porosity and permeability, and change or rearrange the mineral makeup of the rock.

DIAGNOSTIC FOSSIL

A fossil species or genus that is characteristic of a formation, zone, or stratigraphic unit. The fossil either appears only in the subject formation or is especially abundant; key fossil; characteristic fossil; index fossil; guide fossil.

DIAGONAL OFFSET WELL

A direct offset well is located due north, south, east, or west of the original well. A diagonal offset is a well located diagonal to the original well. A common well-spacing arrangement is one well per 40 acres. If there are 20-acre plots, the wells cannot be direct offsets but must be diagonal offsets—about a 45° angle from the well being offset. This pattern places the wells far enough apart to satisfy the distance requirements.

Diamond bit

DIAMOND BIT

See Bit, Diamond.

DIAMOND SHEAR BIT

See Bit, Diamond Shear.

DIAPIR

An anticlinal fold or dome in which the overlying rocks have been broken by the movement (upward) of softer, plastic core material. In sedimentary strata, diapirs usually contain cores of salt or sometimes shale; a piercement dome.

DIATOMITE

A soft, friable, siliceous sedimentary rock consisting mainly of siliceous cell walls (frustules) of the diatom, a one-celled aquatic plant closely related to algae. Some diatomite is found in lake deposits, but the largest deposits are in the salt water of oceans. Diatomite, because of its absorptive capacity and chemical stability, is used as filter material and as a filler or extender in paints, rubber, and plastic.

DIE

A replaceable, hardened steel piece; an insert for a wrench or set of tongs that bites into the pipe as the tool is closed on the pipe; a tong key. Also, in the plural, dies are cutters for making threads on a bolt or pipe.

Cathead and diesel drilling engines

DIESEL, RUDOLPH

The German mechanical engineer who invented the internal-combustion engine that bears his name.

DIESEL ENGINE

A four-stroke cycle internal-combustion engine that operates by igniting a mixture of fuel and air by the heat of compression and without the use of an electrical ignition system.

DIESEL FUEL
A fuel made of the light gas-oil range of refinery products. Diesel fuel and furnace oil are virtually the same product. Self-ignition is an important property of diesel fuel, as the diesel engine has no spark plugs; the fuel is ignited by the heat of compression within the engine's cylinders. *See* Diesel Engine; *also* Cetane Number.

DIESEL FUEL & CLEAN AIR ACT, 1990
Section 211 (i) of U.S. 1990 Clean Air Act requires the petroleum industry to differentiate clearly off-highway diesel from highway-use diesel employing a dye-marker system. This stiff regulation became effective in October 1993. *See* Blue 8, Automate.

DIESELING
The tendency of some gasoline engines to continue running after the ignition has been shut off. This is often caused by improper fuel or carbon deposits in the combustion chamber hot enough to ignite the gasoline sucked into the engine as it makes a few revolutions after being turned off.

DIFFERENTIAL PRESSURE
Any difference in pressure between upstream and downstream where there is a restriction in the flow of a liquid or a gas. For example, the difference in pressure between that at the bottom of a well and at the surface or the wellhead; the difference in the pressures in a pipeline on the two sides of a gate valve or orifice.

DIFFERENTIAL-PRESSURE STICKING
Another name for wall sticking, a condition downhole when a section of drillpipe becomes stuck or hung up in the deposit of filter cake on the wall of the borehole.

DIGGER
One who digs or drills a well; a driller.

DIGGING TOOLS
Hand tools used in digging a ditch, i.e., shovels, picks, mattocks, spades.

DIKE
Volcanic rock in a molten state (magma) that has pushed its way to the surface and filled crevices in other kinds of rocks then cooled and solidified. *See* Sill.

DIP
The angle that a geological stratum makes with a horizontal plane (the horizon); the inclination downward or upward of a stratum or bed.

DIP, FAULT-PLANE
The angle a fault plane makes with the horizontal. Fault-plane dips can vary from vertical to nearly horizontal.

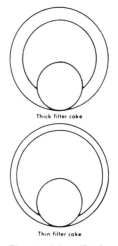

Thick filter cake

Thin filter cake

Pipe stuck in wall cake

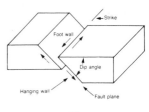

Dip

DIP SLIP
See Slip.

DIRECT CONNECTED
Refers to engines and pumps or electric motors and pumps or compressors that do not have mechanism between them to enhance or reduce the speed, the rpms; the prime mover and the driven machine have identical speeds.

DIRECTIONAL DRILLING
The technique of drilling at an angle from the vertical by deflecting the drill bit. Directional wells are drilled for a number of reasons: to develop an offshore lease from one drilling platform; to reach a pay zone beneath land where drilling cannot be done, e.g., beneath a railroad, cemetery, or lake; and to reach the production zone of a burning well to flood the formation. *See* Killer Well.

Directional drilling

DIRTY CARGO
Bunker fuel and other black residual oils.

DISCHARGE VALVE
See Valve, Discharge.

DISCONFORMITY
An unconformity in which the bedding planes above and below the break (an interruption in a sequence of sedimentary rocks) are parallel or nearly so. This indicates to the geologist that there occurred a long interval of erosion or an interval when no deposits were laid down. A parallel unconformity; erosional unconformity.

DISCONNECT, HOSTILE CONDITIONS
A disconnect is when an offshore drilling vessel, semisubmersible or drillship, must cut loose from its moorings and disconnect from the riser and move off location because of high ocean currents, a hurricane, or other severe weather conditions. After disconnecting from moorings and the large-diameter riser—through which the well is being drilled and is the drilling mud channel—the vessel is moved off and waits out the storm, giving way before the angry weather as little as necessary for safety. When the storm abates, the drilling vessel moves back on location and reconnects and mixes a new batch of mud or receives it *ex situ*. This complicated and often dangerous operation is referred to as Driveoff/Disconnect.

DISCOVERY WELL
An exploratory well that encounters a new and previously untapped petroleum deposit; a successful wildcat well. A discovery well may also open a new horizon in an established field.

DISCOVERY-WELL ALLOWABLE
An allowable above that of wells in a settled field. Some states allow the operators of a discovery well to produce at the maximum efficiency rate (M.E.R.) until the costs of the well have been recovered in oil or gas.

DISCRETIONARY GAS
Gas not committed by short- or long-term contracts; shut-in gas awaiting more favorable market conditions (better price). Discretionary or surplus gas is often sold on the spot market, on a one-time basis.

DISPATCHER
One who directs the movement of crude oil or product in a pipeline system. He receives reports of pumping rates and line pressures and monitors the movement of batches of oil; he may also operate remote, unmanned stations.

DISPOSABLE WELLS
This is not a misprint; a disposable well is a slim-hole well (6 inches in diameter or less) drilled for informational purposes only. Once the downhole intelligence is gathered, recorded, and put to use, the well is plugged and abandoned, P.&A.

DISPOSAL WELL
A well used for the disposal of salt water. The water is pumped into a subsurface formation sealed off from other formations by impervious strata of rock; a service well.

DISSOLVED GAS
Gas contained in solution with the crude oil in the reservoir. *See* Solution Gas.

DISSOLVED-GAS DRIVE
The force of expanding gas dissolved in the crude oil in the formation that drives the oil to the well bore and up to the surface through the production string.

DISTILLATE
Liquid hydrocarbons, usually water-white or pale straw color and of high A.P.I. gravity (above 60°), recovered from wet gas by a separator that condenses the liquid out of the gas stream. (Distillate is an older term for the liquid; today, it is called condensate or natural gasoline.) *See* Condensate.

DISTILLATE FUEL OILS
See Middle Distillates.

DISTILLATE FUEL SYSTEM
A small packaged refinery of 200 to 500 barrels per day capacity or throughput for making diesel fuel, naphtha, and heavy fuel oil. Used

Distilling columns *(Courtesy Shell)*

Distillation system

Steam-powered early-day ditching machine *(Courtesy Lone Star Gas Co.)*

Modern ditching machine

in remote locations (drilling site, pipeline station, production facility) as a source of fuel where a supply of sweet, sulfur-free crude oil is available.

DISTILLATION
The refining process of separating crude oil components by heating and subsequent condensing of the fractions by cooling.

DISTILLATION, DESTRUCTIVE
Distillation is heating to drive off the volatile components of a substance and condensing the gases to a liquid. In destructive distillation there is nothing left of the original substance except an ash, almost pure carbon, after driving off all gaseous components.

DISTILLATION COLUMN
A tall, cylindrical vessel at a refinery or fractionating plant where liquid hydrocarbon feedstocks are separated into component fractions, rare gases, and liquid products of progressively lower gravity and higher viscosity.

DISTILLATION SYSTEM
A small, temporary "refinery" (200 to 1,000 b/d) set up at a remote drilling site to make diesel fuel and low-grade gasoline from available crude oil for the drilling engines and auxiliary equipment.

DISTILLATION UNIT, BASIC
Although they may vary plant-to-plant, a modern basic distillation unit separates crude oil into five raw product streams: naphtha, kerosene, light gas-oil, heavy gas-oil, and reduced crude or residue.

DITCH GAS
Gas encountered downhole at shallow depths and in small amounts.

DITCHING
Making a running trench with a trencher or ditching machine for a pipeline or electric cable or for drainage. *See* Trencher.

DITCHING MACHINE
See Trencher.

DIVERTER SPOOL
An element in a subsea diverter stack located above the choke and kill valves and below a shear ram. The spool has two or more outlet nozzles with integral valving that can be opened simultaneously in the event of a gas kick. This would divert the gas subsea away from the wellhead and floating drilling platform. Gas from a severe kick, if not diverted away from the platform, can aerate the water, reducing its density so that it will not support the floating equipment; the platform and attending vessels sink.

DIVERTER SYSTEM

An assembly of nipples and air-actuated valves welded to a well's surface or conductor casing for venting a gas kick encountered in relatively shallow offshore wells. In shallow wells there is often insufficient overburden pressure around the base of the conductor casing to prevent the gas of a substantial kick from blowing out around the casing. When a kick occurs, the blowout preventer is closed and the valves of the diverter system are opened to vent the gas harmlessly to the atmosphere.

DIVESTITURE

Specifically as it relates to the industry, to break up, to fragment an integrated oil company into individual, separate companies, each to be permitted to operate within only a single phase of the oil business: exploration, production, transportation, refining, or marketing.

DIVISION ORDER

A contract of sale to the buyer of crude oil or gas directing the buyer to pay for the product in the proportions set forth in the contract. Certain amounts of payment go to the operator of the producing property, the royalty owners, and others having an interest in the production. The purchaser prepares the division order after determining the basis of ownership and then requires that the several owners of the oil being purchased execute the division order before payment for the oil commences.

D.M.E.: DIMETHYLETHER

Natural gas or methane from coalbeds are building blocks for producing oxygenates for use in the manufacture of cleaner burning gasolines. Air Products & Chemical Co., working under a grant from the Department of Energy (D.O.E.), developed a one-step approach to synthesizing dimethyl ether. The new one-step method reduces the cost of the oxygenate.

D.M.W.D. or M.W.D.

See Downhole Measurement While Drilling.

DOCTOR SWEET

A term used to describe certain petroleum products that have been treated to remove sulfur compounds and mercaptans that are the sources of unpleasant odors. A product that has been so treated is said to be "sweet to the doctor test."

DOCTOR TEST

A qualitative method of testing light fuel oils for the presence of sulfur compounds and mercaptans, substances that are potentially corrosive and impart an objectionable odor to the fuel when burned.

Doghouse

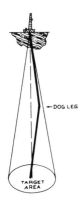

Doglegged hole

D.O.E.'S NEW DIRECTION
The Department of Energy stated in mid-1993 that the agency is redirecting its research and development priorities. The Secretary said, "By re-inventing D.O.E., we are placing the emphasis on energy conservation and efficiency, renewable energy sources, natural gas utilization, and transfer of technologies from the labs to boost competitiveness, as well as several other beneficial research and science projects."

DOGHOUSE
A portable, one-room shelter (usually made of light tank iron) at a wellsite for the convenience and protection of the drilling crew, geologist, and others. The doghouse serves as lunchroom, change house, dormitory, and a room for keeping small supplies and records.

DOG IT
To do less than one's share of work; to hang back; to drag one's feet.

DOGLEG
A deviation in the direction of a ditch or the borehole of a well; a sharp bend in a joint of pipe. *See* Keyseat.

DOG ROBBER
A loyal aid or underling who does disagreeable or slightly unorthodox (shady) jobs for his boss; a master of the "midnight requisition."

DOLLY
Metal rollers fixed in a frame and used to support large-diameter pipe as it is being turned for welding; a small, low platform with rollers or casters used for moving heavy objects.

DOLOMITE
(1) A sedimentary carbonate rock; a variety of limestone or marble rich in magnesium carbonate. Dolomite is closely related to limestone and often occurs interbedded with it. In these instances it is assumed that the dolomite, which is about 90 percent magnesium carbonate, replaced the limestone after the limestone was deposited. (2) A common rock-forming mineral, magnesium carbonate: $CaMg (CO_3)^2$. The mineral is white, colorless, pink, yellow, brown, or gray in color, has perfect rhombohedral cleavage and exhibits a pearly luster. Dolomite is found in extensive beds as dolomite rock and as the material found in veins, cracks, and fissures in other type rocks.

DOLOMITE RESERVOIRS
As the shrinking of the solid volume of limestone takes place during the transformation to dolomite, the resulting rock exhibits higher porosity than the original limestone. A limestone that has been incompletely

dolomitized, with some dolomite mineral in its make up but more conspicuous amounts of calcite, the stone is called "dolomitic limestone."

DOLOMITIC CONGLOMERATE
A conglomerate stone consisting of limestone pebbles cemented together with the mineral dolomite.

DOLOMITIZATION
The process by which limestone is wholly or partly converted to dolomite rock or dolomitic limestone by the replacement of the original calcium (calcite) by magnesium carbonate (mineral dolomite), usually through the action of magnesium-bearing water (seawater or percolating meteoric water). It may occur shortly after the deposition of the limestone or during lithification (turning to stone of the limey sediment) at a later period.

DOLOSTONE
A word that has been around since 1948 when it was proposed by geologist and author, R.R. Shrock, to differentiate dolomite the rock from dolomite the mineral. As careful readers will have ascertained, dolostone is the rock.

DOME
An incursion of one underground formation into the one above it; an upthrust, as a salt dome, that penetrates overlying strata.

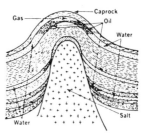

Dome

DOME, NONPIERCEMENT
A rounded, upthrust knoll on the surface caused by the incursion of one underground formation into the one above it. This causes the upward bulge that may be evident on the surface. The nonpiercement lifting of an underground formation by an upthrust or thickening, as in a salt bed, results in an anticline. *See* Salt Dome; *also* Dome, Piercement.

DOME, PIERCEMENT
In a piercement dome, one underground formation has thrust upward into the formation above it, piercing and causing faulting in the general configuration of an anticline. A nonpiercement dome is caused by an underlying formation shouldering into the one above it, also causing an anticline with its characteristic dome-like structure. *See* Salt Dome; *also* Dome, Nonpiercement.

DONKEY PUMP
See Pump, Donkey.

DOODLE BUG
A humorous term, slightly derogatory, for a device for a no-nonsense, direct way to find oil; a black box, usually a magnetometer or a gravimeter.

DOOR MATS

Colloquial term used in the early days to describe small tracts, one-twentieth of an acre, just large enough to accommodate an oil derrick. The concept of pooling had not yet been accepted.

DOPE

Any of various viscous materials used on pipe or tubing threads as a lubricant and corrosion preventive; a tar-base coating for pipelines to prevent corrosion.

DOPE CHOPPER

A machine for removing tar and asphalt coatings from line pipe. The pipe is placed in the chopper where guillotine-like blades cut through the dope but do not damage the pipe. The chunks of coating fall onto a conveyor belt and are carried away from the job.

DOPE GANG

Workers who clean and apply a coat of enamel primer to a pipeline in preparation for coating with a tar-base anticorrosion material and wrapping with tough paper bandage. On large pipeline projects, a machine rides the pipe, cleaning it with rotating metal brushes and then spraying on a primer. A second machine, also riding the pipe, coats and wraps the line in one operation.

DOPE KETTLE

A box-like vessel for heating coal tar or asphalt to liquefy it so it may be used as pipe coating before the pipe is put in the ditch and covered. Dope kettles are mounted on wheels and have an integral fire box beneath the kettle. Dope kettles are now almost obsolete except for small jobs on small lines. *See* Coating and Wrapping.

DOUBLE-ACTION PUMP

See Pump, Double-Action.

DOUBLE-DISPLACEMENT PUMP

See Pump, Double-Displacement.

DOUBLE-HULL REQUIREMENTS

Under the Oil Pollution Act of 1990, all tankers entering U.S. waters after the year 2010 must be double-hulled.

DOUBLE-JOINTING OF LINE PIPE

See Doubling Yard.

DOUBLES

Drillpipe and tubing pulled from the well two joints at a time. The two joints make a stand of pipe that is set back and racked in the derrick. Three-joint stands are called "thribbles," fours are "fourbles."

DOUBLING YARD
An area convenient to a large pipeline construction project where line pipe is welded together in 2-joint lengths preparatory to being transported to the job and strung along the right-of-way.

DOWNCOMER
A pipe in which the flow of liquid or gas is downward.

DOWNHOLE
A term to describe tools, equipment, and instruments used in the well bore; also, conditions or techniques applying to the well bore.

Pipe-handling in doubling yard

DOWNHOLE ASSEMBLIES
There are four downhole assemblies in general use for different drilling conditions. (1) Slick or smooth assembly consisting of regular drillpipe, drill collars at the lower end of the string, and the drill bit. It's called "slick" because it has no reamers or hole openers protruding from the string of drillpipe. (2) Packed-hole assembly is a drillstring consisting of drillpipe, a number of reamers (three- and six-point cutting elements respectively), drill collars and then the bit. (3) Pendulum assembly: Drillpipe, a three-point reamer, several heavy drill collars, and the drill bit. (4) Downhole motor assembly: Drillpipe, several heavy drill collars, a bent sub, and the downhole motor to which the bit is attached.

DOWNHOLE HAZARDS TO WELL CONTROL
Shallow gas; abnormal temperature and pressure gradients; extensive reservoir with high permeability and deliverability; hazardous fluids (gases) H_2S, CO_2; lost circulation.

Doubling yard (Courtesy Reading & Bates)

DOWNHOLE MEASUREMENT WHILE DRILLING
A downhole "real time" data gathering and transmitting system that sends information from the drill bit to the surface by one of several means. The data transmitted by some form of telemetry—hardwire or electronics or hydraulic impulse—includes drilling angle and rate, temperature, type formation, and condition of the bit. The M.W.D. system is the most advanced yet developed to keep the driller and geologist informed on conditions several thousand feet downhole; D.M.W.D.

DOWN IN THE BIG HOLE
A slang expression meaning to shift down into the lowest gear.

DOWNSTREAM
Refers to facilities or operations performed after those at the point of reference. For example, refining is downstream from production operations; marketing is downstream from refining.

Bulldozer

Drag line

DOWNTHROW

The amount of vertical displacement downward of a fault; also, the downward or downthrown side of a fault.

DOWSING RODS

See Doodlebug.

DOZER

Bulldozer; a crawler-type tractor equipped with a hydraulically operated blade for excavating and grading.

DRAG BIT

See Bit, Drag.

DRAG LINE

A type of large excavating machine made with a long boom over which a line runs down to a clamshell bucket. The bucket at the end of the line is swung into position and is then dragged into the material to be moved or dug out.

DRAG-LINE BUCKET

A clamshell bucket; the bucket at the end of the drag line.

DRAG REDUCER

See Flow Improver.

DRAG UP

To draw the wages one has coming and quit the job; an expression used in the oil fields by pipeline construction workers and temporary or day laborers.

DRAINAGE

Migration of oil or gas in a reservoir owing to a pressure reduction caused by production from wells drilled into the reservoir. Local drainage is the movement of oil and gas toward the well bore of a producing well.

DRAINAGE AREA

The area from which a single well can produce oil or gas from a reservoir. The area of drainage depends a great deal on the permeability of the formation the well is producing from. Good permeability of the rock ensures drainage from a larger area than if the formation rock is tight.

DRAINAGE REGIONS

Separate, isolated areas in a subsurface reservoir. Their isolation precludes drainage from the other areas when only one is penetrated by a vertical well. A horizontal well, however, can drill into or across the barriers and receive drainage from all the mini-reservoirs.

DRAINAGE TRACT
A lease or tract of land, usually offshore, immediately adjacent to a tract with proven production; an offshore Federal lease contiguous to producing property whose subsurface geologic structure is a continuation of the producing acreage and therefore more or less valuable as a source of additional oil or gas.

DRAINAGE UNIT
The maximum area in an oil pool or field that may be drained efficiently by one well so as to produce the maximum amount of recoverable oil or gas in such an area.

DRAIN HOLE
The borehole of a horizontal well is the drain hole which can extend laterally through the pay sand for several thousand feet. When cased and perforated, the horizontal borehole is an ideal drain hole.

DRAKE, "COLONEL" EDWIN L.
The man who drilled the country's first oil well near Titusville, Pennsylvania, in 1859 to a depth of 69 1/2 feet using crude cable-tool equipment.

DRAKE'S DRILLER, "COLONEL" EDWIN L.
"Colonel" Drake's head driller was "Uncle" Billie Smith of Titusville, Pennsylvania.

DRAPE FOLD
A fold in layered rocks by an underlying brittle block, at a high angle to the layered rock.

A modern drawworks
(Courtesy National)

DRAPING
The general concordance or agreement of warped strata, lying above any hard, mandrel-like core, e.g., a limestone reef, to the upper surface of the core, owing to initial dip or compaction.

DRAWING THE FIRES
Shutting down a refinery unit in preparation for a turnaround.

DRAWWORKS
The collective name for the hoisting drum, cable, shaft, clutches, power-take-off, brakes, and other machinery used on a drilling rig. Drawworks are located on one side of the derrick floor and serve as a power-control center for the hoisting gear and rotary elements of the drill column.

Spindletop's drawworks

Dresser sleeves

DRESSER SLEEVE

A slip-type collar that connects two lengths of plain-end (threadless) pipe. This type of sleeve connection is used on small-diameter, low-pressure lines.

DRESS-UP CREW

A right-of-way gang that cleans up after the construction crews have completed their work. The dress-up crew smooths the land, plants trees and grass, and builds fences and gates.

DRIFTABILITY

Refers to the concentricity of tubular goods: drillpipe, casing, well tubing. A driftable pipe's inside diameter (I.D.) does not vary from dimension; the drift mandrel can be run through without encountering any kinks, bends, or flat places. Driftability is an important attribute of tubulars because downhole tools—packers, plugs, liners, fishing tools—must be run through the casing or tubing and must be able to move freely.

DRIFT DIAMETER

The inside diameter (I.D.) of a joint of casing, tubing, or line pipe. *See* Drifting the Pipe.

DRIFTING THE PIPE

Testing casing or tubing for roundness; making certain there are no kinks, bends, or flat places in the pipe by use of a drift mandrel or jack rabbit. Pipe must be of proper diameter throughout to be able to run downhole tools such as packers, plugs, etc.

DRIFT MANDREL

A device used to check the size of casing and tubing before it is run. The drift mandrel (jack rabbit) is put through each joint of casing and tubing to make certain the inside diameters are sizes specified for the particular job.

DRILLABLE TOOLS

Packers and downhole cementing tools—float collars, cementers, and plugs—that can be drilled out with polycrystalline diamond bits (P.C.D.).

DRILL BIT, ANTI-WHIRL

A specially designed polycrystalline diamond compact bit with cutting elements positioned to "create a net imbalance force." Such a force, the makers aver, holds or thrusts the bit against the side of the borehole as it rotates, thus creating a stable rotating condition that prevents backward whirling or "dancing."

DRILL BIT, PILOT

A bit sometimes used to drill a deviated hole when necessary to sidetrack an object in the borehole. In drilling a rathole with a whipstock, to change direction, a pilot bit and reamer are used. Such a bit has a projection below the cutting face, with an inverted conical configuration like a carpenter's countersinking bit, and is smaller in diameter than the borehole; the following reamer brings the hole to gauge.

DRILL-BIT METAMORPHISM

A phenomenon where certain rocks or minerals in a normal sedimentary sequence by the action of the drill bit are altered. The extreme localized heat and pressure of thousands of pounds per square inch exerted by the bit on these rocks results in a type of metamorphism that has been termed "bit metamorphism." The phenomenon has been investigated employing petrographic methods, X-ray diffraction, chemical analysis, and scanning electron microscopy, which resulted in the original suspicions that material from the normal sedimentary section downhole was fundamentally altered by the drilling process (the extremes of temperature and pressure) to result in the anomalous cuttings.

DRILL BITS, T.S.D.

Thermally-stable diamond drill bits representing an advance in rock breaking or cutting in deep, harsh environment (hot) gas wells. On high-power motors (high speed: 300 to 1,000 rpm), the T.S.D. bits reduce drilling time, making gauge hole considerably faster than the polycrystalline diamond compact (P.D.C.) or roller cones because of their thermal stability. Conventional rotary drill-bit speed—50 to 200 rpm.

DRILL COLLAR

A heavy, tubular connector between drillpipe and bit. Originally, the drill collar was a means of attaching the drill bit to the drillpipe and to strengthen the lower end of the drill column, which is subject to extreme compression, torsion, and bending stresses. Now the drill collar is used to concentrate a heavy mass of metal near the lower end of the drill column. Drill collars were once a few feet long and weighed 400 or 500 pounds. Today, because of the increased bit pressure and rapid rotation, collars are made up in 1,000-foot lengths and weigh 50 to 100 tons.

DRILL COLLAR SLIPS

Special wedge-shaped steel inserts that fit around the pipe and are held by a bushing in the rotary table. Drill collar slips are made to hold the different sizes and shapes of the heavy tubulars, which are sometimes of smaller outside diameter and in some cases are square in cross section. Also, drill collars are made of "heavy metal," spent uranium or tungsten, to give them added weight. This fact and the different configurations make it necessary to have special slips for drill collars.

DRILL COLLAR, SQUARE

A type of drill collar whose cross section is square instead of circular as in a more conventional collar. Square drill collars are used to prevent or minimize the chances of becoming hung up or stuck in a dogleg downhole. The square corners on the collar, which is located just above the drill bit in the string, act as a reamer and tend to keep the hole passable for the drillpipe.

DRILL COLUMN
The drillstring; the assembly of drilling tools and drillpipe in the borehole, from kelly joint at the rig floor to the drill bit at the bottom of the hole. The drill column may include drill collars and stabilizers that are screwed onto the drillpipe, the drill collars just above the bit and the stabilizers above the collars.

DRILLED-IN PILING
See Piling, Drilled-In.

DRILLED SOLIDS
Rock particles broken and pulverized by the bit and picked up by the drilling mud as it circulates. If the minute rock particles do not drop out in the mud pits or are not removed by surface equipment, they add to the mud's density. This condition can cause serious drilling and circulation problems. *See* Drilling-Mud Density.

DRILLER
One who operates a drilling rig; the person in charge of drilling operations and who supervises the drilling crew.

Driller's console

DRILLER'S CONSOLE
See Information Console.

DRILLER'S LOG
See Log, Driller's.

DRILLHOLE SWEEP
Cleaning the borehole, cased and uncased (open hole), of cuttings, rock chips, sand, and silt by putting all available equipment pumping down the drillpipe and forcing the mud and cuttings up the annulus into the mud tanks or mud pits. On offshore wells, it is imperative to keep the riser free of cuttings. If they accumulate because of the size of the riser pipe and the large volume of drill fluid, additional sweep energy is brought to bear by pumping through the choke or kill line at the bottom of the riser. On certain wells, this purging of the hole and pipes must be done frequently. When hole cleaning is not adequate without continual sweeping, it is advisable to increase the low-shear rate, viscosity better to transport the well cuttings up and out of the hole.

Air drilling rig

DRILLING, AIR
The use of air as a drilling fluid. In certain types of formations, air drilling is considered a better medium than conventional drilling mud.

It is more economical (mud is expensive and the preparation of the slurry and maintaining its condition is time consuming), drilling rates are higher, penetration is faster, and bit life is longer. Although air does a good job of cooling the bit and bringing out the pulverized rock, it has severe limitations. With air drilling, water in the subsurface formations and downhole gas pressure cannot be controlled. When drilling in an area where these two types of intrusions may occur, a mud system must be on standby to avert possible trouble.

DRILLING, CABLE-TOOL
A method of drilling in which a heavy metal bit, sharpened to a point, is attached to a line which is fastened to a walking beam that provides an up-and-down motion to the line and tool.

DRILLING, DELINEATION
Step-out wells that are drilled to establish the boundaries of the play, the trend or the pool. Such a drilling program is heir to a high percentage of failures, dry holes.

DRILLING, EXTENDED REACH
Extended reach drilling is drilling horizontally from a borehole that is begun as a vertical bore. By the use of angle-building assemblies, the drill gradually assumes a horizontal attitude and drills laterally the productive formation. Extended reach drilling is used principally on offshore platforms to cover a large area of an outer continental shelf (O.C.S.) lease. As many as 60 wells have been drilled from a large platform. With the advances in angle-building techniques, using mud motors, extended reach drilling has made significant progress. Some operators have plugged close-in wells and used the platform drilling slots for extended reach wells.

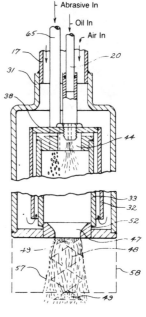

Flame jet drill *(After Browning and Fitzgerald)*

DRILLING, FLAME-JET
A drilling technique that uses jet fuel to burn a hole through rock strata. This leaves a ceramic-like sheath on the walls of the borehole, eliminating the need for casing.

DRILLING, HORIZONTAL
A modern directional-drilling technique using mud motors to begin a well, drilling vertically then diverting the borehole a few degrees from vertical every 50 to 100 feet (angle building) until the well bore is horizontal. The procedure is very effective in producing from thin but porous and permeable formations. To produce from the long axis of a 30 or 60-foot interval is markedly more efficient than vertically across the formation.

DRILLING, INFILL
Wells drilled to fill in between established producing wells on a lease; a drilling program to reduce the spacing between wells in order to increase production from the lease.

DRILLING, KELLY

Drilling, making hole, with the kelly joint turned by the rotary table; the sequence of applied torque: drilling engine, gear box, drive shaft, rotary table, kelly, drillpipe, and drill bit. *See* Mud Motor.

DRILLING, REVERSE CIRCULATION AIR

A method of damage-free drilling in which drilling air is forced down the annulus and up through the drillpipe, bringing out the cuttings. A complete reversal of conventional drilling methods. The advantages cited for the experimental reverse circulation air-drilling are the notably cleaner condition of the borehole; fine-particle dust was vented up the drillpipe and out through the blooie pipe to the pit. Therefore, the skin of any potential productive interval was not clogged or dusted over. Another plus for reverse air-drilling: Small shows or relatively small shows from weak formations can be identified and included in normal completion procedures.

Rotary rig

DRILLING, ROTARY

Drilling a borehole for an oil or gas well with a drill bit attached to joints of hollow drillpipe that are turned by a rotary table on the derrick floor. The rotary table is turned by power from one or more drilling engines. As the bit is turned, boring, cutting, pulverizing the rock, drilling mud is pumped down the hollow drillpipe, out through "eyes" or ports (holes) in the bit, and back up the borehole carrying the rock cuttings to the surface. The drillstring, to which the drill bit is attached, is made up of the kelly joint, lengths of drillpipe, a stabilizer, several heavy tool joints, and then the bit. When the hole is drilled to the producing formation, the borehole is cased and the casing is cemented to prevent any water above the pay zone from entering the hole. After the casing is cemented, it must be perforated to permit the oil from the pay zone to enter the well bore.

DRILLING, SLANT-HOLE

A procedure for drilling at an angle from the vertical by means of special downhole drilling tools to guide the drill assembly in the desired direction. Slant holes are drilled to reach a formation or reservoir under land that cannot be drilled on, such as beneath a town site, a water-supply lake, a cemetery, or industrial property where direct, on-site drilling would be impractical or unsafe. Slant holes are also drilled to flood a formation with water or mud to kill a wild or burning well. *See* Killer Well.

DRILLING, SLIDE

Drilling with a downhole, adjacent-to-the-bit mud motor where nothing in the drillstring rotates except the bit. The bit drills ahead and the drillstring slides along, pushed by the weight on the bit, the weight the driller allows on the string of drillpipe. Slide drilling, a new term since the advent to the mud motor and the nonrotating drillpipe.

DRILLING, TIME

Drilling slowly (with reduced rpms) with a pendulum assembly and a cement plug in the hole in an attempt to straighten a crooked hole or to sidetrack around an obstruction. By the use of a cement plug in a deviated section of the hole, the weighted lower end of the drillstring and bit (using heavy drill collars) has a tendency to stay vertical like a pendulum at rest. With time or slow drilling, the cement plug keeps the bit from following the previously drilled borehole and permits the pendulum assembly to work at drilling straight down.

DRILLING AND BELLING TOOL

A long, large-diameter, cylindrical drill with articulating cutting blades folded into the body of the drill for digging holes for piling-in offshore installations: drilling, producing, or production platforms. The tool, two to 4 feet in diameter, is constructed so that when it reaches the required depth of a few hundred feet the hinged cutting blades are extended to cut out a bell-shaped cavity at the bottom of the borehole. Piles then can be inserted and cemented. This operation is "drilling in" the piling instead of the more common method of driving the piling.

DRILLING BLOCK

An area composed of separate contiguous leaseholds large enough to drill an exploratory well. Before drilling such a well, particularly a deep well, the operator will usually try to acquire a sizeable block of leases surrounding the site of the proposed exploratory well.

DRILLING BREAK

A sudden, incrementally large change in the drilling rate, the rate of penetration, viz., 5 feet per hour to 12 feet per hour, which indicates a softer interval of rock—something not anticipated, a pocket, a lens. Drilling breaks alert the driller to be on the lookout for something not in the well plan, a gas kick or a thief zone.

DRILLING CABLE

A heavy cable, 1 to 2 inches in diameter, made of strands of steel wire.

DRILLING CHOKE

A safety valve in the drillstring to control a gas kick or a threatened blowout up the drillpipe. The automatic valve has a spring-loaded flapper held open by the flow of drilling mud down the drillpipe. If the circulation of the drilling mud is lost for any reason or a pocket of high-pressure gas is drilled into, the flapper valve closes automatically preventing the blowout through the drillpipe. *See* Kelly Cock.

Drilling and belling tool
(Courtesy Calweld)

Drilling cable

DRILLING CLAUSE, EXPRESS
A clause inserted in an oil or gas lease that requires the lessee to drill a well within a certain period of time and to a specified depth. Almost any condition can be inserted into an oil lease, so it is encumbent upon the lessee to read the lease, fine print and all.

DRILLING CONTRACT
An agreement setting forth the items of major concern to both the operator and drilling contractor in the drilling of a designated well at a given location and at a specified time. One standard drilling contract form is the American Petroleum Institute's (A.P.I.); another is the American Association of Oilwell Drilling Contractors' (A.A.O.D.C.). *See* A.P.I. Bid Sheet and Well Specifications.

DRILLING CONTRACTOR
A person or company whose business is drilling wells. Some wells are drilled on a per-foot basis, others are contracted for on a day rate. *See* Turnkey Job.

DRILLING COSTS, INTANGIBLE
See Intangible Drilling Costs.

DRILLING-DELAY RENTAL CLAUSE
The clause in an oil and gas lease giving the lessee the right to maintain or hold the lease from year to year during the primary term either by starting a well, commencing operations, or paying delay rentals. Drilling-delay clauses are put in by the lessee because the courts have ruled that such a clause nullifies any implied covenant to drill a test well on the property. The lessor doesn't care because the clause calls for periodic rental payments. *See* Unless Lease, *also* Or Lease; Delay Rental.

DRILLING FLOOR
See Derrick Floor.

Drilling floor

DRILLING FLUID, BIODEGRADABLE
An environmentally acceptable, biodegradable drilling fluid, developed and patented by Baroid and Henkle HGaA, is used in offshore drilling. The fluid's base ester is derived from vegetable oil, and it exhibits the lubricity and shale stabilization qualities of a mineral oil-based drilling fluid. The base ester is synthesized in the laboratory.

DRILLING FLUIDS DESIGN
Drilling fluids perform several important functions in a drilling well. In high-angle and horizontal drilling, muds are even more critical and are designed to consider all of the following functions: To clean and lubricate the bit and downhole tools; balance formation pressure; control gas

kicks; stabilize sensitive formations; clean the hole of rock cuttings; suspend solids; and drive downhole turbo-drilling motors. The person responsible for designing muds to accomplish these various functions is the rheologist, the mud or fluids engineer.

DRILLING FLUID SPECIALIST
A mud engineer.

DRILLING FUND
A general term describing a variety of organizations established to attract venture capital to oil and gas exploration and development. Usually the fund is established as a joint venture or limited partnership with minimum investments of $5,000–$10,000. Such funds attract high-bracket persons who will receive certain tax benefits.

DRILLING HEAD, ROTATING
A heavy casting bolted to the top of the blowout preventers on the casing through which air or gas drilling is done. The kelly joint fits in the rotating element of the drilling head. Compressed air, as the drilling fluid, enters the drillstring through a flexible hose attached to the kelly. As the bit pulverizes the rock, the chips are brought back up the annulus by the force of the high-pressure air and are vented through the blooie pipe to the burn pit.

DRILLING IN
The procedure of drilling into a producing formation, or the pay zone. After the casing is run and landed just above the oil- or gas-bearing formation, the hole is drilled into the formation, tapping the productive zone. On high-pressure wells the casing is run and securely cemented before drilling in so as to have the well under control, i.e., to be able to control the pressure within the wellhead valves, the blowout-preventer stack, that are attached to the top of the casing at the surface.

Drilling and production island

DRILLING ISLAND
A man-made island constructed in water 10 to 50 feet deep by dredging up the lake or bay bottom to make a foundation from which to drill wells. This procedure is used for development drilling, rarely in wildcatting.

DRILLING JARS
See Jars.

DRILLING JARS, HYDRAULIC
See Jars, Hydraulic.

DRILLING LOG
See Log, Driller's.

Drilling mast

DRILLING MAST

A type of derrick consisting of two parallel legs, in contrast to the conventional four-legged derrick in the form of a pyramid. The mast is held upright by guy wires. This type mast is generally used on shallow wells or for reconditioning work. An advanced type of deep drilling rig employs a mast-like derrick of two principal members with a base as an integral part of the mast.

DRILLING MOTORS, DOWNHOLE

See Mud Motor; Mud Motor, Positive-Displacement; Mud Motor, Turbine.

DRILLING MUD

A special mixture of clay, water, and chemical additives pumped downhole through the drillpipe and drill bit. The mud cools the rapidly rotating bit; lubricates the drillpipe as it turns in the well bore; carries rock cuttings to the surface; and serves as a plaster to prevent the wall of the borehole from crumbling or collapsing. Drilling mud also provides the weight or hydrostatic head to prevent extraneous fluids from entering the well bore and to control downhole pressures that may be encountered.

DRILLING MUD, OIL-BASE

See Oil-Base Mud.

DRILLING-MUD DENSITY

The weight of drilling mud expressed in pounds per U.S. gallon or in pounds per cubic foot. Density of mud is important because it determines the hydrostatic pressure the mud will exert at any particular depth in the well. In the industry, mud weight is synonymous with mud density. To "heavy up on the mud" is to increase its density.

DRILLING OPERATIONS CLAUSE

A saving clause that operates to keep a lease alive after the expiration of the primary term despite the failure to obtain production by that time if drilling operations are being pursued. There is limited authority and very little chance that without such a clause a lease may be kept alive by drilling begun before the expiration of the primary term, even though production results shortly after the expiration date.

DRILLING PERMIT

In states that regulate well spacing, a drilling permit is the authorization to drill at a specified location; a well permit.

DRILLING PLATFORM

An offshore structure with legs anchored to the sea bottom. The platform, built on a large-diameter pipe frame, supports the drilling of a number of wells from the location. As many as 60 wells have been drilled from one large offshore platform.

Drilling platform under construction showing drilling slots

DRILLING PLATFORM, MAT-SUPPORTED
See Mat-Supported Drilling Platform.

DRILLING PLATFORM, MONOPOD
See Monopod Platform.

DRILLING RIG, FOX HOLE
See Fox Hole Drilling Rig.

DRILLING SLOTS
Positions on an offshore drilling platform for additional wells. When a successful well is drilled offshore, other wells are put down slanted out at an angle from the platform by directional drilling. On large offshore platforms, there may be as many as 40, even 60, wells drilled into the reservoir. If all of the multiple wells are successful and the total daily production warrants, the drilling platform will be converted to a producing platform. Drilling equipment is removed, a manifold of well-control valves is built, and pumping equipment installed to move the crude to a production platform where the oil is separated from the produced water, treated with chemicals (if necessary), measured, and pumped to a shore station. (Note the difference between producing and production installations; they are related but quite different in function.)

DRILLING SPOOL
The part of the drawworks that holds the drilling line; the drum of drilling cable on which is spooled the wireline that is threaded over the crownblock sheaves and attached to the traveling block.

DRILLING TENDER
A barge-like vessel that acts as a supply ship for a small, offshore drilling platform. The tender carries pipe, mud, cement, spare parts, and, in some instances, crew quarters.

DRILLING TIME
The time required for the drill bit to drill or penetrate 1 foot of rock; the time required to drill a well exclusive of down time.

DRILLING WITH STANDS
Drilling with stands, as with a top drive system, is drilling with three joints of drillpipe at a time. In rotary-table drilling, one, 30-foot joint is added at a time. After 30 feet of hole, another joint is added to the string. Drilling with stands means the floor men add a stand, three, 30-foot joints to the drill string each time 90 feet of hole is drilled. Stands, 90 feet of pipe at a time, are pulled out of the hole and set back, stood up in the derrick when tripping out; which reduces breaking out time (unscrewing the joints) by as much as one-half. In top-drive rigs, the torque (the

rotary force) is in the power swivel attached to the huge traveling block hanging in the derrick.

Drillship

DRILLPIPE

Heavy, thick-walled steel pipe used in rotary drilling to turn the drill bit and to provide a conduit for the drilling mud. Joints of drillpipe are about 30 feet long.

DRILLSHIP

A self-propelled vessel; a ship equipped with a derrick amid ships for drilling wells in deep water. A drillship is self-contained, carrying all of the supplies and equipment needed to drill and complete a well.

DRILLSTEM

The drillpipe. In rotary drilling, the bit is attached to the drillstem or drill column, which rotates to dig the hole.

DRILLSTEM TEST (D.S.T.)

A method of obtaining a sample of fluid from a formation using a formation-tester tool attached to the drillstem. The tool consists of a packer to isolate the section to be tested and a chamber to collect a sample of fluid. If the formation pressure is sufficient, fluid flows into the tester and up the drillpipe to the surface.

DRILL STRAIGHT UP, TO

An operator who finances his own well without the participation of out-side investors is said to drill straight up.

DRILLSTRING

The kelly joint, drillpipe, drill collars, stabilizer, and drill bit make up a drilling string or, in more common usage, the drillstring or string of tools. *See* Packed-Hole Assembly.

DRILLWELL

See Moonpool.

DRIP

A small in-line tank or condensing chamber in a pipeline to collect the liquids that condense out of the gas stream. Drips are installed in low places in the line and must be blown or emptied periodically.

DRIP GASOLINE

Natural gasoline recovered at the surface from a gas well as the result of the separation of the liquid hydrocarbons dissolved in the gas in the formation; gasoline recovered from a drip in a field gas line; casinghead gasoline.

Hydraulic fracturing a gas well

Offshore drilling platform; note the escape capsule and commuting helicopter

DRIP OILER
See Wick Oiler.

DRIVE
The energy or force present in an oil reservoir that causes the fluid to move toward the well's borehole and up to the surface when the reservoir is penetrated by the drill. A reservoir is very much like a pressure vessel; when a well is drilled into the reservoir, it is as if a valve were opened to vent the pressure. There are several kinds of reservoir drives: gas cap, solution gas, water, and artificial.

DRIVEOFFS, FOUL-WEATHER
See Disconnects.

DRIVE PIPE
A metal casing driven into the borehole of a well to prevent caving of the walls and to shut off surface water. The drive pipe, first used in an oil well by Colonel Drake, was the forerunner of the modern conductor or surface casing.

DRIVE THE HOOPS
To tighten the staves of a wooden stock tank by driving the metal bands or hoops down evenly around the circumference of the tank. Early-day lease tanks were made of redwood in the shape of a truncated cone (nearly cylindrical). Metal bands like those on a wooden barrel held the staves together. Once a year or so, the hoops had to be driven to tighten the seams between the staves to prevent leaks. Today wooden tanks are used on leases to handle salt water and other corrosive liquids. Their staves are held together with steel rods equipped with turnbuckles for keeping the tank watertight.

DROWNING
A colloquial term for the encroachment of water at the well bore into a formation that once produced oil but now produces more and more water.

DRUM
A 55-gallon metal barrel; a standard container used for shipping lubricating oil and other petroleum products.

Drums

DRUM-AND-CABLE PUMPING UNIT
See Pump, Grable Oil-Well.

DRY GAS
Natural gas from the well free of liquid hydrocarbons; gas that has been treated to remove all liquids; pipeline gas.

DRY HOLE
An unsuccessful well; a well drilled to a certain depth without finding oil; a duster.

DRY-HOLE LETTER
A document similar to a bottom-hole letter except that a dry-hole letter creates an obligation to pay the operator a specified amount upon drilling a dry hole. The agreement usually is in the form of a letter that both parties sign. Dry-hole letters are used to help finance a well by providing collateral for a loan for drilling expenses.

DRY-HOLE MONEY
Money paid by one or more interested parties (those owning land or a lease nearby) to an operator who drills a well that is a dry hole. The well, whether successful or dry, serves to "prove their land," providing useful information. Before the well is drilled, the operator solicits dry-hole "contributions" and in return for financial assistance agrees to furnish certain information to the contributors.

DRY-HOLE PLUG
A plug inserted in a well that is dry to seal off the formations that were penetrated by the borehole. This treatment prevents salt water, often encountered in dry holes, from contaminating other formations. *See* Plugging a Well.

DRY OIL
See Wet Oil.

DRY TREE
A Christmas tree installed on land or above water as distinguished from a wet tree, one installed on the sea bed or under water.

D.S.T.
Drillstem test.

DUAL COMPLETION
The completion of a well in two separate producing formations, each at different depths. Wells sometimes are completed in three or even four separate formations with four strings of tubing inserted in the casing. This is accomplished with packers that seal off all formations except the one to be produced by a particular string of tubing.

DUAL DISCOVERY
A well drilled into two commercial pay zones, two separate producing formations, each at a different depth.

DUAL-FUEL ENGINES
Engines equipped to run on liquid as well as gaseous fuel. Stationary engines in the field have modifications made to their carburetors that

Christmas tree of a triple completion well

permit them to operate either on gasoline or natural gas. In some installations, when the gasoline supply is used up, the engine is switched to natural gas automatically.

DUAL-LEG HORIZONTAL WELL
Two horizontal well bores drilled from the same vertical-hole section. For example, a well is taken down to 8,000 feet and there it is kicked off on a lateral leg; the vertical hole is taken on down, and at 9,500 feet the procedure is repeated; the second lateral well is drilled.

DUAL PUMPERS
Two well-pumping units operating through the same wellhead but each pumping through its own well tubing with its own string of rods. A dual completion on the pump. *See* Dual Completion.

DUBAI STORAGE TANK
A specially designed underwater storage tank the shape of an inverted funnel, built by Chicago Bridge & Iron for Dubai Petroleum Company. The tanks have no bottoms and rest on the seafloor supported on their rims. Oil from fields onshore is pumped into the top of the tanks under pressure, forcing the seawater out the bottom. The offshore tanks, which are more than 100 feet tall, also serve as single-point moorings for tankers taking on crude.

DUBBS, CARBON PETROLEUM
Dubbs, a petroleum chemist, developed a cracking process that found wide acceptance in the 1920s and was almost as popular as the Burton still, which was developed earlier for Standard Oil Company of Indiana.

DUCK'S NEST
Colloquial term for a standby drilling mud tank or pit used to hold extra mud, or as an overflow in the event of a gas kick.

DULLS
Badly worn drill bits with lost inserts or broken teeth.

Worn inset bit

DUMP BOX
A heavy wooden or metal box where the contents of a cable-tool well's bailer is emptied. The end of the bailer is lowered into the box which pushes the dart upward, unseating the ball valve and permitting the water, mud, and rock cuttings to empty into the box and slush-pit launder.

DUMP FLOODING
An unusual secondary-recovery technique that uses water from a shallow water bed above the producing pay to flood the oil-producing interval. The water from the aquifer enters the injection string by its own pressure. The weight of the hydrostatic column (water column) produces the necessary force for it to penetrate the oil formation, pushing the oil ahead of it to the producing wells in the field.

Duplex pump *(Courtesy Gaso)*

DUMP GAS

Gas delivered under a dump gas contract, i.e., a gas purchase and delivery contract that does not call for the delivery of a specified amount of gas but that does call for delivery of surplus gas after meeting the terms of a firm gas contract.

DUMP WELL

A disposal well; a well into which produced water (salt water) is pumped. The waste water is pumped downhole into a formation isolated from all others by an impervious layer of rock above and below.

DUPLEX PUMP

See Pump, Duplex.

DUSTER

A completely dry hole; a well that encounters neither gas nor liquid at total depth.

DUTCHMAN

The threaded portion of a length of pipe or nipple twisted or broken off inside a collar or other threaded fitting. Threads thus "lost" in a fitting have to be cut out with a chisel or cutting torch.

D.W.T.

Deadweight ton; a designation for the size or displacement of a ship, e.g., 100,000-dwt crude oil tanker.

DYNAMIC STATIONING

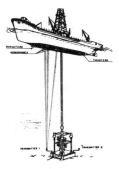

Dynamic stationing

A method of keeping a drillship or semisubmersible drilling platform on target over the hole during drilling operations where the water is too deep for the use of anchors. This is accomplished by the use of thrusters activated by underwater sensing devices that signal when the vessel has moved a few degrees off its drilling station.

E

EARNEST MONEY
A sum of money paid to bind a financial transaction prior to the signing of a contract; hand money.

EASY DIGGING
A soft job; an assignment of work that can be handled without much exertion.

ECCENTRIC
Something that is off center. An eccentric may be a piece of machinery, e.g., a cam, a wheel, or a shaft, whose axis is off center; excentric (derivation).

ECONOMIC DEPLETION
The reduction in the value of a wasting asset by removing or producing the minerals.

ECONOMICS OF A PLAY, FULL-SCALE
The economic analysis of an exploration program. This includes all investments, seismic and geophysical expenses, land costs, staff and overhead, drilling and production costs, even abandonment liabilities.

EDGE WATER
Free water in permeable reservoir rock on the perimeter of the reservoir trap that contains oil and gas. This type of free water in the rock formation surrounding the hydrocarbon deposits is edge water, as distinguished from bottom water.

E.D.M.S.
Electronic Document Management Service. An imaging service that puts all documents, maps, drawings, schematics, general letters of instructions, and engineering specs used by a refinery or a pipeline department on a central computer to be instantly accessed by any company workstation in need of a particular bit of intelligence. An electronic archive, prudently backed up, that makes possible the elimination of voluminous paper files.

EDUCTOR

A form of suction pump; a device using a high-pressure jet of water to create a partial vacuum at an intake opening to draw liquid from a sump.

EFFECTIVE POROSITY

The percent of the total volume of rock that consists of connecting pores or interstices. The part of a rock that is capable of holding a fluid—oil, water, or gas—is the effective porosity.

EFFLUENT

The discharge or outflow from a manufacturing or processing plant; outfall, drainage. *See* Influent.

E.I.A.

Energy Information Agency. A government entity that gathers petroleum industry information, and makes it available on production reserves, and forecasts of markets, both foreign and domestic.

EKOFISK TANK

The first concrete offshore platform installed in the Norwegian quadrant of the North Sea in 1973. The appellation "tank" is from its cylindrical shape. The drilling, production, and accommodation facilities are supported on the truncated concrete cylinder, which has banks and rows of impact-breaking splash holes around its circumferential flanks at and above the splash zone.

ELASTOMER

Any of various elastic materials or substances resembling rubber. The petrochemical industry has produced many types of elastomers that are used for gasket material, guides, swab cups, valve seats, machinery vibration-absorber mounts, etc. Elastomers are highly resistant to chemical decomposition (hydrolysis) in the presence of hydrocarbons, which makes them desirable for use in the petroleum industry—much more so than natural rubber.

ELBOW

A pipe fitting or fabricated short length of pipe in the shape of an L; a 90° turn in a pipe.

ELECTRICALLY OPERATED CONTROL VALVE

See Valve, Electrically Operated Control.

ELECTRICAL RESTIVITY

The electrical resistivity of any material is related to the resistance by the following equation: R = (r x a) L, where *a* is area in square meters of rock material exposed to the current flow; *L*, the length of the material in meters; *r* is electrical resistance in ohms; and *R* is electrical resistivity expressed in Ohm-Meters.

Electrically operated control valve *(Courtesy Magnatrol)*

ELECTRIC LOG
See Log, Electric.

ELECTRIC RIG
See Rig, Electric.

ELECTRIC SERVO MOTORS
See Servomechanism.

ELECTRIC TRACING
See Heat Tracing.

ELECTROLYTICAL TREATMENT OF SEA WATER, THE
The electrolytical treatment of sea water for cooling in the process loop and the use of soft iron anodes reduces the incidence of corrosion. The soft iron anodes set up a weak electrical current that flows to the metals to be protected. The reverse current flow counters the usual flow that is associated with the corrosion of iron and steel. *See* Anode.

ELECTRONIC ARCHIVE
See E.D.M.S.: Electronic Document Management Service.

ELECTROSTATIC PRECIPITATORS
Huge hoppers that filter or electrically attract and capture the dust and other particulate matter in flue gas in fluid catalytic cracking units (F.C.C.U.) and other refinery flues and stacks. Baffle elements in the hoppers through which flue gas travels bear an electric charge which attracts the dust with an opposite charge—the elemental attraction of opposites. It happens even in smoke stacks.

An electrostatic oil treater
(Courtesy C-E Natco)

ELECTROSTATIC TREATER
See Treater, Electrostatic.

ELEPHANT (OILFIELD)
A term used to denote the world's giant oilfields. Prudhoe Bay, in 1970, with an estimated 10-billion-barrel reserve, was an elephant.

ELEVATOR
A heavy, hinged clamp attached to the hook and traveling block by bail-like arms, used for lifting drillpipe, casing, and tubing and lowering them into the hole. In hoisting a joint of drillpipe, the elevators are latched on to the pipe just below the tool joint (coupling), which prevents the pipe from slipping through the elevators.

Zip-lift elevator

ELEVATOR LINKS
See Links.

EMBAYMENT

A large indentation of a coastline; a bay. An embayed coastline.

E.M.F.

Electromotive force.

EMINENT DOMAIN

The right of a government body or public utility (common carrier) to take private property for public use by condemnation proceedings.

EMULSION, OIL-WATER

Very small droplets of water suspended in a volume of crude oil, each droplet surrounded or encased in a film of oil. The water, although heavier than oil, cannot settle to the bottom of the tank until, through the application of heat or mixing with a chemical, the surface tension of the film of oil is reduced sufficiently to free the water droplets. When this occurs, the small droplets join others to form larger ones that have enough mass or weight to settle to the bottom.

EMULSION TREATER

Emulsion treater

A tall cylindrical vessel, a type of oil heater for "breaking down" oil-water emulsions with heat and the addition of certain chemicals. Emulsion treaters have a gas-fired furnace at the bottom of the vessel to heat the stream of oil piped through from the well to the stock tanks; a heater treater.

ENCROACHMENT

The movement of edge water or bottom water into a reservoir as the oil and gas are produced, thus reducing both the reservoir pressure and the volume of oil and gas remaining. Encroachment is a worrisome thing for the producer because he realizes that his wells may soon "go to water," leaving most of the oil (perhaps 70 percent or more) still clinging to the pores of the rock in the reservoir. *See* Channeling.

END-O

The command given by one worker to another or to a group to lift together and move an object forward; a signal to "put out" in a big lift.

ENDOTHERMIC

Refers to a process or chemical reaction that requires the addition of heat to keep it going. Exothermic is the reverse; a process or reaction that once begun gives off heat.

END POINT

The point at which a product or a fraction (hydrocarbon) has totally boiled off or been completely vaporized. The initial boiling point is the temperature at which a product being distilled starts to boil. These two points are called the product's cut points. *See* Boiling Point; *also* Cut Point.

END PRODUCT
Material, substances, goods for consumer use; finished products.

END PRODUCT GAS
See Product Gas.

END USE
Ultimate use; consumption of a product by a commercial or industrial customer.

ENERGY PETROLEUM ALLOCATION ACT OF 1973
This act (E.P.A.A.) was passed in response to the 1973 O.P.E.C. embargo. It provided for mandatory pricing and allocation of crude oil.

ENERGY PRODUCTION AND CONSERVATION ACT
The E.P.C.A. passed in 1975 to amend the Emergency Petroleum Allocation Act (E.P.A.A.). This act established new price controls over various tiers or categories of oil: old oil, new oil, stripper oil, etc.

ENERGY RESEARCH AND DEVELOPMENT AGENCY
This agency was formerly the A.E.C., then in 1977, the E.R.D.A. merged with other agencies to become the U.S. Department of Energy (D.O.E.).

ENERGY SOURCES
Petroleum, coal, hydropower, nuclear, geothermal, synthetic fuels, tides, solar, wind.

ENERGY VALUE OF PETROLEUM AND PRODUCTS
Million Btu per barrel: crude oil, 5.6; distillate fuel oil, 5.8; residual fuel oil, 6.3; gasoline, 5.3; kerosene, 5.7; petroleum coke, 6.0; and asphalt, 6.6. Btu per cubic foot: dry natural gas, 1,031; wet gas, 1,100.

ENGINE, HOT-PLUG
See Hot-Plug Engine.

ENGINE HOUSE
On a cable-tool rig the engine house held the steam-powered drilling engine. Attached to the engine house was the belt hall, which housed the wide, fabric belt that transmitted power from the engine to the bandwheel.

ENHANCED OIL RECOVERY (E.R.O.)
Recovery of crude oil after a well's original rate of production has diminished. This is accomplished, often quite successfully, by secondary

and tertiary recovery methods, such as waterflooding, CO_2 flooding, steam flooding, and steam soak. *See* Five-Spot Waterflood.

ENRICHED-GAS INJECTION

A secondary-recovery method involving the injection of gas rich in intermediate hydrocarbons or enriched by addition of propane, butane, or pentane on the surface or in the well bore as the gas is injected.

ENTITLEMENT PROGRAM PHASEOUT

With the elimination of government price controls on crude oil, the entitlements program ceased to be necessary.

ENTITLEMENTS PROGRAM

A program instituted in 1974 by the federal government to equalize the access to domestic crude by all U.S. refiners—crude oil that was price controlled substantially below world price. The reasoning was that disproportionate access to inexpensive domestic crude would give an unfair advantage to some refiners, those with a large supply of price-controlled oil. The program's aim was to make available to each refiner the same fraction of low-priced oil. Refiners with more price-controlled oil than a calculated national average were required to buy entitlements. Refiners with a lower-than-average amount of price-controlled oil could sell entitlements. The buying and selling of entitlements was between traditional suppliers and purchasers. For example, if Gulf Oil were the traditional supplier of crude oil to Bradford Refining Co. and Gulf had available a larger percent of price-controlled oil than the national average and Bradford had less than the average, Bradford could sell its entitlements for crude-cost equalization to Gulf and Gulf would be required to buy them. In effect, Gulf, the traditional supplier, would pay to Bradford Refining a certain amount of money for each barrel of uncontrolled crude oil it had to buy in the world market.

Degasser

ENTRAINED GAS

Gas that has been picked up or absorbed by a stream of liquid or has entered the stream by pressure and is being carried along. Gas from a formation being drilled through becomes entrained in the drilling fluid and must be removed. *See* Degassing Drilling Mud.

ENTRAINED OIL

Oil occurring as part of the gas stream but as a relatively small percentage of total flow. Special separators are used to remove the liquid from the gas stream.

ENTRAINMENT

The picking up and carrying along as when a current of water collects sediment and moves it along to where it is finally deposited when the stream or current slows down; also, the trapping of gas bubbles in a cement slurry, or water droplets or mist in a gas stream.

ENTRY POSITION
A starting job with a company usually sought by a young man or woman just out of school who wishes to get into the business at whatever level—with the expectation of becoming president in due time.

EOLIAN
Having to do with the wind; may refer to windblown or wind-deposited loess or dune sand; also, erosion and deposition by the wind. From the god of the wind, Aeolus. Has nothing to do with politics, but it seems possible. Also, Aeolian.

E.O.R.
See Enhanced Oil Recovery.

E.P.A.
Environmental Protection Agency.

EPISODE
A term used for a distinctive and significant event or series of events in the geologic history of a region, without implying any particular time limitations. For example, volcanic episode or glacial episode.

E.P. LUBRICANTS
Extreme-pressure lubricating oils and greases that contain substances added to prevent metal to metal contact in highly loaded gears and turntables.

EQUITY CRUDE
In cases where a concession is owned jointly by a host government and an oil company, the crude produced that belongs to the oil company is known as equity crude, as opposed to buy-back (participation) crude. The cost of equity crude is calculated according to the posted price. *See* Buy-Back Crude.

E.R.D. WELLS
Extended reach, high-angle wells. Usually such wells, at angles from the vertical of 60° to 90°, are drilled from offshore platforms with numerous drilling slots. As many as 60 have been recorded.

ERGONOMICS
The science of equipment design with the goal of increased productivity by reducing human fatigue and discomfort on the job; also referred to as Human Engineering or Biotechnology.

EROSIONAL UNCONFORMITY
See Disconformity.

E.R.W. PIPE
Electric resistance welded pipe; (H.F.)E.R.W., high-frequency E.R.W. welded; also induction welded; and D.C. welded. These are some of the methods of welding skelp (flat steel rolls or plate) into all kinds of tubular goods.

ESCAPE BOOMS
Devices used on offshore drilling or production platforms for emergency escape of personnel in the event of a fire or explosion. They consist of counterweighted arms supporting a buoyant head. When the arms are snapped loose from the platform, they fall outward, the head descending to the water. The workers then slide down a lifeline to the floating head.

ESCROW MONEY
See Suspense Money.

ESTUARINE
Pertains to or formed in an estuary, the funnel-shaped mouth of a river where it empties into the sea; the deposits and the biological or sedimentary environment of an estuary.

ET ALS
And others; unnamed participants or interest holders in a deal or a contract; *et al* made plural and used colloquially by oilmen.

ETHANE
A simple hydrocarbon associated with petroleum. Ethane is a gas at ordinary atmospheric conditions.

ETHANOL
Alcohol; one component of gasohol.

EVAPORATION PIT
A man-made excavation to hold salt water to allow the water to evaporate, leaving the salt behind. Such pits are common in stripper-well fields to handle small amounts of brine. But if a large volume of water must be disposed of, disposal wells are drilled as a place to pump the corrosive salt water; a burn pit.

EVAPORITE
A nonclastic sedimentary rock consisting mainly of minerals precipitated from a saline (salty) solution that, over the centuries, has evaporated. Examples of evaporites are gypsum, anhydrite, primary dolomite, and halite (rock salt). A saline deposit.

Deethanizer vessel
(Courtesy Motherwell Bridge)

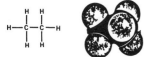

Ethone Molecule

EXCENTRIC
See Eccentric.

EXCESSIVE-PRESSURE WELDING
See Welding, Hyperbaric.

EXOTHERMIC
Refers to a process or chemical reaction that gives off heat. Endothermic is the reverse; a process or reaction that requires the addition of heat to keep it going.

EXPANSION FIT
See Shrink Fit.

EXPANSION JOINT
A section of piping constructed in such a way as to allow for expansion and contraction of the pipe connections without damaging the joints. Specially fabricated, accordion-like fittings are used as expansion joints in certain in-plant hookups where there are severe temperature changes.

Expansion joint or bellows

EXPANSION LOOP
A circular loop (360° bend) put in a pipeline to absorb expansion and contraction caused by heating and cooling without exerting a strain on pipe or valve connections.

EXPANSION-ROOF TANK
A storage or working tank with a roof made like a slip joint. As the vapor above the crude oil or volatile product expands with the heat of the day, the roof-and-apron section of the tank moves upward, permitting the gas to expand without any loss to the atmosphere. The telescoping roof, as it moves up and down, maintains a gas-tight seal with the inner wall of the tank.

Expansion loop

EXPLOITATION WELL
A development well; a well drilled in an area proven to be productive. *See* Infill Drilling.

EXPLORATION ACTIVITIES
The search for oil and gas. Exploration activities include aerial surveying, geological studies, geophysical surveying, coring, and drilling of wildcat wells.

EXPLORATION VESSEL
A seagoing, sophisticated research ship equipped with seismic, gravity, and magnetic systems for gathering data on undersea geologic structures. On the more advanced vessels of this type there are onboard pro-

cessing and interpretation capabilities for the information gathered as the vessel cruises on the waters of the Outer Continental Shelf around the world.

EXPLORATORY WELLS

Wells drilled to find the limits of an oil-bearing formation, often referred to as a pool, only partly developed. *See* Step-Out Well.

EXPLOSION-PROOF MOTOR

A totally enclosed electric motor with no outside air in contact with the motor windings; an enclosed brushless motor. Cooling is by conduction through the frame and housing.

EXPLOSIVE FRACTURING

Using an explosive charge in the bottom of the well to fracture the formation to increase the flow of oil or gas. *See* Well Shooter.

EXPLOSIVE WELDING

See Welding, Explosive.

EX SITU

The opposite of *in situ*. Off the premises; in refinery parlance, *ex situ* usually refers to the practice of using the services of a vendor to do a job or make a repair viz., reactivating a supply of spent or coated catalyst.

EXTENSION TEST

See Outpost Well.

EXTERNAL CASING PACKER

A device used on the outside of a well's casing to seal off formations or to protect certain zones. Often used downhole in conjunction with cementing. The packer is run on the casing and, when at the proper depth, it may be expanded against the wall of the borehole hydraulically or by fluid pressure from the well.

EYEBALL

To align pipe connections or a temporary construction with the eye only; to inspect carefully.

FABRICATED VALVE
See Valve, Fabricated.

FACIES
See Rock Facies.

FACING MACHINE
A device for beveling or putting a machined face on the ends of large-diameter line pipe. The facing machine essentially is a revolving disc chuck holding a number of cutting tools. The chuck is held in alignment against the pipe end by a hydraulically actuated mandrel inserted into the pipe similar to internal lineup clamps used to align pipe for welding. The facing machine is transported and brought into position by a modified boom cat.

Pipe facing machine

FAIL-SAFE
Said of equipment or a system so constructed that, in the event of failure or malfunction of any part of the system, devices are automatically activated to stabilize or secure the safety of the operation.

FAIRLEAD
A guide for ropes or lines on a ship to prevent chafing; a sheave supported by a bracket protruding from the cellar deck of a semisubmersible drilling platform over which an anchor cable runs. Some large floating platforms have anchor lines made up of lengths of chain and cable.

Fairlead

FAIRWAY
A shipping lane established by the U.S. Coast Guard in Federal offshore waters. Permanent structures such as drilling and production platforms are prohibited in a fairway, which significantly curtails oil activity in some offshore areas.

FALL PIPE
A work ship or boat's surface-to-sea floor cylindrical guide used to spot accurately cobbles or gravel on the sea floor for whatever the reason: ballasting, construction, or for scour protection of buried pipelines. The cobbles are dumped over the side into the fall pipe, and are guided to the selected place on the sea floor.

FANNING THE BOTTOM (OF THE BOREHOLE)

Drilling with very little weight on the drill bit in the hope of preventing the bit from drifting from the vertical and drilling a crooked hole. Fanning the bottom, however, is considered detrimental to the drillstring by some authorities as reduced weight on the bit causes more tension on the drillpipe, resulting in pipe and collar fatigue.

FARM BOSS

A foreman who supervises the operations of one or more oil-producing leases.

FARMER'S OIL

An expression meaning the landowner's share of the oil from a well drilled on his property; royalty is traditionally one-eighth of the produced oil free of any expense to the landowner.

FARMER'S SAND

A colloquial term for "the elusive oil-bearing stratum which many landowners believe lies beneath their land, regardless of the results of exploratory wells."

FARMIN, A

An arrangement whereby one oil operator buys in or acquires an interest in a lease or concession owned by another operator on which oil or gas has been discovered or is being produced. Often farmins are negotiated to assist the original owner with development costs and to secure for the buyer a source of crude or natural gas. *See* Farmout Agreement.

FARMOR

Farmor, not farmer, is legalese for the person granting a farmout; grantor, lessor, farmor.

FARMOUT

The name applied to a leasehold held under a farmout agreement.

FARMOUT AGREEMENT

A form of agreement between oil operators whereby the owner of a lease who is not interested in drilling at the time agrees to assign the lease or a portion of it to another operator who wishes to drill the acreage. The assignor may or may not retain an interest (royalty or production payment) in the production.

FAST BREAK

Rapid drilling by the bit, which usually indicates a soft or very porous section of rock. *See* Break.

FAST LINE
On a drilling rig, the fast line is the cable spooled off or on the hoisting drum of the drawworks; the line from the hoisting drum that runs up through the derrick to the first sheave in the crown block. *See also* Dead Line.

Fast line on the drum

FATIGUE LIFE
The working or operational life of pipe or equipment, subject to bending, tensile, pressure or temperature stresses. Not life of fatigue, but life of objects under fatigue. Fatigue life—industry shorthand.

FAT OIL
The absorbent oil enriched by gasoline fractions in an absorption plant. After absorbing the gasoline fractions, the gasoline is removed by distillation, leaving the oil "lean" and ready for further use to absorb more gasoline fractions from the natural gas stream.

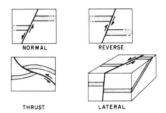
Four common faults

FAULT
A fracture in the earth's crust accompanied by a shifting of one side of the fracture with respect to the other side; the point at which a geological strata "breaks off" or is sheared off by the dropping of a section of the strata by settling.

FAULT, CROSS
A minor fault that intersects a major fault.

FAULT, NORMAL
When the hanging wall moves down in relation to the footwall.

FAULT, OVERLAP
A fault structure in which one layer or crustal unit overrides the other. This can be caused by horizontal pressure or compression rather than by vertical movement or displacement. One side of the broken or fractured bed slides over the other, overlapping as with shingles on a roof. A thrust fault.

FAULT, REVERSE
This phenomenon results when the hanging wall is moved up relative to the footwall. A fault with a dip of 45° or less and where the hanging wall has moved upward relative to the footwall is called a thrust fault. In this instance, horizontal compression rather than vertical displacement is the probable cause.

FAULT, STRIKE-SLIP
A fault where one fault plane moves laterally—endways, rather than up or down on a horizontal plane.

FAULT, THRUST

A fault with a fairly shallow dip (less than 45°) over much of its extent in which the hanging wall has moved upward, or appears so, relative to the footwall. In a thrust fault, horizontal compression or thrust rather than vertical movement or displacement is the cause that identifies it. The rock layer has been subjected to horizontal pressures so that at some point it fractures or faults and one side overrides the other side; an overthrust.

FAULTING, BLOCK

See Block Faulting.

FAULT-PLANE DIP

See Dip, Fault-Plane.

FAULT SCARP

The visible offset of a section of rock and earth caused by faulting of the rock layers beneath the surface. Scarps are often formed when, as the result of an earthquake, there is severe faulting. A scarp may be described as a low, fault-induced, cliff-like structure.

FAULT TRAP

A trap for oil or gas in which the closure, forming the trap, results from the presence of one or more faults.

F.C.C.U., PARTICULATE CONTROL IN

Electrostatic dust removal; Wet scrubbers; Bag house.

F.E.A.

Federal Energy Agency.

FEDERAL ENERGY AGENCY

The government agency that administers the Federal Energy Law whose regulations and directives govern the activities of the oil and gas industry.

FEDERAL ENERGY REGULATORY COMMISSION (F.E.R.C.)

An agency within the Department of Energy (D.O.E.) that replaced the Federal Power Commission in 1977 when the D.O.E. was created. The new commission has essentially the same regulatory powers plus the added responsibility of regulating pipelines, a function previously performed by the Interstate Commerce Commission (I.C.C.).

FEDERAL LEASE

See Lease, Federal.

FEE

The title or ownership of land; short for "owned in fee." The owner of the fee holds title to the land.

FEED
Crude oil or other hydrocarbons that are the basic materials for a refining or manufacturing process; feedstock.

FEEDER LINE
A pipeline; a gathering line tied into a trunk line.

FEEDSTOCK
The raw or semifinished material that is processed in a refinery or other processing plant; charge stock; to charge a still or other processing unit is to pump in a charge of feedstock to be treated or further refined; feed.

FEE ROYALTY
See Royalty, Landowner's.

FEE SIMPLE
Land or an estate held by a person in his own right without restrictions.

FELDSPAR
The name of a group of minerals rather than a single mineral such as quartz. The family of feldspars is made up of combinations of potassium or sodium and calcium with oxygen, aluminum, and silicon. The feldspars are by far the most abundant of the rock-forming minerals and are the most important constituent of the lithosphere (Earth's crust), making up 50 percent or more of its mass—many times more than the runner-up, quartz.

FEMALE CONNECTION
A pipe, rod, or tubing coupling with the threads on the inside.

Female connection

F.E.M.W.D.
Formation evaluation M.W.D. (F.E.M.W.D.); formation evaluation logging, an adjunct to wire-line logging.

FERRAL'S LAW
The statement that the centrifugal force produced by the rotation of the Earth (Coriolis force) causes a rotational deflection of the currents of water and air to the right in the Northern Hemisphere and to the left in the Southern Hemisphere. (Visible evidence when draining the bathtub, in Tulsa, ok, then in Auckland, New Zealand.)

FERROMAGNESIAN
Containing iron (Fe) and magnesium (Mg); may refer to mafic igneous rock, which contains dark-colored ferromagnesian minerals.

F.E.R.C.
Federal Energy Regulatory Commission.

FIBERGLASS PIPING
A type of plastic piping used to handle corrosive liquids that would soon destroy conventional steel piping. Among fiberglass's several advantages are its light weight, ease in making joint connections, and ability to be cut and fit on the job.

FIBER-OPTIC SENSOR
A downhole sensor of formation temperature and pressure using fiber-optics technology. The permanently installed sensor system's developers' (nine interested oil and service companies) target is to achieve downhole measurements with long-term accuracy at pressures above 10,000/lbs. and temperatures above 392°F (200°C). The sensor is for use in remote, permanent subsea locations as far away as 15 miles (25 km) from the readout instrumentation.

FIELD
The area encompassing a group of producing oil and gas wells; a pool.

FIELD BUTANES
A raw mix of natural gas liquids; the product of gas processing plants in the field. Raw mix streams are sent to fractionating plants where the various components—butane, propane, hexane, and others—are separated. Some refineries are capable of using field butanes as 10 to 15 percent of charge stock.

FIELD COMPRESSION TEST
A test to determine the gasoline content of casinghead or wet gas.

FIELD GATHERING LINES
See Gathering Lines.

FIELD GAUGER
See Gauger.

FIELD POTENTIAL
The producing capacity of a field during a 24-hour period.

FIELD TANKS
A battery of two or more 100- to 500-barrel tanks on a lease that receive the production from the wells on the lease; stock tanks.

FILLER PASS
See Welding, Pipeline.

Gauger

Field tank

FILLET WELD
See Weld, Fillet.

FILTER, DEEP-BED GRANULAR MEDIA
This type of heavy-duty filters are effective in removing suspended solids and insoluble hydrocarbons from effluent water.

FILTER CAKE
A plastic-like coating of the borehole resulting from the solids in the drilling fluid adhering and building up on the wall of the hole; the buildup of cake can cause serious drilling problems including the sticking of the drillpipe. *See* Differential-Pressure Sticking.

FILTRATE
The solid material in drilling mud. When filtrate is deposited on the wall of the borehole of the well forming a thick, restrictive layer, it is referred to as filter cake.

FILTRATE SWEEP
A method of diminishing vertical gas migration in preparation for cementing a well's casing. To pump high spurt-loss (low-viscosity) fluid radially into a formation to alter, at least temporarily, the permeability of an interval through which gas is migrating, through small fissures and micro-annuli, into the well's annulus. After the sweep, which impedes the percolating gas, the well's casing can be cemented with specially formulated cement to achieve satisfactory bonding with the walls of the borehole and the casing.

FILTRATION-LOSS QUALITY OF MUD
A drilling-mud quality measured by putting a water-base mud through a filter cell. The mud solids deposited on the filter is filter cake and is a measure of the water-loss quality of the drilling mud. Mud mixtures with low water loss are desirable for most drilling operations.

FINAL PASS/CAPPING PASS
See Welding, Pipeline.

FINES (sed)
Very small particles, especially those smaller than the average, in a mixture of various sized particles, e.g., the silt and clay particles in glacial drift. An engineering term for the silt and clay passing through U.S. standard sieve No. 200. A single fine is as small as the Biblical mote.

FINGER BIT
See Bit, Finger.

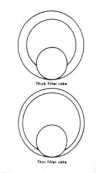

Thick filter cake

Thin filter cake

Pipe stuck in filter cake

Finger board

Tanker loading at finger pier

Finned-tube heat exchanger

FINGER BOARD

A rack high in the derrick made to support, in orderly fashion, the upper ends of the tubing stands that are pulled from the well and set back in the derrick. A hay rake.

FINGERING

Rivulet-like infiltration of water or gas into an oil-bearing formation as a result of failure to maintain reservoir pressure or as the result of taking oil in excess of maximum efficiency rates (M.E.R.). *See* Channel.

FINGER PIER

A jetty or bridge-type structure extending from the shore out into a body of water to permit access to tankers and other vessels where water depth is not sufficient to allow docking at the shore.

FINGERPRINTING

A chemical process that can identify fuel sources and how the fuel was burned. For example, fingerprinting can identify various kinds of fuel oil and the manner in which it was used—burned in an industrial boiler or diesel engine. Environmentalists use this technique to pinpoint sources of pollution and analyze the particular problem being created.

FINNED-TUBE HEAT EXCHANGER

See Heat Exchanger, Finned-Tube.

FIRE FLOODING

See In Situ Combustion.

FIRE PROTECTION, OFFSHORE, PASSIVE

Because of the proximity of all elements on an offshore platform, fire protection must be designed into the structure: fire walls, bulkheads, and other fire- and heat-isolating elements are all parts of acceptable designs.

FIRE TRIMMED

A designation for valves, flanges, and other fittings made to withstand an accidental fire in a plant or process unit. Fire-trimmed valves, when subjected to fire from whatever cause, will not be damaged to the extent that they will leak and thereby add to the emergency. Such valves and flanges have metal gaskets and stuffing boxes with specially formulated fire-resistant packing, or are fitted with metal-to-metal seals.

FIRE WALL

An earthen dike built around an oil tank to contain the oil in the event the tank ruptures or catches fire.

FIRE-WATER SUPPLY

A pond or tank containing water used exclusively in fire fighting.

FIRING LINE
In pipeline construction, the part of the project where the welding is being done.

FIRM GAS
Gas required to be delivered and taken under the terms of a firm gas purchase contract. Firm gas is priced higher than dump gas.

FISH
Anything lost downhole; the object being sought downhole by the fishing tools.

The firing line on a large gas pipeline project

FISHING BID
A small bonus bid in a competitive lease sale. The bid may be made on the basis of very limited geologic and seismic data with no real hope of success if other bids are filed, but with the chance of catching a lease for a modest sum if no others are filed for the particular parcel. A dreamer's bid.

FISHING JOB
The effort to recover tools, cable, pipe, or other objects from the well bore that may have become detached while in the well or been dropped accidentally into the hole. Many special and ingeniously designed fishing tools are used to recover objects lost downhole.

FISHING OPERATION, A
An attempt to remove all pipe and downhole tools from the borehole when the string has parted in more than one place at unknown depths. A far bigger project than a simple fishing job for a lost tool or twisted-off drillpipe.

FISHING TOOLS
Special instruments or tools used in recovering objects lost in a well. Although there are scores of standard tools used in fishing jobs, some are specially designed to retrieve particular objects.

FISHPROOF
Describes an item of equipment used in or over the well's borehole without parts—screws, lugs, wedges, or dogs—that can come loose, fall into the well, and have to be fished out. *See* Captured Bolt.

FISHTAIL BIT
See Bit, Fishtail.

FISSILE
Capable of being split along closely spaced or stacked planes; said of bedding planes that consist of laminae (layers) less than 1/4 inch in thickness.

Fittings (bushing, union, ell, tee plug)

FISSURE
A fracture or crack in a rock in which there is definite and distinct separation, and which is often filled with mineral-bearing material.

FISSURE THEORY OF MIGRATION
An early held theory, now somewhat discredited, that oil and gas migrate extensively through fissures resulting from the arching of sedimentary beds into anticlines.

FITTINGS
Small pipes: nipples, couplings, elbows, unions, tees, and swages used to make up a system of piping.

FITTINGS, TRANSITION
See Transition Fittings.

FIVE-SPOT WATERFLOOD PROGRAM
A secondary-recovery operation where four input or injection wells are located in a square pattern with the production well in the center, a layout similar to a five-of-spades playing card. The water from the four injection wells moves through the formation, flooding the oil toward the production well.

FIXED-BED CATALYST
A catalyst in a reactor vessel through which the liquid being treated drips or percolates through the bed of catalyst material. In other methods, the catalyst is mixed thoroughly with the feedstock as it is pumped into the reactor vessel.

FIXED-RATE ROYALTY
See Royalty, Fixed-Rate.

FIXER'S ROYALTY
See Royalty, Innovator's.

FLAG THE LINE
To tie pieces of cloth on the swab line at measured intervals to be able to tell how much line is in the hole when coming out with the line.

FLAMBEAU LIGHT
A torch used in the field for the disposal of casinghead gas produced with oil when the gas is without a market or is of such small quantity as to make it impractical to gather for commercial use. The use of flambeau lights is now regulated under state conservation laws.

FLAME ARRESTER
A safety device installed on a vent line of a storage or stock tank that, in the event of lightning or other ignition of the venting vapor, will prevent the flame from flashing to the vapors inside the tank.

FLAME-JET DRILLING
See Drilling, Flame-Jet.

FLAME SNUFFER
An attachment to a tank's vent line that can be manually operated to snuff out a flame at the mouth of the vent line; a metal clapper-like valve that may be closed by pulling on an attached line.

FLAMMABLE
Term describing material that can be easily ignited. Petroleum products with a flash point of 80°F or lower are classed as flammable.

FLANGE
(1) A type of pipe coupling made in two halves. Each half is screwed or welded to a length of pipe and the two halves are then bolted together, joining the two lengths of pipe. (2) A rim extending out from an object to provide strength, or for attaching another object.

FLANGED CONNECTION
See Tapped or Flanged Connection.

FLANGE UP
To finish a job; to bring to completion.

FLARE
(1) To burn unwanted gas through a pipe or stack. (Under conservation laws, the flaring of natural gas is illegal.) (2) The flame from a flare; the pipe or the stack itself.

FLARE, SMOKELESS
See Smokeless Flare.

FLARE BOOM
Booms to support flares are incorporated in offshore drilling platforms. They hold the flare tip 40 to 60 feet in the air—well away and canted out from the platform deck.

FLARE-BOOM STACK
A boom jutting up and out, 100 feet or more, from an offshore drilling platform is the support for the flare stack. (On a land well, a blooie pipe serves as a flare stack.) At sea it is vital to vent gas or other flammable substance encountered in drilling as far from the platform as practical. *See* Diverter System.

FLARE GAS
Gaseous hydrocarbons discharged from safety relief valves on process units in a refinery or chemical plant. Should a unit go down from an electrical or cooling water failure, making it necessary to dump a batch

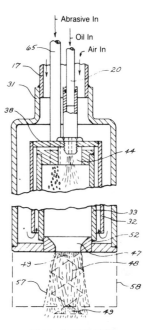

Flame jet drill *(After Browning & Fitzgerald)*

A bolted, high-pressure flange

of liquid feed or product, the flare stack is equipped to handle such an emergency. If it were impossible to dump both gases and liquids in an emergency, the plant personnel and the operating units would be in danger. With the recovery equipment larger plants are installing, flare gases as well as the dumped process fluid are recovered. The gases are used as fuel; the liquids are reprocessed.

FLARE STACK
See Smokeless Flare.

FLARE TIP TO SPLASH ZONE
The complete offshore drilling rig; everything above the water line.

FLASH CHAMBER
A refinery vessel into which a process stream is charged or pumped and where lighter products flash off or vaporize and are drawn off at the top. The remaining heavier fractions are drawn off at the bottom of the vessel.

FLASH OFF
To vaporize from heated charge stock; to distill.

FLASH POINT
The temperature at which a given substance will ignite.

FLASH SET
The very rapid setting of cement downhole; when under high pressure, the slurry's liquid bleeds off into the porous formation leaving the solids on the surface of the rock; a condition very similar to the formation of filter cake by the loss of the fluid in drilling mud.

Flexible coupling between electric motor and centrifugal pump

FLEXIBLE COUPLING
A connecting link between two shafts that allows for a certain amount of misalignment between the driving and driven shaft without damage to bearings. Flexible couplings dampen vibration and provide a way to make quick hookups of engines and pumps, which is useful in field operations. *See* Pipe Coupling, Flexible.

FLINT
A rock almost identical to chert; the synonym for chert; black chert. Flint, the mineral, is a very hard, somewhat impure variety of chalcedony, usually black or shades of gray. It breaks with a conchoidal fracture and is hard enough to spark fire when struck with steel. The stone in a flintlock musket.

Float switch *(Courtesy Jo-Bell Products)*

FLOAT
(1) A long, flatbed trailer, the front end of which rests on a truck, the rear end on two dual-wheel axles. Floats are used in the oil fields for trans-

porting long, heavy equipment. (2) The buoyant element of a fluid-level shutoff or control apparatus. An airtight canister or sphere that floats on liquids and is attached to an arm that moves up and down, actuating other devices as the liquid level rises and falls.

FLOAT EQUIPMENT, WELL-CEMENTING

Float equipment is used in cementing the casing string firmly and gas-tight to the rock wall of the borehole. The equipment consists of a nonreturn (check) valve mounted or held in a steel shell compatible with the casing string. The valve (flapper, ball or plunger) is made from frangible material so it may readily be drilled out, pulverized. Float equipment provides a circulation path for well fluids from casing to annulus; prevents flow back of cement from annulus into casing; provides blowout protection; and floats the casing into the well bore. *See* Floating the Casing.

Floater

FLOATER

(1) A barge-like drilling platform used in relatively shallow offshore work. (2) Any offshore drilling platform without a fixed base, e.g., semi-submersibles—drillships or drill barges. (3) A floating-roof tank.

FLOATING-ROOF TANK

A storage tank with a flat roof that floats on the surface of the oil thus reducing evaporation to a minimum. The roof rests on a series of pontoons whose buoyancy supports the roof proper; a floater.

Floating roof tank

FLOATING STORAGE

A large, converted, permanently moored oil tanker that holds production from offshore wells for transfer to seagoing oil transport vessels or to lighters for transport to shore stations.

FLOATING STORAGE UNIT

A specially designed vessel for transporting crude oil from an offshore production platform. The loaded, floating storage is towed (it has no means of locomotion) from the platform to a shore docking and off-loading facility.

FLOATING THE CASING

A method of lowering casing into very deep boreholes when there exists the danger of the casing joints separating because of the extreme weight or tension on the upper joints. In floating, the hole is filled with fluid and the casing is plugged before being lowered into the hole. The buoyant effect of the hollow column of casing displacing the fluid reduces the weight and the tension on the upper joints. When the casing is in place, the plug is drilled out.

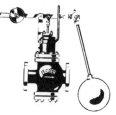
Float-actuated valve
(Courtesy Fisher)

FLOAT VALVE

See Valve, Float.

Floor men

Small flow meter for products *(Courtesy Smith Geosource Inc.)*

Solar Powered Flow Meter (Courtesy Ozzie's Pipeline)

FLOCCULATION
The bunching up or coming together of the particles of a precipitate into tufts or lumps, owing to mutual attraction.

FLOODING
The use of water injected into a production formation or reservoir to increase oil recovery. *See* Secondary Recovery.

FLOODING, DUMP
See Dump Flooding.

FLOOR MEN
Members of the drilling crew (usually two) who work on the derrick floor.

FLOW, PLASTIC
See Plastic Flow; also Turbulent Flow.

FLOW BEAN
The flow-restricting element in a wellhead choke. Beans are interchangeable and come with holes drilled in 64th-inch increments. When the well is to be flowed through a choke for a test, any one of a number of beans may be used—for example, 16/64, 22/64, or 32/64—by inserting one into the choke body. A choke has the general configuration of a conventional gate valve.

FLOW CHART
A replaceable paper chart on which flow rates are recorded by an actuated arm and pen. *See also* Pressure Chart.

FLOW IMPROVER
A patented chemical compound that, when added in relatively small amounts to crude oil being pumped through a pipeline, increases the flow of oil significantly (as much as 50 percent). Pioneer in the field of flow improvers, Conoco Chemicals Co. describes its product, C.D.R., as a "high molecular weight polymer in a hydrocarbon solvent." In other words, a substance that reduces friction in a pipeline.

FLOWING FORMATIONS
Formations that swell or are pushed into areas of lower resistance. Shale swells from drilling fluids and flows into the borehole. Salt beds are another example of formations that flow.

FLOWING TUBING PRESSURE
After fracing and well clean up, the fraced formation often begins flowing up through the production tubing. The flowing tubing pressure is a good indication of the success of the frac job and the vitality of the reservoir.

FLOW LINE, DRILLING WELL

The line at the top of the casing that directs the returning drilling fluid to the mud-gas separator, if there is gas contamination, and then to the mud pit. The returning mud also passes over the shale shaker which strains out and catches a large percentage of the rock cutting from the borehole; the remainder of the cuttings drop out in the mud pits. The shale shaker is where the geologist gets samples of the formation rock being cut.

FLOW NIPPLE

A choke; a heavy steel nipple put in the production string of tubing that restricts the flow of oil to the size of the orifice in the nipple. It is usual to report a new well's production as a flow of a certain number of barrels per day through a choke of a certain size, e.g., 16/64 in., 9/64 in., etc.; a flow plug.

FLOW REGIMEN

A schedule worked out for the flowing or pumping of a well, in light of good production practice. Pumping wells, after a flush production period that may last for a few months or a year, are pumped so many hours a day (depending upon the individual well's capabilities) and rest the remainder of the 24 hours. Flowing wells are kept on a choke; they are not permitted to flow unimpeded. To do so would be wasteful reservoir energy. *See* Gas-Cap Drive.

FLOW SHEET

A diagrammatic drawing showing the sequence of refining or manufacturing operations in a plant.

FLOW STRING

The string of casing or tubing through which oil from a well flows to the surface. *See also* Oil String; Pay String; Capital String; Production String.

FLOW TANK

A lease tank into which produced oil is run after having gas or water removed; production tank.

FLOW TREATER

A single unit that acts as an oil and gas separator, an oil heater, and an oil and water treater.

FLOW VALVE

See Valve, Flow.

FLUID CATALYTIC CRACKING UNIT

A large refinery vessel for processing reduced crude, naphthas, or other intermediates in the presence of a catalyst. Catalytic cracking is regarded

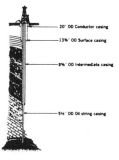

Flow string

Fluid catalytic cracking
unit

as the successor to thermal cracking as it produces less gas and highly volatile material; it provides a motor spirit of 10 to 15 octane numbers higher than that of thermally cracked product. The process is also more effective in producing isoparaffins and aromatics that are of high anti-knock value.

FLUID END

The end of the pump body where the valves (suction and discharge) and the pump cylinders are located. The fluid end of a reciprocating pump is accessible by removing the cylinder heads, which exposes the pistons or pump plungers. The cylinders or liners in most pumps are removable and can be changed for others with larger or smaller internal diameters. Inserting smaller liners and pistons permits pumping at higher pressure but at a reduced volume.

FLUIDICS

Pertains to the use of fluids (and air) in instrumentation. Fluidics is defined as "engineering science pertaining to the use of fluid-dynamic phenomena to sense, control, process information, and actuate." Fluidics provides a reliable system far less expensive than explosion-proof installations required with electrical instrumentation on offshore rigs.

FLUID LEVEL

The distance between the wellhead and the point to which the fluid rises in the well.

FLUID LOSS

A condition downhole in which a water-base drilling mud loses water in a highly permeable zone, causing the solids in the drilling fluid to build up on the wall of the borehole. This buildup of mud solids can result in stuck pipe, which often arises when the hydrostatic head or mud pressure is considerably higher than the formation pressure.

FLUID-LOSS ADDITIVES

Silica flour and very fine, 100-mesh sand are two of several kinds of fluid-loss additives that can be spotted or added to the drilling fluid to clog the pores of a thief zone into which the liquid of the drilling mud is escaping.

FLUID PHASES

Refers to the two kinds of fluid—liquids and gases; liquid phase and gaseous phase. Both are capable of flowing, so they are fluids, although gases are commonly not thought of as fluids. Geologists customarily refer to "multiple fluid phases," meaning oil, condensate, and water as well as gases: natural gas (CH_4), carbon dioxide (CO_2), and hydrogen sulfide(H_2S).

FLUID SPOTTING
Placing a measured quantity of acid downhole in a particular location or spot opposite the interval to be acidized. This is done most effectively by the use of coiled tubing run inside the casing or the production tubing.

FLUSHED-ZONE PROCESS
A technique to check, to forestall, gas migration near the well bore area, in order to effect a gas-tight casing cement job. An inhibited drilling fluid (filtrate), one specially formulated to preclude damage to water-sensitive shales, is pumped under pressure into the formation, thus radially bullheading the migrating gas back away from the borehole. The well, taken on down to T.D., is then cemented free of the maverick gas. An elaborate process, but in numerous field tests the process has exhibited a satisfactory success rate.

FLUSHING OILS
Oils or compounds formulated for the purpose of removing used oil, decomposed matter, metal cuttings, and sludge from lubricating passages and engine parts.

FLUSH PRODUCTION
The high rate of flow of a good well immediately after it is brought in.

FLUVIAL DEPOSITS
A sedimentary deposit of eroded material held suspended, carried along, and laid down by a stream.

FLUVIAL DOMINATED
A term having to do with the action of rivers transporting sediment; as the flow diminishes or slows, the sediment drops out. "Dominated" means the stream is the principal source of the sediment, as in a deltaic reservoir formed at the mouth of a principal river, viz., Mississippi, Orinoco, and Amazon.

FLUXING
To soften a substance with heat so that it will flow; to lower a substance's fusing point.

FLY ASH
Particulate matter; minute particles of ash that escape up the chimneys and smokestacks from coal fires or from furnaces or incinerators burning waste material.

FLYWHEEL
A heavy wheel mounted on the main shaft (the crankshaft) outside the engine that absorbs the torque of the crankshaft, the power strokes, and smoothes them out, while at the same time storing power from one

stroke of the engine to the next. Flywheels smooth the power output of an internal-combustion engine. The momentum of the heavy wheel— sometimes two, one on each side of the engine— does the trick.

FOAMED CEMENT
See Cement, Foamed.

FOAMED OIL
A commonly used frac fluid; a special oil treated with an emulsifier, increasing its viscosity so it can carry a load of proppant material, in suspension, into the formation's newly opened cracks and crevices.

FOAMED-STEAM FLOOD
The injection of a surfactant followed by steam and nitrogen to form a steam foam for the purpose of reducing the loss of the flooding fluid to down dips in the formation being flooded and to fractured, more openly porous intervals or steam paths that developed from previous sweep-flooding activity.

FOCUSED-CURRENT LOG
See Log, Focused-Current.

FOLDING

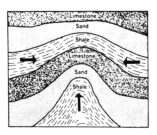

Folding *(Courtesy Petex)*

Folding in layered rock consists of deformation, bending, or curving of the strata without faulting or actually breaking. Folds are caused by draping over basement rock blocks, compression, or compaction (weight of overburden). Folds are classified according to morphology (structure), origin, and type of deformation. *See* Anticline; *also* Syncline.

FOLIATED
Refers to the condition of rock in which the mineral grains are flattened, owing to geothermal action (heat) and immense pressure of the overburden. The grains are not only recrystallized and flattened but have a leaf-like or layered texture, hence the term *foliated*. This condition is especially prevalent in metamorphic rock. Examples of foliated metamorphic rocks are slate, schist, and gneiss.

FOLLOWER
An adjustable piece that fits into the gland to compress the packing against the moving part that can be screwed into the gland or forced into the gland by nuts on stud bolts.

FOOTAGE CONTRACT
A contract for the drilling of a well in which the drilling contractor is paid on a footage basis as the well is taken down. Sometimes the price per foot changes as the well progresses and different formations are encountered.

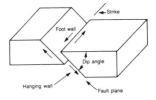

Foot wall

FOOT-POUND
A unit of energy or work equal to the work done in raising 1 pound to the height of 1 foot against the force of gravity.

FOOT VALVE
See Valve, Foot.

FOOTWALL
The underlying side of a fault or break in the strata. *See* Hanging Wall.

FORCED-DRAFT BURNER
Crude-oil disposal equipment on offshore platforms. The burner, mounted on a boom or an extension of the deck, burns crude oil during testing operations. Gas, air, and water manifolded with the test crude stream result in complete combustion of the oil. A platform burner.

FORCE MAJEURE CLAUSE
A lease clause providing that cessation or failure of production shall not cause automatic termination of the leasehold and that the performance of the lessee's covenants shall be excused when the failure of production or performance of covenants is owing to causes set forth in the clause. Such clauses usually list acts of God; adverse weather; compliance with federal, state, or municipal laws; wars; strikes; and other contingencies over which the lessee has no control.

Forced-draft burner

FORCE PUMP
See Pump, Force.

FOREIGN TAX CREDIT
Taxes paid a foreign government by a U.S. company on its overseas oil operations that are creditable against taxes owed the U.S. government. Production sharing by a U.S. company with a foreign government or one of its agencies represents oil royalty payments, not taxes creditable in the United States, according to the Internal Revenue Service.

FOREIGN TRADE ZONE
An area in the United States where imported oil, reduced crude, or intermediates are processed.

FORGIVING CONSTRUCTION
Equipment built in such a way that it can withstand degrees of overloading, misalignment, vibration, or other forms of abuse without being damaged or destroyed.

FORMATION
A strata of rock that is recognizable from adjacent strata by consisting mainly of a certain type of rock or combination of rock types. Thickness

may range from less than 2 feet to hundreds of feet. Formations are the fundamental units of classification of the whole stratigraphic column, the geologic column, anywhere in the world; often combined into groups or subdivided into members.

FORMATION PRESSURE
See Reservoir Pressure.

FORMATION SAMPLE
A rock sample taken from the formation the drill bit has just penetrated; a sample of rock taken with a core bit. A core sample.

FORMULATION
The product of a formula, i.e., plastic, blended oils, gasolines; any material with two or more components or ingredients.

FOSSIL
The remains of once-living plants and animals (flora and fauna) that, as they died, were covered by sand, silt, mud, or lime. Most of the organic matter their bodies contained gradually was replaced by inorganic matter, silica for example, as in petrified wood. Probably the most characteristic property of sedimentary rocks is the presence of fossils—shells, bones, plant fragments, and other more durable elements. In some cases the entire rock may consist of organic matter; coal, for instance, is made up of plant fragments. *See* Coquina.

FOSSIL ENERGY
Energy derived from crude oil, natural gas, or coal.

FOSSIL FUEL
See Fossil Energy.

Four-cycle gas engine

FOURBLE
A stand of drillpipe or tubing consisting of four joints. In pulling pipe from the well, every fourth joint is unscrewed and the four-joint stand is set back and racked in the derrick. This is not a common practice; the usual stands are of two and three joints.

FOUR-CYCLE ENGINE
An internal-combustion engine in which the piston completes four strokes—intake, compression, power, and exhaust—for each complete cycle. The Otto-cycle engine; four-stroke cycle engine.

FOUR-STRING COMPLETIONS
See Dual Completion.

FOURTH-GENERATION RIG
The designation for a large, advanced-design, offshore, seagoing drilling rig. A self-propelled, dynamically positioned rig, capable of drilling in 6,000 foot water depths, in harsh environments, designed to handle high-pressure, high-temperature wells. In 1993, such a rig was so-called fourth generation; the last word in a self-contained, floating, drilling community; ponderous, seaworthy, and costly.

FOXHOLE DRILLING RIG
A top-drive, mechanized pipe-handling rig with enclosed driller's cabin (the foxhole). The cabin has an array of at-hand controls for all pipe handling, tripping, and other necessary rig-floor operations. Developed and patented by A.P. Moeller group.

F.P.C.
Federal Power Commission, an agency of the Federal government; a regulatory body having to do with oil and gas matters such as pricing and trade practices.

F.P.S.
Floating production systems. Such systems can be several types of existing mobile offshore drilling units (M.O.D.U.S.) converted to floating production systems. For marginal offshore fields in deep water, converting existing floating facilities is considerably cheaper than building a new floater.

FRAC FLUID EFFICIENCY
This phrase should be "frac job efficiency." As calculated by the frac team, it is arrived at by computing the amount of frac fluid entering the rock strata, expressed as a percent of the total volume pumped downhole. The volume of frac fluid and proppant present in the fracture at shut in is divided by the total volume pumped down.

FRACING (FRACKING)
See Hydraulic Fracturing.

FRAC JOB
See Hydraulic Fracturing.

FRACKING FLUID
A slurry or foam that carries proppant material in suspension downhole under very high pressure to fracture and prop open the small cracks and fissures made in the producing formation by the intense pressure. After the proppant material (sand grains or microscopic beads) is in place, the pumping of the fracking fluid is discontinued, allowing the fluid to drain out of the formation, leaving the proppant behind to hold open the small cracks.

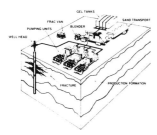

Hydraulic fracturing equipment *(Courtesy Halliburton)*

Bubble tower or fractionator (left) and heater

FRACTION

A separate, identifiable part of crude oil; the product of a refining or distillation process.

FRACTIONATION

The separating of hydrocarbons into fractional components in a fractionating tower by the action of heating to drive off the light ends, the light gases, then progressively heavier fractions, and condensing those fractions or cuts by cooling. *See* Fractionator.

FRACTIONATION TOWER

See Fractionator.

FRACTIONATOR

A tall, cylindrical refining vessel where liquid feedstocks are separated into various components or fractions.

FRAC TRUCKS

Heavy duty trucks equipped with high-pressure reciprocating hydraulic pumps used in fracturing downhole formations. The trucks pump frac fluid, under thousands of pounds pressure and carrying proppant material, down to the face of the formation. A big frac job will involve as many as two dozen trucks linked together in a network of high-pressure piping (10,000 pounds/sq.in. is not unusual).

FRACTURE

A general term for a break in a rock, whether or not there was movement or displacement in the rock. A fracture includes faults, joints, and cracks.

FRACTURE POROSITY

A type of rock that results from openings caused by shattering or breaking a rock that, in its solid state, was more or less impervious with few, if any, cracks or openings.

FRACTURING PRESSURE

The pressure required to overcome the pore pressure of the subject formation. In new wells with high-pore pressure, the hydrostatic pressure of the column of drilling mud does not intrude. In swept, depleted zones, however, producing unwanted quantities of water with pore pressures that have diminished to a minimum, it is often not economically practical to cement squeeze to shut off the water coning. The cement may invade the formation and also shut off the remaining oil flow (trickle drainage).

FRAGMENTAL ROCK

Clastic rock; rocks made up of fragments of older rock or of shell fragments, as in certain limestones; a bioclastic rock.

FRASCH, HERMAN

A Canadian chemist who developed a process for the use of sour crude for making kerosene. The Frasch process opened the market for sour crude from Ohio and Canada just when it was thought the production from Pennsylvania and West Virginia fields had peaked and the country was running low on sweet crude for kerosene and gasoline.

FREE GAS

(1) Natural gas produced by itself, not in the same stream as crude oil or condensate; gas-well gas. (2) Gas that may be used by the lessor (landowner) for lighting and heating if there is surplus gas produced on the lease. The landowner may do so free of charge and at his own risk and expense.

FREE MARKET PRICES

Oil prices not subject to controls by the government; world prices. *See* Posted Price.

FREE RIDE

An interest in a well's oil and gas production free of any expense of that production; a royalty interest; an override.

FREEZE BOX

An enclosure for a water-pipe riser that is exposed to the weather. The freeze box or frost box surrounding the pipe is filled with sawdust, manure, or other insulating material.

FREON

A trademark applied to a group of halogenated hydrocarbons having one or more fluorine atoms in the molecule; a refrigerant.

FRESNO SLIP

A type of horse-drawn earthmoving or cutting scoop with curved runners or supports on the sides and a single long handle used to guide the scoop blade into the earth or material being moved.

FRIABLE

Refers to a rock or mineral that crumbles naturally when weathered, or one that is easily broken or pulverized.

FRICTION CLUTCH

A mechanism in the power train for transferring the power generated by an engine to a pump, compressor, generator, or vehicle. Friction clutches may have disks that are mechanically engaged (brought together) to transfer the power or may have bands made to grip a powered shaft. In a properly adjusted friction clutch there is very little friction because the disks and bands grip almost instantaneously, obviating any heat-generating slippage or friction.

FRICTION WELDING
See Welding, Friction.

FROM PERFORATIONS
A common phrasing of a well-test report is to say, "The Bluebell No. 1 drilled to a total depth of 8,041 feet and was completed from perforations in the Prue zone at 7,410 to 7,436 feet." This indicates that the well was cased (pipe was set) through the producing interval and the casing perforated for 26 vertical feet from 7,410 to 7,436. The well is then produced from the 26-foot interval, or pay zone.

FROST BOX
See Freeze Box.

FROST HEAVE
The action of successive freezing and thawing of the soil which squeezes farmers' fence posts, field stones, and shallowly buried pipelines out of the Earth a little each winter until posts, rocks, and pipelines are completely uncovered. The pipeline quite often ruptured; the farmer's fence down and cows out.

FROST RIVING or WEATHERING
The mechanical splitting or breakup of rocks or soil because of the extremes of pressure by the freezing of water in the cracks or pores.

FROST UP
Icing of pipes and flow equipment at the wellhead of a high-pressure gas well. The cooling effect of the expanding gas as pressure is reduced causes moisture in the atmosphere to condense and freeze on the pipes.

F.S.U.
Former Soviet Union, the Republics.

FUEL-AIR RATIO
The ratio of fuel to air by weight in an explosive mixture that is controlled by the carburetor in an internal-combustion engine.

FUEL OIL
Any liquid or liquefiable petroleum product burned for the generation of heat in a furnace or for the generation of power in an engine, exclusive of oils with a flash point below 100°F.

FUEL-OIL ATOMIZER
See Atomizer, Fuel-Oil.

FUEL SENSING
A device affixed to an internal-combustion engine's carburetor permitting the engine to operate on a mixture of gasoline and from 25 percent

to 85 percent ethanol. The sensing device automatically adjusts the carburetor to the mix of fuel.

FULL BORE
Designation for a valve, ram, or other fitting whose opening is as large in cross section as the pipe, casing, or tubing it is mounted on.

FULLER'S EARTH
A fine, clay-like substance used in certain types of oil filters.

FULL-TERM WORKING INTEREST
See Working Interest, Full-Term.

FULLY INTEGRATED
Said of a company engaged in all phases of the oil business, i.e., production, transportation, refining, marketing. *See* Integrated Oil Company.

FUNGIBLE
Products that are or can be commingled for the purpose of being moved by product pipeline; interchangeable.

FUNNEL VISCOSITY
The number of seconds required for a quart of drilling mud to run through a mud funnel. The funnel viscosity has only one function: to spot a change in the density of the drilling fluid. A change in density, higher or lower, tells the experienced driller and mud men a lot about drilling conditions downhole.

FURFURAL
An extractive solvent of extremely pungent odor used extensively for refining a wide range of lubricating oils and diesel fuels; a liquid aldehyde.

FURNACE OIL
No. 2 heating oil; light gas oil that can be used as diesel fuel and for residential heating; Two oil; distillate fuel.

FUSIBLE PLUG
A fail-safe device; a plug in a service line equipped with a seal that will melt at a predetermined temperature, releasing pressure that actuates shutdown devices; a meltable plug.

A replica of Drake's derrick and engine house *(Courtesy Pennsylvania Historical and Museum Comm.)*

A 6-in. pig trap on a pipeline; note stock tanks in the background

G

GABBRO
A group of dark-colored, igneous, intrusive rocks very much like basalt. The name is thought to be from the town of Gabbro in Tuscany, Italy.

GABIONS
Anti-scour and soil stabilizing devices; wire baskets filled with 3- to 8-inch cobbles and placed on newly buried pipelines to prevent washing or scouring.

GAIN (seis)
A seismic-processing technique used to compensate for loss of energy owing to attenuation naturally resulting from the distance traveled by the wavelets in the acoustic medium. Higher frequency wavelets suffer the greatest attenuation and thus require greater compensatory processing.

GALL, TO
To damage or destroy a finished metal surface, as a shaft journal, by moving contact with a bearing without sufficient lubrication. To chafe by friction and heat as two pieces of metal are forcibly rubbed together in the absence of lubrication.

GALVANIC CORROSION
Electrochemical action that attacks metal installations—tanks, buildings and pipelines—causing corrosion. In pipelines, galvanic corrosion causes pitting, the eating away of the outer surface in small, crater-like depressions that in time become holes in the pipe. *See* Anode.

GAMMA RAY
Minute quantities of radiation emitted by substances that are radioactive. Subsurface rock formations emit radiation quantum that can be detected by well-logging devices and that indicate the relative densities of the surrounding rock.

GAMMA-RAY LOG
See Log, Gamma-Ray.

GANG PUSHER
A pipeline foreman; the man who runs a pipeline or a connection gang; a pusher.

GANG TRUCK

A light or medium-size flatbed truck carrying a portable doghouse or man rack where the pipeline repair crew rides to and from the job. The pipeliners' tools are carried in compartments beneath the bed of the truck.

GAS

"Any fluid, combustible or noncombustible, which is produced in a natural state from the earth and which maintains a gaseous or rarified state at ordinary temperature and pressure conditions." Code of Federal Regulations, Title 30, Mineral Resources, Chap. II, Geological Survey, 221.2.

GAS, ARTIFICIAL

Manufactured gas, made from coal by the Lurgi process, a German coal-gasification process. *See* Coal-Seam Gas.

GAS, COAL-BED

See Unconventional Gas.

GAS, INTERRUPTIBLE

See Interruptible Gas.

GAS, LIQUEFIED PETROLEUM

See Liquefied Petroleum Gas.

GAS, NATIVE

Gas originally in place in an underground structure as opposed to injected gas.

GAS, PENALTY

Quantities of gas provided for a period of time in excess of the amount contracted. Penalty gas customarily brings a high price than gas supplied according to the contract.

GAS, SOLUTION

See Solution Gas.

GAS, SYNTHESIS

A mixture of equal parts of hydrogen and carbon monoxide formed by reacting steam with hot coal or char. The mixture is similar to coal gas, but the main use is in the production of methane or synthetic natural gas. Synthesis gas, also known as "water gas," has an energy content of 980 to 1035 Btu per cubic foot, roughly comparable to the Btu content of natural gas.

GAS, UNASSOCIATED

See Unassociated Gas; *also* Associated Gas.

GAS, UNCONVENTIONAL NATURAL
See Unconventional Natural Gas.

GAS ANCHOR
A device for the bottom-hole separation of oil and gas in a pumping well. The gas anchor (a length of tubing about 5 feet long) is inside a larger pipe with perforations at the upper end. Oil in the annulus between the well's casing and tubing enters through the perforations and is picked up by the pump; the gas goes out through the casing to the wellhead.

GAS BALANCING AGREEMENT
An agreement or a clause in a gas purchase contract wherein the parties set forth the conditions for balancing the underproduction and over-production which in some instances results from split-stream gas sales. Such a clause obviates the need to resort to court rulings based on common law or equitable principles, i.e., a gas-balancing agreement may provide that the party in arrears may make up its underproduction out of 75 percent of the total well production.

GAS BEHIND THE PIPE
See Oil Behind the Pipe.

GASBOIL
A sudden, explosive boil or release of natural gas from a gas kick or blowout upward through the water; it can so aerate the water with gas bubbles that the submersible or drillship will sink because the critical decrease in water density will not support the vessel; the floater will no longer float. In drilling in known high-pressure, shallow-gas areas, operators usually opt for a subsea diverter system, called a "diverter stack." So equipped, the floater is able to contain the gas kick and simultaneously divert the pressure, the gas, horizontally through outlet nozzles. The rush of gas to the surface is on each side of the vessel, vented harmlessly to the atmosphere.

Gas box or degasser

GAS BOTTLES
The cylindrical containers of oxygen and acetylene used in oxyacetylene welding. Oxygen bottles are tall and slender with a tapered top; acetylene bottles are shorter and somewhat larger in diameter.

GAS BOX
Colloquial term for a mud-gas separator at a drilling well. *See* Degassing Drilling Mud.

GAS BUSTER
A drilling-mud/gas separator; a surge chamber on the mud-flow line where entrained gas breaks out and is vented to a flare line; the gas-free mud is returned to the mud tanks.

GAS CAP

The portion of an oil-producing reservoir occupied by free gas; gas in a free state above an oil zone.

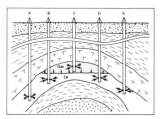

Gas-cap expansion

GAS-CAP ALLOWABLE

A production allowable granted an operator who shuts in a well producing from a gas cap of an oil-producing reservoir. The allowable is transferable to another lease in the same field. The shutting in of the gas-cap producer preserves the reservoir pressure, which is essential to good production practice.

GAS-CAP DRIVE

The energy derived from the expansion of gas in a free state above the oil zone, which is used in the production of oil. Wells drilled into the oil zone cause a release of pressure, which allows the compressed gas in the cap to expand and move downward, forcing the oil into the well bores of the producing wells.

GAS-CAP FIELD

A gas-expansion reservoir in which some of the gas occurs as free gas rather than in solution. The free gas will occupy the highest portion of the reservoir. When wells are drilled to lower points on the structure, the gas will expand, forcing the oil downdip and into the well bores.

GAS CHROMATOGRAPHY

The process of separating gases from one another by passing them over a solid (gas-solid chromatography) or liquid (gas-liquid chromatography) phase. The gases are repeatedly adsorbed and released at different rates resulting in separation of their components.

GAS COLUMN

An interval, a sedimentary section in the well bore containing gas in commercial quantities.

GAS CONDENSATE

Liquid hydrocarbons present in the casinghead gas that condense upon being brought to the surface; formerly distillate, now condensate. Also casinghead gasoline; white oil.

GAS CONDENSATE, RETROGRADE

See Retrograde Gas Condensate.

GAS-CUT MUD

Drilling mud aerated or charged with gas from formations downhole. The gas forms bubbles in the drilling fluid, seriously affecting drilling operations and sometimes causing loss of circulation.

GAS DISTILLATE
See Distillate.

GAS DRILLING
The use of gas as a drilling fluid. *See* Drilling, Air.

GAS DRIVE
See Gas-Cap Drive.

GAS EMISSIONS, FUGITIVE
Fugitive emissions are those that escape confinement, from tank, hose, vents, or open vessels. The new and stricter Environmental Protection Agency regulations make necessary the capture of all fugitive emissions and prevention of future escape, whether from filling the fuel tank of an automobile or storing fuel in tanks without proper vapor traps. The federal regulations have spawned businesses that take entrepreneurial advantage of clean air, clean water, and clean top soil requirements. Advanced-technology gas sniffers, water purification, and soil remedial or restoration work.

One-cylinder gas engine
(Courtesy Arrow Specialties Co.)

GAS ENGINE
A two- or four-cycle internal-combustion engine that runs on natural gas; a stationary field engine.

GAS GRADIENT
The presence of gas in a sedimentary section that may vary in saturation, porosity, and permeability from top to bottom, thus establishing a gas gradient. Part of the gas column may be tight, a part very permeable, and the remainder in between.

GAS HYDRATES
See Hydrate.

GASIFICATION
Converting a solid or a liquid to a gas; converting a solid hydrocarbon such as coal or oil shale to commercial gas; the manufacture of synthetic gas from other hydrocarbons. *See* Synthetic Natural Gas.

GAS INJECTION
Natural gas injected under high pressure into a producing reservoir through an input or injection well as part of a pressure-maintenance, secondary-recovery, or recycling operation.

GAS INJECTION WELL
A well through which gas under high pressure is injected into a producing formation to maintain reservoir pressure.

GASKET
Thin, fibrous material used to make the union of two metal parts pressure-tight. Ready-made gaskets are often sheathed in very thin, soft metal, or they may be made exclusively of metal, or of specially formulated rubber.

GAS KICK
See Kick.

GAS LIFT
A method of lifting oil from the bottom of a well to the surface by the use of compressed gas. The gas is pumped into the hole and at the lower end of the tubing it becomes a part of the fluid in the well. As the gas expands, it lifts the oil to the surface.

GAS-LIFT GAS
Natural gas used in a gas-lift program of oil production. Lift gas is usually first stripped of liquid hydrocarbons before it is injected into the well. And because it is a "working gas" as opposed to commercial gas, its cost per thousand cubic feet (Mcf) is considerably less. Gas-lift and commercial gas commingle when produced, so when the combined gas is stripped of petroleum liquids only the formation gas is credited with the recovered liquids. This is necessary for oil and gas royalty purposes.

GAS LIQUIDS
See Liquefied Petroleum Gas.

GAS LOCK
A condition that can exist in an oil pipeline when elevated sections of the line are filled with gas. The gas, because of its compressibility and penchant for collecting in high places in the line, effectively blocks the gravity flow of oil. Gas lock can also occur in suction chambers of reciprocating pumps. The gas prevents the oil from flowing into chambers and must be vented or bled off.

GAS LOG
See Log, Gas.

GAS MEASUREMENT, STANDARD
A method of measuring volumes of natural gas by the use of conversion factors of standard pressure and temperature. The standard pressure is 14.65 pounds per square inch; the standard temperature is 60°F. One standard cubic foot of gas is the amount of gas contained in one cubic foot of space at a pressure of 14.65 psia at a temperature of 60°F. Using the conversion table, natural gas at any temperature and pressure can be

Gas measuring manifold; note glycol heaters in background

converted to standard cubic feet, the measurement by which gas is usually bought, sold, and transported.

GAS METER, MASS-FLOW
See Mass-Flow Gas Meter.

Gas regulator control valve

GASOHOL
A mixture of 90 percent gasoline and 10 percent alcohol; a motor fuel. Gasohol was first marketed in the late 1970s as a way to stretch available gasoline stocks by using surplus agricultural products to make ethanol or grain alcohol.

GAS OIL
A refined fraction of crude oil somewhat heavier than kerosene, often used as diesel fuel.

GAS-OIL CONTACT
The boundary surface, the interface, between an accumulation of oil and an accumulation of natural gas. The gas, by its nature, is always on top, forming a gas cap. *See* Gas-Cap Drive.

GAS-OIL RATIO
The number of cubic feet of natural gas produced with a barrel of oil; G.O.R.

GASOLINE
Motor gasoline is a blend of different cuts or fractions in the gasoline range. Gasoline is blended for the seasons (winter starting and summer heat), mileage, and antiknock or ping.

GASOLINE, MARINE WHITE
See Marine White Gasoline.

GASOLINE, RAW
See Raw Gasoline.

GASOLINE, STRAIGHT-RUN
The gasoline-range fraction distilled from crude oil. Virgin naphtha.

GASOLINE INDICATOR PASTE
See Indicator Paste, Gasoline.

Gasoline plant

GASOLINE PLANT
A compressor plant where natural gas is stripped of the liquid hydrocarbons usually present in wellhead gas.

GAS or OIL, CLASSIFIED
A gas-oil ratio is frequently used by regulatory agencies to classify a well as either an oil well or a gas well for regulatory purposes.

GAS PURCHASE CONTRACT—TAKE OR PAY
A contract that provides that the purchaser will pay the gas producer for the gas not yet produced if the purchaser, for some valid reason, does not actually take the gas. After such payment, the purchaser is entitled to gas at no cost at some time in the future.

GAS SALES LINE
See Line, Gas Sales.

GASSER
A good gas well with small amounts, if any, of oil or entrained water.

GAS SNIFFER
A colloquial term for a sensitive electronic device that detects the presence of gas or other hydrocarbons in the stream of drilling mud returning from downhole.

GAS SWEETENING
The process of removing hydrogen sulfide (H_2S), carbon dioxide (CO_2), and other impurities from sour gas. Hydrogen sulfide is very corrosive and is deadly to humans, even in small concentrations. Carbon dioxide is somewhat less corrosive and in small concentrations is not harmful. Both gases, however, are contaminants that must be removed from sour gas to make it marketable.

Gas turbine

GAS TURBINE
An engine in which vapor (other than steam) is directed against a series of turbine blades. The energy contained in the rapidly expanding vapors or gases turns the rotors to produce the engine's power.

GAS WELDING
See Welding, Gas.

GAS WELL, SPLIT-STREAM
A gas well with more than one owner and from which each person with an interest receives a share of product, in money or in kind.

GATE
Short for gate valve; common term for all pipeline valves.

Early gas welding rig

GATE, BACKFLOW
A type of swing-check valve made so the clapper's position may be changed from open to closed by an externally mounted handle. The handle is attached to the clapper's fulcrum shaft which protrudes through the side of the valve body. When the clapper is closed (resting on its seat in a normal position), fluid can flow in one direction only; when open (raised from its seat by the handle), fluid can flow in the opposite direction.

Gate valve

Oil inlet line and gauge hatch

Gauging a tank

GATE VALVE
See Valve, Gate.

GATHERING FACILITIES
Pipelines and pumping units used to bring oil or gas from production leases by separate lines to a central point, i.e., a tank farm or a trunk pipeline.

GATHERING LINES
Small-diameter (2 to 4 inch) pipelines that connect the tank batteries on a lease with a booster station in the field. The gathering lines are hooked up to the lease tanks; the oil is pumped to a gathering station or, if the terrain permits, it goes by gravity to the station tanks.

GATHERING SYSTEM
See Gathering Facilities.

GAUGE, DOWNHOLE TOOLS
Gauge has to do with the diameter of a drill bit, reamer, or stabilizer. Boreholes must be kept up to gauge, must maintain hole size for the pipe and downhole tools to follow. When cutting tools wear and are not full gauge, they must be built up by hard banding or discarded.

GAUGE, STRAIN
See Strain Gauge.

GAUGE HATCH
An opening in the roof of a stock or storage tank, fitted with a hinged lid, through which the tank may be gauged and oil samples taken; hatch.

GAUGE HOLE
A gauge hatch.

GAUGING
See Gauger.

GAUGE LINE
A reel of steel measuring tape, with a bob attached, held in a frame equipped with handle and winding crank used in gauging the liquid level in tanks. To prevent striking sparks, the bob is made of brass or other nonsparking material or is sheathed in a durable plastic. The tip of the bob is point zero on the gauge column.

GAUGER
A person who measures the oil in a stock or lease tank, records the temperature, checks the sediment content, makes out a run ticket, and turns the oil into the pipeline. A gauger is the pipeline company's agent and, in effect, buys the tank of oil for his company.

GAUGER, ROYALTY
See Royalty Gauger.

GAUGE ROW (OF CUTTERS)
The outside, the peripheral row of cutters on a drill bit that cuts the borehole to gauge.

GAUGE TANK
A tank in which the production from a well or a lease is measured.

Automatic tank gauge

GAUGE TAPE
Gauge line.

GAUGE TICKET
A run ticket.

G.B. STRUCTURE
A gravity-base structure; a very heavy offshore drilling, producing, and storage platform, with the general configuration of a cylindrical tower weighing as much as 400,000 metric tons. The giant structure is moved from the construction yard unballasted, with ballast tanks full of air. When on location it is ballasted with sea water and settles ponderously to the seabottom, held there firmly by gravity.

Gear box with cover off

GEAR BOX
The enclosure or case containing a gear train or assembly of reduction gears; the case containing a pump's pinion and ring gears.

GEAR PUMP
See Pump, Gear.

GEL
A viscous substance, a jelly-like material, used in well stimulation and formation fracturing to suspend sand or other proppants in the fracturing medium. Gelling agents are mixed with water or light oil to form an emulsion that will carry a quantity of sand for various well workover procedures.

Cutaway of gear pump

GEL BREAKER
A chemical additive to gelled frac fluid that, after a delay of one to three hours, acts to break down the gel which permits the frac fluid, either water or oil, back to its original viscosity to be pumped out of the borehole and the well cleaned up.

GELLED WATER/OIL
Frac fluid made by treating water and certain oils with a gelling agent, a gel activator, to condition either water or oil so they are able to carry one-half to three-quarters of a pound of sand or other material per gallon of fluid. An example of gelled oil and its load is an instance where

183

47,000 gallons of gelled oil transported 34,000 pounds of 20/40-mesh sand into fractures, and broke. Then reverting to its original viscosity, it drained out of the formation, and the well was cleaned up. *See* Gel Breaker.

GEOCHEMICAL EXPLORATION
The search for economic mineral and petroleum deposits by the detection of abnormal concentrations of elements or traces of hydrocarbons in the near-surface materials. This is accomplished by close observation and cursory field tests.

GEOCHEMICAL PROSPECTING
Exploratory methods that involve the chemical analysis of rocks.

GEOCHEMISTRY
The science of chemistry applied to oil and gas exploration. By analyzing the contents of subsurface water for the presence of organic matter associated with oil deposits, geochemistry has proved to be an important adjunct to geology and geophysics in exploratory work.

GEODE
A hollow or partly hollow spherical body, from a few inches to perhaps 2 feet in diameter, found in certain limestone beds and sometimes in shales. It has a relatively thin crust enclosing a cavity partly filled with inward projecting crystals, often perfectly formed and composed of quartz, calcite, or other minerals. The crystals are deposited from solution on the cavity walls. A geode also can be a rock cavity with its lining of crystals that is not a separate unit from the enclosing rock. When the crystals are quartz and are colored violet or pale purple from iron compounds, the crystals are called amethysts and are highly prized as semiprecious stones.

GEODESY
The branch of science concerned with the determination of the size and shape of the Earth and the location of points on its surface; also the gravitational field of the Earth and the study of its tides, polar motion, and rotation.

GEOGRAPHICALLY DISCREET FORMATION
A formation running true to form; maintaining its identification, neither intruded nor intrusive.

GEOLITH CUBE
A depiction, a visual representation of subsurface geology on 3-D cross sections, horizontal and vertical, between wells in a field or to verify the discontinuity or porosity pinchouts of a reservoir under study.

GEOLOGIC AGE
The age of a particular geologic event or feature in relation to the geologic time scale and expressed in centuries (absolute age) or in comparison with the surrounding area (relative age). Geologic age is also used to stress the long-ago periods of geologic history as compared to purely historic times that refer to human history, which is very short.

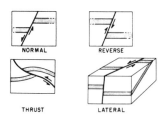

Faulting

GEOLOGICAL STRUCTURE
Layers of sedimentary rocks that have been displaced from their normal horizontal position by the forces of nature.

GEOLOGIC COLUMN
The sequence of formations or layers of rock, sedimentary and metamorphic, from the igneous bedrock to the surface of the earth.

GEOLOGIC DRILLING, HORIZONTAL
Drilling interactively, guiding the bit into and through a hydrocarbon reservoir geologically, by the use of near-bit sensors. Geometric or geologic drilling; two schools of thought on the matter of hitting the target formation, and once there, staying in bounds.

GEOLOGIC ERAS
See Geologic Time Scale.

GEOLOGIC MAP
See Map, Geologic.

GEOLOGIC PERIODS
See Geologic Time Scale.

GEOLOGIC TIME SCALE
According to authorities in the study of the Earth, the Paleozoic Era represents the oldest rocks whose ages are 225 to 600 million years old; the next era is the Mesozoic with rocks 75 to 225 million years old; the most recent era, the Cenozoic, has rocks from the present to 75 million years old. Dividing these eras are periods. Beginning with the oldest, they are: Cambrian, Ordovician, Silurian, Devonian, Mississippian, Pennsylvanian, and Permian, all in the Paleozoic Era. In the Mesozoic Era are the Triassic, Jurassic, and Cretaceous Periods. In the youngest era, Cenozoic, are the Tertiary and Quaternary. Geologists have divided the Quaternary Period, identifying the earlier epoch of the Quaternary Period as Pleistocene.

GEOLOGIST
A person trained in the study of the Earth's crust. A petroleum geologist, in contrast to a hard-rock geologist, is primarily concerned with

Geologist

sedimentary rocks where most of the world's oil has been found. In general, the work of a petroleum geologist consists in searching for structural traps favorable to the accumulation of oil and gas. In addition to deciding on locations to drill, he may supervise the drilling, particularly with regard to coring, logging, and running tests.

GEOLOGIST, GENERATING
A geologist who actively seeks out good prospects for his company, his employer. He uses every means at his command: seismic data, well logs, surface indications, and remote sensing.

GEOLOGIST, HORSEBACK
An amateur, rock-loving person who spots good drilling prospects by observing the Earth's surface features; a humorous, gently pejorative term.

GEOLOGIST, HYDRODYNAMICS
A geologist specializing in the study of the mechanics of fluids in underground formations. His work involves analysis of the test data, the interpretation of fluid pressures from drilling wells and well logs, and applying his findings to the solution of problems associated with oil- and gas-well exploration and development.

GEOLOGRAPH
A device on a drilling rig to record the drilling rate or rate of penetration during each 8-hour tour.

GEOLOGY
The science of the history of the Earth and its life as recorded in rocks.

GEOLOGY, REGIONAL
The geology of any relatively large area treated broadly and mainly from the viewpoint of the distribution and position of stratigraphic units, structural features, and surface forms; areal geology.

GEOMETRIC DRILLING, HORIZONTAL
Drilling automatically to a predetermined target, without deviating from the planned trajectory using a geosteering system. *See* Geologic Drilling, Horizontal.

GEOMETRY OF A RESERVOIR
A phrase used by petroleum and reservoir engineers meaning the shape of a reservoir of oil or gas.

GEOMORPHIC
Pertains to the Earth or, more specifically, to its surface features, its landforms.

GEOMORPHOLOGY
The science that concerns itself with the general features of the Earth's

surface; specifically, the study of the classification, description, origin, and development of present day landforms and their relationship to underlying, subsurface structures.

GEOPHONES
Sensitive sound-detecting instruments used in conducting seismic surveys. A series of geophones is placed on the ground at intervals to detect and transmit to an amplifier-recording system the reflected sound waves created by explosions set off in the course of seismic exploration work.

Geophone

GEOPHYSICAL CAMP
Temporary headquarters established in the field for geophysical teams working the area. In addition to providing living quarters and a store of supplies, the camp has facilities for processing geophysical data gathered on the field trips.

GEOPHYSICAL MODELING
Devising a 3-D (three-dimensional) representation of subsurface geological formations: faults, pinchouts, overthrusts, etc.

GEOPHYSICAL PROSPECTING
The use of geophysical techniques to detect Earth currents (gamma rays), gravity readings, heat flow, and radio activity indicators; also geochemical techniques in the search for oil and gas.

GEOPHYSICAL SHIP
An oceangoing vessel equipped with a full complement of seismographic and geophysical instruments that has seismologists, geophysicists, and other scientists aboard. Geophysical ships conduct seismic and other surveys (geochemical) offshore in waters of the Continental Shelf of all six continents. *See* Seismic Sea Streamer.

GEOPHYSICAL TEAM
A group of specialists working together to gather geophysical data. Their work consists of drilling shot holes, placing explosive charges, setting out or stringing geophones, detonating shot charges, and reading and interpreting the results of the seismic shocks set off by the explosive charges.

GEOPHYSICS
The application of certain familiar physical principles—magnetic attraction, gravitational pull, speed of sound waves, the behavior of electric currents—to the science of geology.

GEOPOLITICS
A pejorative term suggesting organizational manipulations within government, society, or university circles.

GEOPRESSURIZED GAS
Natural gas that exists at great depths in the geopressurized aquifer along the Gulf coast. The gas, because of the extreme pressure in the aquifer, is in solution in the salt water. Geopressurized is a term meaning the fluid pressure in the aquifer is higher than normal hydrostatic pressure of 0.464 pounds per foot of depth.

GEOSCIENCE DATA
Information taken from 3-D visualizations. Geologists, geophysicists, and others with a clear picture of the geoscience process are able to make more accurate analyses than ever before of reservoirs, and where and how to drill a particular formation.

GEOSTATIC PRESSURE
Pressure of the overburden; the loose soil, silt, gravel, and other unconsolidated material overlying bedrock.

GEOSTEERING
A system that uses and integrates directional drilling, drilling mechanics, and petrophysical data measured within a few feet of the bit, and transmits the data to the surface, all in real time. The downhole information is then monitored, interpreted, and converted into visual displays for use by the well-site decision-making technicians and engineers. Geosteering allows the driller to compensate for downhole shocks and vibrations, drillstring washouts, gas influx, cone locking, and other unforeseen events that can become problems.

GEOSTEERING, OTHER ASPECTS OF
Recent advances in geosteering incorporate near-bit sensors that provide azimuthal, restivity, gamma ray, rpm, and inclination measurements, and a wire telemetry system that passes all data from the bit to the M.W.D. system.

GEOSYNCLINE
See Syncline.

Geothermal Power
Plant

GEOTHERMAL GRADIENT
The temperature gradient in well drilling, as the borehole progresses downward into the Earth, is 18°F per 100 feet. In deep wells, 15,000 feet or so, the bottom-hole temperature is 350°F or more. Water-base drilling slurries (mud) cannot be used; the operator must switch to oil-base muds after reaching depths where water in the slurries flashes off, turns to vapor.

GEOTHERMAL POWER GENERATION
The use of underground natural heat sources, i.e., superheated water from deep in the Earth, to generate steam to power turboelectric generators.

GEOTHERMAL PROSPECTING
Exploring for accessible, underground sources of geothermal energy, super-heated water from deep in the Earth.

G.E.R.G.
Geochemical & Environmental Research Group.

GILSONITE
A solid hydrocarbon with the general appearance of coal; uintaite; a black, lustrous form of asphalt that, when treated and refined, yields gasoline, fuel oil, and coke. Found in deposits in Utah.

GIN POLE
(1) An A-frame made of sections of pipe mounted on the rear of a truck bed that is used as a support or fixed point for the truck's winch line when lifting or hoisting. (2) A vertical frame on the top of the derrick, spanning the crown block, providing a support for hoisting. (3) A mast.

Gin pole

GIRBITOL PROCESS
A process used to "sweeten" sour gas by removing the hydrogen sulfide (H_2S).

GIRT
One of the braces between the legs of a derrick; a supporting member.

GIRTH WELD
See Weld, Girth.

GLACIAL TILL
The deposition of sand, rock, and debris left behind when the glaciers melted and receded.

GLOBE VALVE
See Valve, Globe.

GLYCOL DEHYDRATOR
A facility for removing minute particles of water from natural gas not removed by the separator.

Glycol dehydrator

G.M.P.
Gallons of gasoline per 1,000 cubic feet of natural gas produced.

GO-DEVIL
A pipeline scraper; a cylindrical, plug-like device equipped with scraper blades, rollers, and wire brushes used to clean the inside of a pipeline of accumulations of wax, sand, rust, and water. When inserted in the line, the go-devil is pushed along by the oil pressure.

Gooseneck

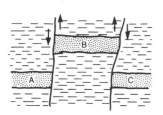

Horst (B) and graben

GONE TO WATER

A well in which the production of oil has decreased and the production of water has increased to the point where the well is no longer profitable to operate.

GOOSENECK

A nipple in the shape of an inverted U attached to the top of the swivel and to which the mud hose is attached.

GOOSING GRASS

Cutting grass and weeds around the lease or tank farm; shaving the grass off the ground with a sharp hoe-like tool, leaving the ground clean.

G.O.R.

Gas-oil ratio.

G.P.M.

Gallons per minute.

GRABEN

A *graben*, from the German, is an elongated block of the earth's crust between faults that has subsided, forming a valley. A famous set of graben is the fault-bordered trough running the length of central Africa, known as the Rift Valley. A horst is the reverse of a graben. It is an elongated crustal block between faults on both sides that is pushed up or elevated relative to the surrounding countryside and forms a plateau.

GRABLE OIL-WELL PUMP

See Pump, Grable Oil-Well.

GRADE

A degree of inclination or the rate of ascent or decent of an inclined surface in relation to the horizontal. A surface that slopes up or down with respect to level. Synonymous with gradient.

GRADED BEDDING

The deposition of successively smaller grain-size sediment from the bottom to the top of a sedimentary unit or sequence. This occurs when a sediment-carrying stream slows down and loses the heavier suspended material first, then next heaviest, and so on until the finest, lightest material drops out.

GRADIENT

The rates of increase or decrease of temperature or pressure are defined as gradients; the rate of regular or graded ascent or descent.

GRADUATE

A calibrated cylindrical vessel, a beaker. A measuring vessel marked in cubic centimeters (cc) and fractions thereof. A graduated cylinder.

GRAINSTONE
A mud-free, grain-supported, carbonate sedimentary rock; a clean rock with sand-size particles making up the matrix.

GRANDFATHERED
To receive permission or approval to carry on an undertaking that was initiated or begun before a current restrictive law or regulation was enacted. Also, not to be penalized for a fait accompli which does not comply with current regulations but was entirely legitimate years before.

GRANITE WASH
A field term for the eroded material from outcrops of granite rock that has been redeposited to form a rock of approximately the same mineral makeup. *See* Basic Wash.

GRANITIC ROCK
A term generally applied to plutonic rock that is coarse grained, containing quartz along with feldspar and mafic (ferromagnesian), dark-colored minerals; an igneous rock; granite.

GRANNY BOARD
See Lazy Board.

Granny board

GRANNY HOLE
The lowest, most powerful gear on a truck.

GRANNY KNOT
A knot tied in such a way as to defy untying; an improperly tied square knot; a hatchet knot.

GRANTEE
The one receiving the money, property, or oil and gas lease; the lessee.

GRANTOR
The person who grants something to another, the grantee. It may be money, property, land, or the right to drill for oil and gas (a lease); a lessor.

GRAPHITE
A mineral; a naturally occurring crystalline form of carbon. It is greasy to the touch and black or dark gray in color. It occurs as crystals, flakes, or grains in thin veins, in bedded masses, or scattered through metamorphic rock. Graphite is used as a lubricant, in paint, and as "lead" in pencils. Among other things, it is made into electrodes, since it is a good conductor of electricity.

A small, modern grass-roots refinery

GRASSROOTS
Said of a refinery or other installation built from the ground up as contrasted to a plant merely enlarged or modernized.

Gravel drilling island

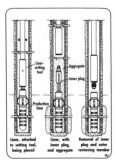

Gravel pack tool

Gravimeter - Underwater

GRASSROOTS DRILLING

Basic drilling: no frills, no slim-hole testing, no M.W.D., no geosteering, just experienced drilling in mature, well-tested fields.

GRAVEL ISLAND

In some locations (in shallow water near shore) in the Arctic, gravel islands, 40 to 50 feet in diameter, are constructed to make a foundation area from which to drill exploratory wells. Gravel is dredged from the sea bottom or transported from a nearby river or delta and dumped into holes cut in the ice. *See* Drilling Island; *also* Ice Platform.

GRAVEL PACKING

Using gravel to fill the cavity created around a well bore as it passes through the producing zone to prevent caving or the incursion of sand and to facilitate the flow of oil into the well bore.

GRAVEYARD SHIFT

A tour of work beginning at midnight and ending at 8 a.m. In pipeline operations, the graveyard shift is customarily from 11 p.m. to 7 a.m. Hoot-owl shift.

GRAVIMETER

A geophysical instrument used to measure the minute variations in the Earth's gravitational pull at different locations. To the geophysicist, these variations indicate certain facts about subsurface formations. Gravitymeter.

GRAVING DOCK

A dry dock; a dock that can accept ships into an enclosure. When the water is pumped out, the ship is left high and dry for repairs.

GRAVITY

(1) The attraction of the Earth's mass for bodies or objects at or near the surface. (2) Short for specific gravity; A.P.I. gravity. (3) To flow through a pipeline without the aid of a pump; to be pulled by the force of gravity.

GRAVITY DRIVE

A natural drive occurring where a well is drilled at a point lower than surrounding areas of producing formations, causing the oil to drain downhill into the well bore. If the reservoir rock is highly permeable and dips sharply toward the well, there is usually good oil recovery.

GRAVITY LINE

A pipeline that carries oil from a lease tank to pumping station without the use of mechanical means; a line that transports liquid from one elevation to a lower elevation by the force of gravity alone.

GRAVITY MAP
See Map, Gravity.

GRAVITYMETER
See Gravimeter.

GRAVITY PUMP
See Pump, Sight.

GRAVITY RETURN HEAT PIPE
See Heat Pipe, Gravity Return.

GRAVITY SEGREGATION
The separation of water from oil or heavy from lighter hydrocarbons by the force of gravity, either in the producing zone or by gravity in the separators after production; the stratification of gas, oil, and water according to their densities.

GRAVITY STRUCTURE
An offshore drilling and production platform made of concrete and of such tremendous weight that it is held securely on the ocean bottom without the need for piling or anchors.

Gravity structure

GRAVITY SURVEY
A survey of subsurface rock densities by the use of a gravimeter. Gravity (gravitational pull) surveys are made at numerous points in an area to determine the location and probable depth of rock masses of lower density than the surrounding formations. The less-dense rocks indicate porous rocks that may be a sedimentary trap or reservoir.

GREASE
(1) A lubricating substance (solid or semisolid) made from lubricating oil and a thickening agent. The lube oils may be very light or heavy cylinder oils; the thickening agent (usually soaps) may be any material that when mixed with oil will produce a grease structure. (2) Colloquial for crude oil.

GREASE CUP
A lubricating device in the shape of a small cup with a spring-loaded lid to hold cup grease to lubricate a shaft or other slow-moving engine or pump part. The top of the grease cup has a spring-loaded cap that keeps pressure on the grease so it will feed onto the moving part gradually; pressure cup.

GREEK FIRE
A mixture of petroleum and lime which, when moistened, would catch

fire. Used from 7th century forward by the Byzantines as a formidable weapon. *Oleum incendiarum.*

GREENHOUSE EFFECT

The widely accepted theory that the combustion (burning) of fossil fuels (coal, oil, and gas) leads to increased carbon dioxide in the atmosphere. As CO_2 is an efficient absorber of heat radiation, the radiation emitted by the Earth will be increasingly absorbed, resulting in higher atmosphere temperatures.

GREEN MAIL

Payment at an inflated price for shares held by a stockholder who is threatening a hostile takeover.

GREEN OIL

A paraffin-base crude oil. Asphalt-base crudes are sometimes referred to as black oil.

GRIEF STEM

Grief stem

Kelly joint; the top joint of the rotary drillstring that works through the square hole in the rotary table. As the rotary table is turned by the drilling engines, the grief stem and the drillpipe are rotated. Grief stems are heavy, thick-walled tubular pieces with squared shoulders that are made to fit into the hole in the rotary table.

GRIND OUT

Colloquial for centrifuge; to test samples of crude oil or other liquid for suspended material—water, emulsion, sand—by use of a centrifuge machine.

GROOVED-END PIPING

A patented type of usually small diameter (2 to 6 inch) pipe with a circumferential, square-sided groove at each end of the nipple or joint of pipe. Groove-ended pipe is joined by inserting a rubber ring in the groove of each segment to be joined and putting on a two-bolt turtle-back clamp. Victaulic is one such brand of groove-end pipe.

GROSS PRODUCTION TAX

A severance tax; a tax usually imposed by a state at a certain sum per unit of mineral removed (barrels of oil, thousands or millions of cubic feet of gas).

GROUND ROLL

A term in seismology for surface waves that are recorded by geophones and must be separated from those waves that are reflected and refracted from the deeper targeted zone or bedding planes. The effect of the surface waves must be eliminated or compensated for to avoid distortion of the seismic trace.

GROUNDMASS
See Matrix.

GROUND-SEAT UNION
A pipe coupling made in two parts: one half is convex, the other half concave in shape, and both are ground to fit. A threaded ring holds the halves together, pressure tight. Used on small-diameter piping.

GROUNDWATER BASIN
An aquifer, not necessarily in the shape of a basin, that has roughly defined boundaries and more-or-less definite charging and discharging systems.

GROUP
Two or more contiguous or closely associated rock formations that have prominent lithologic features (types of rocks) in common. A group is the next lithostratigraphic unit or classification above a formation. To delve more deeply into "group," there is supergroup, subgroup, and synthetic group, all having to do with combinations or divisions of groups according to lithologic features.

GROUP SHOT
Geophysical exploration performed for several individuals or companies on a cost-sharing basis. The companies share the information as well as the cost. This type arrangement is usually for offshore seismic surveys in which several companies are planning to submit bids for offshore leases offered in a government lease sale.

GROUT
(1) A concrete mixture used to fill in around piling, caissons, heavy machinery beds, and foundation work. (2) To stabilize and make permanent. Grout is usually a thin mixture that can be worked into crevices and beneath and around structural forms.

GROWLER BOARD
See Lazy Board.

Growler board

GRUB-STAKE AGREEMENT
An agreement whereby one person undertakes to prospect for oil and agrees to hand over to the person who furnishes money or supplies a certain proportionate interest in the oil discovered. This type of agreement is common for solid minerals but is not often used in oil prospecting.

G.S.A.
Geological Society of America.

GUARANTEED ROYALTY
See Royalty, Guaranteed.

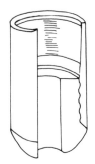

Guide shoe

GUIDE FOSSIL
See Diagnostic Fossil.

GUIDESHOE
A section of casing belled out to permit a downhole tool, such as a whipstock, to be retrieved through the bottom of the casing without hanging up, after the operation is completed.

GULF OIL CORPORATION
One of the giant U.S. fully integrated petroleum companies; responsible for many significant discoveries in the United States and as a pioneer in the Middle East. In 1983, merged with Chevron Oil Company (Standard of California). *See* Seven Sisters.

GUMBO
A heavy, sticky mud formed downhole by certain shales when they become wet from the drilling fluid.

GUM BOOTS
Rubber boots, the kind you pull on like cowboy boots; Wellingtons (Brit.).

GUN BARREL
A colloquial term for a tall, small-diameter tank (a gun barrel) into which oil and water from stripper wells are pumped to separate the two liquids. In this type of homemade separator, the oil works its way to the top of the tank where it is drawn off into the stock tanks. The water goes to the evaporation pit.

Lease tanks (l) with gun barrel (r)

GUNK
The collection of dirt, paraffin, oil, mill scale, rust, and other debris that is cleaned out of a pipeline when a scraper or a pig is put through the line.

GUN PERFORATION
A method of putting holes in a well's casing downhole in which explosive charges lowered into the hole propel steel projectiles through the casing wall. (Casing is perforated to permit the oil from the formation to enter the well.)

An early-day gusher

GUSHER
A well that comes in with such great pressure that the oil blows out of the wellhead and up into the derrick, like a geyser. With improved drilling technology, especially the use of drilling mud to control downhole pressures, gushers are rare today. *See* Blowout.

GUY WIRE
A cable or heavy wire used to hold a pole or mast upright. The end of the guy wire is attached to a stake or a deadman.

GYP
Boiler scale; a residue or deposit from hard water, water with high concentrations of minerals. Pipe and vessels handling hard water become gyped up as the minerals form a hard, rock-like layer on the inner surfaces.

GYPSUM
A soft, usually white mineral consisting of hydrous calcium sulfate ($CaSO_4 \cdot 2H_2O$). It is widely distributed and is the most common sulfate mineral. Gypsum is often found with halite (salt) and anhydrite, forming thick beds interspersed with limestone, clay, and shale.

H

H₂S

H_2S

Hydrogen sulfide.

HABENDUM CLAUSE

The clause in a lease setting forth the duration of the lessee's interest in the property. A habendum clause might read, "It is agreed that this lease shall remain in force for a term of five years from this date and for as long thereafter as oil or gas, of whatever kind, or either of them is produced or drilling operations are continued as hereinafter provided." The primary term in this case is five years. *See* Primary Term.

HAIRPIN HEAT EXCHANGER

See Heat Exchanger, Hairpin.

HALF SOLE

A metal patch for a corroded section of pipeline. The patch is cut from a length of pipe of the same diameter as the one to be repaired. Half soles can be from 6 to 12 feet in length and are placed over the pitted or corroded section of the pipe and welded in place with a bead around the entire perimeter of the half sole.

HALITE

A mineral (NaCL); a natural salt that occurs in massive, granular, and cubic-crystalline forms. Its symbol is HI; rock salt.

HAND LENS

A small magnifying glass with a power of 6× to 10× used in the field for examining rock samples, fossils, and crystals.

HAND MONEY

See Earnest Money.

HANDY

Hand-tight; a pipe connection or nut that can be unscrewed by hand.

HANG A WELL OFF

To disconnect the pull-rod line from a pumping jack or pumping unit being operated from a central power.

Hanging a large rising-stem valve

HANGER, ROD
See Rod Hanger.

HANGING IRON
A colloquial expression for the job of assembling a high-pressure, heavy-duty blowout-preventer stack or production tree. Some of the valve assemblies weigh thousands of pounds or more, so they must be hoisted into place, aligned, and bolted to their mating piece.

HANGING WALL
The hanging wall of a fault is located above the surface of the fault and bears upon the fault surface—the footwall. The footwall is below the fault whether the hanging wall has moved up or down along the fault plane.

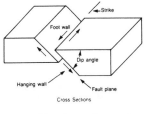

Hanging wall

HANG THE RODS
To pull the pump rods out of the well and hang them in the derrick on rod hangers. On portable pulling units, the rods are hung outside the derrick.

HARD-BANDING
Laying on a coating or surface of superhard metal on a softer metal part at a point or on an area of severe wear or abrasion. Putting a hard surface on a softer metal by welding or other metallurgical process. Where it may be impractical or prohibited by structural constraints to manufacture a part from the harder metal, a coating or hard-surfacing of the part is a practical solution. Also called hard-facing.

HARD-FACING
See Hard-Banding.

HARD MINERAL
A mineral that is as hard or harder than quartz, ranking 7 or higher on the Mohs scale.

HARDPAN
A widely used term for a fairly hard, often clayey layer of earth lying just below the surface. Hardpan is formed by precipitation of relatively insoluble material such as silica, calcium carbonate, iron oxide, and organic matter. It is not noticeably softened by moisture.

HARD ROCK
A term used loosely for igneous and metamorphic rock as distinguished from sedimentary rock, which by comparison is softer and easier to drill. Hard-rock geologists are mining geologists; soft-rock geologists are concerned principally with sedimentary rocks and are usually petroleum geologists.

HARD-ROCK GEOLOGIST

A geologist who specializes in the study of hard rock—igneous and metamorphic—as distinguished from sedimentary or soft rock; a mining geologist.

HARDWARE

(1) Electronic and mechanical components of a computer system, e.g., storage drums, scanners, printers, computers. (2) Mechanical equipment, parts, tools.

HARDWARE CLOTH

A type of galvanized metal screen that can be bought in hole sizes, e.g., 1/8, 1/4, 1/2 inch, etc. The holes are square.

HARDWIRE (TELEMETRY)

Describes a system of communication or information transmission using electric wire from point to point instead of electronic or wireless transmission.

Oil inlet line and hatch

HATCH

See Gauge Hatch.

HATCHET KNOT

A knot that defies untying and so must be cut; a granny knot.

HAT-TYPE FOUNDATION

A metal base or foundation in the shape of an inverted, rectangular cake pan. Hat-type foundations are used for small pumps and engines or other installations not requiring solid, permanent foundations.

HAUL ASS

An inelegant term meaning to leave a place with all haste; vamoose; split.

HAWSER

A large-diameter hemp or nylon rope for towing, mooring, or securing a ship or barge.

HAYFORK PULLEY

A simple pulley; a 4- to 6-inch sheave held in a steel shell or frame with an integral swivel eye for attaching the pulley to a boom or A-frame.

Hay rake

HAY RAKE

Another name for the finger board in the derrick of an oil well.

HEAD

See Hydrostatic Head.

HEADACHE!
A warning cry given by a fellow worker when anything is accidentally dropped or falls from overhead toward another worker.

HEADACHE POST
A frame over a truck cab that prevents pipe or other material being hauled from falling on the cab; a timber set under the walking beam to prevent it from falling on the drilling crew when it is disconnected from the crank and pitman.

Headache post

HEADER
A large-diameter pipe into which a number of smaller pipes are perpendicularly welded or screwed; a collection point for oil or gas gathering lines. *See* Manifold.

HEADING
An intermittent or unsteady flow of oil from a well. This type of flow is often caused by a lack of gas to produce a steady flow thus allowing the well's tubing to load up with oil until enough gas accumulates to force the oil out.

HEAD WELL
A well that makes its best production by being pumped or flowed intermittently.

HEATER
(1) An installation used to heat the stream from high-pressure gas and condensate wells (especially in winter) to prevent the formation of hydrates, a residue that interferes with the operation of the separator. (2) A refinery furnace.

A header or manifold

HEATER TREATER
See Emulsion Treater.

HEAT EXCHANGER, FINNED-TUBE
Small-diameter pipe or tubing with metal fins attached to the outer circumference for cooling water and other liquids or gases. Finned-tube exchangers cool by giving up heat from the surface of the fins to the atmosphere in a manner similar to an automobile radiator. Heat exchangers are not only for cooling but for heat-recovery systems as well. In some plants, finned-tube exchangers are built in ductwork through which the exhaust gas of a turbine flows at 800°F. Oil or process liquids are pumped through the exchanger tubes to use the waste heat to heat the process stream or to make steam.

HEAT EXCHANGER, HAIRPIN
A type of shell-and-tube exchanger with tubes inside a 12- to 18-inch diameter shell that may extend twenty or thirty feet and then double

Crude oil heater (*Courtesy Born Inc.*)

Plate heat exchanger

Battery of vertical heat exchangers

back the same distance like a hairpin. Hairpin exchangers may have bare or finned tubes inside the shell.

HEAT EXCHANGER, PLATE

A relatively low-pressure heat exchanger that uses thin-walled plates as its heat-transfer elements. Because of its thin walls, plate exchangers exhibit a much higher heat transfer coefficient than the more conventional shell-and-tube exchangers. However, because of their less-sturdy construction there are pressure limits to their use.

HEAT EXCHANGER, SHELL-AND-TUBE

A common type of industrial heat exchanger with a bundle of small-diameter pipes (tubes) inside a long, cylindrical steel shell. The tubes (50 to 100 in small units, several hundred in larger ones) run parallel to the shell and are supported, equidistant, by perforated steel end plates. The space inside the shell not filled with tubes carries the cooling water or other liquid. The liquid to be cooled is pumped through the tubes. Heat exchangers act not only as a cooling apparatus but are often used as a waste-heat-recovery system. Heat normally lost to a cooling medium can be used to heat a process stream.

HEAT PIPE, GRAVITY RETURN

A type of passive heat exchanger (requiring no external energy source) that draws heat from a heat source and gives up heat to a heat sink, the atmosphere in most cases. In its basic form, a heat pipe consists of a closed tube (the shell) 2 to 6 inches in diameter and as long as need be. The shell has a porous wick made of fine metal mesh on the inside circumference extending from top to bottom. The shell also contains a quantity of working liquid that may be anhydrous ammonia, liquid metals, glycerine, methanol, or acetone. Heat taken in or absorbed at the lower end of the heat pipe, the end in contact with the heat source, causes the liquid to evaporate and move up the pipe as a vapor. The dissipation of the heat at the upper end condenses the vapor that, as a liquid, moves back down the pipe in the wick by gravity or capillary action. The continuous cycle of vaporization and condensation within the closed pipe makes the heat pipe an efficient, natural-convection heat-transfer loop. On the trans-Alaska pipeline, thousands of heat pipes were installed along the big line to maintain the frozen soil around the vertical support members.

HEAT SINK

An agent or condition that dissipates heat—from the Earth, from a process, or from a heat exchanger. The heat sink for an automobile radiator is the atmosphere, the air passing over the fins. In the case of a refinery heat exchanger, the heat sink may be a liquid pumped through the exchanger's tubes that absorbs the heat.

HEAT TAPE
An electrical heating element made in the form of an insulated wire or tape used as a tracer line to provide heat to a pipeline or instrument piping. The heat tape is held in direct contact with the piping by a covering of insulation.

HEAT TRACING
The paralleling of instrumentation, product, or heavy crude oil lines with small-diameter steam piping or electrical heat tape to keep the lines from freezing or to warm the product or instrument fluid sufficiently to keep them flowing freely. Heat tracing lines, whether steam or electrical tape, are attached parallel to the host piping and both are covered with insulation.

Applying heat tape

HEAT VALUE IN CRUDE OIL
A barrel of crude oil is equal to about 5.8 million Btus.

HEAVE
See Throw.

HEAVE COMPENSATOR
A type of snubber/shock absorber on a floating drilling platform or drillship that maintains the desired weight on the drill bit as the unstable platform heaves on ocean swells. Some compensators are made with massive counterweights; others have hydraulic systems to keep the proper weight on the bit constant. Without compensators, the bit would be lifted off bottom as the platform rose on each swell.

HEAVY BOTTOMS
A thick, black residue left over from the refining process after all lighter fractions are distilled off. Heavy bottoms are used for residual fuel and for asphalt.

HEAVY CRUDE OIL
Crude oil of 20° A.P.I. gravity or less. There are perhaps billions of barrels of heavy oil still in place in the U.S. that require special production techniques, notably steam injection or steam soak, to extract them from the underground formations. Because heavy crude oil is more costly to produce, it and other types of oil are eligible for free-market or world prices.

HEAVY ENDS
In refinery parlance, heavy ends are the heavier fractions of refined oil—fuel oil, lubes, paraffin, and asphalt—remaining after the lighter fractions have been distilled off. *See* Light Ends.

HEAVY FUEL OIL
A residue of crude-oil-refining processes. The products remaining after the lighter fractions—gasoline, kerosene, lubricating oils, wax, and distillate fuels—have been extracted from the crude; residual fuel oil.

HEAVY METAL

Spent uranium or tungsten. Heavy metal is used to make drill tools to add weight to the drill assembly. Drill collars made of heavy metal weigh twice as much as those made of steel and are used to stabilize the bit and to force it to make a straighter hole with less deviation from the vertical.

HEAVY-OIL PROCESS (H.O.P.)

A steam-injection process developed by a subsidiary of Barber Oil Corporation in which steam is injected through horizontal lines into subsurface oil sands containing heavy oil, oil of 20° A.P.I. gravity or less. Conventional steam flooding employs vertical holes through which steam is injected. In the horizontal method, a 7-foot-diameter shaft is drilled into the relatively shallow formation. After it is cased, workmen construct a concrete cavern 25 feet in diameter and 20 feet high. From this work area, lateral holes are drilled several hundred feet in all directions. Perforated pipe is inserted in the drilled holes to carry steam. The steam, injected under pressure, soaks the formation, causing the highly viscous oil to separate from the sand and flow into the laterals after the steam injection is halted.

HEAVY UP

To thicken by adding weighting material to the drilling mud; to increase the weight of a gallon of mud from 9 pounds to 13 pounds, for example. Drilling mud is heavied up when gas is encountered downhole while drilling. The additional weight of several thousand pounds per square inch at the bottom of the hole is sufficient for hole control or to control the pocket of high-pressure gas until it can be cased off.

HEAVY-WALL DRILLPIPE

Drillpipe with thicker walls than regular drillpipe. Heavy-wall is sometimes used in the drillstring to reduce the number of larger diameter and stiffer drill collars. This is true in directional drilling and even in straight holes in certain areas of the country. Some of the advantages of heavy-wall pipe over drill collars are that it can be handled at the rig floor by regular drillpipe elevators and slips, and can be racked in the rig like regular pipe.

HECTARE

In the metric system of measurement, a hectare is 10,000 square meters of land. One hectare is equal to 2.47 acres; an acre is .4049 of a hectare.

HELD BY PRODUCTION

A lease held in force by production from one or more wells on the lease.

HELICAL STRAKE

See Strake, Helical.

HELIDECK

A landing area on an offshore platform, drilling, production or processing, for helicopters that ferry crew members, and carry small parts and supplies from shore stations to seaward platforms.

HELIX, TO

Owing to the spooling, on and off, of coiled tubing, it has a tendency to helix around the well bore. In addition, as it is slide drilling (the tubing is not rotating as in conventional rotary-table drilling), coiled tubing drilling tends to build angle (to deviate from the planned well path). Both of these factors limit the horizontal distance that coiled tubing with its bit and bottom-hole assembly will slide.

HEPTANE

A liquid hydrocarbon of the paraffin series. Although heptane is a liquid at ordinary atmospheric conditions, it is sometimes present in small amounts in natural gas.

HERTZ

A measurement used in seismic applications. A unit of wave frequency equal to one cycle per second, named for Heinrich Hertz, German physicist. Abbreviated, Hz, e.g., 50 Hz = 50 cycles per second. Thus, the higher the frequency, the greater the number of hertz. And the higher the frequency, the shorter the wave length.

HEXANE

A hydrocarbon fraction of the paraffin series. At ordinary atmospheric conditions hexane is a liquid, but often occurs in small amounts of natural gas.

H.H.P.

Hydraulic horsepower.

HIDE THE THREADS

To make up (tighten) a joint of screw pipe until all threads on the end of the joint are screwed into the collar, hiding the threads and making a leakproof connection.

HIGH

A geological term for the uppermost part of an inclined structure where the likelihood of finding oil is considered to be the greatest. *See* Structural High.

HIGH BOTTOM

A condition in a field stock tank when B.S.&W. (basic sediment and water) has accumulated at the bottom of the tank to a depth making it impossible to draw out the crude oil without taking some of the sediment with it into the pipeline. When this condition occurs the pumper must have the tank cleaned before the pipeline company will run the tank of oil.

HIGHBOY
A skid-mounted or wheeled tank with a hand-operated pump mounted on top used to dispense kerosene, gasoline, or lubricating oil to small shops and garages.

HIGH-COLLAPSE CASING
See Casing, High-Collapse.

HIGH-GRAVITY OIL
Crude oil with an A.P.I. gravity in the high 30s to high 40s. Low-gravity or heavy crudes are in the low 20s, even lower: 22–11° A.P.I.

HIGH-PRESSURE GAS INJECTION
Introduction of gas into a reservoir in quantities exceeding the volumes produced in order to maintain reservoir pressure high enough to achieve mixing between the gas and reservoir oil. *See* Solution Gas.

HIGH-PRICED GAS
Natural gas found below 15,000 feet. Because of the costs of drilling and completing wells at these depths, the gas is permitted a special incentive price under the provisions of N.G.P.A., higher than shallow gas. *See also* Tight Gas.

HIGH SEVERITIES
Refers to processes involving high pressures and temperatures.

HIGH-VELOCITY ROCK
Refers not to the speed of the rock, but to its density. Extremely dense rocks such as lavas or metamorphic material are high velocity; in seismic work, sound waves move through them easily and with great speed.

HINGE
The location of maximum curvature or bending in a folded strata of rock, as in a folded or anticlinal structure. It is customary to speak of the hinge as the hinge line because the hinge has lateral dimension. *See* Hinge Line.

HINGE LINE
A line connecting the points of maximum curvature or bending of the bedding planes of a fold. A hinge line is similar to the roof line or ridge line of a house.

HISTORICAL GEOLOGY
A major branch of geology concerned with the evolution of the Earth and its various life forms, from its origins to the present.

HISTORY OF A WELL
A written account of all aspects of the drilling, completion, and operation of a well. (Well histories are required in some states.)

HITTING THE HOOKS
Working on a pipeline, screwing in joints of pipe using pipe tongs; an expression used by the tong crew of a pipeline gang. The tong crews on large-diameter screw pipelines (up to about 12-inch pipe) hit the hooks in perfect rhythms. With three sets of tongs on the joint being screwed in, each large tong, run by two or three men, made a stroke every third beat of the collar pecker's hammer until the joint was nearly screwed in. Then the three tongs, with all six or nine men, hit together to "hide the threads," to tighten the joint the final and most difficult round.

HOISTING DRUM
A powered reel holding rope or cable for hoisting and pulling; a winch. *See* Drawworks.

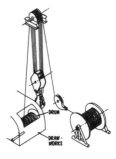
Hoisting drum

HOLD DOWN/HOLD UP
Oscillating anchoring devices or supports for a shackle-rod line to hold the rod line to the contour of the land it traverses. The devices are timbers or lengths of pipe hinged to a deadman or overhead support at one end, the other end attached to and supporting the moving rod line.

HOLDING VESSEL
A converted tanker, in some instances, for holding crude oil produced from offshore wells. The crude is then loaded onto smaller vessels (lighters) to bring the oil to onshore stations. Holding vessels are used to hold and transfer oil while offshore production platforms and undersea pipelines are built. This temporary arrangement allows an earlier cash income from the sale of the oil than would be the case if the producing wells were shut in, awaiting pipeline connections.

HOLE CONTROL
Controlling the hole by what is put in the hole: the drilling fluid. Specifically, it means conditioning shale formations to retard swelling and sloughing; treating drilling mud with various additives (thinners or thickeners, viscosifiers) to maintain proper weight or density and chemical balance for maximum penetration (drilling) rates and for controlling downhole pressures. Hole control also means controlling fluid loss into porous formations and maintaining the best possible circulation rates.

Hole opener *(Courtesy Grant)*

HOLE OPENER
A type of reamer used to increase the diameter of the well bore below the casing. The special tool is equipped with cutter arms that are expanded against the wall of the hole and by rotary action reams a larger-diameter hole.

Repairing holidays

Hook or crane block
(Courtesy Crosby)

HOLIDAYS

Breaks or flaws in the protective coating of a joint of line pipe. Holidays are detected by electronic testing devices as the pipe is being laid. When detected, the breaks are manually coated. *See* Jeeping.

HOLLOW SUCKER ROD

See Sucker Rod, Hollow.

HOME-HEATING OIL

A light gas oil that is similar to diesel fuel in gravity, viscosity, and other properties. No. 2 heating oil is by far the most widely used for domestic furnaces. Two Oil. *See* Gas Oil.

HOMESTEAD

The residence of the head of the family. Many states have laws forbidding the sale or lease of the homestead without the concurrence of both husband and wife. Moreover, the sale of the homestead under execution or other legal process is quite often forbidden.

HOOK

The hook attached to the frame of the rig's traveling block that engages the bail of the swivel in drilling operations. *See* Hook-Load Capacity.

HOOK BLOCK

One or more sheaves or pulleys in a steel frame with a hook attached. *See* Traveling Block.

HOOK-LOAD CAPACITY

The maximum weight or pull a derrick and its lines, blocks, and hook are designed to handle. A rating specification for a drilling rig. (Some large rigs have a hook-load capacity of 2 million pounds.)

HOOKS

Pipelaying tongs named for the shape of the pipe-gripping head of the scissors-like wrench.

HOOK UP

To make a pipeline connection to a tank, pump, or well. The arrangement of pipes, nipples, flanges, and valves in such a connection.

HOOK-WALL TEST

As test of a formation for oil through the drillstring with the use of a hook-wall packer to isolate the formation, forcing what oil there is from the formation into the drillpipe tester. *See* Drillstem Test.

HOOT-OWL SHIFT

The work shift from 11 p.m. to 7 a.m.; the graveyard shift. This work period is so disagreeable and difficult for most workers to handle for any length of time that it commands differential pay, a higher hourly rate than day or evening tours.

H.O.P.
See Heavy-Oil Process.

HORIZON
A zone of a particular formation; that part of a formation of sufficient porosity and permeability to form a petroleum reservoir. *See* Pay Zone.

HORIZONTAL ASSIGNMENT
The assignment of an interest in oil or gas above or below a certain depth in a well; an assignment can specify a particular formation.

HORIZONTAL-CUT OIL PAYMENT
The assignment of a part of an oil payment, providing for a payout, first to the assignee before the assignor receives payment. For example, if A owns an oil payment for the first $100,000 out of an 1/8 of the 7/8 working interest, he may assign a horizontal-cut oil payment to B by transferring the first $50,000 out of 1/8 of the 7/8.

HORIZONTAL DIRECTIONAL DRILLING
Drilling with a specially designed slant rig at an angle from the horizontal beneath a stream, canal, or ship channel. This type of directional drilling has been perfected and is used to make pipeline crossings where dredging a trench across a waterway is too costly or too disruptive of ship traffic and a bridge or A-frame-supported line is prohibited by the authorities.

Horizontal directional drilling *(Courtesy Reading & Bates)*

HORIZONTAL INTEGRATION
Refers to the condition in which a diversified company has resources or investments other than its principal business and from which it makes a profit. Specifically, an oil company is said to be horizontally integrated when, besides oil and gas holdings, it owns coal deposits or is into nuclear energy, oil shale, or geothermal energy. *See* Vertical Integration.

HORIZONTAL SEVERENCE
A conveyance of some portion or all of the minerals above, below, or between certain specified depths in a well, or at a given interval or stratum, e.g., a conveyance of the minerals in the Wilcox formation, or a conveyance of all minerals at a greater or a lesser depth than 5,500 feet.

HORNBLENDE
A dark-colored, rock-forming mineral that occurs in several forms: crystalline, columnar, fibrous, and granular. It is the main constituent of several kinds of igneous rock and is a metamorphic mineral in gneiss and

schist. Hornblende is the most common mineral in the amphibole group. The term *hornblende* is an old German word for any dark, prismatic crystal found with metallic ores but containing no metal. *Blende* indicates something that deceives.

HORSEBACK
A ridge of foreign material in a coal seam.

HORSE FEED
An old oilfield term for unexplainable expense account items in the days of the teamster and line rider who were given an allowance for horse feed. Expenses that needed to be masked in anonymity were simply listed "horse feed."

Horsehead on a pumping
jack

HORSEHEAD
The curved guide or head piece on the well end of a pumping jack's walking beam. The metal guide holds the short loop of cable (the bridle) attached to the well's pump rods.

HORSEPOWER
A unit of power equivalent to 33,000 foot-pounds a minute or 745.7 watts of electricity.

HORSEPOWER, INDICATED AND BRAKE
See Indicated Horsepower; *also* Brake Horsepower.

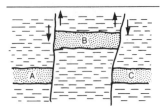

Horst (B) and graben

HORST
See Graben.

HORTONSPHERE
A spherical tank for the storage, under pressure, of volatile petroleum products, e.g., gasoline and L.P.-gases; also Hortonspheroid, a flattened spherical tank, somewhat resembling a tangerine in shape.

HOST COMPUTER
The master computer to whom satellite computers, monitoring and controlling separate operations in the field, report, sending individual station reports to home base. For pumping wells, connected to the network, each well-station might send running time, downtime, temperatures, pressures, and once a month a digitalized dynamometer card.

Hortonsphere

HOSTILE ENVIRONMENT
ABOVE GROUND: One that is not user-friendly; in fact, unfriendly, for tools, equipment, and personnel: Extremes of cold, heat, ocean storms, high altitudes, isolation in remote areas. DOWNHOLE: Heat, high-pressure pockets of gas; H_2S, crooked-hole country, thirsty shales.

HOT FILL/THIRD PASS
See Welding, Pipeline.

HOT-FLUID INJECTION
A method of thermal oil recovery in which hot fluid (water, gas, or steam) is injected into a formation to increase the flow of low-gravity crude to production wells. *See* Hot Footing.

HOT FOOTING
Installing a heater at the bottom of an input well to increase the flow of heavy crude oil from the production wells. *See* Hot-Fluid Injection.

HOTHEAD
A hot-plug engine; a semidiesel.

HOT-NOSE ANOMALY
A colloquial expression for the forward movement of the frac front in one area more rapidly than others so that the profile on the seismic reflection resembles a huge nose.

HOT OIL
(1) Oil produced in violation of state regulations or transported interstate in violation of federal regulations. (2) (Foreign) A term applied to oil produced by a host country after the host country confiscates the assets of a foreign oil company.

HOT PASS
A term describing a bead or course of molten metal laid down in welding a pipeline. The hot pass is the course laid down on top of the stringer bead, which is the first course in welding a pipeline. *See* Pipeline Welding.

HOT-PLUG ENGINE
A stationary diesel-cycle engine that is started by first heating an alloy metal plug in the cylinder head that protrudes into the firing chamber. The hot plug assists in the initial ignition of the diesel fuel until the engine reaches operating speed and temperature. Afterwards the plug remains hot, helping to provide heat for ignition. Hot-tube engine; hot head. *See* Semidiesel.

Hot tapping

HOT TAPPING
Making repairs or modifications on a tank, pipeline, or other installation without shutting down operations. *See* Tapping and Plugging Machine.

HOUDRY, EUGENE J.
A pioneer in developing the use of catalysts in cracking crude oil. Houdry, a wealthy Frenchman, was a World War I hero and auto racer. It is said his interest in cars led him to experiment with more efficient methods of refining and to work with various catalysts until he perfected the catalytic cracking process that bears his name. Although there are several cracking processes in use today, Houdry's work is credited with ushering in the era of catalytic cracking. *See* Hydrocracking.

Hovercraft

HOUSE BRAND (GASOLINE)
An oil company's regular gasoline; a gasoline bearing the company's name.

HOUSING
The covering of a compressor or centrifugal pump, for example. In most cases the housing is integral to the working parts of the mechanism, not just a removable cover.

HOVERCRAFT
See Air-Cushion Transport.

HUFF-AND-PUFF PROJECT
A colloquial reference to a steam-flooding or steam-soak project in which high-pressure steam is injected intermittently into underground formations containing heavy oil (20° A.P.I. or less). *See* Heavy Crude Oil.

HUGOTEN BASIN
A relatively shallow geological basin in southwest Kansas, extending into the Oklahoma and Texas Panhandles. Hugoten has produced significant amounts of gas, and substantial gas reserves still remain.

HUMPHREYS, R.E.
A petroleum chemist who worked with W. M. Burton in developing the first commercially successful petroleum-cracking process using heat and pressure.

HUNDRED-YEAR STORM CONDITIONS
A specification for certain types of offshore installations—production and drilling platforms, moorings, and offshore storage facilities—is that they be built to withstand winds of 125 miles an hour and "hundred-year storm conditions," the biggest blow on record, a storm so intense it occurs only once every 100 years.

HURRY-UP STICK
The name given to the length of board with a hole in one end that the cable-tool driller uses to turn the T-screw at the end of the temper screw when the walking beam is in motion. This enables the driller to perform the job of letting out the drilling line easily and rapidly.

HUYGEN'S PRINCIPLE
A physics principle relating to seismology whereby every point of an advancing wave front is a new center of disturbance from which emanates independent wavelets which create a new wave front.

HYDRATE
(1) A mineral compound in which water is part of the chemical composition, produced by hydration; the transfer of H_2O from the fluid phase into the structure of a mineral. *See* Hydration Reaction. (2) Crystalline compounds of water (H_2O) and methane (CH_4): natural gas. The water freezes, forming ice lattices (microscopic cages) that contain gas molecules. Hydrates resemble wet, melting snow in appearance and are formed frequently in permafrost formations at 90 to 100 feet if natural gas is present. They present problems in pipelines, particularly in the northern regions.

HYDRATION REACTION
A reaction (metamorphic) that results in the transfer of water (H_2O) from the fluid state into the structure of a mineral. *See* Dehydration Reaction.

HYDRAULIC
The term that pertains to movement or action by water or, more commonly, to any fluid, especially liquids in motion and doing work. *See* Hydraulic Fracturing.

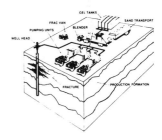

HYDRAULIC FRACTURING
A method of stimulating production from a formation of low permeability, inducing fractures and fissures in the formation by applying very high fluid pressure to the face of the formation, forcing the strata apart. Various patented techniques, using the same principle, are employed by oilfield service companies.

Hydraulic fracturing
equipment on location
(Courtesy Halliburton)

HYDRAULIC HORSEPOWER
A designation for a type of very high-pressure plunger pump used in downhole operations such as cementing, hydrofracturing, and acidizing.

HYDRAULIC WORKOVER UNIT
A type of workover unit that is used on high-pressure wells where it may be necessary to snub the pipe out of and back into the hole when the workover is completed.

HYDROCARBONS
Organic chemical compounds of hydrogen and carbon atoms. There are a vast number of these compounds, and they form the basis of all petroleum products. They may exist as gases, liquids, or solids. An example of each is methane, hexane, and asphalt.

HYDROCRACKATE
The main product from the hydrocracking process; gasoline blending components.

Hydraulic workover unit
(Courtesy Otis)

HYDROCRACKING

A refining process for converting middle-boiling or residual material to high-octane gasoline, reformer charge stock, jet fuel, and/or high-grade fuel oil. Hydrocracking is an efficient, relatively low-temperature process using hydrogen and a catalyst. The process is considered by some refiners as a supplement to the basic catalytic cracking process.

HYDROCYCLONE

A type of in-line centrifuge to separate oil from produced water. Water, as the heavier of the two liquids, is readily separated out of a commingled oil/water stream. Ridding offshore production of produced water is an urgent and ever-present necessity on most production platforms. Recent advances in handling the produced-water problem has reduced both weight and the time it takes to get the water out and back into the sea, clean and nonpolluting.

HYDRODYNAMICS

A branch of science that deals with the cause and effect of regional subsurface migration of fluids.

A portable H_2S monitor
(*Courtesy Ecolyzer*)

HYDROGEN SULFIDE (H_2S)

An odorous and noxious compound of sulfur found in "sour" gas. *See* Sour Gas.

HYDROLOGIC CYCLE

The process of evaporation, convection (circulation of air at different temperatures), advection (the horizontal movement of an air mass), precipitation, runoff, and reevaporation. The cycle of the Earth's water supply.

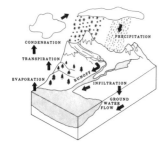

Hydrologic Cycle

HYDROMETER

An instrument designed to measure the specific gravity of liquids; a glass tube with a weighted lower tip that causes the tube to float partially submerged. The A.P.I. gravity of a liquid is read on a graduated stem at the point intersected by the liquid; an areometer.

HYDROPHONES

Sound-detecting instruments used in underwater seismic exploration activities. Hydrophones are attached to a cable towed by the seismic vessel. Sound waves generated by blasts from an air gun reflect from formations below the seafloor and are picked up by the hydrophones and transmitted to the mother ship.

HYDROSTATIC COLUMN

See Hydrostatic Head.

HYDROSTATIC HEAD
The height of a column of liquid; the difference in height between two points in a body of liquid.

HYDROSTATIC PRESSURE
The pressure exerted by a column of water or other fluid. A column of water 1 foot high exerts a pressure of 0.465 pounds per square inch (psi); a column 1,000 feet tall would have a pressure reading at the bottom of the column of 465 psi; in a deep well of 10,000 feet, the hydrostatic pressure at the bottom of the hole would exceed 5,000 psi.

HYDROSTATIC TESTING
Filling a pipeline or tank with water under pressure to test for tensile strength, its ability to hold a certain pressure without rupturing.

HYDROTREATER
A vessel where refinery feed is exposed to or is treated with hydrogen; a part of the train of refining processes.

HYPERBARIC WELDING
See Welding, Hyperbaric.

I.A.D.C.
International Association of Drilling Contractors.

I.C.C.
Interstate Commerce Commission.

ICE PLATFORM
A man-made, thick platform of ice for drilling in the high Arctic. Seawater is pumped onto the normal ocean ice, itself quite thick, where it freezes in the –30° to –40°C temperatures. The platform is built up a few inches at a time with successive pumpings and freezing of the water until the ice is calculated to be thick enough to support drilling operations with a 1,000 to 1,500-ton rig and auxiliary equipment. Ice platform technology was pioneered by Panarctic Oil Ltd., a company with a great deal of experience in Arctic exploration.

ICE PLUGS
Temporary pipeline plugs made of fresh or treated sea water frozen by liquid nitrogen at –196°C (–321°F) to form pressure-tight plugs in order to make repairs, tie-ins, or valve change outs without having to drain long sections of the line.

I.D.
Inside diameter of a pipe or tube; initials used in specifying pipe sizes, e.g., 3 1/2-inch I.D.; also O.D., outside diameter, e.g., 5-inch O.D.

IDIOT STICK
A shovel or other digging tool not requiring a great deal of training to operate.

IDLER GEAR OR WHEEL
A gear so called because it is usually located between a driving gear and a driven gear, transmitting the power from one to the other. It also transmits the direction of rotation of the driving to the driven gear. Without the idler or the intermediate gear, the driving gear by directly meshing with the driven gear reverses the direction of rotation. Idler wheels or pulleys are also used for tightening belts or chains or to maintain a uniform tension on them.

A complete offshore operation with a reel barge, drilling platform, producing platform, and tender.

A rig floor; note the drawworks, the kelly joint, the bushing, and the slips.

IGNEOUS ROCK
Rocks that have solidified from a molten state deep in the earth. Those rocks that have reached the surface while still molten are called lavas; they can form volcanic cones or spread out in flows or sheets, they can be forcibly thrust up between beds of other kinds of rocks in what are called sills, or they can fill crevices and then solidify as "dikes." Rocks that have solidified deep beneath the earth's crust are referred to as plutonic, from the Greek god of the lower regions, Pluto. Granite is an example of plutonic rock.

IGNITION MAGNETO
An electric-current generator used on stationary engines in the field. A magneto is geared to the engine; once the engine is started either by hand cranking or by a battery starter, the magneto continues to supply electric current for the ignition system. Current is produced by an armature rotating in a magnetic field created by permanent magnets.

IGNORANT END
The heaviest end of a tool or piece of equipment to be carried or operated.

I.H.P.
Indicated horsepower.

IMAGING
A general term covering several methods of obtaining information about the Earth's crust without physical contact, without actually digging or touching. Imaging may be done by the use of Landsat and high-altitude observations; by photography, radiated or reflected electromagnetic energy; by infrared detectors, microwave frequency recorders, and radar systems.

IMMATURE ROCK
Rock in the early stages of maturity; clastic sediment that has evolved from the parent rock over a relatively short time (geologic time, that is) and is characterized by unstable minerals and weatherable material, such as clay. An example is immature sandstone, which may contain as much as 10 percent clay. Immature may also refer to argillaceous sedimentary material (clayey) halfway between a clay and a shale. Incomplete or unstable rock; adolescent rock.

IMMISCIBLE
Incapable of mixing; said of two fluids that under normal conditions or in their normal states cannot dissolve completely in one another. *See* Miscible.

IMMUNOASSAY TESTING OF GROUNDWATER
The immunoassay technology, sensitive chemical testing, is useful in identifying contaminant plumes fanning out into lakes and oceans, can

aid in the placement of monitoring wells, and in random sampling and testing of water samples for minute quantities (a few parts per million) of petroleum hydrocarbons.

IMPACT WRENCH
An air-operated wrench for use on nuts and bolts of large engines, valves, and pumps. Impact wrenches have taken the place of heavy end wrenches and sledgehammers in tightening and loosening large nuts. A small version of the impact wrench is the air-operated automobile lug wrench used at modern service stations and garages.

IMPELLER
The wheel-like fan inside a centrifugal pump that impels or propels the fluid forward and out of the discharge opening. As the impeller turns at high speed, it pulls in the fluid at the suction opening of the pump and, with numerous vanes (small fixed blades) on the impeller, impels the liquid around the circular body or housing of the pump, into the discharge opening, and on to the pipeline.

IMPERIAL GALLON (CANADIAN)
One U.S. Gallon = 0.832672 Canadian. One 42-gallon U.S. Barrel = 34.9722 Imperial Gallons.

IMPLIED COVENANT
See Covenant, Implied.

I.M.S.
International M.W.D. Society. An organization "to pursue and communicate all aspects of M.W.D. (Measuring While Drilling) technology."

INACTIVE CATALYST
See Spent Catalyst.

INCENTIVE PRICING
Pricing above the going market price for a product that may be more costly to produce. For example, gas found at great depths, geopressurized or coal-seam gas, may receive incentive pricing if it qualifies under the National Gas Policy Act of 1978 and the Federal Energy Regulatory Commission regulations. Incentive pricing is often the difference between producing a natural resource and not producing because of the high cost of production.

INCLINOMETER
An instrument used downhole to determine the degree of deviation from the vertical of a well bore at different depths. *See* Acid-Bottle Inclinometer.

INCOHERENT ROCK
A rock or a deposit that is loose, unpacked, or unconsolidated. *See* Coherent.

INCOME FUNDS
Funds formed to acquire producing oil and gas properties. Income funds are deemed less risky and more dependable than drilling funds. There are also completion funds and royalty funds.

INCOMPETENT ROCK
A volume or mass of rock that under certain specific conditions and at a specific time is not able to sustain a tectonic force, i.e., some movement of the earth's crust, without cracking or shattering.

INCORPOREAL RIGHTS
Having no material body or form. Said of easements, bonds, or patents. Rights that have no physical existence but that authorize certain activities or interests.

INDEPENDENT PRODUCER
(1) A person or corporation that produces oil for the market, having no pipeline system or refinery. (2) An oil-country entrepreneur who secures financial backing and drills his own well.

INDEX FOSSIL
See Diagnostic Fossil.

INDEX MAP
See Map, Index.

INDICATED HORSEPOWER (I.H.P.)
Calculated horsepower; the power developed within the cylinder of an engine that is greater than the power delivered at the drive shaft by the amount of mechanical friction that must be overcome. *See* Brake Horsepower.

INDICATOR PASTE, GASOLINE
A viscous material applied to a steel gauge line or gauge pole that changes color when it comes in contact with gasoline, making it easy for the gauger to read the height of gasoline in the tank.

INDICATOR PASTE, WATER
A paste material applied to a steel gauge line or wooden gauge pole that changes color when immersed in water. It is used to detect the presence of water in a tank of oil.

INDURATION
The hardening of rock material by heat, pressure, or cementing material; the hardening of soil by chemical action to form hardpan.

INDUSTRIAL GAS
Gas purchased for resale to industrial users.

INERT GAS
Any one of six gases that, under normal conditions, are not inclined to react with any of the other elements. The inert or inactive gases are neon, helium, argon, krypton, xenon, and radon.

INFERENTIAL CONTROLS
Controls on temperatures by noting top and bottom temperatures in a fractionating tower, for example. These can indicate, to experienced operators, the product's composition and progress of the process. Skillful boilerhousing.

INFILL DRILLING
See Drilling, Infill.

INFLUENT
The flow of liquids or gas into a vessel or equipment. *See* Effluent.

INFORMATION CONSOLE
A bank of indicators, counters, and display dials showing weight of the drillstring, weight on the drill bit, mud pump speed, mud pressure, engine speed, etc., to keep the driller informed of all aspects of the drilling operation.

Information console

I.N.G.A.A.
Interstate Natural Gas Association of America.

INHIBITORS
A substance that slows down a chemical reaction. An inhibitor's role is the reverse of a catalyst's. Inhibitors are sometimes used to interfere with a chemical reaction somewhere along the process train.

INITIAL BOILING POINT (I.B.P.)
The temperature at which a product (a cut or a fraction) begins to boil. *See* End Point.

INJECTION STRING
A string of tubing run inside the casing along with the production tubing as a well-control feature. In the event the well has to be killed, mud is pumped down the injection/kill line to fill the hole. The line is also used for well stimulation, washing out sand, and acidizing. Kill string.

INJECTION WELL
See Input Well.

INJECTOR HEAD
The mechanism or fitting on top of the lubricator stack in a coiled tubing set up for drilling. The tubing, unreeled from the great spool, is attached to the bottom-hole assembly—turbo mud motor, the guidance unit, the stabi-

lizer, the drill bit. All are lowered into the hole through the lubricator, a series of valves and a stripper that act as a blowout-preventer stack. In case of a gas kick, the valves on the lubricator are closed on the coiled tubing until the drilling mud can be sufficiently heavied up to balance the hydrostatic column of mud to suppress the gas kick. *See* Bullhead the Bubble Back, To.

Submersible drilling barge

IN KIND
See Royalty in Kind.

IN KIND, OIL COUNTRY PAYMENT
Payment in kind is either payment in barrels of crude oil for a part owner of an oil well, or cubic feet of gas if the interest is in a gas well; payment in product rather than in cash.

INLAND BARGE RIG
See Rig, Barge.

IN-LINE EQUIPMENT
Pumps, separators, heat exchanges integral to a process or processing chain; in the line, not auxiliary or only supporting.

INNAGE GAUGE
A measure of the quantity of oil in a tank calculated on the basis of the depth of oil in the tank; the most common method of gauging a tank. *See* Outage Gauge.

INNOVATOR'S ROYALTY
See Royalty, Innovator's.

Input well

INPUT WELL
A well used for injecting water or gas into a formation in a secondary-recovery or pressure-maintenance operation; injection well.

INSERT BEARING
See Bearing, Insert.

INSERT BIT
See Bit, Insert.

IN SITU COMBUSTION
A technique used in some locations for recovering oil of low gravity and high viscosity from a field when other primary methods have failed. Essentially, the method involves the heating of the oil in the formation by igniting the oil (burning it in place) and keeping the combustion alive by pumping air downhole. As the front of burning oil advances, the heat breaks down the oil into coke and light oil. And as the coke burns, the

Insert drill bit

lighter, less viscous oil is pushed ahead to the well bores of the producing wells.

INSPECTION PIG, BIDIRECTIONAL
A specially designed inspection pig for use in an under-sea pipeline. Because of the lack of a pig trap and launcher at the shore end of the line, the pig's travel is reversed after completing its inspection run, and like a 36-inch diameter trolley it is pushed back to its launching site by reversing, for a brief period, the oil flow in the pipeline.

INSPECTION PLATE
A flat metal plate fitted with a gasket and bolted over an opening in the gear box of a pump or the crankcase of an engine. By removing the plate an inspection of the gears or crank and connecting-rod bearings can be made. On large, multicylinder engines, inspection plates are large enough to permit a mechanic to enter the crankcase to inspect or "change out" a bearing.

Instrument air dryers; hydryers *(Courtesy Prichard-ECO Inc.)*

INSTRUMENT AIR
Dehydrated air; air from which all moisture has been removed to prevent any condensation that would harm the delicate mechanism of air-actuated instruments.

INTAKE VALVE
See Valve, Intake.

INTANGIBLE DRILLING COSTS
Expenditures incurred by an operator for labor, fuel, repairs, hauling, and supplies used in drilling and completing a well for production.

INTANGIBLES
Short for intangible drilling costs.

IN-TANK PUMP
See Pump, In-Tank.

INTEGRATED OIL COMPANY
A company engaged in all phases of the oil business, i.e., production, transportation, refining, and marketing; a company that handles its own oil from wellhead to gasoline pump.

INTEGRATION, HORIZONTAL
See Horizontal Integration; *also* Vertical Integration.

INTERBEDDED
Layers of different types of rock that succeed one another or have been laid down one upon the other. For example, sandstone may be overlain by shale and the shale layer followed by a deposit of limestone. Also, rock material deposited in sequence between other beds.

INTEREST IN AN OIL OR GAS WELL
See Operating Interest; *also* Working Interest.

INTERFACE
The point or area where two dissimilar products or grades of crude oil meet in a pipeline as they are pumped, one behind another.

INTERMEDIATE
Refined product in the middle range of a refinery's output, e.g., straight-run gasoline, naphtha, kerosene, and light gas oil. On either side of the intermediates are the light ends (upper end) and heavy gas oil, lube oil, and residuals (the lower end).

INTERMEDIATE STRING
The string of casing between the surface pipe and the production string; the string of casing that reaches from the wellhead to the pay zone. In a 10,000-foot well, the surface pipe may reach to 1,000 feet; the intermediate string may reach from wellhead to 6,800 feet where it too is cemented, and the capital or production string is run from wellhead to 10,000 feet, cemented, and perforated. After the casing program is completed, the next string of pipe to go in the hole is the 2-inch or 2 1/2-inch well tubing through which the well is either flowed or pumped.

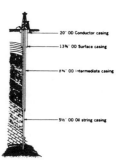

Intermediate string

INTERMITTENT WELL
A stripper or a small-producer well that is pumped on a regular but on a day-off-a-day schedule, when the hole is replenished with oil from the slow but steadily draining formation.

INTERNAL-COMBUSTION ENGINE
A mechanism with internal components: a cylinder in which a piston connected to a piston rod moves up and down; valves or ports to admit a fuel-air mixture and to vent exhaust gases; a sparking element to ignite the fuel-air mixture to cause combustion, which expands the gases in the cylinder and pushes the piston downward, turning the crankshaft to which the piston and rod are attached. When all component parts are properly timed and working smoothly in sequence, the internal-combustion engine is a delight to behold. *See* Otto-Cycle Engine; *also* Diesel Engine; Two Cycle Engine.

INTERRUPTIBLE GAS
A gas supply, usually to industrial plants and large commercial firms, that can be curtailed or interrupted during emergencies or supply shortages in order to maintain service to domestic customers.

INTERSTATE OIL COMPACT
A compact between oil-producing states negotiated and approved by Congress in 1935, the purpose of which is the conservation of oil and gas

by the prevention of waste. The Compact provides no power to coerce but relies on voluntary agreement to accomplish its objectives. Originally, there were six states as members; today, there are nearly thirty.

INTERSTITIAL
Fluid (oil or water) that fills the pores or interstices of rock in underground formations.

INTERVAL
(1) A stratigraphic interval; a body of rock or a strata of rock that occurs between two stratigraphic markers; a recognizable vertical section or unit of rock distinct from that above or below. (2) A section or zone in a formation that may be productive. For example, "The well was completed in the Marchand sand, tested at intervals 10,115 to 10,196, and produced at a rate of 350 bo/d (barrels of oil per day) through a 16/64th-inch choke. Tubing pressure was 1,550 psi."

INTRA-BED MULTIPLE
In seismological parlance, the complex seismic reflection effect whereby a seismic wave bounces repeatedly between intermediate zones before reflecting to the surface. This behavior results in a misleading reading. *See* Peg-Leg Multiple.

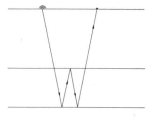

Intra-Bed Multiple

INTRUSION
A large-scale upthrust of a lower formation that forces clay, gypsum, salt, or other relatively plastic sediments into a rounded, plug-like structure, producing an anticline. Such domes, or diapirs, usually contain a core of salt or shale.

INTUMESCENT PIPE COATING
See Pipe Coating, Intumescent.

INVESTMENT CAPITAL
Money an individual has over and above mere discretionary funds that he can afford to invest. Risk capital; money an individual or company can afford to lose.

I.O.C.C.
Interstate Oil Compact Commission.

I.O.S.A.
International Oil Scouts Association.

I.P.A.A.
Independent Petroleum Association of America.

IRON SPONGE
A type of filter for a sour-gas stream containing varying amounts of the deadly H_2S gas. The sour gas passing over or through the iron-sponge

filter is sweetened, as the hydrogen sulfide reacts with the filaments of native iron, Fe.

IRREDUCIBLE WATER

Water that adheres to sand grains in a formation and will not let go. A film of water surrounds each grain, adhering by surface tension. Irreducible or connate water, by taking up pore space reduces the useful porosity of a sedimentary bed. *See* Connate Water.

ISO-

A prefix denoting similarity. Many organic substances, although composed of the same number of the same atoms, appear in two, three, or more varieties or isomers that differ widely in physical and chemical properties. In petroleum fractions there are many substances that are similar, differing only in specific gravity: for example, iso-octane, isobutane, and isopentane.

ISOCHRONAL

Having equal duration; recurring at regular intervals.

ISOMER

Compounds having the same composition and the same molecular weight but differing in properties.

ISOMERIZATION

A refinery process for converting chemical compounds into their isomers, i.e., rearranging the structure of the molecules without changing their size or chemical composition.

ISOPACHOUS MAP

See Map, Isopachous.

ISOPENTANE

A high-octane blending stock for automotive gasoline.

ISOTHERMAL

At constant temperature. When a gas is expanded or compressed at a constant temperature, the expansion or compression is isothermal. Heat must be added to expanding gas and removed from compressing gas to keep it isothermal.

I.T.I.O.

For the antiquarians among you, I.T.I.O. company was the Indian Territory Illuminating Oil company, organized in Oklahoma before it became a state in 1906. Years later, I.T.I.O. was bought out by Prairie Oil & Gas and, subsequently, Prairie became a part of Sinclair Oil and Gas. So much for a fragment of early succession history.

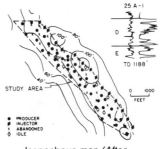

Isopachous map *(After Gates and Brewer)*

J

1890 pipeline gang using pipe tongs and jack board

J-4 FUEL
A designation for highly refined kerosene used as fuel for jet engines.

JACK
An oil-well pumping unit powered by a gasoline engine, electric motor, or rod line from a central power. The pumping jack's walking beam provides the up-and-down motion to the well's pump rods.

JACK BOARD
A wood or metal prop used to support a joint of line pipe while another joint is being screwed into it. Jack boards have metal spikes inserted at intervals to support the pipe at different levels.

JACKET, LEAN-TO
A subordinate jacket connected to the original, more substantial, offshore platform jacket; a lean-to structure is often used to support an auxiliary piece of equipment, such as an additional accommodation or service module.

JACKET, OFFSHORE PLATFORM
See Platform Jacket.

JACKING CYLINDERS
Hydraulic jacking devices for lifting, pushing, and spreading in close places where an ordinary jack cannot be used. Some jacking cylinders are only 4 inches high in a closed position and can exert a powerful thrust for a distance of 3 inches.

Jackknife rig

JACKKNIFE RIG
See Rig, Jackknife.

JACK RABBIT
A device that is put through casing or tubing before it is run to make certain it is the proper size inside and outside; a drift mandrel.

JACK SHAFT
An intermediate shaft in the power train. Jack shafts usually are relatively short and often are splined.

JACKUP BARGE
See Rig, Jackup.

JACKUP RIG
See Rig, Jackup.

JACK-WELL PLANT
A central power pumping from 10 to 25 or more rodline wells. *See* Central Power.

JAM NUT
A nut used to jam and lock another nut securely in place; the second and locking nut on a stud bolt. After the first nut is threaded and tightened on a stud, a second nut is tightened down on the first nut to prevent it from working loose.

Jackup rig; note helipad (top left) *(Courtesy Zapata Off-Shore Co.)*

JAR
A tool for producing a jarring impact in cable-tool drilling, especially when the bit becomes stuck in the hole. Cable-tool jars (part of the drillstring) are essentially a pair of elongated, interlocking steel links with a couple of feet of "play" between the links. When the drilling line is slacked off, the upper link of the jars moves down into the lower link. When the line is suddenly tightened, the upper link moves upward, engaging the lower link with great force that usually frees the stuck bit. *See* Bumper Sub.

JARS, HYDRAULIC
A tool used in the drillstring for imparting an upward or downward jar or jolt to the drillpipe should it get stuck in the hole while drilling or making a trip. The jars' jolting action is initiated either by the weight or tension of the drillstring, which the driller can apply.

Jeeping

JASPER
A type of chert associated with iron ore and containing iron oxide impurities that give the rock its characteristic red color. Yellow, green brown, and black cherts have also been called jasper. Don't be misled, insist on the real thing; jasper is red.

JEEPING
Refers to the operation of inspecting pipe coating with the aid of electronic equipment. An indicator ring is passed over the pipe that carries an electric charge. If there is a break or holiday in the protective coating, a signal is transmitted through the indicator ring to an alarm.

JERKER PUMP
See Pump, Jerker.

Jerker pump *(Courtesy Gaso)*

JERK LINE

A line that connects the bandwheel crank to the drilling cable. As the crank revolves, the drilling line is jerked (pulled up and released suddenly) providing an up-and-down motion to the spudding tools on a cable-tool rig.

JET CHARGES

Shaped explosive charges used in perforating casing downhole to permit oil and gas to enter the well bore from the producing formation. The jet charges in small individual canisters are supported in a light metal frame that can be lowered into the hole on a wireline. The charges are detonated electrically from the surface.

JET FUEL

A specially refined grade of kerosene used in jet-propulsion engines. *See* J-4 Fuel.

Modified jet mixers

JET MIXER (CEMENT)

A device consisting of a hopper to which a water supply under pressure is connected. Sacks of cement are opened and dumped one at a time into the hopper. The high-pressure water is jetted through the lower part of the hopper, mixing with the dry cement to form a slurry for pumping downhole to cement the casing in a well or for a squeeze job. *See* Squeezing a Well.

JET SLED

An underwater trenching machine for burying a pipeline below the seafloor. The patented jet sled straddles the pipeline and scours out the seabed material ahead and beneath the line with a series of high-pressure jets of seawater. The power is supplied by a series of high-pressure pumps aboard an accompanying jet barge. The jetted water, at 1,200 psi, is directed ahead and below the line and literally cuts a ditch in the seafloor into which the line is laid.

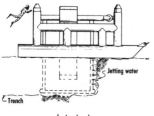

Jet sled

JETTED-PARTICLE DRILLING

A method of drilling in hard rock formations using steel pellets forced at high velocity from openings in the bottom of the drill bit.

JETTING

Injecting gas into a subsurface formation for the purpose of maintaining reservoir pressure.

JETTING THE PITS

A method of removing cuttings from the bottom of the working mud pits. This is done with a suction hose that derives its partial vacuum from a jet-nozzle arrangement, a type of venturi through which clean drilling mud is pumped at high pressure and velocity. At the waist of the venturi, the stream's velocity is increased even more and the pressure at this point is substantially reduced, creating enough vacuum to draw in, through an attached hose, the cuttings from the bottom of the working pit in the manner of a vacuum cleaner. As the chips are drawn into the hose, they are discharged into the reserve pits.

A jetty or pier with oil
lines and loading arms

JETTY
A pier.

JOCKEY
An experienced and proficient driver of large trucks or earth-moving equipment.

JOINT
(1) A length of pipe, casing, or tubing usually from 20 to 30 feet long. On drilling rigs, drillpipe and tubing are run the first time (lowered into the hole) a joint at a time; when pulled out of the hole and stacked in the rig, they are usually pulled two, three, or four at a time, depending upon the height of the derrick. These multiple-joint sections are called stands. (2) A fracture or split in a smooth rock surface (quite often vertical) without any noticeable displacement or movement of the rock. A group of parallel joints is called a set. Joints are caused by weathering in which the rock comes apart along joint planes.

JOINT ADVENTURE
See Joint Venture.

JOINT-INTEREST AGREEMENT
(1) An agreement between two or more concurrent owners who own a tract of land or an oil leasehold and who intend to develop the property. If the joint interest is a joint operating agreement, it is often entered into before there has been any development. If so, one of the owners (often the one with the largest interest) is designated as operator-developer. The parties to the agreement share the expenses and the profits proportionately from the development of the property. (2) An agreement among adjoining landowners or leaseholders to develop a common pool, again sharing expenses and profits. This arrangement is not the same as a unitization agreement.

JOINT OPERATING AGREEMENT
(1) An agreement between two owners or among several concurrent owners for the operation of a leasehold for oil, gas, or other minerals. The agreement calls for the development of the lease or the premises by one of the parties to the agreement, who is designated as operator or unit operator for the joint account. All parties share in the expenses of the operations and in the proceeds resulting from the development.

(2) An agreement among adjoining landowners or leaseholders to develop a common pool, again sharing expenses and profits. This arrangement is not the same as a unitization agreement. *See* Pooling.

JOINT VENTURE
A business or enterprise entered into by two or more partners. Joint venture leasing is a common practice. Usually the partner with the largest interest in the venture will be the operator. *See* Consortium.

JOURNAL
That part of a rotating shaft that rests and turns in the bearing; the weight-bearing segment of the shaft.

JOURNAL BOX
A metal housing that supports and protects a journal bearing. *See* Journal.

J-TUBE
The vertical section of pipe, shaped like the letter J, that connects an offshore production platform's piping to a seabed pipeline. The J-tube is only the guide or mandrel for the subsea pipeline, the end of which is pulled into the curved tube and up to the platform level. This procedure eliminates the need to make underwater connections between the seabed pipeline and a riser pipe. The pipeline is forcibly pulled up through the J-tube to the platform where it can be connected to the platform piping.

JUG
(1) Colloquial term for geophones. (2) Colloquial term for the vertical caverns, shaped like a vinegar jug, leached out of subsurface salt formations for the storage of liquefied petroleum gases and other petroleum products. *See* Salt-Dome Storage.

JUG HUSTLER
One who carries and places geophones in seismic work. Geophones are strung along the ground over an area where seismic shots are to be made by jug hustlers.

Jumbo burner *(Courtesy Otis)*

JUMBO BURNER
A flare used for burning waste gas produced with oil when there is no ready market or the supply is too small or temporary to warrant a pipeline.

JUMBOIZING
A technique used to enlarge an oil tanker's carrying capacity by cutting the vessel in two amidships and inserting a section between the halves.

JUMPER SPOOL
A length of piping fabricated with appropriate connectors to join a subsea production manifold to a sea-floor pipeline. Jumper spools, fabri-

cated on land, after careful measurements of lengths, attitudes, and angles of junction, have made diverless placement of these connections, from a lay barge or work boat, possible and successful.

JUNK BASKET

A type of fishing tool used to retrieve small objects lost in the borehole or down the casing.

JUNK MILL

Drill bit with specially hardened, rough cutting surfaces to grind and pulverize downhole "junk" material or nonretrievable tools or equipment such as millable packers. After the junk has been ground or broken up into small pieces, the pieces can be circulated to the surface by the drilling mud or bypassed by the regular drillstring.

JUVENILE WATER

Water derived from materials deep in the Earth, in magma and plutonic rock. Juvenile water has never been included in the Earthly hydrologic cycle.

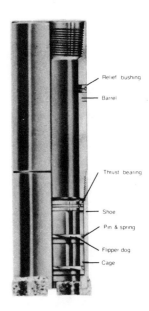

Relief bushing
Barrel
Thrust bearing
Shoe
Pin & spring
Flipper dog
Cage

Junk basket, double catcher

Kelly cock

Kelly joint and bushing

Kelly hose *(Courtesy B.F. Goodrich)*

K

K

The abbreviation for kilo, one thousand. In certain employment ads, notably petroleum industry ads, the letter K is used instead of three zeros in giving salary ranges, e.g., 25K to 60K, also $25K to $60K. To the ad writer this is scientific shorthand meant to catch the eye of the no-nonsense engineer or technical person.

KARST

A type of topography formed in limestone, gypsum, and other erodible rocks, by dissolution; it is known for its sinkholes, small rock-basins, caves, and underground drainage.

KELLY BUSHING

The kelly bushing contains the journal box, which fits in the rotary table and through which the kelly joint passes and is turned when drilling. When the kelly joint is lifted from the hole, the bushing riding on the kelly joint also is withdrawn from the rotary table to make way for the slips, which are inserted in the opening of the rotary table to hold the top joint of the drillstring while the kelly is unscrewed and slipped into the rathole. When going back in the hole, the bushing, riding on the kelly joint, slips into the rotary table and engages the turning mechanism so drilling can resume.

KELLY COCK

A blowout preventer built inside a 3-foot section of steel tubing inserted in the drillstring above the kelly. A kelly cock is also inserted in the string below the kelly joint in some instances.

KELLY HOSE

See Mud, Kelly.

KELLY JOINT

The first and the sturdiest joint of the drill column; the thick-walled, hollow steel forging with two flat sides and two rounded sides that fits into a square hole in the rotary table that rotates the kelly joint and the drill column. Attached to the top of the kelly or grief stem are the swivel and the mudhose.

KELLY SAFETY VALVE
See Kelly Cock.

KELLY SAVER SUB
See Valve, Lower Kelly.

KELLY SPINNER
A mechanism attached to the swivel that spins the kelly joint in and out of the first joint of drillpipe after the kelly has been broken out (unloosened). The spinner saves time in unscrewing and again in screwing in when a joint of drillpipe must be added to the string.

KELLY VALVE, LOWER
See Valve, Lower Kelly.

KELVIN, DEGREES
A unit of temperature equal to 1/273.16 of the Kelvin scale temperature of the triple point of water.

KEROGEN
A bituminous material occurring in certain shales that yield a type of oil when heated. *See* Kerogen Shales.

KEROGEN SHALES
Commonly called oil shales, kerogen shales contain material neither petroleum nor coal but an intermediate bitumen material with some of the properties of both. Small amounts of petroleum are usually associated with kerogen shales, but the bulk of the oil is derived from heating the shale to about 660°F. Kerogen is identified as a pyrobitumen.

KEROGINITE
A type of oil shale found in western Colorado and eastern Utah.

KEROSENE, RAW
Kerosene cut from the distillation of crude oil, not treated or "doctor tested" to improve odor and color.

KEROSINE or KEROSENE SHALE
Any bituminous oil shale.

KEY
(1) A tool used in pulling or running sucker rods of a pumping oil well; a hook-shaped wrench that fits the square shoulder of the rod connection. Rod wrenches are used in pairs; one to hold back-up and the other to break out and unscrew the rod. (2) A slender metal piece used to fasten a pulley wheel or gear onto a shaft. The key fits into slots (keyways) cut in both the hub of the wheel and the shaft.

Kelly spinner *(Courtesy International Tool Co.)*

Kerogen shale semiworks project

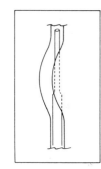

Keyseating drill collars in
crooked hole

Keyseat wiper *(Courtesy
Henderson Tool Co.)*

KEY BED

An easily identified strata or bed of rock that has such distinctive characteristics, lithology (type of rock), or fossils, that it can be used in field mapping or in correlation of subsurface formations; key horizon or marker bed. A signpost for geologists and exploration geologists.

KEY FOSSIL

See Diagnostic Fossil.

KEYSEAT

A section of the well bore deviating abruptly from the vertical, causing drilling tools to hang up; a shoulder in the borehole.

KEYSEATING

A condition downhole when the drill collar or another part of the drillstring becomes wedged in a section of crooked hole, particularly a dogleg, which is an abrupt deviation from the vertical or the general direction of the hole being drilled.

KEYSEAT WIPER

A downhole cutting tool; a type of hole opener that is run on or is a part of the drillstring that wipes or cuts the sides off the borehole where the hole has made a severe turn or deviation from vertical or the general direction of the hole, forming a keyseat. The keyseat wiper cuts away the rock at the point of the abrupt turn in the hole to prevent the drillstring from getting hung up or stuck.

KEYWAY

A groove or slot in a shaft or wheel to hold a key.

KG/CM²

Kilograms per square centimeter; kg/cm^2.

KICK

Pressure from downhole in excess of that exerted by the weight of the drilling mud, causing loss of circulation. If the gas pressure is not controlled by increasing the mud weight, a kick can violently expel the column of drilling mud, resulting in a blowout.

KICK, WATERFLOOD

See Waterflood Kick.

KICKED

A well that is directionally drilled, i.e., at a significant angle (45° or more) from the vertical, is kicked to the producing reservoir. In some instances a well producing from 9,000 feet below the surface may have been kicked horizontally 11,000 feet to reach the reservoir.

KICKING DOWN A WELL

A primitive method of drilling a shallow well using manpower (leg power). In oil's very early days, a pole made from a small tree was used to support the drilling line and bit in the hole. The driller, with his foot in a stirrup attached to the line, would kick downward, causing the pole to bend and the bit to hit the bottom of the hole. The green sapling would spring back, lifting the bit ready for another kick by the driller.

Kicking down a well with spring poles (*Courtesy API*)

KICKOUT

The horizontal or lateral distance from the drillsite reached by a directional well; a well purposely drilled at an angle from the vertical. *See* Kicked; *also* Drilling, Slant-Hole.

KICKOUT CLAUSE

In some purchase contracts for oil and gas, a clause that permits the purchaser, under certain conditions, to renegotiate the contract. Usually the conditions concern pricing or market availability.

KIER, SAMUEL M.

In the early 1850s, Kier was skimming crude oil from the water of his salt wells in Pittsburgh, Pa., and selling it as Kier Rock Oil, a medicinal cure-all. Soon he had more oil than he could peddle in bottles, so he became interested in refining. With the assistance of J.C. Booth, a Philadelphia chemist who designed a crude, coal-fired still, Kier began refining kerosene. By 1859 and the advent of Drake's well, there were nearly a hundred small, one-vessel refineries around the country making kerosene for use in a new lamp that had been invented.

KILL AND CHOKE LINES

Lines connected to the blowout-preventer stack through which drilling mud is circulated when the well has been shut in because excessive pressure downhole has threatened a blowout. Mud is pumped through the kill line and is returned through the choke line, bypassing the closed valves on the B.O.P. When the mud has been heavied up to overcome downhole pressure, drilling can proceed.

KILL A WELL

To overcome downhole pressure in a drilling well by the use of drilling mud or water. One important function of drilling mud is to maintain control over any downhole gas pressures that may be encountered. If gas pressure threatens to cause loss of circulation or a blowout, drilling mud is made heavier (heavied up) by the addition of special clays or other material. *See* Kick.

KILL STRING

See Injection String.

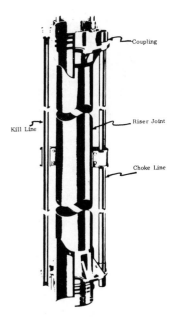

Kill and choke lines attached to marine riser

KILL WATER

Water, at 8.4 lb./gal, used to hold back reservoir fluid in a well's producing formation during workover or other downhole operations requiring a state of equilibrium, vis-á-vis the column of fluid and reservoir pressure. A column of water weighing 0.434 lbs/ft., having a greater density satisfactorily holds at bay reservoir fluids until remedial work is completed or for other reasons. In deep, hot wells water becomes less dense, weighing less per foot and less per gallon: 0.387 lbs/ft and 7.44 lbs/gal., respectively. These approximate values are for well depths of 6,000 to 8,000 feet and temperatures to 400°F. If a well is too deep and too hot for the use of water for the kill, other fluids or mixes are used.

KILLER WELL

A directional well drilled near an out-of-control well to "kill" it by flooding the formation with water or mud. Wells that have blown out and caught fire are often brought under control in this manner if other means fail.

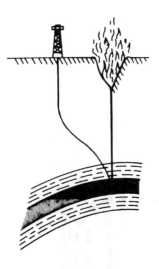

Killer well

KNOCK-OFF POST

A post through which a rod line moves as it operates a pumping jack. When the well is to be hung off (shut down), a block is inserted between the rod-line hook and the knock-off post, which interrupts the line's forward movement, putting slack in the line so that the hook may be disengaged.

KNOCKOUT

A tank or separator vessel used to separate or knock out water from a stream of oil.

KNOCKOUT DRUM, FLARE

A horizontal or vertical tank or vessel for trapping hydrocarbon liquids in a gas-flare system. Separating the liquids out of the gas stream prevents possible damage to refractory linings, also promotes even and more nearly complete burning of the waste gas. To prevent flame propagation into the system (possible flash back into the drum), a liquid seal is used. The seal can also serve to maintain positive (but slight) pressure in the flare header.

KNOWLEDGE BOX

The drilling crew's name for the place the driller keeps his orders and reports; smart box.

KNUCKLE-BUSTER

A wrench so worn or of such poor quality that it will not hold when under the strain of heavy work; a cultivator wrench.

KNUCKLE JOINT
A movable joint connecting two sections of a downhole tool; an integral part of a bottom-hole assembly used to drill at an angle from the vertical; a ball joint. *See* Angle Building.

K.O.C.
Kuwait Oil Company; the country's national oil company.

KORT NOZZLE
A type of ship's propeller that rotates within a cylindrical cowling that concentrates the thrust of the propeller. This produces a nozzle effect as the water is jetted from the cowling. Kort nozzles are installed on some tugboats and drilling-tender vessels because of their maneuverability and response.

K.T.B. SUPERDEEP BOREHOLE
The K.T.B. project is one of the most ambitious geoscientific, deep-hole, exploratory projects ever undertaken by the oil industry—with governmental assistance. The well being drilled in Germany and at this writing, 1994, is 6,000 meters deep, with a target depth of 10,000 M or 32,800 feet, just to see what's down there, and what the bits would have to dig through.

L.A.C.T.
See Lease Automatic Custody Transfer.

LACUSTRINE
Relating to lakes or derived from lakes. Lake deposits are of several different kinds. Some of these can be source beds, while others are reservoir rocks. Lake beds have the same rock types as stream and marine environments, i.e., evaporites, clastics, and carbonates. For example, salt beds, sandstone, and limestone deposits.

LAMINA
The thinnest layer of a sedimentary deposit that is recognizable as a unit of the original material. Laminae are 1/2 to 1 millimeter in thickness and exhibit differences in color, composition, and particle size. Many laminae can make up a bed.

LAMINAR
Composed or arranged in layers like a deck of playing cards; made up of laminae, thin layers or leaves.

LAMINAR FLOW
The movement of a liquid through a pipeline in which the fluid moves in layers as in a nonturbulent stream. The center of the stream moves faster than the edges or bottom, which are slowed down by friction and small obstructions. *See* Plastic Flow.

LAMINATED
Said of a rock such as shale that consists of very thin layers or laminae that can easily be split. Also refers to a sedimentary bed of laminated rock.

LANDED COST (OF OIL)
The cost of a barrel of imported oil offloaded at a U.S. port. Landed cost includes all foreign taxes and royalties plus cost of transportation.

LANDING CASING
Lowering a string of casing into the hole and setting it on a shoulder of rock at a point where the diameter of the borehole has been reduced. The beginning of the smaller-diameter hole forms the shoulder on which the casing is landed.

LANDMAN
A person whose primary duties are managing an oil company's relations with its landowners. Such duties include securing oil and gas leases, lease amendments, and other agreements. A lease hound.

LANDOWNER'S ROYALTY
See Royalty, Landowner's.

LAND PLAY
See Oil Play.

LAP, TO
To hone; to put a smooth finish on a metal surface; also, to fit two mating parts together by polishing one part against the other using a lapping compound, an extremely fine abrasive substance in a grease or grease-like matrix. Example: lapping a valve into its seat.

LAP-WELDED PIPE
Line pipe or casing made from a sheet of steel that is formed on a mandrel. The two edges, tapered to half of normal thickness, are lapped over and welded. *See* Seamless Pipe.

LATERAL LINES
Pipelines that tie into a trunk line; laterals are of smaller diameter and are laid as part of a gathering system or a distribution system. In an oil field, laterals bring oil or gas from individual leases or tank batteries to the booster station and the trunk line.

LATERAL WELLS
High-angle and horizontal wells.

LATERAL WELLS, DUAL
Two wells drilled in opposite directions, kicked off from one vertical borehole, penetrating two separate sands or two longitudinal sections or displacements in the same zone.

Pipelay barge

LAUNDER
See Slush-Pit Launder.

LAY BARGE
A shallow-draft, barge-like vessel used in the construction and laying of underwater pipelines. Joints of line pipe are welded together and paid out over the stern of the barge as it is moved ahead. Lay barges are used in swampy areas, in making river crossings, and laying lines to offshore installations.

LAY-DOWN RACK
A storage area for tubing and drillpipe that are removed from a well and laid down rather than set back and racked vertically in the derrick.

Lay-down rack

LAY DOWN THE TUBING
To pull the tubing from the well, a joint at a time, and remove it from the derrick floor to a nearby horizontal pipe rack. As each joint is unscrewed from the string, the lower end of the joint is placed on a low cart and pulled out to the rack as the driller lowers the pipe which is held up by the elevators.

LAY OFF RISK, TO
To share financial risks; to trade equity position for equipment and services. Laying off some of the risk is a survival technique in a time of high cost in exploration, drilling, completing, and equipping.

LAY TONGS
See Pipe Tongs.

LAZY BENCH
A bench on which workers, when not working, may rest. A perch from which a work operation may be observed by workers or guests.

LAZY BOARD
A stout board with a handle used to support the end of a pipeline while another length of pipe is screwed into it. On small lines, the man operating the lazy board or granny board usually handles the back-up wrench, which holds one joint of pipe firm while another joint is being screwed in. Growler board.

1890 pipeline gang using lazy board and pipe tongs

LB./LB.
Pound per pound (lb/lb). In a refining process, the ratio of ingredients to be mixed or introduced to the process.

LBS.-H₂O/MM
Pounds of water per million standard cubic feet (MMscf) of natural gas. The designation of water content for large volumes of gas. *See* PPM/WT.

L.C.C.V.
Large crude-carrying vessel; tankers from 100,000 to 500,000 deadweight tons capable of transporting 2.5 to 3.5 million barrels of oil in one trip. Cruising speed of L.C.C.V.s is 12 to 18 knots; overall length, about 1,200 feet; draft when fully loaded, more than 80 feet.

LCCV

LEACHING
The dissolving out or separation of soluble material from a rock or ore body by the action of percolating water, rainwater, or irrigation. The removal in solution of nutritive elements from the soil by water washing.

LEAD AND TAIL CEMENT
The first and last stage of a two-stage cement job.

LEAD LINES
Lines through which production from individual wells is run to a lease tank battery.

LEAF CHAIN
A series of links composed of a number of flat metal leaves fastened with steel pins. Each link is made up of 10 to as many as 50 individual flat leaves with a hole in each end through which a connecting pin is inserted to form an "endless" chain.

LEAKOFF, FRAC FLUID
A term used in fracing a downhole formation. When the hydraulic pumps of the fracing crew raise the pressure on the gelled frac fluid to the critical point, the fluid, with the proppant material in suspension, begins to leak off, to gradually invade the formation. Fissures and larger cracks begin to open; the frac job is underway.

LEAN GAS
Natural gas containing little or no liquefiable hydrocarbons. *See* Wet Gas.

LEAN OIL
The absorbent oil in a gasoline absorption plant from which the absorbed gasoline fractions have been removed by distillation. Before distillation to remove gasoline fractions, the oil is referred to as "fat oil."

LEASE
(1) The legal instrument by which a leasehold is created in minerals. A contract that, for a stipulated sum, conveys to an operator the right to drill for oil or gas. The oil lease is not to be confused with the usual lease of land or a building. The interests created by an oil-country lease are quite different from a reality lease. (2) The location of production activity; oil installations and facilities; location of oilfield office, toolhouse, garages.

LEASE, BLOCK
A lease executed by owners of separate tracks or, sometimes, separate leases executed by owners of individual tracts that provide drilling one or more test wells within the combined area or block that satisfy the conditions of the lease for each tract in the block.

LEASE, FEDERAL
An oil or gas lease on federal land issued under the Act of February 25, 1920, and subsequent legislation.

LEASE, NONDRILLING
The nondrilling lease may appear to be contradictory or self-defeating, but such leases are granted under certain circumstances when drilling

on the leased property would be incompatible with the present or proposed use of the land, e.g., municipal water supply (lake), cemetery, industrial area, or national park. A nondrilling lease grants to the lessee the usual rights relative to the oil and gas beneath the property. However, the lease stipulates that no well or supporting surface installations—stock tanks, separators, or pipelines—be placed on the property. Oil and gas production must be from a well drilled on an adjoining property. In some cases a nondrilling lease may be executed to prevent the drilling of a well in the gas cap, which would seriously reduce the reservoir pressure with the result that nominally recoverable oil would be lost. Under these circumstances, adjoining landowners contribute to payments made to the lessor under the terms of the lease in lieu of drilling the well into the gas cap and the payment of customary royalties. When land cannot be drilled on because of the conditions mentioned, slant-hole or deviated drilling is done from adjoining property. With present-day technology, boreholes can be slanted sufficiently to reach a pay zone 2 miles away from the wellsite. *See* Kicked; *also* Drilling, Directional.

LEASE, OR

One of the two most common forms of oil and gas leases; the other is the unless lease. Both types of leases are granted for a primary term, five years for example, and "so long thereafter as oil and gas are produced." In an or lease, the lessee promises to drill on or before the first anniversary date or do something else: pay rental, forfeit the lease, etc. The delay rental clause of an or lease is often written as follows: "Lessee agrees to begin a well on said premises within one year of date hereof or thereafter pay lessor as rental $____ each year in advance to the end of this term or until said well is commenced, or this grant is surrendered as stipulated herein."

LEASE, PAID-UP

A lease that is valid during the entire primary term without payment of delay rentals because the total rentals for the primary term have been paid. With rentals paid up, the delay-rental clause is sometimes deleted. However, some authorities believe it is wiser to leave the delay-rental clause in the lease. Further, if the clause is left in and the rentals are paid up, the lease should say so to protect the lessee in the event the lessor's interest is assigned to a person who would not know the lessee had paid up the lease. The moral: Put everything in writing.

LEASE, SHOOTING

An agreement granting permission to conduct a seismic or geophysical survey. The lease may not give the right to lease the land for oil or gas exploration.

LEASE, STORE
A preprinted lease form (bought at the store) with blanks to be filled in by the parties to the lease.

LEASE, TOP
A lease given by the landowner while his property is still under lease by another person or company. Such a lease becomes effective if and when the existing lease expires or is terminated. The top lease is dated to go into effect the minute the primary term of the existing lease runs out. This is not pleasing to the leaseholder, but it often happens. The person taking the lease and the one giving the lease (landowner) rationalize that the former lessee had three to five years to commence a well but failed to do so; therefore, the land is up for grabs.

LEASE, UNLESS
A type of lease in general use; the other common type is the or lease. There is no single form of the unless lease, but it is known as the unless lease because of the wording of the delay rental clause, which usually takes the following form: "If no well is commenced on said land on or before the date hereof, this lease shall terminate as to both parties unless the lessee on or before that date shall pay or tender to lessor the sum of $____, which shall operate as rental and cover the privilege of deferring the commencement of a well for twelve months from said date."

LEASE, AUTOMATIC CUSTODY TRANSFER
A system of handling on a lease; receiving into tankage, measuring, testing, and turning into a pipeline the crude produced on a lease. Such automatic handling of oil is usually confined to leases with settled production.

LEASE BROKER
A person whose business is securing leases for later sale in the hope of profit. Lease brokers operate in areas where survey or exploration work is being done.

LEASE BURDENS
The lessor's royalty, an overriding royalty, a production payment and similar interests characterized as burdens. Net profits interest, a carried working interest or any other interest payable out of profits are not considered lease burdens.

LEASE CONDENSATE
Liquid hydrocarbons produced with natural gas and separated from the gas at the well or on the lease. *See* Condensate.

LEASEHOLD INSURANCE
A type of insurance offered by some companies that guarantees the estate of the insured in the oil and gas property as specified in the deed.

There is no guarantee as to the value of the reserves (even Lloyds of London would shy away from such assurance), only that the insured has a valid claim to the leasehold. A type of title insurance.

LEASEHOLD INTEREST

An individual's interest as lessee or grantee under an oil, gas, or other mineral lease. Such an interest includes the right to drill, operate, and produce oil and gas, and the grantee is obligated to pay the lessor or grantor a royalty of a percentage of the lease's production, free of any cost to the lessor.

LEASE HOUND

Colloquial term for a person whose job is securing oil and gas leases from landowners for himself or a company for which he works. *See* Landman.

LEASE LINES

Gathering lines on a lease; usually small-diameter (2- to 4-inch) pipelines that carry production from the lease wells to a central tank battery; lead lines.

LEASE PUMPER

See Pumper.

LEASE SALES, COMPETITIVE

The sale of oil and/or gas leases by competitive bidding is usually conducted by submitting sealed bids to the entity: the federal government, the state, or in some cases, a foreign government. Usually, the sealed bid must be accompanied by a good faith cashier's or certified check for a substantial part of the bid.

Lease tank battery

LEASE TANK

A battery of two or more 100- to 500-barrel tanks on a lease that receive the production from the wells on the lease. Pipeline connections are made to the lease tanks for transporting the oil to the trunk line and then to the refinery.

LEASE TANK BATTERY

See Battery.

LEFT-HAND THREAD

A pipe or bolt thread cut to be turned counterclockwise in tightening. Most threads are right-hand, cut to be tightened by turning clockwise. Nipples with one kind of thread on one end, another on the other end, are referred to as bastard nipples.

LEGAL SUBDIVISION

Forty acres; one-sixteenth of a section (square mile).

LEGS, DERRICK
The four corner members of the rig, held together by sway braces and girts.

LENS
A sedimentary deposit of irregular shape surrounded by impervious rock. A lens of porous and permeable sedimentary rock may be an oil-producing area.

LENSING
The thinning out of a strata of rock in one or more directions; the disappearing or pinching out of a formation laterally or horizontally.

LENTICULAR
Said of a body of rock or sedimentary structure whose shape resembles a lens in cross section, especially a double-convex lens like a peach seed. Lenticular pertains to a stratigraphic lens or a lentil, which is a small stratigraphic unit of very limited extent, a part of a much larger formation of different rock or other material, like an almond in a fruitcake.

LESSEE
The person or company entitled, under a lease, to drill and operate an oil or gas well.

L.I.F.O.—F.I.F.O.—F.I.L.O.
Last in, first out; first in, first out; first in, last out. Acronyms that designate the sequence of movement in and out, or the handling of crude oil and products in inventory or held in storage.

LIFT GAS
See Gas-Lift Gas.

LIFTING
(1) Refers to tankers and barges taking on cargoes of oil or refined product at a terminal or transshipment point. (2) Producing an oil well by mechanical means: pump, compressed air, or gas.

LIFTING COSTS
The costs of producing oil from a well or a lease.

LIGHT CRUDE
Crude oil that flows freely at atmospheric temperatures and has an A.P.I. gravity in the high 30s and 40s; a light-colored crude oil. *See* Heavy Crude Oil.

LIGHT ENDS
The more volatile products of petroleum refining, e.g., butane, propane, gasoline.

Oil derrick legs *(Courtesy Parker Drilling)*

Lift pump *(Courtesy KOBE)*

LIGHTER

See Lightering.

LIGHTERING

The use of small, shallow-draft boats in transshipment to shore of oil or other cargo from a large, deep-draft vessel unable to dock at shore facilities because of shallow water. The small boats are called lighters.

LIGHT PLANT

An early-day term for an installation on a lease or at a company camp that provided electricity for lighting and small appliances. The light plant often was simply a belt-driven D.C. generator run off one of the engines at a pipeline pumping station or a pumping well's engine. The lights "surged" with the power strokes of the engines and went out when an engine "went down," but the lights were far better than gas lights—or none at all.

LIGNITE

A brownish-black type of coal that is relatively soft. It is called lignite from the Latin *lignum,* meaning "wood." True to its name, the texture of wood can still be detected in a sample of lignite. Processed lignite is used as an additive in drilling mud, as a dispersant, and as a viscosity-control substance.

LIME

Colloquial for limestone.

LIME PAN

A hard, thin layer, usually below the surface of the ground, that resembles hardpan. Lime pan is cemented mainly with calcium carbonate.

LIMESTONE, OOLITIC

See Oolitic Limestone.

LIMITED PARTNERSHIP, MASTER

A form of organization to finance lease acquisition of both oil and gas drilling operations. In limited partnerships the investor has limited liabilities stemming from the oil operations. Some types of limited partnerships are publicly traded, traded over-the-counter, treated like ordinary equity instruments, and are referred to as Master Limited Partnership (M.L.P.). One reason for the popularity of M.L.P.s, since the mid-1980s, is because the partnership investment is more "liquid," more easily converted to cash than more conventional limited partnerships. One can get in as well as out of M.L.P.s with comparative ease and without ruffling feathers. There are also tax advantages: Distributions, unlike dividends on stock, usually are not subject to corporate taxes. Also, during the early life of the investment, distributions are considered returns of capital and thus are not taxed as ordinary income.

LIMITED WORKING INTEREST
See Working Interest, Limited.

LIMY
Said of soil or rock containing large amounts of lime or limestone; containing calcite; for example, a calcitic dolomite, a limy dolomite.

LINE, GAS SALES
Merchantable natural gas line from a lease or offshore production-processing platform carrying gas that has had water and other impurities removed; a line carrying pipeline gas.

LINE, OIL SALES
Merchantable crude oil line from a lease or offshore, production-processing platform carrying oil that has had water and other impurities removed; a line transporting pipeline oil.

LINE FILL
The amount of gas or oil or product required to fill a new line before deliveries can be made at take-off points or the end of the line.

LINE LIST
Instructions to the pipeline construction crews building a line across the land of many property owners. The instructions list all owners, the length of line across each property, and any special restrictions such as "keep all gates closed and in good repair" and "avoid at all costs damaging large trees." The right-of-way man helps make up the line list.

LINE MILES
Designation of offshore seismic surveying work by ship.

LINE PACK
The volume of gas or barrels of oil maintained in a trunk pipeline at all times in order to maintain pressure and provide uninterrupted flow of gas or oil. There are millions of barrels of oil and billions of cubic feet of gas in the country's pipelines at all times.

LINE-PACK GAS
Gas maintained in a gas-transmission line at all times to maintain pressure and affect uninterrupted flow of gas to customers at take-off points.

Unloading 36-in line pipe

LINE PIPE
Pipeline pipe that is made in even sizes from 2 inches to 48 inches. There are larger sizes in use however; some large gravity loading lines for crude oil tankers are 56 inches. Most line pipe is either lap welded or butt welded. Seamless pipe is usually only for drilling wells. Line pipe, especially the large sizes, has beveled weld ends so the joints can be welded together. Large-diameter screw pipe (12 inches) went out of style in the late 1920s, along with the 200-man pipe-handling and pipelaying crews. Gas welding and then electric welding put them out of business.

LINER
(1) In drilling, a length of casing used downhole to shut off a water or gas formation so drilling can proceed. Liners are also used to case a thief zone where drilling fluid is being lost in a porous formation. (2) A liner is a removable cylinder used in reciprocating pumps and certain types of internal-combustion engines; a sleeve.

LINE RIDER
See Pipeline Rider.

LINERS, FIBERGLASS
Corrosion-resistant fiberglass well bore liners are used in certain older wells in place of more costly, corrosion-prone steel liners. The plastic (fiberglass) liners are not as strong as steel, do not exhibit its compressive strength, so the liners are not set on the bottom. In lowering in and subsequent cementing, extra care is taken to monitor the pumping rates and pressures. Once cemented in place, the fiberglass liners will outlast the old well's steel components.

Internal lineup clamps

LINEUP CLAMP
A device that holds the ends of two joints of pipe together and in perfect alignment for welding. Lineup clamps operate on the outside of the pipe and are used on smaller-diameter line pipe. Large-diameter pipe—20 to 36 inch and over—is aligned by internal, hydraulically operated mandrellike devices.

LINKAGE
A term used to describe an arrangement of interconnecting parts—rods, levers, springs, joints, couplings, pins—that transmit motion and power or exert control.

LINKS
Bail-like arms attached to the elevators. The links engage the hook of the big traveling block that hangs on a number of lines of steel cable from the crown block in the top of the drilling derrick.

LIQUEFIABLE HYDROCARBONS
The light ends separated from crude oil in the refining process that are gaseous at atmospheric pressure. Examples are butane, propane, and pentane. These gaseous fractions are cooled and subjected to pressure, which condenses them to clear liquids. When stored or transported, the liquefied hydrocarbons are stored in pressure vessels. *See* Liquefied Petroleum Gas.

LIQUEFIED ENERGY GASES (L.E.G.)
A term that includes liquefied petroleum gas (L.P.G.) and liquefied natural gas (L.N.G.).

LIQUEFIED NATURAL GAS (L.N.G.)
Natural gas that has been liquefied by severe cooling (–160°C.) for the purpose of shipment and storage in high-pressure cryogenic tanks. To transform the liquid to a usable gas, the pressure is reduced and the liquid is warmed.

LNG carrier

LIQUEFIED NATURAL GAS CARRIER (L.N.G.C.)
A specially designed oceangoing vessel for transporting liquefied natural gas.

LIQUEFIED PETROLEUM GAS (L.P.G.)
Butane, propane, and other light ends separated from natural gasoline or crude oil by fractionation or other processes. At atmospheric pressure, liquefied petroleum gases revert to the gaseous state. L.P.G.; L.P.-Gas.

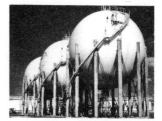

LPG storage tanks

LIQUID HOLDUP
In natural gas lines, particularly offshore, petroleum liquids (condensate) drop out of the gas stream and collect in low places in the line, which causes a blockage or holdup in the line. *See* Slug Catcher.

LIQUID HYDROCARBONS
Petroleum components that are liquid at normal temperatures and atmospheric pressure.

LIQUID TURBINE METER
See Turbine Meter, Liquid.

LITER
A metric unit of volume; 1.057 U.S. quarts; 61.02 cubic inches.

LITHIFY
To change sediment into rock as occurs during diagenesis; to harden into rock.

LITHOFACIES
A lateral, recognizable subdivision of a stratigraphic unit, distinguished from adjacent subdivisions on the basis of lithology. A facies characterized by particular lithographic features.

LITHOFACIES MAP
A facies map based on lithographical characteristics showing area variations in a given stratigraphic unit.

LITHOLOGY
The description of rock specimens and rock showing at outcrops on the basis of color, mineral composition, grain, or crystallization; the study of the physical properties or character of a rock.

LITHOSPHERE
That part of the Earth that consists of solid rock from 30 to 60 miles thick and includes oceanic and continental crusts in the upper sections. Literally, sphere of rock, from the Latin *lithos,* meaning "stone."

LITTLE BIG-INCH PIPELINE
A 20-inch products pipeline built from East Texas to the East Coast during World War II to solve the problem caused by tanker losses as the result of submarine warfare. After the war, the line was sold to a private gas-transmission company.

LIVE OIL
Crude oil that contains dissolved natural gas when produced.

LIVE WELL WORKOVER
Workover (pulling the production tubing and washing the casing free of sand and debris) of a high-pressure well without killing the well with drilling mud. To kill the well with mud requires much preparation and considerable expense, and risks damaging the formation. As an alternative, live well workovers are accomplished by using snubbing operations. *See* Snubbing; *also* Snubber.

L.N.G.
Liquefied natural gas.

L.N.G.C.
Liquefied natural gas carrier.

L.N.G. VAPORIZATION TRAIN
Regasification, odorization (so leaks can be detected by humans), metering, and pressure reduction compatible with the buyer's or receiver's facilities. These four steps are required to offload and deliver to the buyer refrigerated, pressurized, natural gas in the liquid state, arriving

at a sea terminal by ship with its large, spherical, heavily insulated tanks.

LOAD CELLS
An electronic device, a signal system, used on sucker-rod pumps to determine and monitor rod loads. The device is also employed with pump-off controllers to shut off a pumping well as soon as the fluid in the tubing and the well is exhausted—after the allotted or the programmed pumping time.

Loading arms

LOADING ARMS
Vertical standpipes with swivel-jointed arms that extend to a tanker or barge's deck connections for loading crude oil or products.

LOADING RACK
An elevated walkway that supports vertical filling lines and valves for filling tank cars from the top.

LOAD OIL
Oil of any kind put back into a well for any purpose, e.g., hydraulic fracturing, shooting, or swabbing.

LOCAL DRAINAGE
The movement of oil or gas toward the well bore of a producing well. *See* Drainage, *also* Migration, Local.

Loading rack *(Courtesy Diamond Alkali)*

LOCATION
The wellsite; the place where a well is to be drilled or has been drilled; a well-spacing unit, e.g., "Two locations south of the discovery well...."

LOCATION DAMAGES
Compensation paid by an operator to the owner of the land for damages to the surface or to crops during the drilling of a well. Mud pits must be dug, a surface leveled for tanks and rigs, and access roads built, so there are always some location damages to be paid.

LOESS
A nonstratified, fine-grained blanket deposit that covers large areas of Europe, Asia, and North America. It consists of windblown silt, clay particles, and fine sand. It is yellowish-gray in color and is deposited in beds sometimes as thick as 100 feet. Although porous and friable, the mineral grains are held together by calcareous cement, which allows the loess to stand as steep or nearly vertical bluffs along rivers, notably in the U.S. along the Mississippi River. The word loess is from the German *loss,* meaning "loose."

LOG
A record of activity or results of plans or surveys. In drilling a well there are a number of different kinds of logs: downhole density logs taken

6"   16"

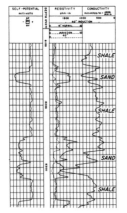

Caliper log

Electric Log
Induction

with various electronic instruments; mud logs that monitor cuttings from the bottom of the hole; driller's logs, a diary-like account of everything that happened during the driller's tour of duty.

LOG, CALIPER
A well log that shows the variations in the diameter of the uncased borehole, from bottom to top. The log is made with a device with springloaded arms that press against the wall of the hole as it is drawn upward, measuring the varying diameters. *See* Drift Mandrel.

LOG, CASING-COLLAR
A well log showing relative magnetic intensity used in cased well bores to identify the threaded joints or casing collars of successive joints of casing. The log is run simultaneously with a gamma-ray log to correlate between the geologic interval and the location of the casing collars for perforating, setting a plug, or other operations. It is hard enough to perforate the casing without having to shoot through a casing collar.

LOG, DENSITY
A well log of induced radioactivity that shows the density of rocks downhole and the fluids they contain. In effect, a density log is a porosity log of the wall-contact type, indicating formation density by recording the back scatter, or reflection, of the induced gamma rays.

LOG, DRILLER'S
A record kept by the driller showing the following: when the well was spudded in, the size of the hole, the bits used, when and at what depth various tools were used, the feet drilled each day, the point at which each string of casing was set, and any unusual drilling condition encountered. In present-day wells, the driller's log is supplemented by electrical well logs.

LOG, ELECTRIC
An electrical survey of a well's borehole before it is cased, which reflects the degree of resistance of the rock strata to electric current. From the results of the survey, geologists are able to determine the nature of the rock penetrated by the drill and some indication of its porosity.

LOG, FOCUSED-CURRENT
The resistivity log curves from a multi-electrode sonde designed to focus the current radially through the rocks in a horizontal disk-shaped pattern. This permits sharp definition of sedimentary bed boundaries and improved measurement of resistivity. Focused-current logs are sold under several trade names, e.g., Laterolog, Guard Log.

LOG, GAMMARAY

A well-logging technique wherein a well's borehole is bombarded with gamma rays from a gamma-ray-emitting device to induce output signals that are then recorded and transmitted to the surface. The gammaray signals thus picked up indicate to the geologist the relative density of the rock formation penetrated by the well bore at different levels.

LOG, GAS

Identifying gas in the drilling mud with the use of a gas chromatograph or other detector. A seat-of-the-pants method is to watch for gas bubbles in the mud. To identify the kind of gas—methane, hydrogen sulfide, carbon dioxide, or helium—scientific instruments are used.

LOG, MAGNETIC

A well log showing relative magnetic intensity in cased well bores to identify the threaded joints or casing collars of successive joints of well casing. The log is made simultaneously with a gamma-ray log to correlate between the geologic interval or formation and the location of the casing collars when the well's casing is to be perforated, when a plug is to be set, or for other downhole operations.

LOG, MUD

A progressive analysis of well bore cuttings washed up from the bottom of the well by the drilling fluid. Rock chips and fragments are retrieved with the aid of the shale shaker and are examined by the geologist. He can tell by the type of rock the bit is cutting what formation is being penetrated.

LOG, NOISE

A sound-detection system inside a logging tool designed to pick up vibrations caused by flowing liquid or gas downhole. The device is used to check the effectiveness of a squeeze job, to estimate the gas flow from perforated formations, etc.

LOG, POROSITY

A generic term for well-log curves whose measurements relate to a formation's porosity or density; a sonic log or density log.

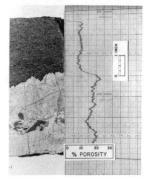

Porosity Log

LOG, SAMPLE

A record of rock cuttings as a well is being drilled, especially in cable-tool drilling. The cuttings, brought to the surface by the bailer, are saved and the depth where obtained is recorded. This record shows the characteristics of the various strata drilled through.

LOG, SONIC

The sonic log is a recording of the time required for a sonic wave to travel a given distance through a formation. This time increment is called the "interval transit time" which is commonly measured in microseconds per foot. Interval travel time is the reciprocal of the sound velocity expressed in feet per second. The relation between these two expressions of compression and propagation rate are used by the seismologist-interpreter in his calculations. No logging tool measures porosity; it must be arrived at from other sources of seismic information. Sonic logs provide one such source of these data.

LOG, TEMPERATURE

A well log, usually run in a cased hole, that indicates temperature variations from the top to the bottom of the hole. It is often used to identify the top of the cement that has been pumped into the hole and is curing (setting up) between the casing and the wall of the borehole; a thermal log.

LOG, THERMAL

See Log, Temperature.

LOGGING UNIT

A wireline truck or skid-mounted unit with a spool of wireline. The spool on which the wire is wound is powered by a small engine to reel in the thousands of feet of wire lowered into the hole with the logging tool.

LOG ROAD

See Cord Road.

LOGS, INSURANCE

M.W.D. logs, although costly, are often run as insurance in the event the operator is unable to run a wireline log.

LONG STRING

See Production String.

L.O.O.P.

Louisiana Offshore Oil Port (L.O.O.P.); the United State's first superport; a facility for offloading supertankers and V.L.C.C.s too large for shallow, nearshore installations. L.O.O.P., 19 miles out in the Gulf of Mexico, is designed to offload crude oil and pump to shore stations 1.4 million barrels a day. The facilities are such that they can unload tankers in a few hours, thus reducing the vessels' turnaround time, a big saving in demurrage.

LOOPING A LINE

The construction of a pipeline parallel to an existing line, usually in the same right-of-way, to increase the throughput capacity of the system;

Logging unit minus
wireline spool

Log road

doubling a pipeline over part of its length, with the new section tied into the original line.

LOOSE-VALVE TREE

The designation for a Christmas tree or production tree nippled up or made up with individual valves as contrasted to solid-block tree valves, i.e., two or more valves made in one compact steel block. A stacked, loose-valve tree.

Loose-valve tree

LOSE RETURNS

Refers to a condition in which less drilling mud is being returned from downhole than is being pumped in at the top. This indicates that mud is being lost in porous formations, crevices, or a cavern.

LOSS OF CIRCULATION

A condition that exists when drilling mud pumped into the well through the drillpipe does not return to the surface. This serious condition results from the mud being lost in porous formations, a crevice, or a cavern penetrated by the drill.

LOST-CIRCULATION MATERIAL

Fibrous material, and high viscosity pills to increase the viscosity of the drilling fluid. The fibrous material to plug fissures and cracks in the formation that is "stealing" the drilling fluid. In efforts to plug holes in the thief zone, many unlikely things have been pumped downhole: pecan shells, burlap scraps, cotton seed hulls, and shreds of oakum—anything that can be pumped and might gang up to form a plug.

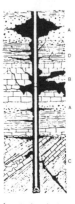

Lost circulation

LOW

See Structural Low.

LOWBOY

A low-profile, flatbed trailer with multiple axels (6 to 10) for transporting extra-heavy loads over relatively short distances. The many wheels and axels spread the weight of the trailer and its load over a large area to avoid damaging streets and highways. The low bed makes it easier to load and unload the heavy equipment it was designed to move.

LOWER IN

To put a completed pipeline in the ditch. This is done with sideboom tractors that lift the pipe in slings and carefully lower it into the ditch.

LOWER KELLY VALVE

See Valve, Lower Kelly.

Lowboy *(Courtesy Elder-Oilfield)*

LOW-GRAVITY OIL
Oil with an A.P.I. gravity in the mid to high 20s; heavy oil is designated as oil 20° A.P.I. or lower.

L.P.-GAS
See Liquefied Petroleum Gas.

L.P.-GAS DRIVE
The injection of high-pressure enriched gas or an L.P.G. slug to affect the miscible displacement of oil. *See* Tertiary Recovery.

LUBE OIL
Short for lubricating oil or lubricant. Also lube and lubes.

LUBRICATING OIL, MULTIGRADE
Specially formulated lubricating oil that flows freely in cold weather and in the heat of engine operation maintains sufficient viscosity or "body" to properly lubricate the engine, e.g., 10–30 S.A.E.

LUBRICATION SYSTEM, GRAVITY SPLASH
A type of lubrication system for relatively slow-moving machinery. The crankcase of a pump, for example, contains the lube oil. As the crank-shaft turns, the crank throws and connecting rods splash through the reservoir of oil, creating a "storm" of lubrication for all bearings inside the crankcase.

LUBRICATOR, MUD
A temporary hookup of pipes and valves for introducing additional, heavy drilling mud into the well bore to control gas pressure. Through one or two joints of large-diameter casing attached atop the wellhead, the heavy mud is fed into the well bore, against pressure, as through a lubricator.

Oil lubricator *(Courtesy Lincoln)*

LUBRICATOR, OIL
A small, box-like reservoir containing a number of gear-operated pumps. The individual pumps, working in oil, measure out a few drops at a time into small, copper lines that distribute the lubricant to the bearings.

LUBRICITY
Slippery. In the context of drilling and completing a well, lubricity has to do with reducing drag and torque on drillpipe, casing, and downhole tools. The same rationale as greasing a pig at the county fair.

LUCAS, CAPT. ANTHONY F.
It was Capt. Lucas' Spindletop gusher in 1901 (75,000 b/d) that ushered in the modern oil age of large oil companies. John H. Galey and James M. Guffey owned the Spindletop gusher located near Beaumont, Texas.

LURGI PROCESS

A process for the commercial gasification of coal that originated in Germany.

L.X.T. UNIT

A low-temperature separator; a mechanical separator that uses refrigeration obtained by expansion of gas from high pressure to low pressure to increase recovery of gas-entrained liquids.

M³
Cubic meter.

MACARONI TUBING
Small-diameter tubing (3/4–1 1/4 inch O.D.) used in slim-hole wells or for certain well workover procedures.

MACROSEEP
Refers to a big oil or gas seep on the deep ocean floor. Oil seeps on shore, in draws and creek beds, were the first good indicators of where the oil was in America's early days of exploration, 1859. Ocean floor seeps are still good indicators of the upward migration of hydrocarbons, oil and gas.

MAFIC
A type of igneous rock composed mainly of one or more ferromagnesian, dark-colored minerals. The term *mafic* is made up of Magnesium and Ferrous plus *ic*.

MAGNETIC LOG
See Log, Magnetic.

MAGNETIC SURVEY
A geophysical survey of possible oil-bearing rocks using a magnetometer for measuring the relative intensity of the earth's magnetic effect or gravitational pull. When there are recordable differences in the magnetic pull at different points in an area, they may indicate an anomaly, an underground formation favorable to the accumulation of oil. *See* Magnetometer.

MAGNETO
See Ignition Magneto.

MAGNETOMETER
An instrument for measuring the relative intensity of the earth's magnetic effect. Used to detect rock formations below the surface; an instrument used by geophysicists in oil exploration work.

MAIN LINE
Trunk line; a large-diameter pipeline into which smaller lines connect; a line that runs from an oil-producing area to a refinery.

MAKE A HAND
To get with the work to be done; to look alive; to use your head, too.

MAKE UP
To screw a pipe or other threaded connection tight by the use of a wrench.

MAKEUP GAS
(1) Gas that has been paid for previously, but not taken under a take-or-pay clause in a gas sales contract. The gas can be taken later, free of additional cost. (2) Gas made available to keep in equitable balance the proportionate share amounts of gas in pooled or unitized leases. If one party to the agreement is underproduced, does not take or market his share during a specified time, an overproduced party gives a certain percent of his share to the underproduced party. This arrangement is known as *balancing* and is used in most split-stream gas sales agreements. The underproduced party also can be paid in cash, instead of in kind, to make up the imbalance.

Making up a connection

MAKEUP TORQUE
The power necessary to screw a joint of pipe into another sufficiently tight to hold and not loosen under working conditions.

MAKING HOLE
Progress in drilling a well, literally.

MALE CONNECTION
A pipe, rod, or coupling with threads on the outside circumference.

Male connection

MANHOLE
A hole in the side of a tank or other vessel through which a man can enter. Manholes have fitted covers with gaskets that are kept bolted in place when the tank is in use.

MANIFEST
A document issued by a shipper covering oil or products to be transported by truck.

MANIFOLD
An area where pipelines entering and leaving a pumping station or tank farm converge and that contains all valves for controlling the incoming and outgoing streams.

Automated production
manifold

MANIFOLD, UNDERSEA TRAWLABLE

A subsea production manifold with a protective structure over its valves and connectors which are keenly vulnerable to trawlers' lines and seins, or other lines or cables dragged across the seafloor.

MAN RACK

A portable doghouse or cab mounted on a flatbed truck for transporting pipeline workers to and from the job.

MAN-RIDING WINCH

The designation for a type of winch used offshore for man-riding applications, such as diving bells, divers' baskets, and "bird cages." The winches, some air-operated, have a built-in safety factor of 8:1 with special steel cable and dual braking systems. *See* Bird Cage.

MANTLE

That part of the Earth that lies beneath the crust and the lithosphere extending downward about 1,800 miles to the outer core, and which is approximately 4,200 miles in diameter. Iron and magnesium-bearing minerals comprise the mantle, which is much denser than the overlying lithosphere. Since the mantle has greater density, seismic waves travel through it at greater velocity than through the lithosphere.

MAP, AREAL

A geologic map showing the lateral or horizontal extent and location of rock units on the surface of a given area.

MAP, BASE

A map that contains latitude and longitude lines, land and political boundaries, rivers, lakes, and major cities.

MAP, CONTOUR

A map showing land-surface elevations by the use of contour lines. Structure-contour maps are used by geologists and geophysicists to depict subsurface conditions or formations. *See* Map, Isopachous.

MAP, GEOLOGIC

A map that records geologic information such as the distribution, nature, and age of rock units, and the occurrence of structural features such as faults, folds, and joints. It may also show mineral deposits, if any, and fossil areas. The map may indicate specific geologic structures by means of outcrops, noting the direction and dip angles. These are shown by conventional symbols, hieroglyphs to the layman.

MAP, GRAVITY

Results of reconnaissance gravity surveys; display of gravity measurements taken in an area. *See* Gravimeter.

MAP, INDEX

A small-scale map that shows the location of, or specific information regarding, a small area on a larger map and pinpoints special features. For example, the small-scale map, often superimposed in the corner of a larger map, may show an area of exploration and drilling activity in relation to the main surface features, i.e., streams, roads, lakes, and villages.

MAP, ISOPACHOUS

A map that shows the thickness of a formation, a bed, or other layered body of rock in a geographic area by means of isopaches (points of equal thickness) connected by lines on the map; a kind of subterranean contour map readily understood by geologists.

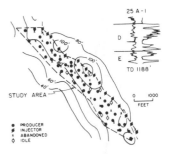

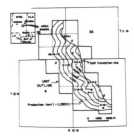

Isopachous map *(After Gates and Brewer)*

MAP, RELIEF

A model of an area in which variation in the surface is shown in relief; a three-dimensional model of a surface area.

MAP, STRUCTURE-CONTOUR

A map that shows subsurface configurations by means of contour lines similar to a surface-contour map; tectonic map.

MAP, SURVEY

A map which contains geologic information of the surface and/or the subsurface.

MAP, TECTONIC

A map that shows the architecture of the earth's crust. It is similar to a structure-contour map, but in addition to showing faults, folds, and other displaced formations, a tectonic map also shows the kinds of rocks the structures are made of, and some indication of their ages and presumed development.

Structure-contour map
(Courtesy Mace)

MAP, TOPOGRAPHIC

A map that shows, in detail, the physical features of an area of land, including rivers, lakes, streams, roads.

MARBLE

A metamorphic rock consisting of fine- to coarse-grained recrystallized calcite or dolomite whose crystals are of uniform size and, to the naked eye, look like grains of sugar in a sugar cube. In common usage, any crystallized carbonate rock that will take a polish and can be used in construction or ornamental stone work.

Map, Topographic

MARGINAL STRIKE

A discovery well on the borderline between what is considered a commercial and a noncommercial well; a step-out well that may have overreached the pool boundary.

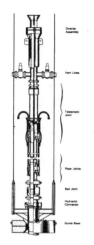

Marine riser system

MARGINAL WELL

A low-producing well, usually not subject to allowable regulations.

MARINE OIL

Petroleum found by wells offshore or on the continental shelf.

MARINE RISER

A string of specially designed steel pipe that extends down from a drillship or floating platform to the subsea wellhead. Marine risers are used to provide a return fluid-flow conductor between the well bore and the drill vessel and to guide the drillstring to the wellhead on the ocean floor. The riser is made up of several sections, including flexible joints and a telescoping joint to absorb the vertical motion of the ship caused by wave action.

MARINE WHITE GASOLINE

Gasoline made for camp stoves, lanterns, blow torches, boat motors. Marine white contains no tetraethyl lead or other additives that could clog the needle valves of gasoline appliances.

MARKER BED

See Key Bed.

MARKER CRUDE

Top-quality crude oil against which all other crudes in a country or region are measured for pricing and quality. For example, Arabian Light crude is a marker crude oil in the Middle East.

MARKET-OUT PROVISIONS

Clauses in a gas contract that permit renegotiating the contract terms by producers or gas pipelines when the market conditions or costs of production change substantially. Gas contracts are usually long-term; without such provisions for review and modification, the supplier or transmission company may be locked into an unsatisfactory arrangement for years.

MARL

A general term for a variety of materials occurring as loose, earthy deposits consisting mainly of a mixture of clay and calcium carbonate. Marl is usually gray but is sometimes white. It is used as an additive or dressing for soils deficient in lime; a clayey limestone.

MARSH BUGGY

A tractor-like vehicle whose wheels are fitted with extra-large rubber tires inflated with air for use in swamps. The great, balloon-like tires are 10 or 12 feet high and 2 or 3 feet wide, providing buoyancy as well as traction in marshland. The marsh buggy is indispensable in exploration work in swampy terrain.

Marsh buggy

MASSAGING

The evaluation and reevaluation, the mulling over of data or information relative to a drilling prospect; for example, when a decision cannot or will not be made. Massaging is often construed as stalling, waiting for even more information with which to make a judgment. A pejorative term.

MASS-FLOW GAS METER

A gas meter that registers the quantity of gas in pounds, which is then converted to cubic feet. Mass-flow meters, which are somewhat more accurate than orifice meters, are used in many refineries where large volumes of gas are consumed.

Drilling mast

MAST

A simple derrick made of timbers or pipe held upright by guy wires; a sturdy A-frame used for drilling shallow wells or for workover; a gin pole.

MASTER BUSHING

The large bushing that fits into the rotary table of a drilling rig into which the kelly bushing fits. When the kelly bushing is lifted out of the master bushing, tapered slips are then inserted around the drillpipe to hold it securely while another joint is added to the drillstring.

MASTER GATE VALVE

See Valve, Master Gate.

Master bushing with kelly bushing above

MATING PARTS

Two or more machine or equipment parts made to fit and/or work together, e.g., piston and cylinder, pump plungers and liners, or sucker rod box and pin.

MATRIX

The fine-grained material filling the interstices (spaces) between the larger particles or grains in a sedimentary rock; the natural material surrounding a sedimentary particle. A tasty example: a raisin-nut cake. The cake is the matrix; the raisins and nuts are the "sedimentary" particles surrounded by the matrix.

MATRIX ACIDIZING

Treating the (usually) fine-grained material surrounding the larger grains of clastic material with acid to enlarge pores, create fissures, and enlarge indigenous cracks.

MATRIX POROSITY

The porosity of the finer part of a carbonate rock as opposed to the porosity of the coarser constituents of the rock.

MATS, CONCRETE ANTI-SCOUR

Anti-scour installations laid down over buried pipeline for scour-erosion protection. Interlocking or wire-connected blocks of concrete have proven effective over buried subsea lines as well as for stabilizing river banks at pipeline crossings.

MAT STRUCTURE

The steel platform placed on the seafloor as a rigid foundation to support the legs of a jackup drilling platform.

MAT-SUPPORTED DRILLING PLATFORM

A self-elevating (jackup) offshore drilling platform whose legs are attached to a metal mat or substructure that rests on the seafloor when the legs are extended.

MATTOCK

A tool for digging in hard earth or rock. The head has two sharpened steel blades; one is in the shape of a pick, the other the shape of a heavy adz.

MATURE FORMATION

A formation from which mobile oxides (such as soda) and leachable, weatherable material have disappeared, e.g., sandstone from which the clays are absent; said also of clastic sediment evolved from its parent rock by processes acting over a very long time and characterized by stable minerals such as quartz.

MAVERICK GAS

Shallow, high-pressure gas that is hard to corral because it is often drilled into with little warning, and at shallow depths there is insufficient head of drilling mud (hydrostatic head) to prevent the initial, often destructive, kick. *See* Diverter System.

MAXIMUM EFFICIENT RATE (M.E.R.)

Taking crude oil and natural gas from a field at a rate consistent with "good production practice," i.e., maintaining reservoir pressure, controlling water, etc.; also the rate of production from a field established by a state regulatory agency.

MAZUT

The Russian word for a heavy fuel oil, for factory boilers, etc.

M.C.F. AND M.M.C.F.

Thousand cubic feet (Mcf); the standard unit for measuring volumes of natural gas. MMcf is one million cubic feet.

M.E.A.

Short for monoethanolamine, an organic base used in refining operations to absorb acidic gases in process streams. Also D.E.A., diethanolamine, another common organic base with an uncommon name.

Refinery coolers for a stream of DEA

MEASURE, UNITS OF

LENGTH

1 centimeter	= 0.3937 inches	= 0.0328 feet
1 meter	= 39.37 inches	= 1.0936 yards
1 kilometer	= 0.6213 miles	= 3,280 feet
1 foot	= 0.3048 meters	
1 inch	= 2.54 centimeters	
1 mil	= 0.001 inch	

SQUARE MEASURE

1 sq centimeter	= 0.1550 sq inches	
1 sq meter	= 1.196 sq yards	= 10.784 sq feet
1 sq kilometer	= 0.386 sq miles	
1 sq foot	= 929.03 sq centimeters	
1 sq mile	= 2.59 sq kilometers	
1 sq inch	= 1 million sq mils	

MEASURED DEPTH

The depth of the borehole of a directional well, a well purposely drilled at an angle from the vertical. Measured depth of a well may exceed true vertical depth by 20 to 30 percent. *See* Drilling, Horizontal.

MEASUREMENT WHILE DRILLING

See Downhole Measurement While Drilling.

MECHANICALLY UNSTABLE FORMATIONS

Friable, easily broken, mechanically weak rock. Said of unconsolidated material.

MECHANICAL RIG

See Rig, Mechanical.

MECHANICAL WEATHERING

See Weathering.

MEDIA BED

A filter bed; the filtering material through which a fluid gravitates or is pumped to remove impurities or suspended material. Filter beds can consist of sand, charcoal, walnut shells, or special clays.

MEGGER

Colloquial shorthand for megaohm meter, a meter for checking the insulation on a submersible electric motor when it is undergoing routine maintenance. The test can readily indicate whether or not the motor (100 hp, usually) is wet. If wet, it is sent to a dryout area.

MEMBER

A subordinate layer or deposit in a formation, e.g., a gravelly member or a clay member of a certain formation. *See* Formation.

M.E.R.

Maximum efficient rate (of production).

MERCAPTANS

Chemical compounds containing sulfur, present in certain refined products that impart objectionable odor to the product.

MERCHANTABLE OIL

Oil (crude) of a quality as to be acceptable by a pipeline system or other purchaser; crude oil containing no more than 1 percent B.S.&W.

MERCURY NUMBER

A measure of the free sulfur in a sample of naphtha. Mercury is mixed with a sample and shaken, and the degree of discoloration in the sample is compared with a standard to determine the mercury number.

METAMORPHIC ROCK

Rocks formed by the metamorphosis of other rocks. When either igneous or sedimentary rocks are subjected to enough heat, pressure, and chemical action, their character and appearance are changed. These factors act to cause recrystallization of the minerals of the rock. Granite may become gneisses or schists; sandstones become quartzites; shales become slates; limestone becomes marble.

METAMORPHISM

Changes in rock induced by pressure, heat, and the action of water that results in a more compact and highly crystalline condition.

METEORIC WATER

Part of the hydrologic cycle. It comes from rain and snowmelt, percolates into the subsurface through permeable soil and rock, and then returns to the surface in springs and seeps.

METER CHART

A replaceable paper chart for recording pressure or flow for a 24-hour period. As the chart revolves on its spindle, an inked pen traces the variations in pressure or volume.

Meter chart *(Courtesy Acco)*

METES AND BOUNDS
The boundaries of a tract of land, especially the boundaries of irregular tracts of land—claims, grants, or reservations—in which the direction and length of each boundary line are stated and in which the lines are located by reference to natural or man-made features of the landscape, such as a large rock, stream, road, or fence.

METHANE
A colorless, odorless, flammable gas (CH_4). Methane is the main constituent of natural gas, which is produced as free gas, and also associated with crude oil as it comes from the well. The simplest saturated hydrocarbon.

METHANE PROFILER, SUBSEA
Sea water at various depths are analyzed with gas-purging methods to detect a natural gas or methane source beneath the seafloor. This passive exploration tool was time consuming and subject to a high degree of error. A new method, developed by a German firm, is much more accurate because it continuously samples the seawater for evidence of hydrocarbons. Other uses for the profiler are detecting gas leaks around subsea wellheads, pipeline manifolds, and risers.

METHANE-RICH GAS PROCESS
See M.R.G. Process.

METHANOL
Methyl alcohol; a colorless, flammable liquid derived from methane (natural gas).

METRIC SYSTEM CONVERSION

inches	× 0.0254	= meters
feet	× 0.305	= meters
miles	× 1609	= meters
miles	× 1.609	= kilometers
millimeters	× 0.03937	= inches
centimeters	× 0.3937	= inches
meters	× 39.37	= inches
meters	× 3.281	= feet
kilometers	× 0.621	= miles
sq centimeters	× 0.155	= sq inches
sq meters	× 10.764	= sq feet
cu centimeters	× 0.061	= cu inches
liters	× 0.2642	= gallons
gallons	× 3.78	= liters

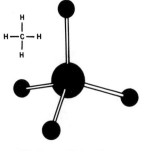

Methane Molecule

METRIC SYSTEM PREFIXES

micro	= one-millionth	hecto	= one hundred
milli	= one one-thousandth	kilo	= one thousand
centi	= one-hundredth	myria	= ten thousand
deci	= one-tenth	mega	= one million
deca	= ten		

METRIC TON

A unit of weight equal to 1,000 kilograms or 2,204.6 pounds. A metric ton of oil is 6.5 to 8.5 barrels, depending upon the oil's gravity. A good approximation is 7.5 barrels of oil is one metric ton. In Europe and the Middle East production and refining, throughput figures are expressed in tons of crude or products instead of barrels as in the United States.

MICA AND LIQUID CASING

Two kinds of lost-circulation material used in the drilling fluid or spotted at the offending interval to retard or, ideally, to stop the loss of drilling fluid into cracks, fractures or crevices.

MICELLAR-POLYMER FLOODING

See Micellar-Surfactant Flooding.

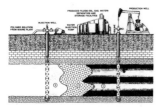

Polymer Flooding

MICELLAR-SURFACTANT FLOODING

A tertiary-recovery technique; a method of recovering additional crude oil from a field depleted by conventional means including repressuring and waterflooding. Micellar-surfactant drive or flooding involves injecting water mixed with certain chemicals into the producing formation. The chemical solution reduces the surface tension of the oil clinging or adhering to the porous rock, thus "setting the oil free" to be pumped out with the flooding solution. Such a project may have various names, e.g., micellar, micellar-polymer, soluble-oil, petroleum-sulfonate.

MICRO-ANNULI

Minute crevasses and invisible pores in near-surface unconsolidated formations or material that permit vertically migrating gas to reach the surface. *See* Filtrate Sweep.

MICROBALLOONS

A foam blanket that floats on the liquid in storage tanks to reduce losses from evaporation. The blanket is composed of billions of hollow, balloon-like plastic spheres containing a sealed-in gas—usually nitrogen. The spheres are almost microscopic in size. When poured in sufficient quantity on top of crude oil or refined products in a tank, they spread across the surface, forming a dense layer that is effective in reducing evaporation.

MICROBAL ENHANCED
Refers to fluids into which microorganisms have been introduced to bi-
ograde hydrocarbon substances. Such enhanced fluids are formulated
and used to clean out clogged production tubing, casing, and the well
bore itself. The microbes literally eat the solid substances, principally
paraffins and sludge, and the remaining converted material is easily
drained or washed out.

MICROEMULSION
An emulsion whose droplets or colloidal are smaller in size; the system
of a mixture of microdroplets in an immiscible liquid.

MICROEMULSION FLOODING
In certain secondary-recovery missions, a microemulsion is used to
flood a tight formation. The immiscible droplets in a normal emulsion
are too large to flush the tight formation with low permeability.

MICRON
A unit of measure equal to one-thousandth of a millimeter. Fines and
other low-gravity solids in drilling mud are described as being so many
microns in size (10 microns, for example) and must be removed from the
circulating mud by the use of a desilting device.

MICRO PALEONTOLOGIST
A scientist who studies the smallest specimens of plant and animal life
in past geologic ages. His subjects are not observable to the naked eye
so he studies them under the microscope.

MIDCONTINENT AREA
Generally, the area lying between the Rocky Mountains and the
Mississippi River, the Canadian border and the Gulf of Mexico, or more
specifically, the greater part of Kansas and Oklahoma, all of Texas except the coastal belt, southeastern New Mexico, northern Louisiana,
southern Arkansas, and western Missouri.

MIDCONTINENT CRUDE
Oil produced principally in Kansas, Oklahoma, and North Texas.

MIDCONTINENT LEASE
A term descriptive, generally, of an Unless Lease or Producer's 88 Lease,
but not descriptive of any particular lease. *See* Lease, Unless; *also*
Producer's 88 Lease.

MIDDLE DISTILLATES
The term applied to hydrocarbons in the so-called middle range of refinery
distillation, e.g., kerosene, light diesel oil, heating oil, and heavy diesel oil.

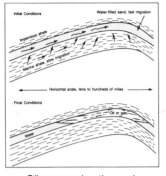

Oil or gas migration and accumulation

Taper mill

Pilot mill

Skirted junk mill

MIDNIGHT REQUISITION
Obtaining material without proper authority; borrowing unbeknown to the "lender"; swiping for a "good" cause.

MIGRATION (seis)
A seismic-processing technique whereby an adjustment is made for naturally occurring distortions in the seismic section or map. After migration, apparent structures are often reduced in size, but the angle of the reflecting planes are increased.

MIGRATION, LOCAL
The underground movement of oil and gas as the result of a difference in pressure in the producing reservoir. When drilling occurs in one part of the reservoir reducing the reservoir pressure, oil and gas tend to move or migrate toward the area of reduced pressure—the boreholes. Local drainage.

MIGRATION, PRIMARY
The movement of hydrocarbons upward from the source beds or source rocks where the oil and gas were formed. The gases and liquids percolate upward to permeable reservoir rocks where they are trapped by impermeable layers, a cap rock. There they remain until discovered by some intrepid wildcatter.

MIGRATION, SECONDARY
The movement of hydrocarbons within the porous and permeable reservoir rocks that results in the segregation of the oil and gas in different parts of the formation. Lighter hydrocarbon fractions (gas) break out or separate from the liquids (oil) to form gas caps or gas reservoirs. If the formation pressure is extremely high, the gas may not be able to break out of solution. In this case, the gas remains in solution until the reservoir pressure is reduced by drilling. An example of gas remaining in solution until the pressure is released is the opening of a carbonated soft drink bottle; the fizzing is the CO_2 escaping or breaking out of the solution.

MILL
To grind up; to pulverize with a milling tool.

MILLABLE
Said of material used downhole, i.e., packers, bridges, and plugs "soft" enough to be bored out or pulverized with a milling tool.

MILL A WINDOW, TO
To mill or cut a hole in the casing with a drillstring-conveyed milling tool. Milling a window usually signifies a kickoff point for a deviated, extended reach or horizontal well. To begin the necessary angle build-

ing, a retrievable whipstock is run to shunt the drill bit from vertical to a sail angle of a few degrees.

MILLIDARCY
A unit of permeability of a rock formation; one-thousandth of a darcy. *See* Darcy.

MILLING
Cutting a "window" in a well's casing with a tool lowered into the hole on the drillstring.

MILLING TOOL
A grinding or cutting tool used on the end of the drill column to pulverize a piece of downhole equipment or to cut the casing.

Milling tools

MILL SCALE
A thin layer or incrustation of oxide that forms on the surface of iron and steel when it is heated during processing. Pipelines must be cleaned of mill scale before being put in service carrying crude oil, gas, or products. This is done by running steel-bristle pigs and scrapers.

MILL-TOOTH BIT
See Bit, Mill-Tooth.

Mill-tooth drill bit

MINERALOGY
The study of minerals: their formation, occurrence, properties, composition, and classification.

MINERAL SPIRITS
Common term for naphthas (solvents) used for dry cleaning and paint thinners.

MINIMUM TENDER
The smallest amount of oil or products a pipeline will accept for shipment. Regulations set minimum tender amounts a common carrier pipeline is required to take into its system and pump to destination.

MINI-SEMI
A scaled-down semisubmersible drilling platform built for service in relatively shallow water.

MISCIBLE
Mixable; fluids that are capable of dissolving in one another.

MISCIBLE FLOOD
A secondary or tertiary oil recovery method in which two or more formation-flooding fluids are used, one behind the other. For example,

carbon dioxide may be injected into the formation followed by water-flooding. *See* Tertiary Recovery.

MIST
Small, almost microscopic droplets of water entrained in natural gas. Such gas must be treated to remove the water before it will be accepted by a gas-transmission pipeline.

MIXER, TANK
See Tank Mixer.

M.M. BTU, $
So many dollars per million Btu; a pricing formula in some gas purchase contracts that is tied directly to formulas involving prices paid for No. 2 fuel oil at specific locations in the U.S.; $/M.M. Btu.

M.M. BTU/HR
Million Btu (British thermal units) per hour; rating used for large industrial heaters and other large thermal installations such as furnaces and boilers.

M.M.S. MINERALS MANAGEMENT SERVICE
An arm of the Department of Energy that monitors federal and Indian land leases with regard to royalty owed the U.S. government.

MOBILE PLATFORM
A self-contained, offshore drilling platform with the means for self-propulsion. Some of the larger semisubmersible drilling platforms are capable of moving in the open sea at 5–7 knots.

MOBILE PRODUCTION UNITS
More-or-less temporary production units offshore that serve as oil- and gas-processing stations with oil storage and loading facilities. Mobile units are often used to handle the production from subsea and platform wells. There are several reasons why temporary mobile production units may prove to be economically desirable: (1) To obtain production data to justify permanent facilities; (2) To reduce field hookup time; (3) To generate cash flow (get some money coming in) while permanent structures are built; (4) To provide an acceptable return on marginal or otherwise noneconomic discoveries.

MOCK-UP
A full-sized structural model built accurately to scale for study and testing of an installation to be used or operated commercially. For deep-water offshore work, mock-ups are made to simulate conditions in subsea wellhead chambers and seafloor work areas.

MODULE

An assembly that is functional as a unit and can be joined with other units for increasing or enlarging the function; for example, a gas-compressor module; an electronic or hydraulic module.

MOGAS

A not-so-euphonious contraction of motor gasoline.

MOHS SCALE

A standard by which the hardness of minerals can be rated. The scale includes 10 minerals listed from softest to hardest: talc, gypsum, calcite, fluorite, apatite, orthoclase feldspar, quartz, topaz, corundum, and diamond.

MOINEAU MOTOR

A positive-displacement mud motor.

MOLDIC POROSITY

Porosity resulting from the removal, usually by solution, of an individual constituent or component of a rock.

MOLECULAR SIEVE

A bed of desiccant material (a drying agent) that absorbs water from a refinery or hydrocarbon recovery plant's feedstock. The superabsorbent material (usually in pellet form) is in layers in a bed. The feedstock is passed through the sieve, at which time it gives up its molecules of water. In a number of refinery processes, notably cryogenic processes, water in the process feedstock can cause untold difficulties. Specifically, the molecular sieve is the desiccant material itself.

MOMENTUM KILL OF BLOWOUT

In certain instances, a well blowout can be stifled and finally killed by a concerted effort of the rig crew after evaluating the blowout, its volume, density, temperature, and the general condition of the well. A kill fluid is pumped through an improvised stinger—a few joints of pipe made up and inserted in the well at the wellhead. The weighted drill fluid is pumped through the stinger at a high rate. The large volume of mud contacts the gas/water or gas/oil flow, slowing it and finally choking it off—a practical application of fluid dynamics. If, however, the blowout is too big, too powerful, a momentum kill will not be successful. A killer well may have to be drilled to flood the formation, choking off the flow, that obviously is more than just a kick. *See* Bullhead the Gas Back, To.

M.O.N.

See Motor Octane Number.

Monkey board

MONEY LEFT ON THE TABLE
A phrase referring to the difference between the highest and the second-highest bid made by operators or companies when bidding on federal or state oil leases. For example: high bid, $1,000,000; second-highest bid, $750,000. Money left on the table, $250,000.

MONKEY BOARD
A colloquial and humorous reference to the tubing board high in the derrick.

MONKEY WRENCH
An adjustable, square-jawed wrench whose adjusting screw-collar is located on the handle and whose head can be used as a hammer; a crude wrench suitable for mechanical work of the roughest kind.

MONOBORE WELL COMPLETIONS
Drilling one-size hole, top to bottom, and running one-size casing, and in some instances, no tubing. The advantages: the chances of damaging the formation are reduced; larger perforating guns can be used; workovers can be done by wireline without shutting in the well. Also, a drilling unit is not required for intervention, and a minimum of equipment is left in the hole; a full bore helps to keep well pressure up, and pipe wall scaling is much less. Monobore for shallow and medium-depth wells.

MONOCLINE
A geological term for rock strata that dip in one direction. When the crest of an anticline is eroded away, a partial cross section of the strata making up the fold is exposed at the earth's surface and the undisturbed lower flanks form what are called monoclines.

MONOLITH
A large, upstanding mass of rock, sometimes a volcanic spire; a sky-scraper rock; also, a piece of exposed bedrock.

MONOPOD PLATFORM
A type of offshore drilling platform with a single supporting leg. The design of the monopod makes it effective in arctic regions where thick, moving bodies of ice present serious problems for more conventional platforms.

MONUMENT
In surveying, a natural structure (large exposed rock) or man-made structure (road, fence, or cairn) that marks the location or a corner, or other survey reference point.

MOONPOOL
The opening in a drillship through which drilling operations are conducted; the moonpool or drillwell is usually located amidship, with the derrick rising above.

MOORING SYSTEM, MULTIBUOY
See Multibuoy Mooring System; *also* S.B.M.; Single-Buoy Mooring.

MOOSE AND GOOSE MEN
A humorous and somewhat sarcastic term for conservation (Environmental Protection Agency) people who, by law, can shut down a drilling well or a construction project to allow a rare or endangered species of bird to incubate her eggs unmolested, or migrating or mating moose to go about their important business without being disturbed.

Multibuoy mooring
system

MOPE POLE
A lever; a pry pole usually made by cutting a small tree; used on pipeline construction as an adjunct to the jack board and in lowering the pipeline into the ditch.

MORAINE
A ridge, mound, or other distinct accumulation of loose, unconsolidated, unstratified glacial drift (clay, sand, gravel, stones, and boulders) deposited by a glacier when it retreated (melted away) thousands of years ago; a heap of earth and stony debris transported and left behind by a glacier during and after the Ice Age, or Glacial Epoch.

MORMON BOARD
A broad, reinforced, sled-like board with eye bolts on each end and a handle in the center. Used to backfill a pipeline ditch using a team of horses or a tractor pulling the board forward and a workman pulling it back into position for another bite.

MORNING REPORT
The report the toolpusher or drilling supervisor makes each morning after assembling the drilling reports of the drillers under his supervision. The report includes depths reached at the end of each tour, footage drilled, mud records, formations penetrated, bit weights, rotary speeds, cores taken, pump speeds and pressures, and other pertinent information of the past 24 hours of operation.

MOTHER ROCK
See Country Rock.

MOTION COMPENSATOR

A hydraulic tensioner system for offshore drilling from a "floater," a drillship or semisubmersible, that maintains constant tension or support for the drillstring in the hole. As the platform moves up and down on ocean swells, the motion-compensator mechanism slacks off or takes up on the drillstring, thus maintaining a constant and proper weight on the drill bit at the bottom of the hole.

MOTOR OCTANE NUMBER

The measure of a gasoline's antiknock qualities, whether or not it will knock or ping in an engine with a given compression ratio. Motor octane number of a gasoline is determined by test engines run under simulated conditions of load and speed. *See* Octane Rating.

MOTOR SPIRIT

A highly volatile fraction in petroleum refining; an ingredient of motor gasoline.

Mousehole

MOUSEHOLE

A hole drilled to the side of the well bore to hold the next joint of drillpipe to be used. When this joint is pulled out and screwed onto the drillstring, another joint of drillpipe is made ready and slipped into the mousehole to await its turn. *See* Rathole.

M.R.G. PROCESS

Methane-rich gas process. M.R.G. is a patented process (Japan Gasoline Co.) to make synthetic natural gas from propane. Liquid propane is hydrodesulfurized and gasified with steam at temperatures between 900° and 1000°F. The resulting gas mixture is methanated, scrubbed to remove CO_2, dried, cooled, and fed to distribution lines.

M.S.C.F.

Thousand standard cubic feet of gas (Mscf). A standard cubic foot of gas is gas volume corrected for standard temperature and pressure. *See* Gas Measurement, Standard.

M.T.B.E.

Methyl tertiary butyl ether, one of the important oxygenates for use in reformulating gasoline to reduce noxious emissions to meet new, stiff regulations of the Clean Air Act, C.A.A.

MUD

See Drilling Mud.

MUD, OIL-BASE

See Oil-Base Mud.

MUD, OIL-CUT
See Oil-Cut Mud.

MUD AGITATOR
An electric motor-driven impeller for use in drilling-mud tanks to keep the weighting material in suspension and thus maintain uniform mud weight throughout the tank. *See* Tank Mixer.

MUD BARREL
A small bailer used to retrieve cuttings from the bottom of a cable-tool drilling well.

MUD-BASE CEMENTS
The conversion of drilling muds into cement suitable for oilfield use had been an area of interest for decades. In the past few years, some progress has been made converting certain drilling mud formulations to practical cement by introducing a hardening agent. Shell Oil has developed a patented mud solidification formula to overcome limitations of other conversion methods. The method employs high-quality blast furnace slag (B.F.S.) as the cementitious or hardening agent. Certain drilling muds can be solidified by the addition of 40–50 lbs./bbl. of B.F.S. to a water-base mud. Thickening time, flow qualities, and compressive strength can be controlled by the addition of common mud additives such as sodium hydroxide lignosulfate thinners, and sodium carbonate. Mud solidification makes possible significant saving in cost and an environmental plus by the convenient and timely reuse of drilling fluids downhole.

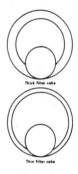

Thick filter cake

Thin filter cake

Mud cake

MUD CAKE
See Filter Cake.

MUD COLUMN
The height of the column of drilling mud standing in the borehole of a well; the column of drilling mud from bottom to top.

MUD-COOLING TOWER
In drilling in or near a geothermal reservoir, the drilling mud becomes superheated and must be cooled to avoid flashing or vaporizing of the liquid (water or oil) in the mud stream at the surface. Cooling also reduces the thermal stress on the drillstring.

MUD CUP
A device for measuring drilling mud density or weight; a funnel-shaped cup into which a measured quantity of mud is poured and allowed to run through, against time.

Mud hog

Mud hose *(Courtesy B.F. Goodrich)*

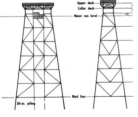

Mud line

MUDDED OFF

Said of a well whose producing formation has been blocked or plugged by the buildup of filter cake or the choking off of the permeability of the reservoir formation by the drilling mud being forced into the formation. This is always a threat because the column of mud exerts thousands of pounds of pressure per square inch on the face of the porous producing formation. As each foot of the mud column exerts more than one-half pound per square inch, a 5,000-foot column of mud in the hole exerts about 3,000 pounds per square inch on the face of the porous, producing formation.

MUD ENGINEER

One who supervises the preparation of the drilling mud, tests the physical and chemical properties of the slurry, and prepares reports detailing the mud weight and additives used. A drilling fluid specialist.

MUD HOG

A mud pump; a pump to circulate drilling mud in rotary drilling; slush pump.

MUD HOPPER

A drilling mud-mixing device consisting of a vessel, a hopper, in the general configuration of an inverted, truncated pyramid. The lower end (the small end) of the hopper is attached to and opens into a tube or pipe through which water or other liquid is pumped at high pressure. As the dry ingredients are dumped into the hopper, they are drawn into the jetting stream of liquid and mixed into a slurry.

MUD HOSE

The flexible, steel-reinforced rubber hose connecting the standpipe with the swivel and kelly joint on a drilling rig. Mud is pumped through the mud hose to the swivel and down through the kelly joint and drillpipe to the bottom of the well.

MUD LINE

The sea or lake bottom; the interface between a body of water and the Earth.

MUD LOG

See Log, Mud.

MUD LUBRICATOR

See Lubricator, Mud.

MUD MOTOR

A downhole drilling motor that derives its power from the force of the drilling mud forced through it by the mud pumps at the surface. Mud motors (turbine and positive displacement) are located at the lower end of the drillstring just above the drill bit. In mud-motor drilling, the drillstring does not turn; only the drill bit rotates.

MUD MOTOR, POSITIVE-DISPLACEMENT
A downhole drilling motor that turns the drill bit through the force of the drilling mud pumped at high pressure through the motor. Based on the Moineau principle, positive-displacement mud motors convert the flow of mud into rotational power, and this power rotates the drill. The motor consists of a long, eccentric rotor inside a close-fitting tube of equal length (the stator). The mud motor is attached to the lower end of the drillpipe just above the bit. The drillpipe does not turn, serving only as support for the motor and bit and as a conduit for the drilling mud.

MUD MOTOR, TURBINE
A type of mud motor in which the drilling mud, upon entering the pump, exerts force on the multiple blades or vanes attached to an axial spindle. As the mud is pumped at high pressure though the motor, the vanes and spindle rotate turning the drill bit. *See* Mud Motor, Positive-Displacement.

MUD-MOTOR DRILLING
See Turbodrilling.

MUD PIPE
A string of casing set through layers of mud and unconsolidated sediment in an offshore location to serve as support and as a stiffening element for the surface pipe that is run inside the mud string. Whenever a couple of hundred feet of soft mud is encountered at a well location, a mud string is run through the soft material down to a supporting formation and cemented in. Then the surface pipe to which the blowout preventers and control valves are connected is run inside the mud string. The mud pipe is, in effect, an outside liner for the surface pipe.

Mud pit

MUD PITS
Excavations near the rig into which drilling mud is circulated. Mud pumps withdraw the mud from one end of a pit as the circulated mud, bearing rock chips from the borehole, flows in at the other end. As the mud moves to the suction line, the cuttings drop out, leaving the mud "clean" and ready for another trip to the bottom of the borehole. *See* Reserve Pit.

MUD PUMP
See Pump, Mud.

MUD SCOW
A portable drilling-mud tank in the shape of a small barge or scow used in cable-tool drilling when relatively small amounts of mud were

Mud pump *(Courtesy Gardner's Denver)*

needed or in a location when a mud pit was not practical. Also, a conveyance, a kind of large sled for transporting pipe and equipment into a marshy location. The mud scow is pulled by a crawler-type tractor that would not bog down as would a wheeled vehicle.

MUDSTONE
A well-packed, hard mud with the texture and composition of shale but lacking shale's fine layering and ability to split along definite planes; a fine-grained sedimentary rock consisting of clay and silt, about in equal parts. Mudstone, as the name suggests, will turn into a gritty mud when wet with water. Mudstones are usually gray to very dark gray in color.

Mud tanks

MUD TANK
Portable metal tank to hold drilling mud. Mud tanks are used where it is impractical to dig mud pits at the wellsite.

MUD THINNERS
Chemical agents or additives to freshwater-based drilling muds that prevent or retard flocculation of the mud solids.

MUD UP
In the early days of rotary drilling and before the advent of accurate well logging, producible formations could be mudded up (plastered over) by the sheer weight of the column of drilling mud, so said the cable-tool men who were skeptical of the newfangled drilling method. Mudding up also occurs in pumping wells. The mud may be from shaly portions of the producing formation, from sections of uncased hole, or from the residue of drilling mud.

Mud mixer in drilling mud tank

MUD VALVE, AUTOMATIC
See Valve, Lower Kelly.

MUD WEIGHT
Weight is an important property of drilling mud because the weight of the mud has much to do with drilling rates, proper circulation, and well or hole control. Drilling mud weight is determined by the amount of weighting material added to the slurry. Eight pounds to the gallon is light, while 16 pounds to the gallon is very heavy mud. *See* Mud Engineer.

Mule skinner and team

MULE SKINNER
Forerunner to the truck driver; a driver of a team or span of horses or mules hitched to an oilfield wagon. Unhitched from the wagon, the team was used to pull, hoist, and do earthwork with a slip or Fresno. The skinner got his name from the ability to skin the hair off a mule's rump with a crack of the long reins he used, appropriately called butt lines.

MULLET
Humorous and patronizing reference to an investor with money to put into the drilling of an oil well with the expectation of getting rich; a sucker; a person who knows nothing about the oil business or the operator with whom he proposes to deal.

MULTI-BRANCH LATERAL WELL
A vertical well is taken down to a target zone. Then, at a kickoff point just above the target, one lateral takes off to penetrate the sedimentary section, another lateral is put through in the opposite direction, 180° away, to investigate the same interval. If Saturation, Porosity, and Permeability, the Three Muskateers of all good wells, are in evidence, the multibranch horizontal will be a success.

MULTIBUOY MOORING SYSTEM
A tanker loading facility with five or seven mooring buoys to which the vessel is moored as it takes on cargo or bunkers from submerged hoses that are lifted from the seabottom. Submarine pipelines connect the pipeline-end manifold to the shore.

Multibuoy mooring system

MULTIPAY WELL
See Multiple Completion.

MULTIPLE COMMUNITY LEASE
A term that describes the effect of the execution by the owners of separate parcels of land, each counterparts of a lease describing a larger area than that owned by any individual lessor and providing that the interests of all lessors in the described area shall be considered pooled and unitized.

MULTIPLE COMPLETION
The completion of a well in more than one producing formation. Each production zone will have its own tubing installed, extending up to the Christmas tree. From there the oil may be piped to separate tankage. *See* Dual Completion.

MULTIPLE-ORIFICE VALVE
See Valve, Multiple-Orifice.

Christmas tree of a triple completion well

MULTIPLIER
A device or linkage for increasing (or decreasing) the length of the stroke or travel of a rod line furnishing power for pumping wells on a lease. A beam that oscillates on a fulcrum and bearing to which is attached the rod line from the power source (central power) and a rod line to the pumping well. By varying the distance from the fulcrum of the two rod-line connections, the travel of the well's rod line can be lengthened or shortened to match the stroke of the well's pump.

M.W.D.
See Downhole Measurement While Drilling.

M.W.D. SENSORS, DOWNHOLE
Measurements While Drilling technology continues to evolve, and the downhole sensor combinations available to M.W.D. users continue to grow. At this writing (1994), most M.W.D. tools offer sensors to monitor inclination, azimuth, gamma ray, formation resistivity, density, neutron porosity, weight-on-bit, downhole torque, and annular temperature.

M.W.D. SURFACE SYSTEM
Surface components of a typical M.W.D. system include pressure transducers for signal detection, electronic signal decoding equipment, and various analog and digital readouts and plotters.

M.W.D. TOOL, THE
The M.W.D. tool is designed to generate power, acquire data, and transmit this data uphole to the surface receiving system. Advanced, sophisticated sensors accomplish these things, and equipment at the surface receives and makes intelligible the downhole coded information.

N.A.C.E.
National Association of Corrosion Engineers.

N.A.F.T.A.
The 1992 North American Free Trade Agreement—among the nations of Mexico, the United States, and Canada.

NAMEPLATE RATING
The manufacturers' ratings as to speed (rpm), working pressure, horsepower, type of fuel, voltage requirement, etc., printed or stamped on the makers' nameplates attached to pumps, engines, compressors, or electric motors. To ensure proper and lasting performance of machines and equipment, nameplate ratings are always heeded.

NAMING A WELL
See Well Naming.

NANNOFOSSIL
An extremely small fossil, smaller than a microfossil, which can be studied only under a powerful microscope. Nanno = 1 billionth part.

NANOFILTER
A patented, superfine, selective filter capable of removing certain ions from seawater. The advanced filtration system is used to treat injection water in large-volume, offshore water-flood programs. *See* Nanofiltration.

NANOFILTRATION
An advanced, patented process of removing sulfate ions from seawater to be injected in a water-flood program. The filtration eliminates barium sulfate (barium) precipitation in the flood water at the point of entrance or breakthrough.

NANT
Scandinavian for small stream.

NAPHTHA
A volatile, colorless liquid obtained from petroleum distillation used as a solvent in the manufacture of paint, as dry-cleaning fluid, and for blending with casinghead gasoline in producing motor gasoline.

NAPHTHENE-BASE CRUDE OIL
See Asphalt-Base Crude.

NATIONAL ENERGY BOARD, CANADIAN
Canada's version of the U.S.'s F.E.R.C.

NATIONAL PETROLEUM RESERVE—ALASKA
An area west of Prudhoe Bay field and south of Point Barrow containing millions of acres set aside in 1980 and held in reserve for national security purposes. *See* Naval Petroleum Reserves.

NATIVE GAS
See Gas, Native.

NATURAL CONDENSATE CRUDE OIL
Hydrocarbons that are liquid at atmospheric conditions, but under the original conditions of the reservoir—the pressure and temperature—were in a gaseous state.

NATURAL GAMMA RAY LOGGING
See Log, Gamma Ray.

Natural gas well

NATURAL GAS
Gaseous forms of petroleum consisting of mixtures of hydrocarbon gases and vapors, the more important of which are methane, ethane, propane, butane, pentane, and hexane; gas produced from a gas well.

NATURAL GAS, COMPRESSED
A source of automotive fuel that, in the early 1990s, was gaining acceptance. Automakers were building "dedicated vehicles," i.e., engines designed and built to efficiently burn C.N.G. Other companies were offering conversion services to convert gasoline engines to the new fuel, compressed natural gas.

NATURAL GAS, UNCONVENTIONAL
See Unconventional Natural Gas.

NATURAL GAS DEHYDRATOR
See Dehydrator.

A pair of in-line natural
gas dehydrators

NATURAL GASOLINE
Drip gasoline; a light, volatile liquid hydrocarbon mixture recovered from natural gas. A water-white liquid similar to motor gasoline, but with a lower octane number. Natural gasoline, the product of a compressor plant or gasoline plant, is much more volatile and unstable than commercial gasoline because it still contains many lighter fractions that have not been removed.

NAVAL PETROLEUM RESERVES
Areas containing proven oil reserves that were set aside for national defense purposes by Congress in 1923. The reserves, estimated to contain billions of barrels of crude oil, are located in Elk Hills and Buena Vista, Ca.; Teapot Dome, Wy.; and on the North Slope in Alaska. The term Naval Reserves is too restrictive today, so the reserves are now called National Petroleum Reserves.

NEAT
Refers to a substance or a product that is pure or nearly so; unadulterated; clean. Also, near-neat, i.e., a product containing desirable additives.

NEEDLE COKE
See Coke, Needle.

NEEDLE VALVE
See Valve, Needle.

Needle valves. Cutaway shows needle and seat *(Courtesy Markad Service Co.)*

NEGATIVE-GRAVITY ANOMALY
See Anomaly, Negative-Gravity.

NEGATIVE NOMINATION
The term to describe arguments submitted by state and federal officials that certain offshore tracts should not be open for leasing because of unspecified risks, sensitive ecological conditions, or hazards that make drilling and production unwise from a safety or environmental standpoint.

NEOGENESIS
The formation of new minerals as by diagenesis—deposition, compaction, cementation—lithification, or metamorphism.

NEOPRENE
A rubber-like product derived from petroleum and compounded with natural rubber to produce a substance highly resistant to chemicals and oils. Neoprene, first called polychloroprene, was discovered by W. Carothers, Ira Williams, A. Collins, and J. Kirby of the DuPont research laboratory.

NET-BACK PRICING
A method of determining the wellhead price of oil and gas by deducting from the downstream price, the destination price, the transportation, and other charges that arise between wellhead and point of sale.

NET-BACK TRANSACTION
An arrangement; a transaction whereby the crude seller is paid based on the price of the refined product rather than on a predetermined price for

the crude. The refiner, however, would be guaranteed a fixed profit per barrel with the balance of the net going back to the producer. This method of payment for all parties involved was used in the early 1980s by Saudi Arabia in dealing with Aramco, the Arabian American Oil Co.

NET OIL ANALYZER
See Oil Analyzer, Net.

NET POSITIVE SUCTION HEAD
See Suction Head, Net Positive.

NET PROFITS BIDDING
Competitive bidding for a lease wherein the lease is awarded to the person, the lessee, agreeing to pay the largest share of net profits to the lessor.

NET PROFITS INTEREST
A share of gross production from a property, measured by the net profits from the operation of the property. Such an interest is carved out of the working interest and represents an economic interest in the oil and gas produced from the property. Sometimes referred to as net royalty.

NET REVENUE INTEREST
A fractional share of the working interest not required to contribute to, nor liable for, any part of the expense of drilling and completing the first well on the property or lease. Net revenue is income from a property after all costs including taxes, royalties, and other assessments have been paid.

NET ROYALTY
See Net Profits Interest.

NEUTRAL STOCK
Lubricating oil stock that has been dewaxed and impurities removed and can be blended with bright stock to make good lube oil; one of the many fractions of crude oil that, owing to special properties, is ideal as a blending stock for making high-quality lube oil.

NEUTRON-GAMMA RAY LOGGING
See Neutron Logging.

NEUTRON LOGGING
A process whereby formations bored through by the drill, the walls of the borehole, are bombarded with neutrons. The logging is performed by lowering a neutron-emitting device, a source, along with a detector. The detector produces output signals indicating the radiation emitted from the bombarded formations. The output signals are transmitted to

the surface where a record of the downhole radiation, correlated with the different depths of the detector, is made. A neutron log indicates whether there is fluid in the formation, but it cannot differentiate between oil and water. The geologist and the owner are forced to make an educated guess; an oxymoron, to be sure. As a result of neutron bombardment, a formation may emit either neutrons or gamma rays, or both. When gamma rays are detected, the resulting neutron log is called a *neutron-gamma ray log*. When neutrons are detected, the log is described as a *neutron-neutron log*.

NEUTRON-NEUTRON LOGGING
See Neutron Logging

NEW OIL
For the purposes of price regulation under the Emergency Petroleum Allocation Act of 1973, new oil is the production from a property in excess of production in 1972; all subsequent production from a property producing in 1972. *See* Old Oil.

NEWTONIAN FLOW
Viscous flow: flow of a liquid in which the shear strain is directly proportional to the shear stress.

NEWTONIAN LIQUID
A flowable substance in which the rate of shear strain is directly proportional to the shear stress. This constant ratio is the viscosity of the liquid.

N.G.A.
Natural Gas Act. An act of Congress that empowers the Federal Power Commission to set prices and regulate the transportation of natural gas.

N.G.P.A.
(1) Natural Gas Policy Act of 1978. (2) Natural Gas Processors Association, successor to the Natural Gasoline Association of America.

NIGERIAN NATIONAL OIL COMPANY (N.N.O.C.)
The Nigerian National Petroleum company founded in 1971.

N.I.O.C.
The National Iranian Oil Company, the state-owned energy company of Iran, established in 1951. In 1913, Iran took over virtually full ownership of all petroleum-related activities.

NIPPLE
A short length of pipe with threads on both ends, or with weld ends.

Lift nipples

A typical nippled-up gas well

NIPPLE CHASER
The material man who serves the drilling rig; the person who makes certain all supplies needed are on hand.

NIPPLE UP
To put together fittings in making a hookup; to assemble a system of pipe, valves, and nipples as in a Christmas tree.

NITROGEN GAS INJECTION
An inert fluid (gas) that is injected into a reservoir for gas-cap displacement, for displacement of carbon dioxide slugs, and for gravity drainage, which is forcing oil downward into lower wells on an updip formation. An enhanced oil recovery technique.

NOBLE METAL (CATALYST)
A metal used in petroleum refining processes that is chemically inactive with respect to oxygen.

NOISE LOG
See Log, Noise.

NOMINAL
(1) Very small; not worth mentioning; a nominal service charge. (2) In name only.

NOMINATIONS
(1) The amount of oil a purchaser expects to take from a field as reported to a regulatory agency that has to do with state proration. (2) Information given to the proper agency of the federal government or a state relative to tracts of offshore acreage a person or company would like to see put up for bid at a lease sale.

NOMOGRAPH
A device used by engineers and scientists for making rapid calculations; a graph that enables one, with the aid of a straightedge, to find the value of a dependent variable when the values of two or more independent variables are given.

Nomograph

NONASSOCIATED
Free gas; gas not in contact with the crude oil in the reservoir.

NONBRANDED INDEPENDENT MARKETER
One who is engaged in marketing or distributing refined petroleum products but who is not a refiner, is not affiliated with nor is controlled by a refiner, and is not a branded independent marketer. A branded marketer is similar, but in addition has an agreement with a refiner for the

use of the refiner's brand name, logo, or other identifying marks. Flying Horse; Orange Disc; Chevron; B.P.; Total; 66; Block T; *ad infinitum.*

NONCLASTIC ROCK
Rocks inorganically and organically precipitated from sea or lake waters and those formed by the accumulation of organic material. Inorganically precipitated rocks are made as saturated sea and lake water deposits material in solution. Limestone, chert, and evaporite are formed in this way. Evaporite deposits are formed as water evaporates, increasing the concentration of minerals. The minerals then precipitate out and are deposited. Examples of evaporites are rock salt, gypsum, potassium, and magnesium salts. Organically precipitated nonclastic rocks are formed as the result of the life processes of plants and animals. Marine flora extract carbon dioxide from seawater and cause the precipitation of calcium carbonate. Marine fauna extract calcium carbonate from seawater in the process of forming their shells and exoskeletons. Reefs develop as the skeletal remains accumulate on the seafloor. This type of rock usually has good porosity and permeability, and thus is a very good reservoir for petroleum.

NONCONFORMITY
An unconformity between the layers of sedimentary rock and older igneous or metamorphic rock that was exposed to erosion before the overlying sedimentary rock covered it. A term once widely used for angular unconformity where the older rock was tilted or folded; angular discordance.

NONCONSENT PENALTY
A penalty against a party to a pooling or unitization agreement who chooses not to participate in the cost of drilling a particular well by the operator or by one of the members. The penalty may be in terms of acreage, production, or cash. If the well is productive, industry practice calls for a penalty of 200 to 300 percent of the nonparticipant's proportional drilling and completion costs for development wells, 300 percent or more for wildcats, and as much as 1000 percent in the case of an offshore productive well.

NONCONSENT WELL
A well drilled in a unitized area by the operator of the area or by one of the participating members, to which one member in interest has not consented. The nonconsenting member is not liable for any of the costs if the well is dry; if it is productive, he/she will be entitled to share in the proceeds only on the basis of a nonconsent penalty.

NONDRILLING LEASE
See Lease, Nondrilling.

NONEXCLUSIVE INFORMATION

In the oil patch, nonexclusive info usually means geophysical reports available to everybody, all companies alike. Such reports are important bodies of information for the industry, but individual companies seldom act on such information alone. It's secret, exclusive, "hot" info that gets the oilman's adrenaline pumping and the landman on the way to the acreage to take the lease. *See* Proprietary Information.

NONFERROUS

Containing no iron; nonferrous tools, valves, or rods are made of other metal or combinations of metals, e.g., brass, copper, bronze, spent uranium, or tungsten. Nonsparking tools are made of nonferrous metals, usually brass or bronze, because they are softer and will not give off sparks when struck against another piece of metal.

NONOPERATING INTEREST

An interest in an oil or gas well bearing no cost of development or operation; the landowner's interest; landowner's royalty.

NONOPERATOR

The working-interest owner or owners other than the one designated as operator of the property; a "silent" working-interest owner.

NONPARTICIPATING ROYALTY

See Royalty, Nonparticipating.

NONPIERCEMENT DOME

See Dome, Nonpiercement.

NONSPARKING TOOLS

Hand tools made of bronze or other nonferrous alloys for use in areas where flammable oil or gas vapors may be present.

N.O.R.M.

Natural Occurring Radioactive Materials. Two types of N.O.R.M. contamination concern regulators: radium contamination of oil production piping and attendant facilities, and radon contamination of natural gas processing works. Oilfield N.O.R.M. is brought to the surface in production water as dissolved radionuclides. The saltier the water, the greater the N.O.R.M. concentration. High levels of radiation are found in scale formed on the interior of tubulars as precipitates of carbonates and sulfates of calcium, barium, and strontium. Radon gas is formed by the radioactive decay of radium-226. It is chemically unreactive, but chemically it is similar to ethane and propane so it tends to follow those gas streams during processing. Quantifying radiation is done with three units: the Curie, a measure of total radiation emitted from a radioactive substance; the rad, a unit of absorbed dose; and rem, a measure of potential harm from radiation to the human body.

NORMAL MOVE OUT (N.M.O.)
A term in seismic application referring to the results from the variation in distance from the shot point to each individual geophone in the string or the array. Seismic reflections from the same zone or bedded plain arrive later at the most distant geophone, and so must be compensated for, or a misleading or false reading will result.

NO-TERM LEASE
A lease that may be kept alive indefinitely by the payment of delay rentals. This is an old-type lease agreement. Any contemporary no-term lease probably was conceived in error, someone misread the store-bought form and did not fill in the blanks properly.

N.P.R.
Naval Petroleum Reserve.

N.P.R.–A.
National Petroleum Reserve-Alaska.

N.P.R.A.
National Petroleum Refiners Association.

N.P.S.H.
Net positive suction head.

N.P.T.
National pipe thread; denotes standard pipe thread.

"N" STAMP
Designates equipment qualified for use in nuclear installations: pipe, fittings, pumps, valves, etc.

NUMBER 2 FUEL OIL
Furnace oil; also Two Oil, distillate fuel. No. 2 Oil.

NUMBER 6 OIL
A heavy, low-gravity road oil, including residual asphaltic oils, used as treatment for dirt roads. Residual oils generally are produced in six grades, from zero (the most fluid at atmospheric temperatures) to six (the most viscous).

NUTATING DISK METER
A type of positive displacement flow-meter that has a nutating disk in

the throat of the meter. The disk resembles a butterfly valve on trunions (pin axles). It lies horizontal, parallel to the line of flow of the liquids being pumped and loads and unloads causing the disk to oscillate or nod on its axle. The axle or stem is connected to a counter, which is attached to a numerical register or pulse counter.

NUT CUTTING, DOWN TO THE
The crucial point; the vital move or decision; a "this-is-it" situation.

O.&S.
See Over and Short.

O.A.P.E.C.
Organization of Arab Petroleum Exporting Countries; seven of O.P.E.C.'s thirteen countries in 1968 joined to further the aspirations of the Arab world and to demonstrate unity.

OBJECTIVE DEPTH
The depth to which a well is to be drilled. Drilling contracts often state that the hole shall be drilled to a specified depth or to a certain identifiable formation, whichever comes first, e.g., "to 5,500 feet or the Skinner sand the objective depth."

OBLIGATION WELL
See Well, Obligation.

OBLIQUE SLIP
See Slip.

O.B.O. VESSEL
A specially designed vessel for carrying ore and crude, both in bulk form. The first oil and bulk ore tanker/carrier was launched in 1966 and used in handling relatively small cargoes of oil and ore.

O.C.A.W.
Oil, Chemical, and Atomic Workers Union, a labor organization representing a large number of the industry's refinery and other hourly workers.

O.C.S.
Outer Continental Shelf.

OCTANE RATING
A performance rating of gasoline in terms of antiknock qualities. The higher the octane number, the greater the antiknock quality; e.g., 94 octane gasoline is superior in antiknock qualities to a gasoline of 84 octane.

O.D.
Outside diameter of pipe; O.D. and I.D. (inside diameter) are initials used in specifying pipe sizes, e.g., 4 1/2-inch O.D.; 8 5/8-inch I.D.

An LNG carrier, the *Lachmer Louisiana*.

A tank farm manifold.

ODORANT
A chemical compound added to natural gas to produce a detectable, unpleasant odor to alert householders should they have even a small leak in the house piping. Odorants are used also in liquids or gases being stored or transported to detect leaks.

OFFLOADING
Another name for unloading; offloading refers more specifically to liquid cargo crude oil and refined products.

OFFSET, DIAGONAL
See Diagonal Offset Well.

OFFSET ROYALTY
See Royalty, Offset.

OFFSET WELL
(1) A well drilled on the next location to the original well. The distance from the first well to the offset well depends upon spacing regulations and whether the original well produces oil or gas. (2) A well drilled on one tract of land to prevent the drainage of oil or gas to an adjoining tract where a well is being drilled or is already producing.

OFFSHORE "WELL NO. 1"
The first offshore well (out of sight of land) was drilled on November 14, 1947, in the Gulf of Mexico, 43 miles south of Morgan City, Louisiana. By 1976, more than 18,000 wells had been drilled offshore.

OFFSITES
A general term for facilities built off the immediate site of a refinery, chemical, or processing plant but that are necessary to the efficient operation of the plant. Examples of offsites are tankage, rail spurs, material sheds, fire-water ponds, etc.

OFF-SPEC
Off specification. Refers to oil, gas, or petroleum products not up to specification, e.g., too much water or emulsion in crude oil; water, sulfur, or condensate in a gas stream; or products in a products pipeline that are mixed. Off-spec products being pumped in a pipeline are switched into a slop tank and sold as off-spec or pumped back to the refinery for redistillation.

OFF-SYSTEM SALES
Gas sold off the system or from a natural gas pipeline system that has more gas than it can dispose of through its regular contract channels. This may occur when a gas transmission company has signed a take-or-

pay contract with its supplier. When gas demand slacks off, the pipeline company must take the stipulated volume of gas, or if it cannot take the gas into the system, it must pay for it anyway. To get rid of its surplus gas, the transmission company sells to a smaller and often local gas pipeline company.

OFF THE SHELF
Said of a product or equipment that is ready and waiting at a supplier's warehouse and can be taken "off the shelf" and shipped immediately. Refers also to techniques and procedures that have been perfected and are ready to be employed on some job.

O.G.J.—*THE OIL & GAS JOURNAL*
The highly respected weekly magazine, published in Tulsa, Oklahoma, that represents and reports on all segments of the petroleum industry in the United States, as well as on significant events in the oil world abroad.

OHM
A unit of electrical resistance equal to that of a conductor in which a current of one ampere is produced by a potential of one volt across its terminals; the unit of electrical resistance was named for Georg Simon Ohm, an 18th Century German physicist.

OGJ — Oil and
Gas Journal

O.I.C.
Oil Information Committee of the American Petroleum Institute (A.P.I.).

OIL
Crude petroleum (oil) and other hydrocarbons produced at the wellhead in liquid form; includes distillates or condensate recovered or extracted from natural gas.

OIL, ATTIC
See Attic Oil.

OIL ANALYZER, NET
A well-testing installation that separates the oil flow and water content of individual wells on a lease. The analyzer automatically determines net oil and net water in a liquid stream. This information is important on leases where the production of individual wells (perhaps with different royalty owners) is to be commingled in the lease tanks or the pipeline gathering system.

OIL-BASE MUD
Drilling mud whose liquid component is an oil rather than water, which is the most common fluid used to mix with the various clays to

make drilling mud. Oil-base muds are used in very deep wells where the bottom-hole temperatures of 300° to 400°F preclude the use of water-base muds. Also, oil-base muds are often used when drilling through clay formations, which have a tendency to absorb the water from water-base muds and swell to the extent that the drillpipe becomes stuck.

OIL BEHIND THE PIPE
Refers to oil and gas sands or formations knowingly passed through, never produced. Such formations usually were of low permeability (tight formations) that, say 20 years ago, were uneconomical to produce when oil was around $5 or less a barrel. Other times formations would be purposely ignored because the operator was going deeper for bigger game, so the less-spectacular, plain-Jane sands were cased off. When the price of crude oil reached $30 per barrel, the bypassed formations looked pretty good and were opened up and produced.

OIL BONUS
A payment in oil to a lessor (usually the landowner) in addition to the cash bonus and royalty payment he is entitled to receive. *See also* Bonus.

OIL BROKER
One who acts as a go-between in the domestic or international crude-oil market. A broker will find a market for a quantity of crude or product not committed by long-term contract. Just as readily, he will come up with oil for someone who wishes to buy. Brokers perform a useful function in the oil business by being knowledgeable about the industry's supply and demand situation. They are the unobtrusive link between buyer and seller, independent producer and small refiner. For their services, the brokers receive either a flat fee or a percentage of the deal they help consummate.

OIL COLUMN
The pay zone; the producing interval of a well. "The drillship cut through an oil column of 600 feet" means the drill bit bored through 600 feet of one or more subsurface formations capable of producing oil or gas.

OIL COUNTRY TUBULAR GOODS
Well casing, tubing, drillpipe, drill collars, and line pipe.

OIL-CUT MUD
Drilling mud with which crude oil has been unintentionally mixed. This may occur when drilling into or through an oil-bearing formation whose pressure is sufficient to overcome the pressure or weight of the column

of mud in the hole. Oil also may become mixed with the drilling mud when a drillstem test is taken. *See* Gas-Cut Mud; *also* Oil-Base Mud.

OILER
The third man at a pumping station in the old days. The normal shift crew on a large gathering or mainline station was the station engineer, the telegraph operator-assistant engineer, and the oiler, whose job included feeling the engine and pump bearings, keeping the wick oilers full and dripping properly, and cleaning and mopping the station floors.

OIL FINDER
A wry reference to a petroleum geologist.

OIL HOUSE
The facility at a refinery where lubricating oils and greases are barreled and packaged. In most cases, the oil house is where the automated canning line is located.

OIL IMPORT TICKET
A license issued by an agency of the federal government to refiners to buy certain amounts of crude oil shipped in from abroad.

OIL IN PLACE
Crude oil estimated to exist in a field or a reservoir; oil in the formation not yet produced.

OIL IN PLACE, ORIGINAL
The estimated number of stock-tank barrels of crude oil in a known reservoir before any production takes place. Known reservoirs include those that are being produced, those with proven reserves but yet to be tapped, and those that have been depleted. These designations are from the American Petroleum Institute, Division of Statistics.

OIL KITCHEN
An imaginative designation for hydrocarbon activity in a province: the occurrence of tar balls on a nearby beach; migrant oil or gas shows in exploratory wells; geo-survey anomalies; gas-sniffer evidence in marine runs; and ultraviolet fluorescence in water seeps. *See* Tar Balls.

OIL LUBRICATOR
See Lubricator, Oil.

OIL-MIST SYSTEM
A lubricating system that pneumatically conveys droplets of a special oil from a central source to the points of application. An oil-mist system

is economical in its use of lubricant and efficient on many types of anti-friction applications.

OIL PATCH

A term referring broadly to the oil field, to areas of exploration, production, and pipelining.

OIL PAYMENT

A share of the oil produced from a well or a lease, free of the costs of production.

OIL PLAY

Interest and activities in leasing and drilling in an area. "Play" means oil-related things are going on in an area: landsmen are leasing, seismic work is being done, and wells are being drilled.

OIL POOL

An underground reservoir or trap containing oil. A pool is a single, separate reservoir with its own pressure system so that wells drilled in any part of the pool affect the reservoir pressure throughout the pool. An oil field may contain one or more pools.

OIL PROPERTY

According to the U.S. Treasury regulations, an oil property is a geological deposit (of oil or gas) on or in a parcel of land or a lease owned by a taxpayer. For tax purposes each separate geological deposit (producing formation) constitutes an oil property unit. (A dual-completion well from two separate and distinct formations represents two oil property units.) Geophysical and geological costs incurred to discover and produce the property must be allocated to the property unit and included in the depletable basis of the unit. Intangible drilling costs are deducted on each unit. Also, depletion—percentage depletion or cost depletion—is calculated separately for each property unit. The oil property concept is not easy to grasp at first glance. It, the concept, might be said to consist of three phases or parts: geological, geophysical, and legal. If there is a geological deposit being produced on a particular plot of land, those two facts make an oil property, and by law the taxpayer owes several kinds of taxes. Two of the more prominent taxes are the production tax or severance tax and income tax. Then there are others: state and county, and maybe even township taxes.

OIL RING

A metal ring that runs on a horizontal line shaft in the bearing well, which has a supply of lube oil. As the ring slowly rotates through the well of oil, it deposits oil on the shaft. Oil rings are generally made of brass and are used on relatively slow-moving shafts.

OIL ROCKS

See Oil Column; *also* Sedimentary Basin.

OIL ROYALTY
See Royalty, Oil.

OIL RUN
(1) The production of oil during a specified period of time. (2) In pipeline parlance, a tank of oil gauged, tested, and put on the line; a run. *See* Run Ticket.

OILS, VAPOR PHASE
Special oils that release vapors that coat machinery and other equipment to protect them from corrosion and damage from moisture. Drilling equipment, especially in offshore environments, are particularly vulnerable to corrosion and are routinely protected by vapor-phase oils and then covered. This is part of the mothballing or cocooning of laid-up jackup rigs and other floaters.

OIL SALES LINE
See Line, Oil Sales.

OIL SANDS BITUMEN
A heavy, petroleum-like substance found in certain consolidated sand formations at the surface of the Earth or at relatively shallow depths where it can be surface mined after the removal of a few feet of overburden. The extraction process is complicated but basically it involves the heating of the oil sands to separate the oil. The oil is floated off and undergoes treatment before it is piped to a refinery.

Oil shale semiworks
project

OIL SHALE
Kerogen shale.

OIL SKIMMER, DRUM
A rotating drum mounted on the bow of a skimming vessel, the barrel, covered with absorbent felt. As the vessel moves into the oil spill, the slowly rotating fuzzy barrel soaks up a quantity of oil and a pressure roller squeezes the oil, and some water, into a catch basin on board.

OIL SLICK
An oil spill on water. A small amount of oil can spread into a sizeable and alarming slick. Oil companies have emergency clean-up procedures that are ready in coastal areas or where ever the danger of spills is present.

OIL SPILL
A mishap that permits oil to escape from a tank, an offshore well, an oil tanker, or a pipeline. Oil spill has come to mean oil on a body of water where even small amounts of oil spread and become highly visible.

Oil spill boom

OIL-SPILL BOOM
Any of various devices or contraptions to contain and prevent the fur-

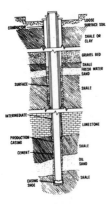

Oil string

Oil-well pumping unit

ther spread of oil spilled on water until it can be picked up. A curtain-like device deployed around or across the path of a drifting oil spill. The curtain is weighted on the bottom edge to hold it a foot or two below the surface and has floats on the upper edge to hold the curtain a foot or more above the surface. Once surrounded, the oil is sucked up by a vacuum cleaner-like suction pump.

OIL STRING
See Production String.

OIL-WATER CONTACT
The interface between the accumulation of oil in a reservoir and the bottom water underlying the oil. *See* Water Drive; *also* Water Coning.

OIL-WATER EMULSION
See Emulsion, Oil-Water.

OIL-WELL PUMP
See Pumping Unit.

OIL-WELL PUMP, GRABLE
See Pump, Grable Oil-Well.

OLD OIL
For the purposes of price regulation, under the Emergency Petroleum Allocation Act of 1973, old oil is production from a property up to the 1972 level of production. Any production in excess of this amount from a property is new oil.

ON ARRIVAL PRICING OF CRUDE OIL
See Crude Oil Pricing, On Arrival.

ON BOTTOM, TURNING TO THE RIGHT
On a rotary rig, this expression means that drilling is proceeding normally.

102 GAS
A short, easy-to-remember method of referring to the various classifications and ceiling prices of natural gas is to refer to the sections in the Natural Gas Policy Act of 1978, numbered 102 through 109. To begin, 102 gas is new gas and gas from newly discovered reservoirs on old Outer Continental Shelf leases; 103 gas is gas from new onshore gas wells; 104 gas is gas previously dedicated to interstate gas sales; 105 gas is gas sold under existing intrastate contracts; 106 gas is gas sold under rollover gas sales contracts; 107 gas is high-cost gas; 108 gas is gas from stripper wells; and 109 gas is an omnibus classification applying to several types of gas supplies not covered by other sections of N.G.P.A.

ONE, FOUR DIAALKYLUMINOANTHROQUINONE
See Blue 8, Automate—a dye marker concentrate for off-highway diesel fuel. E.P.A. Sec. 2ll (i). October 1993.

ONE-THIRD FOR ONE-QUARTER
A term used by independent oil operators who are selling interests in a well they propose to drill. An investor who agrees to a one-third for one quarter deal will pay one-third of the cost of the well to casing point and receive one-fourth of the well's net production.

ON-LEASE GAS
Gas produced and consumed on the same lease.

ON-LINE PLANT
Gas processing plant located on or near a gas transmission line that takes gas from the trunk line for processing—stripping, scrubbing, drying—and returns the residue gas to the line.

ON STREAM
Term used for a processing plant, a refinery, or pumping station that is operating.

ON THE BEAM
Refers to a well on the pump operated by a walking beam instead of a pumping jack.

ON THE BIG SPROCKET
Said of a person who is moving in influential circles or has suddenly gone from a small job to one of considerably larger responsibility; a big operator, often used pejoratively.

ON THE LINE
(1) Said of a tank of oil whose pipeline valve has been opened and the oil is running into the line. (2) A pumping unit that has been started and is pumping on the pipeline.

ON THE PUMP
A well that is not capable of flowing and is produced by means of a pump.

O.O.G.
Office of Oil and Gas, Department of the Interior.

O.O.I.P.
Original oil in place.

OOLITIC LIMESTONE
This type of limestone consists of small, round grains of calcium car-

bonate cemented together. The small grains, called oolites, are formed by the precipitation of calcium carbonate from warm, shallow sea water and deposited on sand grains or shell fragments. Ocean currents continually roll the fragments as the concretions build up, causing the oolites to become spherical. As the oolites accumulate into beds, they form good reservoir rock.

O.P.A. 90
The Oil Pollution Act of 1990. The surety bond requirement for oil spill cleanup and end-of-production cleanup is June 1993, $50 m. The proposal by Mineral Management Service (M.M.S.) is to increase the bond level to $150 m, an increase of 300 percent. The oil industry believes this places a heavy burden on independent operators in the Gulf of Mexico and will severely limit their participation in future leasing and drilling operations.

OP. CO.
Acronym for operating company.

OP. DRILLING SERVICE
Optimization drilling; a consulting service first developed by American Oil Company that makes available to operators of drilling rigs technical, geological, and engineering information gathered from wells drilled in the same area. Included is advice on mud programs, bits, drill speed, and pressures as well as consultation with drilling experts.

O.P.E.C.
See Organization of Petroleum Exporting Countries.

OPEN FLOW
The production of oil or gas under wide-open conditions; the flow of production from a well without any restrictions (valves or chokes) on the rate of flow. Open flow is permitted only for testing or cleanout. Good production practice nowadays is to produce a well under maximum efficient rate conditions.

OPEN FLOW PRESSURE
The natural reservoir pressure when oil or gas is being produced in open flow with no artificial restrictions such as chokes or pinched valves.

OPEN HOLE
An uncased well bore; the section of the well bore below the casing; a well in which there is no protective string of pipe.

OPEN-HOLE LOGGING
Logging operations in an uncased well bore. The well is logged below the relatively shallow surface pipe.

OPERATING AGREEMENT, JOINT
See Joint Operating Agreement; *also* Joint Venture.

OPERATING INTEREST
An interest in oil and gas that bears the costs of development and operation of the property; the mineral interest less the royalty. *See* Working Interest.

OPERATOR
(1) An actuating device; a mechanism for the remote operation and/or control of units of a processing plant. Operators usually are air or hydraulically actuated. Their main use is for opening and closing stops and valves. (2) A person who works a shift at a processing plant, refinery, or pumping station. One who is in charge of and responsible for a unit in the plant or station; a skilled hourly worker. *See* Plant Operator.

Operator or actuator
(Courtesy Fisher)

ORCUTT, W.W.
One of the first geologists actually hired to look for oil. He was hired by the Union Oil Co. of California in 1889. At that time there was still great skepticism in the oil fraternity as to the value of geology.

ORGANIC REEF
A type of structural trap for oil and gas; a former coral reef now buried under sediment deposited at a much later time. Reefs are of limestone and, if porous and permeable enough, are good sources of petroleum. Reefs are usually long and narrow. *See* Reef Reservoir.

ORGANIC ROCK
A sedimentary rock, a kind of rock consisting primarily of the remains of plants and animals, the material that originally was a part of the skeleton of an animal or plant; biogenetic rock.

ORGANIC SUBSTANCE
A material that is or has been part of a living organism. Oil, although classified as a mineral, is an organic substance derived from living organisms.

ORGANIZATION OF PETROLEUM EXPORTING COUNTRIES (O.P.E.C.)
Oil producing and exporting countries in the Middle East, Africa, and South America that have organized for the purpose of negotiating with oil companies on matters of oil production, prices, and future concession rights. O.P.E.C., in 1984, had 13 members: Algeria, Ecuador, Gabon, Indonesia, Iraq, Iran, Libya, Kuwait, Nigeria, Qatar, Saudi Arabia, United Arab Emirates, and Venezuela. The organization was created in November 1960.

Orifice meter

ORIFICE METER
A measuring instrument that records the flow rate of gas, enabling the volume of gas delivered or produced to be computed.

ORIGINAL OIL IN PLACE
See Oil in Place, Original.

ORIMULSION
A heavy petroleum and water mix, an inverse emulsion, that originated in Venezuela, the Orinoco river basin. The Venezuela state oil company developed Orimulsion in order to use and to market the heavy, low-gravity crude oil for use in industrial and commercial boilers.

ORIMULSION "LIQUID COAL" HEATING OIL
A blend of Orinoco basin, extra-heavy crude oil, water, and chemicals. The blend can be used in place of coal under utility and industrial boilers. The Venezuelan, state-owned Lagoven, S.A., developers of the economically important oil/water-chemical blend, measures the country's output in tons instead of barrels, which exempts it from the country's production quota under O.P.E.C.

O-RING
A circular rubber gasket used in flanges, valves, and other equipment for making a joint pressuretight. O-rings in cross section are circular and solid.

OR LEASE
See Lease, Or.

OROGENIC BELT
A region subjected to folding and other deformation during a geologic or tectonic cycle.

O.S.H.A.
Occupational Safety & Health Administration.

OSMOSIS
The diffusion of a solvent through a semipermeable membrane separating the solvent and a solution; or equalizing a dilute solution and a more concentrated one until the solutions are of equal concentration.

OSMOTIC TRANSPORT
A phrase used to indicate the movement of water from a water-base drilling fluid into the clays in a shale sequence. The osmosis of water from drilling fluids into the severely dehydrated clays and shales (owing to the effect of the overburden for eons of time and the complete absence of water in the formation) is an ever present problem when

drilling with a water-base drilling fluid. Switching to an oil-base mud prevents this condition from arising.

OTTO-CYCLE ENGINE
A four-stroke cycle gas engine; the conventional automobile engine is an Otto-cycle engine, invented in 1862 by Beau de Rochas and applied by Dr. Otto in 1877 as the first commercially successful internal-combustion engine. The four strokes of the Otto cycle are intake, compression, power, and exhaust.

OUTAGE GAUGE
A measure of the oil in a tank by finding the distance between the top of the oil and the top of the tank and subtracting this measurement from the tank height.

Otto-cycle engine

OUTBOARD BEARING
See Bearing, Outboard.

OUTCROP
A subsurface rock layer or formation that, owing to geological conditions, appears on the surface in certain locations. That part of a strata of rock that comes to the surface.

OUTER CONTINENTAL SHELF (O.C.S.)
"All submerged lands (1) which lie seaward and outside the area of lands beneath the navigable waters as defined in the Submerged Lands Act (67 Stat. 29) and (2) of which the subsoil and seabed appertain to the U.S. and are subject to its jurisdiction and control."

OUTFALL
See Effluent.

OUTPOST WELL
An oil or gas well drilled some distance from an oil pool partly developed in the expectation of extending the limits of the pool. An outpost well is far enough away from proven production to make the outcome uncertain but not far enough away to be considered a wildcat; a step-out well; an extension well.

OVER AND SHORT (O.&S.)
In a pipeline gathering system O.&S. refers to the perennial imbalance between calculated oil on hand and the actual oil on hand. This is owing to contraction, evaporation, improper measuring of lease tanks, and losses through undetected leaks. Oil is paid for on the basis of the amount shown in the lease tanks. By the time this oil is received at the central gathering station, the amounts invariably are short, which represents a loss to the pipeline system.

OVERBURDEN
In strip or surface mining, the earth (rocks, sand, shale) overlying the seam of mineral deposited. If there is too much overburden to be removed—80 feet or more—it is uneconomical to mine the deposit by surface means.

OVERHEAD
A product or products taken from a processing unit in the form of a vapor or a gas; a product of a distillation column.

OVERKILL TECHNIQUES
Solutions or remedies for a persistent and troublesome problem that, upon mature reflection, were too drastic, costly, and inappropriate.

OVERLAP FAULT
See Fault, Overlap.

OVERPRESSURED
Said of a formation whose pore pressure is such that it must be controlled by mud weight in the hole at least equal to the formation pressure. Overpressured drilling (excessive hydrostatic pressure on the formation) could do it irreparable damage by mudding off or clogging the face, the pores of the formation.

OVERPRESSURIZED ZONE
An extremely high-pressure gas formation often drilled into without prior knowledge of the potential for damage or even disaster. Overpressurized zones usually are small areas caused, geologists believe, by tectonic stress, compression by movement of the surrounding formation. Such zones are a driller's nightmare because there usually is no warning before the gas pocket is drilled into, and then a dangerous blowout can result.

OVERPRODUCED
In gas well terminology, the condition of having produced and sold more than one's proportionate share of a well's gas in a split-stream arrangement. *See* Gas Balancing Agreement, Gas Well, Split-Stream.

OVERRIDE
See Royalty, Overriding.

OVERRIDE SYSTEM
A backup system; controls that take over should the primary system of controls fail or be taken out for adjustment or repair; a redundancy built in for safety and operational efficiency.

OVERRIDING ROYALTY
See Royalty, Overriding.

OVERSHOT

A fishing tool; a specially designed barrel with gripping lugs on the inside that can be slipped over the end of tubing or drillpipe lost in the hole. An overshot tool is screwed to a string of drillpipe and lowered into the hole over the upper end of the lost pipe. The lugs take a friction grip on the pipe, which can then be retrieved.

OVER-THE-DITCH COATING

Coating and wrapping line pipe above the ditch just before it is lowered in. Most line pipe is coated and wrapped in the pipe yard and then transported to the right-of-way and strung. Over-the-ditch coating has the advantage of minimizing scuffing or other damage to the coating suffered through moving and handling.

OVERTHRUST

A large-scale thrust fault of low angle with displacement (relative movement of the two sides of the fault) usually measured in miles.

OVERTURNED FOLD

Refers to a fold or branch of a fold that has been tilted degrees beyond perpendicular; in some instances, as much as 45°. The normal bottom-to-top sequence then appears reversed. Also refers to an over-fold.

OXIDES

Mineral compounds characterized by the linkage of oxygen with one or more metallic elements such as cuprite, CU_2O or spinel, $MgAl_2O_4$

OXYACETYLENE WELDING

See Welding, Oxyacetylene.

OXYGENATES

Additives for motor gasoline to promote cleaner burning in the engine and thus reducing polluting emissions, unburned hydrocarbons, and carbon monoxide. *See* M.T.B.E. and T.A.M.E.

Top sub

Type A packer

Bowl

Spiral grapple

Grapple control

Guide

Circulating overshot with spiral grapple

Spiral grapple

Wall hook guide

Circulating overshot with spiral grapple

Coating and wrapping over the ditch

PACKAGE PLANT

A facility at a refinery where various refined products are put in cartons and boxes ready for shipment. Waxes, greases, and small-volume specialty oils are boxed in a package plant.

PACKED COLUMN

Refers to a fractionating refinery tower where various trays and integral platforms are loaded with packing material, some randomly dumped, other types systematically placed. *See* Bubble Tower.

PACKED-HOLE ASSEMBLY

A drill column containing special tools to stabilize the bit and keep it on a vertical course as it drills. Included among the tools are stabilizer sleeves, square drill collars, and reamers. Packed-hole assemblies are often used in "crooked-hole country."

PACKER

An expanding plug used in a well to seal off certain sections of the tubing or casing when cementing and acidizing or when a production formation is to be isolated. Packers are run on the tubing or the casing and when in position can be expanded mechanically or hydraulically against the pipe wall or the wall of the well bore.

Packer *(Courtesy Otis)*

PACKER, EXTERNAL-CASING

A type of inflatable packer run on the outside of the casing, between the casing and the wall of the borehole. After it is run, the packer expands outward against the wall of the hole isolating the producing zone, the pay zone, from the upper sections of the borehole. External packers are used to prevent gas migration through the cement column or between cement and formation. A hook-wall packer.

PACKER, HOOK WALL

See Packer, External Casing.

PACKER, INFLATABLE

A downhole packer often used in well cementing on squeeze jobs. The packer, run on the casing and then inflated, seals the annulus between the casing and the wall of the well bore; a hook wall packer.

PACKING

Any tough, pliable material—rubber or fiber—used to fill a chamber or "gland" around a moving rod or valve stem to prevent the escape of gas or liquid; any yielding material used to effect a pressure-tight joint. Packing is held in place and compressed against a moving part by a "follower," an adjustable element of the packing gland.

PACKING, RANDOM-DUMPED

Metallic packing in various configurations for refinery vacuum towers. Some metallic packing is in the shape of small, short cylinders 2 1/2 or 3 1/2 inches in diameter with windows or tabs cut in the side walls to facilitate the passage of vapors upward through the 3- to 7-foot-deep bed of randomly dumped packing. The metallic packing, looking very much like a bright, homogeneous scrap pile as it is dumped onto the packing support plate, takes the place of bubble-cap trays or fractionating trays in vacuum towers.

PACKING GLAND

A stuffing box; a chamber that holds packing material firmly around or against a moving rod or valve stem to prevent the escape of gas or liquid.

PACKING OFF

The condition in a borehole, especially a high-angle or horizontal bore, when the hole is not being cleaned sufficiently and fills with rock cuttings or sloughs, and is finally packed off, plugged. This may call for a wiper trip, pulling the drillpipe and drill to clean the hole.

PACKLESS VALVE

See Valve, Packless.

PADDING MACHINE, PIPELINE

A self-contained earth-sifting machine that screens and sifts backfill to eliminate rocks and debris. It then deposits the sifted earth onto the line in the ditch. The sifted earth is called "padding" as it covers the pipe, protecting its coating against damage from rocks that normally are present in backfill. The padding machine obviates the need to haul rock-free earth from off-site to pad the pipeline as the ditch is backfilled in rough, rocky terrain.

PAID-UP LEASE

See Lease, Paid-Up.

PALEONTOLOGY

The science that deals with plant and animal life in past geologic time. The study is based on fossil plants and animals, their relationship to present-day plants and animals, and their environments.

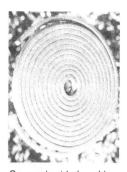
Square-braided packing

Stuffing box or packing gland

Padding Machine Pipeline (Courtesy Soneboz Corporation)

PALYNOLOGY

The science that deals with the study of live and fossil spores and with pollen grains and other microscopic plant structures. As palynology concerns oil prospecting, particularly stratigraphic problems, the science involves age-dating rocks and determining the environment in which sedimentary formations were laid down. This can be observed from well borehole cuttings, cores, and surface outcrop samples; also, microscopic analysis of source rock samples and other basic geochemical studies.

PAPER TRAIL

The mandated record keeping to guarantee or prove compliance with state and federal regulations, for example the Clean Air Act of 1990 and E.P.A. regulations.

PARAFFIN

A white, odorless, tasteless, chemically inert, waxy substance derived from distilling petroleum; a crystalline, flammable substance composed of saturated hydrocarbons.

PARAFFIN-BASE CRUDE

Crude oil containing little or no asphalt materials; a good source of paraffin, quality motor lubricating oil, and high-grade kerosene; usually has lower nonhydrocarbon content than an asphalt-base crude.

PARALLEL UNCONFORMITY

See Disconformity.

PARAMETER

A set of physical properties where values determine the characteristics or behavior of a system; a factor that restricts what is possible or results.

PARASITE STRING

Refers to an injection string of small-diameter tubing installed to a predetermined depth in the borehole to inject nitrogen or other inert gas into the fluid column to maintain an underbalanced well.

PARTICIPATION

A type of joint venture between a host country and an international oil company holding concession rights in that country. Participation may be voluntary on the part of the oil company or as the result of coercion by the host country.

PARTICIPATION CRUDE

A certain percent of the crude oil produced by an oil company, which is the host country's share of production under the terms of a concession or participation agreement. Forty years ago, the percent of participation crude was fairly modest, but beginning in the 1960s the percent began creeping up until it now frequently exceeds 50 percent.

PARTICULATE MATTER
Minute particles of solid matter—cinders and fly ash—contained in stack gases.

PARTNERING VENTURE
A drilling and production joint venture between a state oil company (foreign country) and an international oil company, as operator. This arrangement is a relatively new (1993) avenue to mutual advantage. The major oil company, or an experienced company, brings to the union its know-how, its expertise, its service-company connections. The state oil company supplies the tract leases, some information from previous wells drilled in the immediate area, and sometimes provides a percent of the drilling and completion costs. The drawbacks to this arrangement are the reliability of the well data furnished by the state entity, dealing with bureaucrats, and the question of who will make the decisions. A new member of the major's E.&D. department may have to be drilling superintendent/diplomat. Partnering: a union of nonequals.

PARTNER UP, TO
To pick up a piece of the action; to come in on a deal at the beginning or anywhere along the line from spudding in to well completion.

PASS-THROUGH PROVISION
A provision in a price-control law or other regulation permitting certain increased costs of a product to be "passed through" or passed on to customers by allowable price increases.

PASS-THROUGH ROYALTY
See Royalty, Pass-Through.

PATROL
See Pipeline Patrol.

PATTERN FLOODS
Secondary oil-recovery programs of waterflooding in which the injection wells are placed in one of several patterns, e.g., in line, staggered line, five spot or seven spot. These configurations are intended to sweep the formation of additional oil.

PAWLS
The spring-loaded or gravity operated "dogs" on a ratchet are pawls. The pivoted metal (sometimes wooden) tongue that by falling into place behind each tooth on a ratchet as it is being turned prevents the wheel, as on a windlass, from turning backwards or reversing itself.

PAY
Pay zone; the producing formation in an oil or gas well.

PAYBACK
See Payout.

PAY HORIZON
See Pay Zone.

PAYOUT
The recovery from production of the costs of drilling, completing, and equipping a well. Sometimes included in the costs is a pro rata share of lease costs.

PAYOUT, CARRIED-INTEREST
In a carried-interest arrangement, the payout is the recovery by the carrying party of development and operating costs of the well or lease (more than one well). Concerning a production payment, it is the recovery by the payee (the person being paid) of the stipulated sum due him from the well's production. In general, the payout is the recovery or payback, out of production, of the costs of drilling, completing, and equipping a well for production.

PAY STRING
The pipe through which a well is produced from the pay zone. Also called the "long string" because only the pay string of pipe reaches from the wellhead to the producing zone.

PAY ZONE
The subsurface, geological formation where a deposit of oil or gas is found in commercial quantities.

P.C.V.
Positive Crankcase Ventilation.

P.D.C. BIT
See Bit, Diamond Shear.

P.D.V.S.A
Petroleum de Venezuela, the national energy company that assumed control of petroleum resources in 1976.

PEAK-SHAVING L.N.G. PLANT
A liquefied natural gas plant that supplies gas to a gas pipeline system during peak-use periods. During slack periods the liquefied gas is stored. With the need for additional gas, the liquid product is gasified and fed into the gas pipeline.

PEA PICKER
An inexperienced worker; a green hand; a boll weevil.

PEG-LEG MULTIPLE
A slang term for a seismic reflection that bounces once between intermediate formations before being reflected by a deeper zone. This action results in a false reading for which adjustments must be made.

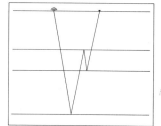

Peg-Leg Multiple

PEMEX (PETROLEOS MEXICANOS)
The national oil company of Mexico, established to take control of all petroleum resources expropriated in 1938.

PENALTY BONUS
See Nonconsent Penalty.

PENCIL ABSTRACT
An informal summary of instruments of record. *See* Bob-Tail Abstract.

PENDULUM DRILL ASSEMBLY
A heavily weighted drill assembly using long drill collars and stabilizers to help control the drift from the vertical of the drill bit. The rationale for the weighted drill assembly is that, like a pendulum at rest, it will resist being moved from the vertical and will tend to drill a straighter hole.

PENNSYLVANIA-GRADE CRUDE OIL
Oil with characteristics similar to the crude oil produced in Pennsylvania from which superior-quality lubricating oils are made. Similar-grade crude oil is also found in West Virginia, eastern Ohio, and southern New York.

PERCENTAGE DEPLETION
A method of computing the allowance for depletion of an oil or gas well, or other mining of minerals, for federal income tax purposes. A provision in the tax law that exempts a certain percent of mineral production from income tax. For an oil well, the percentage depletion rate was 27 1/2 percent, then 22 percent, of the well's gross production, excluding royalty, up to 50 percent of the net income from the property. The exemption is now 15 percent for production up to a certain level.

PERFORATING
To make holes through the casing opposite the producing formation to allow the oil or gas to flow into the well. Shooting steel bullets through the casing walls with a special downhole "gun" is a common method of perforating.

PERFORATING, WIRELINE

Perforating the well's casing by lowering the perforating gun into the downhole casing on a wireline to the desired depth. The other method is to lower the gun on the tubing. Both methods are used in the field; each has its advantages.

PERFORATING GUN

A special tool used downhole for shooting holes in the well's casing opposite the producing formation. The gun, a steel tube of various lengths, has steel projectiles placed at intervals over its outer circumference, perpendicular to the gun's long axis. When lowered into the well's casing on a wireline opposite the formation to be produced, the gun is fired electrically, shooting numerous holes in the casing that permit the oil or gas to flow into the casing.

PERFORATING GUN SYSTEMS

A downhole perforating gun system has a number of components, all of which must operate smoothly when and only when they are programmed to do so. The system consists of a firing device, carrier, shaped charges, explosive detonation components, and gun accessories. A group of majors (major oil companies) organized what is called a Program to Evaluate Gun Systems (P.E.G.S.). Service companies, those who use perforating guns, and whose gun systems were under investigation, were asked to demonstrate thermal and pressure integrity of their equipment to assure their clients that their guns were safe and trustworthy. Guns that accidentally fired prematurely could be deadly destructive.

PERFORATION, SHAPED-CHARGE

See Shaped-Charge Perforation.

PERFORATION BREAKDOWN, PRECISION

A well-completion procedure for isolating individual holes in a well perforation job and injecting acid or other treating fluid into that exposed portion of the formation. Breaking down perforations individually, using a straddle packer, assures that each aperture receives its quota of acid or frac fluid, and that a few holes will not be given a disproportionate share of the fluid as in conventional treatments. The procedure is time consuming but, in most instances, it is very effective.

PERFORATION TUNNELS

The "worm holes" in the wall of the borehole made by the perforation charges from the gun.

PERFS.

Short for perforations, the holes made in a well's casing by a perforating gun lowered into the casing opposite the pay zone, the formation to be produced. *See* Perforating Gun.

PERMAFROST
The permanently frozen layer of earth occurring at variable depths in the Arctic and other frigid regions.

PERMEABILITY
A measure of the resistance offered by rock to the movement of fluids through it. Permeability is one of the important properties of sedimentary rock containing petroleum deposits. The oil contained in the pores of the rock cannot flow into the well bore if the rock in the formation lacks sufficient permeability. Such a formation is referred to as "tight." *See* Porosity; *also* Absolute Permeability.

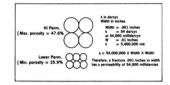
Permeability

PERMEABILITY, ABSOLUTE
A term describing a rock's permeability when only one fluid is present in the pore space.

PERMEABILITY, EFFECTIVE
Effective permeability is less than absolute. This is because the fluid that wets and clings to the solid part of the rock reduces the hydraulic area available to the other fluid (oil) to flow through to the borehole. The ratio of effective-to-absolute is known as "relative permeability."

PERMEABILITY TRAP
A trap for oil and gas formed when the permeability of a reservoir decreases laterally or horizontally. With no permeability, any oil or gas still in the pores of the reservoir rock is unable to move toward the well bore or any other direction.

PERMIAN RED BEDS
See Red Beds.

PERSUADER
An oversize tool for a small job; an extension added to the handle of a wrench to increase the leverage.

PERTAMINA
The national oil company of Indonesia; a fully integrated petroleum company.

PERVIOUS ROCK
Porous rock, rock through which liquids and gases can pass; permeable rock.

P.E.S.A.
Petroleum Equipment Suppliers Association.

PETCOCK
See Stopcock.

PETROBRAS
The acronym for the national oil company of Brazil, established in 1953.

PETROCHEMICALS
Chemicals derived from petroleum; feedstocks for the manufacture of a variety of plastics and synthetic rubber.

PETROCHEMISTRY
The study of the chemical composition of rocks. Petrochemistry is one part or aspect of the broader science of geochemistry and is not to be confused with petroleum chemistry, which is the science of synthesizing substances derived from crude oil, natural gas, and natural gas liquids.

Petrochemical plant

PETROFRACTURING
A type of hydrofracturing of an oil-bearing formation in which a mixture of oil, sand, and chemicals is pumped under high pressure into the face of the formation in an attempt to increase the flow of oil. The mixture is sometimes pumped down the tubing (inside the casing) after setting a packer to isolate the zone to be treated. Another method is to pump the mixture down the casing, followed by a slug of water or drilling mud. *See* Hydraulic Fracturing.

PETROGLYPH
A rock carving or, more specifically, carving upon the surface of a rock made in prehistoric times. A pictograph is a drawing or painting on a rock surface, also of ancient age. The difference is somewhat academic.

PETROLEOS BRASILEIRO, S.A.
Brazil's national oil company.

PETROLEOS del PERU
The national oil company of Peru, S.A.

PETROLEOS de VENEZUELA
The national oil company of Venezuela; P.D.V.S.A.

PETROLEUM
In its broadest sense, the term embraces the whole spectrum of hydrocarbons—gaseous, liquid, and solid. In the popular sense, petroleum means crude oil.

PETROLEUM ARBITRAGE
See Arbitrage, Product.

PETROLEUM COKE
See Coke, Petroleum.

PETROLEUM GEOLOGY
An important branch of geology that concerns itself with the origin, migration, and accumulation of oil and gas deposits in commercial quantities. It involves the application of geochemistry, geophysics, paleontology, structural geology, and stratigraphy to the problems of discovering oil and gas deposits. Petroleum geologists are also intimately involved in the greasy day-to-day work of drilling by advising, identifying, and counseling on handling downhole problems such as lost circulation, acidizing, setting pipe, and hydrofracing.

PETROLEUM PRODUCTS DEMAND
A study by PlanEcon, Inc. of Washington, D.C. and D.R.I./McGraw-Hill, Lexington, Mass. concludes that petroleum products demand in eastern Europe and the former Soviet Union (F.S.U.) will change drastically by the year 2000 as countries in the region embrace free-market economies. Also poised for dramatic changes is the distribution and use of petroleum products in seven eastern European countries and 15 F.S.U. republics.

PETROLEUM RESERVES
Crude oil stored by the U.S. Department of Energy (D.O.E.) as fuel in the event of an emergency or a prolonged oil embargo. Caches of crude oil are located in Louisiana, Texas, and New Mexico in caverns mined in thick salt formations and in abandoned salt mines. *See* Strategic Petroleum Reserves.

PETROLEUM RESERVES, NAVAL
See Naval Petroleum Reserves.

PETROLEUM RESERVES, OCCURRENCE OF
It is estimated by geologists that close to 60 percent of the world's petroleum reserves are in sandstone; the other 40 percent are in limestone, dolomite, and conglomerates. One percent is found in other rock formations, which are sufficiently fractured to provide space for oil and gas accumulation.

PETROLEUM RESERVES, STRATEGIC
See Strategic Petroleum Reserves.

PETROLEUM ROCK
Sandstone, limestone, dolomite, shale, and other porous rock formations where accumulations of oil and gas may be found.

PETROLEUM-SULFONATE FLOODING
See Micellar-Surfactant Flooding.

PETROLEUM TAR SANDS
Native asphalt, solid and semisolid bitumen, including oil-impregnated rock or sands from which oil is recoverable by special treatment.

A tar sand plant

Processes have been developed for extracting the oil, referred to as synthetic oil.

PETROLIFEROUS
Containing or yielding petroleum or hydrocarbons; said of certain types of rock formations.

PETROLOGIST
A specialist in petrology; a geologist who studies the origin, history, occurrence, structure, and chemical composition of sedimentary rocks; also a specialist in the acoustical properties of rocks who often works with geophysicists in determining the presence of oil and gas in sedimentary formations.

PETROLOGY
The science that deals with the origin, history, occurrence, structure, chemical composition, and classification of rocks.

PETROPHYSICS
The study of reservoir rocks and their relation to the surrounding stratigraphy.

PH (pH)
A symbol used in expressing both acidity and alkalinity on a scale whose values run from 0 to 14, with 7 representing neutrality; numbers less than 7, increasing acidity; greater than 7, increasing alkalinity.

PHASE
See Fluid Phases.

PHASE-IN CRUDE
The share of participation crude that a host country may sell and an operating oil company must buy. This often happens when a host country has no outlets yet developed for the crude oil discovered and being produced. The country thus has time to phase in its processing plants or other outlets for its share of crude being produced. *See* Buy-Back Crude.

PHOSPHATE
A rock containing one or more minerals containing phosphorus of sufficient purity and quantity to be used as a source of phosphoric compounds or free phosphorus. Nearly 90 percent of the world's production of this source of fertilizer is sedimentary phosphate rock, which consists of calcium phosphate together with calcium carbonate and other minerals.

PHOTOMETRIC ANALYZER
A device for detecting and analyzing the changes in properties and quantities of a plant's stack gases. The analyzer, through the use of elec-

An electrochemical pH indicator *(Courtesy Foxboro Co.)*

tronic linkage, automatically sounds a warning or effects changes in the stack emissions.

PHYSICAL DEPLETION
The exhausting of a mineral deposit or a petroleum reservoir by extraction or production.

PICKLE
A cylindrical weight (2–4 feet in length) attached to the end of a hoisting cable, just above the hook, for the purpose of keeping the cable hanging straight and thus more manageable for the person using the wireline.

A headache ball or pickle

PICTOGRAPH
See Petroglyph.

PIER
A walkway-like structure built on pilings out from shore a distance over the water for use as a landing place or to tie up boats.

PIERCEMENT DOME
See Dome, Piercement.

PIG
(1) A cylindrical device (3–7 feet long) inserted in a pipeline for the purpose of sweeping the line clean of water, rust, or other foreign matter. When inserted in the line at a "trap," the pressure of the oil stream behind it pushes the pig along the line. Pigs or scrapers are made with tough, pliable discs that fit the internal diameter of the pipe, thus forming a tight seal as they move along cleaning the pipe walls. (2) *verb.* To run or put a pig or scraper through a pipeline; to clean the line of rust, mill scale, corrosion, paraffin, and water.

A pier

PIG LAUNCHER AND RECEIVER
A facility on a pipeline for inserting and launching a pig, scraper, or batching pig. The launcher essentially is a breech-loading cylinder isolated from the pipeline by a series of gate valves. After the pig is loaded into the launching cylinder like a shell into a shotgun, a hinged plug or cap is closed behind it. Then oil under pressure from the pipeline is admitted to the cylinder behind the pig. The pig is launched; it is pushed into the pipeline and moved along at about 3 or 4 miles an hour by the oil pressure behind it. To receive a pig approaching the station manifold, a valve is opened on the bypass line, permitting the pig to be pushed into the receiving cylinder or trap along with the sludge ahead of it. The valves are closed, isolating the pig, at which time the end cap of the receiver is unlatched. The sludge drains into a sump and the pig is removed for cleaning and reconditioning.

Pipeline pig

Pig trap

PIG SIGNALS

A device or mechanism on a pipeline that signals the approach of a pipeline scraper or pig at a pump station or receiving manifold. Signals may be mechanical, in which case an indicator or "flag" is tripped by the passage of the pig at a given point or the pig may actuate an electronic warning device that announces its arrival at the station.

PIG TRAP

A scraper trap.

PILELESS PLATFORM

A concrete offshore drilling platform of sufficient weight to hold the structure firmly in position on the seafloor. Referred to as a "gravity structure," the platform is constructed onshore and then floated and towed to location where it is "sunk" by flooding its compartments. Some platforms of this type have oil storage facilities within the base of the structure. *See* Gravity Structure; *also* Tension-Leg Platform.

PILING, DRILLED-IN

Piling that is inserted into holes drilled by special large-diameter bits. In this operation the piles are cemented in to achieve more stability. Drilled-in piling is often used to secure platform jackets to the ocean floor. *See* Drilling and Belling Tool.

PILL, SPOTTING A

Spotting a pill refers to the remedial action or preventive measure of pumping a measured amount of an oily or chemical substance downhole to a predetermined depth to free a stuck pipe or to lubricate a dog leg or keyseat to prevent the drillpipe from hanging up.

Pillow tanks being filled

PILLOW TANKS

Pliable, synthetic rubber and fabric fuel "tanks" that look like giant pillows. Pillow tanks, first used by the military to store fuel, are now in service at remote locations to store diesel fuel, gasoline, and lube oil until steel tankage can be erected. Easily deployable, the rubber pillows can be filled by tank truck or air shuttle and, when no longer needed, they may be emptied, folded up, and taken to another location.

PILOT HOLE

A small-diameter borehole drilled in an exploratory well to gather subsurface geologic information in a relatively safe manner. Gas kicks, if they are encountered, are easier and cheaper to handle. For example, an 8-inch hole with 6-inch drill collars would provide a good margin of control. The smaller annular clearance provides enough friction pressure to help kill the well if necessary. For exploratory wells in high-pressure gas country or in unknown territory, a pilot hole is the best defense against trouble; a "fraidy hole."

PILOT MILL
A type of junk mill with a tapered center projection below the cutting surface of the bit to guide or pilot the bit into the open end of a piece of junk or a tool to be milled out downhole.

PILOT PLANT
A small model of a future processing plant used to develop and test processes and operating techniques before investing in a full-scale plant.

PILOT VALVE
See Valve, Pilot.

Pilot-operated relief valve

PINCHING A VALVE
Closing a valve partway to reduce the flow of liquid or gas through a pipeline. *See* Cracking a Valve.

PINCHOUT
The disappearance or "wedging out" of a porous, permeable formation between two layers of impervious rock. The gradual, vertical "thinning" of a formation, over a horizontal or near-horizontal distance, until it disappears.

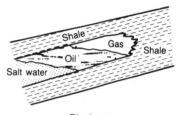

Pinchout

PIPE
Oil country tubular goods: line pipe, well casing, well tubing, and drillpipe. Line pipe (for pipelines) is lap welded or butt welded and comes from the pipe mill either plain ended (for welding) or threaded. Drillpipe is seamless. Well tubing is threaded at one end of the joint and has a tool joint, a connecting collar, threaded on the inside circumference on the other end. Well casing is threaded at one end and has a threaded collar at the other end.

PIPE, PRESSURE DESTRUCTION OF
Three principal types of pressure that cause deformation or destruction of steel pipe, tubing, or casing: tensile (longitudinal stress); burst (pressure from within); collapse (external pressure or impact).

PIPE, SLOTTED FOR THE PAY ZONE
See Slotted Pipe For The Pay Zone.

PIPE COATING, INTUMESCENT
The treatment of offshore and refinery piping with a specially formulated plastic coating that in the event of exposure to extreme heat or fire swells, covering the pipe with a frothy, inert material that protects the pipe.

Flexible coupling between electric motor and centrifugal pump

Pipelay barge

Installing pipeline anchors

PIPE COUPLING, FLEXIBLE

A flange-like coupling made with two elements of the coupling connected by alternate layers of an elastomer and thin sheets of metal. This permits one face or end of the coupling to tilt in relation to the other end like a spool made of rubber. Flexible pipe couplings are used in drilling and production risers in offshore operations. *See* Ball Joint.

PIPE FACING MACHINE

See Facing Machine.

PIPEFITTER

One who installs and repairs piping, usually of small diameter. An "oil patch plumber" according to pipeliners who traditionally work with large-diameter pipe.

PIPE FITTINGS

See Fittings.

PIPE LAX PILL

Special, lubricous oils or formulated fluids that are spotted downhole to free drillpipe or casing stuck in a dog leg, key seat, or just because of excessive borehole drag.

PIPELAY BARGE

See Lay Barge.

PIPELINE, COAXIAL

A multiwall pipe, a pipeline within a pipeline, a larger pipe with a smaller one inside surrounded by a matrix of grout of cement or plastic resin. Coaxial lines are laid from shore station to a subsea manifold or production platform, and they are the corrosion engineer's answer to pipeline integrity in critical and highly corrosive locations.

PIPELINE ANCHOR

Pipelines laid in swampy areas, across rivers, or offshore must be held down in the ditch even when backfilled. One method used for most pipelines is to screw auger-shaped steel pieces (anchors) into the ground, one on each side of the pipe in the ditch, and connect the tops of the two anchors across the pipe. This secures the pipe like a giant staple in a fence post. Anchors are placed every 25–100 feet, depending upon the type of soil or whether the pipe is submerged in water. Buoyancy of a pipeline transporting gas is enough so that even when buried it has a tendency to work its way upward out of the ditch. Another method of holding a line in the ditch is weighting the pipe with large concrete blocks at intervals along the line.

PIPELINE BLINDS

A type of in-line flange with removable blinding or sealing disks that are held pressure tight against the pipe ends. The top of the flange is re-movable for inspecting for leaks. Pipeline blinds are for low-pressure installations.

PIPELINE CAT
A tough, experienced pipeline construction worker who stays on the job until it is flanged up and then disappears—until the next pipeline job. A hard-working, temporary construction hand; a boomer.

PIPELINE DELUMPER
A motor-driven chopping machine that is flanged into a pipeline to break up any solid material that may have found its way into the fluid stream. The electric motor furnishes power for the chopper blades. Delumpers are used for the most part on coal slurry pipelines.

Pipeline delumper

PIPELINE GAS
Gas under sufficient pressure to enter the high-pressure gas lines of a purchaser; gas sufficiently dry so that liquid hydrocarbons—natural gasoline, butane, and other gas liquids usually present in natural gas—will not condense or drop out in the transmission lines.

PIPELINE GAUGER
See Gauger.

PIPELINE HEATER
See Heater.

PIPELINE INSPECTION SPHERE
A manned bathysphere for inspecting offshore pipelines or to investi-gate the underwater terrain, the seafloor, for a proposed route for laying a pipeline. The diving sphere is lowered to the seafloor by a boom and tackle extending from the deck of a work boat or diving tender equipped with support systems.

A manned bathysphere

PIPELINE OIL
Clean oil; oil free of water and other impurities so as to be acceptable by a pipeline system.

PIPELINE PADDING
Fine, rock-free earth used as the initial backfill for coated and wrapped pipelines. The covering of screened soil is essential in rocky terrain where the backfill material contains rock fragments large enough to pierce or otherwise damage the pipe's anticorrosion coating.

PIPELINE PATROL
The inspection of a pipeline for leaks, washouts, and other unusual con-

Pipeline

Pipeline sling

Pipeline spread

ditions by the use of light, low-flying aircraft. The pilot reports by radio to ground stations on any unusual condition on the line.

PIPELINE PRORATIONING
The refusal by a purchasing company or a pipeline to take more oil than it needs from the producer by limiting pipeline runs from the producer's lease, an informal practice in the days of overproduction when market conditions were unsatisfactory or when the pipeline system lacked storage space. Also referred to as purchaser prorationing.

PIPELINER
One who does pipeline construction or repair work: welders, ditching machine operators, cat drivers, coating-and-wrapping machine operators, connection men; broadly, anyone who is involved in the building, maintenance, and operation of a pipeline system.

PIPELINE RIDER
One who covers a pipeline by horseback, looking for leaks in the line or washed-out sections of the right-of-way. The line rider has been replaced by the pipeline patrol using light planes or, for short local lines, by the pickup truck and the man on foot.

PIPELINE SCRAPER
See Scraper.

PIPELINE SLING
See Sling, Pipeline.

PIPELINE SPREAD
See Spread.

PIPELINE TARIFF
See Tariff.

PIPELINE WELDING
See Welding, Pipeline.

PIPE MILL, PORTABLE
See Portable Pipe Mill.

PIPE RAM
See Ram.

PIPE RAMP
A ramp opposite the door of the drilling rig used to skid joints of casing, drillpipe, and tubing up to the rig floor from the pipe rack.

PIPE STRAIGHTENER
A heavy pipeyard press equipped with hydraulically powered man-

drels for taking the kinks and bends out of pipe. The replaceable mandrels come in sizes from 2–12 inches.

PIPE TONGS
Long-handled wrenches that grip the pipe with a scissors-like action used in laying a screw pipeline. The head (called the butt) is shaped like a parrot's beak and uses one corner of a square "tong key," held in a slot in the head, to bite into the surface of the pipe in turning it.

Pipe handler in large
pipe yard

PIPE YARD
An area set aside for the cleaning, straightening, coating and wrapping, and storing of line pipe. Pipe yards usually are near the field headquarters of a pipeline company. Or, if a new line is being laid, pipe yards are located at central points along the route of the line.

PISTON PIN
See Wrist Pin.

PITCH
Asphalt; a dark brown to black bituminous material found in natural beds; also produced as a black, heavy residue in oil refining. *See* Brea.

PITCHER PUMP
See Pump, Pitcher.

PIT LINERS
Specially formulated plastic sheeting for lining earthen or leaking concrete pits to prevent seepage of oil or water into the ground.

Pitman

PITMAN
The connecting piece between the crank on a shaft and another working part. On cable-tool rigs, the pitman transmits the power from the bandwheel crank to the walking beam.

PITOT TUBE
A measuring device for determining the gas flow rates during tests. The device consists of a tube with a 1/8-inch inside diameter inserted in a gas line horizontal to the line's long axis. The impact pressure of the gas flow at the end of the tube compared to the static pressure in the stream is used in determining the flow rate.

PITTED PIPE
Line pipe corroded in such a manner as to cause the surface to be covered with minute, crater-like holes or pits.

Corroded pipe showing
deep pits

PLAIN-END PIPE
Pipe that has not been threaded at the pipe mill; pipe to be used in a

welded pipeline. Plain-end pipe must have the ends beveled before it is ready for welding.

PLANAR
Lying in a plane or a succession of planes, usually parallel as in the bedding of sediment; planar as in the cleavage of a rock or the layering of shale.

PLANT OPERATOR
An employee who runs plant equipment, makes minor adjustments and repairs, and keeps the necessary operating records.

PLANT TURNAROUND
See Turnaround

PLASTIC FLOW
The flow of liquid (through a pipeline) in which the liquid moves as a column; flowing as a river with the center of the stream moving at a greater rate than the edges, which are retarded by the friction of the banks (or pipe wall). *See* Turbulent Flow.

PLAT
A map of land plots laid out according to surveys made by the Government Land Office showing section, township, and range; a grid-like representation of land areas showing their relationship to other areas in a state or county.

PLAT BOOK
A book containing maps of land plots arranged according to township and range for counties within a state. *See* Plat.

Plate heat exchanger

PLATE HEAT EXCHANGER
See Heat Exchanger, Plate.

PLATFORMATE
High-octane gasoline blending stock produced in a catalytic reforming unit, commonly known as a platformer.

PLATFORM AVERAGE
Said of a well drilled and completed within the average cost of an off-shore, platform-supported well.

Platform burner

PLATFORM BURNER
See Forced-Draft Burner.

PLATFORMER
A catalytic reforming unit that converts low-quality, straight-chain paraffins or naphthenes to low-boiling, branched-chain paraffins or aromatics of higher octane; a refinery unit that produces high-octane blending stock for the manufacture of gasoline.

PLATFORM JACKET
A supporting structure for an offshore platform consisting of large-diameter pipe welded together with pipe braces to form a four-legged stool-like structure (stool without a seat). The jacket is secured to the seafloor with pilings driven through the legs. The four-legged offshore platform is then slipped into legs of the jacket and secured with pins and by the weight of the platform and equipment.

PLATFORM TREE
A production Christmas tree on an offshore platform; an assembly of control and production valves used on offshore platforms through which wells are produced.

Jacket piling

PLATFORM WELL
An offshore well drilled from a fixed platform secured to the sea floor by pilings.

PLAY, OIL
New activity in oil country, as an extension of settled production or in a promising area somewhat removed from the established field. Seismic activity (shooting), leasing, and wildcatting in or on a trend. Lease play: a flurry of leasing by the landmen.

P.L.E.M.
Pipeline-end manifold; an offshore, submerged manifold connected to the shore by pipelines that serve a tanker loading station of the multibuoy-mooring type.

PLENUM
A room or enclosed area where the atmosphere is maintained at a pressure greater than the outside air. Central control rooms at refineries are usually kept at pressures of a few ounces above the surrounding atmosphere to prevent potentially explosive gases from seeping into the building and being ignited by electrical equipment. Some offshore drilling and production platforms are provided with plenums as a safety measure. *See* Acoustic Plenum.

PLUG
To fill a well's borehole with cement or other impervious material to prevent the flow of water, gas, or oil from one strata to another when a well is abandoned; to screw a metal plug into a pipeline to shut off drainage or to divert the stream of oil to a connecting line; to stop the flow of oil or gas.

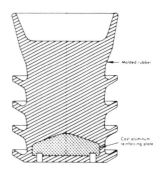
Top plug

PLUG, BRIDGE
See Bridge Plug.

Bottom plug

PLUG, BULL

See Bull Plug.

PLUG BACK

Plugging back means closing off sections of the borehole and exploring a shallower formation. An 8,000-foot well may be plugged back 1,000 feet to 7,000 feet; and the porous formation, previously drilled through and cased but which was not very promising, will now be tested. To plug back, the borehole is filled with cement to the projected level. If the hole is cased to 8,000 feet, the casing may be pulled up in the hole sufficiently to save the 1,000 feet of good pipe. The borehole is then filled with cement up to the new shallower zone. The casing protecting the new interval is cemented, perforated, and the formation tested and, with luck, produced.

PLUGGING A WELL

To fill up the borehole of an abandoned well with mud and cement to prevent the flow of water or oil from one strata to another or to the surface. In the industry's early years, wells were often improperly plugged or left open. Modern practice requires that an abandoned well be properly and securely plugged.

PLUG VALVE

See Valve, Plug.

PLUNGER

The piston in the fluid end of a reciprocating pump. *See* Pump, Plunger.

PLUNGER PUMP

See Pump, Plunger.

PLUTONIC ROCK

See Igneous Rock.

PNEUMATIC HOIST

See Air Hoist.

POCK MARKS

Crater-like depressions on the seafloor found in areas of the North Sea near the Norwegian trench, roughly paralleling the Norwegian coast. The pocks are believed, by geologists, to be caused by natural gas migration from underground pockets which gradually erodes the soft seabed to form the small craters that range in size from 5–50 feet deep, and 30 to several hundred feet wide. The pocks were discovered during a survey of a pipeline route from offshore wells to the Norwegian coast. *See* Terrasic Forms.

POGO PLAN

A plan for financing oil and gas exploration developed for offshore exploration. The form of the plan is usually corporate, the investors receiving shares of stock in the corporation and other securities.

POINT MAN
The member of a pipeline tong crew who handles the tips (the points) of heavy pipelaying tongs. He is the "brains" of the crew as he keeps his men pulling and "hitting" in unison and in time with the other tong crews working on the same joint of screw pipe.

POISON PILL
A financial tactic (as increasing indebtedness) used by a company to ward off a hostile takeover.

POLISHED ROD
A smooth brass or steel rod that works through the stuffing box or packing gland of a pumping well; the uppermost section of the string of sucker rods attached to the walking beam of the pumping jack.

POLYCHLOROPRENE
See Neoprene.

Polycrystalline diamond bit *(Courtesy Strata Bit Co.)*

POLYCRYSTALLINE DIAMOND COMPACT BIT
See Bit, Polycrystalline Diamond Compact.

POLYETHYLENE
A petroleum-derived plastic material used for packaging, plastic housewares, and toys. The main ingredient of polyethylene is the petrochemical gas ethylene.

POLYETHYLENE PIPE
A tough, plastic pipe made to handle corrosive liquids in the oilfield. It is lightweight, flexible, and not affected by acids, brines, or other corrosive forces.

POLYMERIZATION
A refining process of combining two or more molecules to form a single heavier molecule; the union of light olefins to form hydrocarbons of higher molecular weight. Polymerization is used to produce high-octane gasoline blending stock from cracked gases.

POLYVINYL CHLORIDE
See P.V.C.

PONTOONS
The elements of a floating roof tank that provide buoyance; airtight metal tanks that float on the fluid and support the movable deck structure of the roof.

PONY RODS
Sucker rod made in short lengths of 2–8 feet for use in making up a

string of pumping rods of the correct length to connect to the polished rod of the pumping jack. Pony rods are screwed into the top of the string just below the polished rod.

POOH & LAYDOWN
Oilfield parlance for pull out of the hole and lay down rods and tubing from a pumping well.

POOL
See Oil Pool.

POOL, SALT-DOME/SALT-PLUG
See Salt-Dome/Salt-Plug Pool.

POOLING
The bringing together of small, contiguous tracts, resulting in a parcel of land large enough for granting a well permit under applicable spacing regulations. Pooling is often erroneously used for unitization. Unitization describes a joint operation of all or some significant portion of a producing reservoir.

POOLING, THEORY OF CONTRACT
A theory concerning the nature of interests arising from a pooling or unitizing agreement. Under this theory, the relationship of the parties to the agreement is treated as one, as an entity, arising from a contract without a transfer of interests in land.

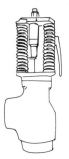

Pop-off valve

POOL OCTANE
The term "pool octane" is the weighted average octane of all gasoline components in a blend. Finished motor gasoline is a precise blend of many components. Modern refiners tailor their gasolines for climate, altitude, and other conditions that affect automobile engine performance.

POP-OFF VALVE
See Valve, Relief.

POPPET VALVE
See Valve, Poppet.

POPPING
The discharge of natural gas into the atmosphere; a common practice in the 1920s and 1930s, especially with respect to sour gas and casinghead gas. After the liquid hydrocarbons were extracted, the gas was "wasted" as there was no ready market for it.

PORCUPINE
A cylindrical steel drum with steel bristles protruding from the surface; a super pipe-cleaning pig for swabbing a sediment-laden pipeline.

PORE
The minute or microscopic voids in porous rock; rock that is porous is able to hold oil, gas, and water. If the pores are interconnected, the rock is permeable and a good reservoir rock.

PORE FILL
What fills the pores of a sedimentary formation: oil, water, gas, or a mixture of all three. In very permeable rock, drilling fluid under great hydrostatic pressure can force its way into the rock's pores, plugging them permanently if the reservoir pressure is insufficient to open up the minute passages when the mud is circulated out and the well is completed. *See* Pore Pressure.

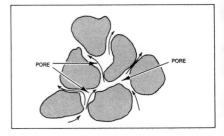

Pore Fill

PORE PRESSURE
In subsurface rock formations, normal pore pressure is hydrostatic pressure due to the head of formation water filling the pores of the rock in communication with the water table or the ocean. This is defined as an "open system." In a "closed system," abnormal pore pressure may arise because of the lack of communication between the hydrostatic head and a venting system.

PORE SPACE
See Pore.

POROSITY
The state or quality of being porous; the volume of the pore space expressed as a percent of the total volume of the rock mass; an important property of oil-bearing formations. Good porosity indicates an ability to hold large amounts of oil in the rock. And with good permeability the quality of a rock allows liquids to flow through it readily; a well penetrating the formation should be a producer.

POROSITY, ABSOLUTE
The percentage of the total volume of a rock sample or core composed of voids or pore spaces, interstices.

POROSITY, EFFECTIVE
See Effective Porosity.

POROSITY, FRACTURE
See Fracture Porosity.

POROSITY, SECONDARY
The term refers to natural fracturing, solution channeling, and dolomitization of the sedimentary reservoir.

POROSITY LOG
See Log, Porosity.

POROSITY POD

Relatively small areas (large enough for three or four wells) where the porosity and permeability of the underground formation (a sand body or sand lens) is higher than average for the remainder of the formation. A local area of better-than-average porosity in a sandstone formation, which can mean increased saturation of oil and gas.

POROSITY TRAP

A trap for petroleum formed by lateral variation of porosity of the reservoir rock, e.g., as a result of cementation, the presence of clay minerals or a decrease in matrix grain size; a stratigraphic trap.

PORTABLE PIPE MILL

A very large, self-propelled factory-on-wheels that forms, welds, and lays line pipe in one continuous operation. The pipe is made from rolls of sheet steel (skelp) shaped into a cylindrical form, electric welded, tested, and strung out behind the machine as it moves forward.

P.O.S.C.—PETROCHEMICAL OPEN SOFTWARE CORP

P.O.S.C. was organized by five leading, major corporations—TEXACO, MOBILE, ELF ACQUITAINE, CHEVRON, and BRITISH PETROLEUM to establish industry standards in data systems and application within the industry. "To define, develop, and deliver, through an open process, an industry standard open system, software integration platform for petroleum upstream technical computing applications." In short, let's standardize.

POSITIVE CRANKCASE VENTILATION SYSTEM

A system installed on automobiles manufactured after 1968 to reduce emissions from the engine's crankcase. The emissions oil and unburned gasoline vapors are directed into the intake manifold and from there they mix with the gasoline to be burned.

POSITIVE-DISPLACEMENT MUD MOTOR

See Mud Motor, Positive-Displacement.

POSITIVE-DISPLACEMENT PUMP

See Pump, Positive-Displacement.

Possum belly

POSSUM BELLY

A metal box built underneath a truck bed to hold pipeline repair tools—shovels, bars, tongs, chains, and wrenches.

POSTED PRICE

The price an oil purchaser will pay for crude oil of a certain A.P.I. gravity and from a particular field or area. Once literally posted in the field, the announced price is now published in area newspapers. With government control affecting almost all aspects of the industry, prices of oil and gas are not permitted to be set by the industry's supply-and-demand requirements as they once were.

POSTULATED OIL COLUMN
A phrase employed by geologists when hypothesizing (making educated assumptions) concerning the presence of oil in a particular sedimentary section downhole that is under study.

POT
See Valve Pots.

POT STILL
See Still, Pot.

POT STRAINER
See Strainer, Pot.

POUNDS PER SQUARE INCH GAUGE
Pressure as observed on a gauge; psig.

POUR POINT
The temperature at which a liquid ceases to flow or at which it congeals.

POUR-POINT DEPRESSANT
A chemical agent added to oil to keep it flowing at low temperatures.

POWDER MONKEYS
Workers who handle dynamite on pipeline construction jobs. They follow the rock-drill crew that have drilled spaced holes in the rock section of a pipeline ditch. The powder monkeys insert sticks of dynamite in the holes, cutting sticks in half, if necessary, to fill the hole including the detonating cap. Breathing the fumes, the vapor that arises from working with the explosive, produces severe headaches. To circumvent this hazard, experienced workers put small amounts of dynamite on their tongues, that will fit on the tip of a knife blade, and eat it. It is not an unpleasant taste, similar to creme of tartar. (The author, in his younger days, was a powder monkey and ingested his share of 40 percent dynamite; it works.)

Old central power plant

POWER
See Central Power.

POWER HOUSE, THE
See Central Power.

POWER SKID
A portable frame or sturdy foundation holding a prime mover—engine or motor—to form an engine-pump, engine-compressor, or engine-

generator unit or skid. A flatbed winch truck can winch the portable skid onto the truck to carry it to a work location.

POWER SLIPS
Patented casing-and-tubing slips that fit into a companion master bushing of the rotary table where a joint of drillpipe or casing is held for making up or breaking out, screwing in or screwing out, a joint. The driller actuates the mechanism so the slips drop into place in the bushing around the pipe.

POWER SWIVEL
A swivel that not only supports the drillstring, but also turns the drillpipe. On a rig equipped with a power swivel (hanging from the traveling block and hook), there is no conventional rotary table located on the rig floor. The torquing power is provided by the air or hydraulic motor in the swivel.

POWER SYSTEMS, DRILLING RIG
See Rig, Electric; *also* Rig, Mechanical.

POWER-TAKE-OFF
A wheel, hub, or sheave that derives its power from a shaft or other driving mechanism connected to an engine or electric motor; the end of a power shaft designed to take a pulley.

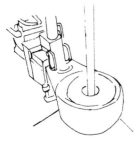

Power tongs

POWER TONGS
An air or hydraulically powered mechanism for making up and breaking out joints of drillpipe, casing, or tubing. After a joint is stabbed, the power tongs are latched onto the pipe, which is screwed in and tightened to a predetermined torque.

POWER TRAIN
The connecting mechanical elements that transmit the power generated by an engine to the driven item of equipment, i.e., pump, generator, feed mill, automobile. The power train may include crankshaft, transmission, clutch, drive shaft, differential, and axles.

P.P.G.
The symbol for pounds per gallon, the weight of drilling fluid. A drilling well's mud engineer's job is to adjust the weight of the mud to the well bore circumstances, to heavy up (to add weighting material) or to water back (to dilute the mud with injections of water).

P.P.M.
Parts per million; a measure of the concentration of foreign matter in air or a liquid.

P.P.M./VOL.
Parts per million (of water) in a given volume of natural gas. *See* Lbs-H_2O/MMscf.

P.P.M./WT.
Parts per million (of water) in a given weight of gas; used to express water content in a small amount of gas. *See* Lbs-H_2O/MMscf.

PRAIRIE-DOG PLANT
A small, basic refinery located in a remote area.

PRESSURE CHART
A circular paper chart on which pressure variations on a pressure gauge are recorded by an inked pen. The pen is attached to a moveable arm actuated by the gauge mechanism. Pressure charts usually record for 24 hours; then a fresh chart is put on the recording gauge by the lease pumper or plant operator.

PRESSURE CUP
See Grease Cup.

PRESSURE DIFFERENTIAL
A difference in pressure at two points in a closed system; the pressure difference between that on the upstream side of a gate valve and the downstream side.

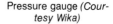

PRESSURE GAUGE
Used on gas or liquid lines to make instantly visible the pressure in the lines. Some gauges have damping devices to protect the delicate mechanisms from the transient pulsations of line pressure.

Pressure gauge *(Courtesy Wika)*

PRESSURE GAUGE, BLOWOUT-BACK
A type of high-pressure gauge for air or liquids that, for safety's sake, is made with a solid steel front (where the dial and pointer or hands are located) and a back plate that will blow out if the Bourdon tube in the gauge ruptures owing to excess pressure or material failure.

PRESSURE GRADIENT
See Gradient.

PRESSURE MAINTAINENCE
An oilfield term describing a most important activity. It means keeping up the reservoir pressure by any available means: reinjection (recycling) of produced gas, water flood, repressuring with carbon dioxide (CO_2), and shutting in flared gas.

PRESSURE SNUBBER
See Pulsation Dampener.

Pressure snubber

PRESSURE VESSEL
A cylindrical or spherical tank so constructed as to hold a gas or a liq-

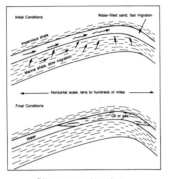

Oil or gas migration

uid under pressure. Pressure vessels are used to hold air for air-actuated valves, air-starting of engines, and other pneumatic applications. In a refinery or chemical plant, pressure vessels are integral parts of the processing chain where feedstock is subjected to both heat and pressure as part of the refining process.

PRIMARY MIGRATION
See Migration, Primary.

PRIMARY PRODUCTION
Primary recovery; production from a reservoir by natural energy (gas cap, solution gas, or water drive) that results in flowing wells, or wells on the pump with the oil flowing freely by gravity to the well bore. *See* Secondary Recovery.

PRIMARY RECOVERY
See Primary Production.

PRIMARY TERM
The period of time an oil and gas lease is to run or to be valid. When a lease's primary term expires, the lease must be renewed, if possible, or the interest in the property reverts automatically to the lessor or landowner. *See* Lease, Or; *also* Delay Rental.

PRIME A PUMP
To prime is to establish suction in a pump by filling it with liquid (water or oil) that acts to seal all clearances and voids, displacing the air. As the impeller (centrifugal pump) turns or the plunger (reciprocating pump) moves, expelling the prime charge, vacuum or suction is created, drawing in the liquid to be pumped.

PRIME MOVER
The term describes any source of motion; in the oil field it refers to engines and electric motors; the power source. Prime mover is also applied to large four-wheel-drive trucks or tractors.

PRIVATE BRAND DEALER
A gasoline dealer who does not buy gasoline from a "major" supplier but retails under the brand name of an independent supplier or his own brand name.

PROCESS STEAM
Steam produced in a refinery's or chemical plant's boilers to heat a process stream or for use in a refining process itself. Of the energy used in the United States, a large percentage is consumed in the production of process steam. Petrochemical plants are important users of superheated, high-pressure steam.

PROCESS STREAM

The charge or stream of liquids or gases moving through the different processes in a refinery or fractionating plant.

PROCESS WATER

Fresh water used in refining and other processing plants. A 100,000/bbl/day refinery uses from 6–9 million gallons of water a day so recycling and reuse of water is essential. The main uses of water (H_2O) are make up, cooling, boiler feed, and wash down.

PROCESSING PLATFORM

A production platform.

PRODUCED WATER

Water, usually salt water or brine, produced with oil in a pumping well. Small amounts of salt water can be separated out at the well site and put in an earthen evaporation pit. Large volumes must be dealt with by pumping it back into disposal wells, which force the super-salt brine into a porous formation isolated by impervious strata above and below.

PRODUCER'S 88 LEASE FORM

Generally, Producer's 88 lease forms contain an "unless" clause rather than an "or" clause, and they are executed for a number of years "and so long after" as oil or gas is produced. In short, a certain structural pattern is usually indicated by the phrase Producer's 88 lease, but in no way indicates the details of the leasing agreement. The ambiguity of this printed lease form was recognized by the courts as far back as 1933. A judicial opinion at that time states in part, "As we see it, the reference to Producer's 88 form lease is as incapable of definite application as if the term 'oil and gas lease form' had been used."

PRODUCING PLATFORM

An offshore structure with a platform raised above the water to support a number of producing wells. In offshore operations, as many as 60 wells are drilled from a single large platform by slanting the holes at an angle from the vertical away from the platform. When the wells are completed, the drilling equipment is removed and the platform is given over to operation of the producing wells.

PRODUCING SAND

See Sand.

PRODUCING ZONE

The interval or section of a porous and permeable formation in a well that has hydrocarbons (oil, gas, or condensate) present and being produced; pay zone; pay horizon.

Vertical heat exchangers on a process stream

Processing platform

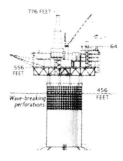

Producing platform

PRODUCT ARBITRAGE
See Arbitrage, Product.

PRODUCT GAS
End product gas; gas resulting from a special manufacturing process; synthetic natural gas.

PRODUCT IMPORT TICKET
A license issued by an agency of the federal government to a refiner or marketer to import product from abroad.

PRODUCTION CHOKE
Flow-control device that limits the flow of oil to a predetermined rate by the size of the aperture in the bean, the interchangeable orifice in the choke body. There are positive chokes with a fixed aperture and adjustable chokes whose flow rate can be changed by operating the handwheel as on a gate valve. Adjustable chokes usually are for relatively low-pressure wells. *See* Wellhead Choke.

PRODUCTION FLOATERS
Permanently moored large tankers that serve as production storage for offshore fields. In addition to storage, the tankers have facilities aboard to process the oil and to offload into smaller vessels. The production floaters are a relatively inexpensive method of handling crude oil from small fields or from one being developed. Fixed platforms are much more expensive, do not have the capacity of a tanker, and take much longer to build and bring on stream.

Production island *(Courtesy Conoco)*

PRODUCTION ISLAND
An island made by dredging up material from the bottom of a lake or the ocean bottom to support one or more producing wells. Production islands are constructed in shallow water, close to shore, and are usually cheaper to build than steel production platforms. And with a lower profile, the islands are less offensive to the esthetic eye. Also, an island can be landscaped to hide the pumping wells and other equipment from view.

PRODUCTION PACKER
An expandable plug-like device for sealing off the annular space between the well's tubing and the casing. The production packer is run as part of the tubing string, inside the casing; when lowered to the proper depth, the packer is mechanically or hydraulically expanded and "set" firmly against the casing wall, isolating the production formation from the upper casing while permitting the oil or gas to flow up the tubing.

PRODUCTION PAYMENT
A share of the oil or gas produced from a lease, free of the cost of production, that terminates when a specified sum of money has been real-

ized; an oil payment. Oil payments may be reserved by a lessor or carved out by the owner of the working interest. An interesting thing about a production payment is that there is a duty to deliver oil or gas under the agreement "when, if, and as" it is produced; there is no personal liability on the part of the grantor of the production payment.

PRODUCTION PAYMENT LOAN
A loan that is to be repaid out of the production of a well.

PRODUCTION PLATFORM
An offshore structure built for the purpose of providing a central receiving point for oil produced in an area of the offshore. The production platform supports receiving tanks, treaters, separators, and pumping units for moving the oil to shore through a submarine pipeline.

Production platform

PRODUCTION RATIO
One measure of a well's productivity is the gas/oil ratio (G.O.R.). If the ratio of gas-to-oil is too high, too much gas, too little oil, the well is neither fish nor fowl, neither a good gas well nor a good oil well, but can be acceptable.

PRODUCTION SKID
A prefabricated oil and gas production unit assembled on a base or skid onshore and transported to an offshore platform by one or more derrick barges. After the skid has been lifted into position and secured to the platform, it is connected to the flow lines of the offshore wells it is to serve and begins its function of receiving, separating, treating, storing, and pumping the oil and gas to shore stations. *See* Production Platform.

PRODUCTION STRING
The casing set just above or through the producing zone of a well. The production string is the longest yet smallest-diameter casing run in a well. It reaches from the pay zone to the surface.

PRODUCTION SYSTEM, OFFSHORE
A system of this nature includes one or more wells, a subsea gathering system, storage and tanker-loading facilities, and a shuttle tanker or tankers to transport the crude to a shore station.

PRODUCTION TAX
See Gross Production Tax.

PRODUCTION TREE
See Christmas Tree.

Production tree

PRODUCT LUBRICATED
Describes a pump whose bearings are lubricated by the liquid it is pumping. The pump is constructed with channels and wells that fill

with product and in which the bearings or other moving parts run. Product lubricated equipment, needless to say, handles only clean liquids, i.e., various kinds of oils with lubricating qualities.

PRODUCTS PIPELINE
A large-diameter pipeline that transports, from refinery to distributor's terminal, various refined products, e.g., gasoline, kerosene, heating oil and diesel fuel. Products in the line are separated by inflated synthetic rubber spheres. In the case of fungible products, diesel fuel, and light gas oil or heating oil, no mechanical separators are used. As the products are pumped down the line, one behind the other, there is very little mixing at the interface.

PRODUCT YIELD
From a 42-gallon barrel of crude oil the average yield is as follows: gasoline, 49.6 percent; jet fuel and kerosene, 6.6 percent; gas oil and distillates, 21.2 percent; residual fuel oil, 9.3 percent; lubricating oil, 7.0 percent; other products, 6.3 percent. With modern-day refining methods, these product percentages can be changed, depending upon market demand.

PROFIT-SHARING BIDS
A type of bidding for federal and sometimes state oil leases in which a relatively small cash bonus is paid for the lease acreage plus a share in the net profits should the lease prove to be commercially productive. In some instances, bidders have offered a 75 to 90 percent share in net profits for an especially promising parcel. This type of bidding substantially reduces the front-end cost for an operator but extends the payout time for his wells.

PROGRESSING-CAVITY PUMP
See Pump, Progressing Cavity.

PROJECT WELLS
A Canadian term for wells involved in specialized drilling as part of oil sands *in situ* projects for the production of raw bitumen. Canada has vast deposits of heavy-oil or bitumen that are exploited by *in situ* combustion programs. When oil prices are high, the oil sands drilling, production is viable and profitable. When prices slump, these programs are shut down as uneconomical.

PROPAGATING FRACTURES
Fractures or tears in a ruptured, high-pressure pipeline that travel almost instantaneously longitudinally (along the length of the pipe) driven by the pressure of the gas or liquid escaping through the initial rupture or break. The metal of the pipe tears, literally, along the

pipe until the stress of the escaping fluid diminishes to the point where the pipe metal resists the stress. Propagating fractures are dangerous and destructive. One method that has been used to arrest the runaway tearing of ruptured pipe is to put steel bands on the pipe at intervals that reinforce the pipe wall at that point and stop the tearing by containing or overcoming the stress. This would confine the pipe blowout to a relative short span, thus saving pipe and protecting life and limb.

PROPANE
A petroleum fraction; a hydrocarbon, gaseous at ordinary atmospheric conditions, but readily converted to a liquid. When in a liquid state, propane must be stored in a high-pressure metal container. Propane is odorless, colorless, and highly volatile. It is used as a household fuel beyond the gas mains.

PROPPANT PACK
The proppant material left in the cracks, fissures, and fractures of the formation after being forced there under extremely high pressure, transported by the viscous frac fluid. As the frac fluid thins and leaks off, retreats, it leaves its proppant pack in place, holding open the fractures.

Variety of propping agents *(Courtesy Pan American Petroleum)*

PROPPANTS
Material used in hydraulic fracturing for holding open the cracks made in the formation by the extremely high pressure applied in the treatment; the sand grains, beads, or other miniature pellets suspended in the fracturing fluid that are forced into the formation and remain to prop open the cracks and crevices permitting the oil to flow more freely.

PROPPANT TRACER
A radioactive material, Iridium 192, for example, along with the sand or other proppant material (ceramic beads) makes possible keeping track of the proppants. The operators can observe on the monitor how the frac and the proppant are doing, whether the high-pressure gel has opened a single large fissure with the resultant fluid loss or is proceeding equally on a wide front through the formation.

Proppants, close-up view *(Courtesy Amoco)*

PROPPED FRACTURE
A pressure-induced formation fracture that has been invaded by the viscous frac fluid (gelled water or gelled oil) carrying proppant material. When the frac fluid breaks, becomes less viscous, it drains out of the fractured rock, leaving behind the proppant material, holding open the minute cracks and fissures.

PROPRIETARY DATA
Information on subsurface geological formations gathered or purchased from a supplier of such data by an operator and kept secret; land and offshore reconnaissance surveys from seismic, magnetic, and gravity studies that are privately owned.

PROPRIETARY INFORMATION
Information that is yours, exclusively. In the oil business, seismic work which you commission results in certain information as to geologic formations, favorable or unfavorable areas to drill. This is proprietary information. Some geophysical companies will run seismic surveys and sell the information to more than one client. This type of report is labelled nonexclusive. It may be useful; but as others have it too, it is of no particular advantage to your company.

PRORATIONING
Restriction of oil and gas production by a state regulatory commission, usually on the basis of market demand. Prorationing involves allowables that are assigned to fields, and from fields to leases, and then allocated to individual wells. *See* Pipeline Prorationing.

PROTECTIVE STRING
A string of casing used in very deep wells and run on the inside of the outermost casing to protect against the collapsing of the outer string from high gas pressures encountered.

PROVED RESERVES
Oil that has been discovered and determined to be recoverable but is still in the ground.

PROVINCE, GEOLOGIC
An area of geologic interest; the loosely defined location of related depositional systems or stratigraphy; province, larger than an area, but smaller than a region. How big is big?

P.S.I.A.
Pounds per square inch absolute (psia), which includes atmospheric pressure.

PUDDING STONE
A conglomerate; a stone that, in looks, reminds you of a pudding with raisins and nuts.

PUGH CLAUSE
A type of pooling clause that provides that drilling operations on a product from a pooled unit or units shall hold the lease in force only on

the lands included within such unit—not on land outside the voluntary pooled unit. This clause purportedly was originated by and took its name from L.C. Pugh of Crawly, Louisiana in 1947, although other oil states had similar clauses in leases prior to this date.

PULLED THREADS
Stripped threads; threads on pipe or tubing damaged beyond use by too much torque or force used in making up the joint.

PULLER
A portable, workover rig; a rig used for shallow- or medium-depth drilling or workovers.

PULLING CASING
Removing the well's casing from the borehole. Most leases permit the lessee to pull the casing at any time. Casing is pulled when a well is abandoned and is to be plugged according to regulations. If the casing has not been cemented, the operator may be able to salvage one-half to three-fourths of the footage. The surface pipe is not salvageable, but the intermediate and long string may be saved to be used another day.

PULLING MACHINE
A pulling unit.

PULLING RODS
The operation of removing the pumping or sucker rods from a well in the course of bringing up the bottom-hole pump for repairs or replacement. Rods must also be pulled if they have parted downhole. The rods above the break are pulled in a normal manner; the lower section must first be retrieved with a fishing tool.

Pulling rods

PULLING TOOLS
Taking the drillpipe and bit out of the hole. If the tools are to be run again (put back in the hole), the drillpipe is unscrewed in two or three-joint sections (stands) and stacked in the derrick.

PULLING UNIT
A portable, truck-mounted mast equipped with winch, wirelines, and sheaves, used for pulling rods or well workovers.

PULL ONE GREEN
To pull a drill bit from the hole before it is worn out; to pull a bit before it is necessary.

PULL ROD
See Shackle Rod.

Pulling unit

Pulsation dampener

Centrifugal pump

A feeder pump (center foreground) feeding chemicals to heater treater

PULL-ROD LINE
See Shackle Rod.

PULSATION DAMPENER
Various devices for absorbing the transient, rhythmic surges in pressure that occur when fluid is pumped by reciprocating pumps. On such pumps air chambers are installed on discharge lines, which act as air cushions. To protect pressure gauges and other instruments from the incessant pounding, fine-mesh, sieve-like disks are placed in the small tubing or piping to which the gauge is attached; this arrangement filters out much of the surging that can damage delicate gauges.

PUMP, BOTTOM-HOLE
A pump located in the bottom of the well and not operated by sucker rods and surface power unit. Bottom-hole pumps are compact, high-volume units driven by an electric motor, or are hydraulically operated.

PUMP, BUBBLE-POINT
A type of downhole oil pump very sensitive to gas. When the saturation pressure is reached, gas is released, which gas-locks the pump until pressure is again built up by the oil flowing into the well bore. This type of pump regulates, in effect, oil production from a reservoir with a gas drive.

PUMP, CASING
A sucker-rod pump designed to pump oil up through the casing instead of the more common method of pumping through tubing. A casing pump is run into the well on the sucker rods; a packer on the top or bottom of the pump barrel provides packoff or seal between the pump and the wall of the casing at any desired depth. Oil is discharged from the pump into the casing and out the wellhead.

PUMP, CENTRIFUGAL
A pump made with blades or impellers in a close-fitting case. The liquid is pushed forward by the impellers as they rotate at high speed. Centrifugal pumps, because of their high speed, are able to handle large volumes of liquid.

PUMP, CHEMICAL FEEDER
A small-volume pump used on oil leases to inject chemicals into flow lines. The pump may be located at the wellhead and be actuated by the motion of the pumping jack. The chemical is used to break down water/oil emulsions that may be contained in the crude oil stream.

348

PUMP, CHEMICAL INJECTION
A small-volume, high-pressure pump for injecting chemicals into producing wells or pipelines. Chemicals are injected into oil streams to reduce any emulsified oil to free oil and water. When the droplets of water are freed of their film of oil, the water will drop out, settle out of the
oil stream, and can be drawn off.

PUMP, DONKEY
Any small pumping device used on construction jobs or other temporary operations.

PUMP, DOUBLE-ACTION
A reciprocating pump (plunger pump) with two sets of suction and discharge valves permitting it to pump fluid during the forward and backward movement of each plunger.

PUMP, DOUBLE-DISPLACEMENT
A type of downhole, rod pump that has plungers placed in tandem and
operated simultaneously by the pump rods.

Duplex pump *(Courtesy Gaso)*

PUMP, DUPLEX
A two-cylinder reciprocating plunger pump.

PUMP, FORCE
A barrel pump; a portable, hand-operated, one-cylinder pump for moving limited amounts of liquid, pumping out sump pits, or transferring
oil or water from one small tank to another. The pump has one horizontal barrel and a plunger attached to a vertical handle. When moved
back and forth, the handle, attached to a fulcrum at the base of the
pump, actuates the plunger.

PUMP, GEAR
A type of rotary pump made with two sets of meshing gears. When rotated on their shafts in the pump housing, fluid is taken in the suction
port and forced out the discharge port. As the gears rotate, they mesh in
a rolling action like an old-fashioned clothes wringer. Gear pumps, like
other rotary pumps, efficiently handle small volumes of fluid, often of
high viscosity, at high pressures.

Cutaway of gear pump

PUMP, GRABLE OIL-WELL
A patented, drum-and-cable pumping unit that can be installed in a
wellhead cellar. The unit raises and lowers the pumping rods by winding and unwinding cable on a drum or spool. The low profile of the
pumping unit makes it ideal for use in populated areas and to protect
the beauty of the landscape.

PUMP, GRAVITY
See Pump, Sight.

PUMP, IN-TANK
A type of vertical submersible pump used to remove oil from a storage tank over the top instead of through the pipeline connection in the side of the tank a foot or two from the bottom. In-tank pumps are used when the depth of sediment in the bottom of the tank prohibits pumping from the tank connection.

PUMP, JERKER
A single-barrel, small-volume plunger pump actuated by the to-and-fro motion of a shackle-rod line and an attached counterweight. The jerker pumps on the pull stroke of the rod line; it takes in fluid (the suction stroke) as the counterweight pulls the plunger back from the pumping stroke. Jerkers pump small volumes but can buck high pressure.

Jerker pump *(Courtesy Gaso)*

PUMP, LUBE OIL
See Lubricator, Oil.

PUMP, METERING
A small-volume rotary or reciprocating pump for injecting measured amounts of additives into process lines or into crude oil lines at the well-head. Metering pumps, as the name implies, inject small, metered amounts of chemical additives against high pressure, often as high as 5,000 psi. A chemical feeder-pump.

PUMP, MUD
A large, reciprocating pump that circulates drilling mud in rotary drilling. The duplex (two-cylinder) or triplex (three-cylinder) pump draws mud from the suction mud pit and pumps the slurry downhole through the drillpipe and bit and back up the borehole to the mud settling pits. After the rock cuttings drop out in the settling pit, the clean mud gravitates into the suction pit where it is picked up by the pump's suction line. In rotary drilling there are at least two mud pumps, sometimes more. In case of a breakdown or other necessary stoppages, another pump can be immediately put on line.

Mud pump

PUMP, PITCHER
A small hand pump for very shallow water wells. Looking much like a large, cast-iron cream pitcher, the pitcher pump is built on the order of the "old town pump" with one exception. The pitcher pump's handle, working on a fulcrum, does not have a string of pump rods attached. Water is pumped by the suction created by a leather cup and valve arrangement in the throat or lower body of the pump together with a

foot valve 20 feet or so down in the tubing. A simple and elegantly fundamental pumping machine.

PUMP, PLUNGER
A reciprocating pump in which plungers or pistons moving forward and backward or up and down in cylinders draw in a volume of liquid and, as a valve closes, push the fluid out into a discharge line.

PUMP, POSITIVE-DISPLACEMENT
A pump that displaces or moves a measured volume of liquid on each stroke or revolution; a pump with no significant slippage; a plunger or rotary pump.

Reciprocating pump
(Courtesy Partek Pumps)

PUMP, PROGRESSING CAVITY
A positive-displacement pump that is a modified screw pump. A single revolving rotor in a close-fitting case moves the fluid along in a continuous stream. This type pump is used for viscous liquids, heavy oils or corrosive slurries.

PUMP, RECIPROCATING
A pump with cylinders and pistons or plungers for moving liquids through a pipeline; a plunger pump. The pistons or plungers move forward and backward, alternately drawing fluid into the cylinders through the suction valves and discharging the liquid through discharge valves into a pipeline. Reciprocating pumps are used extensively in the field and at refineries for moving crude oil and products. They handle relatively small volumes but do so at high pressures. Large volumes of oil moved in trunk or main lines are pumped with large high-speed centrifugal pumps.

Rod-line pump

PUMP, ROD
A class of downhole pumps in which the barrel, plunger, and standing valve are assembled and lowered into the well through the tubing. When lowered to its pumping position, the pump is locked to the tubing to permit relative motion between plunger and barrel. The locking device is a holddown and consists either of cups or a mechanical, metal-to-metal seal.

PUMP, ROD-LINE
An oil-well pump operated by a shackle-rod line; a pumping jack. *See* Rocker.

PUMP, ROTARY
A positive-displacement pump consisting of rotary elements—cams, screws, gears, or vanes—enclosed in a case; employed, usually, in handling small volumes of liquid at either high or low pressures. Because of

Rotary pump *(Courtesy Roper Pump Co.)*

Screw pump

Screw Pump (for
well bore)

the close tolerances in the meshing of the gears or cams, rotary pumps cannot handle liquids contaminated with grit or abrasive material without suffering excessive wear or outright damage.

PUMP, SAND

A cylinder with a plunger and valve arrangement used for sucking up the pulverized rock, sand, and water from the bottom of the well bore. More effective than a simple bailer. Shell pump; sludger.

PUMP, SCREW

A rotary pump made with one, two, or three screws or spiral members. When rotated on their shafts, the screws closely mesh and take in fluid at the suction end of the pump and force it out the discharge port in a continuous stream. Screw pumps, which are small, usually are driven by electric motors but can be hooked up to gas engines. Screw pumps and other types of rotary pumps are used in refineries and chemical plants to handle highly viscous fluids and as transfer pumps for small volumes of liquid at high pressures.

PUMP, SIGHT

An antique gasoline-dispensing system in which the gasoline was pumped by hand into a 10-gallon glass tank atop the pump in plain sight of the customer. When the glass cylinder had been pumped full, the attendant opened the valve on the filling hose, which permitted the gasoline to gravitate into the vehicle's tank. Gravity pump.

PUMP, SIMPLEX

A one-cylinder steam pump used in refineries and processing plants where extra or excess steam is available. Simplex pumps are simple, direct-acting pumps with the steam piston connected directly to the pump's fluid plunger.

PUMP, SINGLE-ACTION

Single-action pumps discharge on the forward stroke and draw in fluid on the return stroke.

PUMP, SPLIT-CASE, TWO-STAGE

A small suction pump in a two-section but integral case with two back-to-back impellers. The two-stage pump is used for boiler feed, hydraulic cleaning, and other applications where high pressure is required. (Developed by Worthington Inc., Dresser Industries.)

PUMP, STEAM

A reciprocating pump that receives its power from high-pressure steam. Steam is piped into the pump's steam chest and from there it is admitted to the power cylinder where it acts upon the pump's power pistons, driving them to and fro as the steam valves open and close. The fluid end of the pump is driven by the steam pistons. *See* Pump, Simplex.

PUMP, SUBMERSIBLE
A bottom-hole pump for use in an oil well when a large volume of fluid is to be lifted. Submersible pumps are run by electricity and, as the name implies, operate below the fluid level in the well.

PUMP, SUCKER-ROD
A rod pump.

PUMP, TRAVELING-BARREL
A downhole pump, operated by rods, in which the barrel moves up and down over the plunger instead of the plunger reciprocating in the barrel as in more conventional pumping devices.

PUMP, TRIPLEX
A reciprocating pump with three plungers or pistons working in three cylinders. The triplex pump discharges fluid more evenly than a duplex or two-plunger pump, as it has a power stroke every one-third of a revolution of the crankshaft compared to every half revolution for the duplex pump.

Triplex pump *(Courtesy Gaso)*

PUMP, TUBING
A class of downhole pumps in which the barrel of the pump is an integral part of the tubing string. The barrel is installed on the bottom of the string of tubing and is run into the well on the tubing string. The plunger assembly is lowered into the pump barrel on the string of pump rods.

PUMP, TURBINE
A type of centrifugal pump driven by a direct-connected electric motor; commonly used to aerate large settling ponds.

PUMP, VACUUM
A pump to create a partial vacuum on the well's casing and thereby on the pay zone in the expectation of increasing production—moderately successful.

Turbine pump

PUMP, VANE
A type of rotary pump designed to handle relatively small volumes of liquid products: gasoline and light oils as well as highly viscous fluids.

PUMP, VERTICAL
A type of submerged, centrifugal pump used to aerate large settling basins or pump out mine water or tanks that must be emptied from the top down, e.g., the underwater storage tanks of Dubai Petroleum Co. in the Arabian Gulf. Vertical pumps are directly connected to powerful electric motors.

PUMP AROUND

A refinery maneuver or practice of drawing off fluid from a process vessel and running it through that stage of the process again; or drawing down a quantity of feed at a certain stage of the process and pumping it back into the process loop farther downstream.

PUMP A WELL DEAD, TO

To load the borehole; to pump weighted drilling mud into the borehole or the casing (if the hole is cased, pipe is set) until the hydrostatic head kills the well, shuts off the escaping gas or oil.

PUMPDOWN

(1) In cementing the casing in a well or squeezing a well, pumping the cement slurry down the casing and up the borehole between the casing and the wall of the hole. Sometimes the cement is pumped downhole through the well tubing. *See* Stab-In Cementing. (2) Lowering the crude oil or products on hand by pumping down the storage tanks.

PUMPER

A person who operates a well's pumping unit, gauges the lease tanks, and keeps records of production; a lease pumper.

PUMPING, BACKSIDE

An arrangement whereby two adjacent pumping wells, in the same pay zone, may be operated by one engine or electric motor. The double-sided, front-and-back, pumping jack is built to utilize the down-stroke weight of the string of rods of one well to counterbalance and assist in the upstroke, the pumping stroke, of the adjacent well. One drawback to this ingenious arrangement is if one well pumps off or the rods part, both wells are incapacitated.

PUMPING DERRICK

See Derrick, Pumping.

Small Pump Jack

PUMPING JACK

An oil-well pumping unit consisting of a gasoline engine or electric motor that actuates a walking beam in a pumping action. At the well end of the walking beam is a horsehead attached to the string of sucker rods or pump rods. The other end of the beam is weighted with counterweights to balance the string of rods in the well. *See* Pumping Unit, Beam-Balanced; also Pumping Unit, Crank-Balanced.

PUMPING RIG

A pumping derrick.

PUMPING UNIT
A pump connected to a source of power; an oil-well pumping jack; a pipeline pump and engine.

PUMPING UNIT, AIR-BALANCED
An oil-well pumping jack equipped with a piston and rod that work in an air chamber to balance the weight of the string of sucker rods. The device is attached to the well end of the walking beam and, acting as a shock absorber, does away with the need for counterweights on the rear end of the walking beam.

Air-balanced pumping unit

PUMPING UNIT, BEAM-BALANCED
An oil-well pumping unit that carries its well-balancing weights on the walking beam on the end opposite the pump rods. The weights are usually in the form of heavy iron plates added to the walking beam until they balance the pull or weight of the string of pumping rods.

PUMPING UNIT, CRANK-BALANCED
An oil-well pumping unit that carries its counterweights on the two cranks that flank the unit's gear box. The string of pump rods is balanced by adding sufficient extra iron weights to the heavy cranks. The walking beam on this type unit is short and is not used as a balancing member.

Beam-balanced pumping unit

PUMPING UNIT, LOW-PROFILE
A sucker rod pump without the usual horsehead pumping jack. Instead, it is operated by a drum-and-cable mechanism located in the wellhead pit. *See* Pump, Grable Oilwell.

PUMPING WELL
An oil well that has ceased flowing (if it ever did) and is on the pump. Nearly all wells are brought in and immediately put on the pump. A pumping well may produce hundreds of barrels of oil per day (along with hundreds of barrels of water) or it may be a stripper well pumping 10 or fewer barrels a day. Many wells have been on the pump and earning their keep for more than 25 years.

PUMP OFF
To pump a well so rapidly that the oil level falls below the pump's standing valve; to pump a well dry, temporarily.

Crank-balanced unit
(Courtesy Lufkin Industries Inc.)

PUMP RODS
Sucker rods; steel rods about 30 feet long and 5/8 to 1 inch in diameter that are screwed together to form a string of rods that connects the

bottom-hole pump in the well's tubing to the pumping jack at the surface. Steel rods attached to a reciprocating pump's pistons. *See* Pump, Reciprocating.

PUMP SPECIFICATIONS
A plunger pump designated as 6 × 12 duplex is a two-cylinder pump whose cylinders are 6 inches in diameter with a stroke of 12 inches. A pump with replaceable liners (cylinders) may carry a specifications plate that reads: 4–6 × 10. This pump can be fitted with liners and pistons from 4 inches to 6 inches in diameter; its stroke is 10 inches.

PUP JOINT
A joint of pipe shorter than standard length; any short piece of usable line pipe.

PUP JOINTS, A.P.I.
Short sections of well tubing made to American Petroleum Institute standards. Pup joints come in different lengths to make up a string of tubing of the proper length, from the bottom of the well to the tubing hanger in the wellhead. Made at the pipe mill under controlled conditions, the short joints are of the same quality as the rest of the tubing.

PURCHASER PRORATIONING
See Pipeline Prorationing.

PUSH-DOWN RIG
See Rig, Push-down.

PUSHER
See Gang Pusher.

PUT ON THE PUMP
To install a pumping unit on a well. Some wells are pumped from the time they are brought in or completed; others flow for a time (sometimes for many years) and then must be put on the pump.

P.V.C.
Polyvinyl chloride; a commercial resin derived from petroleum; the principal ingredient of P.V.C. is ethylene. The resin can be molded, extruded, or made into a thin, tough film.

PYROBITUMEN
A material that, when heated, yields bitumens, petroleum-like substances: asphalt, kerogen, coal tar. Oil shales are pyrobitumens; they yield kerogen upon being heated. *See* Kerogen Shales.

PYROMETER

An instrument for measuring very high temperatures beyond the range of mercury thermometers. Pyrometers use the generation of electric current in a thermocouple or the intensity of light radiated from an incandescent body to measure temperatures.

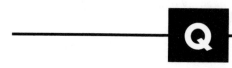

Q

QUAD
A convenient term for a quadrillion Btu, 1 and 15 zeros, when dealing with large amounts of energy. Btu: British Thermal Unit.

QUAD-LATERAL WELL
A horizontal well with two stacked laterals, two laterals one above the other, in separate geological zones, and two stacked laterals in the opposite direction, all four orthogonal (mutually horizontal) to the fracture plane. This technique doubles the formation exposure and reduces the cost per foot of drain hole.

QUANTUM LEAP
An abrupt change; sudden increase; a dramatic advance; also a quantum jump. *Webster's Ninth Collegiate Dictionary.*

QUARTER-TURN VALVE
See Valve, Quarter-Turn.

Quarter-turn ball valve
(Courtesy Marpac)

QUARTZ
An important rock-forming mineral; crystalline silica. Next to feldspar, quartz is the most common mineral that occurs, either in transparent six-sided (hexagonal) crystals or in crystalline masses. It is colorless, or given various colors by impurities forms the major proportion in most sands, and is widely distributed in igneous, metamorphic, and sedimentary rocks. Quartz is very hard, a 7 on the Mohs scale (diamond is 10), and will scratch glass.

QUARTZITE
A very hard, compact sandstone (but not metamorphosed) that consists of quartz grains that have been solidly cemented together with silica (SiO_2). The cementing silica grows around each quartz grain, locking them tightly together and filling the original pore spaces of the sedimentary rock.

QUENCHING
The very rapid cooling of a substance, a charge, in order to preserve certain chemical-physical characteristics of the material in the high-temperature state which would be changed by slow cooling.

QUENCH OIL
A specially refined oil with a high flash point used in steel mills to cool hot metal.

QUICK-ACTING VALVE
See Valve, Quick-Acting.

QUITCLAIM
An instrument or document releasing a certain interest in land owned by a grantor at the time the agreement takes effect. The key phrase of a quitclaim is: "... to release, remise, and forever quitclaim all right, title, and interest in the following described land."

R

Racking board

R.&D.

Research and Development; often used to denote a function up to the stage where the commercial potential of a process or technology can be evaluated. *See* Pilot Plant.

RABBIT

A plug put through lease flow lines for the purpose of clearing the lines of foreign matter and water and to test for obstructions. *See* Pig.

RACKING BOARD

A platform high in the derrick, on well-service rigs, where the derrickman stands when racking tubing is being pulled from the well.

RACK PRICING

Selling to petroleum jobbers or other resellers F.O.B. at the refinery, with the customers picking up pipeline or other transportation charges.

RADIAL BEARING

See Bearing, Radial.

RADIUS PROFILE

What is involved in deviating a vertical borehole from K.O.P. (kick off point) to an eventual deviation of 90°, to horizontal. Involved in this radical change of direction are questions of angle building and the radius of the curve. There are radii that vary from short, to medium, to long. Much depends upon the depth of the hole before kick off is attempted. Shallow sands require short-radius curves; tapping a deep formation, the operator can be more leisurely in coming around to horizontal; the radius can be longer; he can start out and build a continuous curve—no danger of dog legs or keyseats.

RAFFINATE

In solvent-refining practice, raffinate is that portion of the oil being treated that remains undissolved and is not removed by the selective solvent.

RAINBOW

(1) The iridescence (blues, greens, and reds) imparted to the surface of

water by a thin film of crude oil. (2) The only evidence of oil from an unsuccessful well; "Just a rainbow on a bucket of water."

RAM, BLIND
A closure mechanism on a blowout-preventer stack that is operated hydraulically to close in or shut in a well threatened with excessively high pressure. When actuated, the two elements of the closure mechanism come together, sealing off the well bore. It is called blind because the ram does not close on the pipe as does a pipe ram or shear ram. *See* Ram, Shear.

RAM, SHEAR
A closure mechanism on a well's blowout-preventer stack fitted with chisel-like jaws that are hydraulically operated. When the ram is closed on the pipe, the jaws or blades cut the pipe, permitting the upper section to be removed from the B.O.P. stack.

RAMPING SCHEDULE
A fracing operation, with its convocation of hydraulic pumps and high-pressure piping, pumping frac fluid-carrying proppant material into a downhole formation.

RAMS, CHANGE
To change rams means to remove one set of pipe rams from the blowout-preventer stack and replace them with another set of the proper size and type. When the size of the drillpipe is changed for any reason, the size of the rams must be changed so they will fit the pipe they are to close on in the event of a gas kick or a threatened blowout.

RANDOM DUMP PACKING
See Vacuum Tower, Packed.

RANGE OIL
Kerosene-type product used in oil or kerosene stoves or cooking ranges.

RATEABLE TAKE
(1) Production of oil and/or gas in such quantities that each landowner whose property overlies a producing formation will be able to recover an equitable share of the oil or gas originally in place beneath his land. (2) Production in accordance with allowables set by a state regulatory commission. (3) In some states, common carriers and common purchasers of gas and oil are prohibited from discriminating in favor of one supplier over another.

RATHOLE
(1) A slanted hole drilled near the well's borehole to hold the kelly joint when not in use. The kelly is unscrewed from the drillstring and lowered

Ram-type preventer

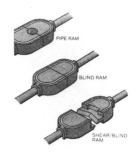

Ram elements

into the rathole as a pistol into a scabbard. (2) The section of the borehole that is purposely deviated from the vertical by the use of a whipstock.

RATHOLE AHEAD
To drill a hole of reduced diameter in the bottom of the regular borehole to facilitate the taking of a drillstem test.

RATHOLE RIG
A light, portable drilling rig often used to spud a well and set conductor pipe, or for preparing the hole for a coiled tubing unit to be moved in to drill the well to T.D., depth or target depth; a spudding rig; and from earlier days, a Forth Worth spudder.

RAW GAS
Gas straight from the well before the extraction of the liquefied hydrocarbons (gasoline, butane); wet gas.

RAW GASOLINE
The untreated gasoline cut from the distillation of crude oil.

RAW MIX
A stream of mixed components: butane, propane, hexane, and others; the product of gas-processing plants that is sent on to fractionating plants for the separation of the various components. *See* Field Butanes.

RAW-MIX STREAM
A mixture of natural gas liquids being pumped through a pipeline; commingled gas liquids.

REACTOR VESSEL
Any of the large vertical vessels at a refinery in which chemical reactions or changes in the feedstock take place. Catalytic crackers, regenerators, and fractionators are, broadly speaking, reactor vessels.

REAL TIME
Current, ongoing, right now; time as ascertained by engineers and scientists concurrent with a project in hand. Testing, measuring, identifying, proving.

REAMER
A tool used to enlarge or straighten a borehole; a milling tool used to cut the casing downhole. Reamers are run on the drillstring and are built with cutting blades or wheels that can be expanded against the walls of the hole.

REBOILER
A refinery heater that reheats or reboils a part of a process stream drawn off a distilling column and then reintroduced to the column as a vapor.

Reboiling is a process of reworking a part of the charge in a distilling column to ensure more complete fractionating.

RECIPROCATING PUMP
See Pump, Reciprocating.

Reciprocating pump

RECLAIMED OIL
Lubricating oil that, after undergoing a period of service, is collected, rerefined, and sold for reuse.

RECTIFIER BED
A source of electric current for protection against corrosion of pipelines, tanks, and other metal installations buried or in contact with the earth. Using a source of A.C. electric current, the rectifier installation converts the A.C. to D.C. and allows the D.C. to flow into the metal to be protected. By reversing the flow of electric current, the corrosion is inhibited. Metal corrosion is a chemical action that produces minute quantities of current that normally flow away from the metal into the ground.

RECUMBENT FOLD
A geologic term for an overturned fold, the broad surface, the axial surface of which is horizontal; a breaking-wave form

RECYCLING (GAS)
Injecting gas back into a formation to maintain reservoir pressure to produce a larger percentage of oil from the formation.

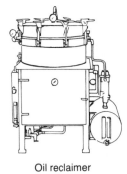

Oil reclaimer

RED BEDS
Sedimentary strata composed mainly of sandstone, siltstone, shale, and in some places, thin layers of conglomerate and limestone that are red owing to the presence of ferric oxide coating each grain. An example is the Permian sedimentary rocks of the southwestern U.S.; Permian red beds.

REDUCED CRUDE OIL
Crude oil that has undergone at least one distillation process to separate some of the lighter hydrocarbons. Reducing crude lowers its A.P.I. gravity.

REDUNDANCY
As it refers to oilfield operations, redundancy is the complete backup or duplication of a power unit or a process unit in a field, refinery or offshore operation. Should the primary source of power fail, a backup or standby source is ready instantly to take over.

REEF RESERVOIR
A type of reservoir trap composed of rocks, usually limestone, made up of the skeletal remains of marine animals. Reef reservoirs are often char-

acterized by high initial production that falls off rapidly, requiring pressure maintenance techniques to sustain production. *See* Organic Reef.

REEFS, ORGANIC
Barrier reefs; pinnacle reefs; fringing reefs; and patch reefs.

REEL BARGE
A pipelaying barge equipped with a gigantic reel on which line pipe up to 12 inches in diameter is spooled at a shore station. To lay the pipe, it is unspooled, run through straightening mandrels, inspected, and paid out over the stern of the barge in the manner of a hawser.

Reel barge

REELED TUBING
A well-service tool used in well workovers. The 1-inch or so flexible tubing is carried on a large spool mounted on a specially equipped truck. The tubing is inserted in the well through the wellhead valves and is used basically for flushing out the well and reestablishing a circulating path.

REENTRY
To reestablish contact with the well's borehole in offshore waters, after having moved off location because of weather or other reasons halting drilling operations. A notable example of reentering was that of the Deep Sea Drilling Program by the Scripps Institution of Oceanography when the crew of the drillship Glomar *Challenger* reentered the hole nine times while drilling in 14,000 feet of water in the Atlantic. *See* Acoustic Reentry.

REEVING A LINE
To string up a tubing or other line in preparation for hoisting; to run a line from the winch up and over a sheave in the crown block and down to the derrick floor.

Refinery fractionating towers

REFINER-MARKETER
A marketer of gasoline and/or heating oils who operates his own refinery.

REFINERY
A modern refinery is a large plant of many diverse processes. A refinery receives its charge stock, or crude oil, from the field via pipeline or from a tanker if the plant is located on a waterway. By processes that include heating, fractionating, pressure, vacuum, reheating in the presence of catalysts, and washing with acids, the crude is divided into hundreds of components: from exotic light gases to volatile liquids down through gasoline, naphtha, kerosene, gas oils, and light and heavy lubricating oil stocks to heavy bunker fuel, residual oil, and finally petroleum coke, the bottom of the barrel.

REFINERY, CONVERSION
A petroleum refinery with a high "bottoms" upgrading capacity. Typical conversion units usually include a needle-grade coker, an anode-grade coker, and a thermal cracker.

REFINERY, SKID-MOUNTED
A small, basic refining unit that is transportable by lowboy trailer to locations where low-grade or straight-run gasoline and diesel fuel are needed and a source of crude oil is available. For example, such a midget "distillation system" can be trucked to a remote drilling site and can supply fuel for diesel drilling engines and gasoline for auxiliary equipment.

REFLECTION SHOOTING
A type of seismic survey in which the travel time of artificially produced seismic waves is measured as they are reflected back to the surface from rock boundaries or surfaces of formations that have different densities. Sound waves are reflected back to recording instruments at the surface much quicker from high-velocity layers—very hard, dense rock—than from soft porous formations that may or may not contain fluids.

REFLECTOR
An interface between media of different elastic qualities that reflect seismic waves.

REFLUX
Return flow; a reflux vessel, a boiler in a refinery or processing plant that reheats certain liquid streams.

REFORMING PROCESSES
The use of heat and catalysts to effect the rearrangement of certain hydrocarbon molecules without altering their composition appreciably; the conversion of low-octane gasoline fractions into higher octane stocks suitable for blending into finished gasoline; also the conversion of naphthas to obtain more volatile product of higher octane number.

Reformer furnace

REFRACTION
The deflection of a ray of light or of an energy wave (such as a seismic wave, a sound wave) owing to its passage from one medium to another of a different density, which changes its velocity.

REGENERATOR
A refinery vessel into which inactive or spent catalyst is pumped to regenerate it, to burn off the coating of carbon or coke. Air at a temperature of 1,100°F is mixed with the spent catalyst, causing the oxidation of the carbon and leaving the catalyst clean and regenerated.

Reformer charge heater

Reid vapor test gauge
(Courtesy Weksler)

Relief valve

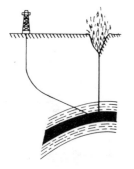

Relief well drilling

REG. NEG.

Abbreviation for Regulations Negotiated on reformulated gasoline vis-á-vis the ethanol vs. gasoline controversy.

REGULATED OUT OF BUSINESS, TO BE

To be unable to comply with new and stiffer regulations because of increased costs or other reasons beyond one's control.

REID VAPOR PRESSURE

A measure of volatility of a fuel, its ability to vaporize. Reid vapor pressure, the specific designation, is named after the man who designed the test apparatus for measuring vapor pressure.

REINVENTING D.O.E.

See D.O.E.'s New Direction.

RELEASED OIL

Under the Emergency Petroleum Allocation Act of 1973, released oil is old oil production equal to any volume of new oil produced. Unlike old oil, released oil could be sold at free market prices.

RELIEF MAP

See Map, Relief.

RELIEF VALVE

See Valve, Relief.

RELIEF WELL

See Well, Relief.

REMOTE SENSING, USING

The use of Landsat and high-altitude imagery is recommended by experienced explorationists, generating geologists, and others as an adjunct to seismic work. Remote sensing and seismic used together significantly improves the chances of discovering a likely prospect. Drilling a well, a wildcat, on an imagery anomaly alone, without any other support or corroboration is assuming more risk than is prudent or necessary at present. For exploration areas, remote sensing is useful in deciding where to run seismic lines or what seismic work already done in the area to buy. Remote sensing is also useful in locating large faults and so called "sweet spots," where fracture and joint patterns intersect. High-altitude imagery is one of the latest exploration tools now available to the hunter of hydrocarbons in virgin, wildcat areas as well as mature basins.

RENTAL, DELAY

See Delay Rental.

REPEATER STATION
An electronic installation, part of a surveillance and control system for offshore or other remote production operations.

REPRESSURE GAS
Gas purchased for injection into an underground formation, a reservoir, for maintaining reservoir pressure. *See* Recycling (Gas).

REPRESSURING OPERATION
The injection of fluid into a reservoir whose pressure has been largely depleted by producing wells in the field. This secondary-recovery technique is used to increase the reservoir pressure in order to recover additional quantities of oil. *See* Service Well.

REREFINED OIL
See Reclaimed Oil.

RESEARCH OCTANE NUMBER
See R.O.N.

RESERVE PIT
An excavation connected to the working mud pits of a drilling well to hold excess or reserve drilling mud; a standby pit containing already mixed drilling mud for use in an emergency when extra mud is needed.

RESERVOIR
A porous, permeable sedimentary rock formation containing quantities of oil and/or gas enclosed or surrounded by layers of less-permeable or impervious rock; a structural trap; a stratigraphic trap.

RESERVOIR, CARBONATE
Reservoirs composed of limestone, dolomite, or carbonitite; sedimentary beds composed of 50 percent or more of carbonate minerals.

RESERVOIR, CLASTIC
A reservoir composed of rocks made up of fragments of pre-existing rocks or minerals that were transported some distance from their place of origin, deposited, compacted, and cemented. Sandstone and shale are the most common and prevalent clastics.

RESERVOIR, GRANITE WASH
In certain areas and at varying depths, oil and gas are found in granite-wash reservoirs *See* Granite Wash.

RESERVOIR ENERGY
See Reservoir Pressure.

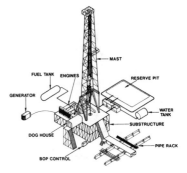

Reserve Pit

RESERVOIR ENGINEER

A petroleum engineer; one who advises production people on matters relating to petroleum reservoirs: estimating and determining effects of reservoir pressure drops, gas and water encroachment, changes in gas-oil ratios, rates of production, and feasibility of secondary and tertiary recovery programs.

RESERVOIR MODELER

A reservoir engineer or geologist who, by various means, simulates petroleum reservoirs. Using data from wells in the area, seismic information, test-hole findings, cores, and rock samples, the modeler projects and expands his information beyond what is known and provable into the realm of the conjectural. This is accomplished with inferences based on an assumed continuity of the data in hand. The work of the reservoir modeler is important in producing a field at the maximum efficient rate (M.E.R.). It is necessary also in projects such as waterflooding, thermal recovery of oil, and hydraulic fracturing.

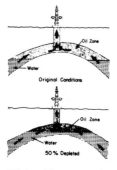

Water-drive reservoir

RESERVOIR PRESSURE

The force that drives oil and gas to the boreholes of the wells drilled in the reservoir. Good production practice dictates that the reservoir pressure be maintained in order to recover as much oil as possible from the field. Reservoir pressure is derived from a gas cap above the oil in the formation, from a water drive, or from gas in solution under pressure (solution-gas drive). It is equal to the shut-in pressure (at the wellhead) plus the weight in pounds of the column of oil in the hole; formation pressure.

RESERVOIR ROCK

Sedimentary formations where nearly all of the world's petroleum has been found. Nearly 60 percent of the world's oil and gas reserves are in sandstone; the other 40 percent are in limestone, dolomite, etc. The most widely held theory is that oil and gas, formed in the source rocks, migrated upward into the porous and more-or-less permeable sedimentary formations that are the reservoirs for petroleum.

RESERVOIRS, HARD TO CATCH

In horizontal drilling, hard-to-catch reservoirs are those not lying entirely or consistently in a single horizontal plane. The bit and borehole, ideally, move in a horizontal plane. The sensors at or near the bit guide the drillstring and, in most instances, keep it in the pay. Thin sections and those that downdip or updip are illusive because there is insufficient vertical latitude for the steering mechanism, as sensitive as it might be, to keep the bit in the thin pay zone. *See* Shoestring Sand.

RESID MARKET

The market for residual oils; black oils market.

RESIDUAL
A term used to describe oils that are "leftovers" in various refining processes; heavy black oils used in ships' boilers and in heating plants.

RESIDUE GAS
Gas that remains after processing in a separator or a plant to remove liquids contained in the gas when produced. *See* Tail Gas.

RESIDUUM
What is left after crude oil has been refined to extinction; a heavy, black, tar-like substance remaining after all usable fractions have been distilled off. The bottom of the barrel, literally.

RESOLUTION, HIGH
The making clear or recognizable a sound or a photograph; high resolution, very clear or magnified so as to make visible the components of an object.

RESOURCES VS. RESERVES
The availability of minerals, including oil, gas, coal, and oil shale, is stated in terms of resources. Resources include all deposits known or believed to exist in such forms that economic extraction is currently or potentially feasible. Reserves are that part of identifiable resources that can be economically extracted with current technology and at current prices.

RETROFITTING
To modify or add to an engine, item of equipment, or operating plant something new for the sake of efficiency, better performance, or increased safety. To retro (go back) and fit or make a change or refinement in the original item of equipment or plant, e.g., "The Ft. Lewis gas plant was retrofitted with automation...." After years of hand operation, the plant was modernized and made more efficient.

RETROGRADE GAS CONDENSATE
A liquid hydrocarbon (condensate) formed in deep formations as the reservoir pressure is reduced through production of natural gas. As the pressure is reduced, the gas condenses to form a liquid instead of the usual pattern of liquid changing to gas. Hence the term "retrograde gas condensate." As liquefaction occurs, the formation rock is "wet" by the condensate, which is then not as recoverable as when it was in a gaseous state.

RETURNS
Colloquial rig-floor term for the drilling mud that comes back up the borehole and flows into the mud pits. Drilling mud is pumped downhole through the hollow drillpipe and drill bit to the bottom of the hole where

it picks up the rock cuttings and brings them back to the surface. On the return trip the drilling mud flows up the annulus, the space between the outside of the drillpipe and the wall of the borehole. Much attention is paid to returns because they can tell a lot about downhole conditions.

REVERSE CIRCULATION

A technique used in fishing for "junk" in the bottom of the well's borehole. A junk basket is lowered into the hole just above the junk to be retrieved, and through ports in the sides of the basket the drilling mud is jetted to the bottom of the hole and back into the open end of the tool, washing the junk back up into the junk basket.

REVERSE FAULT

See Fault, Reverse.

Reverse fault

REVERSE OSMOSIS

A process used in the industry for removing salt and other contaminants from water. The process uses the phenomenon of osmosis, the diffusion through a semipermeable membrane of a solvent, leaving behind the solute or dissolved substance. In reverse osmosis, the solvent (water) diffuses through the man-made membrane, leaving the salt and other contaminants behind.

REVERSE OUT, TO

To expel excess cement from the drillpipe or casing or both after cementing the casing or after a squeeze job. If the cement was pumped down the drillpipe and up the annular space between the casing and the wall of the borehole, when the job is complete the cement left in the drillpipe can be reversed out or expelled by pumping water or drilling mud down the well's casing, thus forcing the cement in the drillpipe back to the surface, and clearing the pipe before the cement has time to set up or harden.

REVERSIONARY INTEREST

As used in the oil field, a legal term to denote an interest that develops after the occurrence of certain events or facts. For instance, in farmout agreements the farmor (party granting the farmout) retains a fractional override and reserves the right to back in, to convert his overriding interest to a percent of the working interest—25 percent, for example—after payout (after the operator has been reimbursed out of the well's production for the costs of the well). The back-in right of the farmor is sometimes referred to as a reversionary interest.

REWORKING A WELL

To restore production where it has fallen off substantially or ceased altogether; cleaning out an accumulation of sand and silt from the bottom of the well.

REWORKING CLAUSE
A clause in an oil and gas lease that permits the lessee to work over a well after the end of the primary term of the lease. When production from the lease has diminished below the break-even point or the profit level, lessee is permitted to plug back or take the well on down in the hope of finding more production. There is, of course, a limit put on reworking time, during which lessee must exhibit due diligence.

RHABDOMANCY
The practice, the art, the science of divination by rods or wands. *See* Doodlebug

RHEOLOGY
A science dealing with the deformation and flow of matter, more specifically the movement of liquids, of slurries; in the drilling industry, the mud specialist is the rheologist who devises mud programs and maintains the proper character and weight of the slurry for optimum drilling and safety conditions. When a gas kick threatens, the mud man heavies up on the slurry, for example, increasing the mud weight from 10 lbs/gal. to 14 lbs/gal—or 18 lbs/gal.

RIBBON ROCK
A rock showing a succession of thin layers of different composition or colors. For example, gray shale may be interspersed with dark brown dolomite and nearly white limestone.

RICH GAS
Natural gas containing significant amounts of liquefiable hydrocarbons, i.e., casinghead gasoline, butane, propane, etc.; wet gas.

RIG
(1) A drilling rig. (2) A large tractor-trailer.

RIG, ACCOMMODATION
A semisubmersible platform with facilities aboard to house and maintain hundreds of oil workers; a floating hotel and supply vessel used as a support for construction projects far from shore, e.g., the North Sea.

RIG, BARGE
A drilling rig mounted on a barge-like vessel for use in shallow water or swampy locations. Barge rigs are not self-propelled and must be towed

Disguised Rig
Urban Setting

Rig

Submersible drilling
barge

Track Mounted
Mobile Rig

Jackknife rig

Rig floor

Jackup rig *(Courtesy Zapata Off-Shore Co.)*

or pushed by a towboat. In addition to all necessary drilling equipment, such barges have crew quarters.

RIG, DEEP

A specially designed drilling derrick built to withstand the extreme hook loads of ultradeep (20,000 to 30,000 feet) wells. Deep rigs, in addition to extra-strong structural members, have massive substructures 25 to 35 feet high to accommodate the large and tall blowout-preventer stacks flanged to the wellhead. Hook loads on deep rigs often exceed 800 tons, 1,600,000 pounds.

RIG, ELECTRIC

A drilling rig that receives its power from a system comprised of diesel engine-D.C. generator-D.C. motor. A typical engine generator-motor rig-up would include four such sets: two for the mud pumps, one for the drawworks and rotary table, and one somewhat smaller set for lighting and auxiliary loads. Another type of electric rig uses the same power-flow system but the generators are A.C., whose current is converted to D.C. current to drive D.C. motors for variable-speed drilling operations.

RIG, HELICOPTER

A drilling rig so constructed that it may be broken down into segments and transported into nearly inaccessible mountainous areas or other rugged terrain. The helicopter picks up one diesel engine, one mud pump at a time, and swings it onto a new location. Such rigs usually are shallow-well rigs or seismic shot-hole rigs.

RIG, JACKKNIFE

A mast-type derrick whose supporting legs are hinged at the base. When the rig is to be moved, it is lowered or laid down intact and transported by truck.

RIG, JACKUP

A barge-like, floating platform with legs at each corner that can be lowered to the seafloor to rise or jack up the platform above the water. Towed to location offshore, the legs of the jackup rig are in a raised position, sticking up high above the platform. When on location, the legs are run down hydraulically or by individual electric motors.

RIG, MECHANICAL

The most common type of drilling rig is the mechanical compound rig. Mechanical rigs are diesel engines coupled directly to the equipment or through compound shafts to drive the rotary, drawworks, and mud pumps. Separate engine-A.C. generator sets provide lighting and power for auxiliary functions. *See* Rig, Electric.

RIG, PUSH-DOWN

A drilling rig that is a modification of rigs used by the mining industry and for drilling water wells. The drillpipe is supported within an A-frame, with the rotary and its pipe-turning mechanism on top of the first joint of drillpipe 30 feet or so up in the A-frame. As the drillpipe and bit are rotated, the pipe is pushed downward hydraulically until the first joint is in the hole and the rotary is at floor level. A second joint is then added, and the rotary is raised to the top to turn and push down on the second joint. A pushdown rig has a conventional mud system, but the rig is practical only for drilling holes to about 3,500 feet in relatively soft formations.

RIG, ROTARY
A derrick equipped with rotary drilling equipment, i.e., drilling engines, drawworks, rotary table, mud pumps, and auxiliary equipment; a modern drilling unit capable of drilling a borehole with a bit attached to a rotating column of steel pipe.

Semiautomatic drilling rig adding a joint of pipe

RIG, SLANT
A drilling derrick designed to drill from offshore platforms at angles of 20° to 35° from the vertical. The slant rig, canted from the vertical, has a companion structure for racking the drillpipe vertically when coming out of the hole on a trip. The rig's traveling equipment—block, hook, swivel, and kelly joint—moves up and down on rails that are an integral part of the derrick. With a slant rig it is possible to reach farther out from a drill platform, particularly in relatively shallow water, than with a conventional rig using directional drilling techniques.

RIG, SPLIT-LEVEL
A land rig design in which the diesel engines, gear compound, and drawworks are at or near ground level, 12–15 feet below and behind the rig floor. On the rig floor are the catworks and the rotary table as on a conventional rig. The power from the high-speed (1,800 rpm) diesel engines is transmitted through clutches and compound to the rotary table through a torque table, rising at about a 45° angle to the gear and chain drive on the rig floor.

Rotary rig

RIG, SS CLASS 2000
The designation for the class of semisubmersible drilling platforms (the largest built to date: 1976) that are of 18,000-ton displacement, 2,000-ton deck-load capacity, and capable of drilling in 2,000 feet of water.

RIG, STANDARD PUMPING
A conventional pumping unit consisting of an engine or electric motor operating a walking beam that raises and lowers the sucker rods in an up-and-down pumping motion.

RIG BUILDER
(1) A person whose job is to build or (in a modern context) to assemble a derrick. Steel derricks are erected by bolting parts together.
(2) Originally, a person who built derricks on the spot out of rig timbers

Split-level drilling rig

and lumber on which he used crosscut saws, augers, axes, hammers, and the adz to fit the wood to his pattern.

RIG-DOWN

To prepare to move the drilling rig and associated tools and equipment to another location or to storage; to stack the tools; to disassemble the mud system, disconnect the engines, lay the derrick down (a jackknife or other portable rig), fill the mud pits, and load up the pipe and fittings and other equipment ready for transport to another wellsite.

RIG FLOOR

See Derrick Floor.

RIG HANDS

A generic term for roughnecks, floor men, drillers, mud engineers, and roustabouts.

Bulldozer clearing right-of-way

RIGHT-OF-WAY

(1) A legal right of passage over another person's land. (2) The strip of land for which permission has been granted to build a pipeline and for normal maintenance thereafter. The usual width of right-of-way for common-carrier pipelines is 50 feet.

RIGHT-OF-WAY GANG

A work crew that clears brush, timber, and other obstructions from the right-of-way. The crew also installs access gates in fenced property. *See* Dress-Up Crew.

RIGHT-OF-WAY MAN

A person who contacts landowners, municipal authorities, and government agency representatives for permission to lay a pipeline through their property or through the political subdivision. He also arranges for permits to cross navigable waterways, railroads, and highways from the proper authorities.

RIG MANAGER

One who supervises all aspects of offshore rig operation. Large semi-submersibles, anchored miles at sea with hundreds of workers, are much like a small town engaged in drilling a well in hundreds of feet of water. The rig manager is the resident boss of this floating microcosm.

RIG REGISTER

A roster of offshore drilling equipment—jackups, semisubmersibles, drillships, platforms, tenders and drilling barges—deployed around the world. The register, a modern Jane's *Fighting Ships* as it were, was introduced by *Petroleum Engineer* magazine. It is kept current and lists the vessel's or platform's depth capability, equipment, whether self-propelled or towed, and other pertinent information.

RIG SET-OFF

Moving a drilling rig from over the well bore so that the well can be completed. Skids and rollers are positioned under the 60 to 80 foot derrick, either wooden or bolted metal, and pulled aside, even to another location by trucks or bulldozers. If to another staked location, the maneuver is referred to as "skidding the rig." *See* Skidding the Rig.

RIG SUPERVISOR

One who directs all aspects of the drilling, testing, and completion of a well. He applies optimization techniques to improve drilling performance; demands safe working practices, and is concerned about the environment and does his best to protect it. He might be called a super tool pusher.

RIG TIMBERS

Large-dimension wooden beams used to support the derrick, drilling engines, or other heavy equipment; heavy, roughcut timbers used in the trade by rig builders when derricks were built rather than assembled.

RIG-TIME WORK

See Day-work Basis.

RIG-UP

To make preparations to drill; to get all equipment in place ready to make hole: dig the cellar and mud pits; set up the derrick, reeve the lines; set engines and pumps; connect the lines of the mud system and set auxiliary equipment. Also have necessary bits, tubular goods, valves, rams, and fittings on hand.

RIM FIRE

A fire occurring at the edge or rim of a floating-roof tank. The rim is the contact between the floating roof and the wall or shell of the tank. As tight as the roof is to the shell, some vapors usually escape and can be ignited by lightning or some other source of sparks.

RING GEAR

A toothed gear, usually a large-diameter circular gear, in a pump that meshes with and is driven by a pinion gear. The pinion gear is a small-diameter gear on the end of a power-driven shaft. A ring and pinion gear are a set.

RIPARIAN

Refers to or is located on the shore or banks of a body of water. Usually *riparian* pertains to a watercourse such as creek or a river. Riparian lands are those along the bank of a river, including land beneath the river, the river bed. Riparian rights are the legal rights regarding a waterway, which belongs to the one who owns the land bordering the watercourse.

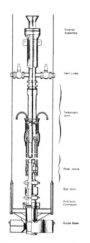

Marine riser system

RIPARIAN LEASE CLAUSE

An excerpt from such a clause: "If this land is riparian to, bounds, or embraces within its boundaries a stream, lake or other body of water then all the lessor's riverbed rights and lands under water, and all area now or hereafter added by accretion, are included and covered by this lease." McQueen, "Development of lease provisions: MidContinent Area." 1965.

RISER

(1) A pipe through which liquid or gas flows upward. (2) In offshore drilling by semisubmersible, jackup, fixed platform, or drillship, a riser is the casing extending from the drilling platform through the water to the sea bed through which drilling is done. *See* Marine Riser.

RISK OF THE HOLE

The chance of some damage or mishap affecting the hole drilled for an oil or gas well. For example, in 1978 a Louisiana court concluded that under the usual Turnkey contract, the risk of the hole shifts to the operator-owner as soon as the last core provided for in the contract is furnished. Control of the hole also shifts at the same time.

RIVER CLAMP

Heavy steel weights made in two halves bolted on screw pipe at each collar to strengthen the joints and keep the line lying securely on the river bottom or in a dredged trench.

RIVER TAMING STRUCTURES

Stabilizing structures and devices—expanded metal matting, tethered concrete slabs (gabions) and plantings—used to secure riverbanks during high water or flood conditions where pipelines cross. Even where pipeline crossings are bored, as has been the practice for the past two decades, stabilizing the banks to prevent scouring is important.

RIVET BUSTER

An air-operated (pneumatic) chisel-like tool for cutting off rivet heads. Used by tankies when tearing down an old tank or other vessel put together with rivets.

ROCK, CLASTIC

See Clastic Rock.

ROCK, IGNEOUS

See Igneous Rock.

ROCK, METAMORPHIC

See Metamorphic Rock.

ROCK A WELL

To agitate a "dead" well by alternately bleeding and shutting in the pressure on the casing or tubing so that the well will start to flow.

ROCK CHUTE
See Fall Pipe.

ROCKER
A counterbalance installed on a shackle-rod line, operating a pumping jack, to pull the rod line back after its power stroke. Rod lines can only pull, so on the return stroke the line is kept taut by a counterbalance. Rockers often are in the shape of a box or crate filled with rocks. One edge of the box is attached to a fulcrum bearing on which it moves back and forth like a rocking chair.

ROCK FACIES
A geologic term meaning the exposed and visible surface of a rock or formation; the look of a rock by which it may be identified.

ROCK HOUND
A geologist; a humorous but affectionate colloquialism for a person who assiduously pursues rock specimens in a search for evidence of oil and gas deposits.

ROCK PRESSURE
An early-day term for a well's shutin or wellhead pressure when all valves are closed and the pressure is observed at the surface.

ROCK SALT
See Halite.

ROD
(1) Sixteen and one-half feet; the unit of measure used in buying certain types of pipeline right-of-way. (2) A sucker rod; an engine's connecting rod; a piston rod.

RODDAGE FEE
The fee paid to a landowner for the easement of a pipeline right-of-way across his property. Right-of-way is measured in rods (16 1/2 feet); hence the term roddage fee. *See* Right-of-Way.

ROD-GUIDE COUPLING, WHEELED
Rod string centralizers/guides with small, built-in wheels that ride on the inner wall of production tubing, thus preventing the abrasive action of sucker rods rubbing on the tubing.

ROD HANGER
A rack with finger-like projections on which rods are hung when pulled from the well; a vertical rack for hanging lengths of pumping rods.

ROD JOB
See Pulling Rods.

Rod job

Rod-line pump

Roller bearing

ROD LINE
See Shackle Rod.

ROD-LINE PUMP
See Pump, Jerker.

ROD-LINE WELL
An oil well that is pumped by power from a rod line or shackle-rod line, which is given its to-and-fro pumping motion by a central power. In old stripper fields, a number of wells (sometimes 20 or more) will be pumped by a central power with shackle rods running to each well's small pumping jack.

ROD PUMP
See Pump, Rod.

ROILY
Muddy, sediment-filled water; turbulent or swirling water.

ROLL A TANK
To agitate a tank of crude oil with air or gas for the purpose of mixing small quantities of chemical with the oil to break up emulsions or to settle out impurities.

ROLLER BEARING
See Bearing, Roller.

ROLLER BIT
See Bit, Roller.

ROLLERS, CASING-AND-TUBING
A steel tubular device for opening up and reconditioning buckled, dented, or collapsed casing and tubing in the hole. The long, steel tool with a tapered end has a series of rollers. The tool is forced into the damaged pipe and, as it is pushed down and rotated by the drillstring, the series of rollers forces the damaged pipe open and restores it to its original diameter and roundness.

ROLL IN
To include the cost of new facilities, service, and supply as part of the overall cost of operating a company for the benefit of all customers served by a pipeline or other common carrier; to roll in the cost of new supplies and facilities for the purpose of arriving at a new rate structure.

ROLLING PIPE
Turning a joint of screw pipe into the coupling of the preceding joint by the use of a rope looped once around the pipe and pulled by a rope

crew. This procedure was used on larger-diameter line pipe—10 and 12 inch—to make up the connection rapidly before the tongs were put on the pipe for the final tightening.

R.O.N.

Research Octane Number; a measure of a gasoline's antiknock quality determined by tests made on engines running under moderate conditions of speed and load. M.O.N., motor octane number, is a measure of a gasoline's antiknock characteristics determined by tests under more severe conditions of load and speed. *See* Octane Rating.

ROOF ROCK

Impervious rock such as shale that acts as a barrier to the upward movement or migration of oil and gas; roof rock overlies a reservoir rock, forming a trap.

ROOT RUN

The first course of metal laid on by a welder in joining two lengths of pipe or other elements of construction; the stringer bead.

R.O.P.

Rate of penetration; how fast the hole is being taken down.

ROPE SOCKET

A device for securing the end of a steel cable into a connecting piece a clevis, hook or chain. A metal cup or socket (like a whip socket) into which the cable end is inserted and which then is filled with molten lead or babbin.

ROTARY BIT

See Bit, Rotary.

ROTARY BUSHING

The metal casting that fits into the master bushing of the rotary table on a drilling well and through which the kelly joint moves downward as drilling proceeds. The kelly bushing is turned by the rotary table and the bushing rotates the kelly and drillstring.

ROTARY DRILLING

See Drilling, Rotary.

ROTARY HEAD

A heavy casting used to close off the annular space around the kelly joint while drilling when there is pressure at the surface. The rotary head rotates with the kelly joint but maintains a pressure-tight seal. This piece of equipment is installed on top of the blowout-preventer (B.O.P.) stack, and it is used principally in air or gas drilling. On the fixed element of

Rotary pump *(Courtesy Roper Pump Co.)*

Rotary rig

Rotary table *(Courtesy U.S. Steel)*

the rotating head, there is a 4-inch pipe, the "blooie pipe," through which gas or air and rock chips from downhole are vented to a burn pit.

ROTARY HOSE
The kelly hose. *See* Mud Hose.

ROTARY-PERCUSSION BIT
See Bit, Rotary-Percussion

ROTARY PUMP
See Pump, Rotary.

ROTARY REAMER
A rock-cutting tool inserted in the drill column just above the drill bit for the purpose of keeping the hole cut to full diameter. Often in drilling deep, hard-rock formations, the bit will become worn or distorted, thus cutting less than a full hole. The following reamer trims the hole wall, maintaining full diameter.

ROTARY RIG
See Rig, Rotary.

ROTARY SLIP
See Slip.

ROTARY TABLE
A heavy, circular casting mounted on a steel platform just above the derrick floor with an opening in the center through which the drillpipe and casing must pass. The table is rotated by power transmitted from the drawworks and the drilling engines. In drilling, the kelly joint fits into the square opening of the table. As the table rotates, the kelly is turned, rotating the drill column and the drill bit.

ROTARY TONGS
The massive, counter-weighted tongs used on the drill floor to screw joints of drillpipe, tubing or casing; the generic term for the heavy wrenches used by the rough necks on the rig floor.

ROTARY VALVES
See Valves, Rotary.

ROTATING DRILLING HEAD
See Drilling Head, Rotating.

ROUGHNECKS
Members of the drilling crew; the driller's assistants who work on the derrick floor and up in the derrick racking pipe, tend the drilling engines and mud pumps, and on "trips," which operate the pipe tongs to break out or unscrew the stands of drillpipe.

Aerial view of the Sweeney refinery.

Well workover unit, pulling tubing. *(Courtesy Kellar Services Inc.)*

ROUND-POINT SHOVEL
A digging tool whose blade is rounded and tapers to a point in the center of the cutting edge. A long-handled shovel, standard equipment for digging ditches by hand.

ROUND TRIP
See Trip.

ROUSTABOUT
A production employee who works on a lease or around a drilling rig doing manual labor.

Making a trip

ROYALTY
See Royalty, Landowner's.

ROYALTY, COMPENSATORY
Payments to royalty owners as compensation for loss of income that they may suffer due to the failure of the operator to develop a lease properly.

ROYALTY, FEE
Landowner's royalty.

ROYALTY, FIXED-RATE
Royalty calculated on the basis of a fixed rate per unit of production, without regard for the actual proceeds from the sale of the production.

ROYALTY, GAS
Royalty paid on natural gas produced from a well or lease. The percentage of royalty—1/8, 3/16 or 1/4—is stated in the oil and gas lease. Gas royalty is based on market or the price at the wellhead. The usual gas royalty does not provide for "taking in kind" as do the terms of an oil royalty.

ROYALTY, GUARANTEED
The minimum amount of royalty income a royalty owner is to receive under the lease agreement, regardless of his share of actual proceeds from the sale of the lease's production.

ROYALTY, INNOVATOR'S
A type of overriding royalty paid to the person instrumental in bringing a company to a concession from a foreign government; British: a fixer's royalty. *See* Royalty, Overriding.

ROYALTY, LANDOWNER'S
A share of gross production of oil and gas, free of all costs of production. Occasionally, the term is used to describe an interest in production created by the landowner outside the lease and distinguished from the conventional lessor's royalty. In this case the landowner's royalty, outside

of the lease, may have any specified duration. In general usage, land-owner's and lessor's royalty are synonymous.

ROYALTY, MINERAL

Occasionally, the term "mineral royalty" is used to describe a non-participating royalty, which means the royalty owner does not share in bonus or rental monies nor does he have the right to execute leases or explore or develop. In Louisiana, such royalty is the right to participate in production of minerals from land owned by an-other, or land subject to mineral servitude owned by another. Unless expressly qualified by the parties, a royalty is a "right to share in gross production, free of mining or drilling or production costs." Louisiana Mineral Code Article. 80 [R.S 31:80 (1975)]. *See* Mineral Servitude.

ROYALTY, MINIMUM (Federal)

A minimum royalty, in the context of federal leases, is the annual payment made to the government, once the lease proves productive, of $1.00 an acre in lieu of rental payments.

ROYALTY, NONPARTICIPATING

Like other royalties, it is an expense-free interest in oil or gas as and when produced. Nonparticipating means the royalty interest does not share in bonus or rental monies nor does it have the right to execute leases or to explore and develop. The amount of the royalty is set by the instrument or document that creates the royalty. For example, a one-sixteenth nonparticipating royalty is one out of every 16 barrels of oil, free of cost, delivered at the surface. If an instrument used the phrase "one-half of all present and future royalties on oil produced," it would also give the nonparticipating royalty owner one out of every 16 barrels if the lease royalty was one-eighth. If it was more or less, the owner would get proportionately more or less. Nonparticipating royalties can run for a fixed term, for as long as oil or gas is produced, or in perpetuity. However, in some states such perpetual interests are held to be invalid.

ROYALTY, OFFSET

Royalty payable in lieu of drilling an offset well. For example, a lessee or operator has a separate lease on two adjacent tracts of land, A and B. If he drills on tract A, which drains tract B, he normally would be required to drill on tract B to protect it from having its oil drain to the well on tract A. But drilling on tract B might be uneconomic because well A is able to recover the oil under both tracts. Under the circumstances, the lessee might elect to pay the royalty owners on tract B an offset royalty (if there is an offset royalty clause) instead of drilling the required offset well. The offset royalty is sometimes a flat annual amount, or it may be measured by estimating what the royalty would have been had the offset well been drilled and put on production.

ROYALTY, OIL

The lessor's or landowners's share of oil produced on his land. The customary 1/8 royalty can be paid in money or in oil. In some instances, another fraction of production is specified as royalty.

ROYALTY, OVERRIDING

An interest in oil and gas produced at the surface free of any cost of production; royalty in addition to the usual landowner's royalty reserved to the lessor. A 1/16 override is not unusual.

ROYALTY, PASSTHROUGH

An unusual type of royalty paid on production from a well drilled on or through one track of land, bottomed out, and producing from an adjacent tract.

ROYALTY, SHUTIN

Payment to royalty owners under the terms of a mineral lease that allows the operator or lessee to defer production from a well that is shut in for lack of a market or pipeline connection.

ROYALTY, SLIDING-SCALE

Royalty paid to the federal government on oil and gas production from a government lease, usually offshore, which varies from the normal 16 percent up to 50 percent of the value of the production. As the value of production increases, the percentage of royalty also increases to a maximum of 50 percent.

ROYALTY, TAPERED

A royalty that begins at a specified rate and declines each year by a given amount or percentage regardless of the rate of production.

ROYALTY, TERM

A royalty interest limited by time or productivity of the lease. Most royalty interests are created for a fixed period of time "and so long thereafter as oil and gas are produced." But there are such interests that run only for a specified, fixed length of time with no qualifying "thereafter" clause.

ROYALTY BIDDING

An uncommon practice of bidding on federal leases by offering a high royalty interest to the government on any production discovered on the tract in lieu of the traditional cash bonus. Royalty interests as high as 70 and 80 percent of gross production have been offered. The advantages to a company bidding royalty interests instead of cash could be a savings in millions of dollars of front money. In case the lease is unproductive, the company is out only the cost of the well and any seismic or other exploratory expenses.

ROYALTY BONUS

An overriding royalty or oil payment reserved by the lessor, the landowner. Usually any consideration received or promised to a lessor on the execution of a lease in excess of the customary one-eighth royalty is called a bonus or royalty bonus. Oil bonus.

ROYALTY DEED

A legal instrument conveying (transferring) a royalty interest from one party to another. The instrument (written document) names the grantor and grantee, describes the land or lease and the size of the interest conveyed, and contains the witnessed signature of the grantor. Royalty deeds are not uniform as to the duration of the interest; they may be forever, for a term of years " and so long thereafter as oil and gas are produced," or for the duration of the oil and gas lease.

ROYALTY FUNDS

Funds organized to invest in oil and gas royalties. This is one of the safest oil investments, particularly on settled production. *See* Income Funds.

ROYALTY GAUGER

An employee or other designated person who gauges the runs of oil or notes the gas meter readings for the owner; the royalty owner's own man who keeps tab on the buyer's gauges and computations.

ROYALTY IN KIND OR IN SPECIE

The right to receive what is due from an oil or gas lease under a royalty interest in specie; that is, oil or gas itself: a barrel of oil or cubic foot of gas to hand.

ROYALTY OIL OR GAS

Oil and gas payments made in kind to lessor under the terms of an oil and gas lease. The usual arrangement is for the operator of the lease to market the product and pay the landowner by royalty check.

ROYALTY POOL

A pooled or unitized area in which owners of royalty in segregated portions of the area share in the royalty on production of all wells located in the area covered by the agreement without regard as to their location.

RUGGEDIZE, TO

A made-up verb form to indicate, by zealous manufacturers, that their good, sturdy tools have been made even stronger; they have been ruggedized.

RULE OF CAPTURE

In common law, regardless of where oil originated or migrated from, the person capturing oil by drilling a well and producing it is entitled to the benefits therefrom.

RUN

A transfer of crude oil from a stock tank on a production lease to a pipeline gathering system, for transportation to the buyer's facilities; running oil from a tank into a pipeline for delivery to a purchaser.

RUNNING THE TOOLS

Putting the drillpipe, with the bit attached, into the hole in preparation for drilling.

RUN OUT OF HOLE

To put so many strings of casing or liners in the borehole that the hole's diameter becomes too small to run production tubing or downhole tools. This condition may occur when unexpected intervals of saltwater, cavey sections, or a pocket of high pressure gas must be cased off, thus reducing the size of the hole for subsequent strings of casing; the hole becomes too small to run a production string or to operate in.

RUN TICKET

A record of the oil run from a lease tank into a connecting pipeline. The ticket is made out in triplicate by the gauger and is witnessed by the lease owner's representative, usually the pumper. The run ticket, an invoice for oil delivered, shows opening and closing gauge, A.P.I. gravity and temperature, tank temperature, and B.S.&W. The original of the ticket goes to the purchaser; copies go to the pumper and one for the gauger.

RUPTURE DISC

A thin, metal plug or membrane in a fitting on a pressure line made so as to blow out or rupture when the pressure exceeds a predetermined level; a safety plug. *See* Soft Plug.

Saddle (Courtesy
Plidco Int'l, Inc.)

Vertical safety valves on
high-pressure gas line

S

SADDLE
A clamp, fitted with a gasket, for stopping the flow of oil or gas from holes or splits in a pipeline; a device for making temporary repairs to a line. The clamp conforms to the curve of the pipe and is held in place by U-bolts that fit around the pipe and extend through the clamp.

SADDLE BEARING
See Bearing, Saddle.

S.A.E.
Society of Automotive Engineers.

S.A.E. NUMBER
A classification of lubricating oils in terms of viscosity only. A standard established by the Society of Automotive Engineers. S.A.E. 20; S.A.E. 10W-30 multiviscosity lubricating oil.

SAFETY PLUG
See Rupture Disc.

SAFETY VALVE
See Valve, Relief.

SAG BASIN
A regional depression forming a basin; a gently sloping discreet land surface forming a fairly symmetrical basin; a sag or downwarp.

SAIL ANGLE
The angle from the vertical in a drilling kickoff, at the start of the deviated or horizontal section of borehole.

SALE LEASE-BACK
A method of freeing equity (raising capital) in a company. Buildings and other real estate are sold and, simultaneously with the sale, leased back under a long-term lease arrangement. Some of the advantages of sale lease-back: it converts real estate into liquid assets, cash; and it can generate a profit on the sale of the properties.

SALT ANTICLINE
A piercement-type geologic structure formed by the upthrust of a salt

dome. Unlike a salt dome, the salt core is elongated rather than roughly cylindrical in shape; a salt wall. *See* Salt Dome.

SALT-BED STORAGE
See Salt-Dome Storage.

SALT DOME
A subsurface mound or dome of salt. Two types of salt domes are recognized: the piercement and nonpiercement. Piercement domes thrust upward into the formations above them, causing faulting; nonpiercement domes are produced by local thickening of the salt beds and merely lift the overlying formations to form an anticline.

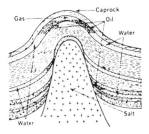

Salt dome

SALT-DOME/SALT-PLUG POOLS
Structural or stratigraphic traps associated with rock-salt intrusions; pools formed by the intrusion of underlying salt formations into overlying porous and permeable sedimentary layers creating traps favorable to the presence of oil and gas.

SALT-DOME STORAGE
Cavities leached out of underground salt formations by the use of superheated water for the storage of petroleum products, especially L.P.-gases.

SALT PILLOW
A low-profile salt dome rising from its source bed at considerable depth beneath the surface; an "embryonic salt dome."

SALT PLUG
The salt core of a salt dome. It is relatively symmetrical, maybe a mile in diameter, and has thrust up through the surrounding sediments from the source bed 3 to 5 miles below the surface.

SALT WALL
See Salt Anticline.

SALTWATER DISPOSAL WELL
A service well; a disposal well.

SAMPLE
Cuttings of a rock formation broken up by the drill bit and brought to the surface by the drilling mud. Rock samples are collected from the shale shaker and are examined by the geologist to identify the formation, the type of rock being drilled.

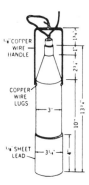

Beaker sample

SAMPLE BAG
A small cotton bag with a drawstring to hold rock-cutting samples. Each bag with its sample is tagged with identifying information: well name, lease, location, depth at which cuttings were taken, etc.

Samson post

SAMPLE LOG
See Log, Sample.

SAMPLER
See Beaker Sampler.

SAMPLE VALVE
See Valve, Sample.

SAMSON POST
A heavy, vertical timber that supports the well's walking beam.

SAND
Short for sandstone; one of the more prolific sedimentary rock formations. In informal usage, other sedimentary rocks are often referred to as "sands."

SAND BODY
A sand or sandstone formation defined by upper and lower layers of impervious rock.

SAND CONTROL
A technique for coping with sand from unconsolidated (loose, unpacked) formations that migrate (drift or wash) into downhole pumping equipment or into the borehole. *See* Gravel Packing.

SANDED UP
A well clogged by sand that has drifted and washed into the well bore from the producing formation by the action of the oil.

SAND-FRAC JOB
Fracturing a formation with fluid under extremes of pressure then introducing fine-grained proppant material (sand, ceramic beads, or other noncrushable grit). The size of a frac job is usually measured by the pounds of sand (generic term for proppant material) injected with the fracing fluid, e.g., a 10,000 lb. frac job; i.e., 5 tons of proppant material forced into the minute cracks and interstities of the fractured formation to prop them open.

SAND LENS
A localized, usually shallow deposit of sandstone, surrounded by shale, in which some hydrocarbons and water accumulate; a sand body.

SAND LINE
A wireline (cable) used on a drilling rig to raise and lower the bailer or sand pump in the well bore. Logging devices and other lightweight equipment are also lowered into the hole on the sand line.

SANDOUT, TO
A sandout occurs when suspended proppant material (sand or ceramic beads, etc.) carried by the frac fluid stack up and clog the face of the formation, the small fractures and the pores, blocking the completion of the frac job.

SAND PUMP
See Pump, Sand.

SAND REEL
A small hoisting drum on which the sand line is spooled and used to run the bailer or sand pump on a cable tool rig. The sand reel is powered by contact with the bandwheel.

SANDS
Common terminology for oil-bearing sandstone formations. (Oil is also found in limestone, shales, dolomite, and other porous rock.) In informal or loose usage, other sedimentary rocks are referred to as sands.

SAND SEPARATOR
A device for removing "drilled solids," pulverized rock and sand, from drilling mud. The sand separator is used in addition to the shale shaker and by removing most of the abrasive material, reduces wear on mud pumps and bits.

Sand separator

SANDSTONE
A clastic rock composed of grains and minute fragments of quartz, usually cemented together by silica, calcium carbonate, iron oxide, etc. Sandstone occurs in many colors, degrees of consolidation, and is a good reservoir rock for the accumulation of oil, gas, and water.

SAND-WASHING A WELL
Sand-washing is ridding a well bore of accumulated sand and other debris. Coiled tubing is used in this operation because of the savings in time and expense. It can be run in and out of the well bore without having to make threaded connections, which makes for faster trip times, and the washing can be done under well pressure.

SATELLITE PLANT
A facility that supports the main processing plant; a plant that derives its feedstock or raw material from the main processing unit.

SATELLITE PLATFORM
Production platform.

Satellite platform

SATS GAS PLANT
A refiner's term for the part of the refinery that processes gas streams carrying saturates to be stripped out of the gases.

SATURATES
Components of refinery-process gas streams: methane, ethane, propane, butanes, and others. Saturates is a synonym for hydrocarbons whose carbon atoms are saturated with hydrogen atoms. These gas streams are further refined in a facility called by refinery engineers the sats gas plant.

SATURATION
(1) The extent to which the pore space in a formation contains hydrocarbons or connate water. (2) The extent to which gas is dissolved in the liquid hydrocarbons in a formation.

SATURATION PRESSURE
See Bubble-Point Pressure.

SAYBOLT SECONDS
See Seconds Saybolt Furol; *also* Seconds Saybolt Universal.

SAYBOLT VISCOSIMETER
See Viscosimeter.

S.B.M. SYSTEM
Single-Buoy Mooring System.

Single-buoy mooring
system

S.B.R.
Initials for Synthetic Butadiene Rubber, the main ingredients of which are derived from petroleum. S.B.R. is used in the manufacture of tires, hose shoes, and other severe-duty products.

SCAB LINER
A casing string, a liner, run inside the casing in a hole that has collapsed or is in danger of being deformed so that downhole tools cannot be run. This often happens when the original casing is run through an unstable interval of a rock formation that "flows" or is forced toward the borehole by the pressure of the rock column or overburden. A notable example of this phenomenon is when drilling through salt beds that flow or are squeezed or forced inward toward the borehole with such force as to collapse the steel casing.

S.C.A.D.A.
The acronym for Supervisory Control and Data Acquisition, a system installed on pipelines to provide real-time information on gathering and main line systems.

S.C.A.D.A. SYSTEM
Computer-based Supervisory Control And Data Acquisition systems monitor and control pumping wells, line heaters, pumps, and shutdown systems. Operators or attendants are equipped with cellular telephones

and carry laptop computers equipped with internal modems to enable them to receive alarms and to take appropriate action promptly.

SCALPING SHAKER
The first cuttings catcher in the line of shakers. The scalping shaker takes out the largest rock chips and other debris from the flow of drilling mud from downhole. From the scalper the flow cascades onto a succession of two or three high-speed finer mesh shakers—then into the desander or desilter cycle. All of this activity is done to keep the drilling fluid free of foreign matter: chips, sand, and silt.

SCARP
See Fault Scarp.

SCAT RIG
A rack that carries welding generators, gas bottles (CO_2), and spools of welding wire along the pipe being welded. The rig is powered by a small diesel engine. Automatic welding heads ("bugs") are moved ahead on the pipe as the joints are welded.

SCAVENGER OIL
A pejorative term for crude oil that's hard to get, to produce. The term usually refers to heavy oil that must be heated by steam flood or steam soak or *in situ* combustion before it will flow into the well bore; refers also to shale oil where the oil shale must be retorted to extract the liquid kerogen which, in turn, must be distilled to yield petroleum fractions. When all the "easy oil" has been exhausted, there will still be hundreds of billions of barrels of what the industry looks upon as problem oils.

SCHEDULER
A person in a dispatch office who plans the future movement of batches of crude oil or product in a pipeline system, keeping batches separated and making arrangements for product input and downstream deliveries. *See* Dispatcher.

SCHEMATIC SECTION
A graphic representation or diagram of a geologic cross section showing the relationships of various strata and the strata with the surface; also the layout of a process plant or an undersea pipeline manifold, drawn to scale.

SCHIST
A distinctly foliated (leaf-like structured) metamorphic rock that can be readily split into very thin flakes or layers owing to the parallel arrangement of most of the minerals in the rock. Mica and hornblende are two minerals often present in large amounts in schist, as is muscovite, a type of mica that commonly occurs in both gneiss and schist.

SCHLUMBERGER (Slumber-jay)

Trade name of a pioneer electrical well-surveying company. In many areas it is common practice to speak of an electric well log as a "slumberjay," even though the log was made by another company.

SCOPING, ENGINEERING

To examine and study in detail for possible improvements in a process or an operation routine.

SCOURING

The erosion or washing away of the sand/clay covering of a buried subsea pipeline. Scouring caused by sea currents is a serious problem for undersea lines. Excessive scouring causes spanning, the hanging of a section of the line one to several feet off bottom. If allowed to go uncorrected, the pipeline welds crack or the pipe ruptures from its unsupported weight. Subsea lines are inspected for scouring and spanning by sidescan sonar devices or by diver inspection.

SCOUT

A person hired by an operator or a company to seek out information about activities of drilling wells in an area, survey data, drilling rates and depths, and well potentials.

SCOUTING A WELL

Gathering information, by all available means, about a competitor's well—the depth, formations encountered, well logs, drilling rates, leasing, and geophysical reports.

SCOUT TICKETS

A written report of wells drilling in the area. The reports contain all pertinent information—all that can be found out by the enterprising oil scout: operator, location, lease, drilling contractor, depth of well, formations encountered, results of drillstem tests, logs, etc. On tight holes the scout is reduced to surreptitious means to get information. Talks to water hauler, to well-service people who may be talkative or landowner's brother-in-law. The bird-dogging scout estimates the drillpipe set-backs for approximate depth; he notes the acid trucks or the shooting (perforating) crew; and through his binoculars, he judges the expressions on the operator's face: happy or disgruntled.

SCRAPER

A pig; a cylindrical, plug-like device equipped with scraper blades, wire brushes, and toothed rollers used to clean accumulations of wax, rust, and other foreign matter from pipelines. The scraper is inserted in the line at a scraper trap and is pushed along by the pressure of the moving column of oil.

Pipeline scraper (Courtesy Oil States Rubber Co.)

SCRAPER TRAP
A facility on a pipeline for inserting and retrieving a scraper or pig. *See* Pig Launcher and Receiver.

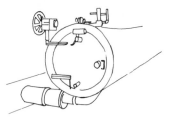

Scraper trap

SCREEN, 100-MESH
A fine screen for sizing sand for proppant material, or a fluid-loss additive to drilling mud.

SCREEN-OUT, FRACTURING
The clogging of the face of the formation by proppant material (sand, ceramic beads, etc.) during hydraulic fracturing; blockage of the fractures by excessive buildup of solids that prevents the flow of the fracturing slurry from the well bore into the minute cracks and crevices of the formation.

SCREW CONVEYOR
A mechanism for moving dry, solid material—pelletized plastics, sulfur, cement, etc.—from one location to another by means of a helix or screw rotating in a cylindrical conduit. Archimedes thought of it first.

SCREW PIPE
Line pipe, pipeline pipe. Screw pipe used almost exclusively for small- and large-diameter pipelines in the 1920s is called threaded pipe. One end of a 30-foot joint has machine-cut threads put on at the pipe mill; the other side has a collar connector with threads on the inside diameter. When laying screw pipe, the threaded end (spigot) was stabbed into the collar (bell) and screwed in with pipe tongs until the joint was pressure-tight. All large-diameter pipelines are now welded together by arc (electric) welding. Other types of screw pipe are well casing and well tubing. Drillpipe is also screwed together, but it is not in the same category as line pipe, casing, and tubing.

Screw conveyor

SCREW PUMP
See Pump, Screw.

SCRUBBING
Purifying a gas by putting it through a water or chemical wash; also the removal of entrained water.

SCRUBBING PLANT
A facility for purifying or treating natural gas for the removal of hydrogen sulfide or other impurities.

SEA LINES
Submarine pipelines; lines laid on the ocean floor from offshore wells to production platform and to receiving stations onshore.

Screw pump

SEALS

Thin strips of metal, imprinted with serial numbers, used to "seal" a valve in an open or closed position. The metal strip has a locking snap on one end into which the free end is inserted, locking it securely. Seals are used on tanks in a battery to prevent the undetected opening or closing of a valve.

SEALS, METAL TO METAL

Seals in pipe joints, flanges, or in compatible machine parts that require no gasket. Metal-to-metal seals are noted for integrity, resisting high pressures, fire damage, and external deformation.

SEAMLESS PIPE

Pipe made without an axial seam; pipe made from a billet or solid cylinder of hot steel and "hot-worked" on a mandrel into a tubular piece without a seam. *See* Lap-Welded Pipe.

SEA TERMINAL

An offshore loading or unloading facility for large, deep-draft tankers. The terminal is served by filling lines from shore or by smaller, shallow-draft vessels.

Sea terminal

SECONDARY MIGRATION

See Migration, Secondary.

SECONDARY POROSITY

Cracks, fissures, and fractures occurring naturally in the reservoir rock.

SECONDARY RECOVERY

The extraction of oil from a field beyond what can be recovered by normal methods of flowing or pumping; the use of waterflooding, gas injection, and other methods to recover additional amounts of oil.

SECONDED, TO BE (Brit.)

To be lent to a company or group that has use for one's talent and expertise, e.g., a geophysicist seconded to a national oil company on a one-year contract by the parent company.

SECONDS SAYBOLT FUROL (S.S.F.)

A measurement of the viscosity of a heavy oil. Sixty cubic centimeters of an oil are put in an instrument known as a Saybolt viscosimeter and are permitted to flow through a standardized orifice in the bottom at a specified temperature. The number of seconds required to flow through is the oil's viscosity, its S.S.F. number. *See* Seconds Saybolt Universal.

SECONDS SAYBOLT UNIVERSAL (S.S.U.)

A measurement of the viscosity of a light oil. A measured quantity of oil—usually 60 cubic centimeters—is put in an instrument known as a

Saybolt viscosimeter and is permitted to flow through an orifice in the bottom at a specified temperature. The number of seconds required for the flowthrough is the oil's S.S.U. number, its viscosity.

SECTION MILL
A downhole cutting tool made with expandable arms used to cut sections out of the casing in the hole. The mill is attached to the end of the drillstring and lowered into the hole to the point where the casing is to be cut. The cutter arms are then expanded, either hydraulically or mechanically, against the casing wall. As the drillpipe is rotated, the cutters do their work.

SECTION OF LAND
One square mile; 640 acres; sixteen 40-acre plots.

SEDIMENTARY BASIN
An extensive area (often covering thousands of square miles) where substantial amounts of unmetamorphosed sediments occur. Most sedimentary basins are geologically depressed areas (shaped like a basin). The sediment is thickest in the interior and tends to thin out at the edges. There are many kinds of such basins, but it is in these formations that all the oil produced throughout the world has been found.

SEDIMENTARY ROCK
Rock formed by the laying down of material in layers compacted by succeeding deposits. (1) Layers made up of clay, sand, or gravel particles that are derived from the decomposition or disintegration of preexisting rock: clastic sedimentary rock. (2) Rock chemically precipitated in water such as evaporating lakes and shallow arms of the sea: rock salt, gypsum, borax. (3) Rock made of organic sediments: coal, oil shales, limestones.

SEDIMENTATION
The process of forming or accumulating sediment of any kind in layers. Broadly speaking, sedimentation also includes the separation of rock particles from the original rock by erosion or abrasion, the transporting of the particles and fragments to where they are to be deposited, and the actual deposition or settling out of the particles; further, the chemical or other digenetic changes that occur to transform the deposited sediment into solid rock.

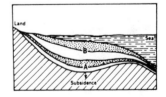

Sedimentation process

SEDIMENTOLOGY
The study of sedimentary beds, their formation, maturation, consolidation, and characteristics.

S.E.G.
The Society of Exploration Geophysicists, a professional organization of geophysicists engaged in exploration for oil and gas.

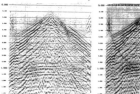

Downhole
Seismic Tool

Drilling a seismic shot
hole

Seismic thumper

Seismic Section

SEEP FINDER, U.V.

An airborne flurosensor used to detect ultraviolet fluorescence from aromatic hydrocarbons in the film on the sea surface. The oily film originates from seepage offshore. To the geologist, the seepage indicates the possibility of hydrocarbons, perhaps in commercial quantities, and usually prompts further investigation into the source.

SEISGUNS

Special vibriosis instruments that, when detonated, produce shock waves that penetrate the Earth and are reflected back to seismograph instruments on the surface. *See* Vibriosis.

SEISMIC SEA STREAMER

A cable, trailed from a geophysical vessel, towing a series of hydrophones along the seafloor recording seismic signals from underwater detonations. As the vessel moves slowly ahead, harmless electronic or air detonations are set off that are reflected from rock formations beneath the seafloor and picked up by the sensitive, sound-detecting hydrophones. *See* Geophone.

SEISMIC SECTION

A map or chart that depicts the combined effect of numerous seismic reflections or traces. *See* Seismic Trace.

SEISMIC SHOT HOLE

See Shot Hole.

SEISMIC SURVEY

See Seismographic Survey.

SEISMIC THUMPER

See Vibrator Vehicle.

SEISMIC TRACE

The record of a single seismic impulse reflected by a subsurface zone to a set of geophones, commonly referred to as jugs. Multiple traces are combined or stacked to get a seismic section.

SEISMOGRAM

The record produced by a seismographic survey.

SEISMOGRAPH

A device that records vibrations from the earth. As used in the exploration for oil and gas, a seismograph records shock waves set off by explosions detonated in shot holes and picked up by geophones.

SEISMOGRAPHIC SURVEY

Geophysical information on subsurface rock formations gathered by means of a seismograph; the investigation of underground strata by recording and analyzing shock waves artificially produced and reflected from subsurface bodies of rock.

SEISMOMETER
A device for receiving and recording shock waves set off by an explosion or other seismic sources and reflected by underground rock formations.

SEISMOMETER, REFERENCE
In seismic prospecting, a detector placed on the surface of the Earth to record successive shots under identical or similar conditions to permit overall time comparisons. It is used when shooting wells for velocity measurements.

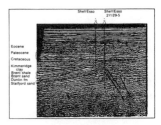
The record made by a seismometer

SEIZE
To stick together, as two pieces of metal that have become hot from excessive friction as one piece moves relative to the other; to bond or adhere, as a piston to a cylinder from heat and pressure.

SELECTIVE STIMULUS TOOL
A patented, straddle injection packer for through-tubing and coiled tubing applications. The downhole tool spots treatment fluids (acids) in predetermined intervals of a producing zone without having to pull the well's production string of tubing. The S.S.T. tool uses two inflatable (and recoverable) packers to isolate the treatment fluid.

SEMIDIESEL
A misnomer for a diesel-cycle engine whose compression is not high enough to create sufficient heat to ignite the injected fuel when starting cold. Semidiesels, or more correctly, hot-head or hot-plug diesels, are equipped with a plug that extends into the firing chamber heated by a torch or by electricity to assist in the ignition of the diesel fuel until the engine is running and up to operating temperature; a small, low-compression diesel engine. *See* Hot-Plug Engine.

Semisubmersible drilling platform

SEMISUBMERSIBLE
A large, floating drilling platform with a buoyant substructure, part of which is beneath the surface of the water. Semisubmersibles are virtually self-contained, carrying on their main and lower decks all supplies and personnel for drilling and completing wells in hundreds of feet of water and miles from shore. Some of the huge platforms are self-propelled and are capable of moving at 6 to 8 knots. As they often drill in waters too deep for conventional chain-and-cable anchors, they maintain their position over the borehole by the use of thrusters, jets, or Kort nozzles controlled by onboard computers. Some of the largest floaters are designated SS-2000 with 18,000-ton displacement, a 2,000-ton deck-

load capacity, and are capable of drilling in 2,000 feet of water in severe weather and sea conditions.

SENECA OIL COMPANY
Organized in the late 1850s in New Haven, Connecticut, Seneca was the first oil company to drill for oil. The man they chose to drill their well was none other than "Colonel" Edwin L. Drake.

SENSORS, GEOSTEERING
The very advanced geosteering systems have near-bit sensors that provide azimuthal, formation resistivity, gamma ray, rpm, and inclinational measurements, plus a telemetry system that passes this potpourri of information from near the bit to the M.W.D. system.

SEPARATOR
A pressure vessel (either horizontal or vertical) used for the purpose of separating well fluids into gaseous and liquid components. Separators segregate oil, gas, and water with the aid, at times, of chemical treatment and the application of heat. *See also* Sand Separator.

SEPARATOR, LOW-TEMPERATURE (GAS)
See L.X.T. Unit.

SEPARATOR, SAND
See Sand Separator; *also* Decanting Centrifuge.

SEPARATOR GAS
Natural gas separated out of the oil by a separator at the well.

SEQUENCE STRATIGRAPHY
A geographically discreet (self-contained, not sprawling) succession of major rock units deposited under related environmental conditions. A succession of rock formations arranged in chronological order and showing their relative position and geologic age.

SERVICE CONTRACT
A performance contract, similar to a Turnkey contract.

SERVICE LOOPS
Flexible umbilical bundles containing electrical, hydraulic, and air lines for use on drilling rigs and on offshore platforms.

SERVICE TOOLS
A variety of downhole equipment used in drilling, completion, and workover of oil and gas wells.

SERVICE WELL
A nonproducing well used for injecting water or gas into the reservoir

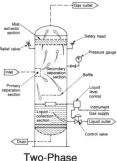

Two-Phase
Separator

Separator

or producing formation in pressure maintenance or secondary-recovery programs; also a salt water disposal well.

SERVO
Short for servomechanism.

SERVOMECHANISM
An automatic device for controlling large amounts of power with a small amount of force. An example of a servomechanism is the power steering on an automobile. A small force on the steering wheel activates a hydraulically powered mechanism that does the real work of turning the wheels.

An engine-generator set
*(Courtesy Allis-
Chalmers)*

SERVOMOTOR
A power-driven mechanism that supplements a primary control operated by a comparatively small force. *See* Servomechanism.

SET
A power unit of an engine and an electric generator, an engine and a pump, or other prime mover and work unit, is a set.

SETBACK
The space on the derrick floor where stands of drillpipe or tubing are "set back" and racked in the derrick. Offshore drilling platforms often list the stand capacity of their setbacks as an indication of their pipe-handling capability and capacity. On transportable, mast-type derricks used on land, setbacks are outside the derrick proper.

Drillpipe set back on the
outside of a portable rig

SET CASING
To cement casing in the hole. The cement is pumped downhole to the bottom of the well and is forced up a certain distance into the annular space between casing and the rock wall of the drill hole. It is then allowed to harden, thus sealing off upper formations that may contain water. The small amount of cement in the casing is drilled out in preparation for perforating to permit the oil to enter the casing. The decision to set casing (or pipe) is an indication that the operator believes he has a commercial well.

SETTLED PRODUCTION
The lower average production rate of a well after the initial flush production tapers off; the production of a well that has ceased flowing and has been put on the pump.

SEVEN SISTERS
A term applied to the seven large international oil companies: Exxon, Texaco, Gulf, Standard of California, and Mobil of the U.S.; and British Petroleum and Royal Dutch Shell, the two overseas sisters. It is said that

these seven companies control a major portion of production and refinery runs in the *free world*. The term was first used by Enrico Mattei, then head of the Italian government oil company Ente Nazionale Idrocarburi.

SEVERANCE TAX
A tax levied by some states on each barrel of oil or each thousand cubic feet of gas produced. Production tax.

SEVERE PROCESS
Subjecting feedstock to high temperatures and pressures.

SEVERE-SERVICE PIPING
Piping used to handle sour gas, gas characterized by hydrogen sulfide, corrosive or abrasive liquids, high temperature, or high pressure. Such piping is installed at wellsites to connect the well to separators, scrubbers, and gas sweeteners and in refineries for handling acids and other corrosive material.

SHACKLE ROD
Jointed steel rods, approximately 25 feet long and 3/4 to 1 inch in diameter, used to connect a central power with a well's pumping unit or pumping jack. Shackle-rod lines are supported on metal posts (usually made of 2-inch line pipe) topped with wooden guide blocks that are lubricated with a heavy grease.

SHAKE OUT
To force the sediment in a sample of oil to the bottom of a test tube by whirling the sample at high speed in a centrifuge. After the sample has been whirled for three to five minutes, the percent of B.S.&W. (sediment and water) is read on the graduated test tube.

SHAKER, SCALPING
See Scalping Shaker.

SHALE
A very fine-grained sedimentary rock formed by the consolidation and compression of clay, silt, or mud. It has a finely laminated or layered structure. Shale breaks easily into thin parallel layers; a thinly laminated siltstone, mudstone, or claystone. Shale is soft but sufficiently hard packed (indurated) so as not to disintegrate upon becoming wet. However, some shales absorb water and swell considerably, causing problems in well drilling. Most shales are compacted and consequently do not contain commercial quantities of oil and gas.

SHALE BREAK
A thin layer of shale that was deposited between harder strata or that occurs within a layer of sandstone or limestone.

SHALE OIL

Oil obtained by treating the hydrocarbon kerogen found in certain kinds of shale deposits. When the shale is heated, the resulting vapors are condensed and then treated in an involved process to form what is called shale oil or synthetic oil.

SHALE OUT, TO

The presence of a shale body in a horizontal stratum of a productive oil sand effectively blocking the horizontal or lineal flow of oil or gas to the well bore. Such an interruption makes water flooding of the formation less than successful.

SHALES, KEROGEN

See Kerogen Shales.

SHALES, REACTIVE

Shales that react to the incursion by or absorption of water, usually from water-base drilling fluid, by swelling, often enough to fill the borehole through the shale section, closing in on the drillpipe, sticking the pipe. Shales with strands of clay in their make up are the worst offenders.

Shale shaker *(Courtesy Sweco)*

SHALE SHAKER

A vibrating screen for sifting out rock cuttings from drilling mud. Drilling mud returning from downhole carrying rock chips in suspension flows over and through the mesh of the shale shaker, leaving small fragments of rocks that are collected and examined by the geologist for information on the formation being drilled.

SHALLOW GAS, DANGER OF

See Maverick Gas; *also* Diverter System.

SHALLOW-GAS STACK

A blowout-preventer stack made up especially to handle shallow gas kicks, dangerous gas kicks from shallow drilling depths offshore. Such stacks differ from more conventional stacks in having diverter spools with outlet nozzles equipped with automatic valves to divert a potentially dangerous kick away from the well.

SHALLOW-WELL CASING

See Casing, Shallow-Well.

SHALY

Composed of or having the characteristics of shale, i.e., easily split along thinly layered bedding planes. Also refers to fine-grained, thinly laminated sandstone that is easily split in layers owing to even thinner layers of shale. The property of splitting easily into thin layers is called *fissility*.

SHAPED-CHARGE PERFORATION
A perforation technique using shaped explosive charges instead of steel projectiles to make holes in casing. Quantities of explosives are made in special configurations and detonated at the bottom of the hole against the casing wall to make the perforations.

SHARPSHOOTER
A spade; a narrow, square-ended shovel used in digging. Sharpshooters are one of the pipeliner's digging tools used for squaring up a ditch or the sides of a bell hole.

SHAVETAILS
A skinner's term for his mules.

SHEAR
The deformation or breaking of an object as the result of sideward stress on one part of the object and an equal sideward stress, but in the opposite direction, on the contiguous part of the object. Breaking an apple in two by holding it in your hands and twisting each half in opposite directions is exerting stress and causes the apple to shear.

SHEAR PIN
A retaining pin, bolt, or screw designed to shear or give way before damage can be done to the item of equipment it is holding in place. A common use for a shear pin is to secure a propeller to a shaft. Should the propeller strike an obstruction, the pin will shear, preventing damage to the shaft or other parts of the power train. In other applications, shear pins or screws are used in downhole tools or equipment to hold a part in position until the tool is landed or in place. Then when thrust or torque is applied, the pin or screw shears, permitting an element of the tool to assume a predetermined attitude.

SHEAR RAM
See Ram, Shear.

Sheave

SHEAVE
A grooved pulley or wheel; part of a pulley block, a sheave can be on a fixed shaft or axle (as in a well's crown block) or in a free block (as in block and tackle).

SHEET DEPOSIT
A mineral deposit that is stratiform (uniform in thickness) and forms a stratum or sheet, and is extensive in area in relation to its thickness, e.g., 3,000 feet thick, covering an area of several square miles.

SHEET IRON
Galvanized, corrugated sheet metal used for roofing, garages, and other more or less temporary buildings. Because a sheet iron or corrugated

iron building is relatively inexpensive and easy to assemble, this kind of construction is common on oil leases.

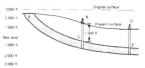

Blanket or sheet sand

SHEET SAND
See Blanket Deposit.

SHELL-AND-TUBE HEAT EXCHANGER
See Heat Exchanger, Shell-and-Tube.

SHELL PUMP
See Pump, Sand.

SHERIFF'S DEED
Conveyance under terms of bankruptcy proceedings; a tax-sale deed.

SHIM
Thin sheet of metal used to adjust the fit of a bearing or to level a unit of equipment on its foundation. For fitting a bearing, a number of very thin (0.001- to 0.030-inch) shims are put between the two halves of the bearing (between the box and cap). Shims are added or removed until the bearing fits properly on the journal.

SHIM STOCK
Thin sheets of metal (brass, tin, or steel) out of which shims for bearing caps and other engine parts that require standoffs of .001 to several thousandths of an inch or mms. Shims can be bought from suppliers, or they can be cut out of thin stock in the shop or field with metal shears or tin snips.

SHIRT TAIL ABSTRACT
A short brief of the instrument.

SHIRTTAILS
Colloquial term for the structural members or shanks of a drill bit that anchor the cutting wheels; the frame of the bit below the threaded pin.

Shock absorber

SHOCK ABSORBER
A spring-loaded slip joint run in the drillstring just above the bit to absorb vibrations and dynamic bit forces while drilling.

SHOESTRING SAND
Narrow strands of saturated formation that pinch out or are bounded by less-permeable strata that contain no oil.

SHOOT A WELL
To detonate an explosive charge in the bottom of a well to fracture a tight formation in an effort to increase the flow of oil. *See* Well Shooter.

SHOOTER

See Well Shooter.

SHOOTING 300 LINE MILES

Refers to a seismic survey, a shoot, on land or sea; line miles or line kilometers. If offshore, the shooting would take place by an exploration vessel trailing seismic streamers. *See* Seismic Sea Streamers.

SHOOTING LEASE

See Lease, Shooting.

SHORESIDE EXTENDED REACH WELLS

Extended reach drilling technology lends itself to exploring near-shore reservoirs that previously were overlooked as being too limited in size to warrant drilling/production platforms. With the advances in extended reach and horizontal drilling technology, these mini-reservoirs can be reached and produced economically from shore locations. Also, in environmentally sensitive areas, e.g., lower California, where the presence of offshore drilling and production facilities (platforms, pipelines and shuttle tankers) with their implicit threat of spills are forbidden by law to operate, land-based directional and horizontal drilling to reach near-shore (1 to 3 miles) reservoirs is now a fait accompli; an example of the innate ingenuity of the industry.

SHORT TANK

A colloquial term for a lease tank not full to the top that the lease owner wants run. Some small independent operators want a gauger to run a short tank at the end of the month to get some revenue from their stripper lease to pay the expenses of the lease. It requires as much time and work to run a short tank as a full one, so gaugers are sometimes reluctant to work a short, half-, or third-full stock tank.

SHORT TRIP

Pulling the drillstring part way out of the hole. Short trips may be necessary to raise the drill up into the protective string of casing to avoid having the drillstring stuck in the hole by a cave-in or sloughing of the wall of the borehole below the protective casing.

SHOT BLASTED

See Shot Peening.

SHOT CHARGE

The explosive charge put in a seismic shot hole. *See* Seismographic Survey; *also* Vibrator Vehicle.

SHOTGUN TANK

A tall, slender tank for separating water and sediment from crude oil. *See* Boot.

SHOT HOLE
A small-diameter hole, usually drilled with a portable, truck-mounted drill, for "planting" explosive charges in seismic operations.

SHOT PEENING
A method of increasing the service life of drillpipe by bombarding the surface of the pipe with ultrahard particles at high velocity, like pounding the pipe all over with a ballpeen hammer. In this treatment surface irregularities are smoothed, microcracks are closed, and the surface of the pipe to a depth of several thousandths of an inch is strengthened. Shot peening has the effect of increasing the metal's apparent hardness and wear resistance. Blacksmiths were onto this technique 100 years ago as they pounded buggy springs, axles, and plowshares with ballpeen hammers to dimple the surface and make the metal stronger.

Drilling a shot hole

SHOT POINT
The shot hole; the point at which a detonation is to be made in a geophysical survey.

SHOW OF OIL
Any sign or indication of petroleum (live oil or gas) that may be detected by odor or seen through a hand lens (observing rock cuttings) or by subjecting the cuttings to ultraviolet light, which causes fluorescence in petroleum. A show can also be detected in the drilling mud by an instrument called a mud sniffer, which can detect natural gas when the mud comes up the borehole. In a tight formation with low permeability, a show of oil may be all an operator gets. The best show, of course, is when the well comes in and flows from a gas or water drive; no need for a hand lens or mud sniffer then.

SHRINK FIT
An extremely tight fit as the result of "shrinking" one metal part around another. A heated part is placed around a companion piece. As the heated part cools, the once-hot piece contracts and a shrink fit results. Conversely, an expansion fit may be made by cooling a part (a valve-seat insert, for example) to extremely low temperature with dry ice and placing the part in position. As it returns to normal temperature, a tight expansion fit will result.

SHUTDOWN WELL
A well is shut down when drilling ceases, which can happen for many reasons: failure of equipment, waiting on pipe, waiting on cement, waiting on orders from the operator, etc. Not to be confused with a Shut-In Well.

Shutin gas well (Courtesy Pan American Petroleum)

High-pressure needle valve on shutin well. Pressure 4,000 psi

Side-boom cats

Sidetrack drilling

SHUT IN

To close the valves at the wellhead so that the well stops flowing or producing; also describes a well on which the valves have been closed.

SHUTIN PRESSURE

Pressure as recorded at the wellhead when the valves are closed and the well is shut in.

SHUTIN ROYALTY

See Royalty, Shutin.

SHUTIN WELL

A well is shut in when its wellhead valves are closed, shutting off production. A shutin well often will be waiting on tankage or a pipeline connection.

S.I.A.L.

An acronym for the silicon-and-aluminum-bearing rocks in the earth's crust. Together, the siatic and the simatic rocks form the earth's crust, which is estimated to range from 3 to 30 miles thick. *See* Sima.

SIDE-BOOM CATS

See Boom Cats.

SIDE-DOOR ELEVATORS

Casing or tubing elevators with a hinged latch that opens on one side to permit it to be fastened around the pipe and secured for hoisting.

SIDELINE WELL

Another name for an offset well drilled to prevent local drainage from the operator's lease to an adjacent property.

SIDE-SCAN SONAR

An electronic device that transmits high-frequency sound waves through water and records the vibrations reflected back from an object on the seafloor. Side-scan sonar is used to map the ocean floor; to discover the mounds, escarpments, or other obstructions when an undersea pipeline is to be built or a drilling/production platform is to be set. As the name suggests, side-scan sonar looks or listens sideways for the echoes from sound waves sent out horizontally rather than vertically. Sonar is a euphonious acronym of Sound Navigation Range.

SIDETRACKING

Drilling another well beside a nonproducing well and using the upper part of the nonproducer. A method of drilling past obstructions in a well, i.e., lost tools, pipe, or other material blocking the hole.

SIDEWALL CORING

A coring technique in which a rock sample is taken from the wall of the borehole in an interval that has already been drilled. A hollow bullet is fired into the wall of the borehole and then retrieved on a flexible steel cable with the rock sample, the core, which can vary in size from 3/4 to 1 1/4 inches in diameter, and from 1 to 4 inches in length. Sidewall samples are taken by operators who wish to inspect a zone or interval which they may have neglected to examine on the way down or to substantiate a conventional core taken earlier.

SIDEWALL TAP OR COCK
A small-diameter valve inserted in the wall of a tank or other vessel for drawing samples or bleeding off pressure.

SIDEWELL CORE
A sidewall core. A core or cores taken from the circumference of the open, uncased borehole.

SIGHT GLASS
A glass tube in which the height of a liquid in a tank or pressure vessel may be observed. The glass tube is supported by fittings that extend through the vessel wall, thus allowing the fluid in the tank to assume a corresponding level in the glass.

SIGHT PUMP
See Pump, Sight.

S.I.G.M.A.
Society of Independent Gasoline Marketers of America.

SIGNATURE BONUS
A sum of money (often quite a substantial sum) paid by the lessee for the execution of a lease by the landowner. Signature bonuses usually are paid by an oil company to a host country for concession rights and an exploration lease. This is often done after oil has been discovered and development has progressed to the point where the country is benefiting significantly and begins to levy heavy taxes.

SIGNATURE OF A BASIN
This can mean several things but usually means securing enough information about a basin to know its extent, its subsurface formations (from successful wells, well records, cores, seismic studies, and delineation wells), its potential, and its chances for further drilling of commercial wells. Major corporations that need to produce and build reserves will wait until a basin is defined before acquiring leases (if there are any left) to drill. Or they may negotiate farmin agreements in a productive basin whose signature seems genuine and convincing.

Silencers on stationary engines

SILENCER
A large cylindrical vessel constructed with an arrangement of baffles,

ports, and acoustical grids to muffle the exhaust noises of stationary engines.

SILICA
Silica is a fairly stable dioxide of the element silicon, SiO_2. It occurs in quartz, sand, diatomite, chert, and flint. When combined in silicates, it forms an essential part of many minerals.

SILICEOUS LIMESTONE
This is usually a thin-bedded, dark, dense limestone that is a mixture of calcium carbonate and chemically precipitated silica; a silicified limestone showing evidence of a replacement of calcite by silica.

SILICEOUS SANDSTONE
A sandstone whose grains are cemented with quartz or cryptocrystalline silica. (Cryptocrystalline means a crystalline structure too small or fine to see with the naked eye or even with a hand lens; indistinctly crystalline.)

SILICEOUS SHALE
A fine-grained, hard rock with a shaly structure containing a large amount (as much as 85 percent) of silica. It may be formed by the silicification of ordinary shale by precipitation or by the accumulation of organic material at the time the shale was deposited. Some geologists believe it is not a true shale and have called it a porcelainite, from porcelain, because of its appearance.

SILT
(1) A rock fragment or particle smaller than a very fine sand and larger than a particle of coarse clay; a loose aggregate of unconsolidated (unlithified) mineral or rock particles of silt size. Silt is most often minute particles of sand and clay. (2) A town in western Colorado.

S.I.M.A.
An acronym for the silicon-and-magnesium-bearing rocks in the Earth's crust. Together the simatic and sial or siatic rocks form the Earth's crust, which is estimated to be 3 to 30 miles thick. *See* Sial.

SIMPLEX PUMP
See Pump, Simplex.

SIMULATION
A technique of getting more production from a downhole formation. Stimulation may involve acidizing, hydraulic fracturing, shooting, or simply cleaning out to get rid of and control sand.

SINGLE-ACTION PUMP
See Pump, Single-Action System.

SINGLE-BUOY MOORING SYSTEM
An offshore floating platform (20 to 35 feet in diameter) connected to pipelines from the shore for loading or unloading tankers. The S.B.M. system is anchored in deep water, thus permitting large tankers to offload or lift cargo in areas where it is impractical to build a loading jetty or the close-in water is too shallow for deep-draft vessels.

SINGLE-POINT MOORING
See Single-Buoy Mooring System.

SINKER BARS
Long, cylindrical weights used in downhole telemetry work to weigh down the cable or cables against well pressure that prevents them and other equipment from being lowered into the well bore.

Single-buoy mooring
system

SIT-OUT
A sit-out describes the position of a carried party under a carried interest arrangement. The carried party sits out while the carrying party drills and completes a well totally at his expense.

SITTING ON A WELL
The vigil of the geologist, the operator, and other interested parties who literally sit waiting for the well's drill to bore into what is expected to be the producing formation. The geologist examines the cuttings brought up by the drilling mud to ascertain just when the pay zone is penetrated. On a "big well," a very good well, everyone knows when the pay is reached; on small or marginal wells, the geologist may be the only one who recognizes it.

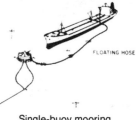

Single-buoy mooring

SIZING SCRAPER
A cylindrical plug-like tool that is pushed or pulled through a length of pipe to test for roundness.

SKANKY
Said of a well site and immediate surroundings that resemble a scrap heap; as the British say, scruffy.

SKELP
Rolls of sheet steel used by a portable pipe mill to form cylindrical line pipe.

SKID
Squared, wooden timbers used to support line pipe while it is being welded; any rough-cut lumber used to move or support a heavy object

Skidding the rig

SKIDDING THE RIG
Moving the derrick from one location to another without dismantling the structure; transporting the rig from a completed well to another lo-

Skid-mounted unit

Skimmer

cation nearby by the use of skids (heavy timbers), rollers, and a truck or tractor. Transportable folding or jackknife rigs are seldom skidded; they are folded down to a horizontal position and moved on a large flatbed

SKID-MOUNTED UNIT

Refers to a pumping unit or other oilfield equipment that has no permanent or fixed foundation but is welded or bolted to metal runners or timber skids. Skid-mounted units are usually readily movable by pulling as a sled or by hoisting onto a truck. *See* Compressor, Skid-Mounted.

SKID TANK

A product-dispensing tank mounted on skids or runners; can be pulled or carried on a truck.

SKIMMER

A type of oil spill clean-up device propelled over the water that sucks or paddles the oil into a tank.

SKIMMING PLANT

(1) A topping plant; (2) A facility built alongside a creek or small stream to catch and skim off oil that, in the early days in some fields, was turned into creeks or accidentally discharged from lease tanks or from broken pipelines.

SKIM OIL

Oil skimmed off or recovered from a salt-water-gathering system. In some oil states, notably Oklahoma and Texas, a monthly report is made of the recovered oil and the final disposition of the saltwater.

SKIM TANK

A vessel for separating trace oils from effluent water or injection water. The skim tank must be large enough to permit entrained oil globules to rise to the top of the water where it can be skimmed off. Under favorable conditions and with a properly designed skim tank, an oil concentration of only 40–50 parts per million (P.P.M.) is possible.

SKIN DAMAGE

Damage to the face of the borehole through the producing formation; a plastering over or clogging of the pay zone, the producing interval, by the action of the perforating gun explosions. With sufficient pressure differential, the damage can be minimized or overcome entirely. The gas or oil pressure, the formation pressure, forcibly clears the damaged area.

SKINNER
See Mule Skinner.

SLAB PATCH
A metal patch made out of a section of pipe welded over a pitted or corroded section of pipeline. *See* Half Sole.

SLAG-DRILLING—MUD CEMENTING
Refers to the practice of using hydraulic blast furnace slag mixed with certain drilling-fluid formulations to cement a well's casing.

Mule skinner and his team

SLANT-HOLE DRILLING
See Drilling, Slant-Hole.

SLANT RIG
See Rig, Slant.

SLANT WELL
A directional well.

SLATE
A fine-grained, compact metamorphic rock that has a slaty cleavage, i.e., can be split in thin sheets or slabs. Most slates were once shales before they were metamorphosed by the action of heat and pressure.

Slant-hole drilling

SLEEVE FITTING
A pipeline repair fitting, a cylindrical piece that is slipped over the ends of two sections of pipe to be joined. A patented sleeve fitting is the Dresser sleeve, which is fitted with rubber gaskets and end-collars that can be tightened by four bolts lying parallel to the sleeve's long axis. Some sleeves may be welded to the pipeline once they have been put in place as a temporary repair.

SLICK LINE
A nonelectric wireline for downhole work of measuring, setting packers, etc.

SLICK STRING
A drillstring consisting of drillpipe with a drill bit at the lower end of the pipe, no stabilizers, hole openers, reamers, or square drill collars, just drillpipe and bit. Slick strings are usually for shallow wells in easy-digging country. *See* Packed-Hole Assembly.

Slant rig

SLIDE VALVE
See Valve, Slide.

Pipeline sling

Motorized loader

Slips inserted into rotary
table housing

SLIDING-SCALE ROYALTY
See Royalty, Sliding-Scale.

SLIM-HOLE DRILLING
A means of reducing the cost of a well by drilling a smaller-diameter hole than is customary for the depth and the types of formations to be drilled through. A slim hole permits the scaling down of all phases of the drilling and completion operations, i.e., smaller bits, less powerful and smaller rigs (engines pumps, drawworks), smaller pipe, and less drilling mud.

SLING, PIPELINE
A wide rubber and fabric sling for lowering in or handling coated and wrapped pipe. The slings, at the end of the boom cat's hoisting lines, are used to minimize scuffing or damaging the pipeline's anticorrosion coating.

SLIP
A horse-drawn, earth-moving scoop. The slip has two handles by which the teamster guides the metal scoop into the ground at a slight angle to skim off a load of earth. Teams and slips were used to dig slush pits and build tank dikes before the days of the bulldozer. A full slip would hold about one-half cubic yard.

SLIP-FAULT PLANES
Fault planes exhibit three general types of movement or slips, one plane on another: (a) Dip slip when one plane or one side of the fault moves vertically, up or down; (b) Oblique slip—when one plane moves both vertically and laterally, up or down, and sideways as well; (c) Strike slip— when one plane moves laterally only with little or no upward or downward movement.

SLIP JOINT
A special sleeve-like section of pipe run in the drillstring to absorb the vertical motion of a floating drilling platform caused by wave action; a heave compensator.

SLIP LOAD
The weight of the string of drillpipe, tubing, or casing suspended in the drill hole by the slips.

SLIPS
Wedge-shaped pieces of metal that fit into a bushing in the rotary table to support the string of tubing or drillpipe.

SLIPSTICK
An engineer's slide rule: a log-log rule; an instrument consisting of a

ruler and a medial slide graduated with logarithmic scales used for rapid calculations.

SLOP TANK
(1) On a products pipeline, a tank where off-specification products or interface mix is stored. (2) At a marine terminal, a tank for holding the oil-water mix from a vessel that has washed down its compartments. (3) Any vessel used for retaining contaminated oil or water until it can be properly disposed of.

SLOTTED LINER
A technique employed in controlling sand incursion and other hole debris in the borehole to insert a slotted liner. This serves to keep the hole clean and to stabilize that section of the pay zone without unduly affecting drainage.

SLOTTED PIPE FOR THE PAY ZONE
Casing or production tubing with longitudinal slots run on the lower end of the string, opposite the producing zone. Slotted pipe, in horizontal boreholes, takes the place of perforated pipe and resists plugging by sands and matrix debris, which is more of a possibility with gun perforating.

SLOW DRILLING
See Drilling, Time.

SLUDGE
An oleo-like substance caused by the oxidation of oil or by contamination with other material; a thick, heavy emulsion containing water, carbon, grit, and oxidized oil.

SLUDGE ACID
See Acid Sludge.

SLUDGER
See Pump, Sand.

SLUG
A measured amount of liquid injected into a pipeline; a batch; a pipeline scraper or pig.

SLUG CATCHER
An arrangement of piping at a gas pipeline terminal made to intercept a slug of liquid in the pipeline and separate it out of the gas stream. In gas lines from offshore wells, petroleum liquids (condensate) accumulate by dropping out or condensing and collecting in low places in the line. When enough condensate (and water) collects to block the flow of gas, the pressure buildup forces the liquids forward through the pipe as a

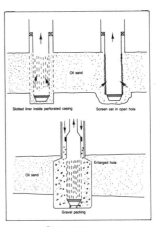

Slotted Liner

Pipeline pig or slug

slug. At the terminal or processing station the slug is caught by a slug catcher and diverted to its own tankage.

SLUG FLOW

Uneven flow in a gas pipeline in which there is heavy condensation and dropout of both water and gas liquids: natural gasoline, butane, etc. The liquids accumulate in low places in the pipeline or at a riser and shut off or block the gas flow until enough pressure builds up to blow the liquid slug out of the line. The pipeline heads up, then blows. *See* Heading; *also* Head Well; Slug Catcher.

SLUGGING

(1) Intermittent flow in a pipeline. When gas and oil are pumped in the same line, the oil will accumulate in low places until sufficient gas pressure builds behind it to push it out forcibly as a slug. (2) A small slug of acid pumped into a pumping well to open up the formation as part of a well workover operation.

SLURRIFY, TO

To make a pumpable slurry out of contaminated soil, well cuttings and other accumulated waste material at a well site. After converting this material into a pumpable slurry, it is pumped into an acceptable formation or into the annulus between casing and the wall of the borehole.

SLURRY

(1) A mixture of water or oil and pulverized solid material that can be poured or pumped in a pipeline. Slurry pipelines for transporting pulverized coal have proved to be an economical and environmentally acceptable way to move coal across the country from mine to power plant. (2) A thin pourable mix of water and cement; drilling mud that is mixed at the wellsite and pumped down the well bore inside the hollow drillpipe.

SLURRY DESIGN

Drilling mud slurries are carefully made up, designed by the mud engineer who is trained in rheology, the characteristics and control of drilling mud systems. The mud engineer, the rheologist, continually monitors his mud system for weight/gallon and volume to check on possible loss of circulation. To control a gas kick he heavies up on the slurry; if the mud is needlessly heavy, he can water back.

SLURRY OIL

A type of residual oil remaining after catalytic cracking of certain feed stocks in the manufacturing of gasoline.

SLURRY PIPELINE

A pipeline whose primary service is carrying a mixture of crushed solids in a water or oil medium. The common use of the term refers to a pipeline carrying pulverized coal in water. A pipeline is the cheapest and most efficient form of transportation for liquids. In recent years the pumping of small-particle solids, notably coal in water, has gained favor with shippers who are attracted by the pipeline's economics and safety, as well as environmental acceptance.

SLUSH-PIT LAUNDER
A wooden or metal square-sided conduit or sluice box where the bailer is dumped, the water, mud, and rock chips flushed down the launder into the slush pit. This device, a cousin to the launder used in washing ore from a mine, is part of a cable-tool drilling scene.

Slush pump

SLUSH PUMP
See Pump, Mud.

SMALL-REFINER BIAS
See Entitlement Program.

SMART BOX
See Knowledge Box.

SMART PIGS
Instrumented internal inspection devices for specially constructed pipelines. Smart pigs can, by passing through a section of the line, detect erosion, pitted areas, out-of-round spots, incipient cracks, and leaching. The instrumented pig is pushed through the pipeline by the forward movement of the crude oil, liquid product or gas in the line in the same way conventional line-cleaning pigs are moved.

A flare

SMOKELESS FLARE
A specially constructed vertical pipe or stack for the safe disposal of hydrocarbon vapors or, in an emergency, process feed that must be disposed of. Smokeless flares are equipped with steam jets at the mouth of the stack to promote the complete combustion of the vented gases. The jets of steam induce greater air flow and cool the flame, resulting in complete combustion without smoke or ash.

SMOKE POINT
One of the specifications on jet engine fuel. Kerosene or jet fuel with a low smoke point is not as desirable as fuel with a high smoke point. Hydrotreating the fuel reduces the smoke or gives it a higher smoke point. This is not a contradiction, as it appears. The high and low smoke points indicate the high and low points on the wick of a testing device made like an old-fashioned kerosene lamp. The higher the wick can be turned up while burning the sample of jet fuel without producing smoke, the cleaner burning it is; thus, the high smoke point.

Snatch block with hook

SNAP GRABBER
A member of a work gang who manages to find easy jobs to keep himself busy while the heavy work is being done by his companions. A fully occupied loafer.

SNATCH BLOCK
A block whose frame can be unlatched to insert a rope or wireline; a single-sheave block used more often for horizontal pulling than for hoisting with A-frame or mast.

S.N.G.
Synthetic natural gas.

SNOW-BANK DIGGING
Colloquial expression for the relatively soft, easy drilling in sand, shales, or gumbo.

SNUB
To check a running line by taking a turn around a post or fixed object; to take up and hold fast the slack in a line; to secure or hold an object from moving with an attached rope turned around an anchoring piece.

SNUBBER
An ingenious rig-up of lines and blocks to push down on joints of pipe that must be put into the well through the blowout-preventer stack against very high well pressure. With a special hookup, the upward pull of the rig's traveling block and hook is transmitted to lines and a yoke that push down on a joint of drillpipe, forcing it by the packing of the rams in the B.O.P. stack while the rams are holding the well pressure leak-tight. After a number of joints of pipe are forced in (the joints are screwed together), their weight equals the upthrust of the well's pressure so the snubbers may be removed and the remainder of the pipe put in through the B.O.P. without being pushed.

SNUBBING
In this procedure, the tubing is pulled out of the well through a stack of blowout-preventer valves and rams to maintain pressure control at all times. Because of high well pressure, the pipe must be snubbed out and then snubbed back in the hole using special hydraulic rig-ups that ease the pipe out and then force it back in the hole against pressure. The actual sand and debris washing is done with small-diameter reeled tubing that is inserted in the well's casing in the same procedure after the tubing is pulled from the well.

SOAPSTONE
A metamorphic rock of massive or interlaced fibrous or flaky texture.

Soapstone is soft and greasy to the touch and is composed chiefly of talc with varying amounts of micas and other minerals. It can be cut easily with a saw; used for table tops and other kinds of useful and decorative purposes.

SOFT FORMATION
Chalk, anhydrites, marl, and mudstone are referred to as soft formations; easy digging.

SOFT MINERAL
A mineral that is softer than quartz, which means less than 7 on the Mohs scale of hardness. A hard mineral is harder than quartz, i.e., 8, 9, or 10 on the Mohs scale.

SOFT PLUG
A safety plug in a steam boiler, soft enough to give way or blow before the boiler does from excessively high pressure; the plug in an engine block that will be pushed out in case the cooling water in the block should freeze, thus preventing the ice from cracking the block.

SOFT-ROCK GEOLOGIST
A colloquial term for a geologist who specializes in the study of sedimentary rocks; petroleum geologist.

SOFT ROPE
Rope made of hemp, sisal, jute, or nylon, as distinguished from wire rope, which is a steel cable.

SOFTWARE
The collection of programs used in a particular application for use in a computer. Tapes, cards, and disk packs containing programs designed for a process or series of processes.

SOIL BINDER
See Binder.

SOL
A fluid (liquid or gas) in which there is a homogeneous suspension or dispersion of colloidal matter. A sol (from solution) is more fluid than a gel.

SOLENOID
An electrical unit consisting of a coil of wire in the shape of a hollow cylinder and a movable core. When energized by an electric current, the coil acts as a bar magnet, instantly drawing in the movable core. A solenoid on an automobile's starting mechanism causes the starter-motor

gear to engage the toothed ring on the vehicle's flywheel, turning the engine. Solenoids are used also for opening and closing quick-acting, plunger-type valves, as those on washing machines and automatic dishwashers.

SOLIDS CONTROL
Refers to keeping the drilling fluid free of rock cuttings, sand, and silt. To this end, the operator and his mud man (rheological engineer or mud engineer) keep watch over the quality, the consistency of the circulating mud. Shale shakers, desilters, and desanding equipment are employed to keep the solids in the mud within efficient operating limits.

SO-LONG-AS CLAUSE
See Thereafter Clause.

SOLUBLE-OIL FLOODING
See Micellar-Surfactant Flooding.

SOLUTION CHANNELS
Tunnels or channels in formations being drilled that often cause loss of circulation. Geologists opine that the channels, from the size of a soda straw to inches in diameter, may have been caused by rock fractures that were leached out into channels by percolating underground water.

SOLUTION GAS
Natural gas dissolved and held under pressure in crude oil in a reservoir. *See* Solution-Gas Drive.

SOLUTION-GAS DRIVE
An oil reservoir deriving its energy for production from the expansion of the natural gas in solution in the oil. As wells are drilled into the reservoir, the gas in solution drives the oil into the well bore and up to the surface.

SOLVENT
A liquid capable of absorbing another liquid, gas, or solid to form a homogeneous mixture; a liquid used to dilute or thin a solution.

SOLVENT OIL
Specially formulated oils that are capable of absorbing another liquid or solid to form a homogeneous mixture; an oil used to dilute or thin a solution. *See* Lean Oil; *also* Fat Oil.

SONIC INTERFACE DETECTOR
A pipeline-sensing probe for detecting the approach of a product interface by identifying the change in sound velocities between the two products being pumped. The electronic device has a probe inserted through the wall of the pipeline, protruding into the fluid stream. The probe picks up the variations in sound velocities and through the proper link-

age, can give an audible alarm or actuate valves when the interface arrives.

SORBENTS, OIL
High-tech materials made into various shapes—booms, pillows, and "snakes"—to absorb oil in the event of a spill. The thirsty, sponge-like material absorbs, even attracts, crude oil lying on water in a thin film, and sucks up, ingests, many times its weight in oil. After it is wrung out and put back on the oily water, its absorbent qualities are still intact, only slightly diminished for having its fibrous matrix coated with oil.

SORTING
The process by which particles of sediment having a particular characteristic such as size, shape, or weight are naturally selected and separated from dissimilar particles by the action of running water; the degree of similarity of sediment in a sedimentary layer; well sorted—large apples in one barrel, small apples in another.

SOUP
Nitroglycerine used in shooting a well. Nitro in its pure form is a heavy, colorless, oily liquid made by treating glycerin with a mixture of nitric and sulfuric acids. It is usually mixed with absorbents for easier handling. Nitro, when used in well shooting, is put in tin "torpedos," 4 to 6 inches in diameter, and lowered into the well on a line. The bottom of each torpedo can is made to nest in the top of the preceding one, so as many cans as necessary for the shot can be lowered in and stacked up. Nitro is measured in quarts; the size of the shot depends upon the thickness and hardness of the formation to be fractured.

SOURCING OF HYDROCARBONS
A term, not in respectable use, except perhaps by postgraduates, for "where the oil and gas comes from." We know how it travels, migrates, seeps and, when injudiciously unearthed by the drill, gushes. But no one really knows the true, primordial source of petroleum. There are guesses (hypotheses), but they are just that.

SOUR GAS
Natural gas containing chemical impurities, notably hydrogen sulfide (H_2S) or other sulfur compounds that make it extremely harmful to breathe even small amounts; a gas with a disagreeable odor resembling that of rotten eggs.

A portable H_2S monitor
(Courtesy Ecolyzer)

SOUR PRODUCTS
Gasoline, naphthas, and refined oils that contain hydrogen sulfide (H_2S) or other sulfur compounds. Sourness is directly connected with odor.

SOUR-SERVICE TRIM
A designation by manufacturers of oilfield fittings and equipment that

their products have finishes resistant to corrosion by hydrogen sulfide (H_2S) and other corrosive agents in "sour" oil and gas. *See* Sour Gas.

SOUR-WATER STRIPPER TOWER
See Stripper Tower, Sour-Water.

SOUR WELL
A well where H_2S, CO_2, and chlorides are present in various concentrations and mixes. H_2S is lethal, even in small doses; all are corrosive.

SPACERS AND WASHERS
Specially formulated fluids for removing drilling mud from a well's borehole just ahead of the cement in a downhole cementing job. It is essential to a good cement job that the mud be removed and the wall of the hole be clean to ensure a good bond between cement and the wall. Spacers are thick fluids that displace the drilling mud ahead of the cement in a slug or piston-like manner, owing to the fluid's high viscosity and weight differential. Washers are much thinner fluids that separate the drilling mud from the cement being pumped downhole and simultaneously remove the coating of mud left on the formations. This is accomplished through a combination of turbulent and surfactant action.

Before well spacing was controlled

SPACING PATTERN
Geographic subdivision established by governmental authority, usually state, defining the number of acres to be allotted to each well drilled in a common reservoir.

SPALLING
The crumbling, scaling, or disintegrating of a metal surface.

SPAN LIMITS
Limits set on a process unit to prevent overfilling with feed stock or, conversely, to cause the unit to run dry.

SPANNING
A condition in which an undersea pipeline, once buried, has been uncovered by the action of sea currents and has several joints unsupported, spanning or swinging free. This is a hazardous condition for a line; if not corrected, the pipe could rupture or pull apart (fail in tension), which means big trouble indeed.

SPEARS
Fishing tools for retrieving pipe or cable lost in the borehole. Some spears resemble harpoons with fixed spurs; others have retractable or releasing-type spurs.

SPECIFICATIONS, PUMP
See Pump Specifications.

SPECIFIC GRAVITY

The ratio between equal volumes of water and another liquid, both at standard temperature of 60°F, where the weight of the water is given the value of 1. In the oil industry the specific gravity of oils is given in A.P.I. gravity. In the case of gas, the ratio is between air and gas, the volume of air being assigned a value of 1. For solids, the measure of the weight of a mineral compared to an equal volume of water at its maximum density which is at 39.2°F or 4°C. At this density, the weight of a volume of water is said to have a value of 1. For example, quartz has a specific gravity of 2.7, which means it weighs 2.7 times as much as an equal volume of water at 39.2°F.

SPEC SHOOT

Geophysical exploration made on a speculative basis, with the intention to sell the seismic data to one or more companies with an interest in the area.

SPECULATIVE SURVEY

A seismic imaging survey undertaken on speculation by one of the geophysical services. They complete a survey then offer it, for a price, to a company with interests in the survey area.

SPENT ACID

An acid that has reacted with another substance to the degree that its molecular makeup has changed; an acid that has reacted itself out of existence, just as a pan of soapsuds becomes spent and ineffective after the suds have washed a sinkful of greasy dishes.

SPENT CATALYST

A catalyst that has become coated with a residue from the reactions it has been a part of or has been promoting. For example, in a catalytic cracking unit (cat cracker), some petroleum coke is formed and ends up as a deposit on the minute grains of the catalyst, which then becomes inactive or spent. The spent catalyst is not discarded (it's expensive stuff) but is made fresh and active again by a treatment in a vessel called a regenerator. Here, the carbon coating is burned off the catalyst by injecting 1,100°F air into the vessel.

Blowout preventer stack with spherical BOP at the top

SPHERICAL BLOWOUT PREVENTER

A large, barrel-shaped well closure mechanism attached to the top of the well's casing. Its purpose is to close around the drillpipe in the event of a severe gas kick or threatened blowout. When the preventer's closing mechanism is hydraulically actuated, pressure is applied to a piston that moves upward, forcing the packing elements to extend into the well bore and around the drillpipe in a pressure-tight seal. Should the spherical preventer be damaged or for some reason not hold pressure, rams in

Spindletop well

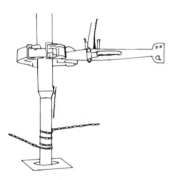

Spinning chain

Spinning wrench *(Courtesy Eckel)*

the B.O.P. stack below can also be closed on the pipe to hold the pressure until the well is killed (the pressure is equalized) by the injection of heavier drilling mud.

SPHEROID

As it applies to the industry, a spheroid is a steel storage tank in the shape of a sphere flattened at both poles designed to store petroleum products, mainly L.P.-gases, under pressure. *See* Hortonsphere.

SPIDERS

The hinged latching device attached to the elevators (the hoisting arms that lift pipe and casing in the derrick). An elevator spider is a unit attached to the traveling-block hook for hoisting pipe, casing, and tubing out of the hole and lowering in. The spider is manually locked around a length of tubing just below the tool joint. Some advanced types of elevator spiders are air-operated.

SPINDLETOP

The name of the gusher brought in by Capt. Anthony Lucas near Beaumont, Texas, in 1901. The well, the first important producer ever drilled with rotary tools, blew-in (literally) and produced at the rate of 75,000 to 100,000 barrels a day.

SPINNER

A flow-rate indicator on small-diameter product lines. The in-line device gets it name, quite simply, from an impeller-counter visible through a glass section of the pipeline.

SPINNING CHAIN

A light chain used by the drilling crew on the derrick floor when running and pulling tubing or drillpipe. After a joint has been "broken" or loosened by the pipe tongs, the spinning chain is given several turns around the pipe and, when the chain is pulled, the pipe is rotated counterclockwise and quickly unscrewed.

SPINNING TONGS

See Spinning Wrench.

SPINNING WRENCH

An air-operated drillpipe or tubing wrench used in place of the spinning chain and the winch-operated wrenches.

SPIN UP

To screw one stand of drillpipe or tubing rapidly into another with a spinning chain. After making up the joint in this manner, the heavy pipe tongs are applied to make the joint tight.

SPIRAL-WALLED CASING

See Casing, Spiral-Walled.

SPLASH ZONE

The area where waves of an ocean or lake strike the support members of offshore platforms and production installations; the water line. The splash zone is particularly subject to corrosion because of the action of both salt water and air.

SPLIT-LEVEL RIG

See Rig, Split-Level.

Split-level drilling rig

SPLIT SLEEVE

A type of pipeline repair clamp made in two halves that bolt together to form a pressure-tight seal over a hole or split in the pipe. Split sleeves also are made to enclose leaking valves and flanges until they can be permanently repaired.

SPLIT-STREAM GAS WELL

One in which owners of a gas well each sell respective shares of the gas to separate gas purchasers. This arrangement usually results in an imbalance in sales, one pipeline often being able to take more gas than the other. To balance the interests, a calculation of "make-up gas" is made. Such quantity is delivered to the under-delivered party's purchaser, he hopes before the well is depleted. Cash balancing is also possible though not common, except as the result of litigation.

SPLIT-STREAM PRODUCTION

The production of gas or oil, usually gas, divided between two or more purchasers. The division may not involve double or multiple pipeline connections.

SPLIT-STREAM SALES (of Gas)

When joint owners of a gas well each sell respective shares of produced gas to different purchasers, and sometimes at different prices, these sales are called split-stream gas sales. *See* Make-Up Gas; *also* Split-Stream Gas Well.

SPLITTER

A fractionator vessel at a refinery or gas reformer plant that "splits" the charge into various usable fractions. *See* Butane Splitter.

S.P.M.

Strokes per minute; indicates the speed or pumping rate of reciprocating pumps.

SPONGE, H₂S SCAVENGER

A patented chemical-grade iron powder—1 1/2 to 50 microns—which is added to drilling mud to react with H_2S to form iron pyrite, FeS_2, thus neutralizing the hydrogen sulfide, a deadly gas.

Casing spool with attached valve

SPONGE COKE
See Coke, Sponge.

SPONGE OIL
A type of lean oil used in refinery absorber columns to absorb light petroleum fractions or a lighter lean oil that has vaporized in an upstream process.

SPONSON
An air chamber along the sides of a barge or small ship to increase buoyancy and stability. Sponsons are used on crane barges for additional buoyancy and to minimize listing when heavy, offside lifts are being made with the crane.

SPOOL, CASING-AND-TUBING
Short-length castings, flanged on both ends, used in Christmas tree assemblies to separate and support the various valves in the stack. Spools act as spacers for the valves in the blowout preventer.

SPOOLABLE WORKOVER SYSTEM
Essentially, a coiled-tubing workover system with preinstalled gas-lift assembly and subsurface safety valve. The patented system reduces work time, provides adequate well control, and reduces substantially the amount of support equipment required.

SPOOL PIECE
A short section of piping specially cut to join the ends of two pipelines lying at unusual attitudes to each other or in tight, difficult-to-reach places. In undersea work, spool pieces are used to connect a seabed flow line to a platform riser or two undersea lines. Spool pieces are difficult to measure and cut because of the pitch and yaw angle of the pipes to be joined. A spool piece may either be a simple nipple with the ends cut at the proper angles, or it may include a valve or other fittings.

SPOT CHARTER TANKER RATES
The cost per ton to move crude oil or product by tanker from one port to another on a one-time basis, as compared to long-term charter rates. Spot charter rates fluctuate widely with demand and availability of tonnage.

SPOT FLOODING
See Five-Spot Waterflood Program.

SPOT MARKET
The market of oil resellers or brokers who supply oil on a one-time basis, often paying more for the crude than posted prices on the world market for the same gravity and type of oil. Spot market sales usually

are made to buyers whose normal supply has been interrupted or who need an extra million barrels or so for special purposes storage, for instance, against a rainy day.

SPOT-MARKET GAS SALES
The marketing of surplus natural gas on the spot market by producing and transmission gas pipeline companies. The surplus gas sales (gas not committed to long-term contracts) are made to the highest bidder, usually in relatively small volumes, and delivery is for short periods of time. The etymology of spot market is lost in antiquity; it probably is derived from a one-time, on-the-spot deal; collect on delivery; C.O.D.

SPOT MARKET SALES
(1) The term applied to sales of crude oil or products on a one-time basis and usually at prices above the going rate or world prices. Often these sales are arranged by an oil broker who can obtain certain quantities of oil for a price and for a one-time sale. (2) Sales of domestic crude oil by major producers to independent refiners from the majors' temporary overproduction or surplus. These spot sales usually are intermittent and often at prices somewhat below the posted prices.

SPOTTING FLUID
Oil-base or other non-oil fluid formulations for freeing stuck pipe in the borehole. The fluid is spotted by pumping a slug or small batch downhole to the point where the pipe is stuck in the built-up filter cake. The spotting fluid penetrates the filter cake and wets the drillpipe, making it easier to free the pipe.

SPRAY-ICE ISLANDS
Drilling islands in the Arctic made of ice by spraying sea water into the air where it loses heat rapidly and falls as coarse, porous-ice pellets. After weeks of spraying sea water, the ice build up is large enough in mass to rest on the sea bottom if the water is no more than 30 to 40 feet deep. When the ice mass reaches the mud line, the sea bottom, continued spraying and build up of ice develops the island with a freeboard of 50 to 70 feet. This ingenious method of island-making is cheaper than a dredged-gravel island.

Pipeline spread

SPREAD
A contractor's men and equipment assembled to do a major construction job. A "spread" may be literal, as the men and equipment are strung out along the right-of-way for several miles. On well workover or other jobs, the spread is a concentration of the equipment for the work.

SPREAD BOSS
The person in charge of men and equipment on a large pipeline or other construction project; the stud duck.

Spring-loaded check
valve

SPRING LOADED

Refers to an item of equipment, machinery, or valve incorporating one or more springs to effect an action or motion. A spring (spiral, coil, or leaf) that, when compressed, exerts a pressure or force against whatever is compressing it equal to the compressive force. This stored-up energy of the compressed spring is used to close a valve after being opened by a momentary greater force (a pop-off or relief valve); a machine's working part to assume its original position after being acted upon for a split instant by a larger force, e.g., the instantaneous closing of an automobile's exhaust and intake valves after being opened by the engine's push rods and rocker arms.

SPUD

To start the actual drilling of a well.

SPUD DATE

The date specified in a farmout or other exploration contract for the spudding in of a well—the actual first penetration of the Earth by the drill bit.

SPUDDER

The name for a small, transportable cable-tool drilling rig. Spudders are used in shallow-well workovers, for spudding in, or for bringing in a rotary-drilled well.

SPUDDING BIT

See Bit, Spudding.

S.P.W.L.A.

Society of Petroleum Well-Log Analysts.

SQUARE DRILL COLLAR

See Drill Collar, Square.

SQUEEZE, HESITATION

A type of low-pressure squeeze cementing in which periods of pumping are alternated with periods of shutdown. During shutdown or non-pumping periods, the hydrostatic head continues to exert a differential pressure on the slurry in the hole. And during these static periods, the slurry develops gel strength, thus increasing resistance to flow when pumping resumes. Pumping relatively small volumes of cement with time out to permit the gelling of each batch has the advantage of filling all voids without the danger of damaging low bottom-hole pressure wells and/or their pay zones.

SQUEEZE, RUNNING

A type of cement squeeze in which the slurry is pumped continuously downhole until all the voids are filled and surface squeeze pressure is obtained. Although surface pressure may be fairly quickly arrived at, all voids may not be filled. The slurry, which is only cement and water, and possibly a retarder or accelerator, may dehydrate so rapidly that the entire area to be squeezed may not be covered and thus a good hydraulic seal is not obtained.

SQUEEZE JOB
See Squeezing a Well.

SQUEEZING A WELL
A technique to seal off with cement a section of the well bore where a leak or incursion of water or gas occurs; forcing cement to the bottom of the casing and up the annular space between the casing and the wall of the borehole to seal off a formation or plug a leak in the casing; a squeeze job.

SQUIB SHOT
A small charge of nitroglycerin set off in the bottom of a well as part of a workover operation. After cleaning out a well, freezing the producing interval of sand and silt, a small explosive charge may be set off to "wake up the well."

S.S. CLASS 2000 RIG
See Rig, S.S. Class 2000.

S.S.F. & S.S.U.
Seconds Saybolt Furol and Seconds Saybolt Universal.

STABBER
(1) A pipeline worker who holds one end of a joint of pipe and aligns it so that it may be screwed into the collar of the preceding joint. Before the days of the welded line, the pipeline stabber worked only half a day because of the exhausting nature of his work. (2) On a pipe-welding crew, the stabber works the lineup clamps or lineup mandrel. (3) On a drilling rig, the floorman (roughneck) who centers the joint of pipe being lowered into the tool joint.

Stabbing a joint of drillpipe

STABBER (DERRICK)
A rig hand who, during the running of casing or tubing, stands on the tubing board high in the derrick and guides the pipe into position so the threaded end can be set in the bell (the pipe collar or coupling) and made up by the floor men with the tongs.

STABBING BOARD
A platform 20 to 40 feet up in the derrick used in running casing. The derrickman stands on the stabbing board and assists in guiding the

threaded end of the casing into the collar of the preceding joint that is hanging in the slips in the rotary table.

STABBING GUIDE
A funnel-shaped device that latches onto the box (female end) of a tool joint or tubing so the next joint can be stabbed without damaging the threads or risking cross-threading.

STABILIZER
A bushing the size of the borehole inserted in the drill column to help maintain a vertical hole, to hold the bit on course. The bushing or sleeve can be the fixed or rotating type with permanent or replaceable wings or lugs. (The lugs protrude from the body of the sleeve, making contact with the wall of the hole.)

STAB-IN CEMENTING
A method of cementing large-diameter casing in the borehole in which cement is pumped down through the drillpipe. The drillpipe is landed in a special casing shoe at the bottom of the casing. When the drillpipe is locked into the casing shoe, pumping cement downhole begins. When the cement works its way up the outside of the casing, filling the annular space, and reaches the surface, cement pumping is stopped and water and drilling mud are started down the pipe behind the cement. This displaces the cement to the bottom of the tubing. Stab-in cementing uses less cement than pumping down the casing and minimizes contamination at the cement-drilling mud interface.

STACK
Smokestack.

STACK GAS
Gases that are vented to the air through various stacks at refineries and power-generating plants. Stack gases contain carbon monoxide, carbon dioxide, sulfur compounds, and particulate matter (small grains of solid material). Today, it is mandatory that industrial stacks have scrubbers, electrostatic precipitators, and other devices to reduce the amount of noxious gases and gritty particulate matter.

STACK THE TOOLS
Pulling the drillpipe and laying it down (stacking outside the derrick) in preparation for skidding or dismantling the derrick. If the rig is transportable, it is folded down and made ready to move.

STALKS
Colloquialism for joints of line pipe, tubing, or drillpipe.

STAND
A section of drillpipe or tubing (one, two, three, or sometimes four

Stands of pipe in the derrick

joints) unscrewed from the string as a unit and racked in the derrick. The height of the derrick determines the number of joints that can be unscrewed in one stand of pipe. *See* Doubles.

STANDARD CUBIC FOOT OF GAS
The volume of gas contained in 1 cubic foot of space at a pressure of 14.65 pounds per square inch absolute and a temperature of 60°F. Volumes of gas are bought and sold corrected to the standard pressure and temperature.

STANDARD GAS MEASUREMENT
See Gas Measurement, Standard.

STANDARD PUMPING RIG
See Rig, Standard Pumping.

STANDARD RIG
See Rig, Cable-Tool.

STANDARD TOOLS
See Cable Tools.

STANDBY PIT
See Reserve Pit.

STANDBY RIG TIME
Payment made during the period of time when the drilling rig is shut down awaiting a decision from the lease owner and other interested parties whether or not drilling is to continue.

STANDPIPE
The pipe that conveys the drilling mud from the mud pump to the swivel. The standpipe extends partway up the derrick and connects to the mud hose, which is connected to the gooseneck (a curved pipe) of the swivel.

Standpipe

STANDUP DRILLING UNIT
When a drilling unit or spacing pattern is established using lines bounding sections of land or regular subdivisions thereof, the spacing pattern will be in the form of a square—40-, 160-, 640-acre units—or in the shape of a rectangle—20- or 80-acre units. A rectangular drilling unit or spacing pattern in which the long axis runs north and south is sometimes described as a standup unit (or spacing). When the long axis runs east and west, the unit may be termed a lie-down, lay-down, prone, or a recumbent unit.

STANDUP SPACING
See Standup Drilling Unit.

STARVE A PUMP

To have insufficient suction head at the pump's intake connection. A pump whose capacity or pumping rate is greater in volume than the volume of fluid being fed into it is being starved, which can cause cavitation, particularly in rotary and centrifugal pumps. *See* Suction Head, Net Positive.

STATIC HEAD

Short for hydrostatic head.

STATIC LINE

A wire or line to drain off or ground static electricity that may have built up from friction in a vehicle or its contents; a grounding line for gasoline transports to prevent arcing of static charges when unloading.

STATION KEEPING

See Dynamic Stationing.

STATION OPERATOR

See Operator.

STATUTE MILE

A measure of distance on land equal to 5,280 feet; 1,760 yards; 1,609.35 meters; 1.61 kilometers; 880 fathoms; 80 chains; or 320 rods. Commonly called a mile.

Mobile steam generators for steam flooding

STEAM FLOODING

A secondary or tertiary oil recovery method in which superheated, high-pressure steam is injected into an oil formation to heat the oil, to reduce its viscosity so it will separate from the oil sand and drain into the well bore. The water from the cooled and condensed steam is pumped out of the well with the oil and separated at the surface. *See* Heavy Oil Process (H.O.P.).

STEAMING PLANT

See Treating Plant.

STEAM INJECTION

A method of recovering very heavy crude oil from underground formations. *See* Steam Flood.

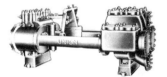

Steam pump *(Courtesy Union Steam Pump)*

STEAM PUMP

See Pump, Steam.

STEAM SALES

Steam from a cogeneration plant is sold by pounds per hour (lbs/hr), e.g., 10,000 pounds per hour is sold to a nearby paper mill.

STEAM SOAK
See Steam Flooding.

STEAM TRAP
A device on a steam line designed to trap air and water condensate and automatically bleed the air and drain the water from the system with a minimum loss of steam pressure.

STEEL REEF
Refers to the artificial reefs formed by the substructures of offshore drilling and production platforms that attract a variety of marine life from barnacles and algae to many kinds of fish.

STEEL STORAGE
Refers to the storage of crude oil and products in aboveground steel tanks.

STEP-OUTS
Wells drilled a location or two from the discovery well, often in each direction, to establish the configuration of the trend or field. *See* Delineation Well.

STICK ELECTRODE WELDING
See Welding, Stick Electrode.

STILE
Steps made for walking up and over a fence or other obstruction. Made in the shape of the letter A, stiles are used on farms and fenced leases to get to the other side without going through a gate. *See* Cattle Guard.

STILL, ATMOSPHERIC
A refining vessel in which crude oil is heated and product is distilled off at the pressure of 1 atmosphere.

STILL, PIPE
A type of distillation unit in which oil to be heated passes through pipes or tubes in the form of a flat coil, similar to certain kinds of heat exchangers. There are two main chambers in a pipe still: one where the oil is preheated by flue gases (the convection chamber); the other where the radiant-heat chamber raises the oil to the required temperature. No distillation or fractionation takes place in the still proper. The hot oil is piped to a bubble tower or fractionation tower where the oil flashes or vaporizes. The vapors are then condensed into a liquid product.

STILL, POT
A closed vessel in which crude was heated and the vapors piped to cooling coils where the gases condensed into products such as kerosene and

Steam trap *(Courtesy Armstrong)*

Storage tank under construction

Stick electrode welding

light oils. The pot still evolved into the shell still that did the same work but on a larger and more sophisticated scale. *See* Still, Shell; *also* Still, Pipe.

STILL, SHELL
The oldest and simplest form of a distillation still; a closed vessel in which crude oil is heated and the resulting vapors conducted away to be condensed into a liquid product.

STILL, TUBE
A pipe still.

STILL, VACUUM
A refining vessel in which crude oil or other feedstock is distilled at less than atmospheric pressure.

STINGER
A few joints of pipe made up and put into a gas or gas-water or gas-oil blowout in order to pump kill fluid (weighted drilling mud) down the hole to stifle and finally bring the well under control by hydrostatic pressure. *See* Momentum Kill of Blowout.

Pipelaying vessel with stinger extending into water

STIRRUP BEARING
See Bearing, Stirrup.

STOCK AND DIES
A device for making threads on the end of a joint of pipe or length of rod; an adjustable frame holding a set of steel dies or cutting teeth that is clamped over the end of the pipe to be threaded. When properly aligned, the dies are rotated clockwise in the frame, cutting away excess metal and leaving a course of threads.

STOCK TANK
(1) A lease tank into which a well's production is run. (2) Colloquial term for a cattle pond, particularly in the Southwest.

Stock tanks

STOP
A common term for a type of plug valve used on lease tanks and low-pressure gravity systems.

STOP-AND-WASTE VALVE
See Valve, Stop-and-Waste.

STOPCOCK
A type of plug valve usually installed on small-diameter piping; petcock.

STOPCOCKING
Shutting in wells periodically to permit a buildup of gas pressure in the formations and then opening the wells for production at intervals.

STOPPEL
A plug inserted in a pipeline to stop the flow of oil while repairs are being made; a specially designed plug inserted in a pipeline through the use of a tapping and plugging machine.

STORAGE, SALT-BED
See Salt-Dome Storage.

STORAGE JUG
The name applied to underground salt cavities for the storage of L.P.-gases and other petroleum products. Jug-shaped cavities are leached or washed out of salt beds using superheated water under pressure. The resulting underground caverns, some are 100 feet in diameter and 900 feet deep, are ideal storage wells for petroleum products. *See* Salt-Dome Storage.

STORE LEASE
See Lease, Store.

STORM CHOKE
A safety valve installed in the well's tubing below the surface to shut the well when the flow of oil reaches a predetermined rate. Primarily used on offshore, bay, or townsite locations, the tubing valve acts as an automatic shutoff in the event there is damage to the control valve or the Christmas tree.

STORM CONDITIONS, HUNDRED-YEAR
See Hundred-Year Storm Conditions.

STOVE OIL
A light fuel oil or kerosene used in certain kinds of wickless-burner stoves.

STOVEPIPE METHOD
Adding one joint at a time (as in building a stovepipe) in laying an offshore pipeline from a weld-and-lay barge. In contrast, reel-barge pipelaying is done by unreeling a spool of pipe over the stern of the reel barge, over the stinger and onto the seafloor. On some of the largest reel barges, 12,000 feet of 10-inch pipe can be carried on the massive reel and payed out like a giant hawser as the barge moves through the water at about a mile an hour. A reel barge can lay as much pipe in an hour or so as can be welded and laid from a conventional lay barge in a day. This

capability is very important where weather windows may be of short duration, as in the North Sea or in the extremely hostile environment of Arctic water.

STRADDLE PACKER
A downhole tool, a packer whose elements are spaced one foot or so apart. A straddle packer is used to isolate a perforation in the casing so treating fluid may be injected into one particular aperture only.

STRADDLE PLANT
See On-Line Plant.

STRAIGHT RUN
Refers to a petroleum product produced by the primary distillation of crude oil; the simple vaporization and condensation of a petroleum fraction without the use of pressure or catalysts.

STRAIGHT-RUN GASOLINE
See Gasoline, Straight-Run.

STRAINER, POT
An inline strainer used to catch and hold debris being pumped through a pipeline in a products line, a refinery, or processing plants. The strainer is flanged and is bolted into a pipeline.

STRAIN GAUGE
Any of various devices that measure the deformation of a structural element, pipe, or cable subject to loads; a tensiometer.

STRAKE, HELICAL
The helical band of metal or durable plastic attached to tall metal smokestacks or vent stacks. The helical strakes, like giant grapevines entwining the stack, reduce the oscillation of the structure caused by the wind. Strakes are designed to protrude from the circumference of the stack a distance equal to about 10 percent of the stack's diameter.

STRAPPING
Measuring a tank with the use of a steel tape to arrive at its volume; strapping involves measuring the circumference at intervals, top to bottom, height and steel thickness, and computing deadwood. Tank tables are made from these measurements.

STRATAPAX
See Bit, Diamond Shear.

STRATEGIC PETROLEUM RESERVES
Crude oil stored in underground formations and sealed caverns as a fuel reserve in the event of a national emergency or a prolonged oil embargo

Strainer *(Courtesy Andale)*

by foreign suppliers. The caches of crude are located in various areas across the country.

STRATIFICATION
The formation or deposition of sedimentary material in definite layers, the arrangement of sedimentary rocks in more-or-less horizontal strata, a structure produced by the deposition of sediment in layers. Stratification may be caused by a difference in texture, hardness, cementation, and structure.

STRATIFORM DEPOSIT
Sedimentation or mineral deposit of uniform thickness. *See* Sheet Deposit.

STRATIGRAPHIC BREAK
See Break.

STRATIGRAPHIC COLUMN
See Geologic Column.

STRATIGRAPHIC SEQUENCE
A chronological succession of sedimentary rocks, from older rocks below to younger ones above, without significant interruption; a sequence of bedded rocks.

STRATIGRAPHIC SYSTEM
A major chrono-stratigraphic unit of worldwide importance; the rock laid down or formed during a period of geologic time. In the United States there are 12 recognized systems ranging in age from Quaternary, the youngest, to Cambrian, the oldest.

STRATIGRAPHIC TEST
A test well drilled to obtain information on the thickness, lithology, porosity, and permeability of the rock layers drilled through or to locate a key bed. Such wells are often drilled to evaluate a potentially productive pay zone.

STRATIGRAPHIC TRAP
A type of reservoir capable of holding oil or gas, formed by a change in the characteristics of the formation—loss of porosity and permeability, or a break in its continuity—which forms the trap or reservoir.

STRATIGRAPHIC TEST HOLE
A hole drilled to gather information about a stratigraphic formation, the general character of the rocks, their porosity and permeability.

STRATIGRAPHIC TRAP FIELD
A porous, permeable formation cut off or hemmed in by a porosity/permeability pinch out. *See* Pinch Out.

STRATIGRAPHY

Geology that deals with the origin, composition, distribution, and succession of rock strata.

STRAW IN THE CIDER BARREL

To have a well in a producing reservoir or to have an interest in a well in a producing field.

STREAM

A stream—whether oil, gas, or product—is what is being pumped through a pipeline, moved from one process unit to another. Onstream means a plant or refinery is operating and handling streams of charge stock or plain stock.

STREAM DAY

An operating day on a process unit as opposed to a calendar day. Stream day includes an allowance for regular downtime.

STRESS CORROSION OF STEEL

Stress-corrosion cracking is defined as the beginning and subsequent growth of cracks of an alloy, a metal alloy, as the result of corrosion and tensile stress. Corrosion is so insidious and destructive that cracking can occur in steel piping and equipment well below the yield point of the metal.

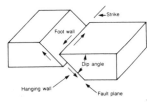

Strike

STRIKE

(1) The direction or trend taken by a structural surface, e.g., a bedding plane or fault plane, as it intersects the horizontal. The strike of a fault plane can be shown by the line of intersection between the fault plane and a horizontal surface. Since the strike line is always horizontal and has direction, it can be measured by azimuth or bearing. For example, N30°W or N60°E. The strike of a layer of rock that has dipped is the direction (angle measured from north) taken by the line of intersection of a horizontal and an inclined plane. The inclined plane is the surface of the dipping layer of rock (the fault plane); the horizontal plane is an imaginary surface determined by a level or surveying instrument. (2) A good well; to make a strike is to find oil in commercial quantities; a hit.

STRIKE SLIP

See Slip.

STRIKING WENCH

A heavy-duty end wrench for tightening large hex nuts by striking the wrench handle with a sledgehammer. Striking wrenches are used for tightening nuts on large flanges, head bolts, etc.

STRING

(1) A succession of joints of tubing makes a string of tubing. Drillpipe,

drill collars, stabilizers, and a drill bit screwed together or made up and lowered into the borehole is a drillstring. (2) A string of dry holes is three or more and is bad news indeed.

STRING DESIGN
A phrase that refers to the make up of the string of tools to be lowered into the hole to drill. For example in crooked hole country, areas where it is difficult to keep the bit drilling on a vertical course, the drillstring might be made up of stiff, heavy components such as heavy square drill collars, stabilizers, and reamers screwed onto the regular drillpipe; a packed-hole assembly. In easy digging, shallow wells, a slick string would be used consisting of nothing more than drillpipe and bit. Nothing exotic like stabilizers, reamers, or keyseat wipers would be required.

STRINGER BEAD
A welding term that refers to the first bead or course of molten metal put on by the welder as two joints of line pipe are welded together. *See* Welding, Pipeline.

STRINGERS
Thin streaks or strands of harder material, often very abrasive, in a downhole section of sedimentary rock.

STRINGING PIPE
Placing joints of line pipe end to end along a pipeline right-of-way in preparation for laying; i.e., screwing or welding the joints together to form a pipeline.

Stringing pipe

STRIP
To disassemble; to dismantle for the purpose of inspection and repair; to remove liquid components from a gas stream. *See also* Stripping the Pipe.

STRIP CHART
A meter chart in the form of a roll of paper towels. The more common meter charts are circular sheets of ruled and calibrated paper with a hole in the center so they may be fitted on the meter's spindle. Strip charts may record gas flow or pressure fluctuations for a week as opposed to most circular charts' 24-hour run.

STRIPPER
An oil well in the final stages of production; a well producing less than 10 barrels a day. Most stripper wells are pumped only a few hours a day. In 1982 there were more than 400,000 stripper wells in the United States producing 20 percent of the country's oil.

Chart, Circular type
(Courtesy
Sonceboz
Corporation)

Stripping job

Stripping plant

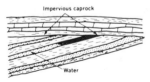

Oil trapped under a cap rock

STRIPPER COLUMN
See Fractionator.

STRIPPER TOWER, SOUR-WATER
A refinery vessel; a tower for the physical removal of contaminants from "sour water"—water from knockout drums, condensates from accumulators and other processing units—before it undergoes biological treatment or is discharged in the plant's waste-water system.

STRIPPER WHEEL
A handwheel that is attached to the upper rod in a string of sucker rods in the well to unscrew them.

STRIPPING JOB
See Pulling Rods.

STRIPPING PLANT
See Gasoline Plant.

STRIPPING THE PIPE
The job of removing drillpipe or tubing from a well under pressure while maintaining control of the well. The pipe is stripped by withdrawing it, a stand at a time, through a wellhead plug equipped with a hydraulic closure mechanism (ram) that maintains pressure contact with the pipe being withdrawn.

STRIPPING THE WELL
To pull the rods and tubing from the well at the same time. The tubing must be stripped over the rods, a joint at a time.

STRUCTURAL FEATURE
A feature or condition caused by the deformation or displacement of rocks, such as a fold or a fault. A more common term used in the field for such features is simply structure.

STRUCTURAL HIGH
A general term for geologic features that include anticline, dome, and crest; in general usage, a high.

STRUCTURAL LOW
A term that includes a structural basin, syncline, saddle, or sag; in general usage, a low.

STRUCTURAL TRAP
A type of reservoir containing oil and/or gas formed by movements of the Earth's crust that seal off the oil and gas accumulation in the reservoir, forming a trap. Anticlines, salt domes, and faulting of different kinds form structural traps. *See* Stratigraphic Trap.

STRUCTURE
The general arrangement, disposition, or relative position of rock masses of an area. As used in petroleum geology, a structure means the physical arrangement of rock formations, such as an anticline, syncline, or reef, that indicates a possible trap or accumulation of oil and gas.

STRUCTURE CONTOUR MAPS
Contour maps drawn by geophysicists and geologists to show underground rock strata.

STRUCTURE PIPE
In deep, offshore wells drilled from floating platforms (semisubmersibles or drillship), the first pipe sunk to about 1,000 feet. B.M.L. (below mud line) is a large-diameter, cassion-like pipe as a stabilizing element. Inside the structure pipe is the conductor pipe, down 1,500 to 2,000 feet; next is the surface casing at about 3,000 ft. B.M.L. These depths will vary with the type of seabed and subsea formations to be drilled.

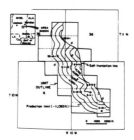

Structure contour map
(Courtesy Mace)

STUB IN
To attach a line (usually of smaller diameter) to an existing line, manifold, or vessel and make the connection by cutting a hole in the existing installation and welding on a nipple or other fitting.

STUB LINE
An auxiliary line attached to an existing line by use of a tap saddle or by welding on a nipple or other fitting.

STUCK PIPE
Refers to drillpipe stuck in the well's borehole from one or more of several possible causes: the formation above the drill bit caves or sloughs off, filling an interval of the hole; keyseating; junk or a foreign object wedged against the pipe; sand or shale packed around the tool joints; or a section of shale, absorbing water from the drilling mud, swelling sufficiently to reduce the size of the hole.

STUD DRIVER
A mechanical device for driving or screwing stud bolts into a bored and threaded hole; a wrench-like device attached to one threaded end of a stud bolt without damaging the threads. When torque is applied, the other end of the bolt screws into the hole. Simple stud drivers are hand-held but for large jobs they can be adapted for impact wrenches, drill-press, or air motors.

STUD DUCK
Top man; the big boss.

Stuffing box on pumping well

Submersible barge platform

Submersible manipulator
(Courtesy Ocean Engineering)

STUFFING BOX
A packing gland; a chamber or "box" to hold packing material compressed around a moving pump rod or valve stem by a "follower" to prevent the escape of gas or liquid.

SUB
A short length of tubing containing a special tool to be used downhole; a section of steel pipe used to connect parts of the drill column that, because of difference in thread design, size or other reason, cannot be screwed together; an adapter.

SUB, BENT
A short section of drillpipe in the drillstring, located just above the drill bit and mud motor, that is an angle builder for beginning or continuing the curve that converts the vertical hole to horizontal. The sub's axis is bent or off center only slightly—1° to 3°.

SUBLEASE
An assignment of an oil and gas lease wherein the assignor transfers to another party an interest of a lesser amount than is contained in the original lease; transfers 80 acres of an original 160 acre-lease, for example.

SUBLIMATION
The natural processes by which a solid substance vaporizes without going through a liquid state, e.g., snow and ice.

SUBMERSIBLE
A small "submarine," usually manned by two operators, that is used for subsea observation of seabottom conditions when an underwater pipeline is to be laid, for pipeline inspections, etc.

SUBMERSIBLE BARGE PLATFORM
A type of drilling rig mounted on a barge-like vessel used in shallow coastal waters. When on location, the vessel's hull is submerged by flooding its compartments, leaving the derrick and its equipment well above the water line.

SUBMERSIBLE DRILLING BARGE
A barge-like vessel capable of drilling in deeper water than the smaller and simpler barge platform. The submersible drilling barge has a drilling deck separate from the barge element proper. When floated into position offshore in water as deep as 100 feet, the barge hull is flooded. As it slowly sinks, the drilling platform is simultaneously raised on jacking legs at each corner of the barge, keeping the drill platform well above the water surface.

SUBMERSIBLE PUMP
See Pump, Submersible.

SUBSALT EXPLORATION

Drilling in 400 to 800 feet of water depth, then through a horizontal salt sheet as thick as 3,000 feet, then on down another 14,000 to 18,000 feet to the prospective pay zone in the Gulf of Mexico. The presence of subsalt formations has been suspected for years by geologists. Only recently (1992–1993), through the use of advanced 3-D seismology, have these suspicions been proven to be correct. There are hydrocarbons down there; geologists now are convinced. But with the subject formations 15,000 to 18,000 feet below the mudline, they are accessible only to major corporations or well-heeled independents.

Subsea inspection vessel
(Courtesy Comex)

SUBSEA COMPLETION SYSTEM

A self-contained unit resembling a bathysphere to carry men to the ocean bottom to install, repair, or adjust wellhead connections. One type of modular unit is lowered from a tender and fastened to a special steel wellhead cellar. Men work in a dry, normal atmosphere. The underwater wellhead system was developed by Lockheed Petroleum Services Ltd. in cooperation with Shell Oil Company.

SUBSTITUTE GAS

See Synthetic Natural Gas.

SUBSTRUCTURE

The sturdy platform upon which the derrick is erected. Substructures are from 10 to 30 feet high and provide space beneath the derrick floor for the blowout-preventer valves.

Substructure

SUBSURFACE GEOLOGY

The examination, identification, and correlation of rock formations, structures, and other features beneath the surface of the land or oceans. The study is made by examining rock samples brought to the surface by exploratory drilling of boreholes, geophysical methods, and inference.

SUCKER ROD

Steel rods that are screwed together to form a "string" that connects the pump inside a well's tubing downhole to the pumping jack on the surface; pumping rods.

SUCKER ROD, HOLLOW

In certain applications, such as slim-hole pumping, hollow sucker rods are used, serving the dual purpose of rod and production tubing in the same string. Traveling-barrel pumps are most often used with hollow-rod pumping. The rods are attached to the cage or pull tube (traveling barrel); the pump is installed in the seating nipple, or a packer-type pump anchor is used.

SUCKER-ROD GUIDES

Small washer-like devices attached to a pumping well's sucker rods to

Sucker rod, pin end

center the rods in the tubing as the rods move up and down. The guides prevent excessive wear of the tubing and the rods as well.

SUCKER-ROD PUMP
See Pump, Rod.

SUCKER-ROD SCRAPERS
Perforated disks attached to the string of sucker rods of a pumping well to prevent the buildup of paraffin on the inside of the tubing. As the rods move up and down, the perforated disks (several to each rod) scrape off the paraffin attempting to coat and then build up on the tubing, reducing the amount of oil that can be pumped from the well.

SUCTION HEAD, NET POSITIVE
The hydrostatic head; the height of the column of liquid required to ensure that the liquid is above its bubble-point pressure at the impeller eye of a centrifugal pump. If a pump requires 10 feet of net positive suction head to fill properly and prevent cavitation, then the minimum liquid level above the pump's immediate intake connection should be 12 feet. The additional 2 feet of liquid level are needed to overcome the friction of connecting piping.

SUCTION VALVE
See Valve, Suction.

SUITCASE ROCK
Any formation that indicates further drilling is impractical. Upon hitting such a formation, drilling crews traditionally pack their suitcases and move on to another site.

SUPERCHARGE
To supply air to an engine's intake or suction valves at a pressure higher than the surrounding atmosphere. *See* Supercharger.

SUPERCHARGER
A mechanism such as a blower or compressor for increasing the volume of air charge to an engine over that which can normally be drawn into the cylinders through the action of the pistons on the suction strokes. Superchargers are operated or powered by an exhaust-gas turbine in the engine's exhaust stream.

SUPERFUND
Money collected from the oil industry and others to clean up chemical waste sites. Comprehensive Environmental Response, Compensation and Liability Act of 1982. The superfund act taxes gasoline and the chemicals that gasoline contains for funds to clean up hazardous waste sites. The act also imposes a tax on crude oil and on imported petroleum products.

SUPERPORT
A terminal or oil-handling facility located offshore in water deep enough to accommodate the largest, deep-draft oil tankers.

SUPERPOSITION
Refers to the geologic fact that in a normal sequence of layered rock, the youngest rock is on top and the oldest on the bottom, having been deposited first. A sequence of sedimentary rock is assumed to have been deposited or laid down horizontally. When a well is drilled, the youngest rock is encountered first and on down successively older rock is penetrated by the drill bit—unless, of course, the orderly sequence has been interrupted or overturned by a major deformation caused by a shifting of the Earth's crust.

312,000-dwt supertanker, the *Universe Ireland*

SUPERTANKER
The largest crude oil carrier yet designed.

SUPPLY-BOAT MOORING SYSTEM (S.B.M.S.)
A type of sea terminal for tankers and supply boats featuring a single leg securely fixed to the ocean floor with a truss-like yoke that attaches to the bow of the vessel being loaded. Loading lines are supported by the yoke, which is hinged to the boat allowing free articulation to accommodate any kind of sea condition during loading. The leg of the mooring system is equipped with a universal joint and is able to rotate as the ship weathervanes.

SUPPORT AGREEMENTS
These agreements are contracts entered into to help in the financing of drilling operations, to lend support in cases where outside financial help is needed. Support agreements are often called "contribution agreements." Three of the most common support or contribution agreements are Dry Hole Agreement, Bottomhole Contribution Agreement, and Acreage Contribution Agreement.

SURETY REQUIREMENTS
The U.S. government requires surety bonds to be posted to cover end-of-production clean up for wells off the continental coasts. Wells must be properly plugged with cement. Fifty thousand to 3 million dollars are required to cover the clean up and final removal of platforms, mobile rigs, pipelines, and other seabed installations situated on government offshore tract leases. *See* O.P.A. 90.

SURFACE GEOLOGY
The examination, identification, and correlation or classification of rocks, outcrops, and other formations as seen on the surface by the geologist.

Large-diameter well casing; surface pipe

SURFACE PIPE

The first casing put in a well, which is cemented into place and serves to shut out shallow water formations, and also as a foundation or anchor for all subsequent drilling activity. *See* Production String; *also* Anchor String.

SURFACE PIT

A dug pit for the disposal of oilfield brine by evaporation or seepage, or in early days, overflow by fresh water runoff after a good downpour.

SURFACE RIGHT

The right of an owner of an oil and gas lease, the lessee, to use as much of the surface of the land as may reasonably be necessary to conduct the operations under the lease. Most applicable state laws require that compensation be paid by the operator for damage to the surface or for any resulting crop loss.

SURFACE TENSION

A property of liquids in which the surface tends to contract to the smallest possible area, as in the formation of spherical raindrops, a phenomenon attributed to the attractive force or cohesion between the molecules of a liquid.

SURFACE WATER

Water that falls as rain or snow and stands on the ground until it evaporates or soaks into the earth and becomes groundwater. *See* Meteoric Water; *also* Juvenile Water.

SURFACTANT FLOODING

See Micellar-Surfactant Flooding.

SURGE CHAMBER

See Surge Tank.

SURGE TANK

A vessel on a flow line whose function is to receive and neutralize sudden, transient rises or surges in the stream of liquid. Surge tanks often are used on systems where fluids flow by heads owing to entrained gas.

SURPLUS BOTTOMS

A phrase that refers to oil tankers, to the number and capacity of tankers: one bottom equals one tanker. Surplus bottoms: surplus tankers.

SURRENDER CLAUSE

A lease clause authorizing the lessee to surrender all or part of his leased property. It follows: "Lessee has the right at any time without the lessor's consent to surrender all or any part of the leased premises and to be relieved of all obligations concerning the acreage released."

SURVEY, GRAVITY
See Gravity Survey.

SURVEY, SEISMOGRAPHIC
See Seismographic Survey.

SURVEY MAP
See Map, Survey.

SURVEY STAKES
Wooden markers driven into the earth by a survey crew identifying the boundaries of a right-of-way, the route of a pipeline or a well location. Survey stakes may bear notations indicating elevation or location.

SURVEYOR'S CHAIN
A measuring instrument; a chain of 100 links, each link equaling 7.92 inches.

SURVEYOR'S TRANSIT
See Transit.

SUSPENDED DISCOVERY
An oil or gas field that has been identified by a discovery well but is yet to be developed.

SUSPENSE MONEY
The term applied to revenue or money collected by a regulated gas pipeline company after filing a rate increase, which is subject to an obligation to refund the money to purchasers if the regulatory agency controlling such increases fails to approve the increase. Escrow money.

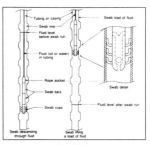

Swabbing

SWAB, TO
(1) To clean out the borehole of a well with a special tool attached to a wireline. Swabbing a well is often done to start it flowing. By evacuating the fluid contents of the hole, the hydrostatic head is reduced sufficiently to permit the oil in the formation to flow into the borehole, and if there is enough gas in solution the well may flow for a time. (2) To glean as much information as possible from a person.

SWAG
A downward bend put in a pipeline to conform to a dip in the surface of the right-of-way or to the contours of a ravine or creek; a sag.

SWAGE
A heavy, steel tool, tapered at one end, used to force open casing that has collapsed downhole in a well.

Sway braces *(Courtesy Parker Drilling)*

SWAGE NIPPLE

An adapter; a short pipe fitting, a nipple that is a different size on each end, e.g., 2-inch to 3-inch; 2-inch to 4-inch.

SWAMPER

A helper; the person who helps a truck driver load and unload and helps take care of the vehicle.

SWAY BRACES

The diagonal support braces on a rig structure. Along with the horizontal girts, sway braces hold the legs (the corner members) of the rig in place.

SWEEP EFFICIENCY

Refers to secondary oil recovery drives using various fluids with additives under pressure.

SWEET

Having a good odor; a product testing negative to the doctor test—free of sulfur compounds.

SWEET CRUDE

Crude oil containing very little sulfur and having a good odor.

SWEETENING

See Gas Sweetening.

SWEET GAS

Natural gas free of significant amounts of hydrogen sulfide (H_2S) when produced.

SWEET SPOT

An area where two geologic features meet; where two different sets of fracture and joint patterns intersect.

SWELLING CLAY

Clay that is capable of absorbing large quantities of water, which increases its volume up to eight times. A good example is bentonite, a soft, plastic, very porous, light-colored rock that got its name from deposits found in the Benton formation, Rock Creek district, of eastern Wyoming. Bentonite and other clays similar in composition are used to thicken drilling muds.

SWEPT-FREQUENCY EXPLOSION

A type of controlled explosion used in seismic work in which a string of small detonations are set off in sequence instead of the more conventional single large explosion. In oil and gas exploration, swept-frequency explosions are a vibration or shock source in conducting seismographic surveys.

SWING CHECK
A check valve.

SWING JOINT
A combination of pipe fittings that permits a limited amount of movement in the connection without straining the lines, flanges, and valves.

SWING LINE
A suction line inside a storage or working tank that can be raised or lowered by a wireline attached to a hand winch mounted on the outside of the tank. By raising the swing line above the level of water and sediment in the tank, only the clean oil is pumped out.

SWING MAN
One whose job is working in place of other employees on their days off. In a refinery or pump station operating 24 hours a day, there are three shifts of workers and a swing shift. The swing shift covers the days off of the other workers, so the swing man works two day shifts, two evening shifts, and one graveyard or hoot-owl shift. The other graveyard shift is worked either by another swing man or another plant worker who is not a regular shift worker.

SWING SHIFT
See Swing Man.

SWITCHER
A person who works on an oil lease overseeing the filling of lease stock tanks. When a tank is full he switches valves, turning the production into other tanks. A switcher works on a lease with flowing production; if the lease had only pumping wells, he would be called a pumper.

SWITCHES, LIMIT
A pneumatic or manual switch installed on a rotary valve (ball or butterfly) that makes electrical contact in both open and shut positions, indicating its position (O or S), on a remote control panel.

SWIVEL
A part of the well-drilling system; a heavy, steel casting equipped with a bail—held by the hook of the traveling block—containing the wash pipe, gooseneck, and bearings on which the kelly joint hangs and rotates; the heavy link between the hook and the drillstring onto which the mud hose is attached; an item of traveling equipment.

SYNCLINE
A bowl-shaped geological structure usually not favorable to the accumulation of oil and gas. Stratigraphic traps are sometimes encountered in synclines. *See* Anticline.

Swivel *(Courtesy B.F. Goodrich)*

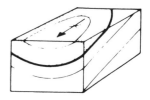
Plunging syncline *(Courtesy Petex)*

SYNFUEL
Short for synthetic gas or oil.

SYNTHANE PLANT
A coal-to-gas pilot plant operated by the Energy Research and Development Administration in Pennsylvania. Designed to produce 1.2 MMcfd of pipeline gas, designated as synthane, synthetic methane.

SYNTHESIS GAS
A mixture of equal parts of hydrogen and carbon monoxide formed by reacting steam with hot coal or char. The resulting mixture can be burned as fuel, and is similar to coal gas. The primary use for the gas is in the manufacture of methane or synthetic natural gas. Synthesis gas, also know as water gas, has an energy content of 980 to 1035 Btu per standard cubic foot, roughly the same value as natural gas.

SYNTHETIC NATURAL GAS
Commercial gas made by the reduction or gasification of solid hydrocarbons: coal, oil shale and tar sand. Syngas; substitute gas. *See* Gasification.

SYNTHETIC OIL
A term applied to oil recovered from coal, oil shales, and tar sands.

SYSTEM
A major chronostratigraphic unit recognized in much of the world that represents the fundamental unit of Phanerozoic rocks (rock in which there is evidence of plant and/or animal life as observed from fossils). From the Greek meaning "visible life." The systems recognized in the United States are, in order of increasing age, Quaternary, Tertiary, Cretaceous, Jurassic, Silurian, Ordovician, and Cambrian.

TACK WELD
Spot welds temporarily joining two joints of pipe to hold them in position for complete welding.

TAIL
To carry the light end of a load; to extricate a vehicle from a ditch or mud.

TAIL CHAIN
The short length of chain, with a hook attached, on the end of a winch line.

TAIL ENDS
In a distillation column at a refinery, tail ends are the overlapping ends of the distillation curves of two products. For example, when naphtha and kerosene are being distilled, the end point of naphtha is about 325°F but the initial boiling point of kerosene is about 305°F. Before naphtha reaches its end point, kerosene has begun to boil or vaporize. This unavoidable overlap results in tail ends; the high end of one product and low end of a closely related product.

TAIL GAS
Residue gas from a sulfur-recovery unit; any gas from a processing unit treated as residue.

TAILGATE (OF THE PLANT)
The place or the area where a product leaves a refinery or processing plant. If the product is a liquid or a gas, the product leaves the plant by pipeline.

TAILING
Leftovers from a refining process; refuse material separated as residue.

TAILING-OUT RODS
Unscrewing and stacking rods horizontally outside the derrick. As a rod is unscrewed, a worker takes the free end and, as the elevators holding the other end are slacked off, he "walks" the rod to a rack where it is laid down.

TAKE AND PAY CLAUSE
A clause in a gas purchase contract requiring the buyer to take a specified quantity of gas and to pay for it. In the event the buyer fails to take the specified quantity, the contract measure of damages is applicable. In contrast is the Take Or Pay Clause which requires the buyer to pay the purchase price for the specified quantity, whether taken or not.

TAKE-OR-PAY CLAUSE
A contractual clause specifying how many years later "free" makeup gas will be available to a buyer. *See* Makeup Gas.

TAKEOVER PROVISION
A provision in most farmout agreements that allows the farmor (sic) to take over the job of completing, reworking, or operating a well. In this case, the farmee or the operator elects not to do the work or is unable to do so.

TALLY THE PIPE
In setting pipe or casing a well, it is important to keep tab on the footage of pipe run in the hole; before lowering a joint, it is carefully tallied (measured) so the operator, by counting the number of joints run, knows to the foot where the bottom of the casing is downhole.

Tallying the pipe

TALUS
Rock fragments at the base of a cliff, sometimes forming a slope of chips and larger fragments one-fourth to one-third the way up the face of the disintegrating rock cliff.

T.A.M.E.
Tertiary amyl methyl ether, an oxygenate that added to motor gasoline promotes cleaner burning engine performance. *Also* M.T.B.E.

TANDEM
A heavy-duty flatbed truck with two closely coupled pairs of axles in the rear; a 10-wheeler.

TANK
(1) Cylindrical vessel for holding, measuring, or transporting liquids. (2) Colloquial for small pond; stock tank.

TANK, BULLET
See Bullet Tank.

TANK BATTERY
See Battery.

Tank battery *(Courtesy Cities Service)*

TANK BLANKETING
Covering the crude oil or refined product in a storage tank with chemi-

cally inert microballoons which form a nonevaporation blanket over the contents, thus reducing the costly evaporation. *See* Microballoons.

TANK BOTTOMS
Oil-water emulsion mixed with free water and other foreign matter that collect in the bottoms of stock tanks and large crude storage tanks. Periodically, tank bottoms are cleaned out by physically removing the material or by the use of chemicals that separate oil from water, permitting both to be pumped out.

Tanker terminal

TANK DECKS
So called box girders, in the fabrication of offshore platforms, that are sealed as rectangular tanks and used for oil storage or seawater ballast.

TANK DIKE
A mound of earth surrounding an oil tank to contain the oil in the event of a rupture in the tank, a fire, or the tank running over.

TANKER PIRATES
Hijacking of crude oil tankers have, in the late 1980s and early 1990s, taken their toll in Indonesian waters—Malaysia, Singapore, and the Malacca Straight. Pirates board the vulnerable ships (low freeboard, slow speed, small crews) tie up and rob the crew, and leave the tanker to sail on out of control. The chances of collision and resulting oil spill and pollution are extremely high.

Tank farm

TANKER TERMINAL
A jetty or pier equipped to load and unload oil tankers. *See* Sea Terminal.

TANK FARM
A group of large riveted or welded tanks for storage of crude oil or product. Large tank farms cover several square miles.

Tankies

TANKIES
The name given to workers who build tanks of all sizes, shapes, and materials—iron, steel, wood. Some lease tanks are erected by bolting together precut and bored sections of tank iron (1/4-inch steel sheets); larger tanks are put together from heavier steel sheets and are welded together to make the floor, walls, and roof. In the old days large storage tanks were riveted, and the 55,000-barrel tank was about the largest size. Today's tanks can hold up to 1 million barrels!

TANK MIXER
Motor-driven propeller installed on the shell of a storage tank to stir up and mix tank sediments with the crude. The propeller shaft protrudes through the shell, with the motor mounted on the outside. Turbulence

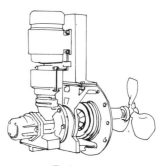

Tank mixer

453

created by the prop thrust causes the B.S.&W. to remain suspended in the oil as it is pumped out.

TANK TABLE

A printed table showing the capacity in barrels for each one-eighth inch or one-quarter inch of tank height, from bottom to the top gauge point of the tank. Tank tables are made from dimensions furnished by tank strappings. *See* Strapping.

TANK TRAIN

Tank train

A new concept in the rail shipment of crude oil, products, and other liquids developed by General American Transportation (G.A.T.X.). Tank Train tank cars are interconnected, which permits loading and unloading of the entire train of cars from one trackside connection. This arrangement does away with the need for the conventional loading rack, and vapors from the filling operation can be more easily contained. *See* Densmore, Amos.

TANK TURNS

Refers to the cycles of filling and emptying of product tankage at proprietary or reseller terminals. A record of this activity can be a measure of profitability, also the adequacy of the terminal's tankage.

TANKWAGON

The word for the old-fashioned tank mounted on a horse-drawn wagon; the tankwagon that carried crude oil from lease tank battery to pipeline booster station; and gasoline to the service stations with hand-operated sight pumps. The word tankwagon is now generic, even for tank trucks, e.g., tankwagon prices at the refinery.

TAP

Taper tap

A cutting tool for making threads in a drilled hole in metal, wood, or other hard material. A slightly tapered, bolt-like threaded tool that is forcibly screwed into a bored hole, cutting away some of the metal, forming internal threads. For cutting external threads on a bolt or pipe, dies are used.

TAPER MILL

A type of junk mill; an elongated, tapered, grinding and pulverizing bit (tapered from several inches to 1 to 2 inches in diameter) whose surface has been hard-faced with superhard, durable cutting material.

TAPPED OR FLANGED CONNECTION

Indicates the two types of pump or process unit connections available from suppliers. Tapped is an internally threaded (female) connection into which an externally threaded piece may be screwed; a flanged connection is one furnished with a screw or weld flange.

TAPPING AND PLUGGING MACHINE

A device used for cutting a hole in a pipeline under pressure. A nipple with a full-opening valve attached is welded to the line. The tapping machine is screwed onto the valve and, working through the open valve, bores a hole in the line. The tapping drill is withdrawn, the valve is closed, and the tapping machine is unscrewed from the valve. A connection can then be made to the pipeline at the valve.

T.A.P.S.

Trans-Alaska Pipeline System; a large-diameter pipeline built from the oil-rich North Slope of Alaska to the warm-water port of Valdez on the state's south shore.

TAP SADDLE

A type of pipeline clamp with a threaded hole in one of the two halves of the bolt-on clamp for use when a pipeline is to be tapped, to have a hole made in it for drawing off gas or liquid. Tap saddles are used on fieldlines, 2 to 10 inches; for tapping larger lines, nipples are welded to the pipe and a tapping machine is used.

TAR BALLS

Balls of a tar-like substance found on seashores which indicate oil seeps in the near vicinity or offshore. The tar balls began as accumulations of hydrocarbons which have weathered, having given up their more volatile fractions thus becoming residual or tar-like. Strong winds along the shore roll the patches of residue into lumps, then into balls.

TARIFF

A schedule of rates or charges permitted a common carrier or utility; pipeline tariffs are the charges made by common carrier pipelines for moving crude oil or products.

TARIM BASIN

A geologic basin found in northwest China. This large feature has yet to be explored to any significant extent, but it is considered to have a very high potential.

TARMAT

Very heavy, topped crude verging on tar.

TAR SANDS

See Petroleum Tar Sands.

TATTLETALE

A geolograph; a device to record the drilling rate or rate of penetration during an 8-hour tour.

Tapping and plugging machine

Tar sand plant

T.B.A.

Among marketing department people, T.B.A. stands for tires, batteries, and accessories.

T.C.F.

Trillion cubic feet (of gas).

T.D.

Total depth. Said of a well drilled to the depth intended.

T.D. THE WELL

To cease drilling and making hole because of some difficulty and to call where drilling stopped Total Depth or T.D. This is seldom done except in the pay zone or in a nonproductive interval with the option to plug back to a lesser pay previously drilled through—making the best of a bad situation.

TEAMING CONTRACTOR

A person who furnished teams of horses and mules and oilfield wagons for construction and earthwork in the oil fields. Some large teaming contractors in the early days kept stables with 600 teams (1,200 horses and mules). In the days of dirt roads in the booming oil fields, the horse and wagon was the most dependable mode of transportation.

TEAMSTER

See Mule Skinner.

Teamster and team

TEAPOT DOME

Part of the Naval Petroleum Reserves set aside by Congress in 1923. Teapot Dome in Wyoming was the center of controversy and scandal in the 1920s during the presidency of Warren G. Harding.

TECHNICAL SCALE OF MINERAL HARDNESS

In addition to the Mohs scale, there is a so-called technical scale of hardness that includes 15 minerals, listed softest to hardest: talc, gypsum, calcite, feldspar, apatite, orthoclase, pure silica glass, quartz, topaz, garnet, fused zircon, corundum, silicon, carbide, boron carbide, and diamond.

TECHNOLOGY AT THE BIT

An omnibus phrase alluding to the geosteering system which provides a number of essential bits of information to the surface by wireless telemetry. There, decision-making people—geologists, engineers, and the driller—put the information to use. *See* Geosteering.

TECTONIC BLOCK

A mass of rock that has obviously been moved in relation to the surrounding or adjacent rock by the action of tectonic forces, the movement of the Earth's crust.

TECTONIC BRECCIA
Breccia formed by tectonic forces, some movement of the Earth's crust that has shattered brittle rocks.

TECTONIC MAP
See Map, Tectonic.

TECTONICS
A word derived from the Greek *tektonikos,* "of a builder." A branch of geology that deals with the broad architecture of the earth's crust, the structural and deformational features of a region, particularly with folding and faulting, and other massive displacements of rocks. Tectonics is closely related to structural geology but usually concerns itself with the broader aspects of the Earth's tumbled and deformed crust. Tectonics also may refer to the forces involved in producing such phenomena.

T.E.L.
Tetraethyl lead.

TELEGRAPH
A device for the remote control of a steam drilling engine on a cable-tool rig. The "telegraph" consisted of a wire or a small cable running between the pulleys—one at the driller's stand, the other mounted on the steam valve of the engine. By turning his wheel, the driller regulated the speed of the engine by opening or closing the steam valve.

TELEGRAPHER'S BUG
An automatic Morse-code-sending machine operated by pressure from the telegrapher's thumb and forefinger. The advantage of the bug is that it makes dots in rapid succession by a slight pressure on the thumb lever; dashes are made one at a time with the forefinger. A popular, patented bug is the Vibroplex, which has a beetle on the nameplate, hence the name.

TELEGRAPH KEY
A Morse-code-sending instrument made with a spring-loaded lever on a fulcrum. When the lever is depressed, the brass lever or key makes contact with a fixed terminal, closing the electric circuit which energizes two small coils into magnets. The magnets draw down a small bar on the telegraph sounder, making a dot or a dash sound depending upon the length of time (split seconds) contact is made by the telegrapher. Dots are short, dashes are longer.

TEMPERATURE BOMB
A device used downhole to measure bottom-hole and circulating temperatures on a drilling well. One technique involves attaching a temperature-sensitive probe in a protective sleeve attached to a carrier mounted on the drillpipe.

TEMPERATURE CONVERSION

°F to °C: °C = 5/9 (°F – 32); °C to °F: °F = 9/5 (°C + 32).

TEMPERATURE DEDUCTION

A conversion of the metered volume of oil or gas produced on the basis of a standard or agreed temperature base, customarily 60°F. Crude oil in the reservoir at high temperature and pressure may contain quantities of gas in solution which increases the volume. After the oil is produced and temperature and pressure reduced, the dissolved gas comes out of solution. This causes the crude oil to shrink in volume. In much the same way, the volume of gas from a gas well changes as it is brought to surface conditions.

TEMPERATURE GRADIENT

See Thermal Gradient.

TEMPER SCREW

A device on the cable of a string of cable tools that permits the driller to adjust the tension on the drilling line. A temper screw is made in the general form of a turnbuckle.

TEMPLATE PLATFORM

An offshore platform whose supporting legs fit into a frame previously constructed and anchored to the seafloor. The platform, constructed on-shore, is taken out to location by a crane barge where it is set into the frame.

TENDER

(1) A permit issued by a regulatory body or agency for the transportation of oil or gas. (2) A barge or small ship serving as a supply and storage facility for an offshore drilling rig; a supply ship. (3) A quantity of crude oil or refined product delivered to a pipeline for transportation. Regulations set the minimum amount of oil that will be accepted for transportation.

TENSILE LOADING

A term that describes a buildup of longitudinal stress in a pipe or cable. In a pipeline or string of other tubulars—drillpipe, tubing or casing—tensile loading can adversely affect not only the threaded joints, causing leakage or failure, but also the pipe itself.

TENSILE STRENGTH

The resistance to being pulled apart laterally; tensile, from tension; resistance to lengthwise stress. Hydrostatic testing of pipeline pipe is testing for tensile strength or resistance to bursting.

TENSIOMETER

A gauge attached to a cable or wire rope to detect the tension being ap-

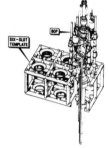

Floater-type subsea template

Jackup rig (left) and drilling tender (right)

plied. From two positions on a section of the cable a sensitive gauge measures the stretch and twist of the cable under load, indicating the tension on a scale; a strain gauge.

TENSIONER SYSTEMS
Tensioner systems are installed on deepwater floating-drilling platforms to maintain a constant tension on the marine riser. Two types of systems are used: the deadweight system and the pneumatic system. Tensioning systems serve the dual purpose of compensating for the vertical motion of the drilling vessel or platform and maintaining a constant tension or lifting force on the riser.

Tensioner system

TENSION-LEG PLATFORM
A semisubmersible drilling platform held in position by multiple cables anchored to the ocean floor. The constant tension of the cables makes the platform immune to heave, pitch, and roll caused by wave action and conditions that affect conventional semisubmersibles.

TERM CLAUSE
The clause in a lease, mineral deed, or royalty deed that fixes the duration of the interest granted, assigned, or reserved.

TERM ROYALTY OR MINERAL INTEREST
A mineral or royalty interest of limited duration. Most such interests are created to endure for a fixed period and so long thereafter as oil, gas, or other minerals are produced. Some interests remain in effect for a fixed and definite period, 20 years, say, without the thereafter clause. In the interest of clarity, the latter can be referred to as Fixed Term Interests and those with a thereafter clause Defeasible (Voidable) Term Interests.

TERM-LIMIT PRICING
An agreement on price between a supplier and a wholesaler or jobber that runs for a specified length of time.

TERMINATION OF LEASE
Expiration, cancellation, or forfeiture of the lessee's interest in leased property. A lease may be terminated by the lessee's exercise of a surrender clause in the lease; and under most forms of leases, failure of production in commercial quantities at or after the expiration of the primary term will terminate the lease—unless it can be rescued and preserved by some lease-saving clause in the lease such as the Shut-In Gas Well clause, Drilling Operations clause, Continuous Drilling Operations clause, etc. *See* Drilling Operations Clause.

TERRASTIC FORMS
A geological term for landforms or Earth forms found in offshore areas of the continental shelf. The undersea landforms may be terraces,

mounds, oval, crater-like depressions or a succession of parallel ridges. In laying underseas pipelines, these various forms are often encountered and always present problems for the pipeline engineers. *See* Pock Marks.

TERRIGEOUS

A geologic adjective denoting that the formation or the fill was derived from the land, a wash-in of top or shallow, unconsolidated clastic (sometimes) material to form a fill or a deltaic fan (the beginnings of a delta).

TERTIARY INCENTIVE OIL (T.I.O.)

Oil produced under a tertiary oil recovery program that has tax and pricing incentives to compensate for the higher costs involved in producing such oil.

TERTIARY RECOVERY

The third major phase of crude oil recovery. The primary phase is flowing and finally pumping down the reservoir until it is "depleted" or no longer economical to operate. Secondary recovery usually involves re-pressuring or simple waterflooding. The third or tertiary phase employs more sophisticated techniques of altering one or more of the properties of crude oil, e.g., reducing surface tension. This is accomplished by flooding the formations with water mixed with certain chemicals that "free" the oil adhering to the porous rock so it may be taken into solution and pumped to the surface. *See* Micellar-Surfactant Flooding.

TEST COUPONS

Small samples of materials—metals, alloys, coatings, plastics, and ceramics—that are subjected to heat, cold, pressure, humidity, and other conditions of stress to test durability and performance under simulated operating conditions.

TESTING, HYDROSTATIC

See Hydrostatic Testing.

TEST LOOP

A configuration of piping set up, usually out of doors, to test meters, valves, corrosion inhibitors, strainers, flanges, and other components of process units, pipeline manifolds, and line pipe under simulated operating conditions. A test loop may be a large spread extending several hundred feet in both directions from the observation site, or it may be not much larger than a laboratory test bench. Both kinds are used in the petroleum industry's persistent and aggressive testing of designs, methods, and materials. Much of the industry's testing is under the auspices of, or by the American Petroleum Institute, the A.P.I.

TEST WELL

An exploratory well; a wildcat well.

TETRAETHYL LEAD (T.E.L.)
A lead compound added, in small amounts, to gasoline to improve its antiknock quality. Tetraethyl lead is manufactured from ethyl chloride, which is derived from ethylene, a petrochemical gas.

TEXAS DECK
The top deck of a large semisubmersible drilling platform. The upper deck of any offshore drilling rig that has two or more platform levels.

TEXAS TOWER
A radar or microwave platform supported on caissons anchored to the ocean floor. The tower resembles an offshore drilling platform in the Texas Gulf, hence the name.

TEXTURE, ROCK
The general appearance or physical characteristics of a rock, its crystals, their arrangement, size, and shape; the graininess or other elements of its composition. The term structure is sometimes used erroneously for texture. Structure refers to the larger features, e.g., blocky fracture, columnar structure, platy or banded aspects. Structure is best observed in outcrops rather than in hand specimens.

THE LAST, LONGEST, AND SMALLEST
Colloquial term for the production string of pipe to go in the well. *See* Capital String; *also* Pay String.

THEODOLITE
See Transit.

THEREAFTER CLAUSE
The clause in an oil or gas lease which provides for the continuation of a lease beyond the primary term.

THERMAL CRACKING
A refining process in which heat and pressure are used to break down, rearrange, or combine hydrocarbon molecules. Thermal cracking is used to increase the yield of gasoline obtainable from crude oil.

THERMAL GRADIENT
The measured change in temperature, increase or decrease. In drilling, the deeper the borehole the higher the bottom-hole temperature.

THERMAL LOG
See Log, Temperature.

THERMAL OXIDIZERS
A large, cylindrical furnace, with refractory lining and banks of burners at various levels, for burning refinery gases before they are vented to the flare tower.

Thermowell thermocouple *(Courtesy Honeywell)*

THERMAL SYPHON

A method of cooling a slow-moving, stationary gas engine not equipped with a water pump. The water in the engine's water jacket is heated to boiling (212°F) as the engine runs. The steam that forms in the water jacket expels the hot water in slugs into a large cooling tank connected to the engine's cooling system. As the hot water is expelled, cool water from the tank gravitates into the cooling system to repeat the cycle. The hydrostatic head in the cooling tank keeps the engine's water jacket full. The water loss by the formation of steam is made up by adding water to the tank.

THERMOCOUPLE

A pyrometer; a temperature-measuring device used extensively in refining. The thermocouple is based upon the principle that a small electric current will flow through two dissimilar wires properly welded together at the ends when one junction is at a higher temperature than the other. The welded ends are known as the "hot junction," which is placed where the temperature is to be measured. The two free ends are carried through leads to the electromotive force detector, known as the "cold junction." When the hot junction is heated, the millivolts can be measured on a temperature scale.

THERMOMETRIC HYDROMETER

A hydrometer that has a thermometer as an integral part of the instrument to show the temperature of the liquid. This is of first importance as the density or A.P.I. gravity varies with the temperature. Hydrometers used by pipeline gaugers are thermometric hydrometers.

THICKS AND THINS

Thicks and thins sometimes characterize a sedimentary formation laid down on an eroded, unconformable surface. The eroded surface may have hills and valleys, high and low places. When the sedimentary material was deposited, it filled in the low places, resulting in thicks, and was shallower over the high places, causing thins.

THIEF

A metal or glass cylinder with a spring-actuated closing device that is lowered into a tank to obtain a sample of oil or to the bottom of the tank to take a column of heavy sediment. The thief is lowered into the tank on a line that, when jerked, will trip the spring valve, enabling the operator to obtain a sample at any desired level.

Thief

THIEF FORMATION

A rock sequence into which fluid is lost during drilling operations; a thief zone.

THIEF HATCH
An opening in the top of a tank large enough to admit a thief and other oil-sampling equipment.

THIEFING A TANK
Taking samples of oil from different levels in a tank of crude oil and from the bottom to determine the presence of sediment and water with the use of a thief.

THIEF ZONE
A very porous formation downhole into which drilling mud is lost. Thief zones, which also include crevices and caverns, must be sealed off with a liner or plugged with special cements or fibrous clogging agents before drilling can resume.

Oil inlet line and thief hatch

THIN OUT
See Pinch Out.

THIN WELL
A well drilled into a producing interval that is relatively, and disheart-eningly, narrow or thin.

THINNERS
See Mud Thinners.

THINS
See Thicks and Thins.

THIRD-FOR-A-QUARTER DEAL
A joint venture for drilling a well; a standard arrangement among partners in the oil patch for financing a well. It follows: The operator transfers three-quarters of the leasehold interest in a prospect (a well to be drilled) to three investors in return for payment of 100 percent of the cost of drilling and, if the well hits, completing the well. To continue: As the deal includes three people, plus the operator, each person (other than the operator) has put up one-third of the money for drilling the well and will receive one-quarter interest in the well. The operator's one-quarter interest is his reward for searching for, identifying, and leasing the prospect, as well as supervising the drilling and completion of the well. The classic Third-for-a-Quarter Deal.

THIRD-GENERATION HARDWARE
Equipment developed from earlier, less-sophisticated models or prototypes; the latest in the evolution of specialized equipment.

THIRTY DAY-SIXTY DAY CLAUSE
A provision, common in an Unless Lease form, intended to keep the lease alive if a dry hole is drilled during the primary term, if production

Thread protectors on casing

ceases during the primary term or thereafter, or if there is no production at the end of the term but drilling operations are then underway. The clause provides that under these circumstances (and certain others), the lease will remain in force if the lessee pursues with all diligence drilling or reworking operations as specified. While there are a number of variations in the wording of the clause, the foregoing captures the intent.

THIXOTROPIC
The property of certain specially formulated cement slurries—used in cementing jobs downhole—that causes them to set, become rigid, when pumping ceases. But when force is applied again (pumping is resumed) the cement again becomes a pumpable slurry. This procedure may be repeated until the predetermined setting time of the cement is reached.

THREAD PROTECTOR
A threaded cap or lightweight collar screwed onto the ends of tubular goods (pipe, casing, and tubing) to protect the threads from damage as the pipe is being handled.

3-D SEISMIC PROGRAM
Seismic surveys shot from surface stations to map underground stratigraphy; to profile the underlying strata in search of updips, down dips, faults, and other promising anomalies.

Thribbles racked outside a deep rig

THRIBBLES
Drillpipe and tubing pulled from the well three joints at a time. The three joints make a stand of pipe that is racked in the derrick. Two-joint stands are doubles; four-joint stands are fourbles.

THROTTLING
Adjusting the flow of oil or gas on a flowing well. Wells are never permitted to flow at an unrestricted rate. To do so is not considered good production practice. In an oil well, unrestricted flow except for short test periods. To do so would deplete prematurely the reservoir drive, whether gas cap, solution gas, or water drive. The result: leaving much oil behind to be pumped, if the producer is lucky.

THROUGH-CONDUIT VALVE
See Valve, Through-Conduit.

THROW
The vertical movement of a fault's displacement. *See* Heave.

Through-conduit valve with rising stem

THROWING THE CHAIN
Wrapping the spinning chain around the drillpipe in preparation for running the pipe up or backing it out. Crew members become proficient at throwing the chain in such a way as to put several wraps on the pipe with one deft motion.

THRUST BEARING
See Bearing, Thrust.

THRUSTERS
Jets or propellers on large tankers, drillships, and deepwater drilling platforms that provide a means to move the vessel sideways—at right angles to the ship's normal line of travel—when docking or in maintaining position in water too deep for conventional anchors. *See* Dynamic Stationing.

Thrust fault

THRUST FAULT
A fault with a dip of 45° or less over most of its extent, on which the hanging wall appears to have moved upward relative to the foot wall. Compression horizontally rather than vertically is a characteristic feature of the thrust fault. *See* Hanging Wall.

THRUST SHEET
A body of rock above a large thrust fault.

THUMPER
See Vibrator Vehicle.

Thumper

TIDELANDS
Land submerged during high tide. The term also refers to that portion of the continental shelf between the shore and the boundaries claimed by the states. The federal government now has the right to produce oil and gas from this area of the continental shelf.

TIE-BACK STRING (Of casing)
Well casing that is run from the top of the liner back to the surface, the wellhead. A tie-back string often serves as the production string or production casing.

TIE-IN
An operation in pipeline construction in which two sections of line are connected: a loop tied into the main line, a lateral line to a trunk line.

Tie-in

TIGHT FORMATION
Tight sand. *See* Tight Gas.

TIGHT GAS
Natural gas produced from a tight formation, one that will not give up its gas readily or in large volumes. The production of tight gas is more costly and therefore less attractive to producers owing to the need for fracturing, acidizing, and other expensive treatments to free the gas from the relatively impermeable formations. In view of these constraints, such gas has been given an incentive price of 150 percent of the

price of gas from new, conventional onshore gas wells by the Natural Gas Policy Act of 1978.

TIGHT-GAS DESIGNATION

The certification by a government regulatory agency that gas being produced is from a tight, low-permeability formation and is therefore eligible for incentive pricing under the Natural Gas Policy Act of 1978. *See* Tight Gas; *also* Deep Gas.

TIGHT GAS SAND

The Federal Energy Regulatory Commission (F.E.R.C.) has set high wellhead price ceilings on gas from gas sands at depths below 15,000 feet and with permeability of 0.01 millidarcy or less. The tight gas sands are designated by F.E.R.C. and state regulatory agencies. Spacing is one well on 320 acres for wells less than 12,000 feet deep, and one well per 640 acres for deeper wells.

TIGHT HOLE

A drilling well about which all information—depth, formations encountered, drilling rate, logs—is kept secret by the operator.

TIGHT SAND

A sandstone or limestone formation whose porosity and permeability are such that they will not readily give up their oil and gas. The formation may have good saturation, but because of low permeability the hydrocarbons are unable to flow freely from one pore space to another. As a result, producing from a tight sand is often not economically feasible. Modern hydrofracturing methods can often make such tight formations producible by fracturing the formation or acidizing.

TILT-BED FIELD

Tilted beds of coal that are drilled parallel to the bedding to produce methane gas. *See* Coal-Seam Gas.

TILTING-DISC CHECK VALVE

See Valve, Tilting-Disc Check.

TIME CUSHION

Factored in spare time for large field projects with a planned completion

TIME DRILLING

See Drilling, Time.

TIME FRAME

A period of time, within boundaries, with respect to some action or project.

TIN HAT
The metal, derby-like safety hat worn by all workers in the oil fields, refineries, and plants to protect their heads.

TIN SNIPS
Metal shears for cutting shims, corrugated roofing, gasket material, or screen.

T.L.P.
Term-limit pricing.

TO FARM INTO A LICENSE
To buy into, to take an equity in, a field or blocks of leases from the original lessee or consignee from a foreign government, to drill, produce, and develop.

TO FLOW BY HEADS
A condition of a flowing well in which the hole or the production tubing loads up with oil; and when sufficient gas pressure builds under the column of oil, it is blown up the hole in intermittent, uneven gushes. When this condition exists, it is evidence that neither the production of oil or gas is sufficient for an even, consistent flow. Also called Heading.

TOE BOARD
The enclosure at toe height around a platform or on a catwalk to prevent tools or other objects on the platforms from being kicked off accidentally.

TOLUENE
An aromatic hydrocarbon resembling benzene but less volatile and flammable. It is used as a solvent and as an antiknock agent in gasoline.

TOMOGRAPHY
A technique for making detailed x-rays of a predetermined plane section of a solid object while blurring out the images of other planes.

TONG GANG
In the days of screw pipe the pipeliners who manned the heavy hooks (tongs) to screw together the joints of 8- and 12-inch line pipe were called the tong gang (circa 1920).

TONG JERK LINE
See Jerk Line.

TONG LINE
The line (hemp or manila rope) attached to the tubing or drillpipe tongs and looped a couple of turns around the cathead. When the driller tight-

Tin hat

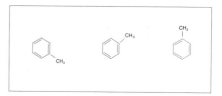

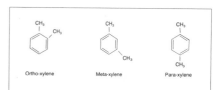

Ortho-xylene   Meta-xylene   Para-xylene

Toluene Molecule

Tubing tongs

Drillpipe tongs *(Courtesy Peter Bawden Drilling Ltd.)*

Power casing tongs *(Courtesy Hughes Tool Co.)*

Tool house

Tool joints

·ens the loops on the spinning cathead, the tong line pulls the tongs to break out the pipe joint or make it up.

TONG KEY

A rectangular bar of case-hardened steel inserted in the key slot of a pair of pipelay tongs, used in the 1920s and '30s. One edge of the tong key bit into the outer circumference of the pipe as the tongs were moved in a thread-tightening motion. When one edge of the key became worn and smooth, it could easily be removed and turned 90° to expose another sharp edge. When all four edges were worn smooth, the key was replaced with a new one.

TONGS

See Pipe Tongs.

TON OF CRUDE OIL

A ton of crude oil is 6.5 to 8.5 barrels, depending on the oil's specific gravity. For rough approximation, 7.5 barrels equals a metric ton or long ton, 1,000 kilograms or 2,204.6 pounds.

TOOLDRESSER

In cable-tool drilling, a worker who puts a new cutting edge on a drill bit that is worn or blunted. Like a blacksmith, the tooldresser heats the bit in a charcoal fire and, using a hammer, draws out the metal into a sharp, chisel-like cutting edge.

TOOLHOUSE

A shed or temporary sheet-metal building where tools and engine/pump parts are stored for future use on the lease or by a pipeline repair gang. *See* Doghouse.

TOOLIE

A tooldresser on a cable-tool rig.

TOOL JOINT

Heavy-duty threaded joint specially designed to couple and uncouple drillpipe into stands of such length that they can be racked in the derrick. Intermediate couplings between the tool joints are made with regular drillpipe collars.

TOOL JOINT LEAK DETECTOR

A hydraulic testing device that is clamped around a tool joint after it is made up tight in the drillstring and before it is lowered in the hole. The leak detector puts a 1,000-psi pressure or more on the outside circumference of the joint and holds the pressure for a few seconds. The smallest leak in the connection is indicated on a gauge by a drop in pressure.

TOOLPUSHER

A supervisor of drilling operations in the field. A toolpusher may have one drilling well or several under his direct supervision. Drillers are directed in their work by the toolpusher.

TOP

In petroleum geology, a term for the uppermost surface of a formation when it is drilled into. It is usually identified by the first appearance of a distinctive feature, such as a marked change in lithology, a break, or the occurrence of an index or guide fossil. The top is often shown by a distinctive configuration on an electric log and is used in correlation studies and structure-contour mapping.

T.O.P. (TAKE OR PAY)

Take or Pay refers to clauses in natural gas contracts which specify that a gas pipeline or distributor (the buyers) must take the quantity of gas the contract calls for, or if unable to take said quantity, the pipeline or the distributor must pay for it.

TOP-DRIVE DRILLING

Drilling with stands instead of with singles is one of the obvious advantages of top-drive (with power swivel) drilling which eliminates the rotary table and kelly joint. *See* Power Swivel.

TOP HOLE

The topmost section of hole in drilling a well, the hole for the conductor casing that is of larger diameter than the remaining well bore. Top-hole drilling includes the rathole and the mousehole. Special bits are used for top holes and are weighted by adding a few drill collars for the relatively shallow, large-diameter holes.

TOP LEASE

See Lease, Top.

TOPOGRAPHIC MAP

See Map, Topographic.

TOPOGRAPHY, BASKET OF EGGS

A landscape of many side-by-side drumlins (from Irish Gaelic drum hill or ridge—low, rounded hillocks of glacial drift) with marshy areas in between. (Quite possibly named by an Irish geologist after a pint or two at Quinlan's).

TOP OUT

To finish filling a tank; to put in an additional amount that will fill the tank to the top.

TOPPED CRUDE OIL

Oil from which the light ends have been removed by a simple refining process.

TOPPING PLANT
An oil refinery designed to remove and finish only the lighter constituents of crude oil, such as gasoline and kerosene. In such a plant, the oil remaining after these products are taken off is usually sold as fuel oil.

TOPS
The fractions or products distilled or flashed off at the top of a tower or distillation unit in a refinery.

TORPEDO
An explosive device used in shooting a well. The well-shooting torpedo was invented and used by Colonel E.A.L. Roberts, a Civil War veteran, in 1865. The first torpedoes used black powder as an explosive; later, nitroglycerin was substituted for the powder.

TORQUE
A turning or twisting force; a force that produces a rotation or torsion, or tends to.

TORQUE AND DRAG
The two conditions that, in excess, seriously impede downhole drilling progress. On occasion torque and drag cause serious, time-consuming and costly trouble: torque, the twisting or turning force can result in twist-offs if too great a turning force is applied to a drillstring bound up in a dogleg or keyseat; excess drag results in stuck pipe, and the inability to run casing to depth.

TORQUE CONVERTER
An item of hydraulic equipment that is installed between the prime movers (drilling engines, for example) and the driven components (mud pumps and rotary) to transmit a smooth, continuous flow of power. The torque converter absorbs or cushions the pulsations, the transient, uneven surges of torque in the power train.

TORQUE REDUCER
A chemical additive for water-base drilling fluids that reduce the friction or drag of the drillpipe against the sides of the borehole. The chemical (there are several patented products that reduce drag) is also used in drilling mud preparation, for setting pipe, running casing, and lubricating all tight places downhole such as dog legs and keyseats. (At this writing, there are no torque reducers for use in oil-base muds.) *See* Flow Improver.

TORQUE TUBE
On split-level rigs the torque tube transmits torque or power from the drilling engines, at ground level, to the rig floor 15 to 20 feet above. The torque tube is part of the power train.

TORQUE TUBE

The driveshaft on a split-level rig that transmits power from the diesel engines at ground level to the rotary table and catworks on the rig floor. (Strictly speaking, the torque tube proper is the covering [for safety] for the steel drive shaft which provides the torque.) *See* Rig, Split Level.

TORQUE UP, TO

To become progressively more difficult to turn, as in screwing a joint of pipe into a collar or valve. Drillpipe begins to torque up when drilling through a section or interval of shale that absorbs water and swells into the borehole, thus closing on the drillpipe. Often the necessary torque to turn the drill increases disastrously, twisting off the pipe.

TORQUE WRENCH

A tool for applying a turning or twisting motion to nuts, bolts, pipe, or anything to be turned and that is equipped with a gauge to indicate the force or torque being applied. Torque wrenches are useful in tightening a series of bolts or nuts with equal tension, as on a flange or engine head.

Torque wrench

TORR

A unit of pressure equal to 1 millimeter (mm) of mercury (HG) at 0° Centigrade or 1/760 of an atmosphere. The torr is named for the 17th century Italian physicist, Evangelista Torricelli, who invented the barometer. This very small unit of pressure is used in refineries in certain calibrations of vacuum tower pressure variations.

TORREY CANYON

An oil tanker that ran aground March 18, 1967, off the coast of England, causing the largest, most costly, and most publicized oil spill up to that time. The mishap touched off reactions that put oil-spill pollution in the international spotlight.

TORSION BALANCE

A delicate instrument used by early-day geophysical crews to measure the minute variations in magnetic attraction of subsurface rock formations. As the differences in attraction of the subsurface features were plotted over a wide area, the geophysicist had some idea as to where sedimentary formations that might contain oil were located in relation to nonsedimentary rocks. The torsion balance has been superceded by the less-complicated (to use) gravity meter or gravimeter. *See* Gravimeter; *also* Magnetometer.

TOTAL DEPTH (T.D.)

The depth of a well when drilling is completed. Total depth of a well is the vertical distance from the rig floor to the bottom of the hole. A 10,000-foot well may take 11,300 feet of casing to complete the well because the well bore has drifted several degrees from vertical, adding 1,300 feet to the depth of the hole, not the depth of the well.

TOTAL VERTICAL DEPTH
See T.V.D.

TOUR
A work period; a shift of work, usually eight hours, performed by drilling crews, pump station operators, and other oilfield personnel. In the field "tour" is pronounced tower, to rhyme with sour; a trick.

TOWER, MUD-COOLING
See Mud-Cooling Tower.

TOWER HAND
A member of the drilling crew who works up in the derrick; derrickman.

TOWER INTERNALS
Refers to the material, the irregular-shaped metal pieces, piled randomly on fractionator trays, tower trays. There are at least two kinds in general use: structured and random packing. The packing provides configuration and surfaces to aid in the aeration of the hot-liquid charge as the hot gases move upward through the tower trays.

TOWN GAS
See Coal Gas.

TRACER LINES
Small-diameter tubing paralleling and in contact with process or instrumentation piping in a refinery or other plant to provide heat or cooling for the fluid or gases in transit. More often tracer lines carry steam. In the field, larger-diameter tracer lines are used to heat low-gravity, viscous crude oils so they may be pumped. *See* Heat Tape.

TRACT BOOK
A record book maintained by the district land offices of the Bureau of Land Management (B.L.M.), listing all entries affecting described land.

TRACTOR FUEL
A low-octane fuel, less volatile than motor gasoline, used in low-compression farm tractors.

TRADER
One who deals in bulk petroleum or products both domestic and foreign; one who operates in the international oil market, arranging for supplies and trading surpluses of one product for others; an oil broker.

Pressure transmitter

TRANSDUCER
A device or instrument actuated by power from one kind of system that

472

in turn supplies power to another system. A classic example of a transducer is the telephone receiver that is actuated by electric power and supplies acoustic power to the atmosphere. A form of transducer is an air or hydraulic system that will actuate an electric system by pressure on a contact switch. Another and true form of transducer is the thermocouple, wherein heat on two dissimilar pieces of metal will create a small, measurable electric current.

TRANSIENT PRODUCTION
The production of a new well before the flow rate stabilizes; also referred to a flush production. Most wells stabilize production within a few weeks or sooner.

TRANSIT
A telescope mounted on a calibrated base, a tripod, for measuring horizontal as well as vertical angles; a theodolite. A transit is commonly used by surveyors for running levels.

TRANSITE PIPE
A patented composition pipe for handling corrosive liquids and salt water.

TRANSITIONAL DRILLING
Drilling with a bit that is designed and built to drill rapidly in soft formations and also can handle hard, abrasive formations as well. Drilling with a multipurpose bit.

TRANS-PANAMA PIPELINE
An 80-mile, 40-inch and 36-inch line built across the Isthmus of Panama as a joint venture by the Republic of Panama, Northville Industries Corp., and the Chicago Bridge and Iron Group. The line, completed in 1982, transports Prudhoe Bay or North Slope crude from the terminal at Puerto Armuelles on the Pacific to Chirique Grande on the Atlantic. The new line shifts the movement of 800,000 barrels of oil daily from the Panama Canal and numerous transshipment tankers.

TRANSITION FITTINGS
When using plastic pipe in the field or at a plant, it is usually necessary to make connection with steel tank fittings or a pipeline. If so, special transition fittings, made with one end acceptable to the plastic pipe and the other end a standard thread end or weld end, are installed.

TRANSPONDER
A radio or other electronic device that, upon receiving a designated signal, emits a signal of its own.

TRANSSHIPMENT TERMINAL
A large deepwater terminal where crude oil and products are delivered

Fault trap

Traveling block *(Courtesy B.F. Goodrich)*

by supertankers (L.C.C.V.) and transshipment of product is by smaller tankers. Such terminals have large storage capacities and high-volume unloading facilities to accommodate the mammoth vessels that carry more than two million barrels of oil each trip.

TRAP

A type of geological structure that retards the free migration of oil and concentrates the oil in a limited space. A mass of porous, permeable rock that is sealed on top and down both flanks by nonporous, impermeable rock, thus forming a trap. *See* Anticline.

TRAVELING BARREL

The barrel of a downhole tubing pump in which the barrel moves up and down. The barrel, connected to the pump rods, moves up and down over the plunger instead of the plunger moving in the pump barrel, as in other downhole rod pumps or tubing pumps. *See* Pump, Traveling-Barrel.

TRAVELING-BARREL PUMP

See Pump, Traveling-Barrel.

TRAVELING BLOCK

The large, heavy-duty block hanging in the derrick and to which the hook is attached. The traveling block supports the drill column and "travels" up and down as it hoists the pipe out of the hole and lowers it in. The traveling block may contain from three to six sheaves depending upon the loads to be handled and the mechanical advantage necessary. The wireline from the hoisting drum on the drawworks runs to the derrick's crown block and down to the traveling block's sheaves.

TRAVEL-TIME CURVE

In seismology, a plot of wave-train traveltime against corresponding distance along the Earth's surface from the source to the point of observation; a time-distance curve.

TRAYED COLUMNS

Any of several kinds of vertical, cylindrical refining or processing columns fitted with internal horizontal trays or baffles over which charge stock flows from top to bottom in a vaporization or absorption process. *See* Bubble-Cap Trays.

TREATER, ELECTROSTATIC

An oil treater that uses A.C. and D.C. electrical force fields to cause the water droplets in the oil-water emulsion to come together, coalesce, and then drop out by gravity. The patented dehydrator uses some heat in its process, particularly on low-gravity crude oils.

An electrostatic oil treater

TREATING PLANT
A facility for heating oil containing water, emulsions, and other impurities and with the addition of chemicals, causing the water and oil to separate. The water and other foreign matter settle to the bottom of the tank and are then drawn off.

TREE SAVER
A patented, mandrel-like piping made to slip into and through the valves and connecting spools of a Christmas tree when a well is to be stimulated, acidized, or hydrofraced under high pressure. The mandrel or inner sleeve takes the pressure, protecting the tree both from the high pressure and any corrosive or abrasive fluids during the stimulation operation.

Trencher

TRENCHER
A ditching machine; a large, self-propelled machine with digging buckets fixed to an "endless" chain belt or circular frame that, when rotated, scoops out a ditch to predetermined width and depth.

TREND
A trend may encompass one or more oil pools, a field or a basin. But a trend is an area of exploration, drilling, and production which has exhibited or is exhibiting direction as the field develops. There may be dry holes on its right flank and on its left flank, but the trend of successful wells, like Tennyson's *Light Brigade*, moves forward. *See* Oil Play.

Ferris-wheel-type
trencher

TRIANGULATION
A trigonometric operation for finding the directions and distances to and the coordinates of a point by means of bearings from two fixed points a known distance apart; a method of surveying in which the stations are points on the ground at the vertices of a chain or network of triangles, whose angles are measured instrumentally, and whose sides are derived by computation from selected sides or baselines the lengths of which are obtained by direct measurement on the ground or by computation from other triangles. Triangulation is generally used where the area surveyed is large and requires the use of geodetic methods. Also, triangulation is the network or system of triangles into which any part of the Earth's surface is divided in a trigonometric survey.

TRICK
See Tour.

TRICKLE-CHARGED BATTERY
A storage battery, usually for standby emergency service, kept charged by a small amount of current from a primary electrical source. Should the main source of power fail the battery, fully charged, is ready for use.

Making a trip

Triplex pump *(Courtesy Gaso Pump)*

TRIM, EQUIPMENT
Special noncorroding finish; finishes or construction details that mark the product as ideally fitted for certain specific service, e.g., certain valves have special-service internals: silencing and anticavitating, for example.

TRIM, SOUR-SERVICE
See Sour-Service Trim.

TRIP
Pulling the drillpipe from the hole to change the bit and running the drillpipe and new bit back in the hole. On deep wells, round trips, as trips are sometimes called, may take 24 hours, three 8-hour shifts.

TRIP GAS
High-pressure gas encountered in drilling deep wells that can cause serious control problems when the tools are pulled out of the hole while making a trip. The driller must exercise extreme care to prevent loss of control or a blowout. Sufficient mud must be in the hole to provide the hydrostatic head necessary to contain the downhole gas pressure. Sometimes, in order to come out of the hole under high-pressure conditions, the crew must resort to stripping the pipe, i.e., removing the drill-string through the well's stack of control valves, the blowout preventer, on the wellhead.

TRIPLE POINT OF WATER
The condition of temperature and pressure in which the three phases of a substance—liquid, gaseous, solid—can exist in equilibrium. *See* Kelvin Degrees.

TRIPLEX INJECTION SKID
A skid-mounted triplex pump for injecting slurrified well cuttings and other debris back into an acceptable formation or, in some instances, into the annulus between casing and the wall of the borehole.

TRIPLEX PUMP
See Pump, Triplex.

TRIPPED OUT
Refers to drillpipe, casing, or tubing that has been removed from the well bore (the hole) by withdrawing the pipe a stand (two or three joints) at a time. *See* Tripping the Bit

TRIPPING DOUBLES
An expression meaning pulling the drillpipe out of the hole (or going in) in two-joint stands. Tripping doubles requires one-third more pipe connections to make up and break out by the floor men than if they were

tripping "thribbles," three-joint stands. Handling thribbles calls for a large, tall derrick, as three-joint stands are 90 feet high and can present problems in windy areas.

TRIPPING THE BIT
Removing the bit from the hole and running it in again. (In removing the bit, the drillpipe must be pulled a stand at a time in order to reach the bit.) *See* Trip.

TRUNK LINE
Main line.

TRUNNION VALVE
See Valve, Trunnion.

TUBE BUNDLE
The name given to the tubes in the core of a heat exchanger. The tubes or pipes, all the same length, are spaced equidistant in parallel rows and are supported by perforated end plates, thus forming a bundle.

Tube bundle

TUBE STILL
See Still, Pipe.

TUBE TURN
A weld or flanged fitting in the shape of a U used in the construction of manifolds, exchanger bundles, and other close pipe work.

TUBING
(1) Pipe, 2 to 4 inches in diameter, that is put into a well to be flowed or pumped. The tubing goes inside the well's casing and is the longest string of pipe, reaching from the surface valves to the pay zones. In a pumping well the bottom-hole pump is screwed onto the bottom joint of tubing (the first joint run). (2) Small-diameter (1/2 to 1 inch) flexible piping; copper tubing; plastic tubing; glass tubing. *See* Pump, Tubing.

TUBING ANCHOR
A downhole, packer-like device run in a string of tubing that clamps against the wall of the casing. The tubing anchor prevents the "breathing" of the tubing, the cyclic up-and-down movement of the lower section of tubing, as the well is pumped by a rod pump.

TUBING AND CASING ROLLERS
A downhole tool for reconditioning buckled, dented, or collapsed well tubing or casing. The tool is lowered into the hole, entering the small, deformed diameter of the damaged pipe. As the cylindrical tool is forced lower and rotated, it pushes out dents and restores the pipe to its original diameter.

Tubing board

TUBING BOARD

A small platform high in the derrick where a derrickman (a member of the drilling crew who is not affected with acrophobia) stands to rack drillpipe or tubing as it is being pulled and set back.

TUBING-CONVEYED PERFORATING

A method of lowering in perforating charges on a string of tubing as opposed to the wireline method. There are advantages to the tubing-conveyed technique: larger charges for deeper penetration or for larger entry holes; the ability to perforate an interval of almost any length on a single trip; and shot density to increase the flow area and production rate in unconsolidated formations. The charge holder or gun on certain patented equipment can be disengaged from the tubing and dropped so that further work such as testing, remedial slick-line work, or well stimulation can be performed without pulling the tubing.

TUBING HEAD

The top of the string of tubing with control and flow valves attached. Similar in design and function to the casinghead, the tubing head supports the string of tubing in the well, seals off pressure between casing and the inside of the tubing, and provides connections at the surface to control the production of gas or oil.

TUBING PUMP

See Pump, Tubing.

TUBING SPINNER

An air or hydraulically operated device for spinning out or spinning up a joint of well tubing or drillpipe. Spinners are much faster at screwing pipe and are safer than using a rope or spinning chain. The tool latches onto the pipe and then torque is applied. The joint is turned by an inner mechanism actuated by air or hydraulic pressure.

Tubing spool *(Courtesy Cameron Iron Works)*

TUBING SPOOL

A heavy, forged-steel fitting that is flanged to the casinghead and into which the tubing hangers fit; an element of the aboveground well completion hookup.

TUBULAR GOODS

See Oil Country Tubular Goods.

TUFA

A carbonate rock formed of limy deposits from spring waters saturated with calcium carbonate. Calcareous tufa, as it is sometimes referred to, is distinguished from volcanic tufa. It is easily cut, takes a high polish and, like travertine, is used as decorative stone for bank lobbies and railway stations.

TUFF

Volcanic ash that has been compressed or compacted into a layered rock. It has been widely used as a building material in regions of volcanic activity. Greek and Roman cities were built largely of tuff.

TUNDRA

A vast area in the Arctic lying between the permanent ice cap and the more southerly forested region. Even in the warmest months of summer, the subsoil remains frozen, the top few inches supporting only limited vegetation.

TURBIDITIC FORMATION

A formation where the matrix is characterized by turbulent, swirling action of the sediment at the time of the major depositions.

Turbine meter *(Courtesy Rockwell)*

TURBIDITY LIMESTONE

A limestone indicating resedimentation by turbidity currents in fast-moving, sediment-laden streams; also turbidity sands.

TURBINE METER, LIQUID

A mechanism inserted into a liquid flow line that measures volumetric flow rate and total flow. The meter is constructed with vanes on a spindle inside a housing that can be flanged into a flow line. The movement of liquid through the meter exerts a force on the curved vanes, causing the spindle to turn, as on a water wheel. The spindle is connected to a counter and readout mechanism, showing rate of flow and total daily or monthly through put.

Motor-driven turbine pump

TURBINE MUD MOTOR

See Mud Motor, Turbine.

TURBINE PUMP

See Pump, Turbine.

TURBOCHARGER

A centrifugal blower driven by an engine's exhaust-gas turbine to supercharge the engine. To supercharge is to supply air to the intake of an engine at a pressure higher than the surrounding atmosphere.

TURBODRILLING

A type of rotary drilling in which a fluid-drive turbine (motor) is placed in the drillstring just above the drill bit. The mud pressure from the pumps at the surface pumping mud down through the drillpipe turns the turbine that rotates the drill bit. The drillpipe does not rotate as in conventional drilling; hence, there is no kelly joint being turned by the rotary table.

Turbodrill

TURBOEXPANDER

A mechanical device, a turbine that converts the energy of a high-pressure gas stream into force or motion. When a gas pipeline brings gas under pressure to a refinery or processing plant, the line pressure is much higher than can be used. Some plants, instead of installing pressure-reducing piping, direct the stream into and through a turbo-expander. This produces power to generate electricity, for example, at the same time reducing the incoming gas to useable pressure levels.

TURBULENT FLOW
The movement of liquid through a pipeline in eddies and swirls that tend to keep the column of liquid together, rather than running like a river with the center of the stream moving faster than the edges. *See* Plastic Flow.

TURNAROUND
The planned, periodic inspection and overhaul of the units of a refinery or processing plant; the preventive maintenance and safety check requiring the shutting down of a refinery and the cleaning, inspection, and repair of piping and process vessels.

Hook-and-eye turnbuckle
(Courtesy Crosby)

TURNBUCKLE
A link with a screw thread at one end and a swivel at the other; a right-and-left screw link used for tightening a rod, a guy wire, or stay.

TURNKEY
To perform a complete job as under a turnkey contract, to take over and perform all necessary work of planning, procurement, construction, completion, and testing of a project before turning it over to the owner for operation.

TURNKEY CONTRACT
A contract in which a drilling contractor agrees to furnish all materials and labor and do all that is required to drill and complete a well in a workmanlike manner. When on production, he delivers it to the owner ready to "turn the key" and start the oil running into the lease tank, all for an amount specified in the contract.

TURNKEY WELL
A well drilled under a turnkey contract.

Turntable *(Courtesy U.S. Steel)*

TURNTABLE, ROTARY DRILLING
See Rotary Table.

TURRET-MOORED ICE-DRILLING BARGE
A drilling barge of new concept developed by Dome Petroleum Ltd. for use in Arctic waters where floating or moving ice is a danger to conventional drillships or barges. The new barge has a 16-anchor mooring system attached to a swivel directly beneath the drilling derrick. At the approach of advancing ice on the barge's beam, the vessel weathervanes

until its bow is headed into the ice flow. This maneuver reduces the tension on the mooring lines to a small fraction of that on a vessel moored in a fixed position.

TURTLEBACK

A two-part clamp for joining lengths of shackle rod. The connector is in the general configuration of an English walnut; the two halves are held together by a bolt and nut.

T.V.D.

Total vertical depth. The T.V.D. of a well is almost always less than the M.D. or measured depth. This is true because very few boreholes are exactly vertical. On deviated wells, the average well, it is safe to say has a measured depth from a few hundred to a 1,000 feet or more greater than the true vertical depth.

TWIST A TAIL

To bring pressure to bear in order to speed up a job or to get action from someone who is suspected of dragging his feet.

TWIST-OFFS

Drillpipe affected by excess torque, the turning force, until it fails laterally; it shears—twists off.

TWO-CYCLE ENGINE

An internal-combustion engine that produces one power stroke for each revolution of the crankshaft. Intake, compression, ignition, and power stroke are accomplished in one revolution.

TWO OIL

Colloquial term for No. 2 heating oil; home heating oil; furnace oil.

TWO-PHASE FLOW

The transportation of a liquid and a gas by a single pipeline. In some instances, notably in an offshore environment, crude oil and natural gas are moved to shore stations through the same line. In this type of transportation, there can be different flow patterns depending upon a number of parameters, such as flow rates of the two phases, liquid and gaseous, line characteristics, and physical properties of the liquid phase. *See* Plastic Flow; *also* Turbulent Flow.

TWO-PHASE PIPELINE

Two-phase pipeline is one capable of carrying a liquid and a gas stream simultaneously.

TWO-STAGE COMPRESSOR

Two-stage identifies a type of compressor that intakes gas and compresses or raises the pressure in the first chamber of the compressor and passes the gas into the second-stage chamber where it is further compressed, raising the pressure to the required level.

TWO-STEP PIPE THREADS
A pipe thread configuration used on well tubing and casing in which there are two independent sets of threads in a terrace arrangement with a shoulder between them. The threads are on both pin and box. When the joint is stabbed, the first or starting thread of both sets of threads is engaged simultaneously.

U

U-BOLT

A bolt in the shape of a U, both ends of which are threaded. A follower or saddle piece fits over the threaded ends and is held in place by nuts. U-bolts or U-clamps are used to hold two ends of wirelines together or to make a loop in a length of wire cable by turning back the running part (the loose end) on the standing part of the cable and clamping them together.

UINTAITE (GILSONITE)

Gilsonite, a solid hydrocarbon with the general appearance of coal. The mineral was named for Samuel H. Gilson, the first person to be seriously interested in mining and marketing the mineral. The name uintaite was taken from Utah's Uinta basin where the mineral was discovered in commercial quantities. Uin-ta-tite.

ULLAGE

The reserve space in a storage tank between the top of the oil and the top of the tank. This space or ullage allows for expansion of the oil when it warms up from the sun or artificial heating.

ULTRASONIC ATOMIZER

A development in burners for heating oils in which high-frequency sound waves are focused on the stream of fuel, forming a spray of microscopic fuel droplets. The resulting intimate mixture of fuel and air makes for greater combustion efficiency.

UMBILICALS, CONTROL

Lines of communication, life lines: electric, air, or hydraulic lines from a control or supply center to dependent satellites.

UNASSOCIATED GAS

Natural gas occurring alone, not in solution or as free gas with oil or condensate. *See* Associated Gas.

Unassociated gas well

UNBRANDED GASOLINE

Gasoline sold by major refiners to jobbers and other large distributors without bearing the name of the refiner.

UNCONFORMITY

The surface that separates two rock units. If the rock layers on either side (top and bottom) are parallel, it is a parallel unconformity; if they lie at an angle to each other, it is an angular unconformity. For example, a layer of sandstone is lying on a layer of limestone. Where the two dissimilar formations touch or meet, this surface is an unconformity; the upper layer does not conform to the lower layer or vice versa.

UNCONFORMITY, ANGULAR

An unconformity between two groups of rocks whose bedding planes are not parallel or where the older rocks dip at a steeper angle than the younger, overlying rocks. Also, an unconformity in which the younger sediments are deposited on the eroded surface of a tilted or folded section of older rock.

UNCONSOLIDATED

Loosely packed, often porous, near-surface formations of sand, gravelly sand, certain shales, and mudstone.

UNCONVENTIONAL GAS

Coal-bed gas, methane produced from gaseous beds or layers of coal from brown coal (lignite) to bituminous. A relatively new source of methane. Coal bed wells are drilled and usually completed open-hole as opposed to cased and perforated. Shallow beds are more productive because there is less overburden, a freer and looser formation. Horizontal boreholes have proven effective in tapping the bedded coal. Concomitant with drilling in known coal beds is drilling tight-gas formations.

UNCONVENTIONAL NATURAL GAS

The term applied to natural gas so difficult and expensive to find and produce that the sources have been bypassed in favor of more easily obtainable supplies. Such sources can be found in tight sandstone reservoirs in the western and southwestern states, in certain shales in the Appalachian Basin, and in geopressurized reservoirs along the Gulf Coast. Geologists have known of these sources for many years but, because of the low prices even for more cheaply producible conventional gas, the unconventional gas supplies have remained virtually untouched. Also, to get at these marginal sources, advances in technology have to be made and at great cost.

UNDERBALANCED WELL

A well in which the hydrostatic head, the pressure exerted by the hole full of drilling mud, is greater than the reservoir pressure. This is a damage-sensitive condition wherein drilling fluid, the slurry, leaks off into the formation and can partially plug the pores of the formation, blocking the in-flow of oil to the borehole.

UNDERGROUND STORAGE
In certain areas of the country where there are underground caverns, petroleum and products are stored for future use. All caverns are not suitable; some are not naturally sealed and would permit the stored oil to leak into subsurface water sources. *See* Salt-Dome Storage.

UNDER PRODUCED.
In gas-well terminology, the condition of having produced and sold less than your proportionate share of the gas produced by a well in a split-stream arrangement. The producer is generally entitled to gas balancing or, if reserves are not sufficient, to cash balancing. *See* Gas-Balancing; *also* Split Stream Production.

UNDERREAM
To enlarge the size of the borehole of the well by the use of an under-reamer, a tool with expanding arms or lugs that, when lowered into the hole, can be released at any depth to ream the hole with steel or insert cutters.

UNDERREAMER
A type of drilling tool used to enlarge the diameter of the borehole in certain downhole intervals. The underreamer is made with expandable arms fitted with cutters. When in position, the expandable arms are released and the cutters chew away the rock to enlarge the hole. When the reamer is pulled from the hole, the arms fold in toward the body of the tool.

Cavern hewn for oil storage (*Courtesy Shell*)

UNDIVIDED-INTEREST PIPELINE
A large pipeline in which each of several owners has an undivided interest as a tenant-in-common in the line. One of the owners is the operator or agent. Each of the owners accepts shipments of oil or gas, controls tariffs and tenders, collects his revenue, and pays his share of the cost of operating the line. This type of pipeline may be thought of as one large line with a number of smaller but different-sized lines inside, each line with a different owner. The owners operate separately with the economies inherent in a large pipeline.

UNIBOLT COUPLING
A patented coupling or flange for joining two lengths of pipe. The two mating halves of the coupling have tapered shoulders. When torque is applied to the two halves by a single bolt, drawing the bolt lugs together, the coupling is tightened. Unibolt couplings are for medium-diameter piping and take up less space than conventional multibolt flanges.

UNIDIRECTIONAL FRACTURE TREND
A rock structure, laid down in such manner and sequence that when submitted to great pressures as in fracing, its lines of fracture are unidirectional, they move in one direction only. Also, a fracture that propagates or progresses in one direction only. The fracture once begun does not deviate from its general heading.

UNION
A pipe connector or coupling made of two mating pieces, one for each end of the two pipes to be joined. The two halves of the union are held together by a threaded ring. Unions join small-diameter pipe, 2 inches to 4 inches. *See* Flange.

UNION, GROUND SEAT
See Ground Seat Union.

UNION, WING
A coupling for small-diameter pipe. The two mating halves of which are held together by a winged, threaded ring that can be made up gas tight by striking on of the wings or lugs with a hammer; a hammer connection.

UNIT OPERATOR
Head well puller; the man in charge of the pulling unit crew that does routine subsurface work on producing wells, e.g., cleaning out, changing pumps, pulling rods, and tubing.

UNITIZATION
A term denoting the joint operation of separately owned producing leases in a pool or reservoir. Unitization makes it economically feasible to undertake cycling, pressure maintenance, or secondary-recovery programs.

UNIVERSAL JOINT
A shaft coupling able to transmit rotation to another shaft not directly in line with the first shaft; a movable coupling for transmitting power from one shaft to another when one shaft is at an angle to the other's long axis.

UNLESS CLAUSE
The clause in an unless lease that provides for the termination of the lease interest unless the lessee commences drilling or pays rental during the primary term of the lease. *See* Delay Rental.

UNLESS LEASE
See Lease, Unless.

UNLOADING THE HOLE
A colloquial term referring to a gas kick blowing the mud and tools out of the hole—when the well is out of control.

UNMANNED STATION
A pipeline pumping station that is started, stopped, and monitored by remote control. Through telecommunication systems, most intermediate booster stations on large trunk lines are unmanned and remotely controlled from the dispatcher's office.

Unmanned station

UPDIP WELL
A well located high on a structure where the oil-bearing formation is found at a shallower depth.

UPHILL WELDER
A bellhole welder; an accomplished welder capable of welding on work in any position.

UPSET TUBING
Tubular goods made thicker in the area of the threads in order to compensate for the metal cut away in making threads. In the manufacture of casing and drillpipe, the additional metal is usually put on the inside. In well tubing, especially the smaller sizes, the thickening is on the outside. This is known as exterior-upset tubing.

UPSTREAM
Facilities or operations performed before those at the point of reference. Oil production is upstream from pipeline transportation, and transportation is upstream from refining. *See* Downstream.

U.S.G.S.
U.S. Geological Survey; an agency of the federal government that, among its many services and duties, regulates the placement of wells in federal offshore lands.

U.S. SYNTHETIC FUELS CORP.
A corporation established by the Energy Security Act of 1980 with the power to make loans to any company for development of synthetic fuel projects, to make loan guarantees on obligations to provide funds for such projects, and to make purchase agreements for synfuel production (guarantee prices received by a company for the production from a synfuel project).

VACUUM DISTILLATION

Distillation under reduced pressure (less than atmospheric) that lowers the boiling temperature of the liquid being distilled. This technique, with its relatively low temperatures, prevents cracking or decomposition of the charge stock.

VACUUM FLASHER

A refinery vessel; a large-diameter column where charge stock is distilled at less than atmospheric pressure. The pressure in some flasher vessels is less than one-third atmospheric—4 or 5 pounds per square inch. At this reduced pressure, lighter fractions of the heavy charge stock will flash off or vaporize. The lower the pressure, the lower the boiling point for all liquids.

VACUUM STILL

See Still, Vacuum.

VACUUM TAR

See Asphalt.

VACUUM TOWER, PACKED

A fractionating refinery tower that has random-dumped packing in place of trays and bubble caps. The packing in a vacuum tower is a quantity of metallic cylinders, 2 to 3 inches in diameter with tabs cut in the sides and bent inward to increase the surface area and to slow the flow of the hot liquid and gases moving upward in the tower. *See* Vacuum Flasher.

VALVE, AUTOMATIC MUD

See Valve, Lower Kelly.

VALVE, BALL

A type of quick-opening valve with a spherical core, a ball with a full-bore port, that fits and turns in a mating cavity in the valve body. Like plug valves, ball valves open or close by a quarter turn, 90°, of the valve handle attached to the spherical core.

VALVE, BALL AND CAGE

A ball-and-seat valve in which the ball, as it lifts off the seat and then

Vacuum still *(Courtesy ARCO)*

Ball valve

488

drops and reseats, is held in a vertical plane by a surrounding metal cage. *See* Valve, Ball and Seat.

VALVE, BALL AND SEAT

The discharge and suction valves in a reciprocating pump. The ball, a metal sphere, is nested in a circular seat, and when pressure is exerted on the ball during the discharge strokes of the pump, the ball fits pressure-tight in the ground seat. During the suction stroke, the ball is lifted out of the seat by the inrushing liquid (oil) until the end of the stroke. Then on the discharge stroke the ball seats securely in the seat.

Packless or bellows valves

VALVE, BELLOWS

A small packless valve used on chemical, caustic liquid, and steam lines. Instead of packing in the conventional sense, the valve stem is attached to a thin-metal bellows (like a small concertina), that with the stem in any position packs off, prevents leakage. *See* Packless Valve.

VALVE, BELLOWS-SEALED

A packless valve; a valve without conventional packing around the valve stem. The stem, as it moves up and down (when opened and closed), is kept pressure-tight by an attached metal bellows that expands and contracts as the valve stem moves. This type valve is for severe-service piping, i.e., steam, chemicals, or other corrosive liquids.

Block valve

VALVE, BLEEDER

A small valve on a pipeline, pump, or tank from which samples are drawn or to vent air or oil; sample valve.

VALVE, BLOCK

A large heavy-duty valve on a crude oil or products trunk line placed on each side of a pipeline river crossing to isolate possible leaks at the crossing.

VALVE, BLOCK-AND-BLEED

A heavy-duty mainline valve made to hold bubble-tight against high pressure. The valve is made with a small bleeder line and valve that are tapped into the block valve's bonnet. When the block valve is closed, its effectiveness may be checked by opening the bleeder valve for evidence of any leakage from the upstream or high-pressure side.

VALVE, BUTTERFLY

A type of quick-opening valve whose orifice is opened and closed by a disk that pivots on a shaft in the throat of the valve.

VALVE, BYPASS

A valve by which the flow of liquid or gas in a system may be shunted past a part of the system through which it normally flows; a valve that controls an alternate route for liquid or gas.

Butterfly valve with manual actuator (*Courtesy Fisher*)

VALVE, CHECK

A valve with a free-swinging tongue or clapper that permits fluid in a pipeline to flow in one direction only; back-pressure valve.

VALVE, CRYOGENIC

An advanced design of a ball valve. The V.O.R.C. (variable orifice rotary control) cryogenic block valve is made to the most exacting specifications as it handles liquids ranging in temperatures from –173°C to –265°C (100°K to 8°K), and must shut off gas tight. In the cryogenic industry—liquefaction, separation, and purification of a list of gases—there are three separate categories of valves: (1) warm valves, those which are a part of the operating process but are located upstream of the cold box; (2) cryogenic isolation (block) valves which are fully open or completely closed; and (3) cryogenic process valves which either are operated pneumatically or electrically.

VALVE, DISCHARGE

One of two sets of valves in a reciprocating pump. The other set is the suction or intake valves.

VALVE, DUMP

Any of various quick-acting valves for dumping or emptying rapidly the contents of a tank or other vessel. Also, a quick-acting exhaust valve actuated by a solenoid that will instantly evacuate the air chamber of a pneumatic valve to effect its rapid closure.

Electrically operated control valve *(Courtesy Magnatrol)*

VALVE, ELECTRICALLY OPERATED CONTROL

A small-diameter valve used in process piping that is opened or closed by a small electric motor.

VALVE, FABRICATED

A type of valve or other fitting that is built and welded together from wrought iron and forged steel pieces to make a particularly strong high-pressure valve. Most valves are steel castings with bodies, bonnets, and packing glands cast separately and assembled.

VALVE, FLOAT

A valve whose stem is actuated by an arm attached to a float; an automatic valve operated, through linkage to a float mechanism, by the change in liquid level in a tank or other vessel.

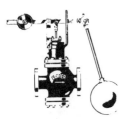

Float-actuated valve *(Courtesy Fisher)*

VALVE, FLOW

An adjustable valve on the production tubing that controls the flow of the well; an adjustable choke. Some wells are completed with a bottom-hole choke, a type of flow valve on the first joint of tubing at the bottom of the well.

VALVE, FOOT
A type of check valve used on the foot or lower end of a suction-pipe riser to maintain the column of liquid in the riser when the liquid is being drawn upward by a pump.

VALVE, GATE
A pipeline valve made with a wedge-shaped disk or "tongue" that is moved from open to closed position (up to down) by the action of the threaded valve stem.

VALVE, GLOBE
A type of pipeline valve that shuts off as the stem, rotated by the hand wheel, moves a mating part downward onto a ground seat that is integral to the valve body.

Gate valve

VALVE, INTAKE
The valve in a reciprocating pump or a four-stroke-cycle internal-combustion engine through which a charge of fluid or a mixture of fuel and air is drawn into pump or engine. The intake valves are one-half of the set of pump or engine valves; the other half of the set are the discharge valves.

VALVE, LOWER KELLY
An automatic valve attached to the lower end of the kelly joint that opens and closes by mud pump pressure. The purpose of the valve is to prevent the mud in the kelly joint from pouring out on the derrick floor each time the kelly is disconnected from the drillpipe. When the mud pump is stopped, the kelly valve automatically closes. After a joint of drillpipe is added to the string and the kelly is made tight, the pumps are started, the mud pressure opens the kelly valve, and drilling resumes. The automatic valve saves valuable mud, keeps the rig floor dry, and speeds up the job of making a connection.

VALVE, MASTER GATE
A large valve on the wellhead used to shut in a well if it should become necessary.

VALVE, MULTIPLE-ORIFICE
A patented orifice valve with two orifice plates or disks in pressure-tight contact. One disk can be rotated through 90°. For full flow through the valve, the orifices in the two disks are in perfect alignment. To reduce the flow, the movable disk is rotated a certain number of degrees, which partially covers the orifice in the fixed disk, thus restricting the flow through the valve.

VALVE, NEEDLE
A valve used on small, high-pressure piping where accurate control of

Needle valves. Cutaway shows needle and seat (*Courtesy Markad Service Co.*)

small amounts of liquid or gas is desired. The "tongue" of the valve is a rod that tapers to a point and fits into a seat that permits fine adjustments as when used with pressure gauges.

Pilot-operated relief valve

VALVE, NONRETURN
See Check Valve.

VALVE, PACKLESS
A special kind of valve that uses a welded bellows rather than soft packing around the valve stem. The stem of the packless valve does not rotate; it is raised and lowered into the valve body by a connecting stem outside the fluid cavity. Packless or packingless valves are usually for small-diameter piping (one-quarter to 2-inch) and are used on piping carrying hazardous or toxic fluids or gases and for high-pressure steam.

VALVE, PILOT
A small relief valve that, through a linkage of pressure piping, controls the opening of a larger relief or safety valve. A pilot valve is usually employed to modulate or dampen the action of a larger valve as it opens, to relieve the system pressure.

VALVE, PLUG
A type of quick-opening pipeline valve constructed with a central core or "plug." The valve can be opened or closed with one-quarter turn of the plug; a stop.

Pop-off valve

VALVE, POPPET
A type of check valve installed in a riser or a downhole packer to prevent fluid from rising vertically in the pipe or the well bore. A spring-loaded vertical valve that permits downward flow as fluid pressure opens the valve. Pressure from below moving upward is blocked by the valve's clapper.

VALVE, POP-OFF
See Valve, Relief.

VALVE, QUARTER-TURN
A plug valve, ball valve, or butterfly valve. A valve made with a plug or sphere with a full-bore opening on the horizontal axis that can be opened or closed with a quarter or 90° turn of the handle. A bufferfly valve with its disk that rotates on a shaft or trunion in the valve body also is opened and closed with a quarter turn of the handle.

Quarter-turn ball valve
(Courtesy Marpac)

VALVE, QUICK-ACTING
A quarter-turn valve; butterfly, ball, and plug valves are called quick-acting valves because a 90° turn of the handle closes or opens them.

VALVE, RELIEF

A valve that is set to open when pressure on a liquid or gas line reaches a predetermined level; a pop-off valve.

VALVE, RISING-STEM

A large-diameter pipeline valve whose stem is fixed to the valve's tongue or gate. When the valve is fully opened, the stem of the valve rises to a height equal to the depth of the valve, the diameter of the valve's throat, so on a 24-inch, rising-stem valve, the stem will rise through the stuffing box. Most rising-stem valves are electric-motor driven; and in a large manifold, each valve wears a number or letter identifying marker so open or closed valves may be easily identified from a distance from the pipeline or tank farm office.

VALVE, ROTARY

See Valve, Plug.

VALVE, SAFETY

See Valve, Relief.

VALVE, SAMPLE

A small-diameter valve in the wall of a tank from which samples of oil or a refined product are taken; a bleeder valve.

VALVE, SLIDE

Very large, box-like valves for flues and stacks. Made from sheet steel, the valves are mechanically or hydraulically operated.

VALVE, STOP-AND-WASTE

A type of plug valve that, when in a closed position, drains the piping above or beyond it. When the valve is turned a quarter turn to shut it off, a small port or hole in the valve body is uncovered, permitting water above the valve to drain out, preventing a freeze-up in cold weather. Stop-and-waste valves are used mainly on small-diameter water piping.

VALVE, STORM

A hurricane season control valve that is run with a packer on tubing or drillpipe several joints below the rig floor or the surface to control the well pressure when the drilling well is to be shut down and made secure at the approach of a severe storm. It is customary to circulate to kill the well, pull the drillpipe up into the casing, run the storm valve and packer. The pipe above the packer and valve assembly is backed off (unscrewed) at the safety joint and pulled. The master valve or blind rams on the wellhead can be closed, completely securing the well.

VALVE, SUCTION

One of two sets of valves in a reciprocating pump, e.g., a mud pump, triplex pump, etc. The other set of pump valves is the discharge valves.

Through-conduit valve
with rising stem

VALVE, THROUGH-CONDUIT

A class of gate valve whose valve body is made so that the gate or tongue of the valve and its seating element extend down through the fluid passageway or conduit of the valve. In an ordinary gate valve, the gate seats at the bottom of the conduit and does not extend through. Through-conduit valves, when in an open and closed position, seal off the body of the valve from the fluid pumped through the pipeline. This is an important feature when the fluid being handled is corrosive.

VALVE, TILTING-DISC CHECK

A type of check valve, usually for large-diameter pipelines, with the disc mounted on trunnions instead of a hinge as in more conventional check valves. One advantage of the tilting disc is its quiet operation, the absence of "slam" as in other types of check valves.

VALVE, TRUNNION

A butterfly valve whose orifice is opened or closed by a disk in the throat of the valve that rotates on trunnions or pins seated in the body of the valve. Trunnion valves are opened and closed by a quarter turn of the valve handle, a 90° turn.

VALVE, WEDGE-PLUG

A patented type of quarter-turn valve, a plug valve, for use at severe temperatures, –140° to 800°C. When the valve is opened or shut, a 90° turn of the handwheel lifts the plug as it is turned and reseats the plug or core of the valve in its new position. This slight lifting and turning overcomes one of the negative features of a plug valve in severe service: it sticks and is difficult to turn without unseating.

Valve and seats (Courtesy Woolley)

VALVE AND SEAT

On a reciprocating pump, the seat is firmly held in the body of the pump, the suction and discharge cavity; the valve, held in place and guided by its stem, moves up on the suction stroke of the pump. On the discharge stroke, it closes or seats itself.

VALVE PACKING

See Packing.

VALVE POTS

The wells in the body of a reciprocating (plunger) pump where the suction and discharge valves are located. Valve pots are on the fluid end of the pump and are covered and sealed by heavy, threaded plugs or metal caps bolted over the top of the pots.

VALVE RATINGS

One of the commonly used ratings for valves is the flow coefficient (Cv), which is defined as the amount of water in gallons per minute that can flow through the valve at 1 psi pressure drop.

VANE PUMP
See Pump, Vane.

VAN SYCKEL, SAMUEL
The man who invented and successfully operated the first crude oil pipeline. The line was 2 inches and ran from Pithole City, Pa., to a railhead 5 miles away. It pumped 81 barrels the first day, thus sounding the knell for the teamster and his wagonload of oil barrels.

VAPOR LOCK
A condition that exists when a volatile fuel vaporizes in an engine's fuel line or carburetor, preventing the normal flow of liquid fuel to the engine. To handle gas lock or vapor lock, the gas must be bled off the system by removing a line or loosening a connection, or the lines and carburetor must be cooled sufficiently to condense the gas back to a liquid.

VAPOR PRESSURE
The pressure exerted by a vapor held in equilibrium with its liquid state. Stated inversely, it is the pressure required to prevent a liquid from changing to a vapor. The vapor pressure of volatile liquids is commonly expressed in pounds per square inch absolute at a temperature of 100°F. For example, butane has a vapor pressure of 52 psia at 100°F.

VAPOR RECOVERY UNIT
A facility for collecting and condensing vapors of volatile products being loaded into open tanks at refineries, terminals, and service stations. The vapors are drawn into a collecting tank and, by pressure and cooling, are condensed to a liquid. V.R. units significantly reduce air pollution by petroleum vapors.

VAPOR TENSION
See Vapor Pressure.

VARIETAL MINERAL
A mineral either found in considerable amounts in a rock or a distinctive characteristic of a rock; a mineral that distinguishes one rock from another.

VARNISH
A dark shiny coating composed of iron oxide and some manganese oxide formed on the surface of boulders and rock ledges after very long exposure to the weather. The thin, dark coating is thought to have formed as the rocks exuded mineralized solutions that, as they evaporated, left the "desert varnish."

V-belt and engine power
take-off

V-BELT
A type of "endless" V-shaped belt used in transmitting power from an engine's grooved drive-pulley to the grooved sheave of a pump, com-

pressor, or other equipment. The V-belt, a bigger and tougher version of the automobile fan belt, is used in sets of from two to twenty belts, depending upon the size of the drive pulley.

V-DOOR
The opening in the derrick opposite the drawworks used for bringing in drillpipe and casing from the nearby pipe racks.

Drillpipe going in V-door

VENDER DOCUMENTS
The results of testing or private research done in a specialized field (theirs) by service or supply companies which are published as advertisements or as useful information brochures for the industry which they serve. Progress in metallurgy by manufacturers of oilfield tubular goods; the latest in drill-bit technology; an improved cement for squeeze jobs or new surfactant mixes for waterflood projects are examples of useful vender documents. Good service companies know many of the industry's problems and, to their credit, work toward being part of the solution, for a profit.

VENT LINE
(1) A horizontal 4- to 6-inch pipe that vents gases from oil storage tanks; a vent line lets the tank breathe as the oil is warmed during the day and cooled at night. (2) Blooie pipe. (3) A line at a pumping well that vents gas from the well, in the event it is blown, to clean the well.

VENTURI EFFECT
The decrease in pressure which results from the increased velocity of a fluid as it is pumped through a constricted section of pipe. The reduced pressure occurs at the location of the smaller diameter of the pipe. The venturi principle is used in measuring fluid flow and in creating a vacuum for certain instrumentation and for drawing gasoline into an engine's carburetor. The venturi tube bears the name of the Italian physicist, G.B. Venturi, who discovered the phenomenon in the 18th century.

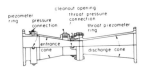

Venturi tube *(Courtesy Honeywell)*

VENTURI METER
An instrument for measuring the volume of flowing gases and liquids. It consists of two parts—the tube through which the fluid flows and a set of indicators that show the pressures, rate of flow, or quantity discharged. The tube, in the shape of an elongated hourglass, is flanged into a pipeline carrying the fluid. The effect of the tube is to increase the velocity and decrease the pressure at the point where the tube's diameter is reduced. The relationship between the line pressure and the pressure at the narrow "waist" of the tube is used in computing the rate of flow.

VERTICAL INTEGRATION
Refers to the condition in which a company produces raw material,

transports it, refines or processes it, and markets the product, all as one integrated operation. Specifically, an oil company is said to be vertically integrated when it finds and produces oil and gas, transports it in its own pipelines, refines it, and markets its products under its brand name. According to the critics of the industry, this is not in the country's best interest. *See* Horizontal Integration.

VERTICAL MIGRATION (OF OIL & GAS)
The upward movement of hydrocarbons from very deep Mesozoic rocks, for example, to shallower, porous, sedimentary formations where they are trapped by impervious layers—and await the wildcatter's drill.

VERTICAL-MOORED PLATFORM (V.M.P.)
A buoyant drilling-producing platform moored to the seafloor by flexible risers cemented into the seabed. Wells are drilled through the risers by conventional methods and completed at the platform deck. When all wells are drilled and completed, the V.M.P. becomes a producing platform. The buoyancy of the platform exerts sufficient tension on its mooring systems to stabilize it in all kinds of weather.

VERTICAL SEISMIC PROFILE
A seismic technique whereby the impulse source or shot is lowered into an existing borehole. The geophones can be located on the surface or in the borehole of an adjacent well. Surface-related distortions are then minimized or eliminated.

VESSEL, EXPLORATION
See Exploration Vessel.

VIBRATOR VEHICLE
A specially designed tractor-like vehicle used to produce shock waves for geophysical and seismic surveys. The vehicle incorporates a hydraulically operated hammer or thumper that strikes the ground, setting off shock waves that are reflected from subsurface rock formations and recorded by seismic instruments at the surface.

VIBRIOSIS VEHICLE
See Vibrator Vehicle.

VIBROPLEX
A patented automatic telegraph-sending machine; a telegrapher's bug.

VIBROSEIS
Producing seismic shockwaves by the use of "thumpers" or vibrator vehicles.

VICTAULIC COUPLING
A patented pipe coupling made in two halves that fit the grooved ends of two lengths of pipe and are forced together by bolts. Before the halves

Seismic vibrator vehicle

Vibrator in Action

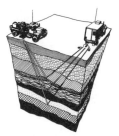

Vibroseis

Victaulic coupling

of the coupling are put in position, a rubber ring is placed over the junction of the two lengths of pipe. When the coupling is tightened with the two bolts, the rubber is compressed, making a pressure-tight connection.

VINTAGE (of Gas Sales Contracts)

A term borrowed, we suspect, from the vintners, that denotes the time a gas sales contract was made. Vintaging is a method or system of determining the price to be charged for natural gas subject to price control regulations under which the price was based on the vintage of the gas as new or old gas. The determination of whether the gas is new or old is based solely on the date of sale of the subject gas.

VINTAGING

See Vintage (Gas Sales Contracts).

VISCOSIFERS

Chemicals or other additives that increase the viscosity of drilling muds, frac fluids, or enhanced recovery (E.O.R.) liquids for surfactant flooding of fraced formations.

VISCOSIMETER

A device or apparatus for measuring the viscosity of liquids by determining the rate of flow through a small, calibrated hole at a standard temperature. *See* Viscosity.

VISCOSITY INDEX

An arbitrary scale used to show the changes in the viscosity of lubricating oils with changes in temperatures.

VISCOSITY

One of the physical properties of a liquid, i.e., its ability to flow. It happens that the more viscous an oil, for example, the less readily it will flow. So the term has an inverse meaning—the lower the viscosity, the faster the oil will flow. Motor oil with a viscosity of S.A.E. 10 flows more readily than oil with a viscosity of S.A.E. 20. *See* Seconds Saybolt.

V.L.C.C.

Very large crude carrier; a crude oil tanker of 160,000 deadweight tons or larger, capable of transporting 1 million barrels or more.

V.L.P.C.

Very large product carriers (oceangoing tankers).

VOLATILE-LADEN CRUDE

See Crude Oil, Volatile-Laden.

VOLATILITY

The extent to which gasoline or oil vaporizes; the ease with which a liquid is converted into a gaseous state.

VOLUME 90 (O.G.J.)
In publication jargon, volume 90 on the *Oil & Gas Journal* signifies the 90th year of publication. Begun as the *Oil Investor's Journal,* its ownership and name changed eight years later, and frequency of publication went from bi-monthly to weekly. It has remained so since 1910. From cable tools to geosteering, 3-D, and horizontal bores. Well done!

VOLUME TANK
A small cylindrical vessel connected to a gas line in the oil field to provide an even flow of gas to an engine and to trap liquids that may have condensed in the gas line.

VOLUNTARY POOLING AND UNITIZING
See Multiple Community Lease.

V.R.U.
Vapor recovery unit, one of the process units at a refinery, specifically as part of an F.C.C.U., fluid catalytic cracking unit.

VUGS
Large pits or cavities found in certain types of sedimentary rock.

VUGULAR-TYPE ROCK
Rock with large pits or cavities in its structure. Limestone, which often contains pits and cavities, is an example of a vugular-type sedimentary rock.

VULTURE FUND
A name, bordering on the macabre, for an investment fund put together by speculators to buy up distressed property, oil and gas leases, or small gas plants or refineries whose owners are in a financial pinch and are being foreclosed; also such property repossessed and held by banks or loan companies.

Pneumatic rock drills supported by a side-boom cat

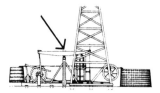

Walking beam

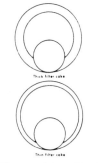

Pipe stuck in wall cake

WACKE

A sandstone consisting of a variety of angular, unsorted mineral and rock fragments and lots of clay and silt as a matrix; a dirty sandstone containing 10 percent or more of argillaceous or clayey material.

WAGON DRILLS

A battery of pneumatic rock drills mounted on a wagon. A type of trailer pulled along a pipeline right of way to drill holes in rock for the placement of dynamite charges to break up the rock so it can be removed by trenching equipment or bulldozers.

WAIT AND WEIGHT

A well-killing procedure in which the well is shut in and the weight of the mud is increased to the point required to kill the well, to control the kick. The heavyweight mud is then circulated into the well, and the kick fluids (gas or oil or both) are circulated out. This method requires the patience of Job and steady nerves as one waits for the mud to be weighted sufficiently to control what could be a blowout.

WALKING BEAM

A heavy timber or steel beam supported by the Samson post that transmits power from the drilling engine via the bandwheel, crank, and pitman to the drilling tools. The walking beam rocks or oscillates on the Samson post, imparting an up-and-down motion to the drilling line or to the pump rods of a well.

WALL CAKE

See Filter Cake.

WALL CLEANER

A scraping or cutting device attached to the lower joints of casing in the string for the purpose of cleaning the wall of the borehole in preparation for cementing. There are numerous types of scratching, raking, and cutting devices designed to remove the clay sheet or "filter cake" deposited by the circulating drilling mud. Mechanically cleaning the walls frees the production formation from the caked mud and also enlarges the hole diameter through the production zone making for more efficient oil flow into the well bore.

WALL SCRAPER
See Wall Cleaner.

WALL STICKING
A condition downhole when a section of the drillstring becomes stuck or hung up in the deposit of filter cake on the wall of the borehole. Also referred to as differential sticking, an engineer's term.

WALL-STUCK PIPE
See Differential-Pressure Sticking

WARM UP
To hammer a pipe coupling so as to loosen the threaded connection. Repeated pounding with a hammer literally warms the connection as well as "shocking" corroded threads so that they can be unscrewed.

WASHERS AND SPACERS
See Spacers and Washers.

WASHOUT
A leak of high-pressure drilling fluid at a drill collar or tool joint that becomes progressively worse, wearing away at the metal until the pipe at the threads is weakened, and under continuous torque and vibration twists off, parts. Washouts can be detected by closely monitoring the circulating drilling fluid pressure and the returns.

WASHOUT (Drill String)
A drilling-mud leak at a collar or tool joint that, under high pressure from the mud pumps, soon cuts the threads until circulation is affected or worse—the pipe parts.

WASHOVER PIPE
A type of drillpipe large enough in diameter to be run over the well's drillstring. A washover shoe, a tough milling tool, is attached to the washover pipe to cut away obstructions in the well's annulus.

Washover pipe

WASHOVER SHOE
A sharp-toothed milling tool used downhole to clear the annulus of obstructions that may be sticking the drillpipe. The washover shoe is run on the string of washover pipe, which is slipped over the drillpipe and lowered into the hole. The washover pipe with shoe attached is turned by the kelly joint and rotary table. As the shoe, a torus-shaped cutting head pointed with tungsten carbide teeth, moves downward it cuts and grinds away any and all obstructions, metal or rock.

WASTE
Wiping material; cotton waste; tangled skein of cotton thread from a textile mill used in engine rooms or on dirty jobs to wipe up oil and grease.

WASTE LIQUIDS
Oilfield brine, cut oil, tank-bottom sediment, concentrated sulfur water, and acid waters.

WASTE-WATER PLUME
The visible trace of discolored waste or chemically polluted water as it leaves the waste pipe of a processing plant and pours into a body of fresh or clear salt water.

WASTING ASSET
A material (usually mineral property) whose use results in depletion; a nonreplaceable mineral asset; oil, gas, coal, uranium, and sand.

WATER, CONNATE
See Connate Water.

WATER, JUVENILE
See Juvenile Water.

WATER, METEORIC
See Meteoric Water.

WATER BACK, TO
To reduce the density or weight of drilling mud by adding water which thins and lightens the slurry, the drilling fluid.

WATER-BASE DRILLING MUD
Conventional drilling mud mixtures or slurries have a water base, either fresh or salt water. Only in special cases do drilling muds have an oil base. Using a light oil about the grade of diesel fuel is, of course, more costly, and the drilling or penetration rate is not as good as a water-base mud.

WATER-BATH HEATERS, INDIRECT-FIRED
A type of fluid heater using hot water to heat gas and oil or to vaporize propane or cryogenic liquids. Conventional heaters are direct-fired, like a teakettle heats water. Indirect-fired heaters warm a circulating liquid that in turn heats the charge.

WATER COLUMN
In engineers' parlance, a body of water, top to bottom, under consideration or study; a hypothetical cylinder or stand of water; waterdepth.

WATER CONING
The encroachment of water in a well bore in a water-drive reservoir owing to an excessive rate of production. The water below the oil moves upward to the well bore through channels, fissures, and permeable streaks, leaving the oil sidetracked and bypassed.

WATER CYCLE
See Hydrologic Cycle.

WATER DRIVE
The force of water under immense pressure below the oil formation that, when the pressure is released by drilling, drives the oil to the surface through the well bore.

WATER-DRIVE RESERVOIR
An oil reservoir or field in which the primary natural energy for the production of oil is from edge or bottom water in the reservoir. Although water is only slightly compressible, the expansion of vast volumes of it beneath the oil in the reservoir will force the oil to the well bore.

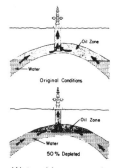

Water-drive reservoir

WATERED-OUT RESERVOIR
Such reservoirs usually are gas reservoirs with strong water drives. As the gas is produced and depleted, more and more salt water is produced until it becomes so costly to dispose of the water in disposal wells or to recycle that the operation is no longer profitable. This can happen also in oil reservoirs, but in a different way. *See* Water Coning.

WATERFLOODING
One method of secondary recovery in which water is injected into an oil reservoir to force additional oil out of the reservoir rock and into the well bores of producing wells.

WATERFLOOD KICK
The first indication of increased crude oil production as the result of a waterflood project. In such an operation, the massive and forcible injection of water into a reservoir, it may be a year or longer before there is a kick, a measurable increase in the field's production rate.

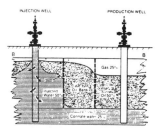

Waterflooding process
(After Clark, courtesy
SPE)

WATERFLOOD PROGRAM, FIVE-SPOT
See Five-Spot Waterflood Program.

WATERGLASS TREATMENT
Sodium silicate (waterglass), a viscous substance that, before the advent of refrigeration on the farm, was used to preserve eggs and other items of food, and in tunnel construction is a water-incursion stopper. In oil well drilling, water glass is used for the same purpose: to seal off certain water-bearing formations.

WATER-HAMMER
Water-hammer is a common side effect of most valve-closure mechanisms that depend on reverse-flow forces to aid in the closure. The phenomenon of water-hammer is caused by the sudden stoppage of a flowing fluid which converts the kinetic energy of the flow into a pressure

Waterflood facilities

pulse that sometimes can be very high and damaging to a formation. Flapper-type valves are most likely to cause hammering because rapid back-flows, when suddenly reversed, have built up significant velocities; and once valve closure begins, it happens almost instantaneously with a resultant sudden pressure build; the effect on the flapper of smart hammer blows.

WATER INDICATOR PASTE
See Indicator Paste, Water.

WATER LOGGING
A condition arising in waterflooding when the injected water fails to flush out the oil in the oil-bearing formation owing to insufficient movement of the fluids. The flooded formation turns out to be not as permeable to water as to oil. As a result, the well becomes water logged.

WATER LOSS IN DRILLING MUD
See Filtration-Loss Quality of Mud.

WATER PAD
A quantity of treated water pumped downhole ahead of a measured volume of gelled acid in a frac job. In certain treatments a slug of water, a pad of water, goes first, followed by the gelled acid, then another pad of treated water, all pumped to the face of the formation at extremely high pressure. After reaching the critical pressure, the fluids break into the formation. The gelled acid carrying in suspension the proppant material which, after a time (1 to 3 hrs.), is deposited as the gel breaks, losing its viscosity and ability to hold the proppant. Then the fluids, treated water and spent acid, drain back into the borehole and are circulated out or pumped out to clean the hole.

WATER PRESSURE
Pounds per square inch = height of the water column in feet × 0.434; e.g., 10-foot column of water × 0.434 = 4.34 pounds per square inch of pressure.

WATERSHED
A water parting; or the ridge or line or high ground separating two drainage basins. The Continental Divide, the backbone of the lower 48 states is a classic and large-scale water shed: west into the Pacific and Gulf of Mexico; east into the Mississippi River and the Gulf.

WATER STRING
The casing used downhole to shut off a water-bearing formation. To seal off the water interval, the casing is cemented. *See also* Liner.

WATER TABLE
The level of ground water in the earth; the surface below which all pores of the ground are filled with water.

WAX

Paraffin. In processing lubricating oil, one step is removing the paraffin, which retards the oil's ability to flow at low temperatures. In pumping wells the tubing becomes so coated with wax or paraffin on the inside that the flow of oil is blocked. When this happens, the tubing is pulled and heated to melt the paraffin or it is scraped out. In fields producing paraffin-based oil, this problem of plugged tubing is costly and time consuming. *See* Paraffin.

WEAK ZONE

A porous formation or one full of capillary channels or crevices that cause loss of circulation, loss of drilling fluid into the formation, while drilling.

WEATHERED CRUDE

Crude oil that has lost an appreciable quantity of its volatile components owing to natural causes (evaporation) during storage and handling.

WEATHERING

(1) The physical and chemical forces that break rock. Wind, rain, and the freezing-thawing cycle, which often work in concert, degrade even the hardest exposed rock. This inexorable work is called mechanical and chemical weathering. When rocks are broken into fragments, it is from mechanical weathering. Chemical weathering can break down certain rocks by dissolving or decomposing the minerals, making the exposed portions more susceptible to further breakdown by other forces. (2) The old practice of allowing highly volatile products such as natural gasoline to stand in tanks vented to the atmosphere to "weather off," to lose some of the more volatile fractions before being pumped into a pipeline. This wasteful procedure is no longer permitted.

WEATHER WINDOW

The period of time between storms when relatively calm weather prevails. In offshore work, particularly in the stormy North Sea, weather windows are often of short duration so work must be planned to take advantage of the brief intervals of good weather. This is true also for exploration work and supply operations in the Arctic.

WEDGE-PLUG VALVE

See Valve, Wedge-Plug.

WEDGING OUT, A

A formation that laterally becomes thinner and thinner, less porous, less permeable, and finally like a wedge, ends; wedges out.

WEEVIL-PROOF

Refers to tools and equipment simple to operate or to assemble; fittings and equipment parts impossible for a boll weevil, a green hand, to put together improperly.

WEIGHBRIDGE
A facility to measure the contents of rail tank cars loading L.P.-gas at a refinery or terminal. The rail cars are moved onto the scales and loading is done while the cars are on the weighbridge (scales). When the tank car is filled, the flow is automatically shut off and a ticket for the net weight is printed simultaneously.

WEIGHT COATING
Coating line pipe with a heavy overcoat of steel-mesh reinforced concrete to weigh it down and secure it to the bottom of a river or the ocean. Weight coating of a section of pipeline always includes anticorrosion coatings.

Weight indicator and driller's control console

WEIGHT INDICATOR
A large scale-like instrument suspended a few feet above the derrick floor that constantly displays the total weight of the drillstring in the hole as the well is being drilled. By observing the indicator, the driller can tell at a glance the weight of the string and, just as importantly, the weight or downward pressure on the drill bit.

WEIGHTING MATERIAL
Small pellets or particles of inert, nonabrasive material such as barite (barium sulfate) added to drilling mud to increase its unit weight per gallon.

WELD, FILLET
An electric or oxyacetylene weld joining two pieces of metal whose ends overlap; a weld that fills in the angular or concave junction of two overlapping pieces; a strip weld.

WELD, GIRTH
A pipeline weld on the outer circumference of the pipe made as two joints are welded together.

Weld-end fittings (swage, ell, flange, tee)

WELD, WET
A weld made underwater, "in the wet," without the use of a dry box as in hyperbaric (extreme pressure) welding.

WELD-END FITTINGS
Nipples, flanges, valves, and plugs without threads but with plain, beveled ends that can be welded to nonthreaded, plain-end pipe. For proper welding, the ends of both fittings and pipe are beveled to provide a V-shaped groove for the courses of welding metal.

WELDER, DOWNHILL

A beginner welder who is not capable of welding in the bellhole, upside down, as it were. Downhill welders can weld on horizontal surfaces, and to a limited degree on vertical, but they cannot manage welding above their heads as uphill welders can.

WELDER'S HELPER

A person who assists the welder. His most important job is to brush the weld with a wire brush to dislodge rust and scale. He also keeps the welder supplied with rods and holds or turns the piece being welded.

Welder's helper with emery wheel

WELDING, AUTOMATIC

An electric welding unit that rides on a track-like guide on the outer circumference of the line pipe being welded. When set in motion, the welding head or "bug" lays on a bead, a course of metal from its weld-wire electrode. *See* Welding "Bug."

WELDING, CO$_2$-SHIELDED

A semiautomatic technique of field welding that has the advantage of making welding a hydrogen-free operation, thus eliminating hydrogen cracking of the weld metal; an inert gas-shielded welding process; electric welding in which the molten metal being laid down is blanketed by CO_2 to protect it from active gases making contact with the molten surface.

WELDING, EXPLOSIVE

A method of welding in which a shaped explosive charge is used to "fast-expand" the end of a section of pipe into the bore of a special steel sleeve to form a solid bond. The shaped charge is inserted into the end of the pipe over which the sleeve is placed. When the charge is detonated, the force expands the pipe's outer circumference forcibly to the sleeve's inner circumference, making a secure, pressure-tight bond. This welding technique creates little heat, which for certain jobs is more desirable than fusion welding in which both pieces of metal must be heated to a high temperature.

Automatic welding machine

WELDING, FRICTION

The fusing of two pieces of metal by the heat of friction between the two. In welding tool joints to drillpipe, this method is used by Nippon Kokan, a Japanese steel company that pioneered the procedure. The tool joint is spun against the pipe end at high speed until the joint and the pipe are red hot or white hot, at which time the two pieces of steel fuse, forming a uniform, durable weld.

WELDING, GAS

Welding with oxygen and acetylene or with oxygen and another gas. *See* Welding, Oxyacetylene.

CO$_2$-shielded welding

Pipeline welding; note goggles and hood

Wire welding machine

Welding-bottle gauges

Welding bug

WELDING, HYPERBARIC

Welding on the seafloor "in the dry" but under many atmospheres of pressure (compression). In hyperbaric welding of undersea pipelines, a large frame is lowered into the water and clamped to the pipeline. Then an open-bottomed, box-like enclosure is placed in the center of the frame over the pipe. Power lines and life-support umbilicals are connected to the box. The seawater is displaced with breathing-gas mixtures for the diver-welders, permitting them to work in the dry but high-pressure atmosphere. Excessive-pressure welding.

WELDING, OXYACETYLENE

The use of a mixture of oxygen and acetylene in heating and joining two pieces of metal. When the weld edges of the two pieces are molten, metal from a welding rod is melted onto the molten puddle as the welder holds the tip of the rod in the flame of the torch. Oxygen and acetylene are also used in cutting through metal. The intense heat generated at the tip of the cutting torch (about 3,500°F), literally melts away the metal in the area touched by the flame. *See* Welding Torch.

WELDING, PIPELINE

In pipeline welding, the bevelled ends of two joints are brought together and aligned with clamps. Welders then lay on courses of weld metal called passes or beads designated as (1) stringer bead, (2) hot pass, (3) third pass or hot fill (for heavy-wall pipe), (4) filler pass, and (5) final or capping pass.

WELDING, STICK ELECTRODE

Electric-arc welding in which the welding rod or electrode is hand-held as compared to automatic welding.

WELDING, WIRE

Electric welding with a continuous wire electrode instead of the more common hand-held electrode or welding rod. Wire welding is used in automatic electric welding on pipelines to lay down filler beads or hot passes after the joints of pipe have been joined by tack welds and stringer beads with the use of hand-held welding rods.

WELDING BOTTLE

Steel cylinders of oxygen and acetylene gas used in oxyacetylene or gas welding. The oxygen bottle is green and taller and smaller in diameter than the black acetylene bottles.

WELDING-BOTTLE GAUGE

A type of small, adjustable-flow regulator screwed onto oxygen and acetylene gas bottles to regulate the flow of gases to the welding torch.

WELDING "BUG"

An automatic electric welding unit; specifically, the welding head that contains the welding-wire electrode and moves on the pipe's circumference on an aligned track-like guide. Used in welding large-diameter line pipe.

WELDING GOGGLES

Dark safety glasses used by oxyacetylene welders and welders' helpers to protect their eyes from the intense light of the welding process and from the flying sparks. *See* Welding Hood.

WELDING HOOD

A wraparound face and head shield used by electric welders. The hood has a dark glass window in the face shield; the hood tilts up, pivoting on the headband.

WELDING MACHINE

A self-contained electric generating unit composed of a gasoline engine direct-connected to a D.C. generator that develops current for electric welding. Welding machines or units are skid-mounted and transportable by dragging or by truck.

WELDING MACHINE, PIPELINE

After two joints of pipe are joined by tack welds, automatic wire-welding machines are used to put on the filler beads.

WELDING TORCH

An instrument used to produce a hot flame for welding; a hand-held, tubular device connected by hoses to a supply of oxygen and acetylene and equipped with valves for regulating the flow of gases to the tip or welding nozzle. By opening the valves to permit a flow of the two gases from the tip, the torch is ignited and then, with adjustments of the valves, a hot (3,500°F) flame results.

Cutting torches in a holder for beveling pipe ends

WELL

A hole drilled or bored into the earth, usually cased with metal pipe, for the production of gas or oil. Also, a hole for the injection under pressure of water or gas into a subsurface rock formation. *See* Service Well.

WELL, DEEPEND

A producing well that has a highly permeable section below the main pay zone. These permeable sections can be thief zones and, if so, they are squeezed off with a cement job.

WELL, OBLIGATION

A well which must be drilled under the terms of various agreements or regulations. For example, a farmout agreement, lease agreement, government regulations, or other binding contract. Failure to drill such a well will result in forfeiture of the lease or farmout assignment. Neither payment of compensatory royalty nor negotiations will discharge this obligation. Only a well drilled with reasonable diligence will do.

WELL, RELIEF

A directional well drilled near an out-of-control or burning well to kill the well by flooding the formation with water or drilling mud; a well drilled as close as possible or prudent to an out-of-control well and into the same formation in order to vent off or relieve the flowing pressure of the blowout so that the wild well may be brought under control. In some instances, more than one relief well is drilled to reduce the flow of the blowing well. A killer well.

WELL CLEAN UP

Refers to ridding the borehole of spent frac fluid after a frac job. With the use of a viscosity breaker, the gelled frac fluid, after cracking or fracturing the formation and leaving behind its load of proppant material, returns to its original state and drains back into the borehole where it is pumped or circulated out—leaving the hole clean.

WELL COMPLETION

The work of preparing a newly drilled well for production. This is a costly procedure and includes setting and cementing the casing, perforating the casing, running production tubing, hanging the control valves (nippling up the production tree, i.e., Christmas tree), connecting the flow lines, and erecting the flow tanks or lease tanks.

WELL COMPLETIONS, PRIMARY

Open hole; acidizing; fracture stimulation; cementing; and gravel packing.

WELL COST, AVERAGE

In 1983 the average cost to drill, case, and complete a well was $410,000. Included in this average were many deep, multimillion-dollar wells—the 15,000 to 25,000-foot gas wells—and many thousands of shallower wells costing $200,000, or even less.

WELLHEAD

The top of the casing and the attached control and flow valves. The wellhead is where control valves, testing equipment, and take-off piping are located.

WELLHEAD CELLAR

An airtight submarine chamber enclosing an underwater wellhead large enough to permit work to be carried on in a dry and normal atmosphere. The wellhead cellar is a piece of equipment that has been developed for well completions and other work in a deepwater environment.

Wellhead cellar

WELLHEAD CHOKE

A type of control valve through which a well is flowed when testing or regulating flow. Most wellhead chokes come in an assortment of flow beans that are calibrated in 64th-inch openings or orifices. The beans are interchangeable in the valve body, so a well can be flowed through a 16/64-inch or a 22/64-inch choke, etc.

WELLHEAD PLATFORM

A large drilling platform with 36 to 44 drilling slots for extended reach and high-angle wells. Each successful well means one wellhead. When the drilling program is complete, the platform then becomes a wellhead platform.

WELLHEAD PRESSURE

The pressure exerted by a well's oil or gas at the casinghead or the well-head when all valves have been closed for a period of time, usually 24 hours. The pressure is shown on a gauge on the wellhead. *See* Reservoir Pressure.

4,000 psi wellhead pressure

WELLHEAD PRICING

The price of gas or oil—usually gas—at the wellhead, as produced, free of the cost of transportation. A version of F.O.B.; Freight On Board.

WELL JACKET

A structure built around a completed offshore well to protect its production tree (the valves and piping protruding above the surface of the water) from damage by boats or floating debris. The structure is equipped with navigational warning lights and other devices to signal its position at night or in a fog.

WELL LOGGING

Gathering and recording information about subsurface formation, the nature and extent of the various downhole rock layers. Also included are records kept by the driller, the record of cuttings, core analysis drill-stem tests, and electric, acoustic, and radioactivity logs. Any pertinent information about a well, written and saved, is a log—from sailing ship days.

WELL NAMING

The naming of a well follows a longstanding, logical practice. First is the name of the operator or operators drilling the well; then the landowner from whom the lease was obtained; and last the number of the well on the lease or the block. For example Gulf Oil drills on a lease acquired from Dorothy Doe, and it is the first well on the lease. The name will appear in the trade journals as Gulf Doe No. 1, sometimes Gulf 1 Doe. For a lease from the State of Oklahoma by Gulf, the name would appear on a sign at the well as: Gulf State No. 1 and perhaps followed by its location—SW NW 2-29s-13w. (SW quarter of NW quarter [40 acres] of Section 2, Township 29 south, Range 13 west.)

WELL PERMIT
The authorization to drill a well issued by a state regulatory agency.

WELL PLATFORM
An offshore structure with a platform above the surface of the water that supports the producing well's surface controls and flow piping. Well platforms often have primary separators and well-testing equipment. *See* Producing Platform.

Well platform

WELL PROGRAM
The step-by-step procedure for drilling, casing, and cementing a well. A well program includes all data necessary for the toolpusher and drilling crews to know: formations to be encountered, approximate depth to be drilled, hole sizes, bit types, sampling and coring instructions, casing sizes, and methods of completion—or abandonment if the well is dry.

WELLS, EXPANDABLE EXPLORATION
Expandable wells are slim-hole or coiled-tubing wells of 3-inch diameter or less. If a producing formation is encountered, production is discovered, a conventional rig is brought in to drill a larger hole, to case, cement and perforate the casing. If the slim hole is dry, it is easier and cheaper to P.&A., plug and abandon.

WELLS, FIELD
Wells drilled in an established field, a recognized play or pool. Field wells are development wells as opposed to exploratory wells.

WELLS, HOT
Wells whose drilled depth, whose boreholes, have reached a temperature gradient of 200 to 400°F. Water-base drilling fluids must be abandoned and oil-base slurries substituted. *See* Temperature Gradient.

WELLS, STEAM
Wells drilled for the purpose of injecting high-pressure, high-temperature steam into a heavy-oil formation in a steam-flood program.

WELL SHOOTER
A person who uses nitroglycerin and other explosives to shoot a well, to fracture a subsurface rock formation into which a well has been drilled. The shooter lowers the explosive into the well bore on a wireline. When the explosive charge has been landed at the proper depth, it is detonated electrically.

WELL SORTED SEDIMENTARY MATERIAL
The phrase refers to a collection of different shapes and sizes of grains in

a sedimentary bed. When grains of sand are spherical and all one size, porosity of rock is at its maximum, regardless of the size of the grains. Strange but true. And porosity diminishes as the grains become more angular because such grains of sediment pack together more closely.

WELL SOUNDER
An electronic device to determine from the surface, the fluid level in a well. Spotting the fluid level (oil or water) can indicate bottom-hole pressure (B.H.P.), a measure of the well productivity index and the efficiency of the downhole pump.

WELL STIMULATION
See Stimulation.

WELL SYMBOLS
Symbols used on a map to indicate the kind of well and its condition—successful, dry, abandoned, etc.

WET GAS
Natural gas containing significant amounts of liquefiable hydrocarbons.

WET JOB
Pulling tubing full of oil or water. As each joint or stand is unscrewed, the contents of the pipe empties onto the derrick floor, drenching the roughnecks. The tubing is standing full of fluid because the pump valve on the bottom of the tubing is holding and will not permit the fluid to drain out as it is being hoisted out of the hole.

WET OIL
Oil as it comes from a well, with produced water and all. Dry oil, so-called, is oil free of produced water; it has been dewatered, put through an oil/water separator, dried out as it were.

WET STRING
See Wet Job.

WETTING THE GAS CAP
A phenomenon occurring in a gas-cap field when the pressure of the gas in the cap is depleted faster than that of the oil zone. The pressure differential that results forces an upward movement of solution gas and oil into the gas cap. Capillary action causes the pore spaces of the rock to retain the fluids (oil). As a result, some of the oil from the oil zone that was moved upward by the diminished pressure in the gas cap becomes unrecoverable.

WET TREE
An underwater Christmas tree; a stack of blowout preventers, other control and production valves (sales valves) on the sea floor controlling one well. Wet trees are used for subsea completions where there is no per-

API STANDARD OIL-MAPPING SYMBOLS	
Location	○
Abandoned location	erase symbol
Dry hole	-○-
Oil well	●
Abandoned oil well	-●-
Gas well	○
Abandoned gas well	-○-
Distillate well	◐
Abandoned distillate well	-◐-
Dual completion – oil	◉
Dual completion – gas	◎
Drilled water-input well	⌀w
Converted water-input well	●w
Drilled gas-input well	⌀G
Converted gas-input well	●G
Bottom-hole location (× indicates bottom of hole. Changes in well status should be indicated as in symbols above.)	○---··
Salt-water disposal well	⊕SWD

Standard well symbols
(Courtesy API)

Wet job

513

manent, stable production platform above the water. The undersea valves are controlled hydraulically or pneumatically.

WET WELD
See Weld, Wet.

WET WELDING
The kind that is done underwater, sometimes very deep with specially designed equipment.

WHALE PASTURE
A whimsical and imaginative description of the vast offshore.

WHEELED ROD GUIDE COUPLINGS
Sucker rod couplings that act as centralizers for the string of rods in a pumping well. The cylindrical couplings are made with several small wheels integral to the coupling that roll free as rods move up and down in the production tubing, thus reducing friction on both rods and tubing.

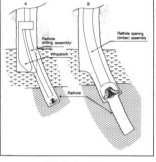

Whipstocking

WHIPSTOCK
A tool used at the bottom of the borehole to change the direction of the drilling bit. The whipstock is, essentially, a wedge that crowds the bit to the side of the hole, causing it to drill at an angle to the vertical.

WHIPSTOCK ANCHOR
A downhole tool used to prevent the downward movement of a whip-stock during milling or sidetracking. Essentially, the tool is a plug run and set permanently in the casing just below the projected window to be cut in the casing wall. The anchor isolates the casing and the formations below the window and keeps the whipstock from moving, forcing the milling tool to cut the casing at the intended point. Inside tracking, the anchor directs the drill bit at an angle from the vertical to sidetrack or drill a new hole around an obstruction, a fish, or to correct the direction of the bit in a severely deviated hole.

Whirley crane barge
(Courtesy R. Reece Bar-rett Assoc.)

WHIRLEY
The name applied to a full-revolving crane for offshore duty. Other barge-mounted cranes revolve 180°—over the stern and over both sides of the vessel.

WHISPERING INTEREST
A very small interest in a well, an interest one doesn't talk about in the company of those with substantial interests of 10, 15, or 20 percent, say. The same cautions are operative for a person with a 30-acre ranch when in the company of Texas cattlemen, some of whose spreads are the size of Rhode Island.

WHITE, DR. ISRAEL CHARLES
The "father of petroleum geology." Dr. White brought about the transi-

tion from superstition and "creekology" to scientific geological methods. He was a poor West Virginia boy who grew up to become world famous as the discoverer of the anticlinal or structural theory of oil accumulation.

Whittaker survival capsule on display *(Courtesy Whittaker Co.)*

WHITE CARGO
Clean cargo; a term to describe distillate—gasoline, kerosene, heating oils—carried by tankers.

WHITE OIL
A colloquial term for condensate, gas condensate, casinghead gasoline; liquid hydrocarbons produced with natural gas. *See* Condensate.

WHITTAKER SYSTEM
A patented system for protection of crews working on offshore drilling platforms, semisubmersibles, and other structures. The heart of the system is a survival capsule into which offshore crew members can retreat in the event of fire or explosion or other disaster and lower themselves to the water. The capsule is self-propelled and provides food, water, and first-aid supplies for 28 persons. Large offshore structures have several survival capsules that hang from davits at various locations on the platform. *See* Brucker Survival Capsule.

WICK OILER
A lubricator for large, slow-moving crank bearings. Oil is fed from a small canister, a drop at a time, onto a felt pad or "wick." As the crank turns beneath the wick, a scraper on the crank makes contact with the wick, taking a small amount of oil.

Wildcat well

WILDCATTER
A person or company that drills a wildcat well; a person held in high esteem by the industry, if he is otherwise worthy; an entrepreneur to whom taking financial risks to find oil is the name of the game.

WILDCAT WELL
A well drilled in an unproved area, far from a producing well; an exploratory well in the truest sense of the word; a well drilled out where the wildcats prowl and "the hoot owls mate with the chickens."

WINCH
A device used for pulling or hoisting by winding rope or cable around a power-driven drum or spool.

WINDFALL PROFIT
For the purpose of the crude oil Windfall Profit Tax Act of 1980, the term means the excess of the removal price of a barrel of crude oil over the sum of: the adjusted base price of the crude plus the amount of the sev-

Rope mooring winch

erance tax adjustment allowed by the Internal Revenue Service (I.R.S. Code 4988).

WINDLASS
A winch; a steam or electric-driven drum with a vertical or horizontal shaft for raising a ship's anchor.

WIND-LOAD RATING
Drilling derricks not only are rated as to lifting or load capacity (up to two million pounds) but for wind load, resistance to the wind trying to blow them over when their racks are full of drillpipe or tubing. The biggest sturdiest rigs are designed to withstand 100 to 130 mph wind gusts.

WINDOW (EXPLORATION)
An area in a formation, a sedimentary section, that is promising as to production; a source of hydrocarbons.

WINDOW (GEOLOGY)
An eroded area on a thrust sheet revealing the rocks beneath it, the stratum immediately below.

WINDOW (LEASE INTEREST)
A term that refers to an unsigned interest affecting a pooled or unitized interest. The window, the unsigned portion of the agreement, is an opportunity to join the pooled or unitized interests.

WINDOW (WEATHER)
A time, usually of limited duration, when the weather is suitable for oil operations: moving platforms, building offshore pipelines, or conducting seismic surveys, etc. In hazardous or very hostile environments—the Arctic, the North Sea; also, the Equatorial zones, between the 20th parallels during the monsoons—weather windows are foretold as accurately as possible and used to advantage. Oil men have an enviable record of being remarkably tough, resilient, and persevering in the face of the obstacles nature and man devise to hamper their world-wide operations: permafrost, ocean storms, desert sand, unfriendly natives, whiteouts, floods, hurricanes, and, now environmentalists and even legislatures. But oil and gas will be found.

WING UNION
See Union, Wing.

WIPER TRIP
See Wiping the Pipe.

WIPING THE PIPE
Pulling the drillpipe to clean the borehole of cuttings or sloughs. A

debris-filled annular space increases torque necessary to rotate the bit and can, in some instances, cause differential sticking or hang up of the drillstring. Wiper trips are time-consuming and expensive, but are considered prudent moves in view of the alternatives of stuck pipe or even a twist off.

WIRELINE
A cable made of strands of steel wire; the "lines" used on a drilling rig.

WIRELINE, SLICK
A nonelectric wireline for downhole work of measuring, setting packers, and lowering in bailers and sand pumps.

Wireline trailer (*Courtesy Mathey*)

WIRELINE TOOLS
Special tools or equipment made to be lowered into the well's borehole on a wireline (small-diameter steel cable), e.g., logging tools, packers, swabs, measuring devices, etc.

WIRELINE TRUCK
A service vehicle on which the spool of wireline is mounted for use in downhole wireline work.

WIRE ROPE
A rope made of braided or twisted strands of steel wire; a cable. *See* Soft Rope.

WIRE WELDING
See Welding, Wire.

WITCHING
See Doodlebug.

W.O.B. (WEIGHT ON BIT)
The weight on the drill bit is an important consideration in the drill's rate of penetration (R.O.P.); so is rpm, and whether the bit is turned by the rotary table on the rig floor or by a mud motor at the bottom of the borehole. When a disparity arises between the weight on the bit as shown by the weight gauge at the surface and weight on the bit itself, the condition may indicate crowding of a section of the well bore by the intrusion, the swelling of water-absorbing clay. Or the disparity of weight could be caused by the inadequate cleaning of the hole, the accumulation of rock cutting not being carried away to the surface by the circulating mud. The development of a dog leg or a keyseat and the resultant drag can also affect the weight of the string getting to the drill where it belongs.

Skid-mounted wireline unit

W.O.C. TIME
Waiting-on-cement time; the period between the completion of the ac-

tual cementing operations and the drilling out of the hardened cement plug in the casing at the bottom of the well.

W.O.G.M.
Water-oil-gas mud; a designation for a type of valve used in well-control operations that will handle all four fluids or a combination of them.

WON'T STAND UP
A colloquial phrase referring to unconsolidated formations being drilled through that slough or cave into the borehole; the sides of the hole won't stand up.

WOODCASE THERMOMETER
A thermometer used by gaugers in taking the tank temperature; the temperature of the oil in the tank as contrasted to the temperature of the sample oil to be tested. The thermometer is encased in a wood frame to which a line may be attached for lowering the thermometer into the oil.

Work boat

WORK-BACK MARKET VALUE
See Net-Back Pricing

WORK BOAT
A boat or self-propelled barge used to carry supplies, tools, and equipment to job sites offshore. Work boats have large areas of clear deck space which enable them to carry a variety of loads. *See* Drilling Tender.

WORKING DATE
A term used in some lease forms referring to last working day of the period during which the lessee must begin drilling operations, pay rental, or quit-claim and give up the lease.

WORKING FOR STREET AND WALKER
A sardonic reference to looking for a job.

WORKING GAS
Gas stripped of all liquid hydrocarbons and used in gas-lift production of crude oil; lift gas.

WORKING INTEREST
The operating interest under an oil and gas lease. The usual working interest consists of seven-eighths of the production subject to all the costs of drilling, completion, and operation of the lease. The other one-eighth of production is reserved for the lessor or landowner. *See* Royalty, Landowner's.

WORKING INTEREST, FULL-TERM
A working interest that lasts as long as the well or the lease is produc-

tive; as long as oil and gas are produced in quantities that make the well economic to operate. *See* Working Interest, Limited.

WORKING INTEREST, LIMITED
A working interest that terminates at some time prior to the depletion or exhaustion of the oil and gas reserves. The termination of the interest may be after a certain length of time, after a specified amount of production, or when a definite amount of money has been realized from production. *See* Working Interest, Full-Term.

WORKING PIT
The drilling mud pit at the wellsite that is being pumped out of and into which the returns flow. Besides the working pit that is used while drilling and circulating, there is the reserve pit that holds a supply of mud for use in an emergency, such as loss of returns or a kick.

WORKING PRESSURE
The pressure at which a system or item of equipment is designed to operate. Normal pressure for a particular operation.

WORKING TANK
A terminal or main-line tank pumped into and out of regularly; a tank that is "worked" as contrasted to a storage tank not regularly filled and emptied.

WORKOVER
Operations on a producing well to restore or increase production. Tubing is pulled and the casing at the bottom of the well is pumped or washed free of sand that may have accumulated.

Portable workover rig

WORKOVER FLUIDS
A specially formulated drilling mud used to keep a well under control while it is being worked over, acidized, cleaned out, or fraced. The workover fluid is carefully formulated: not too lightweight to lose control, not too heavy so as to damage the producing formation.

WORK STRING
A string either of drillpipe or tubing, suspended in the well, to which a special tool is attached to do a specific job: fishing, cutting, reaming, squeeze cementing, etc. A work string is usually made up differently from a drillstring. A drillstring, with a bit on the lower end, would be made up to keep the hole straight, put weight on the bit with drill collars; stabilizers, and reamers to keep the hole to size.

WORLDSCALE
Said of refineries, chemical plants, or operations that are as large as any in the world and that are equally or more advanced in design, operating efficiency, and throughput. One of the big boys on the block.

WORM

An inexperienced worker; a green hand; a close relative of the boll weevil.

WORM GEAR

A type of pinion gear mounted on a shaft, the worm gear meshing with a ring gear; a gear in the shape of a continuous spiral or with the appearance of a pipe thread; often used to transmit power at right angles to the power shaft.

WRENCH FLAT OR SQUARE

The flat area on an otherwise round or cylindrical fitting to which an end wrench can be applied to make up or break out a connection. A good example is the wrench square of a sucker rod, at both the box and pin ends, also called "shoulders."

WRINKLE PIPE

To cut threads on a piece of pipe in order to make a connection.

WRIST PIN

The steel cylinder or pin connecting the rod to the engine's or pump's piston. The wrist pin is held in the apron or lower part of the piston by a friction fit and a circular spring clip. The upper end of the connecting rod is fitted with a lubricated bushing that permits the rod to move on the pin. A piston pin.

WRITE DOWN, TO

To reduce the present value of a commodity or a resource by lowering arbitrarily the value, as carried on a company's books. A reduction of the book value of a company's oil reserves, for example, from $30/bbl to $15/bbl automatically reduces the shareholders' equity by half, and at the same time causes a reduction in the company's borrowing base. On the plus side, writing down has tax advantages.

XYLENE
An aromatic hydrocarbon; one of a group of organic hydrocarbons (benzene, toluene, and xylene) that forms the basis for many synthetic organic chemicals. *See* B.T.X.

YIELD
See Product Yield.

YIELD STRENGTH
A measure of the force required to deform drillpipe, tubing, casing, or other tubular goods to the degree that they are permanently distorted.

YIELD TAX
See Severance Tax

ZEOLITIC CATALYST
Catalyst formulations that contain zeolite (any of various hydrous silicates, a mineral) for use in catalytic cracking units.

ZERO DISCHARGE UNIT
An offshore, closed water treatment system consisting of water catching, retention, treatment, and monitoring. This type of environmentally friendly system, is being built into new offshore platforms; the older ones are being retrofitted.

ZONE
An interval of a subsurface formation containing one or more reservoirs; that portion of a formation of sufficient porosity and permeability to form an oil or gas reservoir.

ZONE ISOLATION
A method of sealing off, temporarily, a producing formation while the hole is being deepened. A special substance is forced into the formation where it hardens, allowing time for the well bore to be taken on down. After a certain length of time, the substance again turns to a liquid, unblocking the producing formation.

ZONE OF LOST CIRCULATION
An interval in a subsurface formation so porous or cut with crevices and

Zone of lost circulation
(Courtesy Dresser Mag-cobar)

fissures that the drilling mud is lost in the pores, cracks, or even a cavern, leaving none to circulate back to the surface.

ZONES, PROBLEM

Thief zones, zones of water and near-surface gas incursion, sloughing intervals, and hydrophilic, shaley sections that absorb the water from water-base drilling muds and swell, filling the borehole and sticking the drillpipe. These are some of the troublesome areas that demand special attention, delaying the well and adding to the cost. (No one said drilling a well was easy or cheap.)

ZONES, SWEPT

Swept zones are those that have been produced, waterflooded, and subjected to tertiary oil recovery programs. Depleted zones whose pore pressures have declined to near zero.

1. Accumulator
2. A-frame
3. Air compressor
4. Annular (bag) preventer
5. Annulus
6. Base
7. Bell nipple
8. BOP control
9. Bit (drill)
10. Bradenhead
11. Burning pit
12. Casing-hanger spool
13. Cathead
14. Cat line
15. Catwalk
16. Cellar
17. Centrifuge
18. Chemical barrel
19. Choke line
20. Choke manifold
21. Choke manifold control
22. Compound
23. Conductor casing
24. Crown block
25. Cyclone desander desilter
26. Dead line
27. Degasser
28. Discharge line
29. Doghouse
30. Drawworks
31. Drill collars
32. Driller's console
33. Drilling line
34. Drillpipe
35. Drill tool storage (junk box)
36. Dynamatic or hydromatic
37. Elevators
38. Engines
39. Fast line
40. Fill-up line
41. Flow line
42. Fuel line
43. Fuel tank
44. Generating unit (light plant)
45. Gin pole
46. Hoisting line
47. Hook
48. Intermediate casing
49. Kelly
50. Kelly bushing
51. Kelly (rotary) hose

52. Kill line
53. Ladder
54. Line guide
55. Mast
56. Mast lifting line
57. Mixing (mud) pit
58. Monkey board
59. Mousehole
60. Mud
61. Mud-gas separator (gas buster)
62. Mud gun (submerged)
63. Mud gun (surface)
64. Mud hopper
65. Mud line
66. Mud logging unit
67. Mud (paddle) mixer
68. Mud-mixing plant
69. Oil and grease storage
70. Pipe rack (floor)
71. Pipe racks
72. Pressure (mud) gauge
73. Preventer control lines
74. Preventer (BOP) ram type
75. Production casing
76. Pump drive
77. Pump, mud mixing
78. Pumps, mud
79. Ram wheel
80. Ramp
81. Rathole
82. Reserve drilling line
83. Reserve (mud) pit

Detail for 113

Drilling rig schematic

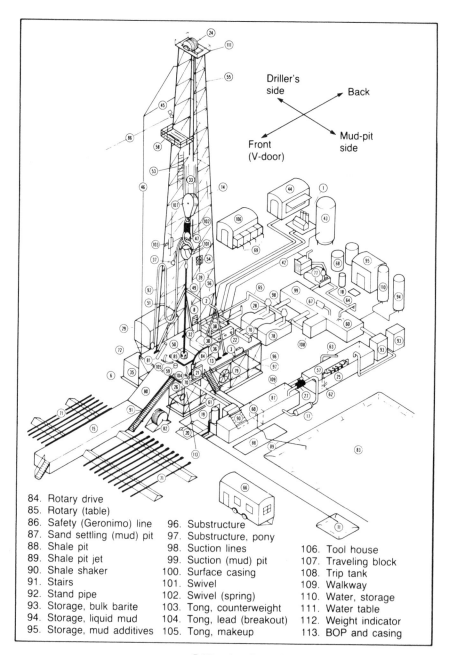

Driller's side
Back
Mud-pit side
Front (V-door)

84. Rotary drive
85. Rotary (table)
86. Safety (Geronimo) line
87. Sand settling (mud) pit
88. Shale pit
89. Shale pit jet
90. Shale shaker
91. Stairs
92. Stand pipe
93. Storage, bulk barite
94. Storage, liquid mud
95. Storage, mud additives

96. Substructure
97. Substructure, pony
98. Suction lines
99. Suction (mud) pit
100. Surface casing
101. Swivel
102. Swivel (spring)
103. Tong, counterweight
104. Tong, lead (breakout)
105. Tong, makeup

106. Tool house
107. Traveling block
108. Trip tank
109. Walkway
110. Water, storage
111. Water table
112. Weight indicator
113. BOP and casing

Drilling rig schematic

Bibliography

American Petroleum Institute, Division of Refining. *Glossary of Terms Used in Petroleum Refining.* Washington, D.C., 1962.

_____. *Petroleum—The Story of an American Industry,* 2 ed. 1935.

_____. *Glossary of Terms Used in Petroleum Refining,* 1953.

Ball, Max W. *This Fascinating Oil Business.* New York: Bobbs–Merrill, 1940.

Bank of Scotland Information Service. *Oil and Gas Industry Glossary of Terms.* Edinburgh: Bank of Scotland Oil Div., 1974.

Bates, Robert L., and Julie A. Jackson. *Glossary of Geology,* 2 ed. Falls Church, Virginia: American Geological Institute, 1980.

Bland, W., and Davidson. *Petroleum Processing Handbook.* New York, N.Y.: McGraw-Hill Publishing Co., 1967.

Brantly, J.E. *Rotary Drilling Handbook.* New York: Palmer Publications, 1952.

Craft and Hawkins. *Applied Petroleum Reservoir Engineering.* Englewood Cliffs, N.J.: Prentice-Hall, 1959.

Desk and Derrick Clubs. *D & D Standard Oil Abbreviator.* Tulsa: PennWell Publishing Company, 1973.

Fanning, Leonard M. *Men, Money & Oil: The Story of an American Industry.* New York: World Publishing Company, 1966.

Harris, L.M. *Deepwater Floating Drilling Operations.* Tulsa: PennWell Publishing Company, 1972.

Howell, J.K., and E.E. Hogwood. *Electrified Oil Production.* Tulsa: PennWell Publishing Company, 1962.

Hunt, John M. *Petroleum Geochemistry and Geology.* San Francisco: W.H. Freeman & Company, 1979.

Interstate Oil Compact Commission (Engineering Committee). *Oil and Gas Production.* Norman, Oklahoma: University of Oklahoma Press, 1951.

Kretchman, H.F. *The Story of Gilsonite.* Salt Lake City, Utah: American Gilsonite Company, 1957.

Maurer, William C. *Advanced Drilling Techniques.* Tulsa: PennWell Publishing Company, 1980.

Miller, Ernest C. *Pennsylvania History Studies* No. 4. Gettysburg: Pennsylvania Historical Association, 1954.

Miller, Kenneth G. *Oil & Gas Federal Income Taxation.* Chicago: Commerce Clearing House, 1966.

Moore, Preston L. *Drilling Practices Manual.* Tulsa: PennWell Publishing Company, 1974.

McCray, A.W. *Oil Well Drilling Technology.* Norman, Oklahoma: University of Oklahoma Press, 1958.

Porter, Hollis P. *Petroleum Dictionary for Office, Field and Factory,* 4 ed. Houston: Gulf Publishing Company, 1948.

Rocks, Lawrence. *Fuels for Tomorrow.* Tulsa: PennWell Publishing Company, 1980.

Sell, George. *The Petroleum Industry.* London: Oxford University Press, 1963.

Stephens, M.M., and O.F. Spencer. *Petroleum Refining Processes.* University Park, Pennsylvania: Pennsylvania State University Press, 1958.

Stovall, J.W., and Howard E. Brown. *Principles of Historical Geology.* New York: Ginn & Company, 1954.

Uren, L.C. *Petroleum Production Engineering (Development),* 4 ed. New York: McGraw–Hill Book Company, 1956.

Wendland, Ray T. *Petrochemicals— The New World of Synthetics.* New York: Doubleday and Company, 1969.

Wheeler, R.R., and Maurine Whited. *From Prospect to Pipeline.* Houston: Gulf Publishing Company, 1975.

Williams, Howard R. *Oil and Gas Terms: Annotated Manual of Legal, Engineering, and Tax Words and Phrases.* Cleveland: Banks–Baldwin Law Publishing Company, 1957.

Williams and Myers. *Manual of Oil and Gas Terms.* New York: Matthew Bender and Company, 1964.

Williamson, E.H., et al. *The American Petroleum Industry.* Chicago: Northwestern University Press, 1963.

Yergin, Daniel. *The Prize.* New York: Simon & Schuster, 1991.

Zabo, Joseph. *Modern Oil-Well Pumping.* Tulsa: PennWell Publishing Company, 1962.

D&D

Standard Oil Abbreviator

Desk and Derrick Clubs
of North America

A	abstract (i.e., A-10)
A	angstrom unit
A-Cem	acoustic cement
A&A	adjustments and allowances
A/	acidized with
A/C	air conditioning
A/CLR	air cooler
A/R	accounts receivable
AA	after acidizing, as above
AAR	Association of American Railroads
ABC	Audit Bureau of Circulation
abd	abandoned
abd loc	abandoned location
abd-gw	abandoned gas well
abd-ow	abandoned oil well
abdogw	abandoned oil & gas well
ABHL	absolute bottom-hole location
ABM	Atlas Bradford modified
abrsi jet	abrasive jet
abs	absolute
ABS	acrylonitrile butadiene styrene rubber
absrn	absorption
abst	abstract
abt	about
abun	abundant
abv	above
ac	acid
ac	acre(s), acreage
AC	alternating current
AC	Austin chalk
ac-ft	acre-feet
ACC	access
ACCEL	accelerometers
ACCESS	accessory, accessories
acct	account (ing)
accum	accumulative, accumulator
acd	acidize (ed) (ing)
acfr	acid fracture treatment
ac-ft	acre feet
ACLD	air cooled
ACLR	air cooler
ACM	acid-cut mud
acrg	acreage
ACS	American Chemical Society
ACSR	aluminum conductor steel reinforced
ACT	actual
ACT	actuated, actuator
ACT	automatic custody transfer
ACW	acid-cut water
AD	actual drilling
AD	authorized depth
ADC	actual drilling cost
add	additive
ADDC	Association of Desk & Derrick Clubs
addl	additional
ADH	adhesive
adj	adjustable
adm	administration, administrative
ADOM	adomite
ADP	automatic data processing
adpt	adapter
adspn	adsorption
ADT	actual drilling time
advan	advanced
AER	aeration, aerator
AF	acid frac
AF	after fracture
AF/CLR	after cooler
AF/COND	after condenser
AFC	Authorization for Commitment
AFC	Authorized for Construction
AFD	auxiliary flow diagram
AFE	Authorization for Expenditure
affd	affirmed
afft	affidavit
AFIT	after federal income tax
AFP	average flowing pressure
AFRA	average freight rate assessment
AG	agitator
AGGR	aggregate
aglm	agglomerate
AIR	average injection rate
AIR COND	air conditioning
AJT	actual jetting time
AL	aluminum
AL	artificial lift
Alb	Albany

alc	alcoholic
ALCOA	Aluminum Company of America
alg	algae
alg	along
ALIGN	alignment (ing)
alk	alkaline, alkalinity
alkyl	alkylate, alkylation
ALLOW	allowable, allowance
alm	alarm
ALOC	allocation
alt	alternate
ALT	altitude
ALY	alloy
amb	ambient
AMI	area of mutual interest
AMM	ammeter
amor	amorphous
amort	amortization
AMP	American melting point
amp	ampere
amp hr	ampere hour
amph	amphipore
Amph	Amphistegina
AMR	addition or modification reque
amt	amount
an	annulus
anal	analysis, analytical
ANC	Anchor (age)
ang	angle, angular
Angul	Angulogerina
anhy	anhydrite, anhydritic
anhyd	anhydrous
Ann	annulus
ANR	amount not reported
ANS	Alaskan North Slope
ANUB	annubar
ANUC	annunciator
ANYA	allowable not yet available
AOAC	Association of Official Agricultural Chemistry
AOF	absolute open flow potential (gas well)
AP&VMA	American Paint and Varnish Manufacturers Association
APHA	American Public Health Association
API	American Petroleum Institute
app	appears, appearance

APPAR	apparatus
appd	approved
appl	appliance
appl	applied
appllc	application
approx	approximate (ly)
apr	apparent (ly)
APR	average penetration rate
apt	apartment
aq	aqueous
AQCR	air quality control region
AQMA	air quality maintenance area
AR	acid residue
Ara	Arapahoe
arag	aragonite
Arb	Arbuckle
arch	architectural
Archeo	Archeozoic
aren	arenaceous
arg	argillaceous
arg	argillite
ark	arkose(ic)
Arka	Arkadelphia
arm	armature
arnd	around
ARO	after receipt of order (purchasing term)
ARO	at rate of
arom	aromatics
ARR	arrange (ed) (ing) (ment)
AS	after shot
AS	anhydrite stringer
AS&W ga	American Steel & Wire gauge
ASA	American Standards Association
ASAP	as soon as possible
asb	asbestos
asbr	absorber
ASD	abandoned-salvage deferred
asgmt	assignment
Ash	ashern
ASO	acid-soluble oil
asph	asphalt, asphaltic
assgd	assigned
assn	association
assoc	associate (d) (s)
asst	assistant
assy	assembly
ASTM	American Society for Testing & Materials

astn	asphaltic stain
ASW	adjustable spring wedge
AT	acid treat (ment)
AT	after treatment
AT	all thread
ATT	after the tanks
At	Atoka
at	atomic
at wt	atomic weight
ATC	after top center
ATD	approved total depth
ATF	automatic transmission fluid
atm	atmosphere, atmospheric
ATP	Authorization to Proceed
ATP	average treating pressure
ATP	average tubing pressure
ATT	attach (ed) (ing) (ment)
att	attempt(ed)
atty	attorney
aud	auditorium
Aus	Austin
auth	authorized
auto	automatic
auto	automotive
autogas	automotive gasoline
aux	auxiliary
AV	annular velocity
AV	Aux Vases sand
av	aviation
avail	available
AVC	automatic volume control
avg	average
avgas	aviation gasoline
AW	acid water
AWD	award
AWG	American Wire Gauge
awtg	awaiting
AWWA	American Water Works Association
az	azimuth
aztrop	Azeotropic

B	billion
B	bulletin

B & B	bell and bell
B & CB	beaded and center beaded
B & S	bell and spigot
B Hn	Big Horn
B slt	base of the salt
B. In.	Big Injun
B. Ls	Big Lime
B. Riv	Black River
B. Bl	Base Blane
B.E.	bevelled end
BS&W	basic sediment and water
B&B	bent & bowed pipe
B&F	bell and flange
B&S	Brown & Sharpe (gauge)
B/	base
B/	bottom of given formation (i.e., B/Frio)
B/B	back to back
B/B	barrels per barrel
B/D	barrels per day
B/dry	bailed dry
B/hr	barrels per hour
B/JT	ball joint
B/L	bill of lading
B/M	bill of material
b/off	buck-off
b/on	buck-on
B/S	back scuttled
B/S	base salt
B/S	bending schedule
B/S	bill of sale
B/SD	barrels per stream day (refinery)
B/VESS	bulk vessel
B/Vlv	ball valve
BA	barrels of acid
bail	bail (ed)
BAL	balance
Ball	Balltown sand
bar	barite (ic)
Bar	Barlow Lime
bar	barometer, barometric
BAR	barrels acid residue
Bark Crk	Barker Creek
Bart	Bartlesville
base	basement (granite)
bat	battery
BAT	before acid treatment

Bate	Bateman	**BF**	blind flange
BAW	barrels acid water	**bf**	buff
BAWPD	barrels acid water per day	**BFIT**	before federal income tax
BAWPH	barrels acid water per hour	**BFL**	baffle
BAWUL	barrels acid water under load	**BFO**	barrels frac oil
BB	bridged back	**BFPD**	barrels fluid per day
BB fraction	butane-butene fraction	**BFPH**	barrels fluid per hour
BBE	bevel both ends	**BFW**	bailer feed water
bbl	barrel	**BFW**	barrels formation water
BC	barrels of condensate	**BFW**	barrels fresh water
Bcf	billion cubic feet	**BFW**	boiler feed water
Bcfd	billion cubic feet per day	**BH**	bottom hole
BCPD	barrels condensate per day	**BHA**	bottom-hole assembly
BCPH	barrels condensate per hour	**BHC**	bottom-hole choke
BCPMM	barrels condensate per million	**BHCS**	borehole compensated sonic
BD	barrels of distillate	**BHF**	Bradenhead Flange
bd	board	**BHFP**	bottom-hole flowing pressure
BD	budgeted depth	**BHL**	bottom-hole location
bd ft	board foot (feet)	**BHM**	bottom-hole money
BD-MLW	barge deck to mean low water	**BHN**	Brinell hardness number
Bd'A	Bois d'Arc	**BHO**	bottom-hole orientation
BDA	breakdown acid	**BHP**	bottom-hole pressure
BDF	broke (break) down formation	**bhp**	brake horsepower
BDL	bundle	**bhp-hr**	brake horsepower-hour
BDNG	bedding	**BHPC**	bottom-hole pressure closet (see also SIBHP and BHSIP)
BDO	barrels diesel oil	**BHPF**	bottom-hole pressure flowing
BDP	breakdown pressure	**BHPS**	bottom-hole pressure survey
BDPD	barrels distillate per day	**BHSIP**	bottom-hole shutin pressure
BDPH	barrels distillate per hour	**BHT**	bottom-hole temperature
BDT	blow-down test	**BID SUM**	bid summary
Be	Baumé	**Big.**	Bigenerina
Be	Berea	**Big. f.**	Bigenerina floridana
be	box end	**Big. h.**	Bigenerina humblei
Bear Riv	Bear River	**Big. nod.**	Bigenerina nodosaria
bec	becoming	**BIN**	binary
Beck	Beckwith	**bio**	biotite
Bel	Beldon	**bit**	bitumen
Bel C	Belle City	**bit**	bituminous
Bel F	Belle Fourche	**bkdn**	breakdown
Belm	belenmites	**BKFLSH**	back flush
Ben	Benoist (Bethel) sand	**bkr**	breaker
Ben	Benton	**BKWSH**	backwash
Bent	bentonite	**BL**	barrels load
berm	berm, sloped wall to keep out flooding	**bl**	blue
bev	bevel (ed)	**BL&AW**	barrels load & acid water
BF	barrels fluid	**Bl/Cb**	blast cabinet
		BL/JT	blast joint
		BLC	barrels load condensate

BLCPD	barrels load condensate per day	**bndry**	boundary
BLCPH	barrels load condensate per hour	**bnish**	brownish
		BNO	barrels new oil
bld	bailed	**BNW**	barrels of new water
BLD FLG, BF	blind flange	**bnz**	benzene
bldg	bleeding	**BO**	backed out (off)
bldg	bleeding gas	**BO**	barrels oil
bldg	building	**BO**	blew out
bldg drk	building derrick	**BO**	blocked off
bldg rds	building roads	**BO**	free-point back off
bldo	bleeding oil	**BOCD**	barrels oil per calendar day
bldrs	boulders	**BOCS**	Basal Oil Creek sand
BLDWN	blowdown	**BOD**	barrels oil per day
BLE	bevel large end	**BOD**	biochemical oxygen demand
blg	bailing	**Bod**	Bodcaw
Blin	Blinebry	**BOE**	bevel one end
blk	black	**BOE**	blowout equipment
blk	block	**Bol.**	Bolivarensis
Blk Lf	Black Leaf	**Bol. a.**	Bolivina a.
Blk Li	Black Lime	**Bol. flor.**	Bolivina floridana
blk lnr	blank liner	**Bol. p.**	Bolivina perca
BLND	blend (ed) (er) (ing)	**Bonne**	Bonneterre
BLO	barrels load oil	**BOP**	blowout preventer
blo	blow	**BOPD**	barrels oil per day
BLOPD	barrels load oil per day	**BOPE**	blowout preventer equipment
BLOPH	barrels load oil per hour	**BOPH**	barrels oil per hour
BLOR	barrels load oil recovered	**BOPPD**	barrels oil per producing day
Blos	blossom	**BOS**	brown oil stain
BLOTBR	barrels load oil to be recovered	**bot**	bottom
BLOYTR	barrels load oil yet to recover	**BP**	back pressure
BLPD	barrels of liquid per day	**BP**	base Pennsylvanian
blr	bailer	**BP**	Bearpaw
BLR	boiler	**BP**	boiling point
blts	bullets	**BP**	bridge plug
BLW	barrels load water	**BP**	bulk plant
BLWPD	barrels load water per day	**BP**	bull plug
BLWPH	barrels load water per hour	**BP Mix**	butane and propane mix
BLWR	blower	**BP/CLR**	bypass cooler
BLWTR	barrels load water to recover	**BPD**	barrels per day
BM	barrels mud	**BPH**	barrels per hour
BM	benchmark	**BPLO**	barrels of pipeline oil
BM	Black Magic (mud)	**BPLOPD**	barrels of pipeline oil per day
BMEP	brake mean effective pressure	**BPM**	barrels per minute
BMI	black malleable iron	**BPSD**	barrels per stream day
bmpr	bumper	**BPV**	back-pressure valve
bn	brown	**BPWPD**	barrels per well per day
bnd	band (ed)	**BR**	building rig
		BR	building road

brach	brachiopod
BRC	brace (ing) (ed)
brec	breccia
BRFL/V	butterfly valve
brg	bearing
Brid	bridger
brit	brittle
brk	break (broke)
brkn	broken
brkn sd	broken sand
BRKR	breaker
BRKS	brakes
brksh	brackish (water)
brkt(s)	brackets(s)
Brn Li	brown lime
brn or br	brown
brn sh	brown shale
Brom	bromide
brtl	brittle
bry	bryozoa
BS	ball sealers
BS	basic sediment
BS	Bone Spring
BS	bottom sediment
BS	bottom settlings
BS&W	bottom (basic) sediment & water
Bscf	billion standard cubic feet
Bscf/d	billion standard cubic feet per day
BSE	bevel small end
BSFC	brake specific fuel consumption
BSHG	bushing
BSI	British Standards Institution
bskt	basket
bsl	basal
bsmt	basement
BSPL	base plate
BSTR	booster
BSUW	black sulfur water
BSW	barrels salt water
BSWPD	barrels salt water per day
BSWPH	barrels salt water per hour
BT	Benoist (Bethel) sand
BTC	buttress thread coupling
BTDC	before top dead center
BTFL/V	butterfly valve
btm (d)	bottom (ed)

btm chk	bottom choke
btry	battery
BTU	British thermal unit
btw	between
BTWLD	butt weld
BTX	benzene toluenexylene (unit)
bu	bushel
Buck	Buckner
Buckr	buckrange
Bul. text.	Buliminella textularia
Bull W	Bullwaggon
Bum.	bottom-hole pressure bomb
bunr	burner
Burg	Burgess
butt	buttress thread
BUZ	buzzer
BV	block valve
BV/WLD	beveled for welding
BW	barrels of water
BW	boiled water
BW	butt weld
BW ga	Birmingham (or Stubbs) iron wire gauge
BW/D	barrels of water per day
Bwg	Birmingham wire gauge
BWL	barrel water load
BWL	body wall loss
BWOL	barrels water over load
BWPD	barrels water per day
BWPH	barrels water per hour
bx	box (es)
BYP	bypass

C	Celsius
C	center (land description)
C	centigrade
c	coarse (ly)
C	core hole
C & F	cost and freight
C & P	cellar & pits
C & W	coat and wrap (pipe)
C to C	center to center

C to E	center to end
C to F	center to face
C.I.F.	cost insurance and freight
C.O.P.	completed on pump
C&A	compression and absorption plant
C&C	circulate & condition
C&CH	circulated and conditioned hole
C&CM	circulated and conditioned mud
C&R	circulate and reciprocate
C/	contractor (i.e., C/John Doe)
C/A	commission agent
C/BM	crawl beam
C/H	cased hole
C/L	center line
c/o	care of
C/W	complete with
CA	corrosion allowance
CAB	cabinet
CaCl₂	calcium chloride
Cadd	Caddell
CAG	cut across grain
cal	calcite, calcitic
cal	caliche
CAL	caliper log
cal	caliper survey
cal	calorie
Calc	calcareous, calcerenite
Calc	calcium
Calc	calculate (ed), calculation
calc gr	calcium-base grease
calc	calceneous
CALIBR	calibrate, calibration
Calv	Calvin
Camb	Cambrian
Cycl canc.	Cyclamina cancellata
Cane Riv	Cane River
Cany	canyon
CaO	calcium oxcide
CAOF	calculated absolute open flow
cap	capacitor
cap	capacity
Cap	Capitan
Car	Carlile
carb	carbonaceous
carb tet	carbontetrachloride
Carm	Carmel

Casp	Casper
CAT	carburetor air temperature
Cat	Catahoula
CAT	catalog
CAT	catalyst, catalytic
cat ckr	catalytic cracker
Cat Crk	Cat Creek
cath	cathodic
caus	caustic
cav	cavity
CB	changed (ing) bits
CB	continuous blowdown
CB	core barrel
CB	counterbalance (pumping equip.)
CBC	cement dump bailer
CBL	cable (ing)
CBU	circulate bottoms up
CC	calcium chloride
CC	carbon copy
CC	casing cemented (depth)
CC	closed cup
cc	cubic centimeter
C-Cal	contact caliper
ccBU	circulate bottoms up
CCHF	center of casinghead flange
Cck	casing choke
CCL	casing collar locator
CCLGO	cat-cracked light gas oil
CCM	condensate-cut mud
CCP	central compressor plant
CCP	critical compression pressure
CCPR	casing collar perforating record
CCR	Conradson carbon residue
CCR	critical compression ratio
CCS	California Coordinate System
CCS	cast carbon steel
CCS	computer control system
CCU	catalytic cracking unit
ccw	counterclockwise
CD	calendar day
CD	cold drawn
CD	contract depth
CD PL	cadmium plate
CDB	cement dump bailer
CDB	common data base
CDBTF	common data base task force
CDL	cut drilling line

535

CDM	continuous dipmeter survey	CH	casinghead (gas)
CDO	certified drawing outline	ch	chert
CDP	Central Delivery Point	ch	choke
Cdr Mtn	Cedar Mountain	CH	closed hole
CDS	continuous directional service	CH	core hole
cdsr	condenser	CH OP	chain operated
Cdy	Cody (Wyoming)	chal	chalcedony
cell	cellar	CHAM	chamfer
cell	cellular	Chapp	Chappel
cem	cement (ed)	CHAR	characteristics
CEMF	counter electromotive force	Char	Charles
Ceno	Cenozoic	Chatt	Chattanooga shale
cent	centralizers	CHD	closed hydrocarbon drain
centr	centrifugal	chem	chemical, chemist, chemistry
ceph	cephalopod		
Cert. ex.	Ceratobulimina eximia	chem prod	chemical products
CERT	certified	Cher	Cherokee
CET	cement evaluation	Ches	Chester
CF	casing flange	CHF	casinghead flange
CF	clay filled	CHG	casinghead gas
Cf	Cockfield	chng	change (ed) (ing)
CF	cold finished	Chngd DP	changed drillpipe
cf	cubic foot (feet)	chrg	charge (ed) (ing)
CFB & G	companion flange bolt and gasket	Chim H	Chimney Hill
		Chim R	Chimney Rock
CFBO	companion flanges bolted on	Chin	Chinle
cfd	cubic feet per day	chit	chitin (ous)
CFE	contractor furnished equipment	chk	chalk
cfg	cubic feet gas	chk	choke
cfgd	cubic feet gas per day	Chkbd	checkerboard
cfgh	cubic feet gas per hour	chkd	checked
CFM	continuous flowmeter	CHKD PL	checkered plate
cfm	cubic feet per minute	CHKV	check valve
CFOE	companion flange one end	chky	chalky
cfp	cubic feet per pound	chl	chloride (s)
CFR	cement friction reducer	chl	chloritic
CFR	cement friction retarder	chl log	chlorine log
CFRC	Coordinating Fuel Research Committee	CHLR	chlorinator
		CHMBR	chamber
cfs	cubic feet per second	CHNL	channel
CG	center of gravity	Chou	Chouteau lime
cg	centigram	CHP	casinghead pressure
cg	coarse grained	chrm	chairman
cg	coring	chromat	chromatograph
CG	corrected gravity	chrome	chromium
cglt	conglomerate, conglomeritic	cht	chart
C-gr	coarse grained	cht	chert
cgs	centimeter-gram-second system	chty	cherty
		Chug	chugwater

536

CI	cast iron	Cl₂	chlorine
CI	contour interval (map)	cm	centimeter
CI engine	compression-ignition engine	cm/sec	centimeters per second
Cib.	Cibicides	CMA	acoustic caliper
Cib. h.	Cibicides hazzardi	CMC	sodium carboxymethylcellu-
CIBP	cast-iron bridge plug		lose
CIE	crude industrial ethanol	Cmchn	comanchean
Cima	Cimarron	CMPARTR	comparator
CIP	cement in place	CMPD	compound
CIP	closed-in pressure	Cmpt	compact
cir	circle	cmt(d)(g)(r)	cement (ed) (ing) (er)
cir	circuit	CN	cetane number
cir	circular	CN/BD	control building
cir mils	circular mils	cncn	concentric
circ	circulate, circulating, circulation	CND	conduit
Cis	Cisco	CNL	compensated neutron log
ck	cake	CNR	corner
ck	check	cntf	centrifuge
CK Mtn	Cook Mountain	cntl	control (s)
cksn	chicksan	CNTN	containment
CL	carload	cntr	center (ed)
cl	centiliter	cntr	container
CL	class	cntr	controller
Clag	Clagget	CNTWT	counter weight
Claib	Claiborne	Cnty	county
Clarks	Clarksville	cnvr	conveyor
clas	clastic	CO	carbon monoxide
CLASS	classification	CO	carbon oxygen
Clav	Clavalinoides	CO	circulated out
Clay	Clayton	CO	clean out
Clay	Claytonville	CO	cleaning out, cleaned out
Cleve	Cleveland	Co	company
Clfk	Clearfork	CO	crude oil
CLFR	clarifier	CO & S	clean out & shoot
CLG	cooling	Co. Op.	company operated
CLG/TWR	cooling tower	Co. Op. S.S.	company-operated service
Cliff H	Cliff House		stations
CLKG	caulking	co-op	cooperative
CLMP	canvas-lined metal petal bas- ket	COBOL	Common Business-Oriented Language
cln (d) (g)	clean (ed) (ing)	COC	Cleveland open cup
Clov	cloverly	Coco	Coconino
clr	clear, clearance	COD	chemical oxygen demand
CLR	cooler	Cod	Codell
CLR/TWR	cooling tower	coef	coefficient
clrg	clearing	COF	calculated open flow (potential)
clsd	closed	COG	coke oven gas
CLTR	collector	COH	coming out of hole
Clyst	Claystone	COL	collar

COL	colored
COL	column
Col ASTM	Color American Standard Test Method
Cole Jct	Coleman Junction
coll	collect (ed) (ing) (ion)
colr	collar
Com	Comanche
Com	Comatula
com	common
Com Pk	Comanche Peak
comb	combined, combination
COMB	combustion
coml	commercial
comm	commenced
comm	commission
comm	communication
comm	community
commr	commissioner
comp	complete (ed) (tion)
comp nat	completed natural
compnts	components
compr	compressor
compr sta	compressor station
compt	compartment
COMPT	component (s)
COMPTR	computer
COMT	comment
COMUT	commutator
con	consolidated
conc	concentrate
conc	concentric
conc	concrete
conch	conchoidal
concl	conclusion
cond	condensate
cond	condition (ed) (ing)
condr	conductor (pipe)
condt	conductivity
conf	confidential
conf	confirm (ed) (ing)
confl	conflict
cong	conglomerate (itic)
conn	connection
cono	conodonts
consol	consolidated
const	constant
const	construction
consv	conserve, conservation

cont (d)	continue (ed)
contam	contaminated, contamination
contr	contractor
contr resp	contractor responsibility
contrib	contribution
conv	converse
CONVT	convector, convection
COOH	coming out of hole
coord	coordinate
COP	crude oil purchasing
coq	coquina
cor	corner
Corp	corporation
corr	correct (ed) (ion)
corr	corrosion
corr	corrugated
correl	correlation
corres	correspondence
COSH	hyperbolic cosine
COTD	cleaned out to total depth
COTH	hyperbolic cotangent
Cott G	Cottage Grove
Counc G	Council Grove
CO_2	carbon dioxide
CP	casing point
CP	casing pressure
cp	centipoise (s)
cp	chemically pure
CP	correlation point
Cp Colo	Camp Colorado
CPA	certified public accountant
CPC	casing pressure closed
cp'd	cemented through perforations
CPF	casing pressure flowing
CPG	cost per gallon
CPG	cents per gallon
cplg	coupling
CPM	cycles per minute
CPO	confirming telephone order (purchasing term)
CPR	Copper River Meridian (Alaska)
CPS	cycles per second
CPSI	casing pressure shut in
C Riv	Cane River
CR	cold rolled
CR	compression ratio
CR	Cow Run
CR Con	carbon residue (Conradson)

cr moly	chrome molybdenum	ctd	coated
cr (d)(g)	core (ed) (ing)	CTD	corrected total depth
CRA	cased reservoir analysis	ctg(s)	cuttings
CRA	chemically retarded acid	CTHF	center of tubing flange
crbd	crossbedded	Ctlmn	Cattleman
CRC	Coordinating Research Council Inc.	CTM	cable tool measurement
		ctn	carton
CRCMF	circumference	Ctnwd	Cottonwood
crd	cored	CTP	cleaning to pits
CRDL	cradle (s)	ctr	center
cren	crenulated	CTS	cement to surface
Cret	Cretaceous	CTT	consumer transport truck
crg	coring	CTU	coiled tubing unit
Crin	crinoid (al)	CTW	consumer tank wagon
Cris	Cristellaria	CU	clean up
crit	critical	cu	cubic
crk	creek	cu cm	cubic centimeter
crkg	cracking	cu ft	cubic foot
Crkr	cracker	cu ft/bbl	cubic feet per barrel
CRN	crane	cu ft/min	cubic feet per minute
crn blk	crown block	cu ft/sec	cubic feet per second
crnk	crinkled	cu in.	cubic inch
Crom	Cromwell	cu m	cubic meter
crs	coarse (ly)	cu yd	cubic yard
CRS	cold-rolled steel	CUB	cubical
CRS	cross	culv	culvert
CRS	retainer	cum	cumulative
crs-xln	coarse crystalline	Cur	Curtis
CRT	cathode ray tube	cush	cushion
CRV	curve	CUST	customer
crypto-xln	cryptocrystalline	Cut B	cut bank
cryst	crystalline	Cut Oil	cutting oil
CS	carbon steel	Cut Oil Act Sul dk	cutting oil active-sulfurized-dark
CS	casing seat		
CS	cast steel	Cut Oil Act Sul trpt	cutting oil active-sulfurized-transparent
cs	centistokes		
CSA	casing set at	Cut Oil Inact S	cutting oil inactive-sulfurized
CSCH	hyperbolic constant	Cut Oil Sol	cutting oil soluble
cse gr	coarse grained	Cut Oil St Mrl	cutting oil straight mineral
csg	casing	cutbk	cutbank
csg hd	casinghead	Cutl	Cutler
csg press	casing pressure	CV	control valve
csg pt	casing point	CV	Cotton Valley
CSK	countersink	cvg(s)	cavings (s)
CSL	center section line	CVO	confirming telephone order (purchasing term)
CSL	county school lands		
CS_2	carbon disulfide	CVR	cover
CT	cable tools	CVTR	convert (er) (ed)
CT	cooling tower	cw	clockwise
CTC	consumer tank car	CW	continuous weld

CW	cooling water
CWE	cold water equivalent
CWP	cold working pressure
CWR	cooling water return
CWS	cooling water supply
cwt	hundredweight
CX	crossover
Cy Sd	Cypress Sand
Cyc	cyclamina
CYC	cyclone
Cycl	cyclone
cyl	cylinder
cyn	canyon
Cyp	Cypridopsis
Cz	Carrizo

D

D	day
D	development
D	dual
D & A	dry and abandoned
D & B	Dun & Bradstreet
D & C	drill and complete
D & D	Desk and Derrick
D.I.	diesel index
D.O.	division order
d-d-1-s-1-e	dressed dimension one side and one edge
d-d-4-s	dressed dimension four sides
d-1-s	dressed one side
D-2	Diesel No. 2
d-2-s	dressed two sides
d-4-s	dressed four sides
d/b/a	doing business as
D/D	day to day
D/L	density log
D/O	division office
D/P	differential pressure
D/P	drill (ed) (ing) plug
D/S	data sheet
D/T	driller's top
D/T	drilling tender
DA	daily allowable
DA	Dresser Atlas

DA	drift angle
DAIB	daily average injection barrels
Dak	Dakota
Dan	Dantzler
Dar	Darwin
DAR	discovery allowable requested darcy (darcies not abbreviated)
dat	datum
db	decibel
DB	diamond bit
DB	drilling break
DBA	depth bracket allowable
DBL	double
DBO	dark brown oil
DBOS	dark brown oil stains
DC	delayed coker
DC	development well, carbon dioxide
DC	diamond core
DC	digging cellar or digging cellar and slush pits
DC	direct current
DC	drill collar
DC	dually completed
DCB	diamond core bit
DCLSP	digging slush pits
DCM	distillate-cut mud
DCS	distributed control system
DCTR	detector
dd	dead
DD	degree day
DD	deviation degrees
DD	drilling deeper
DD	dyna-drilling
DDD	dry desiccant dehydrator
DDT	dichloro diphenyl-trichloroethane
DE	double end
DEA	diethanolamine
DEA unit	diethanolamine unit
Deadw	Deadwood
deaer	deaerator
deasph	deasphalting
debutzr	debutanizer
dec	decimal
decl	decline
decr	decrease (ed) (ing)

deethzr	deethanizer
defl	deflection
Deg	Degonia
deg	degree (s)
deisobut	deisobutanizer
Del R	Del Rio
Dela	Delaware
delv	delivery (ed) (ability)
delv pt	delivery point
demur	demurrage
dend	dendrite (ic)
DENL	density log
depl	depletion
deprec	depreciation
deprop	depropanizer
dept	department
Des Crk	Desert Creek
Des M	Des Moines
desalt	desalter
desc	description
desorb	desorbent
desulf	desulfurizer
det	detail (s)
det	detector
deterg	detergent
detr	detrital
dev	deviate, deviation
Dev	Devonian
devel	develop (ed) (ment)
dewax	dewaxing
Dext	Dexter
DF	derrick floor
DF	diesel fuel
DF	drill floor
DFE	derrick floor elevation
DFO	datum faulted out
DFP	date of first production
DFT	dry film thickness
dg	decigram
DG	development gas well
DG	draft gauge
DG	dry gas
DGA	diglycolamine
DGTL	digital
DH	development well, helium
DH	double hub
DHC	dry-hole contribution
DHDD	dry hole drilled deeper
DHDS	diesel hydrogen desulfurization

DHM	dry-hole money
DHR	dry hole reentered
dia	diameter
diag	diagonal
diag	diagram
diaph	diaphragm
dichlor	dichloride
diethy	diethylene
diff	different (ial) (ence)
DIFL	dual injection focus log
dilut	diluted
dim	dimension
dim	diminish (ing)
Din	Dinwoody
dir	direct (tion) (tor)
dir drlg	directional drilling
dir sur	directional survey
disc	discharge
Disc	Discorbis
disc	discount
disc	discover (y) (ed) (ing)
Disc. grav.	Discorbis gravelli
Disc. norm.	Discorbis normada
Disc. y.	Discorbis yeguaensis
disch	discharge
dism	disseminated
disman	dismantle
displ	displaced, displacement
dist	distance
dist	distillate, distillation
dist	district
distr	distribute (ed) (ing) (ion)
div	division
dk	dark
Dk Crk	Duck Creek
dl	deciliter
DL	drilling line
DLC	dual lower casing
dlr	dealer
DLS	dogleg severity
DLT	dual lower tubing
DM	datum
dm	decimeter
DM	demand meter
DM	dipmeter
DM	drilling mud
dml	demolition
DMPD	dumped

dmpr	damper
DMS	dimethyl sulfide
dm	cubic decimeter
dm/s	cubic decimeter per second
dn	down
dns	dense
DO	development oil
DO	development oil well
do	ditto
DO	drill (ed) (ing) out
DOBLDG	dock operating building
DOC	diesel oil cement
Doc	Dockum
doc	document
DOC	drilled-out cement
doc-tr	doctor-treating
DOCREQ	documentation
DOD	drilled-out depth
DOE	Department of Energy
dolo	dolomite (ic)
dolst	dolstone
dom	domestic
dom AL	domestic airline
DOM WTR	domestic water
DOP	drilled-out plug
Dorn H	Dornick Hills
Doth	Dothan
Doug	Douglas
doz	dozen
DP	data processing
DP	dewpoint
DP	double pipe
DP	drillpipe
DP SW	double pole switch
DPDB	double pole double base (switch)
DPDT SW	double pole double throw switch
dpg	deepening
DPM	drillpipe measurement
dpn	deepen
DBSB	double pole single base (switch)
DPST SW	double pole single throw switch
DPT	deep pool test
dpt	depth
dpt rec	depth recorder
DPU	drillpipe unloaded
DR	development redrill (sidetrack)

dr	drain
dr	drive
dr	drum
dr	druse
Dr Crk	Dry Creek
DRAPR	Delaware River Area Petroleum Refineries
drk	derrick
DRL	double random lengths
drl	drill
drld	drilled
drlg	drilling
drlr	driller
DRM	drum
drng	drainage
dropd	dropped
drsy	drusy
DRV (R)	drive (ing) (er)
dry	drier, drying
ds	dense
DS	directional survey
DS	drillsite
DS	drillstem
DSF	drillsite facility
dsgn	design
DSI	drilling suspended indefinitely
dsl	diesel (oil)
dsmt (g)	dismantle (ing)
DSO	dead oil show
DSS	days since spudded
DST	drillstem test
DST (Strd)	drillstem test with straddle packers
dstl	distillate
dstn	destination
DSU	development well, sulfur
DSUPHTR	desuperheater
DT	downthrown
DT	drilling time
DTD	driller's total depth
dtr	detrital
DTW	dealer tank wagon
DUC	dual upper casing
Dup	Duperow
dup	duplicate
DUR	duration
DUT	dual upper tubing
Dutch	Dutcher
DV	differential valve (cementing)

DVL	develop
DVT	davit
DWA	drilling with air
DWC	drilling and well completion
DWD	dirty water disposal
DWG	drawing
DWG	drilling with gas
dwks	drawworks
DWM	drilling with mud
DWO	drilling with oil
DWP	dual (double) wall packer
DWSW	drilling with salt water
DWT	deadweight tester
DWT	deadweight tons
DWTR	dewatering
DX	development well workover
dx	duplex
dyn	dynamic

E	east
E	exploratory
E of W/L	east of west line
E.D.	effective depth
e.g.	for example
E.T.D.	estimated total depth
E/BL	east boundary line
E/E	end to end
E/L	east line
E/O	east offset
E/2	east half
E/4	east quarter
ea	each
EA	environmental assessment
EAM	electric accounting machines
Earls	Earlsboro
Eau Clr	Eau Claire
ECC	eccentric
Ech	echinoid
ECM	East Cimarron Meridian (Oklahoma)
Econ	economics, economy, economizer
Ect	Ector (County, TX)

ecx	excavation
Ed lm	Edwards lime
EDC	ethylene dichloride
EDCTN(R)	education (tor)
EDD	expected date of delivery
EDP	electronic data processing
Educ	education
Edw	Edwards
EF	Eagle Ford
eff	effective
eff	efficiency
effl	effluent
EFV	equilibrium flash vaporization
Egl	Eagle
Eglwd	Englewood
EHP	effective horsepwer
EIA	environmental assessment
EIR	environmental impact report
EIS	environmental impact statement
EJ	perforating, enerjet
eject	ejector
EL	elevation (height)
el gr	elevation ground
EL/T	electric log tops
Elb	elbert
ELB	elbow
elec	electric (al)
Elec/MAG	electromagnetic
elem	element, elementary
elev	elevation, elevator
Elg	Elgin
ELIM	eliminate (tor) (ed)
ell(s)	elbow (s)
Ellen	Ellenburger
Elm	Elmont
E'ly	easterly
EM	Eagle Mills
Emb	Embar
emer	emergency
EMF	electromotive force
EMN	electromagnetic
EMNI	electromagnetic induction
EMP	European melting point
empl	employee
EMS	Ellis-Madison contact
emul	emulsion
encl	enclosure
End	Endicott
endo	endothyra

eng	engine
engr (g)	engineer (ing)
enl	enlarged
enml	enamel
Ent	Entrada
ENT	entrance
ent	entry
ENV	envelope
ENVIR	environment
EO	emergency order
Eoc	Eocene
EOF	end of file
EOL	end of line
EOM	end of month
EOQ	end of quarter
EOR	east of Rockies
EOR	enhanced oil recovery
EOY	end of year
EP	end point
EP	extreme pressure
Epon.	Eponides
Epon. y.	Eponides yeguaensis
eq	equal, equalizer
Eq	equation (before a number)
equip	equipment
equiv	equivalent
ERC/MK	erection mark
erect	erection
Eric	Ericson
ERW	electric resistance weld
ESD	emergency shutdown
est	estate
est	estimate (ed) (ing)
et al.	and others
et con.	and husband
et seq.	and the following
et ux.	and wife
et vir.	and husband
ETA	estimated time of arrival
eth	ethane
ethyle	ethylene
EU	Eutaw
EUE	external upset end
euhed	euhedral
EUR	estimated ultimate recovery
ev	electron-volts
ev-sort	even sorted
eval	evaluate
evap	evaporation, evaporate
EW	electric weld

EW	exploratory well
EWT	early well tie-ins
EX	example
ex	except
Ex	Exeter
EX-PRF	explosion proof
EXAM	examination
EXC	excitation
exch	exchanger
excl	excellent
EXEC	executive
exh	exhaust
exh	exhibit
exist	existing
EXP	expansion
exp	expense
EXP JT	expansion joint
exp plg	expendable plug
expir	expire (ed) (ing) (ation)
expl	exploratory, exploration
explos	explosive
exr	executor
Exrx	executrix
exst	existing
ext	external
Ext M/H	extension manhole
ext(n)	extended, extension
extr	exterior
extrac	extraction
EYC	estimated yearly consumption

F & D	faced and drilled
F & D	flanged and dished (heads)
F & F	fuels & fractionation
F & L	fuels & lubricants
F & S	flanged and spigot
F to F	face to face
F.G.	fracture gratient
F.O.E.	fuel oil equivalent
F-D	formation density
F-DIA	flow diagram
f-gr	fine grained
F-MET	flowmeter
F-R oil	fire-resistant oil

F-SHT	flow sheet
F/	flowed, flowing
F/DIA	flow diagram
F/FAB	field fabricated
F/GOR	formation gas-oil ratio
F/O opt	farmout option
F/S	flange & screwed
F/S	front & side
F/SW	flow switch
F/WTR	fire water
f/xln	finely crystalline
fab	fabricate (ed) (tion)
FAB	faint air blow
fac	facet (ed)
FACIL	facility (ies)
FACO	field authorized to commence operations
fail	failure
Fall Riv	Fall River
FAO	finish all over
Farm	Farmington
FARO	flowed (ing) at a rate of
fau	fauna
FB	fresh break
FBH	flowing by heads
FBHP	flowing bottom-hole pressure
FBHPF	final bottom-hole pressure, flowing
FBHPSI	final bottom-hole pressure, shut-in
FBP	final boiling point
fbrgs	fiberglass
FC	filter cake
FC	fixed carbon
FC	float collar
FCC	fluid catalytic cracking
FCL	facility capacity limits
FCO	functional check out
FCP	flowing casing pressure
FCV	flow control valve
FD	feed
FD	floor drain
FD	flow diagram
FD EFF	feed effluent
FD/WTR	feed water
FDC	formation density correlated
FDL	formation density log
fdn	foundation
fdr	feeder

Fe	iron
Fe-st	ironstone
FE/L	from east line
fed	federal
FEIS	Final Environmental Impact Statement
FEL	from east line
FELA	Federal Employers Liability Act
FEM	female
Ferg	Ferguson
ferr	ferruginous
fert	fertilizer
$Fe_2(SO_4)_3$	ferric sulfate
FF	fishing for
FF	flat face
FF	frac finder (log)
FF	full of fluid
FFA	female to female angle
FFA	full freight allowed (purchasing term)
FFG	female to female globe (valve)
FFGU	field fuel gas unit
FFL	final fluid level
FFO	furnace fuel oil
FFP	final flowing pressure
FG	fuel gas
FGIH	finish going in hole
FGIW	finish going in with
FGVV	flanged gate valve
FH	full hole
FHP	final hydrostatic pressure
FI	flow indicator
fib	fibrous
FIC	flow-indicating controller
fig	figure
FIH	finished in hole
FIH	fluid in hole
filt	filtrate
fin	final
fin	finish (ed)
fin drlg	finished drilling
FIN GR	finish grade
FIRC	flow indicating ratio controller
fis	fissure
fish	fishing
fisl	fissile
FIT	formation interval tester

fix	fixture
FJ	flush joint
fkt	fault
FL	flashing
FL	floor
FL	flow line
fl	fluid
FL	fluid level
FL	flush
Fl-COC	flash point, Cleveland Open Cup
fl/	flowed (ing)
FL/BD	flammable liquid building
FLA	Ferry Lake anhydrite
flat	flattened
Flath	Flathead
fld	failed
fld	feldspar (thic)
fld	field
FLD	full-length drift
flex	flexible
flg	flowing
flg (d) (s)	flange (ed) (es)
Flip	Flippen
flk	flaky
FLMB	flammable
flo	flow
FLO	flushing oil
Flor fl	Florence flint
flshd	flushed
flt	float
fltg	floating
fltn	flotation
FLTR	filter
flu	flue
flu	fluid
fluor	fluorescence, fluorescent
flw (d) (g)	flow (ed) (ing)
Flwg Pr.	flowing pressure
Flwrpt	flowerpot
FLXBX	flexibox
fm	formation
FM	frequency meter
FM	frequency modulation
Fm W	formation water
f'man	foreman
fmp	feet per minute
fn	fine
FNEL	from northeast line
FNL	from north line
fnly	finely
FNSH	finish
fnt	faint
FNWL	from northwest line
FO	farmout
FO	faulted out
FO	final open
FO	fuel oil
FO	full opening
FOB	free on board
FOCL	focused log
FOE-WOE	flanged one end, welded one end
FOH	full open head
fol	foliated
FONSI	finding of no significant impact
FOR	fuel oil return
Forak	Foraker
foram	foraminifera
Fort	Fortura
FOS	face of stud
FOS	fuel oil supply
foss	fossiliferous
FOT	flowing on test
Fount	Fountain
Fox H	Fox Hills
FP	final pressure
FP	flowing pressure
FP	freezing point
FPI	free-point indicator
FPO	field purchase order
fprf	fireproof
fps	feet per second
f-p-s	foot-pound-second (system)
FPT	female pipe thread
FPTFD	field pressure test flow diagram
FQG	frosted quartz grains
fr	fair
FR	feed rate
FR	flow rate
FR	flow recorder
fr	fractional
fr	from
fr	front
fr	frosted
fr E/L	from east line

fr N/L	from north line	ft-lb	foot-pound
fr S/L	from south line	ft-lb/hr	foot-pound per hour
fr W/L	from west line	Ft R	Fort Riley
FRA	friction reducing agent	Ft U	Fort Union
frac (d) (s)	fracture, fractured, fractures	Ft W	Fort Worth
fract	fractionation, fractionator, fractional	ft-c	foot-candle
		ft/hr	feet per hour
frag	fragment	ft/min	feet per minute
fran	franchise	ft/sec	feet per second
Franc	Franconia	ftg	fittings
FRC	flow recorder control	ftg	footing, footage
Fred	Fredericksburg	FTP	final (flowing) tubing pressure
Fred	Fredonia	FTS	fluid to surface
freq	frequency	ft²	square feet
FRG	forge (ed) (ing)	ft³	cubic feet
Frgy	froggy	FU	fill up
fri	friable	Full	Fullerton
FRM (G)	frame, framing	furf	furfural
Fron	frontier	furn	furnance
fros	frosted	FURN	furnish (ed)
FRP	fiberglass-reinforced plastic	Furn & fix	furniture and fixtures
FRR	field receiving report	Fus	Fuson
FRR	final report for rig	Fussel	Fusselman
frs	fresh	Fusul	Fusulinid
frt	freight	fut	future
Fruit	Fruitland	FV	funnel viscosity
FRW	final report for well	fvst	favosites
frwk	framework	FW	fillet weld
frzr	freezer	FW	fresh water
FS	feedstock	FWC	field wildcat
FS	float shoe	fwd	forward
FS	flow station	FWD	four-wheel drive
FS	forged steel	FWL	from west line
FS&WLs	from south and west lines	fwtr	fresh water
FSEL	from southeast line	fxd	fixed
fsg	fishing	FYE	fiscal year ending
FSIP	final shutin pressure	FYI	for your information
FSIWA	Federation of Sewage and Industrial Wastes Association		
FSL	from south line		
FSP	flowing surface pressure		**G**
FST	forged steel		
FSTN	fasten (ing) (er)		
FSWL	from southwest line	G	gas
ft	feet, foot	g	gram
FT	formation test	G egg	goose egg
Ft C	Fort Chadborne	g mole	gram molecular weight
Ft H	Fort Hayes	G Rk	gas rock

G.M.	gravity meter	**GE**	grooved ends
G. Riv	Gull River	**gel**	gelled
g-cal	gram calorie	**gel**	jelly-like colloidal suspension
G-N₂	gaseous nitrogen	**gen**	generation, generator
G-O₂	gaseous oxygen	**genl**	general
G&MCO	gas & mud-cut oil	**Geo**	Georgetown
G&O	gas and oil	**geo**	geothermal
G&OCM	gas- and oil-cut mud	**geol**	geology (ist) (ical)
G/L	gathering line	**geop**	geophysics (ical)
G/O	gas-oil ratio	**GFLU**	good fluorescence
G/P	gun perforate	**GFR**	gas-fluid ratio
ga	gauge (ed) (ing)	**gg**	grains per gallon
GA	gallons acid	**GGD**	gas lift gas distribution
GA	general arrangement	**gge**	gauge
GAF	gross acre-feet	**GGW**	gallons gelled water
gal	gallon (s)	**GH**	Greenhorn
gal sol	gallons of solution	**GHO**	gallons heavy oil
gal/Mcf	gallons per thousand cubic feet	**GHSG**	gas-handling study group
		GI	gas injection
gal/min	gallons per minute	**Gib**	Gibson
Gall	Gallatin	**GIH**	going in hole
galv	galvanized	**gil**	gilsonite
gaso	gasoline	**Gilc**	Gilcrease
gast	gastropod	**GIP**	gas in pipe
GB	gun barrel	**GIW**	gas injection well
GBDA	gallons breakdown acid	**GL**	gas lift
GC	gas-cut	**GL**	ground level
GCAW	gas-cut acid water	**glau**	glauconite, glauconitic
GCD	gas-cut distillate	**GLBVV**	globe valve
GCLO	gas-cut load oil	**gld thd**	galled threads
GCLW	gas-cut load water	**Glen**	Glenwood
GCM	gas-cut mud	**Glna**	Galena
GCO	gas-cut oil	**GLO**	gas lift oil
GCPA	gas cap participating area	**GLO**	General Land Office (Texas)
GCPD	gallons condensate per day	**Glob**	Globigerina
GCPH	gallons condensate per hour	**Glor**	Glorieta
GCR	gas-condensate ratio	**GLR**	gas-liquid ratio
GCSW	gas-cut salt water	**gls (y)**	glass, glassy
GCT	guidance continuance tool	**GLT**	gas lift transfer
GCW	gas-cut water	**glyc**	glycol
GD li	Glen Dean lime	**GMC**	General Motors Corporation
gd	good	**gm**	gram
gd o&t	good odor & taste	**GM**	ground measurement (elevation)
GDE	geothermal development, failure	**gm-cal**	gram-calorie
Gdld	Goodland	**GMA**	gallons mud acid
GDR	gas-distillate ratio	**gmy**	gummy
GDS	geothermal development, success	**gnd**	grained (as in fine-grained)
Gdwn	Goodwin		
GE	General Electric Company	**gns**	gneiss

GO	gallons oil	Grayb	Grayburg
GO	gas odor	grdg	grading
GO	grind out	grdg loc	grading location
GOC	gas-oil contact	GRDL	guard log
GODT	gas odor distillate taste	GRG	gas reserve group
Gol	Golconda lime	grn	green
Good L	Goodland	Grn Riv	Green River
GOPD	gallons of oil per day	grn sh	green shale
GOPH	gallons of oil per hour	grnd	ground
GOR	gas-oil ratio	grnlr	granular
Gor	Gorham	GRP	group
Gouldb	Gouldbusk	GRS	gas to surface
gov	governor	grs	gross
govt	government	grt	grant (of land)
GP	gas pay	grtg	grating
GP	gasoline plant	grty	gritty
GPC	gas purchase contract	grv	grooved
GPD	gallons per day	grvt	gravitometer
GPF	granite point field	GRVTY	gravity
GPG	grains per gallon	gry	gray
GPH	gallons per hour	GS	gas show
GPM	gallons per minute	GS	guide shoe
GPM	geophysical investigation map	GSC	gas sales contract
GPS	gallons per second	GSG	good show of gas
GPU	Gas Production Unit	GSI	gas well shutin
GQM	geological quadrangle map	gskt	gasket
GR	gamma ray	GSO	good show oil
GR	gauge ring	GSO&G	good show oil and gas
GR	Glen Rose	GST	gamma spectroscopy tool
gr	grade	GSW	gallons salt water
gr	grain	gsy	greasy
gr	gravity	GT	geothermal
gr	grease	GTS	gas to surface (time)
gr	ground	GTSTM	gas too small to measure
gr API	gravity° API	GTY	gravity
gr roy	green royalty	GU	gas unit
Gr Sd	Gray sand	Guns	Gunsite
gr wt	gross weight	GUS	gusset
GR&DC	Gulf Research and Development Company	GV	gas volume
		gvl	gravel
GRA	gallons regular acid	GVLPK	gravel packet
GRAD	gradiomanometer	GVNM	gas volume not measured
grad	gradual, gradually	GW	gallons water
gran	granite, granular	GW	gas well
gran w	granite wash	GW	geothermal wildcat, failure
Granos	Graneros	GWC	gas-water contact
grap	graptolite	GWD	geothermal wildcat, success
grav	gravity	GWG	gas-well gas
Gray	Grayson	GWPH	gallons of water per hour

gyp	gypsum
gypy	gypsiferous
Gyr.	Gyroidina
Gyr. sc.	Gyroidina scal
gywk	graywacke

H

H & V	heating and ventilating
H H P	hydraulic horsepower
H. O.	hole opener
H-SEL	perforating hyper select
H-VOLT	high voltage
H/C	hydrocracker
Hackb	Hackberry
Hara	Haragan
Hask	Haskell
Haynes	Haynesville
haz	hazardous
HB	house brand (regular grade of gasoline)
Hburg	Hardinsburg sand (local)
HBP	held by production
hbr	harbor
HC	hydrocarbon
HC	hydrocracker
HCDS	hydrocarbon drain system
HCGO	heavy coker gas oil
HCO	heavy cycle oil
HCV	hand-control valve
hd	hard
hd	head
HD	heavy duty
HD	high detergent
HD	hot dry rock development, failure
HD	Hydril
HD	perforating, hyperdome
Hd li	hard lime
hd sd	hard sand
hdl	handle
hdns	hardness
hdr	header
HDS	hot dry rock development, successful

HDS	hydrogen delsulfurization
hdwe	hardware
Heeb	Heebner
hem	hematite
Her	Herington
Herm	Hermosa
het	heterostegina
HEX	heat exchanger
hex	hexagon (al)
hex	hexane
HEX HD	hex head
hfg	hydrofining
HFO	heavy fuel oil
HFO	hole full of oil
HFSW	hole full of salt water
HFW	hole full of water
HGCM	heavily (highly) gas-cut mud
HGCSW	heavily (highly) gas-cut salt water
HGCW	heavily (highly) gas-cut water
HGOR	high gas-oil ratio
hgr	hanger
hgt	height
HH	hand hole
HH	hydrostatic heat
HIA	Hydrologic Investigations Atlas
Hick	Hickory
Hill	Hilliard
hky	hackly
HLDN	hold down
HLSD	high-level shutdown
HND/WHL	handwheel
HNDLG	handling
HO	heating oil
HO	heavy oil
HO	hole opener
HO	home office
HO&GCM	heavily (highly) oil- and gas-cut mud
hock	hockleyensis
HOCM	heavily (highly) oil-cut mud
HOCSW	heavily (highly) oil-cut salt water
HOCW	heavily (highly) oil-cut water
Hog	Hogshooter
Holl	Hollandberg
Home Cr	Home Creek

hop	hopper
horiz	horizontal
Hosp	Hospah
HOT	hot oil tar
Hov	Hoover
Hox	Hoxbar
HP	high pressure
HP	horsepower
HP	hydraulic pump
HP	hydrostatic pressure
hp-hr	horsepower-hour
HPF	holes per foot
HPG	high-pressure gas
HPG	high-pressure gauge
HQ	headquarters
HR	heavy reformate
hr	hour (s)
HR Sul W	hole full of sulfur water
HRD	high-resolution dipmeter
hrs	heirs
HRS	hot-rolled steel
HSD	heavy steel drum
HSE	house (ed) (ing)
HST	hydrostatic test
HT	heat tracing (ed)
HT	heat-treated, heater treater
HT	high temperature
HT	high tension
HTA	heat-treated alloy
htr	heater
HTSD	high-temperature shutdown
HU	hook up
Humb	Humblei
Hump	Humphreys
Hun	Hunton
HUX	heavy hydrocrackate
HV	high viscosity
HVAC	heating ventilating and air conditioning
HVGO	heavy gas oil
HVI	high viscosity index
HVL	high volume lift
hvly	heavily
hvy	heavy
HW	hot dry rock wildcat, failure
HWCM	heavily (highly) water-cut mud
HWD	hot dry rock wildcat, success
HWP	hookwall packer
hwy	highway

HX	heat exchanger
HYD	hydraulic
HYD	Hydril thread
HYDA	Hydril Type A joint
HYDCA	Hydril Type CA joint
HYDCS	Hydril Type CS joint
HYDRO	hydro test
hydtr	hydrotreater
Hyg	hygiene
HYGN	hydrogenation
HYPO	hypotenuse
Hz	Hertz
H_2	hydrogen
H_2S	hydrogen sulfite
H_2SO_4	sulfuric acid

I

I.D. sign	identification sign
I-	miscellaneous investigations series
I-O-M	installation operation & maintenance
I/C	interconnection (ing)
I/CFD	interconnecting flow diagram
I/O	input/output
IAB	initial air blow
IB	impression block
IB	iron body (valve)
IBBC	iron body brass core (valve)
IBBM	iron body brass (bronze) mounted (valve)
IBHP	initial bottom-hole pressure
IBHPF	initial bottom-hole pressure, flowing
IBHPSI	initial bottom-hole pressure, shut in
IBP	initial boiling point
IC	iron case
icfos	microfossil (iferous)
ID	inside diameter
IDENT	identify (ier) (ication)
Idio	Idiomorpha
IES	induction electrical survey
IF	internal flush
IFL	initial fluid level

IFP	initial flowing pressure	Inj Pr	injection pressure
IG	injection gas	inl	inland
Ign	igneous	inl	inlet
IGN	ignition	inlam	interlaminated
IGOR	injection gas-oil ratio	Inoc	Inoceramus
IGV/IBV	inlet gate valve/inlet ball valve	INPE	installing (ed) pumping equipment
IH	in hole	INQ	inquire, inquiry
IHP	indicated horsepower	ins	insulate, insulation
IHP	initial hydrostatic pressure	ins	insurance
IHPHR	indicated horsepower hour	insol	insoluble
II	injection index	insp	inspect (ed) (ing) (tion)
IJ	integral joint	inst(d)(g)(l)	install (ed) (ing) (ation)
ILUM	illuminator (s)	inst	instantaneous
imbd	imbedded	inst	institute
IMF	intermediate manifolds	instr	instrument, instrumentation
immed	immediate (ly)	insul	insulate
Imp	Imperial	INT	integral
IMP	impounding	int	interest
Imp gal	Imperial gallon	int	interior
imperv	impervious	int	internal
IMW	initial mud weight	int	intersection
in.	inch (es)	INTCON	interconnection
in. Hg	inches mercury	inter-gran	intergranular
in.-lb	inch-pound	inter-lam	interlaminated
in./sec	inches per second	inter-xln	intercrystalline
Inbded	inbedded	interbd	interbedded
Inc.	incorporated	intgr	integrator
inc	increment	INTL	internal
incd	incandescent	intl	interstitial
INCIN	incinerator, incineration	intr	intrusion
incl	include (ed) (ing)	ints	intersect
INCLR	intercooler	intv	interval
incls	inclusions	inv	invert (ed)
INCM	income (er) (ing)	inv	invoice
incolr	intercooler	invrtb	invertebrate
incr	increase (ed) (ing)	IO	initial open
ind	induction	IP	initial potential
indic	indicate (s) (tion)	IP	initial pressure
indiv	individual	IP	initial production
indr	indurated	IP	intermediate pressure
indst	indistinct	IPA	initial participating area
Inf. L	inflammable liquid	IPA	isopropyl alcohol
Inf. S	inflammable solid	IPE	install (ing) pumping equipment
info	information		
ingr	intergranular	IPF	initial production flowed (ing)
inhib	inhibitor	IPG	initial production gas lift
init	initial	IPI	initial production on intermitter
inj	injection, injected		

IPL	initial production plunger lift
IPP	initial production pumping
IPR	inflow performance rate
IPS	initial production swabbing
IPS	iron pipe size
IPT	iron pipe thread
IR	infrared
IR	injection rate
Ire	Ireton
irid	iridescent
irreg	irregular
IRS	Internal Revenue Service
Irst	ironstone
IS	inside screw (valve)
ISIP	initial shutin pressure (DST)
ISIP	instantaneous shutin pressure (frac)
ISITP	initial shutin tubing pressure
ISO	isometric
ISO/CKR	isocracker
ISOL	isolate (tor)
isom	isometric
isoth	isothermal
ISS	issue
ITB	invitation to bid
ITC	investment tax credit
ITD	intention to drill
IUE	internal upset ends
Ives	Iverson
IVP	initial vapor pressure
IW	injection water
IW	injection well

J	Joule
J&A	junked and abandoned
J/O	joint operation
jac	jacket
Jack	Jackson
Jasp	jasper (oid)
Jax	Jackson sand
JB	junction box
JB	junk basket
jbr	jobber

JC	job complete
jct	junction
JCUMWE	Joint Committee on Uniformity of Methods of Water Examination
Jdn	Jordan
Jeff	Jefferson
JFA	jet fuel (aviation)
JINO	joint interest nonoperated (property)
JJ	junk joint
jmd	jammed
jnk	junk (ed)
JOA	joint operating agreement
JOP	joint operating provisions
JP	jet perforated
JP fuel	jet propulsion fuel
JP/ft	jet perforations per foot
JSPF	jet shots per foot
jt(s)	joint (s)
JTU	jet treating unit
Jud Riv	Judith River
Jur	Jurassic
juris	jurisdiction
JV	joint venture
Jxn	Jackson

K	Kelvin (temperature scale)
K	thousand (i.e., 13K = 13,000)
Kai	Kaibab
kao	kaolin
Kay	Kayenta
KB	kelly bushing
KBM	kelly bushing measurement
KC	Kansas City
kc	kilocycle
kcal	kilocalorie
KD	kiln dried
KD	Kincaid lime
KD	knock down
KDB	kelly drive bushing
KDB-LDG FLG	kelly drill bushing to landing flange

KDB-MLW	kelly drill bushing to mean low water
KDB-Plat	kelly drill bushing to platform
KDBE	kelly drive bushing elevation
Ke	Keener
Keo-Bur	Keokuk-Burlington
kero	kerosine (kerosene)
ket	ketone
kev	thousand electron-volts
Key	Keystone
kg	kilogram
kg-cal	kilogram-calorie
kg-m	kilogram-meter
KGRA	known geothermal resource area
Khk	Kinderhook
kHz	kilohertz
Kia	Kiamichi
Kib	Kibbey
Kin li	Kincaid lime
kin	kinematic
kip	one thousand pounds
kip-ft	one thousand foot-pounds
Kirt	Kirtland
KIT	kitchen
kl	kiloliter
kld	killed
km	kilometer
KMA	KMA sand
KO	kick off
KO	knock out
Koot	Kootenai
KOP	kickoff point
kPa	kilopascal
Kri	Krider
KTLE	kettle
kv	kilovolt
KV	kinematic viscosity
KV	permeability (vertical direction)
kva	kilovolt-ampere
kvah	kilovolt-ampere-hour
kvar	kilovar; reactive kilovolt-ampere
kvar-hr	kilovar-hour
kvp	kilovolt peak
KW	kill (ed) well
kw	kilowatt

kwh	kilowatt-hour
kwhm	kilowatt-hour meter

L

l	liter
L & P	ladder & platform
L U	lease use (gas)
L.P.	line pipe
L-DK	loading dock
L-VOLT	low voltage
L/	Lower, i.e., L/Gallup
L/Alb	Lower Albany
L/Cret	Lower Cretaceous
L/Tus	Lower Tuscaloosa
LA	level alarm
LA	lightning arrester
LA	load acid
La Mte	La Motte
lab	labor
lab	laboratory
LACT	lease automatic custody transfer
lad	ladder
LAG	lagging
Lak	Lakota
lam	laminated, lamination (s)
Land	Landulina
Lans	Lansing
Lar	Laramie
LAS	lower anhydrite stringer
lat	latitude
Laud	Lauders
Layt	Layton
LB	light barrel
lb	pound
lb-in.	pound-inch
lb/ft	pounds per foot
lb/sq ft	pounds per square foot
LBOS	light brown oil stain
lbr	lumber
LC	lease crude
LC	level controller
LC	long coupling
LC	lost circulation

LC	lower casing	lin ft	linear foot
LC	lug cover type (5-gallon can)	LIP	local injection plants
LCGO	light coker gas oil	liq	liquid
lchd	leached	liqftn	liquefaction
LCL	less-than-carload lot	litho	lithographic
LCL	local	LJ	lap joint
LCM	lost circulation material	lk	leak
LCP	local control panel	lk	lock
LCP	lug cover with pour spout	LKG	leakage
LCV	level control valve	LKR	locker
LD	laid down	LLC	liquid level controller
ld	load	LLG	liquid level gauge
ld(s)	land (s)	lm	lime, limestone
LDC	laid-down cost	LMF	lowermost flange
LDCX	lead drill collar	LMn	Lower Menard
LDDCs	laid (laying) down drill collars	Lmpy	lumpy
		LMTD	log mean temperature difference
LDDP	laid (laying) down drillpipe		
LDF	large-diameter flow line	lmy	limy
LDG	landing	Lmy sh	limy shale
LDG	loading	LN	line
LDR	loader	ln	logarithm (natural)
Le C	Le Comptom	LNG	liquefied natural gas
Leadv	Leadville	lngl	linguloid
LEL	lower explosive limit	lnr	liner
Len	Lennep	lns	lense
len	lenticular	LO	load oil
LFO	light fuel oil	LO	lube oil
lg	large	LOA	length overall
lg	length	loc	located, location
lg	level glass	loc abnd	location abandoned
lg	long	loc gr	location graded
Lg Disc	Large Discorbis	log	logarithm (common)
LGD	Lower Glen Dean	long	longitude (inal)
Lge	league	LOS	lease operations
LH	left hand	Lov	Lovell
LH/RP	long handle/round point	Lov	Lovington
LI	level indicator	low	lower
li	lime, limestone	LOX	liquid oxygen
LIB	light iron barrel	LP	line pressure
LIC	level indicator controller	LP	lodge pole
lic	license	lp	loop
Lieb	Liebuscella	LP	low pressure
lig	lignite, lignitic	LP sep	low-pressure separator
LIGB	light iron grease barrel	LP-Gas	liquefied petroleum gas
LIH	left in hole	LPG	liquefied petroleum gas
lim	limit, limonite	LPG	propane
lin	linear	LPO	local purchase order
lin	liner	LPS	low-pressure separation

LR	level recorder
LR	long radius
LRAP	long-range automation plan
LRC	level recorder controller
lrg	large
LRP	long-range plan
ls	limestone
LS	long string
LSD	legal subdivision (Canada)
LSD	light steel drum
lse	lease
LST/COMPTS	list of components
lstr	lustre
lt	light
LT	lower tubing
LT&C	long threads and coupling
ltd	limited
LTD	log total depth
ltg	lighting
LTL	less than truckload
ltl	little
ltr	letter
LTS Unit	low-temperature separation unit
LTSD	low-temperature shutdown
LTT	long-term tubing test
LTX unit	low-temperature extraction unit
lub	lubricate (ed) (ing) (tion)
Lued	Lueders
LUX	light hytrocrackate
LV	liquid volume
LVI	low viscosity index
lvl	level
Lvnwth	Leavenworth
LW	lapweld
LW	load water
LWL	low water loss
lwr	lower

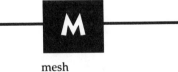

M	mesh
m	meter
m	micron

m	millimicron
M	molar
M	thousand (i.e., 9M = 9,000)
M & R Sta.	measuring and regulating station
M. Tus	Marine Tuscaloosa
m-a	microampere
m-f	microfarad
m-g	microgram
m-gr	medium grained
m-in.	microinch
m-kg	meter-kilogram
m-m	micromicron
m-mf	micro-microfarad
m-v	microvolt
M&F	male and female (joint)
M&FP	maximum & final pressure
M&P	mix and pump
M/	middle
m/l	more or less
M/PLT	masking plate
M/T	marine terminal
M/V	motor vehicle, motor vessel
MA	massive anhydrite
ma	microampere
ma	milliampere
MA	mud acid
MAC	medium amber cut
mach	machine
Mack	Mackhank
Mad	Madison
mag	magnetic, magnetometer
maint	maintenance
maj	major, majority
mall	malleable
man	manifold
man	manual
man op	manually operated
Manit	Manitoban
Mann	Manning
MAOP	maximum allowable operating pressure
Maq	Maquoketa
mar	marine
mar	maroon
March	Marchand
marg	marginal
Marg.	Marginulina
Marg. coco.	Marginulina coco

Marg. fl.	Marginulina flat
Marg. rd	Marginulina round
Marg. tex.	Marginulina texana
margas	marine gasoline
Mark	Markham
Marm	Marmaton
MARSH	Marshal (ling)
mass	massive
Mass. pr.	Massilina pratti
Mat	matter
Math	mathematics
matl	material
MAW	mud acid wash
MAWP	maximum allowable working pressure
max	maximum
May	Maywood
MB	methylene blue
MB	Moody's Branch
MBF/D	thousand barrels fluid per day
MBFPD	thousand barrels fluid per day
Mbl Fls	Marble Falls
Mbo/d	thousand barrels oil per day
mbr	member (geologic)
MBTU	thousand British thermal units
MBW/D	thousand barrels of water per day
mc	megacycle
MC	mud cake
MC	mud cut
MC ls	Moore County lime
MCA	mud cleanout agent
MCA	mud-cut acid
MCB	master circuit board
McC	McClosky lime
MCC	motor control center
McCul	McCullough
McEl	McElroy
Mcfd	thousand cubic feet per day
Mcfgpd	thousand cubic feet of gas per day
MCG	mud-cut gas
mchsm	mechanism
McK	McKee
McL	McLish
McMill	McMillan

MCO	mud-cut oil
MCP	maxium casing pressure
mcr-x	microcrystalline
MCSW	mud-cut salt water
MCT	computer-processed interpretation
MCW	mud-cut water
MD	measured depth
md	millidarcies
MD	Mt. Diablo
md wt	mud weight
MDC	Monel drill collars
MDDO	maximum daily delivery obligation
MDF	market demand factor
mdl	middle
mdse	merchandise
Mdy	muddy
MEA	monoethanolamine
Meak	Meakin
meas	measure (ed) (ment)
mech	mechanic (al), mechanism
Mech DT	mechanical down time
med	median
Med	Medina
med	medium
Med B	Medicine Bow
med FO	medium fuel oil
med gr	medium grained
Medr	Medrano
Meet	Meeteetse
MEG	methane-rich gas
MEK	methylethylketone
memo	memorandum
Men	Menard lime
Mene	Menefee
MEOH	methanol
MEP	mean effective pressure
MER	maximum efficient rate
Mer	Meramec
merc	mercury
mercap	mercaptan
merid	meridian
Meso	Mesozoic
meta	metamorphic
meth	methane
meth-bl	methylene blue
meth-cl	methyl chloride
methol	methanol

methr	methanator
metr	metric
mev	million electron volts
mezz	mezzanine
MF	manifold
MF	mud filtrate
MF-	Miscellaneous Field Studies Map
MFA	male to female angle
mfd	manufactured
MFD	mechanical flow diagram
mfd	microfarad
mfg	manufacturing
MFP	maximum flowing pressure
MFR	manufacture (er)
mg	medium grained
mg	milligram
mg	motor generator
MG	multigrade
MG	thousand gallons
m'gmt	management
mgr	manager
MGS	middle ground shoals
MH	manhole
mh	millihenry
MH	mousehole
mho/m	mhos per meter
MHz	megahertz (megacycles per second)
MI	malleable iron
MI	mile (s)
MI	mineral interest
MI	moving in (equipment)
mica	mica, micaceous
MICR	moving in completion rig
micro-xin	microcrystalline
microsec	microsecond
MICT	moving in cable tools
MICU	moving in completion unit
mid	middle
Mid	Midway
MIDDU	moving (moved) in double drum unit
MIE	moving in equipment
MIK	methylisobutylketone
mil	military
mill	milliotitic
millg	milling
MIM	moving in materials

min	minerals
min	minimum
min	minute(s)
min P	minimum pressure
Minl	Minnelusa
Mio	Miocene
MIPU	moving in pulling unit
MIR	moving in rig
MIRT	moving in rotary tools
MIRU	moving in and rigging up
MIRUSU	moving in rigging up swabbing unit
misc	miscellaneous
Mise	Misener
MISR	moving in service rig
Miss	Mississippian
Miss Cany	Mission Canyon
MIST	moving in standard tools
MIT	moving in tools
MIU	moisture impurities and unsaponifiabales (grease testing)
mix	mixer
MIXG	mixing
MKG	making
mkt	market (ing)
Mkta	Minnekahta
mky	milky
ml	milliliter
ML	mud logger
ml TEL	milliliters tetraethyl lead per gallon
mld	milled
Mle	milled one end
mlg	milling
MLL	master load list
MLU	mud logging unit
MLW	mean low wave
MLW-PLAT	mean low water to platform
Mly	marly
mm	millimeter
MM	million (i.e., 9MM = 9,000,000)
MM	motor medium
mm Hg	millimeters of mercury
MMBLS	millions of barrels
MMBTU	million British thermal units
MMcf	million cubic feet
MMcfd	million cubic feet per day
mmf	magnetomotive force

MMRVB	million reservoir barrels		MPH	miles per hour
MMS	Minerals Management Service		MPL	mechanical properties
			MPT	male pipe thread
MMscfd	million standard cubic feet per day		MPY	miles per year
			MR	marine rig
MNL	manual		MR	meter run
MNR	minor		mr	milliroentgen
mnrl	mineral		MRF	"merf"/main reaction furnace
MO	molybdenum		MRK	marking
MO	motor oil		mrlst	marlstone
MO	moving out		MRQ	memo requesting quotes
mob	mobile		ms	millisecond (s)
MOCT	moving out (off) cable tools		MS	motor severe
MOCU	moving out completion unit		MSA	multiple service acid
mod	model		MSC	mapping subcommittee
mod	moderate (ly)		Mscf	thousand standard cubic feet
mod	modification		Mscf/d	thousand standard cubic feet per day
modu	modular			
MOE	milled other end		Mscf/h	thousand standard cubic feet per hour
MOE	moving out equipment			
Moen	Moenkopi		MSDS	material safety data sheets
mol	molas		msl	mean sea level
mol	mole		MSP	maximum surface pressure
MOL	molecule, molecular		MSR	mud/silt remover
mol	mollusca		mstr	master
mol wt	molecular weight		MSW	muddy salt water
mon	monitor		MSWG	miscible substance group
MON	motor octane number		MT	empty container
Mont	Montoya		MT	magnetic particle examination
Moor	Mooringsport		MT	marcaroni tubing
MOP	maximum operating pressure		Mt. Selm	Mount Selman
Mor	Morrow		MTD	maximum total depth
MOR	moving out rig		MTD	mean temperature difference
Morr	Morrison		MTD	measured total depth
MORT	moving out (off) rotary tools		mtd	mounted
Mos	Mosby		mtg	mounting
mot	motor		mtge	mortgage
mott	mottled		mtl	material
MOU	motor oil units		MTO	material take-off
mov	moving		MTP	maximum top pressure
Mow	Mowry		MTP	maximum tubing pressure
MP	maximum pressure		mtr	meter
MP	melting point		MTR	motor
MP	multipurpose		MTS	mud to surface
mPa	megapascal		Mtx	matrix
MPB	metal petal basket		mud wt	mud weight
MPGH-Lith	multipurpose grease lithium base		mudst	mudstone
			MULTX	multiply, multiplexer
MPGR-Soap	multipurpose grease soapbase		musc	muscovite

MUX	middle hydrocrackate
mv	millivolt
Mvde	Mesaverde
MVFT	motor vehicle fuel tax
MVOP	monthly volume operation plan
MW	megawatt
MW	microwave
MW	mud weight
MW	muddy water
MWD	marine wholesale distributors
MWP	maximum working pressure
MWPE	mill wrapped plain end
Mwy	Midway
mxd	mixed
M1E	milled one end
M2E	milled two ends
m/d	cubic meters per day

N	Newton
N	nonproducer
N	normal (to express concentration)
N	north
N. Cock.	Nonionella Cockfieldensis
N/O	north offset
N/S S/S	nonstandard service station
N/tst	no test
N/2	north half
N/4	north quarter
NA	not applicable
NA	not available
Nac	Nacotoch
nac	nacreous
NaCL	sodium chloride
$NaCO_3$	sodium carbonate
NAG	no appreciable gas
NALRD	Northern Alberta Land Registration District
NaOH	sodium hydroxide
nap	naphtha
NARR	narrative
nat	natural

nat'l	national
Nav	Navajo
Navr	Navarro
NB	new bit
NB	nitrogen blanket
Nbg	Newburg
NC	no change
NC	no core
NC	normally closed
NC	not completed
NCT	national coarse thread
NCT	noncontiguous tract
ND	nippled down
ND	nondetergent
ND	not drilling
NDBOPs	nipple (ed) (ing) down blowout preventers
NDE	not deep enough
NDG	no show gas
Ndl Cr	Noodle Creek
NDT	nipple-down tree
NDT	nondestructive testing
NE	nonemulsifying agent
NE	northeast
NE/4	northeast quarter
NEA	nonemulsion acid
NEC	National Electric Code
NEC	northeast corner
neg	negative
neg	negligible
NEGO	negotiation
NEL	northeast line
NEP	net effective pay
neut	neutral neutralization
Neut. No.	Neutralization Number
New Alb	New Albany shale
Newc	Newcastle
NF	National Fine (thread)
NF	natural flow
NF	no fluid
NF	no fluorescence
NF	no fuel
NFD	new field discovery
NFOC	no fluorescence or cut
NFW	new field wildcat
NG	natural gas
NG	no gauge
NG	no good
NGL	natural gas liquids

NGTS	no gas to surface	NPL UP	nippled up
NHDS	naphtha-hydrogen desulfurization	npne	neoprene
NH_3	ammonia	NPOS	no paint on seams
NH_4Cl	ammonium chloride	NPR	Naval Petroleum Reserve
NIC	not in contract	NPRA	Naval Petroleum Reserve, Alaska
NID	notice of intention to drill	NPS	nominal pipe size
Nig	Niagara	NPSH	net positive suction head
Nine	Ninnescah	NPT	National pipe thread
Niob	Niobrara	NPTF	National pipe thread, female
nip	nipple	NPTM	National pipe thread, male
nitro	nitroglycerine	NPW	new pool wildcat
NL	north line	NPX	new pool exempt (nonprorated)
NL Gas	nonleaded gas		
NLL	neutron lifetime log	NR	new rod
N'ly	northerly	NR	no recovery
NMI	nautical mile	NR	no report, not reported
NO	new oil	NR	nonreturnable, no returns, not reached
NO	Noble-Olson		
NO	normally open	NRC	Nuclear Regulatory Commission
NO	number		
No Inc	no increase	NRI	net revenue interest
no rec	no recovery	NRS	nonrising stem (valve)
No.	number (before a number, i.e., No. 3)	NRSB	nonreturnable steel barrel
		NRSD	nonreturnable steel drum
NOB	not on bottom	NS	no show
nod	nodule, nodular	NSC	not suitable for coating
Nod. blan.	Nodosaria blanpiedi	NSFOC	no show fluorescence or cut
Nod. mex.	Nodosaria mexicana		
NOJV	nonoperated joint ventures	NSO	no show oil
nom	nominal	NSO&G	no show oil and gas
Non	Nonionella	NSPS	new source performance standards
nonf G	nonflammable compressed gas		
NOP	nonoperating property	nstd	nonstandard
NOR	no order required	NT	net tons
nor	normal	NT	no time
NOV	notice of violation	NTD	new total depth
noz	nozzle	NTP	notice to proceed
NP	nameplate	NTS	not to scale
NP	nickel plated	NU	naphfining unit
NP	no production	NU	nippled (ing) up
NP	nonporous	NU	nonupset
NP	not prorated	NUBOPs	nipple (ed) (ing) up blowout preventers
NP	not pumping		
NP	Notary Public	NUE	nonupset ends
NPD	new pool discovery	Nug	nugget
NPDES	National pollution discharge elimination system	num	numerous
		NUT	nipple up tree
NPL	nipple	NUWH	nippling up wellhead

NVP	no visible porosity
NW	no water
NW	northwest
NW/C	northwest corner
NW/4	northwest quarter
NWL	northwest line
NWT	Northwest Territories
NYA	not yet available
NYD	not yet drilled
NYL	nylon
N_2	nitrogen

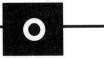

O	oil
O	Osborne
O sd	oil sand
O&G	oil and gas
O&GC SULW	oil and gas-cut sulfur water
O&GCAW	oil and gas-cut acid water
O&GCLW	oil and gas-cut load water
O&GCM	oil and gas-cut mud
O&GCSW	oil and gas-cut salt water
O&GCW	oil and gas-cut water
O&GL	oil and gas lease
O&M	operations and maintenance
O&SW	oil and salt water
O&SWCM	oil and sulfur water-cut mud
O&W	oil and water
O/S	out of service over and short (report)
O/S	out of stock
O/T tk	open-top tank
OA	overall
OAH	overall height
Oakv	Oakville
OAL	overall length
OAW	oil abandoned well
OB	off bottom
obj	object
OBM	oil-based mud
OBMO	outboard motor oil
OBS	observation
OBS	ocean bottom suspension
obsol	obsolete

OBW & RS	optimum bit weight and rotary speed
OC	oil cut
OC	on center
OC	open choke
OC	open cup
OC	operations commenced
OC-	Oil and Gas Investigations Chart
OCB	oil circuit breaker
occ	occasional (ly)
OCM	oil-cut mud
OCS	Outer Continental Shelf
OCSW	oil-cut salt water
oct	octagon, octagonal
oct	octane
OCW	oil-cut water
od	odor
OD	outside diameter
Odel	O'Dell
ODT	oil down to
OE	oil emulsion
OE	open end
OE	overexpenditure
OEB	other end beveled
OEM	oil emulsion mud
OF	open flow
OF	open-file report
off	office, official
off-sh	offshore
OFIC	oil insulated fan-cooled
OFL	overflush (ed)
OFLU	oil fluorescence
OFOE	orifice flange one end
OFP	open flow potential
OFS	offsite
OGJ	Oil & Gas Journal
O'H	O'Hara
OH	open hearth
OH	open hole
OH	overhead
ohm	ohm
ohm-cm	ohm-centimeter
ohm-m	ohmmeter
OI	oil insulated
OIFC	oil insulated, fan cooled
OIH	oil in hole
Oil Cr	Oil Creek
oilfract	oil fractured

OIP	oil in place
OISC	oil insulated, self-cooled
OIT	oil in tanks
OIWC	oil immersed, water cooled
OL	off/on location
OL	open line (no choke)
ole	olefin
Olig	Oligocene
OLN	outline
OMRL	oriented microresistivity
ONR	octane number requirement
ONRI	octane number requirement increase
OO	oil odor
ooc	ooliclastic
OOIP	original oil in place
ool	oolitic
oom	oolimoldic
OP	articles published in outside journals/books
OP	oil pay
OP	outpost
OP	overproduced
op hole	open hole
OPB	old plugback
OPBD	old plugback depth
oper	operate, operations, operator
Operc	Operculinoides
OPI	oil payment interest
opn	open (ed) (ing)
OPO	overseas procurement office
opp	opposite
OPT	official potential test
OPTL	optional
optn to F/O	option to farmout
OR	orange
Or	Oread
Ord	Ordovician
orf	orifice
org	organic
org	organization
ORIAC	Oil Refining Industry Action Committee
ORIENT	orientation
orig	original, originally
Orisk	Oriskany
ORR	overriding royalty
ORRI	overriding royalty interest
ORSANCO	Ohio River Valley Water Sanitation Commission
orth	orthoclase
OS	oil show
Os	Osage
OS	overshot
OS&F	odor stain and fluorescence
OS&Y	outside screw and yoke (valve)
OSA	oil-soluble acid
OSD	operation shutdown
OSF	oil string flange
OSI	oil well shut in
OSIDP	oil standing in drillpipe
Ost	Ostracod
OSTN	oil stain
OSTOIP	original stock tank oil in place
Osw	Oswego
OT	open tubing
OT	overtime
OT&S	odor taste & stain
OTD	old total depth
OTD	original total depth
OTE	oil-powered total energy
otl	outlet
OTS	oil to surface
OTS&F	odor taste stain and fluorescence
OU	oil unit
Our	Ouray
ovhd	overhead
OWC	oil-water contact
OWDD	old well drilled deeper
OWF	oil well flowing
OWFWF	oil well from waterflood
OWG	oil well gas
OWPD	old well plugged back
OWST	old well sidetracked
OWWO	old well worked over
ox	oxidized, oxidation
oxy	oxygen
oz	ounce

P

P	professional paper
P & A	plugged and abandonded
P & ID	process & instrument diagram
P & NG	petroleum and natural gas
P Lar	Post Laramie
P tstg	pump testing
p.	page (before a number, i.e., p. 4)
P.O.	Pin Oak
P.O.	Post Oak
P.P.	present production
P-HDII	perforating Hyperdome II
P-M	Pensky-Martins (flash)
P&C	personal and confidential
P&F	pump and flow
P&IDS	piping and instrument diagrams
P&L	profit and loss
P&P	porosity and permeability
P&P	porous and permeable
P/	pump
P/BLDG	pump building
P/DIA	piping diagram
PA	participating area
Pa	Pascal
PA	pooling agreement
PA	pressure alarm
PA	public address
PAB	per-acre bonus
Padd	Paddock
Paha	Pahasapa
Pal	Paluxy
Paleo	paleontology
Paleo	Paleozoic
Palo P	Palo Pinto
Pan L	Panhandle lime
PAR	per-acre rental
Para	Paradox
Park C	Park City
PART	partial
pat	patent (ed)
patn	pattern
pav	paving

Paw	Pawhuska
PB	plugged back
PB-ADA	report available only through National Technical Information Service
PBD	plugged-back depth
PBE	plain both ends
PBHL	proposed bottom-hole location
pbl (y)	pebble, pebbly
PBP	pulled bid pipe
PBTD	plugged back total depth
PBW	pipe buttweld
PBX	private branch exchange
PBX	switchboard
PC	Paint Creek
pc	piece
PC	poker chipped
pc	port collar
PC	Porter Creek
PCF	pounds per cubic foot
pct	percent
PCV	positive crankcase ventilation
PCV	pressure control valve
PD	geopressure development, failure
pd	paid
PD	per day
PD	plug down
PD	present depth
PD	pressed distillate
PD	proposed depth
PD	pulsation dampener
PD	pumper's depth
PDC	power distribution center
PDC	pressure differential controller
PDET	production department exploratory test
PDI	pressure differential indicator
PDIC	pressure differential indicator controller
PDR	pressure differential recorder
PDRC	pressure differential recorder controller
PDS	geopressure development, success
PDS	power distribution system
pdso	pseudo

pe	pin end	PI	pump in
PE	plain end	PIC	pressure indicator controller
PE	pumping equipment		
PEB	plain end beveled	Pic Cl	Pictured Cliff
PED	pedestal	pinpt	pinpoint
pell	pelletal, pelletoidal	PIP	pump-in pressure
pen	penetration, penetration test	piso	pisolites, pisolitic
Pen A.C.	penetration asphalt cement	pit	pitted
penal	penalty, penalize (ed) (ing)	PJ	pump jack
Penn	Pennsylvanian	PJ	pump job
perco	percolation	pk	pink
perf	perforate (ed) (ing) (or)	pkg (d)	packing, package (ed)
perf csg	perforated casing	pkr	packer
perm	permanent	PL	pipeline
perm	permeable (ability)	PL	plate
Perm	Permian	PL	property line
perp	perpendicular	pl fos	plant fossils
pers	personnel	Plan. palm.	Planulina palmarie
PERT	performance evaluation and review technique	Plan. har.	Planulina harangensis
		plas	plastic
pet	petroleum	PLASR	plaster
Pet	Pettet	platf	platform
Pet sd	Pettus sand	plcy	pelecypod
petrf	petroliferous	pld	pulled
petrochem	petrochemical	PLE	plain large end
Pett	Pettit	Pleist	Pleistocene
PEW	pipe electric weld	plg	plagioclase
pf	per foot	plg	pulling
PF	power factor	plgd	plugged
pfd	preferred	Plio	Pliocene
PFD	process flow diagram	PLMB	plumbing
PFM	power factor meter	pln	plan
PFRACT	prefractionator	plngr	plunger
PFT	pumping for test	PLO	pipeline oil
PG	Pecan Gap	PLO	pumping load oil
Pg	plug	plt	pilot
PGC	Pecan Gap chalk	PLT	pipeline terminal
PGW	producing gas well	plt	plant
pH	acidity or alkalinity	plty	platy
pH	hydrogen ion concentration	PLV	pilot loaded valve
pH	measure of hydrogen potential	PLW	pipe lapweld
Ph	parish	P-M	Pensky Martins
ph	phase	PML	production management
PHC	pipe-handling capacity	pmp (d) (g)	pump (ed) (ing)
Phos	Phosphoria	PN	Performance Number (aviation gas)
PI	penetration index		
PI	Pine Island	pneu	pneumatic
PI	pressure indicator	pnl	panel
PI	productivity index	PNL BD	panel board

PNR	please note and return	**PPI**	Process Performance Index
po	Phrohotite	**ppm**	parts per million
PO	pulled out	**ppn no**	precipitation number
PO	pumps off	**PPP**	pinpoint porosity
PO	purchase order	**ppt**	precipitate
POB	plug on bottom	pr	pair
POB	pump on beam	**PR**	polished rod
POCS	Pacific Outer Continental Shelf	pr	poor
		PR	pressure recorder
POD	plan of development	**PR**	public relations
Pod.	Podbielniak	**PR**	purchasing request
POE	plain one end	**pr op**	present operations
POGW	producing oil and gas well	**PR&T**	pull (ed) rods and tubing
POH	pulled (put) out of hole	**PRC**	pressure recorder control
pois	poison	prcst	precast
pol	polish (ed)	prd	period
poly	polymerization, polymerized	**Pre Camb**	Precambrian
poly cl	polyethylene	**PRECIP**	precipitator
polygas	polymerized gasoline	predom	predominant
polypl	polypropylene	prefab	prefabricated
PONA	parrafins-olefins-napthenes-aromatics	prehtr	preheater
		prelim	preliminary
Pont	Pontotoc	prem	premium
POOH	pull (put) out of hole	**Prep**	prepare, preparing, preparation
POP	putting on pump		
por	porosity, porous	press	pressure
porc	porcelaneous	prest	prestressed
porc	porcion	prev	prevent, preventive
port	portable	**PREV**	previous
pos	position	**PREV DO AVG**	previous daily average
pos	positive	**PRF**	Primary Reference Fuel
poss	possible (ly)		
pot	potential	pri	primary
pot dif	potential difference	prin	principal
POT/WTR	potable water	pris	prism (atic)
pour ASTM	pour point (ASTM method)	priv	privilege
POW	producing oil well	prly	pearly
POWF	producing oil well, flowing	prmt	permit
POWP	producing oil well, pumping	**prncpl lss**	principal lessee (s)
PP	pinpoint	pro	prorated
PP	production payment	prob	probable (ly)
PP	pulled pipe	proc	process
PP	pump pressure	prod	produce (ed) (ing) (tion), product (s)
ppb	parts per billion		
PPB	pounds per barrel	prog	progress
ppd	prepaid	proj	project (ed) (ion)
ppg	piping	**PROP**	property
PPG	pounds per gallon	prop	proportional
PPI	production payment interest	prop	propose (ed)

prot	protection
Protero	Proterozoic
Prov	provincial
PRPT	preparing to take potential test
prtgs	partings
PS	pressure switch
ps	pseudo
PS	pump station
PSA	packer set at
PSB	pressure seal bonnet
PSD	permanently shut down
PSD	prevention of significant deterioration, EPA
PSE	plain small end
psf	pounds per square foot
psi, PSI	pounds per square inch
PSI	profit-sharing interest
psia, PSIA	pounds per square inch absolute
psig, PSIG	pounds per square inch gauge
PSL	pipe sleeve
PSL	Public School Land
PSM	pipe seamless
Psp	prospect
PSV	pressure safety valve
PSW	pipe spiral weld
PT	liquid penetrant examination
pt	part, partly
pt	pint
pt	point
Pt Lkt	Point Lookout
PT	potential test
PTC	permanent type completion
PTD	present total depth
PTD	projected total depth
PTD	proposed total depth
PTF	production test flowed
PTG	pulling tubing
PTN	partition
PTP	production test pumped
PTR	pulling tubing and rods
PTS pot	pipe to soil potential
PTTF	potential test to follow
PU	picked up
PU	pulled up
PU	pumping unit
PUC	project ultimate cost

PUDP	picking up drillpipe
PUIC	pulled up in casing
PULS	pulse (sating) (sation)
PURCH	purchasing
PURF	purification
purp	purple
PV	plastic viscosity
PV	pore volume
PVC	polyvinyl chloride
pvmnt	pavement
PVR	plant (pressure) volume reduction
PVT	pressure-volume-temperature
PW	producing well
PW(15)	present worth at discount rate of 15%
PWHT	postweld heat treatment
PWR	power
PWY	pipeway
PWZ	peripheral wedge zone
Pxy	Paluxy
pyls	pyrolysis
pymt	payment
pyr	pyrite, pyritic
pyrbit	pyrobitumen
pyrclas	pyroclastic

Q. City	Queen City
Q. sd	Queen Sand
QA	quality assurance
QC	quality control
QDA	quality discount allowance
QDRNT	quadrant
qnch	quench
QRC	quick ram change
qry	quarry
qt	quart (s)
qtr	quarter
qty	quantity
qtz	quartz, quartzite, quartzitic
qtzose	quartzose
quad	quadrant, quadrangle, quadruple

QUAL	qualitative
qual	quality
quan	quantity
quest	questionable
quint	quintuplicate

R	radius
R	range
R	rankine (temp. scale)
R	resistivity
r	roentgen
R	rows
R & D	research and development
R & T	rods and tubing
R test	rotary test
R.O.	Red Oak
R(16")	resistivity as recorded from 16" electrode configuration
R-SP	recommended spare part
R&L	road & location
R&LC	road & location complete
R&O	rust and oxidation
R/A	regular acid
R/W	right of way
RA	radioactive
RA	right angle
RACTR	reactor
rad	radical
rad	radiological
rad	radius
RADT	radiant
radtn	radiation
RAGL	raw gas lift
RALOG	running radioactive log
Rang	Ranger
RB	rock bit
RB	rotary bushing
RBLR	reboiler
Rbls	rubber balls
RBM	rotary bushing measurement
RBP	retrievable bridge plug
rbr	rubber
RBSO	rainbow show of oil

RBSOF	rubber ball sand oil frac
RBSWF	rubber ball sand water frac
RC	rapid curing
RC	Red Cave
RC	remote control
RC	reverse circulation
RC	running casing
RCO	returning circulation oil
RCPT	receptacle
RCR	Ramsbottom Carbon Residue
RCR	reverse circulation rig
RCTN	reaction
RCVR	receiver
RCVY	recovery
RCYL	recycle
RD	redrilled
RD	rigged (ing) down
rd	road
rd	round
Rd Bds	red beds
Rd Fk	Red Fork
Rd Pk	Red Peak
rd thd	round thread
RDB	rotary drive bushing
RDB-GD	rotary drive bushing to ground
RDCR	reducer
rdd	rounded
RDMO	rigged down, moved out
RDS	reservoir description service
RDSU	rigged-down swabbing unit
rdtp	round trip
REABS	reabsorber
reacd	reacidize (ed) (ing)
React	reaction (ed)
rebar	reinforcing bar
rec	recommend
rec	record (er) (ing)
rec	recover (ed) (ing), recovery
recd	received
recip	reciprocate (ing)
recirc	recirculate
recomp	recomplete (ed) (ion)
RECOMP	recompressor
recond	recondition (ed)
recp	receptacle
rect	rectangle, rectangular
rect	rectifier
recy	recycle

red	reducing, reducer	**retd**	returned
red bal	reducing balance	**retr**	retrievable
redrld	redrilled	**retr ret**	retrievable retainer
ref	reference	**rev**	reverse (ed)
ref	refine (ed) (er) (ry)	**rev**	revise (ed) (ing) (ion)
refg	refining	**rev**	revolution (s)
refl	reflection	**rev/O**	reversed out
refl	reflux	**RF**	raised face
REFMR	reformer	**RF**	rig floor
REFOO	re-evaluation for overoptimism	**RFFE**	raised face, flanged end
		RFG	roofing
reform	reformate (er) (ing)	**RFG/BD**	refrigeration building
refr	refraction, refractory	**RFLCT**	reflect (ed) (ing) (tion)
REFRIG	refrigerator (rant) (tion)	**RFP**	request for proposal
reg	register	**RFQ**	request for quote
reg	regular, regulator	**RFR**	ready for rig
regen	regenerator	**RFSF**	raised face, smooth finish
reinf	reinforce (ed) (ing) (ment)	**RFSO**	raised face, slip on
reinf conc	reinforced concrete	**RFWN**	raised face, weld neck
rej	reject	**RG**	raw gas
rej'n	rejection	**rg**	ring
Rek	Reklaw	**RG**	ring groove
rel	relay	**Rge**	range
rel	release (ed)	**rgh**	rough
rel	rig released	**RGTR**	register
REL	running electric log	**RH**	rat hole
reloc	relocate (ed)	**RH**	relative humidity
rem	remains	**RH**	right hand
rem	remedial	**RHD**	righthand door
Ren	Renault	**rheo**	rheostat
rent	rental	**RHM**	rat hole mud
Reo. bath.	Reophax bathysiphoni	**RHN**	Rockwell hardness number
rep	repair (ed) (ing) (s)	**RI**	royalty interest
rep	replace (ed)	**Rib**	ribbon sand
rep	report	**Rier**	Rierdon rig
reperf	reperforated	**RIH**	ran in hole
repl	replace (ment)	**RIL**	red indicating lamp
REQ	request	**riv**	rivet
req	requisition	**RIZ**	resistivity invaded zone
reqd	required	**RJ**	ring joint
reqmt	requirement	**RJFE**	ring joint, flanged end
res	research	**RK**	rack
res	reserve (ation)	**rk**	rock
res	resistance, resistivity, resistor	**RKB**	rotary kelly bushing
Res. O. N.	Research Octane Number	**rky**	rocky
resid	residual, residue	**RL**	random lengths
RESIS	resistor (s)	**rlf**	relief
ret	retain (er) (ed) (ing)	**rlg**	railing
ret	return	**rls (d) (ing)**	release (ed) (ing)

rly	relay	RS	rig skidded
rm	ream	RS	rising stem (valve)
Rm	resistivity, mud	RSD	returnable steel drum
rm	room	rsns	resinous
rmd	reamed	RSU	released swab unit
Rmf	resistivity, mud filtrate	rsvr	reservoir
rmg	reaming	RT	radiographic examination
rmn	remains	RT	rig time
RMS	root mean square	RT	rotary table
rmv (l)	remove (al) (able)	RT	rotary tools
rnd	rounded	RT CB	round trip changed bit
rng	running	rtd	retard (ed)
RO	reversed out	RTD	rotary total depth
ro	rose	rtg	rating
Ro	Rosiclare sand	RTG	routing
Rob	Robulus	RTG	running tubing
Rod	Rodessa	RTJ	ring tool joint
ROF	rich oil fractionator	RTJ	ring-type joint
ROGL	rotative gas lift	RTL	Refinery Technology Laboratory
ROI	return on investment		
ROL	rig on location	RTLTM	rate too low to measure
Rok	rock	RTMTR	rotameter
ROM	rough order of magnitude	rtnr	retainer
ROM	run of mine	RTTS	retrievable test treat squeeze (tool)
RON	Research Octane Number		
ROP	rate of penetration	RTU	remote terminal unit
ROR	rate of return	RU	rig (ged) (ging) up
ROS	remote operating system (station)	RU	rotary unit
		rub	rubber
rot	rotary, rotate, rotator	RUCC	rig-up casing crew
ROW	right of way	RUCT	rigging-up cable tools
roy	royalty	RUM	rigging-up machine
RP	rock pressure	RUP	rigging-up pump
RPI	Research Planning Institute	rupt	rupture
rpm	revolutions per minute	RURT	rigging-up rotary tools
rpmn	repairman	RUSR	rigging-up service rig
RPP	retail pump price	RUST	rigging-up standard tools
RPRT	report	RUSU	rigging-up swabbing unit
rps	revolutions per second	RUT	rigging-up tools
rptd	reported	RV	relief valve
RR	railroad	RVP	Reid vapor pressure
RR	Red River	rvs (d)	reverse (ed)
RR	rig released	RVT	rivet
RR	rig repair	Rw	resistivity, water
RR	rigging rotary	Rwa	resistivity, water (apparent)
RR&T	ran (running) rods and tubing	rwk (d)	rework (ed)
		RWTP	returned well to production
RRC	Railroad Commission (Texas)	Rxo	resistivity, flushed zone
RS	rig service		

S

S	seconds
S	south
S	stratigraphic test
s & s	spigot and spigot
S & T	shell and tube
S Bomb	sulfur by bomb method
S O	south offset
S Riv	Seven Rivers
S.L.	sea level
S-T-R	section-township-range
S&F	swab and flow
S&O	stain and odor
s&p	salt and pepper
S/	swabbed
S/C	speed/current
S/E	screwed end
S/FAB	shop fabrication
s/p	shipping point (purchasing term)
S/SR	sliding-scale royalty
S/SW	screwed and socketweld
S/T	sample tops
S/T	speed/torque
S/WTR	sanitary water
S/2	south half
s,t&b	sides, tops, & bottoms
SA	seal assembly
Sab	Sabinetown
sach	saccharoidal
Sad Cr	Saddle Creek
sadl	saddle
saf	safety
SAF/DPT	safety/department
SAFE	surface approximation and formation evaluation
Sal	Salado
sal	salary, salaried
sal	salinity
Sal Bay	Saline Bayou
salv	salvage
samp	sample
SAN	sanitary
San And	San Andres
San Ang	San Angelo

San Raf	San Rafael
Sana	Sanastee
sani	sanitary
sap	saponification
Sap No.	saponification number
Sara	Saratoga
sat	saturated, saturation
Saw	Sawatch
Sawth	Sawtooth
Say Furol	Saybolt furol
SB	sideboom
SB	sleeve bearing
SB	stuffing box
sb	sub
Sb	Sunburst
SBA	secondary butyl alcohol
SBB&M	San Bernardino base and meridian
SBHP	static bottom-hole pressure
sc	scales
SC	self-contained
SC	show condensate
SC DL	slip and cut drill line
SCAF	scaffolding
scatt (d)	scatter (ed)
scf, SCF	standard cubic foot
scfd, SCFD	standard cubic feet per day
scfh, SCFH	standard cubic feet per hour
scfm, SCFM	standard cubic feet per minute
sch	schedule
schem	schematic
SCHL	Schlumberger
scly	securaloy
scolc	scolescodonts
scr	scraper
scr	scratcher
scr	screen
scr (d)	screw (ed)
scrub	scrubber
SCSSV	surface-controlled subsurface safety valve
sctrd	scattered
sd, SD	sand
SD	shut down
sd & sh	sand and shale
SD Ck	side door choke
Sd SG	sand showing gas
Sd SO	sand showing oil

SDA	shut down to acidize
sdd	sanded
SDF	shut down to fracture
sdfract	sandfracked
SDG	siding
SDL	shut down to log
SDO	show of dead oil
SDO	shut down for orders
sdoilfract	sand-oil fracked
SDON	shut down overnight
SDP	set drillpipe
SDPA	shut down to plug and abandon
SDPL	shut down for pipeline
SDR	shut down for repairs
sdtkr	sidetrack (ed) (ing)
SDW	shut down for weather
SDWL	sidewall
SDWO	shut down awaiting orders
sdwtrfract	sand-water fracked
sdy	sandy
sdy li	sandy lime
sdy sh	sandy shale
SE	southeast
SE NA	screw end American National Acme thread
SE NC	screw end American National Coarse thread
SE No.	steam emulsion number
SE NTP	screw end American National Taper Pipe thread
SE/C	southeast corner
SE/4	southeast quarter
Sea	Seabreeze
sec	secant
sec	second (ary)
sec	secretary
sec	section
SECT	section (s) (al) (ing)
sed	sediment (s)
Sedw	Sedwick
SEG	segment
seis	seismograph, seismic
sel	selenite
Sel	Selma
SELECT	selection (tive) (tor)
Sen	Senora
SEO	seal oil
SEP	separate, separator, separation

SEPM	Society of Economic Paleontologists & Mineralogists
sept	septuplicate
seq	sequence
ser	series, serial
Serp	serpentine
Serr	Serratt
serv	service (s)
serv chg	service charge
set	settling
sew	sewer
SEWOP	self-elevating work platform
Sex	sexton
sext	sextuplicate, sextuplet
SF	sandfrac
sfc	surface
SFD	system flow diagram
SFL	starting fluid level
SFLU	slight, weak, or poor fluorescence
SFO	show of free oil
SFP	surface flow pressure
SFT	sample formation tester
sft	soft
sg	specific gravity
SG	show gas
SG	surface geology
SG & W	show gas and water
SG&C	show gas and condensate
SG&D	show gas and distillate
SG&O	show of gas and oil
SGCM	slightly gas-cut mud
SGCO	slightly gas-cut oil
SGCSW	slightly gas-cut salt water
SGCW	slightly gas-cut water
SGCWB	slightly gas-cut water blanket
SGCWC	slightly gas-cut water cushion
sgd	signed
sgl (s)	single (s)
sh	shale
sh	sheet
SH	substructure height
Shan	Shannon
SHDP	slim-hole drillpipe
Shin	Shinarump
shld	shoulder
shls	shells
SHLT	shelter

shly	shaly	Slty	salty
SHM	solar heat medium	slty	silty
shp	shaft horsepower	slur	slurry
shp(g)	ship (ping)	SLV	sleeve
shpt	shipment	S'ly	southerly
shr	shear	SM	Seward Meridian (Alaska)
SHT	straight-hole test	sm	small
SHTG	sheeting	SM	surface measurement
SHTG	shortage	Smithw	Smithwick
shthg	sheathing	Smk	Smackover
SI	shut in	smls	seamless
SIBHP	shutin bottom-hole pressure	smpl	sample
SICP	shutin casing pressure	smth	smooth
SIGW	shutin gas well	SN	seating nipple
SIH	started in hole	SNG	synthetic natural gas
Sil	Silurian	SNUB	snubber, snubbing
silic	silica, siliceous	SNUFF	snuffing
silt	siltstone	SO	shake out
sim	similar	SO	shaled out
Simp	Simpson	SO	show of oil
SIOW	shutin oil well	SO	side opening
SIP	shutin pressure	SO	slip on
Siph. d.	Siphonina davisi	SO&G	show oil and gas
SITP	shutin tubing pressure	SO&GCM	slightly oil- and gas-cut mud
SIWHP	shutin wellhead pressure	SO&W	show oil and water
SIWOP	shutin, waiting on potential	SOC	start of cycle
sk	sack (s)	SOCM	slight oil-cut mud
SK	sketch	SOCSW	slightly oil-cut salt water
Sk Crk	Skull Creek	SOCW	slight oil-cut water
skim	skimmer	SOCWB	slightly oil-cut water blanket
Skn	Skinner	SOCWC	slightly oil-cut water cushion
skr rd	sucker rod	sod gr	sodium-base grease
sks	slickensided	SOE	screwed on one end
skt	socket	SOF	sand-oil fracture
SL	section line	SOH	shot open hole
sl	sleeve	SOH	started out of hole
SL	south line	sol	solenoid
SL	state lease	sol	solids
SLC	steel line correction	soln	solution
sld	sealed	solv	solvent
sli	slight (ly)	som	somastic
Sli	Sligo	somct	somastic coated
sli SO	slight show of oil	SONL	sonic log
slky	silky	SOOH	started out of hole
SLM	steel line measurement	SOOH	strapped out of hole
SLNCR	silencer	SOP	standard operational produre
slnd	solenoid	SOR	start of run
SLPR	sleeper	sort	sorted (ing)
Slt Mtn	Salt Mountain	SOV	solenoid-operated valves

sow	socket weld
SP	self (spontaneous) potential
SP	set plug
sp	shipping point (purchasing term)
sp	shot point
sp	slightly porous
sp	spare
Sp	Sparta
sp	spore
SP	straddle packer
SP	surface pressure
sp	shipping point (purchasing term)
sp gr	specific gravity
sp ht	specific heat
SP SW	single pole switch
sp. vol.	specific volume
SP-DST	straddle packer drillstem test
spcl	special
spcr	spacer
spd	spud (ded) (der)
spdl	spindle
SPDT	single-pole double throw
SPDT SW	single-pole double throw switch
SPE	Society of Petroleum Engineers
spec	specification
speck	speckled
SPF	shot per foot
spf	spearfish
spg	sponge
spg	spring
SPGT	spigot
sph	spherules
Sphaer	Sphaerodina
sphal	sphalerite
spic	spicule (ar)
Spiro. b.	Spiroplectammina barrowi
spkr	sprinkler
spkt	sprocket
spl	sample
spl cham	sample chamber
Spletp	Spindletop
SPLTR	splitter
splty	specialty
splty	splintery
sply	supply
SPM	strokes per minute
Spra	Spraberry
sprf	spirifers
Sprin	Springer
SPST	single-pole single throw
SPST SW	single-pole single throw stitch
SPT	shallower pool (pay) test
sptd	spotted
sptty	spotty
Spud Date	date actually started drilling
SPWY	spillway
sq	square
sq cg	squirrel cage
sq cm	square centimeter
sq ft	square foot
sq in.	square inch
sq km	square kilometer
sq m	square meter
sq mm	square millimeter
sq pkr	squeeze packer
sq yd	square yard (s)
SQRT	square root
sqz	squeeze (ed) (ing)
SR	short radius
SR	swab rate
SR	swab run (s)
SRB	sulfate bacteria
SRG	surge
SRL	single random lengths
SRN	straight-run naphtha
srt	sort (ed) (ing)
SRV	safety relief valve
SS	sandstone
SS	service station
SS	shock sub
SS	short string
SS	single shot
SS	slow set (cement)
SS	small show
SS	stainless steel
SS	string shot
SS	subsea
SS	subsurface
SSA	spot sales agreement
SSG	slight show of gas
SSO	slight show of oil
SSO&G	slight show of oil and gas
SSSV	subsurface safety valve
SSU	Saybolt Seconds Universal

SSUW	salty sulfur water	stnr	strainer
ST	short thread	stoip	stock tank oil in place
st	start	stor	storage
ST (g)	sidetrack (ing)	STP	standard temperature and pressure
St L	Saint Louis lime		
St Ptr	Saint Peter	STPR	strip (per) (ping)
St Gen	Saint Genevieve	stpr (d)	stopper (rd)
ST&C	short threads & coupling	Str	Strawn
sta	station	strat	stratigraphic
Sta Marg	Santa Margarita	strd	straddle
stab	stabilized (er)	strd	strand (ed)
STAG	staggered	strg	storage
Stal	Stalnaker	strg	strong
Stan	Stanley	strg (r)	string (er)
stat	stationary	stri	striated
stat	statistical	strk	streak
State pot	state potential	STROH	strap out of hole
stb, STB	stock tank barrels	strom	stromatoporoid
stb/d, STBPD	stock tank barrels per day	strt	straight
stcky	sticky	strtd	straightened
std	stand (s) (ing)	struc	structure, structural
std (s) (g)	standards	STTD	sidetracked total depth
stdg	standing	STV	stock tank vapor
stdy	steady	stv	stove oil
stdy	study	stwy	stairway
STDZN	standardization	Sty Mtn	Stony Mountain
Stel	Steele	styo	styolite, styolitic
stenp	stenographer	sub	subsidiary
Stens	Stensvad	sub	substance
STG	stage	sub angl	subangular
stging	straightening	Sub Clarks	sub-Clarksville
STH	sidetracked hole	sub rnd	subrounded
STIF	stiffener	subd	subdivision
stip	stippled	SUBST	substitute
stir	stirrup	substa	substation
stk	stock	suc	sucrose, sucrosic
stk	streak (s) (ed)	suct	suction
stk	stuck	sug	sugary
stl	steel	sul	sulfur
stm	steam	sul wtr	sulfur water
STM	steel tape measurement	SULF	sulfur, sulfuric
stm cyl oil	steam cylinder oil	sulf	sulfated
stm eng oil	steam engine oil	SUM	summarize
STM TR	steam trace (ing)	sum	summary
stn (d) (g)	stain (ed) (ing)	Sum	Summerville
Stn Crl	Stone Corral	Sunb	Sunburst
stn/by	stand by	Sund	Sundance
stncl (d) (g)	stencil (ed) (ing)	Sup	Supai
Stnka	Satanka	supl	supply (ied) (ier) (ing)

supp	supplement
suppt	support
suprv	supervisor
supsd	superseded
supt	superintendent
sur	survey
surf	surface
surp	surplus
SUS	Saybolt universal seconds
susp	suspended
SUSP CLG	suspended ceiling
SUV	Saybolt universal viscosity
SV	solenoid valve
svc	service
svcu	service unit
SVI	smoke volatility index
Svry	severy
SW	salt wash
SW	salt water
SW	socket weld
SW	southwest
SW	spiral weld
SW	switch
SW&W	show gas and water
SW/c	southwest corner
SW/4	southwest quarter
Swas	Swastika
SWB	seal-welded bonnet
SWB	swabbed, swabbing
swbd	switchboard
SWC	sidewall cores
SWCM	saltwater-cut mud
SWD	saltwater disposal
swd	swaged
SWDS	saltwater disposal system
SWDW	saltwater disposal well
Swet	sweetening
SWF	saltwater fracture
swgr	switchgear
SWI	saltwater injection
SWION	shut well in overnight
SWLD	seal weld
SWNP	sidewall neutron porosity
SWP	steam working pressure
SWR	statewide rules
SWRK	switchrack
SWS	sidewall samples
swtr	salt water
SWTS	salt water to surface

SWU	swabbing unit
sx	sacks
sxtu	sextuple
Syc	Sycamore
Syl	Sylvan
sym	symbol
sym	symmetrical
syn	synchronous, synchronizing
syn	synthetic
syn conv	synchronous converter
SYNSCP	synchroscope
SYNTH	synthesis
sys	system
sz	size

T	tee
T	tooth, teeth
T	ton (after number—3T)
T	township (as T2N)
T&B	top and bottom
T&C	threaded and coupled
T&C	topping and coking
T.O.	temperature observation
T&BC	top and bottom chokes
T&G	tongue and groove (joint)
T&R	tubing and rods
T&W	tarred and wrapped
T/	top of (a formation)
T/Box	terminal box
T/BRD	terminal board
T/C	tank car
T/C	turbine compressor
T/pay	top of pay
T/S	top salt
T/sd	top of sand
TA	temporarily abandoned
TA	turn around
tab	tabular, tabulating
TACH	tachometer
Tag	Tagliabue
Tal	Tallahatta
Tamp	Tampico
TAN	tangent

Tan	Tansill
Tann	Tannehill
TAPS	Trans-Alaska Pipeline System
Tark	Tarkio
Tay	Taylor
TB	tank battery
TB	thin bedded
tb	tube
TB/BDL	tube bundle
TBA	tertiary butyl alcohol
TBA	tires, batteries, and accessories
TBE	threaded both ends
tbg	tubing
tbg chk	tubing choke
tbg press	tubing pressure
TBP	true boiling point
TC	temperature controller
TC	tool closed
TC	top choke
TC	tubing choke
TCC	tag closed cup (flash)
TCC	thermofor catalytic cracking
TCC	tubing and casing cutter
Tcf, TCF	trillion cubic feet
Tcf/d, TCF/D	trillion cubic feet per day
TCP	tricresyl phosphate
TCV	temperature control valve
TD	time delay
TD	total depth
TDA	temporary dealer allowance
TDI	temperature differential indicator
TDR	temperature differential recorder
TDT	thermal decay time
TE	temporary
tech	technical, technician
TEFC	totally enclosed, fan cooled
tel	telephone, telegraph
TEL	tetraethyl lead
Tel Cr	Telegraph Creek
Temp	temperature
temp	temporary (ily)
Tens	Tensleep
tens str	tensile strength
Tent	tentaculites
tent	tentative
Ter	tertiary
termin	terminate (ed) (ing) (ion)

Tex	Texana
tex	texture
Text. art.	Textularia articulate
Text. d.	Textularia dibollensis
Text. h.	Textularia hockleyensis
Text. w.	Textularia warreni
TFB	trip (ped) for bit
Tfing	Three Finger
Tfks	Three Forks
TFNB	trip for new bit
tfs	tuffaceous
TG	temperature gradient
th	thence
TH	tight hole
Thay	Thaynes
THC	top hole choke
THD	thermal hydrodealkylation
thd	thread, threaded
Ther	Thermopolis
therm	thermometer
therm ckr	thermal cracker
therst	thermostat
THF	tubinghead flange
THFP	top hole flow pressure
thk	thick, thickness
thrling	throttling
thrm	thermal
thru	through
Thur	Thurman
TI	temperature indicator
ti	tight
TIC	temperature indicator controller
TIH	trip in hole
TIM	Timpas
Timpo	Timpoweap
tk	tank
TKF	tank farm
tkg	tankage
tkr	tanker (s)
tl	tool (s)
tl jt	tool joint
TLE	thread large end
TLG	telegraph
TLH	top of liner hanger
TML	tetramethyl lead
tndr	tender
TNS	tight no show
TO	temperature observation

TO	tool open
TOBE	thread on both ends
TOC	tag open cup (flash)
TOC	top of cement
TOCP	top of cement plug
Tod	Todilto
TOE	threaded one end
TOF	top of fish
TOH	trip out of hole
tol	tolerance
TOL	top of liner
tolu	toluene
Tonk	Tonkawa
tons	tons
TOP	testing on pump
topg	topping
topo	topographic, topography
TOPS	turned over to producing section
Tor	Toronto
Toro	Toroweap
TORT	tearing out rotary tools
TOS	top of salt
tot	total
Tow	Towanda
TP	tool pusher
TP	Travis Peak
TP	treating pressure
TP	tubing pressure
TP&A	theoretical production and allocation
TPC	top of cement
TPC	tubing pressure, closed
TPF	threaded pipe flange
TPF	tubing pressure, flowing
tpk	turnpike
Tpka	Topeka
TPSI	tubing pressure shut in
TR	temperature recorder
TR	trace
TR	tract
trans	transfer (ed) (ing)
trans	transformer
trans	transmission
transl	translucent
transp	transparent
transp	transportation
TRC	temperature recorder controller
Tremp	Terempleleau
Tren	Trenton

TRG	to be conditioned for gas
Tri	Triassic
trilo	trilobite
Trin	Trinidad
trip	triplicate
Trip	Tripoli
trip	tripolitic
trip	tripped (ing)
trk	truck
trkg	trackage
Trn	Trenton
TRNDC	transducer
TRO	to be conditioned for oil
TRQ	torque
trt (r)(d)(g)	treat (er) (ed) (ing)
trtr	treater
TRVL	travel (ed) (ing)
TS	tensile strength
TS	topo sheet evaluation
TSD	temporarily shutdown
TSE	thread small end
TSE-WLE	thread small end, weld large end
TSI	temporarily shut in
TSITC	temperature survey indicated top cement at
TSS	Tar Springs sand
tst (r) (g)	test (er) (ing)
tste	taste
TSTM	too small to measure
TT	tank truck
TT	through-tubing
TTC	through-tubing caliper
TTF	test to follow
TTL	total time lost
TTP	through-tubing plug
TTT	through the tanks
TTTT	turned to test tank
Tuck	Tucker
tuf	tuffaceous
Tul Cr	Tulip Creek
tung carb	tungsten carbide
TURB	turbo, turbine
Tus	Tuscaloosa
TV	television
TVA	temporary voluntary allowance
TVD	true vertical depth
TVP	true vapor process

578

TW	tank wagon
Tw Cr	Twin Creek
TWI	techniques of water-resources investigations
twp	township
TWR	tower
twst	townsite
twst off	twisted off
TWTM	too wet (weak) to measure
TWX	teletype
ty	type
typ	typical
tywr	typewriter

U	unclassified
U/	upper (i.e., U/Simpson)
U/C	under construction
U/L	upper and lower
U/W	used with
U/WTR	utility water
UC	upper casing
UCH	use customer's hose
UD	under digging
UFD	utility flow diagram
UG	under gauge
UG	underground
UGL	universal gear lubricant
UHF	ultra-high frequency
ULJ	perforating, Ultrajet
ult	ultimate
UM	Umiat Meridian (Alaska)
UMB	umbrella (s)
un	unit
UNBAL	unbalanced
unbr	unbranded
UNC	unified coarse thread
unconf	unconformity
uncons	unconsolidated
undiff	undifferentiated
unf	unfinished
UNF	unified fine thread
uni	uniform
Univ	university, universal

UNLD	unloading
UNLDR	unloader
UOCO	Union Oil Company
UPS	uninteruptible power supply
UR	underreaming
UR	unsulfonated residue
UR	used rod
USG	United States gauge
UST	ultrasonic test
UT	ultrasonic examination
UT	upper tubing
UT	upthrown
UTL	utility
UTM	universal transverse mercator
UV	ultraviolet
UV	Union Valley
Uvig. lir.	Uvigerina lirettensis

V	valve
V	viscosity
v, V	volt
V	volume
v.	very (as very tight)
v.n.	very noticeable
v.c.	very common
V.P.S.	very poor sample
v.r.	very rare
v-f-gr	very fine-grained
v-HOCM	very heavily oil-cut mud
v-sli	very slight
V/DWG	vendor drawing
V/L	vapor-liquid ratio
V/S	velocity survey
v%	volume-percent
va	Volt-ampere
vac	vacant
vac	vacation
vac	vacuum
Vag. reg.	Vaginuline regina
Val	Valera
Vang	Vanguard
vap (r)	vapor (izor)
var	variable, various

var	volt-ampere reactive	VOLT	voltage
vari	variegated	VP	vapor pressure
VARN	varnish	VR	vapor recovery
VCP	vitrified clay pipe	vrs	varas
vel	velocity	vrtb	vertebrate
vent	ventilator	vrtl	vertical
Ver Cl	Vermillion Cliff	VRU	vapor recovery unit
Verd	Verdigris	vrvd	varved
vert	vertical	VS	velocity survey
ves	vesicular	vs	versus
VESS	vessel	VSC	volumetric subcommittee
vfg	very fine-grain (ed)	VSGCM	very slightly gas-cut mud
VGC	viscosity-gravity constant	VSGCSW	very slightly gas-cut salt water
VHF	very high frequency		
VHGCM	very heavily (highly) gas-cut mud	VSGCW	very slightly gas-cut water
		VSM	vertical support member
VHGCSW	very heavily (highly) gas-cut salt water	VSO&GCM	very slightly oil and gas-cut mud
VHGCW	very heavily (highly) gas-cut water	VSO&GCSW	very slightly oil and gas-cut salt water
VHO&GCM	very heavily (highly) oil and gas-cut mud	VSOCM	very slightly oil-cut mud
		VSOCSW	very slightly oil-cut salt water
VHO&GCSW	very heavily (highly) oil and gas-cut salt water	VSOCW	very slightly oil-cut water
		VSP	very slightly porous
VHO&GCW	very heavily (highly) oil and gas-cut water	VSSG	very slight show of gas
		VSSO	very slight show of oil
VHOCM	very heavily (highly) oil-cut mud	vt	vapor temperature
		vug	vuggy
VHOCSW	very heavily (highly) oil-cut salt water	vug	vugular
VHOCW	very heavily (highly) oil-cut water		
Vi	Viola		
VI	viscosity index		
VIB	vibrate (tor) (ing)		
Virg	Virgelle		

vis	viscosity	W	wall (if used with pipe)
vis	visible	W	water-supply paper
vit	vitreous	w	watt
Vks	Vicksburg	W	west
VLAC	very light amber cut	W	wide
vlv	valve	W Cr	Wall Creek
VM&P Naptha	varnish makers and painters naphtha	w shd	washed
		W.O.B.	weight on bit
VOC	volatile organic compounds	W-F	Washita-Fredericksburg
Vogts	Vogtsberger	w-hr	watt-hour
vol	volume	W&R	wash and ream
vol%	volume-percent	w/	with
vol. eff.	volumetric efficiency	W/CLR	water cooler

W/L	water load	WHIP	wellhead injection pressure
W/O	west offset	whip	whipstock
W/O	without	WHL	wheel
W/SSO	water with slight show of oil	whse	warehouse
W/S TK	welded steel tank	whsle	wholesale
W/sulf O	water with sulfur odor	wht	white
W/2	west half	WI	washing in
w%	weight-percent	WI	water injection
Wab	Wabaunsee	WI	working interest
WAB	weak air blow	WI	wrought iron
WACT	weight averaged catalyst temperature	Wich Alb	Wichita Albany
		Wich.	Wichita
Wad	Waddell	WIH	water in hole
WAG	water-alternating gas (or water and gas)	Willb	Willberne
		Win	Winona
Wap	Wapanucka	Winf	Winfield
War	Warsaw	Wing	Wingate
Was	Wasatch	Winn	Winnipeg
WaSd	Waltersburg sand	WIP	work in place
Wash	Washita	WIW	water injection well
WB	water blanket	wk	weak
WB	wet bulb	wk	week
WB	Woodbine	wkd	worked
WBIH	went back in hole	wkg	working
WC	water closet	wko	workover
WC	water cushion (DST)	wkor	workover rig
WC	water cut	WL	water loss
WC	wildcat	WL	well lines
WC	Wolfe City	WL	west line
WCM	water-cut mud	WL	wireline
WCO	water-cut oil	wlbr	wellbore
WCTS	water cushion to surface	WLC	wireline coring
WD	water depth	wld	welded, welding
WD	water disposal well	WLD/DET	welding detail (s)
WD	wiring diagram	wldr	welder
Wd R	Wind River	WLT	wireline test
Wdfd	Woodford	WLTD	wireline total depth
WE	weld ends	W'ly	westerly
Web	Weber	WN	weldneck
Well	Wellington	WNSO	water not shut off
WF	new field wildcat, dry	WO	waiting on
WF	waterflood	WO	wash oil
WF	wide flange	WO	wash over
WFD	new field wildcat, discovery	wo	washout
WFD	wildcat field, discovery	WO	wildcat outpost, dry
wgt	weight	WO	work order
WH	wellhead	WO	workover
Wh Dol	white dolomite	WOA	waiting on acid
Wh Sd	white sand	WOA	waiting on allowable

WOB	waiting on battery
WOC	wating on cement
WOCR	waiting on completion rig
WOCT	waiting on cable tools or completion tools
WODP	waiting on drillpipe
WOE	successful wildcat outpost
WOG	waiting on geologist
WOG	water oil or gas
Wolfc	Wolfcamp
WOO	waiting on orders
Wood	Woodside
Woodf	Woodford
WOP	waiting on permit
WOP	waiting on pipe
WOP	waiting on plastic
WOP	waiting on pump
WOPE	waiting on production equipment
WOPL	waiting on pipeline
WOPT	waiting on potential test
WOPU	waiting on pumping unit
WOR	waiting on rig or rotary
WOR	water-oil ratio
WORT	waiting on rotary tools
WOS	washover string
WOSP	waiting on state potential
WOST	waiting on standard tools
WOT	waiting on test or tools
WOT&C	waiting on tank and connection
WOW	waiting on weather
WP	new pool wildcat, dry
WP	wash pipe
WP	well pad
WP	working pressure
WPD	new pool wildcat, discovery
WPM	well pad manifolding
wpr	wrapper
WPT	Windfall Profit Tax
WR	White River
Wref	Wreford
WRG	wiring
WRTB	wash and ream to bottom
WS	shallower pool wildcat, dry
WS	water saturation
WS	whipstock
WS	worldscale
WSD	shallower pool wildcat, discovery

WSD	whipstock depth
wsh (g)	wash (ing)
WSIM	water separation index modfied
WSO	water shutoff
WSONG	water shutoff no good
WSOOK	water shutoff OK
WST	waste
WSW	water source wells
WSW	water supply well
WT	wall thickness (pipe)
wt	weight
wt%	weight-percent
WTB	wash to bottom
wtg	waiting
WTH/PRF	weatherproof
wthr(d)	weather (ed)
wtr (y)	water, watery
wtr. cush	water cushion
WTR/PRF	waterproof
WTR/T	watertight
WTS	water to surface
WUT	water-up to
WW	wash water
WW	water well
Wx	Wilcox

X

X	salt
X-bdd(ing)	crossbedded, crossbedding
X-hvy	extra heavy
X-line	extreme line (casing)
X-over	crossover
X-R	X-ray
X-REF	cross reference
X-SECT	cross section
x-stg	extra strong
x/n	crystalline
XFMR	transformer
XHGR	extra-heavy grade pipe
Xing	crossing
Xlam	cross-laminated
Xln	crystalline
XMTR	transmitter
XO	crossover

STANDARD OIL ABBREVIATOR

XO-sub	crossover sub
xtal	crystal
Xtree	Christmas tree
XW	salt water
XX-Hvy	double extra heavy
XX-STR	double extra strong

y	Yates
yd	yard (s)
yel	yellow
YIL	yellow indicating lamp
YMD	your message of date
YMY	your message yesterday
Yoak	Yoakum
YP	yield point
yr	year
Yz	Yazoo

zen	zenith
Zil	Zilpha
ZN	zinc
Zn	zone

10^{12}	trillion
12 GA W.W.S.	12 gauge wire-wrapped screen (in a liner)
3 PH	three phase
3P ST SW	triple pole single throw switch
3P SW	triple pole switch
4P ST SW	four pole single throw switch
4P SW	four pole switch
8rd	eight round pipe
/ft	per foot
/L	line, as in E/L (eastline)
%	percent
°API	degrees, API
°C	degrees Centrigrade, degrees Celsius
°F	degrees Fahrenheit
µg	microgram (s)

abandoned	abd
abandoned, salvage deferred	ASD
abandoned gas well	abd-gw
abandoned location	abd loc
abandoned oil & gas well	abd-ogw
abandoned oil well	abd-ow
about	abt
above	abv
abrasive jet	abrsi jet
absolute	abs
absolute bottom-hole location	ABHL
absolute open flow potential (gas well)	AOF
absorber	absr
absorption	absrn
abstract	abst
abstract (i.e., A-10)	A
abundant	abun
accelerometers	ACCEL
access	ACC
accessory	ACCESS
account (ing)	acct
accounts receivable	A/R
accumulative, accumulator	accum
acid	ac
acid frac	AF
acid fracture treatment	acfr
acid residue	AR
acid-soluble oil	ASO
acid treat (ment)	AT
acid water	AW
acid-cut mud	ACM
acid-cut water	ACW
acidity or alkalinity	pH
acidize (ed) (ing)	acd
acidized with	A/
acoustic caliper	CMA
acoustic cement	A-Cem
acre-feet	ac-ft
acre (s)	ac
acreage	ac, acrg
acrylonitrile butadiene styrene rubber	ABS
actual	ACT

actual drilling	AD
actual drilling cost	ADC
actual drilling time	ADT
actual jetting time	AJT
actuated, actuator	ACT
adapter	adpt
addition or modification request	AMR
additional	addl
additive	add
adhesive	ADH
adjustable	adj
adjustable spring wedge	ASW
adjustments and allowances	A&A
administration, administrative	adm
adomite	ADOM
adsorption	adspn
advanced	advan
aeration, aerator	AER
affidavit	afft
affirmed	affd
after acidizing	AA
after condenser	AF/COND
after cooler	AF/CLR
after federal income tax	AFIT
after fracture	AF
after receipt of order (purchasing term)	ARO
after shot	AS
after the tanks	ATT
after top center	ATC
after treatment	AT
agglomerate	aglm
aggregate	AGGR
agitator	AG
air conditioning	A/C
air cooled	A/CLD
air cooler	A/CLR
air quality control region	AQCR
air quality maintenance area	AQMA
alarm	alm
Alaskan North Slope	ANS
Albany	Alb
alcoholic	alc
algae	alg
alignment (ing)	ALIGN
alkaline, alkalinity	alk
alkalinity or acidity	pH
alkylate, alkylation	alkyl
all thread	AT

allocation	**ALOC**	annubar	**ANUB**
allowable not yet available	**ANYA**	annular velocity	**AV**
allowable, allowance	**ALLOW**	annulus	**an**
alloy	**ALY**	annulus	**Ann**
along	**alg**	annunciator	**ANUC**
alternate	**alt**	apartment	**apt**
alternating current	**AC**	apparatus	**APPAR**
altitude	**ALT**	apparent (ly)	**apr**
aluminum	**AL**	appears, appearance	**app**
aluminum conductor steel	**ACSR**	appliance	**appl**
reinforced		application	**applic**
ambient	**amb**	applied	**appl**
American Chemical Society	**ACS**	approved	**appd**
American melting point	**AMP**	approved total depth	**ATD**
American Petroleum Institute	**API**	approximate (ly)	**approx**
American Public Health	**APHA**	aqueous	**aq**
Association		aragonite	**arag**
American Society for Testing &	**ASTM**	Arapahoe	**Ara**
Materials		Arbuckle	**Arb**
American Standards	**ASA**	Archeozoic	**Archeo**
Association		architectural	**arch**
American Steel & Wire gauge	**AS&W ga**	area of mutual interest	**AMI**
American Water Works	**AWWA**	arenaceous	**aren**
Association		argillaceous	**arg**
American Wire gauge	**AWG**	argillite	**arg**
ammeter	**AMM**	Arkadelphia	**Arka**
ammonia	**NH$_3$**	arkose (ic)	**ark**
ammonium chloride	**NH$_4$Cl**	armature	**arm**
amorphous	**amor**	aromatics	**arom**
amortization	**amort**	around	**arnd**
amount	**amt**	arrange (ed) (ing) (ment)	**ARR**
amount not reported	**ANR**	articles published in outside	**OP**
ampere	**amp**	journals/books	
ampere-hour	**amp-hr**	artificial lift	**AL**
amphipore	**amph**	as soon as possible	**ASAP**
amphistegina	**amph**	asbestos	**asb**
analysis, analytical	**anal**	Ashern	**Ash**
anchor (age)	**ANC**	asphalt, asphaltic	**asph**
and husband	**et con.**	asphaltic stain	**astn**
and husband	**et vir.**	assembly	**assy**
and others	**et al.**	assigned	**assgd**
and the following	**et seq.**	assignment	**asgmt**
and wife	**et ux.**	assistant	**asst**
angle, angular	**ang**	associate (ed) (s)	**assoc**
angstrom unit	**Å**	association	**assn**
angulogerina	**angul**	Association of American	**AAR**
anhydrite stringer	**AS**	Railroads	
anhydrite, anhydritic	**anhy**	Association of Desk & Derrick	**ADDC**
anhydrous	**anhyd**	Clubs	

Association of Official Agricultural Chemists	**AOAC**
at rate of	**ARO**
Atlas Bradford modified	**ABM**
atmosphere, atmospheric	**atm**
Atoka	**At**
atomic	**at**
atomic weight	**at wt**
attach (ed) (ing) (ment)	**ATT**
attempt (ed)	**att**
attorney	**atty**
Audit Bureau of Circulation	**ABC**
auditorium	**aud**
Austin	**Aus**
Austin chalk	**AC**
Authorization for Commitment	**AFC**
Authorization for Expenditure	**AFE**
Authorization to Proceed	**ATP**
authorized	**auth**
authorized depth	**AD**
Authorized for Construction	**AFC**
automatic	**auto**
automatic custody transfer	**ACT**
automatic data processing	**ADP**
automatic transmission fluid	**ATF**
automatic volume control	**AVC**
automotive	**auto**
automotive gasoline	**autogas**
Aux Vases sand	**AV**
auxiliary	**aux**
auxiliary flow diagram	**AFD**
available	**avail**
average	**avg**
average flowing pressure	**AFP**
average freight rate assessment	**AFRA**
average injection rate	**AIR**
average penetration rate	**APR**
average treating pressure	**ATP**
average tubing pressure	**ATP**
aviation	**av**
aviation gasoline	**avgas**
awaiting	**awtg**
award	**AWD**
azeotropic	**aztrop**

B

back flush	**BKFLSH**
back pressure	**BP**
back-pressure valve	**BPV**
back scuttled	**B/S**
back to back	**B/B**
backed out (off)	**BO**
backwash	**BKWSH**
baffle	**BFL**
bailed	**bld**
bailed dry	**B/dry**
bailer	**blr**
bailer feed water	**BFW**
bailing	**blg**
balance	**BAL**
ball joint	**B/JT**
ball sealers	**BS**
ball valve	**B/Vlv**
Balltown sand	**Ball**
band (ed)	**bnd**
barge deck to mean low water	**BD-MLW**
barite (ic)	**bar**
Barker Creek	**Bark Crk**
Barlow lime	**Bar**
barometer, barometric	**bar**
barrel	**bbl**
barrel water load	**BWL**
barrels acid	**BA**
barrels acid residue	**BAR**
barrels acid water	**BAW**
barrels acid water per day	**BAWPD**
barrels acid water per hour	**BAWPH**
barrels acid water under load	**BAWUL**
barrels condensate	**BC**
barrels condensate per day	**BCPD**
barrels condensate per hour	**BCPH**
barrels condensate per million	**BCPMM**
barrels diesel oil	**BDO**
barrels distillate	**BD**
barrels distillate per day	**BDPD**
barrels distillate per hour	**BDPH**
barrels fluid	**BF**
barrels fluid per day	**BFPD**
barrels fluid per hour	**BFPH**
barrels formation water	**BFW**

barrels frac oil	BFO	base of the salt	B slt
barrels fresh water	BFW	Base Pennsylvanian	BP
barrels liquid per day	BLPD	base plate	BSPL
barrels load	BL	base salt	B/S
barrels load & acid water	BL&AW	basement	bsmt
barrels load condensate	BLC	basement (granite)	base
barrels load condensate per day	BLCPD	basic sediment	BS
barrels load condensate per hour	BLCPH	basic sediment & water	BS&W
		basket	bskt
barrels load oil	BLO	Bateman	Bate
barrels load oil per day	BLOPD	battery	btry
barrels load oil per hour	BLOPH	Baume	Be
barrels load oil recovered	BLOR	beaded and center beaded	B & CB
barrels load oil to be recovered	BLOTBR	Bear River	Bear Riv
barrels load oil yet to recover	BLOYTR	bearing	brg
barrels load water	BLW	Bearpaw	BP
barrels load water per day	BLWPD	Beckwith	Beck
barrels load water per hour	BLWPH	becoming	bec
barrels load water to recover	BLWTR	bedding	BDNG
barrels mud	BM	before acid treatment	BAT
barrels new oil	BNO	before federal income tax	BFIT
barrels new water	BNW	before top dead center	BTDC
barrels oil	BO	Beldon	Bel
barrels oil per calendar day	BOPCD	belemnites	Belm
barrels oil per day	BOPD	bell and bell	B & B
barrels oil per hour	BOPH	bell and flange	B & F
barrels oil per producing day	BOPPD	bell and spigot	B & S
barrels per barrel	B/B	Belle City	Bel C
barrels per day	B/D	Belle Fourche	Bel F
barrels per hour	B/hr	benchmark	BM
barrels per minute	B/M	bending schedule	B/S
barrels per stream day	BPSD	Benoist (Bethel) sand	Ben, BT
barrels per stream day (refinery)	B/SD	bent & bowed pipe	B&B
		Benton	Ben
barrels per well per day	BPWPD	bentonite	Bent
barrels pipeline oil	BPLO	benzene	bnz
barrels pipeline oil per day	BPLOPD	benzene toluenexylene (unit)	BTX (unit)
barrels salt water	BSW	Berea	Be
barrels salt water per day	BSWPD	between	btw
barrels salt water per hour	BSWPH	bevel (ed)	bev
barrels water	BW	bevel both ends	BBE
barrels water over load	BWOL	bevel large end	BLE
barrels water per day	BWPD	bevel one end	BOE
barrels water per hour	BWPH	bevel small end	BSE
Bartlesville	Bart	beveled for welding	BV/WLD
basal	bsl	beveled end	B.E.
Basal Oil Creek sand	BOCS	bid summary	BID SUM
base	B/	Big Horn	B. Hn.
Base Blane	B. Bl	Big Injun	B. Inj.

Big Lime	**B. Ls**	blower	**BLWR**
Bigenerina	**Big.**	blowout equipment	**BOE**
Bigenerina floridana	**Big. f.**	blowout preventer	**BOP**
Bigenerina humblei	**Big. h.**	blowout preventer equipment	**BOPE**
Bigenerina nodosaria	**Big. nod.**	blue	**bl**
bill of lading	**B/L**	board	**bd**
bill of material	**B/M**	board-foot; board-feet	**bd ft**
bill of sale	**B/S**	Bodcaw	**Bod**
billion	**B**	body wall loss	**BWL**
billion cubic feet	**BCF,**	boiled water	**BW**
	Bcf	boiler	**BLR**
billion cubic feet per day	**BCFD,**	boiler feed water	**BFW**
	Bcfd	boiling point	**BP**
billion standard cubic feet	**Bscf**	Bois d'Arc	**Bd'A**
billion standard cubic feet per	**Bscf/D,**	Bolivarensis	**Bol.**
day	**Bscfd**	Bolivina a.	**Bol. a.**
binary	**BIN**	Bolivina floridana	**Bol. flor.**
biochemical oxygen demand	**BOD**	Bolivina perca	**Bol. p.**
biotite	**bio**	Bone Spring	**BS**
Birmingham (or Stubbs) iron	**BW ga**	Bonneterre	**Bonne**
wire gauge		booster	**BSTR**
Birmingham wire gauge	**Bwg**	borehole compensated sonic	**BHCS**
bitumen	**bit**	bottom sediment & water	**BS&W**
bituminous	**bit**	bottom (ed)	**btm (d)**
black	**blk**	bottom choke	**btm chk**
Black Leaf	**Blk Lf**	bottom hole	**BH**
Black Lime	**Blk Li**	bottom-hole assembly	**BHA**
Black Magic (mud)	**BM**	bottom-hole choke	**BHC**
black malleable iron	**BMI**	bottom-hole flowing pressure	**BHFP**
Black River	**B. Riv**	bottom-hole location	**BHL**
black sulfur water	**BSUW**	bottom-hole money	**BHM**
blank liner	**blk lnr**	bottom-hole orientation	**BHO**
blast cabinet	**Bl/Cb**	bottom-hole pressure	**BHP**
blast joint	**BL/JT**	bottom-hole pressure bomb	**BHPB**
bleeding	**bldg**	bottom-hole pressure, closed	**BHPC**
bleeding gas	**bldg**	(See SIBHP and BHSIP)	
bleeding oil	**bldo**	bottom-hole pressure, flowing	**BHPF**
blend (ed) (er) (ing)	**BLND**	bottom-hole pressure survey	**BHPS**
blew out	**BO**	bottom-hole shutin pressure	**BHSIP**
blind flange	**BLD FLG,**	bottom-hole temperature	**BHT**
	BF	bottom of given formation (i.e.,	**B/**
Blinebry	**Blin**	B/Frio)	
block	**blk**	bottom sediment	**BS**
block valve	**BV**	bottom settlings	**BS**
blocked off	**BO**	bottoms sediment and water	**B.S.&W.**
Blossom	**Blos**	boulders	**bldrs**
blow	**blo**	boundary	**bndry**
blowdown	**BLDWN**	box end	**be**
blow-down test	**BDT**	box (es)	**bx**

brace (ed) (ing)	**BRC**
brachiopod	**brach**
bracket (s)	**brkt(s)**
brackish (water)	**brksh**
Bradenhead flange	**BHF**
brake horsepower	**bhp**
brake horsepower-hour	**bhp-hr**
brake mean effective pressure	**BMEP**
brake specific fuel consumption	**BSFC**
brakes	**BRKS**
break (broke)	**brk**
breakdown	**bkdn**
breakdown acid	**BDA**
breakdown pressure	**BDP**
breaker	**BRKR**
breccia	**brec**
bridge plug	**BP**
bridged back	**BB**
Bridger	**Brid**
Brinell hardness number	**BHN**
British Standards Institution	**BSI**
British thermal unit	**BTU**
brittle	**brtl, brit**
broke (break) down formation	**BDF**
broken	**brkn**
broken sand	**brkn sd**
bromide	**brom**
brown	**brn or br**
Brown and Sharpe gauge	**B&S ga**
Brown lime	**Brn Li**
brown oil stain	**BOS**
brown shale	**brn sh**
brownish	**bnish**
bryozoa	**bry**
Buckner	**Buck**
buck off	**b/off**
buck on	**b/on**
Buckrange	**Buckr**
budgeted depth	**BD**
buff	**bf**
building	**bldg**
building derrick	**bldg drk**
building rig	**BR**
building roads	**bldg rds**
Buliminella textularia	**Bul. text.**
bulk plant	**BP**
bulk vessel	**B/VESS**
bull plug	**BP**
bullets	**blts**

Bullwaggon	**Bull W**
bumper	**bmpr**
bundle	**BDL**
Burgess	**Burg**
burner	**bunr**
bushel	**bu**
bushing	**BSHG**
butane and propane mix	**BP mix**
butane-butene fraction	**BB fraction**
butt weld	**BW, BTWLD**
butterfly valve	**BRFL/V, BTFL/V**
buttress thread	**butt**
buttress thread coupling	**BTC**
buzzer	**BUZ**
bypass	**BYP**
bypass cooler	**BP/CLR**

cabinet	**CAB**
cable (ing)	**CBL**
cable tool measurement	**CTM**
cable tools	**CT**
Caddell	**Cadd**
cadmium plate	**CD PL**
cake	**ck**
calcareous, calcerenite	**calc**
calceneous	**cale**
calcite, calcitic	**cal**
calcium	**calc**
calcium-base grease	**calc gr**
calcium chloride	**CaCl$_2$**
calcium oxide	**CaO**
calculate (ed), calculation	**Calc**
calculated absolute open flow	**CAOF**
calculated open flow (potential)	**COF**
calendar day	**CD**
calibrate (tion)	**CALIBR**
caliche	**cal**
California Coordinate System	**CCS**
caliper log	**CAL**
caliper survey	**cal**

caulking	**CLKG**	cast iron	**CI**
calorie	**cal**	cast steel	**CS**
Calvin	**Calv**	cast-iron bridge plug	**CIBP**
Cambrian	**Camb**	cat-cracked light gas oil	**CCLGO**
Camp Colorado	**Cp Colo**	Cat Creek	**Cat Crk**
Cane River	**Cane Riv**	Catahoula	**Cat**
canvas-lined metal petal basket	**CLMP**	catalog	**CAT**
canyon	**cany, cyn**	catalyst, catalytic	**CAT**
Canyon Creek	**Cany Crk**	catalytic cracker	**Cat ckr**
capacity, capacitor	**cap**	catalytic cracking unit	**CCU**
Capitan	**Cap**	cathode ray tube	**CRT**
carbon copy	**CC**	cathodic	**cath**
carbon dioxide	**CO₂**	Cattleman	**Ctlmn**
carbon disulfide	**CS₂**	caustic	**caus**
carbon monoxide	**CO**	caving (s)	**cvg(s)**
carbon oxygen	**CO**	cavity	**cav**
carbon residue (Conradson)	**CR Con**	Cedar Mountain	**Cdr Mtn**
carbon steel	**CS**	cellar	**cell**
carbon tetrachloride	**carb tet**	cellar & pits	**C & P**
carbonaceous	**carb**	cellular	**cell**
carburetor air temperature	**CAT**	Celsius	**C**
care of	**c/o**	cement (ed)	**cem**
Carlile	**Car**	cement (ed) (ing)	**cmt (d) (g)**
carload	**CL**	cement dump bailer	**CDB**
Carmel	**Carm**	cement evaluation	**CET**
Carrizo	**Cz**	cement friction reducer	**CFR**
carton	**ctn**	cement friction retarder	**CFR**
cased hole	**C/H**	cement in place	**CIP**
cased reservoir analysis	**CRA**	cement to surface	**CTS**
casing	**csg**	cemented through perforations	**cp's**
casing cemented (depth)	**CC**	cementer	**cmtr**
casing choke	**Cck**	Cenozoic	**Ceno**
casing collar locator	**CCL**	center (ed)	**cntr**
casing collar perforating record	**CCPR**	center (land description)	**C**
casing flange	**CF**	center line	**C/L**
casing point	**csg pt, CP**	center of casinghead flange	**CCHF**
casing pressure	**csg pres, CP**	center of gravity	**CG**
		center of tubing flange	**CTF**
casing pressure, shut in	**CPSI**	center section line	**CSL**
casing pressure, closed	**CPC**	center to center	**C to C**
casing pressure, flowing	**CPF**	center to end	**C to E**
casing seat	**CS**	center to face	**C to F**
casing set at	**CSA**	Centigrade	**C**
casinghead	**csg hd**	centigram	**cg**
casinghead flange	**CHF**	centiliter	**cl**
casinghead gas	**CHG**	centimeter	**cm**
casinghead pressure	**CHP**	centimeter-gram-second system	**cgs**
Casper	**Casp**	centimeters per second	**cm/sec**
cast carbon steel	**CCS**	centipoise	**cp**

centistokes	cs	chloride (s)	chl
central compressor plant	CCP	chlorinator	CHLR
central delivery point	CDP	chlorine	CL$_2$
centralizers	cent	chlorine log	chl log
centrifugal	centr	chloritic	chl
centrifuge	cntf	choke	chk
cephalopod	ceph	Chouteau lime	Chou
Ceratobulimina eximia	Cert. ex.	Christmas tree	Xtree
certified	CERT	chromatograph	chromat
certified drawing outline	CDO	chrome molybdenum	cr moly
certified public accountant	CPA	chromium	chrome
cetane number	CN	Chugwater	Chug
chain operated	CH OP	Cibicides	Cib.
chairman	chrm	Cibicides hazzardi	Cib. h.
chalcedony	chal	Cimarron	Cima
chalk	chk	circle	cir
chalky	chky	circuit	cir
chamber	CHMBR	circular	cir
chamfer	CHAM	circular mils	cir mils
change (ed) (ing)	chng	circulate & condition	C&C
changed (ing) bits	CB	circulate and reciprocate	C&R
changed drillpipe	chngd DP	circulate bottoms up	CBU,
channel	CHNL		ccBU
Chappel	Chapp	circulate (ing) (tion)	circ
characteristics	CHAR	circulated and conditioned mud	C&CM
charge (ed) (ing)	chrg (d) (ing)	circulated and conditioned hole	C&CH
		circulated out	CO
Charles	Char	circumference	CRCMF
chart	cht	Cisco	Cis
Chattanooga shale	Chatt	Clagget	Clag
check	ck	Claiborne	Claib
check valve	CHKV	clarifier	CLFR
checked	chkd	Clarksville	Clarks
checkerboard	Chkbd	class	CL
checkered plate	CHKD PL	classification	CLASS
chemical oxygen demand	COD	clastic	clas
chemical products	chem prod	Clavalinoides	Clav
chemical, chemist, chemistry	chem	clay filled	CF
chemically pure	cp	Claystone	clyst
chemically retarded acid	CRA	Clayton	Clay
Cherokee	Cher	Claytonville	Clay
chert	cht	clean out	CO
cherty	chty	clean out & shoot company	CO & S
Chester	Ches	clean up	CU
chicksan	cksn	clean (ed) (ing)	cln (d) (g)
Chimney Hill	Chim H	cleaned out to total depth	COTD
Chimney Rock	Chim R	cleaning out, cleaned out	CO
Chinle	Chin	cleaning to pits	CTP
chitin (ous)	chit	clear, clearance	clr

Clearfork	**Clfk**	commission agent	**C/A**
clearing	**clrg**	commissioner	**commr**
Cleveland	**Cleve**	common	**com**
Cleveland open cup	**COC**	Common Business-Oriented	**COBOL**
Cliff House	**Cliff H**	Language	
clockwise	**cw**	common data base	**CDB**
closed	**clsd**	command data base	**CDBTF**
closed cup	**CC**	task force	
closed hole	**CH**	communication	**comm**
closed hydrocarbon drain	**CHD**	community	**comm**
closed-in pressure	**CIP**	commutator	**COMUT**
Cloverly	**Clov**	compact	**cmpt**
coarse crystalline	**crs-xln**	companion flange bolt & gasket	**CFB & G**
coarse grained	**C-gr**	companion flange one end	**CFOE**
coarse (ly)	**crs, c**	companion flanges bolted on	**CFBO**
coat and wrap (pipe)	**C & W**	company	**Co**
coated	**ctd**	company operated	**Co. Op.**
Cockfield	**Cf**	company-operated service	**Co. Op. S.S.**
Coconino	**Coco**	stations	
Codell	**Cod**	comparator	**CMPARTR**
Cody (Wyoming)	**Cdy**	compartment	**compt**
coefficient	**coef**	compensated neutron log	**CNL**
coiled tubing unit	**CTU**	complete (ed)(tion)	**comp**
coke oven gas	**COG**	completed natural	**comp nat**
cold drawn	**CD**	completed on pump	**C.O.P.**
cold finished	**CF**	complete with	**C/W**
cold rolled	**CR**	component (s)	**COMPT**
cold rolled steel	**CRS**	components	**compnts**
cold water equivalent	**CWE**	compound	**CMPD**
cold working pressure	**CWP**	compression and absorption plant	**C&A**
Coleman Junction	**Cole Jct**	compression-ignition engine	**CI engine**
collar	**COL**	compression ratio	**CR**
collect (ed) (ing) (tion)	**coll**	compressor	**compr**
collector	**CLTR**	compressor station	**compr sta**
Color American Standard Test	**Col**	computer	**COMPTR**
Method	**ASTM**	computer control system	**CCS**
colored	**COL**	computer-processed	**MCT**
column	**COL**	interpretation	
Comanche	**Com**	concentrate	**conc**
Comanche Peak	**Com Pk**	concentric	**cncn**
Comanchean	**Cmchn**	concentric	**conc**
Comatula	**Com**	conchoidal	**conch**
combined, combination	**comb**	conclusion	**concl**
combustion	**COMB**	concrete	**conc**
coming out of hole	**COOH**	condensate	**cond**
commenced	**comm**	condensate-cut mud	**CCM**
comment	**COMT**	condenser	**cdsr**
commercial	**coml**	condition (ed) (ing)	**cond**
commission	**comm**	conductivity	**condt**

conductor (pipe)	condr	cooling tower	CT
conduit	CND	cooling tower	CLT/TWR
confidential	conf	cooperative	co-op
confirm (ed) (ing)	conf	coordinate	coord
confirming telephone order (purchasing term)	CVO	Coordinating Research Council Inc.	CRC
confirming telephone order (purchasing term)	CPO	Copper River Meridian (Alaska)	CPR
conflict	confl	coquina	coq
conglomerate, conglomeritic	cglt	core barrel	CB
conglomerate (itic)	cong	core hole	C
connection	conn	core (ed) (ing)	cr (d) (g)
conodonts	cono	cored	crd
Conradson carbon residue	CCR	coring	cg
conserve, conservation	consv	corner	cor
consolidated	con	corner	CNR
consolidated	consol	corporation	Corp
constant	const	correct (ed) (ion)	corr
construction	const	corrected gravity	CG
consumer tank car	CTC	corrected total depth	CTD
consumer tank wagon	CTW	correlation	correl
contact caliper	C-Cal	correspondence	corres
container	cntr	corrosion	corr
containment	CNTN	corrosion allowance	CA
contaminated, contamination	contam	corrugated	corr
continue (ed)	cont (d)	cost and freight	C & F
continuous blowdown	CB	cost insurance and freight	C.I.F.
continuous directional service	CDS	cost per gallon	CPG
continuous flowmeter	CFM	Cottage Grove	Cott G
contiuous weld	CW	Cotton Valley	CV
contour interval (map)	CI	Council Grove	Counc G
contract depth	CD	counter electromotive force	CEMF
contractor	contr	counter weight	CNT WT
contractor (i.e., C/John Doe)	C/	counterbalance (pumping equip)	CB
contractor furnished equipment	CFE	counterclockwise	ccw
contractor responsibility	contr resp	county	Cnty
contribution	contrib	county school lands	CSL
control(s)	cntl	coupling	cplg
control building	CN/BD	cover	CVR
control valve	CV	Cow Run	CR
controller	cntr	cracker	Crkr
convector, convection	CONVT	cracking	crkg
converse	conv	cradle (s)	CRDL
convert (er) (ed)	CVTR	crane	CRN
conveyor	cnvr	crawl beam	C/BM
cooler	CLR	creek	crk
cooling water	CW	crenulated	cren
cooling water return	CWR	Cretaceous	Cret
cooling water supply	CWS	crinkled	crnk

crinoid (al)	**Crin**
Cristellaria	**Cris**
critical	**crit**
critical compression pressure	**CCP**
critical compression ratio	**CCR**
Cromwell	**Crom**
cross	**CRS**
crossbedded	**crbd**
crossover	**CX**
crown block	**crn blk**
crude oil	**CO**
crude oil purchasing	**COP**
cryptocrystalline	**crypto-xln**
crystalline	**cryst**
cubic	**cu**
cubic centimeter	**cu cm**
cubic centimeter	**cc**
cubic feet gas	**cfg**
cubic feet gas per day	**cfgd**
cubic feet gas per hour	**cfgh**
cubic feet per barrel	**cu ft/bbl**
cubic feet per day	**cfd**
cubic feet per minute	**cu ft/min, CFM**
cubic feet per pound	**CFP**
cubic feet per second	**cu ft/ sec, CFS**
cubic foot (feet)	**cf**
cubic inch	**cu in.**
cubic meter	**cu m**
cubic meters per day	**m³/d**
cubic yard	**cu yd**
cubical	**CUB**
culvert	**culv**
cumulative	**cum**
Curtis	**Cur**
curve	**CRV**
cushion	**cush**
customer	**CUST**
cut across grain	**CAG**
Cut Bank	**Cut B**
cut drilling line	**CDL**
cutbank	**cutbk**
Cutler	**Cutl**
cutting oil	**Cut Oil**
cutting oil soluble	**Cut Oil Sol**
cutting oil-active-sulfurized dark	**Cut Oil Act Sul-dk**
cutting oil-active-sulfurized transparent	**Cut Oil Act Sul-transpt**
cutting oil-inactive-sulfurized	**Cut Oil Inact Sul**
cutting oil-straight mineral	**Cut Oil St Mrl**
cuttings	**ctg(s)**
Cyclamina	**Cyc.**
Cyclamina cancellata	**Cyc. canc.**
cycles per minute	**cpm**
cycles per second	**cps**
cyclone	**CYC**
cylinder	**cyl**
Cypress sand	**Cy Sd**
cypridopsis	**cyp.**

daily allowable	**DA**
daily average injection barrels	**DAIB**
Dakota	**Dak**
damper	**dmpr**
Dantzler	**Dan**
dark	**dk**
dark brown oil	**DBO**
dark brown oil stains	**DBOS**
Darwin	**Dar**
data processing	**DP**
data sheet	**D/S**
date actually started drilling	**spud date**
date of first production	**DFP**
datum	**dat, DM**
datum faulted out	**DFO**
Davit	**DVT**
day	**D**
day to day	**D/D**
days since spudded	**DSS**
dead	**dd**
dead oil show	**DOS**
deadweight tester	**DWT**
deadweight tons	**DWT**
Deadwood	**Deadw**
deaerator	**deaer**

dealer	dlr	desalter	desalt
dealer tank wagon	DTW	description	desc
deasphalting	deasph	Desert Creek	Des Crk
debutanizer	debutzr	design	dsgn
decibel	db	Desk and Derrick	D & D
decigram	dg	desorbent	desorb
deciliter	dl	destination	dstn
decimal	dec	desulfurizer	desulf
decimeter	dm	desuperheater	DSUPHTR
decline	decl	detail (s)	det
decrease (ed) (ing)	decr	detector	det, DCTR
deep pool test	DPT	detergent	deterg
deepen	dpn	detrital	detr, dtr
deepening	dpg	develop	DVL
deethanizer	deethzr	develop (ed) (ment)	devel
deflection	defl	development	D
Degonia	Deg	development gas well	DG
degree day	DD	development oil	DO
degree (s)	deg	development oil well	DO
degrees API	°API	development redrill (sidetrack)	DR
degrees Centigrade	°C	development well, carbon dioxide	DC
degrees Fahrenheit	°F		
deisobutanizer	deisobut	development well, helium	DH
Del Rio	Del R	development well, sulfur	DSU
Delaware	Dela	development well workover	DX
Delaware River Area Petroleum Refineries	DRAPR	deviate, deviation	dev
		deviation degrees	DD
delayed coker	DC	Devonian	Dev
delivery (ed) (ability)	delv	dew point	DP
delivery point	delv pt	dewatering	DWTR
demand meter	DM	dewaxing	dewax
demolition	dml	Dexter	Dext
demurrage	demur	diagonal	diag
dendrite (ic)	dend	diagram	diag
dense	ds, dns	diameter	dia
density log	D/L, DENL	diamond bit	DB
department	dept	diamond core	DC
Department of Energy	DOE	diamond core bit	DCB
depletion	depl	diaphragm	diaph
depreciation	deprec	dichloride	dichlor
depropanizer	deprop	dichloro-diphenyl-trichloroethane	DDT
depth	dpt		
depth bracket allowable	DBA	diesel (oil)	dsl
depth recorder	dpt rec	diesel fuel	DF
derrick	drk	diesel hydrogen desulfurization	DHDS
derrick floor	DF	diesel index	D.I.
derrick floor elevation	DFE	Diesel No. 2	D-2
Des Moines	Des M	diesel oil cement	DOC

diethanolamine	**DEA**	documentation	**DOCREQ**
diethanolamine unit	**DEA unit**	dogleg severity	**DLS**
diethylene	**diethy**	doing business as	**d/b/a**
different (ial) (ence)	**diff**	dolomite (ic)	**dolo**
differential pressure	**D/P**	dolstone	**dolst**
differential valve (cementing)	**DV**	domestic	**dom**
digging cellar	**DC**	domestic airline	**dom AL**
digging cellar and slush pits	**DCLSP**	domestic water	**DOM WTR**
digging slush pits	**DSP**	Dornick Hills	**Dorn H**
digital	**DGTL**	Dothan	**Doth**
diglycolamine	**DGA**	double	**DBL**
diluted	**dilut**	double end	**DE**
dimension	**dim**	double extra heavy	**XX-Hvy**
dimethyl sulfide	**DMS**	double extra strong	**XX-STR**
diminish (ing)	**dim**	double hub	**DH**
Dinwoody	**Din**	double pole (switch)	**DP**
dipmeter	**DM**	double pole double base (switch)	**DPDB**
direct (tion) (tor)	**dir**	double pole double throw switch	**DPDTSW**
direct current	**DC**		
directional drilling	**dir drlg**	double pole single base (switch)	**DPSB**
directional survey	**dir sur, DS**	double pole single throw switch	**DPST SW**
dirty water disposal	**DWD**	double pole switch	**DP SW**
discharger	**disc, disch**	double random lengths	**DRL**
Discorbis	**Disc.**	Douglas	**Doug**
Discorbis gravelli	**Disc. grav.**	down	**dn**
Discorbis normada	**Disc. norm.**	downthrown	**DT**
Discorbis yeguaensis	**Disc. y.**	dozen	**doz**
discount	**disc**	draft gauge	**DG**
discover (y) (ed) (ing)	**disc**	drain	**dr**
discovery allowable requested	**DAR**	drainage	**drng**
dismantle	**disman**	drawing	**DWG**
dismantle (ing)	**dsmt (g)**	drawworks	**dwks**
displaced, displacement	**displ**	dressed dimension four sides	**d-d-4-s**
disseminated	**dism**	dressed dimension one side and one edge	**d-d-1-s-1-e**
distance	**dist**		
distillate	**dstl**	dressed four sides	**d-4-s**
distillate-cut mud	**DCM**	dressed one side	**d-1-s**
distillate, distillation	**dist**	dressed two sides	**d-2-s**
distribute (ed) (ing) (ion)	**distr**	Dresser Atlas	**DA**
distributed control system	**DCS**	drier, drying	**dry**
district	**dist**	drift angle	**DA**
ditto	**do**	drill	**drl**
division	**div**	drill (ed) (ing) out	**DO**
division office	**D/O**	drill (ed) (ing) plug	**D/P**
division order	**D.O.**	drill and complete	**D & C**
dock operating building	**DOBLDG**	drill collar	**DC**
Dockum	**Doc**	drill floor	**DF**
doctor-treating	**doc-tr**	drillpipe	**DP**
document	**doc**		

drillpipe measurement	**DPM**	dual upper tubing	**DUT**
drillpipe unloaded	**DPU**	dually completed	**DC**
drill site	**DS**	Duck Creek	**Dk Crk**
drillsite facility	**DSF**	dumped	**DMPD**
drillstem	**DS**	Dun & Bradstreet	**D & B**
drillstem test	**DST**	Duperow	**Dup**
drilled	**drld**	duplex	**dx**
drilled-out cement	**DOC**	duplicate	**dup**
drilled-out depth	**DOD**	duration	**DUR**
drilled-out plug	**DOP**	Dutcher	**Dutch**
driller	**drlr**	dyna-drilling	**DD**
driller's top	**D/T**	dynamic	**dyn**
driller's total depth	**DTD**		
drilling	**drlg**		
drilling and well completion	**DWC**		
drilling break	**DB**		
drilling deeper	**DD**		
drilling line	**DL**		
drilling mud	**DM**		
drilling suspended indefinitely	**DSI**	each	**ea**
drilling tender	**D/T**	Eagle	**Egl**
drilling time	**DT**	Eagle Ford	**EF**
drilling with air	**DWA**	Eagle Mills	**EM**
drilling with gas	**DWG**	Earlsboro	**Earls**
drilling with mud	**DWM**	early well tie-ins	**EWT**
drilling with oil	**DWO**	east	**E**
drilling with salt water	**DWSW**	east boundary line	**E/BL**
drillstem test with straddle packers	**DST (Strd)**	East Cimarron Meridian (Oklahoma)	**ECM**
drive (ing) (er)	**DRV (R), dr**	east half	**E/2**
dropped	**dropd**	east line	**E/L**
drum	**dr, DRM**	east of Rockies	**EOR**
druse	**dr**	east of west line	**E of W/L**
drusy	**drsy**	east offset	**E/O**
dry and abandoned	**D & A**	east quarter	**E/4**
Dry Creek	**Dr Crk**	easterly	**E'ly**
dry desiccant dehydrator	**DDD**	Eau Claire	**Eau Clr**
dry film thickness	**DFT**	eccentric	**ECC**
dry gas	**DG**	echinoid	**ech**
dry-hole contribution	**DHC**	economics, (y), economizer	**econ**
dry hole drilled deeper	**DHDD**	Ector (County, TX)	**Ect**
dry-hole money	**DHM**	education (tor)	**Educ (r)**
dry hole reentered	**DHR**	Edwards	**Edw**
dual	**D**	Edwards lime	**Ed lm**
dual (double) wall packer	**DWP**	effective	**eff**
dual injection focus log	**DIFL**	effective depth	**E.D.**
dual lower casing	**DLC**	effective horsepower	**EHP**
dual lower tubing	**DLT**	efficiency	**eff**
dual upper casing	**DUC**	effluent	**effl**

eight round pipe	8rd	envelope	ENV
ejector	eject	environment	ENVIR
Elbert	Elb	environmental assessment	EA
elbow (s)	ell(s), ELB	environmental impact report	EIR
electric accounting machines	EAM	environmental impact statement	EIS
electric log tops	EL/T	Eocene	Eoc
electric resistance weld	ERW	Eponides	Ep.
electric weld	EW	Eponides yeguaensis	Ep. y.
electric (al)	elec	equal, equalizer	eq
electromagnetic	EMN, ELEC/MAG	equation (before a number)	Eq.
		equilibrium flash vaporization	EFV
electromagnetic induction	EMNI	equipment	equip
electromotive force	EMF	equivalent	equiv
electron volts	ev	erection	erect
electronic data processing	EDP	erection mark	ERC/MK
element, elementary	elem	Ericson	Eric
elevation (height)	EL	estate	est
elevation ground	el gr	estimate (ed) (ing)	est
elevation, elevator	elev	estimated time of arrival	ETA
Elgin	Elg	estimated total depth	ETD
eliminate (tor) (ed)	ELIM	estimated ultimate recovery	EUR
Ellenburger	Ellen	estimated yearly consumption	EYC
Ellis-Madison contact	EMS	ethane	eth
Elmont	Elm	ethylene	ethyle
Embar	Emb	ethylene dichloride	EDC
emergency	emer	euhedral	euhed
emergency order	EO	European melting point	EMP
emergency shutdown	ESD	Eutaw	EU
employee	empl	evaluate	eval
empty container	MT	evaporation, evaporate	evap
emulsion	emul	even-sorted	ev-sort
enamel	enml	examination	EXAM
enclosure	encl	example	EX
end of file	EOF	excavation	exc
end of line	EOL	excellent	excl
end of month	EOM	except	ex
end of quarter	EOQ	exchanger	exch
end of year	EOY	excitation	EXC
end point	EP	executive	EXEC
end to end	E/E	executor	Exr
Endicott	End	executrix	Exrx
endothyra	endo	Exeter	Ex
engine	eng	exhaust	exh
engineer (ing)	engr (g)	exhibit	exh
Englewood	Eglwd	existing	exist
enhanced oil recovery	EOR	expansion	exp
enlarged	enl	expansion joint	EXP JT
Entrada	Ent	expected date of delivery	EDD
entrance	ENT	expendable plug	exp plg
entry	ent	expense	exp

expire (ed) (ing) (ation)	**expir**
exploratory	**E**
exploratory well	**EW**
exploratory, exploration	**expl**
explosion proof	**EX-PRF**
explosive	**explos**
extended, extension	**ext (n)**
extension manhole	**Ext M/H**
exterior	**extr**
external	**ext**
external upset end	**EUE**
extra heavy	**X-hvy**
extra heavy grade pipe	**XHGR**
extra strong	**x-stg**
extraction	**extrac**
extreme line (casing)	**X-line**
extreme pressure	**EP**

fabricate (ed) (tion)	**fab**
face of stud	**FOS**
face to face	**F to F**
faced and drilled	**F & D**
facet (ed)	**fac**
facility (ies)	**FACIL**
facility capacity limits	**FCL**
failed	**fld**
failure	**fail**
faint	**fnt**
faint air blow	**FAB**
fair	**fr**
Fall River	**Fall Riv**
Farmington	**Farm**
farmout	**FO**
farmout option	**F/O opt**
fasten (ing) (er)	**FSTN**
fault	**flt**
faulted out	**FO**
fauna	**fau**
favosites	**fvst**
federal	**fed**
Federal Employers Liability Act	**FELA**
Federation of Sewage and Industrial Wastes Association	**FSIWA**

feed	**FD**
feed effluent	**FD EFF**
feed rate	**FR**
feed water	**FD/WTR**
feeder	**fdr**
feedstock	**FS**
feet, foot	**ft**
feet per hour	**ft/hr**
feet per minute	**fpm**
feet per minute	**ft/min**
feet per second	**fps**
feet per second	**ft/sec**
feldspar	**fld**
female	**FEM**
female pipe thread	**FPT**
female to female angle	**FFA**
female to female globe (valve)	**FFG**
Ferguson	**Ferg**
ferric sulfate	$\mathbf{Fe_2(SO_4)_3}$
ferruginous	**ferr**
Ferry Lake anhydrite	**FLA**
fertilizer	**fert**
fiberglass	**fbrgs**
fiberglass-reinforced plastic	**FRP**
fibrous	**fib**
field	**fld**
field authorized to commence operations	**FACO**
field fabricated	**F/FAB**
field fuel gas unit	**FFGU**
field pressure test flow diagram	**FPTFD**
field purchase order	**FPO**
field receiving report	**FRR**
field wildcat	**FWC**
figure	**fig**
fill up	**FU**
fillet weld	**FW**
filter	**FLTR**
filter cake	**FC**
filtrate	**filt**
final	**fin**
final (flowing) tubing pressure	**FTP**
final boiling point	**FBP**
final bottom-hole pressure, flowing	**FBHPF**
final bottom-hole pressure, shut in	**FBHPSI**
Final Environmental Impact Statement	**FEIS**
final flowing pressure	**FFP**

final fluid level	**FFL**	flexibox	**FLXBX**
final hydrostatic pressure	**FHP**	Flippen	**Flip**
final open	**FO**	float	**flt**
final pressure	**FP**	float collar	**FC**
final report for rig	**FRR**	float shoe	**FS**
final report for well	**FRW**	flotation	**fltn**
final shutin pressure	**FSIP**	floating	**fltg**
finding of no significant impact	**FONSI**	floor	**FL**
fine	**fn**	floor drain	**FD**
fine grained	**f-gr**	Florence Flint	**Flor Fl**
finely	**fnly**	flow	**flo**
finely crystalline	**f/xln**	flow (ed) (ing)	**flw (d) (g)**
finish	**FNSH**	flow control valve	**FCV**
finish all over	**FAO**	flow diagram	**F/DIA, FD, F-DIA**
finish going in hole	**FGIH**		
finish going in with	**FGIW**	flow-indicating controller	**FIC**
finish grade	**FIN GR**	flow-indicating ratio controller	**FIRC**
finish (ed)	**fin**	flow indicator	**FI**
finished drilling	**fin drlg**	flow line	**FL**
finished in hole	**FIH**	flowmeter	**F-MET**
fire water	**F/WTR**	flow rate	**FR**
fire-resistant oil	**F-R oil**	flow recorder	**FR**
fireproof	**fprf**	flow recorder control	**FRC**
fiscal year ending	**FYE**	flow sheet	**F-SHT**
fishing	**fish, fsg**	flow station	**FS**
fishing for	**FF**	flow switch	**F/SW**
fissile	**fisl**	flowed (ing) at a rate of	**FARO**
fissure	**fis**	flowed, flowing	**Fl, fl/**
fittings	**ftg**	Flowerpot	**Flwrpt**
fixed	**fxd**	flowing	**flg**
fixed carbon	**FC**	flowing bottom hole pressure	**FBHP**
fixture	**fix**	flowing by heads	**FBH**
flaky	**flk**	flowing casing pressure	**FCP**
flammable	**FLMB**	flowing on test	**FOT**
flammable liquid building	**FL/BD**	flowing pressure	**Flwg Pr, FP**
flange (ed) (es)	**flg (d) (s)**	flowing surface pressure	**FSP**
flanged and dished (heads)	**F & D**	flue	**flu**
flanged and spigot	**F & S**	fluid	**flu**
flanged and screwed	**F&S**	fluid catalytic cracking	**FCC**
flanged gate valve	**FGVV**	fluid in hole	**FIH**
flanged one end, welded one end	**FOE-WOE**	fluid level	**FL**
		fluid to surface	**FTS**
flash point, Cleveland Open Cup	**Fl-COC**	fluorescence, fluorescent	**fluor**
		flush	**FL**
flashing	**FL**	flush joint	**FJ**
flat face	**FF**	flushed	**flshd**
flathead	**Flath**	flushing oil	**FLO**
flattened	**flat**	focused log	**FOCL**
flexible	**flex**	foliated	**fol**

foot-candle	**ft-c**	free point indicator	**FPI**
foot-pound	**ft-lb**	freezer	**frzr**
foot-pound per hour	**ft-lb/hr**	freezing point	**FP**
foot-pound-second (system)	**f-p-s**	freight	**frt**
footing, footage	**ftg**	frequency	**freq**
for example	**e.g.**	frequency meter	**FM**
for your information	**FYI**	frequency modulation	**FM**
Foraker	**Forak**	fresh	**frs**
foraminifera	**foram**	fresh break	**FB**
foreman	**f'man**	fresh water	**FW**
forge (ed) (ing)	**FRG**	fresh water	**fwtr**
forged steel	**FST, FS**	friable	**fri**
formation	**fm**	friction-reducing agent	**FRA**
formation density	**FD**	froggy	**Frgy**
formation density correlated	**FDC**	from	**fr**
formation density log	**FDL**	from east line	**FE/L, FEL,**
formation gas-oil ratio	**F/GOR**		**fr E/L**
formation interval tester	**FIT**	from north line	**FN/L, FNL,**
formation test	**FT**		**fr N/L**
formation water	**Fm W**	from northeast line	**FNEL**
Fort Chadborne	**Ft C**	from northwest line	**FNWL**
Fort Hayes	**Ft H**	from south and west lines	**FS&WLs**
Fort Riley	**Ft R**	from south line	**FS/L, FSL,**
Fort Union	**Ft U**		**fr S/L**
Fort Worth	**Ft W**	from southeast line	**FSEL**
Fortura	**Fort**	from southwest line	**FSWL**
forward	**fwd**	from west line	**FW/L, FWL,**
fossiliferous	**foss**		**fr W/L**
foundation	**fdn**	front	**fr**
Fountain	**Fount**	front & side	**F/S**
four pole single throw switch	**4P ST SW**	frontier	**fron**
four pole switch	**4P SW**	frosted	**fros, fr**
four-wheel drive	**FWD**	frosted quartz grains	**FQG**
Fox Hills	**Fox H**	Fruitland	**Fruit**
frac finder (log)	**FF**	fuel gas	**FG**
fractional	**fr**	fuel oil	**FO**
fractionation, fractionator,	**fract**	fuel oil equivalent	**F.O.E.**
fractional		fuel oil return	**FOR**
fracture gradient	**F.G.**	fuel oil supply	**FOS**
fracture, fractured, fractures	**frac (d) (s)**	fuels & fractionation	**F & F**
fragment	**frag**	fuels & lubricants	**F & L**
frame, framing	**FRM (G)**	full freight allowed (purchasing	**FFA**
framework	**frwk**	term)	
franchise	**fran**	full hole	**FH**
Franconia	**Franc**	full length drift	**FLD**
Fredericksburg	**Fred**	full of fluid	**FF**
Fredonia	**Fred**	full open head	**FOH**
free on board	**FOB**	full opening	**FO**
free point back off	**FPBO**	Fullerton	**Full**

functional check out	**FCO**
funnel viscosity	**FV**
furfural	**furf**
furnace fuel oil	**FFO**
furnace	**furn**
furnish (ed)	**FURN**
furniture and fixtures	**furn & fix**
Fuson	**Fus**
Fusselman	**fussel**
fusulinid	**fusul**
future	**fut**

gauge (ed) (ing)	**ga**
Galena	**Glna**
Gallatin	**Gall**
galled threads	**gld thd**
gallon (s)	**gal**
gallons acid	**GA**
gallons breakdown acid	**GBDA**
gallons condensate per day	**GCPD**
gallons condensate per hour	**GCPH**
gallons gelled water	**GGW**
gallons heavy oil	**GHO**
gallons mud acid	**GMA**
gallons of oil per day	**GOPD**
gallons of oil per hour	**GOPH**
gallons of solution	**gal sol**
gallons of water per hour	**GWPH**
gallons oil	**GO**
gallons per day	**GPD**
gallons per hour	**GPH**
gallons per minute	**gal/min, GPM**
gallons per second	**GPS**
gallons per thousand cubic feet	**gal/Mcf**
gallons regular acid	**GRA**
gallons salt water	**GSW**
gallons water	**GW**
galvanized	**galv**
gamma ray	**GR**
gamma ray spectroscopy tool	**GST**
gas	**G**
gas & mud-cut oil	**G&MCO**

gas and oil	**G&O**
gas and oil-cut mud	**G&OCM**
gas cap participating area	**GCPA**
gas in pipe	**GIP**
gas injection	**GI**
gas injection well	**GIW**
gas lift	**GL**
gas lift gas distribution	**GGD**
gas lift oil	**GLO**
gas lift transfer	**GLT**
gas-liquid ratio	**GLR**
gas odor	**GO**
gas odor distillate taste	**GODT**
gas pay	**GP**
Gas Production Unit	**GPU**
gas purchase contract	**GPC**
gas reserve group	**GRG**
gas rock	**G Rk**
gas sales contract	**GSC**
gas show	**GS**
gas to surface	**GTS**
gas to surface (time)	**GTS**
gas too small to measure	**GTSTM**
gas unit	**GU**
gas volume	**GV**
gas volume not measured	**GVNM**
gas well	**GW**
gas well shut in	**GSI**
gas-condensate ratio	**GCR**
gas-cut	**GC**
gas-cut acid water	**GCAW**
gas-cut distillate	**GCD**
gas-cut load oil	**GCLO**
gas-cut load water	**GCLW**
gas-cut mud	**GCM**
gas-cut oil	**GCO**
gas-cut salt water	**GCSW**
gas-cut water	**GCW**
gas-distillate ratio	**GDR**
gas-fluid ratio	**GFR**
gas-handling study group	**GHSG**
gas-oil contact	**GOC**
gas-oil ratio	**G/O, GOR**
gas-water contact	**GWC**
gas-well gas	**GWG**
gaseous nitrogen	**G-N$_2$**
gaseous oxygen	**G-O$_2$**
gasket	**gskt**
gasoline	**gaso**

gasoline plant	**GP**	Goodwin	**Gdwn**
gastropod	**gast**	goose egg	**G egg**
gathering line	**G/L**	Gorham	**Gor**
gauge	**gge**	Gouldbusk	**Gouldb**
gauge ring	**GR**	government	**govt**
gelled	**gel**	governor	**gov**
general	**genl**	grade	**gr**
general arrangement	**GA**	grading	**grdg**
General Electric Company	**GE**	grading location	**grdg loc**
General Land Office (Texas)	**GLO**	gradiomanometer	**GRAD**
General Motors Corporation	**GM**	gradual, gradually	**grad**
generation, generator	**gen**	grain	**gr**
geological quadrangle map	**GQM**	grained (as in fine grained)	**gnd**
geology (ist) (ical)	**geol**	grains per gallon	**gg, GPG**
geophysical investigation map	**GPM**	gram	**g, gm**
geophysics (ical)	**geop**	gram molecular weight	**g mole**
geopressure development, failure	**PD**	gram-calorie	**gm-cal**
geopressure development, success	**PDS**	Graneros	**Granos**
		Granite Point Field	**GPF**
Georgetown	**Geo**	granite wash	**gran w**
geothermal	**geo, GT**	granite	**gran**
geothermal development, failure	**GD**	grant (of land)	**grt**
geothermal development, success	**GDS**	granular	**grnlr**
		graptolite	**grap**
geothermal wildcat, failure	**GW**	grating	**grtg**
geothermal wildcat, success	**GWS**	gravel	**gvl**
Gibson	**Gib**	gravel packed	**GVLPK**
Gilcrease	**Gilc**	gravitometer	**grvt**
gilsonite	**gil**	gravity	**grav, gr**
glass, glassy	**gls**		**GTY**
glauconite, glauconitic	**glau**		**GRVTY**
Glen Dean lime	**GD Li**	gravity °API	**gr °API**
Glen Rose	**GR**	gravity meter	**G.M.**
Glenwood	**Glen**	gray	**gry**
globe valve	**GLBVV**	Gray sand	**Gr Sd**
Globigerina	**Glob**	Grayburg	**Grayb**
Glorieta	**Glor**	Grayson	**Gray**
glycol	**glyc**	graywacke	**gywk**
gneiss	**gns**	grease	**gr**
going in hole	**GIH**	greasy	**gsy**
Golconda lime	**Gol Li**	green	**grn**
good	**gd**	Green River	**Grn Riv**
good fluorescence	**GFLU**	green royalty	**gr roy**
good odor & taste	**gd o&t**	green shale	**grn sh**
good show of gas	**GSG**	Greenhorn	**GH**
good show oil	**GSO**	grind out	**GO**
good show oil and gas	**GSO&G**	gritty	**grty**
Goodland	**Gdld,**	grooved	**grv**
	Good L	grooved ends	**GE**

gross	grs	head	hd
gross acre-feet	GAF	header	hdr
gross weight	gr wt	headquarters	HQ
ground	gr, grnd	heat exchanger	HX, HTX
ground level	GL	heat tracing (ed)	HT
ground measurement (elevation)	GM	heat-treated alloy	HTA
		heat-treated, heater treater	HT
group	GRP	heater	htr
guard log	GRDL	heating and ventilating	H & V
guidance continuance tool	GCT	heating oil	HO
guide shoe	GS	heating ventilating and air conditioning	HVAC
Gulf Research and Development Company	GR&DC		
		heavily	hvly
Gull River	G. Riv	heavily (highly) gas-cut salt water	HGCSW
gummy	gmy		
gun barrel	GB	heavily (highly) gas-cut water	HGCW
gun perforate	G/P	heavily (highly) oil-cut mud	HOCM
Gunsite	Guns	heavily (highly) oil-cut salt water	HOCSW
gusset	GUS		
gypsiferous	gypy	heavily (highly) oil-cut water	HOCW
gypsum	gyp	heavily (highly) water-cut mud	HWCM
Gyroidina	Gyr.	heavily gas-cut mud	HGCM
Gyroidina scal	Gyr. sc.	heavily oil and gas-cut mud	HO&GCM
		heavy	hvy
		heavy coker gas oil	HCGO
		heavy cycle oil	HCO
		heavy duty	HD
		heavy fuel oil	HFO
		heavy gas oil	HVGO
		heavy hydrocrackate	HUX
Hackberry	Hackb	heavy oil	HO
hackly	hky	heavy reformate	HR
hand hole	HH	heavy steel drum	HSD
hand-control valve	HCV	Heebner	Heeb
handle	hdl	height	hgt
handling	HNDLG	heirs	hrs
handwheel	HND/WHL	held by production	HBP
hanger	hgr	hematite	hem
Haragan	Hara	Herington	Her
harbor	hbr	Hermosa	Herm
hard	hd	Hertz	Hz
hard lime	hd li	heterostegina	het
hard sand	hd sd	hex head	HEX HD
Hardinsburg sand (local)	Hburg	hexagon (al)	hex
hardness	hdns	hexane	hex
hardware	hdwe	Hickory	Hick
Haskell	Hask	high detergent	HD
Haynesville	Haynes	high gas-oil ratio	HGOR
hazardous	haz	high pressure	HP

high temperature	**HT**	hundred weight	**cwt**
high tension	**HT**	Hunton	**Hun**
high viscosity	**HV**	hydraulic	**HYD**
high viscosity index	**HVI**	hydraulic pump	**HP**
high voltage	**H-VOLT**	Hydril	**HYD**
high volume lift	**HVL**	Hydril thread	**HYDT**
high-level shutdown	**HLSD**	Hydril Type A joint	**HYDA**
high-pressure gas	**HPG**	Hydril Type CA joint	**HYDCA**
high-pressure gauge	**HPG**	Hydril Type CS joint	**HYDCS**
high-resolution dipmeter	**HRD**	hydro test	**HYDRO**
high-temperature shutdown	**HTSD**	hydrocarbon	**HC**
highly	**hily**	hydrocarbon drain system	**HCDS**
highway	**hwy**	hydrocracker	**H/C, HC**
Hilliard	**Hill**	hydrodesulfurizer	**HDS**
hockleyensis	**hock**	hydrofining	**hfg**
Hogshooter	**Hog**	hydrogen	**H$_2$**
hold down	**HLDN**	hydrogen delsulfurization	**HDS**
hole full of oil	**HFO**	hydrogen ion concentration	**pH**
hole full of salt water	**HFSW**	hydrogen sulfide	**H$_2$S**
hole full of sulfur water	**HF Sul W**	hydrogenation	**HYGN**
hole full of water	**HFW**	Hydrologic Investigations Atlas	**HIA**
hole opener	**HO, H.O.**	hydrostatic head	**HH**
holes per foot	**HPF**	hydrostatic pressure	**HP**
Hollandberg	**Holl**	hydrostatic test	**HST**
Home Creek	**Home Cr**	hydrotreater	**hydtr**
home office	**HO**	hygiene	**hyg**
hook up	**HU**	hyperbolic constant	**CSCH**
hookwall packer	**HWP**	hyperbolic cotangent	**COTH**
Hoover	**Hov**	hyperbolic cosine	**COSH**
hopper	**hop**	hypotenuse	**HYPO**
horizontal	**horiz**	hyraulic horsepower	**H H P**
horsepower	**HP**		
horsepower-hour	**hp-hr**		
Hospah	**Hosp**		
hot dry rock development, failure	**HD**		

hot dry rock development, successful	**HDS**		
hot dry rock wildcat, failure	**HW**	identification sign	**ID. sign**
hot dry rock wildcat, success	**HWS**	identify (ier) (ication)	**IDENT**
hot oil tar	**HOT**	Idiomorpha	**Idio**
hot-rolled steel	**HRS**	igneous	**ign**
hour (s)	**hr, HRS**	ignition	**IGN**
house (ed) (ing)	**HSE**	illuminator (s)	**ILUM**
house brand (regular grade of gasoline)	**HB**	imbedded	**imbded**
		immediate (ly)	**immed**
Hoxbar	**Hox**	Imperial	**Imp**
Humblei	**Humb**	Imperial gallon	**Imp gal**
Humphreys	**Hump**	impervious	**imperv**

impounding	**IMP**	initial production gas lift	**IPG**
impression block	**IB**	initial production on	**IPI**
in hole	**IH**	intermitter	
inbedded	**inbded**	initial production plunger lift	**IPL**
incandescent	**incd**	initial production pumping	**IPP**
inch (es)	**in.**	initial production swabbing	**IPS**
inch-pound	**in.-lb**	initial shutin tubing pressure	**ISITP**
inches mercury	**in. Hg**	initial shutin pressure (DST)	**ISIP**
inches per second	**in./sec**	initial vapor pressure	**IVP**
incinerator, incineration	**INCIN**	injection gas	**IG**
include (ed) (ing)	**incl**	injection gas-oil ratio	**IGOR**
inclusions	**incls**	injection index	**II**
income (er) (ing)	**INCM**	injection pressure	**Inj Pr**
incorporated	**Inc.**	injection rate	**IR**
increase (ed) (ing)	**incr**	injection water	**IW**
increment	**incr**	injection well	**IW**
indicate (s) (tion)	**indic**	injection, injected	**inj**
indicated horsepower	**IHP**	inland	**inl**
indicated horsepower hour	**IHPHR**	inlet	**inl**
indistinct	**indst**	inlet gate valve/inlet ball valve	**IGV/IBV**
individual	**indiv**	Inoceramus	**Inoc**
induction	**ind**	input/output	**I/O**
induction electrical survey	**IES**	inquire, inquiry	**INQ**
indurated	**indr**	inside diameter	**ID**
inflammable liquid	**Inf. L**	inside screw (valve)	**IS**
inflammable solid	**Inf. S**	insoluble	**insol**
inflow performance rate	**IPR**	inspect (ed) (ing) (tion)	**insp**
information	**info**	install (ed) (ing)	**inst**
infrared	**IR**	install (ing) pumping	**IPE**
inhibitor	**inhib**	equipment	
initial	**init**	installation operation &	**I-O-M**
initial air blow	**IAB**	maintenance	
initial boiling point	**IBP**	installation (s)	**instl**
initial bottom-hole pressure	**IBHP**	installing (ed) pumping	**INPE**
initial bottom-hole pressure,	**IBHPF**	equipment	
flowing		instantaneous	**inst**
initial bottom-hole pressure,	**IBHPSI**	instantaneous shutin pressure	**ISIP**
shut in		(frac)	
initial flowing pressure	**IFP**	institute	**inst**
initial fluid level	**IFL**	instrument, instrumentation	**instr**
initial hydrostatic pressure	**IHP**	insulate	**insul**
initial mud weight	**IMW**	insulate, insulation	**ins**
initial open	**IO**	insurance	**ins**
initial participating area	**IPA**	integral	**INT**
initial potential	**IP**	integral joint	**IJ**
initial pressure	**IP**	integrator	**intgr**
initial production	**IP**	intention to drill	**ITD**
initial production flowed (ing)	**IPF**	interbedded	**interbd**

interconnecting flow diagram	I/CFD
interconnection (ing)	I/C, INTCON
intercooler	incolr, INCLR
intercrystalline	inter-xln
interest	int
intergranular	inter-gran, ingr
interior	int, INTR
interlaminated	irter-lam, intlam
intermediate manifolds	IMF
intermediate pressure	IP
internal	int, INTL
internal flush	IF
Internal Revenue Service	IRS
internal upset ends	IUE
intersect	ints
intersection	int
interstitial	intl
interval	intv
intrusion˙	intr
invert (ed)	inv
invertebrate	invrtb
investment tax credit	ITC
invitation to bid	ITB
invoice	inv
Ireton	Ire
iridescent	irid
iron	Fe
iron body (valve)	IB
iron body brass (bronze) mounted (valve)	IBBM
iron body brass core (valve)	IBBC
iron case	IC
iron pipe size	IPS
iron pipe thread	IPT
ironstone	Fe-st, irst
irregular	irreg
isocracker	ISO/CKR
isolate (tor)	ISOL
isometric	isom, ISO
isopropyl alcohol	IPA
isothermal	isoth
issue	ISS
Iverson	Ives

jacket	jac
Jackson	Jxn, Jack
Jackson sand	Jax sd
jammed	jmd
Jasper (oid)	Jasp
Jefferson	Jeff
jelly-like colloidal suspension	gel
jet fuel (aviation)	JFA
jet perforated	JP
jet perforations per foot	JP/ft
jet propulsion fuel	JP fuel
jet shots per foot	JSPF
jet treating unit	JTU
job complete	JC
jobber	jbr
Joint Committee on Uniformity of Methods of Water Examination	JCUMWE
joint interest nonoperated (property)	JINO
joint operating agreement	JOA
joint operating provisions	JOP
joint operation	J/O
joint venture	JV
joint (s)	jt(s)
Jordan	Jdn
Joule	J
Judith River	Jud Riv
junction	jct
junction box	JB
junk (ed)	jnk
junk basket	JB
junk joint	JJ
junked and abandoned	J&A
Jurassic	Jur
jurisdiction	juris

Kaibab	Kai
Kansas City	KC
kaolin	kao
Kayenta	Kay
Keener	Ke
kelly bushing	KB
kelly bushing measurement	KBM
kelly drill bushing to landing flange	KDB-LDG FLG
kelly drill bushing to mean low water	KDB-MLW
kelly drill bushing to platform	KDB-Plat
kelly drive bushing	KDB
kelly drive bushing elevation	KDBE
Kelvin (temperature scale)	K
Keokuk-Burlington	Keo-Bur
kerosene	kero
ketone	ket
kettle	KTLE
Keystone	Key
Kiamichi	Kia
Kibbey	Kib
kick off	KO
kickoff point	KOP
kill (ed) well	KW
killed	kld
kiln dried	KD
kilocalorie	kcal
kilocycle	kc
kilogram	kg
kilogram-calorie	kg-cal
kilogram-meter	kg-m
kilohertz (See Hz-Hertz)	KHz
kiloliter	kl
kilometer	km
kilopascal	kPa
kilovar-hour	kvar-hr
kilovar, reactive kilovolt-ampere	kvar
kilovolt	kv
kilovolt peak	kvp
kilovolt-ampere	kva
kilovolt-ampere-hour	kvah
kilowatt	kw

kilowatt-hour	kw-h
kilowatt-hour meter	kw-hm
Kincaid lime	Kin Li, KD
Kinderhook	Khk
kinematic	Kin
kinematic viscosity	KV
Kirtland	Kirt
kitchen	KIT
KMA sand	KMA
knock down	KD
knock out	KO
known geothermal resource area	KGRA
Kootenai	Koot
Krider	Kri

La Motte	La Mte
labor	lab
laboratory	lab
ladder	lad
ladder & platform	L & P
lagging	LAG
laid (laying) down drill collars	LDDCs
laid (laying) down drillpipe	LDDP
laid down	LD
laid-down cost	LDC
Lakota	Lak
laminated, lamination (s)	lam
landing	LDG
land (s)	ld (s)
Landulina	Land
Lansing	Lans
lap joint	LJ
lapweld	LW
Laramie	Lar
large	lrg, lg
large-diameter flow line	LDF
Large Discorbis	Lg Disc
latitude	lat
Lauders	Laud
Layton	Layt
Le Comptom	Le C
leached	lchd

lead drill collar	**LDCX**	limit	**lim**
Leadville	**Leadv**	limited	**ltd**
league	**Lge**	line	**LN**
leak	**lk**	line pipe	**L.P.**
leakage	**LKG**	line pressure	**LP**
lease	**lse**	line, as in E/L (east line)	**/L**
lease automatic custody transfer	**LACT**	linear	**lin**
		linear foot	**lin ft**
lease crude	**LC**	liner	**lnr, lin**
lease operations	**LOS**	linguloid	**lngl**
lease use (gas)	**L U**	liquefaction	**liqftn**
Leavenworth	**Lvnwth**	liquefied petroleum gas	**LP-Gas, LPG**
left hand	**LH**		
left in hole	**LIH**	liquid	**liq**
legal subdivision (Canada)	**LSD**	liquid level controller	**LLC**
length	**lg**	liquid level gauge	**LLG**
length overall	**LOA**	liquid oxygen	**LOX**
Lennep	**Len**	liquid penetrant examination	**LPE**
lense	**lns**	liquid volume	**LV**
lenticular	**len**	liquefied natural gas	**LNG**
less than truckload	**LTL**	list of components	**LST/ COMPTS**
less-than-carload lot	**LCL**		
letter	**ltr**	liter	**l**
level	**lvl**	lithographic	**litho**
level alarm	**LA**	little	**ltl**
level control valve	**LCV**	load	**ld**
level controller	**LC**	load acid	**LA**
level glass	**lg**	load oil	**LO**
level indicator	**LI**	load water	**LW**
level indicator controller	**LIC**	loader	**LDR**
level recorder	**LR**	loading	**LDG**
level recorder controller	**LRC**	loading dock	**L-DK**
license	**lic**	local	**LCL**
Liebuscella	**Lieb**	local control panel	**LCP**
light	**lt**	local injection plants	**LIP**
light barrel	**LB**	local purchase order	**LPO**
light brown oil stain	**LBOS**	located, location	**loc**
light coker gas oil	**LCGO**	location abandoned	**loc abnd**
light fuel oil	**LFO**	location graded	**loc gr**
light hydrocrackate	**LUX**	lock	**lk**
light iron barrel	**LIB**	locker	**LKR**
light iron grease barrel	**LIGB**	lodge pole	**LP**
light steel drum	**LSD**	log mean temperature difference	**LMTD**
lighting	**ltg**		
lightning arrester	**LA**	log total depth	**LTD**
lignite, lignitic	**lig**	logarithm (common)	**log**
lime, limestone	**Li, lm, ls**	logarithm (natural)	**ln**
limy	**lmy**	long	**lg**
limy shale	**Lmy sh**	long coupling	**LC**

long handle/round point	LH/RP
long radius	LR
long-range automation plan	LRAP
long-range plan	LRP
long string	LS
long-term tubing test	LTT
long threads and coupling	LT&C
longitude (inal)	long
loop	lp
lost circulation	LC
lost circulation material	LCM
Lovell	Lov
Lovington	Lov
low pressure	LP
low viscosity index	LVI
low voltage	L-VOLT
low water loss	LWL
low-pressure separation	LPS
low-pressure separator	LP sep
low-temperature extraction unit	LTX unit
low-temperature separation unit	LTS unit
low-temperature shutdown	LTSD
lower	lwr, low
Lower Albany	L/Alb
lower anhydrite stringer	LAS
lower casing	LC
Lower Cretaceous	L/Cret
lower explosive limit	LEL
Lower Glen Dean	LGD
Lower Menard	LMn
lowermost flange	LMF
lower tubing	LT
Lower Tuscaloosa	L/Tus
lower, i.e., L/Gallup	L/
lube oil	LO
lubricate (ed) (ing) (tion)	lub
Lueders	Lued
lug cover type (5-gallon can)	LC
lug cover with pour spout	LCP
lumber	lbr
lumpy	lmpy
lustre	lstr

machine	mach
Mackhank	Mack
Madison	Mad
magnetic particle examination	MPE
magnetic, magnetometer	mag
magnetomotive force	mmf
main reaction furnace, "merf"	MRF
maintenance	maint
Manitoban	Manit
major, majority	maj
making	MKG
male and female (joint)	M&F
male pipe thread	MPT
male to female angle	MFA
malleable	mall
malleable iron	MI
management	m'gmt
manager	mgr
manhole	MH
manifold	MF, man
Manning	Mann
manual	man, MNL
manually operated	man op
manufacture (er)	MFR
manufactured	mfd
manufacturing	mfg
mapping subcommittee	MSC
Maquoketa	Maq
Marble Falls	Mbl Fls
macaroni tubing	MT
Marchand	March
marginal	marg
Marginulina	Marg.
Marginulina coco	Marg. coco.
Marginulina flat	Marg. fl.
Marginulina round	Marg. rd.
Marginulina texana	Marg. tex.
marine	mar
marine gasoline	margas
marine rig	MR
marine terminal	M/T
Marine Tuscaloosa	M. Tus
marine wholesale distributors	MWD

market demand factor	MDF	mean low water to platform	MLW-PLAT
market (ing)	mkt	mean low wave	MLW
Markham	Mark	mean sea level	msl
marking	MRK	mean temperature difference	MTD
marlstone	mrlst	measure (ed) (ment)	meas
marly	Mly	measure of hydrogen potential	pH
Marmaton	Marm	measured depth	MD
maroon	mar	measured total depth	MTD
Marshal (ling)	MARSH	measuring & regulating station	M & R Sta.
masking plate	M/PLT	mechanic (al), mechanism	mech
Massilina pratti	Mass. pr.	mechanical down time	Mech DT
massive	mass	mechanical flow diagram	MFD
massive anhydrite	MA	mechanical properties	MPL
master	mstr	mechanism	mchsm
master circuit board	MCB	median	med
master load list	MLL	Medicine Bow	Med B
material	mat'l, mtl	Medina	Med
material safety data sheets	MSDS	medium	med
material take-off	MTO	medium amber cut	MAC
mathematics	Math	medium fuel oil	med FO
matrix	Mtx	medium grained	m-gr, med gr, mg
matter	mat		
maximum	max	Medrano	Medr
maximum & final pressure	M&FP	Meeteetse	Meet
maximum allowable operating pressure	MAOP	megacycle	mc
maximum allowable working pressure	MAWP	megahertz (megacycles per second)	MHz
maximum casing pressure	MCP	megapascal	mPa
maximum daily delivery obligation	MDDO	megawatt	MW
		melting point	MP
maximum efficient rate	MER	member (geologic)	mbr
maximum flowing pressure	MFP	memo requesting quotes	MRQ
maximum operating pressure	MOP	memorandum	memo
maximum pressure	MP	Menard lime	Men li
maximum surface pressure	MSP	Menefee	Mene
maximum top pressure	MTP	Meramec	Mer
maximum total depth	MTD	mercaptan	mercap
maximum tubing pressure	MTP	merchandise	mdse
maximum working pressure	MWP	mercury	merc
Maywood	May	"merf", main reaction furnace	MRF
McClosky lime	McC lm	meridian	merid
McCullough	McCul	Mesaverde	Mvde
McElroy	McEl	mesh	M
McKee	McK	Mesozoic	Meso
McLish	McL	metal petal basket	MPB
McMillan	McMill	metamorphic	meta
Meakin	Meak	meter	m, mtr
mean effective pressure	MEP	meter run	MR
		meter-kilogram	m-kg

methanator	**methr**	millimeter	**mm**
methane	**meth**	millimeters of mercury	**mm Hg**
methane-rich gas	**MEG**	millimicron	**m**
methanol	**methol**	milling	**millg**
methyl chloride	**meth-cl**	milling	**mlg**
methylene blue	**meth-bl**	million (i.e., 9MM = 9,000,000)	**MM**
methylethylketon	**MEK**	million British thermal units	**MMBTU**
methylisobutylketone	**MIK**	million cubic feet	**MMcf**
metric	**metr**	million cubic feet per day	**MMcfd**
mezzanine	**mezz**	million electron volts	**MMev**
mhos per meter	**mho/m**	million reservoir barrels	**MMRVB**
mica, micaceous	**mica**	million standard cubic feet per	**MMscfd**
micro-microfarad	**m-mf**	day	
microampere	**m-a, ma**	millions of barrels	**MMBLS**
microcrystalline	**micro-xin, micro-x**	milliotitic	**mill**
		milliroentgen	**mr**
microfarad	**mfd, m-f**	millisecond (s)	**ms**
microfossil (iferous)	**micfos**	millivolt	**mv**
microgram	**m-g**	mineral	**mnrl**
microgram (millionth)	**µg**	mineral interest	**MI**
microinch	**m-in.**	minerals	**min**
micromicron	**m-m**	Minerals Management Service	**MMS**
micron	**m, µ**	minimium	**min**
microsecond	**microsec. µsec**	minimum pressure	**min P**
		Minnekahta	**Mkta**
microvolt	**µ-v, µv**	Minnelusa	**Minl**
microwave	**MW**	minor	**MNR**
middle	**M/, mdl, mid**	minute (s)	**min**
		Miocene	**Mio**
middle ground shoals	**MGS**	miscellaneous	**misc**
middle hydrocrackate	**MUX**	Miscellaneous Field Studies Map	**MFSM**
Midway	**Mwy**	Miscellaneous Investigations	**MIS**
mile (s)	**mi**	Series	
miles per hour	**MPH**	miscible substance group	**MSG**
miles per year	**MPY**	Misener	**Mise**
military	**mil**	Mission Canyon	**Miss Cany**
milky	**mky**	Mississippian	**Miss**
mill wrapped plain end	**MWPE**	mix and pump	**M&P**
milled	**mld**	mixed	**mxd**
milled one end	**M1E**	mixer	**mix**
milled other end	**MOE**	mixing	**MIXG**
milled two ends	**M2E**	mobile	**mob**
milliampere	**ma**	model	**mod**
millidarcies	**md**	moderate (ly)	**mod**
milligram	**mg**	modification	**mod**
millihenry	**mh**	modular	**modu**
milliliter	**ml**	Moenkopi	**Moen**
milliliters tetraethyl lead per gallon	**ml TEL/G**	moisture impurities and unsaponifiables (grease testing)	**MIU**

molar	M	moving in rotary tools	MIRT
molas	mol	moving in service rig	MISR
mole	mol	moving in standard tools	MIST
molecular weight	mol wt	moving in tools	MIT
molecule, molecular	MOL	moving out	MO
mollusca	mol	moving out (off) cable tools	MOCT
molybdenum	MO	moving out (off) rotary tools	MORT
Monel drill collars	MDC	moving out completion unit	MOCU
monitor	mon	moving out equipment	MOE
monoethanolamine	MEA	moving out rig	MOR
monthly volume operation plan	MVOP	Mowry	Mow
Montoya	Mont	Mt. Diablo	MD
Moody's Branch	MB	mud acid	MA
Moore County line	MC	mud acid wash	MAW
Mooringsport	Moor	mud cake	MC
more or less	m/l	mud cleanout agent	MCA
Morrison	Morr	mud cut	MC
Morrow	Mor	mud filtrate	MF
mortgage	mtge	mud logger	ML
Mosby	Mos	mud logging unit	MLU
motor	MTR, mot	mud to surface	MTS
motor control center	MCC	mud weight	md wt, MW
motor generator	mg	mud-cut acid	MCA
motor medium	MM	mud-cut gas	MCG
motor octane number	MON	mud-cut oil	MCO
motor oil	MO	mud-cut salt water	MCSW
motor oil units	MOU	mud-cut water	MCW
motor severe	MS	mud/silt remover	MSR
motor vehicle fuel tax	MVFT	muddy	Mdy
motor vehicle, motor vessel	M/V	muddy salt water	MSW
mottled	mott	muddy water	MW
Mount Selman	Mt. Selm	mudstone	mudst
mounted	mtd	multigrade	MG
mounting	mtg	multiple service acid	MSA
mousehole	MH	multiply, multiplexer	MULTX
moving	movg	multipurpose	MP
moving (moved) in double drum unit	MIDDU	multipurpose grease lithium base	MPGH-lith
moving in (equipment)	MI	multipurpose grease soap base	MPGR-soap
moving in and rigging up	MIRU		
moving in cable tools	MICT	muscovite	musc
moving in completion rig	MICR		
moving in completion unit	MICU		
moving in equipment	MIE		
moving in materials	MIM		
moving in pulling unit	MIPU		
moving in rig	MIR		
moving in rigging up swabbing unit	MIRUSU		

N

Nacotoch	**Nac**
nacreous	**nac**
nameplate	**NP**
naphfining unit	**NU**
naphtha	**nap**
naphtha-hydrogen desulfurization	**NHDS**
narrative	**NARR**
national	**nat'l**
national coarse thread	**NCT**
National Electric Code	**NEC**
National Fine (thread)	**NF**
National pipe thread	**NPT**
National pipe thread, female	**NPTF**
National pipe thread, male	**NPTM**
national pollution discharge elimination system	**NPDES**
natural	**nat**
natural flow	**NF**
natural gas	**NG**
natural gas liquids	**NGL**
nautical mile	**NMI**
Navajo	**Nav**
Navarro	**Navr**
Naval Petroleum Reserve	**NPR**
Naval Petroleum Reserve Alaska	**NPRA**
negative	**neg**
negligible	**neg**
negotiation	**NEGO**
neoprene	**npne**
net effective pay	**NEP**
net positive suction head	**NPSH**
net revenue interest	**NRI**
net tons	**NT**
neutral, neutralization	**neut**
neutralization number	**Neut. No.**
neutron lifetime log	**NLL**
New Albany shale	**New Alb**
new bit	**NB**
new field, discovery	**NFD**
new field wildcat	**NFW**
new field wildcat, discovery	**WFD**
new field wildcat, dry	**WF**
new oil	**NO**

new pool discovery	**NPD**
new pool exempt (non-operated)	**NPX**
new pool wildcat	**NPW**
new pool wildcat, discovery	**WPD**
new pool wildcat, dry	**WP**
new rod	**NR**
new source performance standards	**NSPS**
new total depth	**NTD**
Newburg	**Nbg**
Newcastle	**Newc**
Newton	**N**
Niagara	**Nig**
nickel plated	**NP**
Ninnescah	**Nine**
Niobrara	**Niob**
nipple	**nip, NPL**
nipple (ed) (ing) down blowout preventers	**NDBOPs**
nipple (ed) (ing) up blowout preventers	**NUBOPs**
nipple-down tree	**NDT**
nipple-up tree	**NUT**
nippled (ing) up	**NU, UP**
nippled down	**ND**
nippling up wellhead	**NUWH**
nitrogen	N_2
nitrogen blanket	**NB**
nitroglycerine	**nitro**
no appreciable gas	**NAG**
no change	**NC**
no core	**NC**
no fluid	**NF**
no fluorescence	**NF**
no fluorescence or cut	**NFOC**
no fuel	**NF**
no gas to surface	**NGTS**
no gauge	**NG**
no good	**NG**
no increase	**No Inc**
no order required	**NOR**
no paint on seams	**NPOS**
no production	**NP**
no recovery	**no rec, NR**
no report, not reported	**NR**
no show	**NS**
no show fluorescence or cut	**NSFOC**
no show gas	**NSG**

no show oil	**NSO**	Northern Alberta Land	**NALRD**
no show oil and gas	**NSO&G**	Registration District	
no test	**N/tst**	northwest	**NW**
no time	**NT**	northwest corner	**NW/C**
no visible porosity	**NVP**	northwest line	**NWL**
no water	**NW**	northwest quarter	**NW/4**
Noble-Olson	**NO**	Northwest Territories	**NWT**
Nodosaria blanpiedi	**Nod. blan.**	not applicable	**NA**
Nodosaria mexicana	**Nod. mex.**	not available	**NA**
nodule, nodular	**nod**	not completed	**NC**
nominal	**nom**	not deep enough	**NDE**
nominal pipe size	**NPS**	not drilling	**ND**
noncontiguous tract	**NCT**	not in contract	**NIC**
nondestructive testing	**NDT**	not on bottom	**NOB**
nondetergent	**ND**	not prorated	**NP**
nonemulsion acid	**NEA**	not pumping	**NP**
nonleaded gas	**NL gas**	not suitable for coating	**NSC**
nonoperated joint ventures	**NOJV**	not to scale	**NTS**
nonoperating property	**NOP**	not yet available	**NYA**
nonporous	**NP**	not yet drilled	**NYD**
nonproducer	**NP**	Notary Public	**NP**
nonreturnable steel barrel	**NRSB**	notice of intention to drill	**NID**
nonreturnable steel drum	**NRSD**	notice of violation	**NOV**
nonreturnable, no returns, not	**NR**	notice to proceed	**NTP**
reached		nozzle	**noz**
nonrising stem (valve)	**NRS**	Nuclear Regulatory Commission	**NRC**
nonstandard	**nstd**	nugget	**Nug**
nonstandard service station	**N/S S/S**	number	**NO.**
nonupset	**NU**	number (before a number)	**No.**
nonupset ends	**NUE**	numerous	**num**
nonemulsifying agent	**NE**	nylon	**NYL**
nonflammable compressed gas	**nonf G**		
Nonionella	**Non**		
Nonionella cockfieldensis	**N. cock.**		
Noodle Creek	**Ndl Cr**		
normal	**nor**		
normal (express concentration)	**N**		
normally closed	**NC**		
normally open	**NO**	Oakville	**Oakv**
north	**N**	object	**obj**
north half	**N/2**	observation	**OBS**
north line	**NL**	obsolete	**obsol**
north offset	**N/O**	occasional (ly)	**occ**
north quarter	**N/4**	ocean bottom suspension	**OBS**
northeast	**NE**	octagon, octagonal	**oct**
northeast corner	**NEC**	octane	**oct**
northeast line	**NEL**	octane number requirement	**ONR**
northeast quarter	**NE/4**	octane number requirement	**ONRI**
northerly	**N'ly**	increase	

O'Dell	**Odel**	oil odor	**OO**
odor	**od**	oil pay	**OP**
odor, stain, and fluorescence	**OS&F**	oil payment interest	**OPI**
odor, taste, & stain	**OT&S**	Oil Refining Industry Action	**ORIAC**
odor, taste, stain, & fluorescence	**OTS&F**	Committee	
		oil sand	**O sd**
off bottom	**OB**	oil show	**OS**
offshore	**offsh**	oil stain	**OSTN**
off/on location	**OL**	oil standing in drillpipe	**OSIDP**
office, official	**off**	oil string flange	**OSF**
official potential test	**OPT**	oil to surface	**OTS**
offsite	**OFS**	oil unit	**OU**
Ohio River Valley Water Sanitation Commission	**ORSANCO**	oil well flowing	**OWF**
		oil well from waterflood	**OWFWF**
ohm	**ohm**	oil well gas	**OWG**
ohm-centimeter	**ohm-cm**	oil well shut in	**OSI**
ohm-meter	**ohm-m**	oil-cut mud	**OCM**
oil	**O**	oil-cut salt water	**OCSW**
oil abandoned well	**OAW**	oil-cut water	**OCW**
oil and gas	**O&G**	oil-powered total energy	**OTE**
oil and gas investigations chart	**OC-**	oil-soluble acid	**OSA**
Oil & Gas Journal	**OGJ**	oil-water contact	**OWC**
oil and gas lease	**O&GL**	old plugback	**OPB**
oil and gas-cut acid water	**O&GCAW**	old plugback depth	**OPBD**
oil and gas-cut load water	**O&GCLW**	old total depth	**OTD**
oil and gas-cut mud	**O&GCM**	old well drilled deeper	**OWDD**
oil and gas-cut salt water	**O&GCSW**	old well plugged back	**OWPB**
oil and gas-cut sulfur water	**O&GC SULW**	old well sidetracked	**OWST**
		old well worked over	**OWWO**
oil and gas-cut water	**O&GCW**	olefin	**ole**
oil and salt water	**O&SW**	Oligocene	**Olig**
oil and sulfur water-cut mud	**O&SWCM**	on center	**OC**
oil and water	**O&W**	one thousand foot-pounds	**kip-ft**
oil-based mud	**OBM**	one thousand pounds	**kip**
oil circuit breaker	**OCB**	ooliclastic	**ooc**
Oil Creek	**Oil Cr**	oolimoldic	**oom**
oil cut	**OC**	oolitic	**ool**
oil down to	**ODT**	open (ed) (ing)	**opn**
oil emulsion	**OE**	open choke	**OC**
oil emulsion mud	**OEM**	open cup	**OC**
oil fluorescence	**OFLU**	open end	**OE**
oil fractured	**oilfract**	open flow	**OF**
oil immersed, water cooled	**OIWC**	open flow potential	**OFP**
oil in hole	**OIH**	open hearth	**OH**
oil in place	**OIP**	open hole	**OH, op hole**
oil in tanks	**OIT**		
oil insulated	**OI**	open line (no choke)	**OL**
oil insulated, fan cooled	**OIFC**	open-top tank	**O/T tk**
oil insulated, self-cooled	**OISC**	open tubing	**OT**

open-file report	OF
operate, operations, operator	oper
operation shut down	OSD
operations and maintenance	O&M
operations commenced	OC
Operculinoides	Operc
opposite	opp
optimum bit weight and rotary speed	OBW & RS
option to farmout	optn to F/O
optional	OPTL
orange	OR
Ordovician	Ord
Oread	Or
organic	org
organization	org
orientation	ORIENT
oriented microresistivity	OMRL
orifice	orf
orifice flange one end	OFOE
original oil in place	OOIP
original stock tank oil in place	OSTOIP
original total depth	OTD
original, originally	orig
Oriskany	Orisk
orthoclase	orth
Osage	Os
Osborne	O
ostracod	ost
Oswego	Osw
other end beveled	OEB
ounce	oz
Ouray	Our
out of service over and short (report)	O/S
out of stock	O/S
outboard motor oil	OBMO
Outer Continental Shelf	OCS
outlet	otl, OUT
outline	OLN
outpost	OP
outside diameter	OD
outside screw and yoke (valve)	OS&Y
overexpenditure	OE
overhead	OH
overproduced	OP
overall	OA
overall height	OAH

overall length	OAL
overflush (ed)	OFL
overhead	ovhd
overriding royalty	ORR
overriding royalty interest	ORRI
overseas procurement office	OPO
overshot	OS
overtime	OT
oxidized, oxidation	ox
oxygen	oxy

Pacific Outer Continental Shelf	POCS
packed	pkd
packer	pkr
packer set at	PSA
packing, package (ed)	pkg (d)
Paddock	Padd
page (before a number)	p.
Pahasapa	Paha
paid	pd
Paint Creek	PC
pair	pr
paleonotology	Paleo
Paleozoic	Paleo
Palo Pinto	Palo P
Paluxy	Pal, Pxy
panel	pnl
panel board	PNL BD
Panhandle lime	Pan L
Paradox	Para
Park City	Park C
parish	Ph
parrafins-olefins-napthenes-aromatics	PONA
part, partly	pt
partial	PART
participating area	PA
partings	prtgs
partition	PTN
partly	ptly
parts per billion	ppb
parts per million	ppm
Pascal	Pa

patent (ed)	**pat**	personnel	**pers**
pattern	**patn**	petrochemical	**petrochem**
pavement	**pvmnt**	petroleum	**pet**
paving	**pav**	petroleum and natural gas	**P & NG**
Pawhuska	**Paw**	petroliferous	**petrf**
payment	**pymt**	Pettet	**Pet**
pearly	**prly**	Pettit	**Pett**
pebble, pebbly	**pbl (y)**	Pettus sand	**Pet. sd**
Pecan Gap	**PG**	phase	**ph**
Pecan Gap chalk	**PGC**	Phosphoria	**Phos**
pedestal	**PED**	phrohotite	**po**
pelecypod	**plcy**	picked up	**PU**
pelletal, pelletoidal	**pell**	picking up drillpipe	**PUDP**
penalty, penalize (ed) (ing)	**penal**	Pictured Cliff	**Pic Cl**
penetration asphalt cement	**Pen A.C.**	piece	**pc**
penetration index	**PI**	pilot	**plt**
penetration, penetration test	**pen**	pilot-loaded valve	**PLV**
Pennsylvanian	**Penn**	pin end	**pe**
Pensky-Martins	**PM**	Pin Oak	**P.O.**
Pensky-Martins (flash)	**P-M**	Pine Island	**PI**
per-acre bonus	**PAB**	pink	**pk**
per-acre rental	**PAR**	pinpoint	**pinpt, PP**
percent	**%, pct**	pinpoint porosity	**PPP**
per day	**PD**	pint	**pt**
per foot	**/ft, pft**	pipe butt weld	**PBW**
per square inch gauge	**psig, PSIG**	pipe electric weld	**PEW**
percolation	**perco**	pipe lap weld	**PLW**
perforate (ed) (ing) (or)	**perf**	pipe, seamless	**PSM**
perforated casing	**perf csg**	pipe sleeve	**PSL**
perforating hyper select	**H-SEL**	pipe spiral weld	**PSW**
perforating, Enerjet	**EJ**	pipe to soil potential	**PTS pot**
perforating, Hyperdome	**HD**	pipe-handling capacity	**PHC**
perforating, Hyperdome II	**P-HDII**	pipeline	**PL**
perforating, Ultrajet	**ULJ**	pipeline oil	**PLO**
performance evaluation and review technique	**PERT**	pipeline terminal	**PLT**
		pipeway	**PWY**
Performance Number (aviation gas)	**PN**	piping	**ppg**
		piping and instrument diagrams	**P&IDS**
period	**prd**	piping diagram	**P/DIA**
peripheral wedge zone	**PWZ**	pisolites, pisolitic	**piso**
permanent	**perm**	pitted	**pit**
permanent-type completion	**PTC**	plagioclase	**plg**
permanently shut down	**PSD**	plain both ends	**PBE**
permeability (vertical direction)	**KV**	plain end	**PE**
permeable (ability)	**perm**	plain end beveled	**PEB**
Permian	**Perm**	plain large end	**PLE**
permit	**prmt**	plain one end	**POE**
perpendicular	**perp**	plain small end	**PSE**
personal and confidential	**P&C**	plan	**pln**

plan of development	POD	port collar	pc
plant	plt	portable	port
plant (pressure) volume reduction	PVR	Porter Creek	PC
		position	pos
plant fossils	pl fos	positive	pos
Planulina harangensis	Plan. hara.	positive crankcase ventilation	PCV
Planulina palmarie	Plan. palm.	possible (ly)	poss
		Post Laramie	P Lar
plaster	PLASR	Post Oak	P.O.
plastic	plas	postweld heat treatment	PWHT
plastic viscosity	PV	potable water	POT/WTR
plate	PL	potential	pot
platform	platf	potential difference	pot dif
platy	plty	potential test	PT
please note and return	PNR	potential test to follow	PTTF
Pleistocene	Pleist	pound	lb
Pliocene	Plio	pound-inch	lb-in.
plug	Pg	pounds per barrel	PPB
plug down	PD	pounds per cubic foot	PCF
plug on bottom	POB	pounds per foot	lb/ft
plugged	plgd	pounds per gallon	PPG
plugged and abandonded	P & A	pounds per square foot	psf, lb/sq ft
plugged back	PB	pounds per square inch	psi, PSI
plugged-back depth	PBD	pounds per square inch absolute	psia, PSIA
plugged-back total depth	PBTD		
plumbing	PLMB	pounds per square inch gauge	psig, PSIG
plunger	plngr	pour point (ASTM method)	pour ASTM
pneumatic	pneu	power	PWR
Podbielniak	Pod.	power distribution center	PDC
point	pt	power distribution system	PDS
Point Lookout	Pk Lkt	power factor	PF
poison	pois	power factor meter	PFM
poker chipped	PC	Precambrian	Pre Camb
polish (ed)	pol	precast	prcst
polished rod	PR	precipitate	ppt
polyethylene	poly cl	precipitation number	ppn No.
polymerization, polymerized	Poly	precipitator	PRECIP
polymerized gasoline	polygas	predominant	predom
polypropylene	polypl	prefabricated	prefab
polyvinyl chloride	PVC	preferred	pfd
Pontotoc	Pont	prefractionator	PFRACT
pooling agreement	PA	preheater	prehtr
poor	pr	preliminary	prelim
porcelaneous	porc	premium	prem
porcion	porc	prepaid	ppd
pore volume	PV	prepare, preparing, preparation	prep
porosity and permeability	P&P		
porosity, porous	por	preparing to take potential test	PRPT
porous and permeable	P&P	present depth	PD

present operations	**pr op**	producing oil well, pumping	**POWP**
present production	**P.P.**	producing well	**PW**
present total depth	**PTD**	production department	**PDET**
present worth at discount rate of 15%	**PW(15)**	exploratory test	
		production payment	**PP**
pressed distillate	**PD**	production payment interest	**PPI**
pressure	**press**	production test flowed	**PTF**
pressure alarm	**PA**	production test pumped	**PTP**
pressure control valve	**PCV**	productivity index	**PI**
pressure differential controller	**PDC**	professional paper	**PP**
pressure differential indicator	**PDI**	profit and loss	**P&L**
pressure differential indicator controller	**PDIC**	profit-sharing interest	**PSI**
		progress	**prog**
pressure differential recorder	**PDR**	project (ed) (ion)	**proj**
pressure differential recorder controller	**PDRC**	project ultimate cost	**PUC**
		projected total depth	**PTD**
pressure indicator	**PI**	propane	**LPG**
pressure indicator controller	**PIC**	property	**PROP**
pressure recorder	**PR**	property line	**PL**
pressure recorder control	**PRC**	proportional	**prop**
pressure safety valve	**PSV**	propose (ed)	**prop**
pressure seal bonnet	**PSB**	proposed bottom-hole location	**PBHL**
pressure switch	**PS**	proposed depth	**PD**
pressure-volume-temperature	**PVT**	proposed total depth	**PTD**
prestressed	**prest**	prorated	**pro**
prevent, preventive	**prev**	prospect	**Psp**
prevention of significant deterioration, EPA	**PSD**	protection	**prot**
		Proterozoic	**Protero**
previous	**PREV**	provincial	**Prov**
previous daily output average	**PREV DO AVG**	pseudo	**pdso, ps**
		public address	**PA**
primary	**pri**	public relations	**PR**
primary reference fuel	**PRF**	Public School Land	**PSL**
principal	**prin**	pull (ed) rods and tubing	**PR&T**
principal lessee (s)	**prncpl lss**	pull (put) out of hole	**POOH**
prism (atic)	**pris**	pulled	**pld**
private branch exchange	**PBX**	pulled (put) out of hole	**POH**
privilege	**priv**	pulled bid pipe	**PBP**
probable (ly)	**prob**	pulled out	**PO**
process	**proc**	pulled pipe	**PP**
process & instrument diagram	**P & ID**	pulled up	**PU**
process flow diagram	**PFD**	pulled up in casing	**PUIC**
Process Performance Index	**PPI**	pulling	**plg**
produce (ed) (ing) (tion), product (s)	**prod**	pulling tubing	**PTG**
		pulling tubing and rods	**PTR**
producing gas well	**PGW**	pulsation dampener	**PD**
producing oil and gas well	**POGW**	pulse (sating) (sation)	**PULS**
producing oil well	**POW**	pump	**P/**
producing oil well, flowing	**POWF**	pump and flow	**P&F**

pump building	**P/BLDG**
pump in	**PI**
pump jack	**PJ**
pump job	**PJ**
pump on beam	**POB**
pump pressure	**PP**
pump station	**PS**
pump testing	**P tstg**
pump-in pressure	**PIP**
pump (ed) (ing)	**pmp (d) (g)**
pumper's depth	**PD**
pumping equipment	**PE**
pumping for test	**PFT**
pumping load oil	**PLO**
pumping unit	**PU**
pumps off	**PO**
purchase order	**PO**
purchasing	**PURCH**
purchasing request	**PR**
purification	**PURF**
purple	**purp**
putting on pump	**POP**
pyrite, pyritic	**pyr**
pyrobitumen	**pyrbit**
pyroclastic	**pyrclas**
pyrolysis	**pyls**

quadrant	**QDRNT**
quadrant (rangle) (ruple)	**quad**
qualitative	**QUAL**
quality	**qual**
quality assurance	**QA**
quality control	**Q.C., QC**
quality discount allowance	**QDA**
quantity	**qty**
quarry	**qry**
quart (s)	**qt**
quarter	**qtr**
quartz, quartzite, quartzitic	**qtz**
quartzose	**qtzose**
Queen City	**Q. City**
Queen Sand	**Q. sd**
quench	**qnch**

questionable	**quest**
quick ram change	**QRC**
quintuplicate	**quint**

R

rack	**RK**
radiant	**RADT**
radiation	**radtn**
radical	**rad**
radioactive	**RA**
radiographic test	**RT**
radiological	**rad**
radius	**R**
radius	**rad**
railing	**rlg**
railroad	**RR**
Railroad Commission (Texas)	**RRC**
rainbow show of oil	**RBSO**
raised face	**RF**
raised face, flanged end	**RFFE**
raised face, slip on	**RFSO**
raised face, smooth finish	**RFSF**
raised face, weld neck	**RFWN**
Ramsbottom Carbon Residue	**RCR**
ran (running) rods and tubing	**RR&T**
ran in hole	**RIH**
random lengths	**RL**
range	**rge**
Ranger	**Rang**
Rankine (temp. scale)	**R**
rapid curing	**RC**
rat hole	**RH**
rat hole mud	**RHM**
rate of penetration	**ROP**
rate of return	**ROR**
rate too low to measure	**RTLTM**
rating	**rtg**
raw gas	**RG**
raw gas lift	**RAGL**
re-evaluation for over- optimism	**REFOO**
reabsorber	**REABS**
reacidize (ed) (ing)	**reacd**
reaction (ed)	**react**

ready for rig	**RFR**	regular, regulator	**reg**
ream	**rm**	Reid vapor pressure	**RVP**
reamed	**rmd**	reinforce (ed) (ing) (ment)	**reinf**
reaming	**rmg**	reinforced concrete	**reinf conc**
reboiler	**RBLR**	reinforcing bar	**rebar**
received	**recd**	reject	**rej**
receiver	**recr**	rejection	**rej'n**
receptacle	**recp, RCPT**	Reklaw	**Rek**
		relative humidity	**RH**
reciprocate (ing)	**recip**	relay	**rly**
recirculate	**recirc**	release (ed) (ing)	**rls (ed) (ing), rel**
recommend	**rec**		
recommended spare part	**R-SP**	released swab unit	**RSU**
recomplete (ed) (ion)	**recomp**	relief	**rlf**
recompressor	**RECOMP**	relief valve	**RV**
recondition (ed)	**recond**	relocate (ed)	**reloc**
record (er) (ing)	**rec**	remains	**rmns, rems**
recover (ed) (ing), recovery	**rec**	remedial	**rem**
recovery	**RCVY**	remote control	**RC**
rectangle, rectangular	**rect**	remote operating system (station)	**ROS**
rectifier	**rect**		
recycle	**recy, RCYL**	remote terminal unit	**RTU**
		remove (al) (able)	**rmv (l)**
red beds	**Rd Bds**	Renault	**Ren**
Red Cave	**RC**	rental	**rent**
Red Fork	**Rd Fk**	Reophax bathysiphoni	**Reo. bath.**
red indicating lamp	**RIL**	repair (ed) (ing) (s)	**rep**
Red Oak	**R.O.**	repairman	**rpmn**
Red Peak	**Rd Pk**	reperforated	**reperf**
Red River	**RR**	replace (ed)	**rep**
redrilled	**redrld, RR**	replace (ment)	**repl**
reducer	**RDCR**	report	**rep, RPRT**
reducing balance	**red bal**	report available only through National Technical Information Service	**PB-ADA**
reducing, reducer	**red**		
reference	**ref**		
refine (ed) (er) (ry)	**ref**	reported	**rptd**
Refinery Technology Laboratory	**RTL**	request	**REQ**
		request for proposal	**RFP**
refining	**refg**	request for quote	**RFQ**
reflect (ed) (ing) (tion)	**refl, RFLCT**	required	**reqd**
reflux	**refl**	requirement	**reqmt**
reformate (er) (ing)	**reform**	requisition	**req**
reformer	**REFMR**	research	**res**
refraction, refractory	**refr**	research and development	**R & D**
refrigerator (rant) (tion)	**REFRIG**	Research Octane Number	**Res. O.N., RON**
refrigeration building	**RFG/BD**		
regenerator	**regen**	Research Planning Institute	**RPI**
register	**reg, RGTR**	reserve (ation)	**res**
regular acid	**R/A**	reservoir	**rsvr**

reservoir description service	**RDS**	rig-up casing crew	**RUCC**
residual, residue	**resid**	rigged (ing) down	**RD**
resinous	**rsns**	rigged down, moved out	**RDMO**
resistance, resistivity, resistor	**res**	rigged-down swabbing unit	**RDSU**
resistivity	**R**	rigging rotary	**RR**
resistivity as recorded from		rigging-up cable tools	**RUCT**
16″ electrode configuration	**R(16″)**	rigging-up machine	**RUM**
resistivity, invaded zone	**RIZ**	rigging-up pump	**RUP**
resistivity, flushed zone	**Rxo**	rigging-up rotary tools	**RURT**
resistivity, mud	**Rm**	rigging-up service rig	**RUSR**
resistivity, mud filtrate	**Rmf**	rigging-up standard tools	**RUST**
resistivity, water	**Rw**	rigging-up swabbing unit	**RUSU**
resistivity, water (apparent)	**Rwa**	rigging-up tools	**RUT**
resistor (s)	**RESIS**	right angle	**RA**
retail pump price	**RPP**	right hand	**RH**
retain (er) (ed) (ing)	**ret, rtnr**	righthand door	**RHD**
retard (ed)	**rtd**	right of way	**ROW**
retrievable	**retr**	ring	**rg**
retrievable bridge plug	**RBP**	ring groove	**RG**
retrievable retainer	**retr ret**	ring joint	**RJ**
retrievable test treat squeeze	**RTTS**	ring joint, flanged end	**RJFE**
(tool)		ring-tool joint	**RTJ**
return	**ret**	ring-type joint	**RTJ**
return on investment	**ROI**	rising stem (valve)	**RS**
returnable steel drum	**RSD**	rivet	**riv. RVT**
returned	**retd**	road (s)	**rd (s)**
returned well to production	**RWTP**	road & location	**R&L**
returning circulation oil	**RCO**	road & location complete	**R&LC**
reverse (ed)	**rvs, rev (d)**	Robulus	**Rob**
reverse circulation	**RC**	rock	**rk**
reverse circulation rig	**RCR**	rock bit	**RB**
reversed out	**rev/O, RO**	rock pressure	**RP**
revise (ed) (ing) (ion)	**rev**	Rockwell hardness number	**RHN**
revolution (s)	**rev**	rocky	**rky**
revolutions per minute	**rpm**	Rodessa	**Rod**
revolutions per second	**rps**	rods and tubing	**R & T**
rework (ed)	**rwk (d)**	roentgen	**r**
rheostat	**rheo**	roofing	**RFG**
ribbon sand	**Rib**	room	**rm**
rich oil fractionator	**ROF**	root mean square	**RMS**
Rierdon rig	**Rier**	rose	**ro**
rig (ged) (ging) up	**RU**	Rosiclare sand	**Ro sd**
rig floor	**RF**	rotameter	**RTMTR**
rig on location	**ROL**	rotary bushing	**RB**
rig released	**RR, R Rel**	rotary bushing measurement	**RBM**
rig repair	**RR**	rotary drive bushing	**RDB**
rig service	**RS**	rotary drive bushing to ground	**RDB-GD**
rig skidded	**RS**	rotary kelly bushing	**RKB**
rig time	**RT**	rotary table	**RT**

rotary test	R test
rotary tools	RT
rotary total depth	RTD
rotary unit	RU
rotary, rotate, rotator	rot
rotative gas lift	ROGL
rough	rgh
rough order of magnitude	ROM
round	rd
round thread	rd thd
round trip	rdtp
round trip changed bit	RT CB
rounded	rdd, rnd
routing	RTG
rows	R
royalty	roy
royalty interest	RI
rubber	rbr, rub
rubber ball sand water frac	RBSWF
rubber ball sand oil frac	RBSOF
rubber balls	Rbls
run of mine	ROM
running	rng
running casing	RC
running electric log	REL
running radioactive log	RALOG
running tubing	RTG
rupture	rupt
rust and oxidation	R&O

Sabinetown	Sab
saccharoidal	sach
sack (s)	sk,sx
saddle	sadl
Saddle Creek	Sad Cr
safety	saf
safety relief valve	SRV
safety/department	SAF/DPT
Saint Genevieve	St Gen
Saint Louis lime	St L
Saint Peter	St Ptr
Salado	Sal
salary, salaried	sal
Saline Bayou	Sal Bay

salinity	sal
salt	X
salt and pepper	s&p
Salt Mountain	Slt Mtn.
salt wash	SW
salt water	SW, swtr, XW
salt water to surface	SWTS
saltwater-cut mud	SWCM
saltwater disposal	SWD
saltwater disposal system	SWDS
saltwater disposal well	SWDW
saltwater fracture	SWF
saltwater injection	SWI
salty	Slty
salty sulfur water	SSUW
salvage	salv
sample	samp, smpl, spl
sample chamber	splcham
sample formation tester	SFT
sample tops	S/T
San Andres	San And
San Angelo	San Ang
San Bernardino base and meridian	SBB&M
San Rafael	San Raf
Sanastee	Sana
sand	SD, sd
sand and shale	sd & sh
sand-oil fracked	sdoilfract
sand-oil fracture	SOF
sand showing gas	Sd SG
sand showing oil	Sd SO
sand-water fracked	sdwtrfract
sanded	sdd
sandfracked	sdfract, SF
sandstone	SS
sandy	sdy
sandy lime	sdy li
sandy shale	sdy sh
sanitary	SAN, sani
sanitary water	S/WTR
Santa Margarita	Sta Marg
saponification	sap
saponification number	Sap No.
Saratoga	Sara
Satanka	Stnka
saturated, saturation	sat
Sawatch	Saw

Sawtooth	**Sawth**	selenite	**sel**
Saybolt furol	**Say furol**	self (spontaneous) potential	**SP**
Saybolt Seconds Universal	**SSU**	self-contained	**SC**
Saybolt universal viscosity	**SUV**	self-elevating work platform	**SEWOP**
scaffolding	**SCAF**	Selma	**Sel**
scales	**sc**	Senora	**Sen**
scatter (ed)	**scatt (d),**	separate, separator, separation	**SEP**
	sctrd	septuplicate	**sept**
schedule	**sch**	sequence	**seq**
schematic	**schem**	series, serial	**ser**
Schlumberger	**SCHL**	serpentine	**serp**
scolescodonts	**scolc**	Serratt	**Serr**
scraper	**scr**	service (s)	**svc, serv**
scratcher	**scr**	service charge	**serv chg**
screen	**scr**	service unit	**svcu**
screw (ed)	**scr (d)**	set drillpipe	**SDP**
screw end American National Acme thread	**SE NA**	set plug	**SP**
		settling	**set**
screw end American National Coarse thread	**SE NC**	Seven Rivers	**S Riv**
		service station	**SS**
screw end American National Taper Pipe thread	**SE NTP**	severy	**Svry**
		Seward Meridian (Alaska)	**SM**
screwed and socketweld	**S/SW**	sewer	**sew**
screwed end	**S/E**	sexton	**Sex**
screwed on one end	**SOE**	sextuple	**sxtu**
scrubber	**scrub**	sextuplicate, sextuplet	**sext**
sea level	**S.L.**	shaft horsepower	**shp**
Seabreeze	**Sea**	shake out	**SO**
seal assembly	**SA**	shale	**sh**
seal oil	**SEO**	shaled out	**SO**
seal weld	**SWLD**	shaly	**shly**
seal-welded bonnet	**SWB**	shallower pool (pay) test	**SPT**
sealed	**sld**	shallower pool wildcat, discovery	**WSD**
seamless	**smls**		
seating nipple	**SN**	shallower pool wildcat, dry	**WS**
secant	**sec**	Shannon	**Shan**
second (ary)	**sec**	shear	**shr**
secondary butyl alcohol	**SBA**	sheathing	**shthg**
seconds	**S, sec**	sheet	**sh**
secretary	**sec**	sheeting	**SHTG**
section	**sec**	shell and tube	**S & T**
section (s) (al) (ing)	**SECT**	shells	**shls**
section line	**SL**	shelter	**SHLT**
section-township-range	**S-T-R**	Shinarump	**Shin**
securaloy	**scly**	ship (ping)	**shp(g)**
sediment (s)	**sed**	shipping point (purchasing term)	**s/p, sp**
Sedwick	**Sedw**		
segment	**SEG**	shipment	**shpt**
seismograph, seismic	**seis**	shock sub	**SS**
selection (tive) (tor)	**SELECT**	shop fabrication	**S/FAB**

short radius	SR	sidetracked hole	STH
short string	SS	sidetracked total depth	STTD
short thread	ST	sidewall	SDWL
short threads & coupling	ST&C	sidewall samples	SWS
shortage	SHTG	siding	SDG
shot open hole	SOH	signed	sgd
shot per foot	SPF	silencer	SLNCR
shot point	sp	silica, siliceous	silic
shoulder	shld	silky	slky
show condensate	SC	siltstone	silt
show gas	SG	silty	slty
show gas and condensate	SG&C	Silurian	Sil
show gas and distillate	SG&D	similar	sim
show gas and water	SG & W	Simpson	Simp
show of dead oil	SDO	single (s)	sgl (s)
show of free oil	SFO	single-pole double throw	SPDT
show of gas and oil	SG&O	single-pole double throw switch	SPDT SW
show of oil	SO	single-pole single throw	SPST
show oil and gas	SO&G	single-pole single throw stitch	SPST SW
show oil and water	SO&W	single-pole switch	SP SW
shut down	SD	single random lengths	SRL
shut down awaiting orders	SDWO	single shot	SS
shut down for orders	SDO	Siphonina davisi	Siph. d.
shut down for pipe line	SDPL	size	sz
shut down for repairs	SDR	sketch	SK
shut down for weather	SDW	skimmer	skim
shut down overnight	SDON	Skinner	Skn
shut down to acidize	SDA	Skull Creek	Sk Crk
shut down to fracture	SDF	sleeper	SLPR
shut down to log	SDL	sleeve	sl, SLV
shut down to plug & abandon	SDPA	sleeve bearing	SB
shut in	SI	slickensided	sks
shutin bottom-hole pressure	SIBHP	sliding-scale royalty	S/SR
shutin casing pressure	SICP	slight (ly)	sli
shutin gas well	SIGW	slight oil-cut mud	SOCM
shutin oil well	SIOW	slight oil-cut water	SOCW
shutin pressure	SIP	slight show of gas	SSG
shutin tubing pressure	SITP	slight show of oil	sli SO, SSO
shutin wellhead pressure	SIWHP	slight show of oil and gas	SSO&G
shutin, waiting on potential	SIWOP	slight, weak, poor fluorescence	SFLU
shut well in overnight	SWION	slightly gas-cut salt water	SGCSW
side door choke	SD Ck	slightly gas-cut water cushion	SGCWC
side opening	SO	slightly gas-cut mud	SGCM
sidewall cores	SWC	slightly gas-cut oil	SGCO
sidewall neutron porosity	SWNP	slightly gas-cut water	SGCW
sideboom	SB	slightly gas-cut water blanket	SGCWB
siderite (ic)	sid	slightly oil- and gas-cut mud	SO&GCM
sides, tops & bottoms	s,t&b	slightly oil-cut salt water	SOCSW
sidetrack (ed) (ing)	sdtkr, ST	slightly oil-cut water blanket	SOCWB

slightly oil-cut water cushion	**SOCWC**	southwest corner	**SW/c**
slightly porous	**sp**	southwest quarter	**SW/4**
Sligo	**Sli**	spacer	**spcr**
slim-hole drillpipe	**SHDP**	spare	**sp**
slip and cut drill line	**SC DL**	Sparta	**Sp**
slip on	**SO**	spearfish	**spf**
slope type of wall to keep out flooding	**berm**	special	**spcl**
		specialty	**splty**
slow set (cement)	**SS**	specific gravity	**sp gr**
slurry	**slur**	specific heat	**sp ht**
Smackover	**Smk,SO**	specific volume	**sp. vol.**
small	**sm**	specification	**spec**
small show	**SS**	speckled	**speck**
Smithwick	**Smithw**	speed/current	**S/C**
Smoke Volatility Index	**SVI**	speed/torque	**S/T**
smooth	**smth**	Sphaerodina	**Sphaer**
snubber, snubbing	**SNUB**	sphalerite	**sphal**
snuffing	**SNUFF**	spherules	**sph**
Society of Economic Paleontologists & Mineralogists	**SEPM**	spicule (ar)	**spic**
		spigot	**SPGT**
Society of Petroleum Engineers	**SPE**	spigot and spigot	**s & s**
socket	**skt**	spillway	**SPWY**
socket weld	**SW**	spindle	**spdl**
sodium-base grease	**sod gr**	Spindletop	**Spletp**
sodium carbonate	**NaCO₃**	spiral weld	**SW**
sodium carboxymethylcellulose	**CMC**	spirifers	**sprf**
sodium chloride	**NaCL**	Spiroplectammina barrowi	**Spiro. b.**
sodium hydroxide	**NaOH**	splintery	**splty**
soft	**sft**	splitter	**SPLTR**
solar heat medium	**SHM**	sponge	**spg**
solenoid	**sol, slnd**	spore	**sp**
solenoid-operated valves	**SOV**	spot sales agreement	**SSA**
solenoid valve	**SV**	spotted	**sptd**
solids	**sol**	spotty	**sptty**
solution	**soln**	Spraberry	**Spra**
solvent	**solv**	spring	**spg**
somastic	**som**	Springer	**Sprin**
somastic coated	**somct**	sprinkler	**spkr**
sonic log	**SONL**	sprocket	**spkt**
sort (ed) (ing)	**srt**	spud (ded) (der)	**spd**
south	**S**	square	**sq**
south half	**S/2**	square centimeter	**sq cm**
south line	**SL**	square foot (feet)	**ft², sq ft**
south offset	**S O**	square inch	**sq in.**
southeast	**SE**	square kilometer	**sq km**
southeast corner	**SE/C**	square meter	**sq m**
southeast quarter	**SE/4**	square millimeter	**sq mm**
southerly	**S'ly**	square root	**SQRT**
southwest	**SW**	square yard (s)	**sq yd**

squeeze (ed) (ing)	**sqz**	steel tape measurement	**STM**
squeeze packer	**sq pkr**	Steele	**Stel**
squeezed	**sq**	stencil (ed) (ing)	**stncl (d) (g)**
squirrel cage	**sq cg**		
stabilized (er)	**stab**	stenographer	**steno**
stage	**STG**	Stensvad	**Stens**
staggered	**STAG**	sticky	**stcky**
stain (ed) (ing)	**stn (d) (g)**	stiffener	**STIF**
stain and odor	**S&O**	stippled	**stip**
stainless steel	**SS**	stirrup	**stir**
stairway	**stwy**	stock	**stk**
Stalnaker	**Stal**	stock tank barrels	**STB, stb**
stand (s) (ing)	**std**	stock tank barrels per day	**STBPD, stb/d**
stand by	**stn/by**		
standard cubic feet per day	**SCFD, scfd**	stock tank oil in place	**st oip**
standard cubic feet per hour	**SCFH, scfh**	stock tank vapor	**STV**
standard cubic feet per minute	**SCFM, scfm**	Stone Corral	**Stn Crl**
standard cubic foot	**SCF, scf**	Stony Mountain	**Sty Mtn**
standard operational procedure	**SOP**	stopper (ed)	**stpr (d)**
standard temperature and pressure	**STP**	storage	**strg**
		stove oil	**stv**
standardization	**STDZN**	straddle	**strd**
standards	**std (s)**	straddle packer	**SP**
standing	**stdg**	straddle packer drillstem test	**SP-DST**
Stanley	**Stan**	straight	**strt**
start	**st**	straight-hole test	**SHT**
start of cycle	**SOC**	straight-run naphtha	**SRN**
start of run	**SOR**	straightened	**strtd**
started in hole	**SIH**	straightening	**stging**
started out of hole	**SOH, SOOH**	strainer	**stnr**
		strand (ed)	**strd**
starting fluid level	**SFL**	strap out of hole	**STROH**
state lease	**SL**	strapped out of hole	**SOOH**
state potential	**State pot**	stratigraphic	**strat**
statewide rules	**SWR**	Strawn	**Str**
static bottom-hole pressure	**SBHP**	streak (s) (ed)	**stk, strk**
station	**sta**	striated	**stri**
stationary	**stat**	string (er)	**strg (r)**
statistical	**stat**	string shot	**SS**
steady	**stdy**	strip (per) (ping)	**STPR**
steam	**stm**	strokes per minute	**SPM**
steam cylinder oil	**stm cyl oil**	stromatoporoid	**strom**
steam emulsion number	**SE No.**	strong	**strg**
steam engine oil	**stm eng oil**	structure, structural	**struc**
steam trace (ing)	**STM TR**	stuck	**stk**
steam working pressure	**SWP**	study	**stdy**
steel	**stl**	stuffing box	**SB**
steel line correction	**SLC**	styolite, styolitic	**styo**
steel line measurement	**SLM**	sub-Clarksville	**Sub Clarks**

subangular	**sub angl**	swab rate	**SR**
subdivision	**subd**	swab run (s)	**SR**
subrounded	**sub rnd**	swabbed	**S/**
subsea	**SS**	swabbed, swabbing	**SWB**
subsidiary	**sub**	swabbing unit	**SWU**
substance	**sub**	swaged	**swd**
substation	**substa**	Swastika	**Swas**
substitute	**SUBST**	sweetening	**Swet**
substructure height	**SH**	switch	**SW**
subsurface	**SS**	switchboard	**PBX, swbd**
subsurface safety valve	**SSSV**	switchgear	**swgr**
successful wildcat outpost	**WOE**	switchrack	**SWRK**
sucker rod	**skr rd**	Sycamore	**Syc**
sucrose, sucrosic	**suc**	Sylvan	**Syl**
suction	**suct**	symbol	**sym**
sugary	**sug**	symmetrical	**sym**
sulfate bacteria	**SRB**	synchronous, synchronizing	**syn**
sulfur by bomb method	**S Bomb**	synchronous converter	**syn conv**
sulfur, sulfuric	**sulf**	synchroscope	**SYNSCP**
sulfuric acid	**H$_2$SO$_4$**	synthesis	**SYNTH**
sulfated	**sulf**	synthetic	**syn**
sulfur	**sul**	synthetic natural gas	**SNG**
sulfur water	**sul wtr**	system	**sys**
summary, summarize	**sum**	system flow diagram	**SFD**
Summerville	**Sumvl**		
Sunburst	**Sb, Sunb**		
Sundance	**Sund**		
Supai	**Sup**		
superintendent	**supt**		
superseded	**supsd**		
supervisor	**suprv**		

supplement	**supp**	tabular, tabulating	**tab**
supply (ied) (ier) (ing)	**supl, sply**	tachometer	**TACH**
support	**sppt**	tag closed cup (flash)	**TCC**
surface	**surf, sfc**	tag open cup (flash)	**TOC**
surface approximation and	**SAFE**	Tagliabue	**Tag**
formation evaluation		Tallahatta	**Tal**
surface-controlled subsurface	**SCSSV**	Tampico	**Tamp**
safety valve		tangent	**tan**
surface flow pressure	**SFP**	tank	**tk**
surface geology	**SG**	tank battery	**TKB**
surface measurement	**SM**	tank car	**T/C**
surface pressure	**SP**	tank farm	**TKF**
surge	**SRG**	tank truck	**TT**
surplus	**surp**	tank wagon	**TKW**
survey	**sur**	tankage	**tkg**
suspended	**susp**	tanker (s)	**tkr**
suspended ceiling	**SUSP CLG**	Tannehill	**Tann**
swab and flow	**S&F**	Tansill	**Tan**

taper pipe thread	TPT	test to follow	TTF
Tar Springs sand	TSS	testing on pump	TOP
Tarkio	Tark	tetraethyl lead	TEL
tarred and wrapped	T&W	tetramethyl lead	TML
taste	tste	Texana	Tex
Taylor	Tay	Textularia articulata	Text. art.
tearing out rotary tools	TORT	Textularia dibollensis	Text. d.
technical, technician	tech	Textularia hockleyensis	Text. h.
techniques of water-resources investigations	TWI	Textularia warreni	Text. w.
		texture	tex
tee	T	Thaynes	Thay
teeth	T	thence	th
telegraph	TLG	theoretical production and allocation	TP&A
Telegraph Creek	Tel Cr		
teletype	TWX	thermal	thrm
television	TV	thermal cracker	therm ckr
telephone, telegraph	tel	thermal decay time	TDT
temperature	Temp	thermal hydrodealkylation	THD
temperature control valve	TCV	thermofor catalytic cracking	TCC
temperature controller	TC	thermometer	therm
temperature differential indicator	TDI	Thermopolis	Ther
		thermostat	therst
temperature differential recorder	TDR	thick, thickness	thk
		thin bedded	TB
temperature gradient	TG	thousand (i.e., 13K = 13,000)	K
temperature indicator	TI	thousand (i.e., 9M = 9,000)	M
temperature indicator controller	TIC	thousand barrels fluid per day	MBF/D, MBFPD
temperature observation	TO		
temperature recorder	TR	thousand barrels of oil per day	MBO/D, MBOPD
temperature recorder controller	TRC		
temperature survey indicated top cement at	TSITC	thousand barrels of water per day	MBW/D, MBWPD
temporarily abandoned	TA	thousand British thermal units	MBtu
temporarily shut down	TSD	thousand cubic feet of gas per day	MCFGPD, Mcfgpd
temporarily shut in	TSI		
temporary (ily)	temp	thousand cubic feet per day	MCFD, Mcfd
temporary dealer allowance	TDA		
temporary voluntary allowance	TVA	thousand electron-volts	kev
tender	tndr	thousand gallons	MG
tensile strength	tens str, TS	thousand standard cubic feet	MSCF, Mscf
Tensleep	Tens		
tentaculites	tent	thousand standard cubic feet per day	MSCF/D, Mscf/d
tentative	tent		
Teremplealeau	Tremp	thousand standard cubic feet per hour	MSCF/H, Mscf/h
terminal board	T/BRD		
terminal box	T/Box	thread large end	TLE
terminate (ed) (ing) (ion)	termin	thread on both ends	TOBE
tertiary	Ter	thread small end	TSE
tertiary butyl alcohol	TBA	thread small end, weld large end	TSE-WLE
test (er) (ing)	tst (r) (g)		

thread, threaded	thd	top of pay	T/pay
threaded and coupled	T & C	top of salt	TOS
threaded both ends	TBE	top of sand	T/sd
threaded one end	TOE	top salt	T/S
threaded pipe flange	TPF	Topeka	Tpka
Three Finger	Tfing	topo sheet evaluation	TS
Three Forks	Tfks	topographic, topography	topo
three-phase	3 PH	topping	tpg
throttling	thrling	topping and coking	T & C
through	thru	Toronto	Tor
through the tanks	TTT	Toroweap	Toro
through-tubing	TT	torque	TRQ
through-tubing caliper	TTC	total	tot
through-tubing plug	TTP	total depth	TD
Thurman	Thur	total time lost	TTL
tight	ti, tite	totally enclosed, fan cooled	TEFC
tight hole	TH	tough	gh
tight no show	TNS	Towanda	Tow
time delay	TD	tower	TWR
Timpas	TIM	township	twp
Timpoweap	Timpo	township (as T2N)	T
tires, batteries, and accessories	TBA	townsite	twst
to be conditioned for gas	TRG	trace	TR
to be conditioned for oil	TRO	trackage	trkg
Todilto	Tod	tract	TR
tolerance	tol	Trans-Alaska Pipeline System	TAPS
toluene	tolu	transducer	TRNDC
ton (after number - 3T)	T	transfer (ed) (ing)	trans
tongue and groove (joint)	T&G	transformer	trans,
Tonkawa	Tonk	translucent	transl
tons	tons	transmission	trans
too small to measure	TSTM	transmitter	XMTR
too wet (weak) to measure	TWTM	transparent	transp
tool (s)	tl	transportation	transp
tool closed	TC	travel (ed) (ing)	TRVL
tool joint	tl jt	Travis Peak	TP
tool open	TO	treat (er) (ed) (ing)	trt (r) (d) (g)
toolpusher	TP		
tooth	T	treater	trtr
top and bottom	T & B	treating pressure	TP
top and bottom chokes	T&BC	Trenton	Tren
top choke	TC	Triassic	Tri
top hole choke	THC	tricresyl phosphate	TCP
top hole flow pressure	THFP	trillion	10^{12}
top of (a formation)	T/	trillion cubic feet	TCF, Tcf
top of cement	TOC	trillion cubic feet per day	TCF/D, Tcf/d
top of cement plug	TOCP		
top of fish	TOF	trilobite	trilo
top of liner	TOL	Trinidad	Trin
top of liner hanger	TLH	trip for new bit	TFNB

trip in hole	**TIH**		
trip out of hole	**TOH**		
trip (ped) for bit	**TFB**		
triple-pole single throw switch	**3P ST SW**		

U

triple-pole switch	**3P SW**	ultimate	**ult**
triplicate	**trip**	ultrahigh frequency	**UHF**
Tripoli	**Trip**	ultrasonic examination	**UT**
tripolitic	**trip**	ultrasonic test	**UST**
tripped (ing)	**trip**	ultraviolet	**UV**
truck	**trk**	umbrella (s)	**UMB**
true boiling point	**TBP**	Umiat Meridian (Alaska)	**UM**
true vapor process	**TVP**	unbalanced	**UNBAL**
true vertical depth	**TVD**	unbranded	**unbr**
tube	**tb**	unclassified	**U**
tube bundle	**TB/BDL**	unconformity	**unconf**
tubing	**tbg**	unconsolidated	**uncons**
tubing and casing cutter	**TCC**	under construction	**U/C**
tubing and rods	**T&R**	under digging	**UD**
tubing choke	**tbg chk, TC**	under gauge	**UG**
		underreaming	**UR**
tubing pressure	**TP, tbg press**	underground	**UG**
		undifferentiated	**undiff**
tubing pressure, shut in	**TPSI**	unfinished	**unf**
tubing pressure, closed	**TPC**	unified coarse thread	**UNC**
tubing pressure, flowing	**TPF**	unified fine thread	**UNF**
tubinghead flange	**THF**	uniform	**uni**
Tucker	**Tuck**	uninterruptible power supply	**UPS**
tuffaceous	**tfs, tuf**	Union Oil Company	**UOCO**
Tulip Creek	**Tul Cr**	Union Valley	**UV**
tungsten carbide	**tung carb**	unit	**un**
turbine compressor	**T/C**	United States gauge	**USG**
turbo, turbine	**TURB**	universal gear lubricant	**UGL**
turn around	**TA**	universal transverse mercator	**UTM**
turned over to producing section	**TOPS**	university, universal	**Univ**
		unloader	**UNLDR**
turned to test tank	**TTTT**	unloading	**UNLD**
turnpike	**tpk**	unsulfonated residue	**UR**
Tuscaloosa	**Tus**	upper (i.e., U/Simpson)	**U/**
12 gauge wire-wrapped screen (in a liner)	**12GA W.W.S**	upper and lower	**U/L**
		upper casing	**UC**
Twin Creek	**Tw Cr**	upper tubing	**UT**
twisted off	**twst off**	upthrown	**UT**
type	**ty**	use customer's hose	**UCH**
typewriter	**tywr**	used rod	**UR**
typical	**typ**	used with	**U/W**
		utility	**UTL**
		utility flow diagram	**UFD**
		utility water	**U/WTR**
		Uvigerina lirettensis	**Uvig. lir.**

vacant	**vac**
vacation	**vac**
vacuum	**vac**
Vaginulina regina	**Vag. reg**
Valera	**Val**
valve	**V, vlv**
Vanguard	**Vang**
vapor pressure	**VP**
vapor recovery	**VR**
vapor recovery unit	**VRU**
vapor temperature	**vt**
vapor (izor)	**vap (r)**
vapor-liquid ratio	**V/L**
varas	**vrs**
variable, various	**var**
variegated	**vari**
varnish	**VARN**
varnish makers and painters naphtha	**VM&P naphtha**
varved	**vrvd**
velocity	**vel**
velocity survey	**V/S, VS**
vendor drawing	**V/DWG**
ventilator	**vent**
Verdigris	**Verd**
Vermillion Cliff	**Ver Cl**
versus	**vs**
vertebrate	**vrtb**
vertical	**vert, vrtl**
vertical support member	**VSM**
very (as very tight)	**v.**
very common	**v.c.**
very fine-grain (ed)	**vfg**
very heavily (highly) gas-cut mud	**VHGCM**
very heavily (highly) gas-cut salt water	**VHGCSW**
very heavily (highly) gas-cut water	**VHGCW**
very heavily (highly) oil- and gas-cut salt water	**VHO& GCSW**
very heavily (highly) oil- and gas-cut mud	**VHO&GCM**
very heavily (highly) oil- and gas-cut water	**VHO&GCW**
very heavily (highly) oil-cut mud	**VHOCM**
very heavily (highly) oil-cut salt water	**VHOCSW**
very heavily (highly) oil-cut water	**VHOCW**
very high frequency	**VHF**
very light amber cut	**VLAC**
very noticeable	**v.n.**
very poor sample	**V.P.S.**
very rare	**v.r.**
very slight	**v-sli**
very slight show of gas	**VSSG**
very slight show of oil	**VSSO**
very slightly gas-cut salt water	**VSGCSW**
very slightly gas-cut water	**VSGCW**
very slightly gas-cut mud	**VSGCM**
very slightly oil- and gas-cut mud	**VSO & GCM**
very slightly oil- and gas-cut salt water	**VSO &GCSW**
very slightly oil-cut mud	**VSOCM**
very slightly oil-cut salt water	**VSOCSW**
very slightly oil-cut water	**VSOCW**
very slightly porous	**VSP**
vesicular	**ves**
vessel	**VESS**
vibrate (tor) (ing)	**VIB**
Vicksburg	**Vks**
Viola	**Vi**
Virgelle	**Virg**
viscosity	**vis, V**
viscosity index	**VI**
viscosity-gravity constant	**VGC**
visible	**vis**
vitreous	**vit**
vitrified clay pipe	**VCP**
Vogtsberger	**Vogts**
volatile organic compounds	**VOC**
volt	**V, v**
volt-ampere	**va**
volt-ampere reactive	**var**
voltage	**VOLT**
volume	**V, vol**
volume-percent	**v%**
volumetric efficiency	**vol. eff.**
volumetric subcommittee	**VSC**
vuggy	**vug**
vugular	**vug**

Wabaunsee	**Wab**
Waddell	**Wad**
waiting	**wtg**
waiting on	**WO**
waiting on acid	**WOA**
waiting on allowable	**WOA**
waiting on battery	**WOB**
waiting on cable tools or completion tools	**WOCT**
waiting on completion rig	**WOCR**
waiting on drillpipe	**WODP**
waiting on geologist	**WOG**
waiting on orders	**WOO**
waiting on permit	**WOP**
waiting on pipe	**WOP**
waiting on pipeline	**WOPL**
waiting on plastic	**WOP**
waiting on potential test	**WOPT**
waiting on production equipment	**WOPE**
waiting on pump	**WOP**
waiting on pumping unit	**WOPU**
waiting on rig or rotary	**WOR**
waiting on rotary tools	**WORT**
waiting on standard tools	**WOST**
waiting on state potential	**WOSP**
waiting on tank and connection	**WOT&C**
waiting on test or tools	**WOT**
waiting on weather	**WOW**
wall (if used with pipe)	**W**
Wall Creek	**W Cr**
wall thickness (pipe)	**WT**
Waltersburg sand	**Wa sd**
Wapanucka	**Wap**
warehouse	**whse**
Warsaw	**War**
Wasatch	**Was**
wash (ing)	**wsh (g)**
wash and ream	**W&R**
wash and ream to bottom	**WRTBw**
wash oil	**WO**
wash over	**WO**
wash pipe	**WP**
wash to bottom	**WTB**

wash water	**WW**
washed	**w shd**
washing in	**WI**
Washita	**Wash**
Washita-Fredericksburg	**W-F**
washout	**wo**
washover string	**WOS**
waste	**WSTw**
water blanket	**WB**
water closet	**WC**
water cooler	**W/CLR**
water cushion (DST)	**wtr. cush, WC**
water cushion to surface	**WCTS**
water cut	**WC**
water depth	**WD**
water disposal well	**WD**
water in hole	**WIH**
water injection	**WI**
water injection well	**WIW**
water load	**W/L**
water loss	**WL**
water not shut off	**WNSO**
water oil or gas	**WOG**
water saturation	**WS**
water separation index modified	**WSIM**
water shutoff no good	**WSONG**
water shutoff OK	**WSOOK**
water shutoff	**WSO**
water source wells	**WST**
water supply well	**WSW**
water to surface	**WTS**
water well	**WW**
water with slight show of oil	**W/SSO**
water with sulfur odor	**W/sulf O**
water-alternating gas (or water and gas)	**WAG**
water-cut mud	**WCM**
water-cut oil	**WCO**
water-oil ratio	**WOR**
water-supply paper	**WSP**
water-up to	**WUT**
water, watery	**wtr (y)**
waterflood	**WF**
waterproof	**WTR/PRF**
watertight	**WTR/T**
wating on cement	**WOC**
watt	**w**
watt-hour	**w-hr**

weak	wk	wildcat	WC
weak air blow	WAB	wildcat field, discovery	WFD
weather	wthr	wildcat outpost, dry	WO
weather (ed)	wthd	Willberne	Willb
weatherproof	WTHR/PRF	Wind River	Wd R
Weber	Web	Windfall Profit Tax	WPT
week	wk	Winfield	Winf
weight	wgt., wt	Wingate	Wing
weight averaged catalyst temperature	WACT	Winnipeg	Winn
		Winona	Win
weight on bit	W.O.B.	wireline	WL
weight-percent	wt%	wireline coring	WLC
welded steel tank	W/S tk	wireline test	WLT
weld ends	WE	wireline total depth	WLTD
weld neck	WN	wiring	WRG
welded, welding	wld	wiring diagram	WD
welder	wldr	with	w/
welding detail (s)	WLD/DET	without	W/O
Welex	Wx	Wolfcamp	Wolfc
wellhead injection pressure	WHIP	Wolfe City	WC
well lines	WL	Woodbine	WB
well pad	WP	Woodford	Wdfd, Woodf
well pad manifolding	WPM		
wellbore	wlbr	Woodside	Woodsd
wellhead	WH	work in place	WIP
Wellington	Well	work order	WO
went back in hole	WBIH	worked	wkd
west	W	working	wkg
west half	W/2	working interest	WI
west line	WL	working pressure	WP
west offset	W/O	workover	WO, wko
westerly	W'ly	workover rig	wkor
wet bulb	WB	worldscale	WS
wheel	WHL	wrapper	wpr
whipstock	whip, WS	Wreford	Wref
whipstock depth	WSD	wrought iron	WI
white	wht		
white dolomite	Wh Dol		
White River	WR		
White Sand	Wh Sd		
wholesale	whsle		
Wichita	Wich.		
Wichita Albany	Wich Alb		
wide	W		
wide flange	WF		
Wilcox	Wx		

X-ray	X-R

yard (s)	**yd**
Yates	**Y**
Yazoo	**Yz**
year	**yr**
yellow	**yel**
yellow indicating lamp	**YIL**
yield point	**YP**
Yoakum	**Yoak**
your message of date	**YMD**
your message yesterday	**YMY**

Z

zenith	**zen**
Zilpha	**Zil**
zinc	**ZN**
zone	**Zn**

ABBREVIATIONS FOR LOGGING
TOOLS AND SERVICES

The appropriate companies and associations have not established standard abbreviations for the logging segment of the oil and gas industry. The following lists by individual companies are supplemented by a Miscellaneous Section, for your convenience.

ATLAS WIRELINE SERVICES

Acoustilog	AC
Acoustilog Caliper Gamma Ray	AC/CAL GR
Acoustilog Caliper Neutron	AC/CAL/ NEU
Acoustilog Caliper Gamma Ray-Neutron	AC/CAL GR/NEU
Acoustic Cement Bond	CBL
Acoustic Cement Bond Gamma Ray	CBL/GR
Acoustic Cement Bond Neutron	CBL/NEU
Acoustic Cement Bond G/R Neutron	CBL/GR/ NEU
Acoustic Signature	AC SIGN
BHC Acoustilog Caliper	AC/CAL
BHC Acoustilog Caliper Gamma Ray	AC/CAL/ GR
BHC Acoustilog Caliper Neutron	AC/NEU
BHC Acoustilog Caliper G/R Neutron	AC/CAL/ GR/NEU
BHC Acoustilog Caliper (Thru Casing)	AC/CAL
BHC Acoustilog Caliper Gamma Ray (Thru Casing)	AC/CAL/ GR
BHC Acoustilog Caliper G/R Neutron (Thru Casing)	AC/CAL/ GR/NEU
Caliper	CAL
Casing Potential Profile	CPP
Chlorinlog	CHL
Chlorinlog-Gamma Ray	CHL GR
Compensated Densilog Caliper	CDL/CAL
Compensated Densilog Caliper Gamma Ray	CDL/ CAL/GR
Compensated Densilog Caliper Neutron	CDL/CAL/ NEU
Compensated Densilog Caliper G/R Neutron	CDL/CAL/ GR/NEU
Compensated Densilog Caliper Minilog	CDL/CAL/ ML
Conductivity Derived Porosity	CDP
Corgun	SWC
Densilog Caliper Gamma Ray Log	CDL/CAL/ GR
Depth Determination	DD
Directional Survey	DIR
Dual Induction Focused Log	DIFL
Dual Induction Focused Log Gamma Ray	DIFL/GR
Electrolog	EL
Formation Tester	FMT
4 Arm High Resolution Diplog	DIP
Frac Log	DIP FRAC
Frac Log-Gamma Ray	DIP FRAC/ GR
Gamma Ray Cased Hole	GR
Gamma Ray/Dual Caliper	GR/CALD
Gamma Ray-Open Hole	GR
Gamma Ray Neutron-Cased Hole	GR/NEU
Gamma Ray Neutron-Open Hole	GR/NEU
Induction Electrolog	IEL
Induction Electrolog Gamma Ray	IEL/GR
Induction Electrolog Neutron	IEL/NEU
Induction Electrolog Gamma Ray Neutron	IEL/GR/ NEU
Induction Log	IEL
Induction Log-Gamma Ray	IEL/GR
Induction Log-Neutron	IEL/NEU
Laterolog	LL
Laterolog-Gamma Ray	LL/GR
Laterolog-Neutron	LL/NEU
Laterolog-Gamma Ray-Neutron	LL/GR/ NEU
Microlaterolog-Caliper	MLL/CAL
Minilog Caliper	ML/CAL
Minilog Caliper Gamma Ray	ML/ CAL/GR
Movable Oil Plot	MOP
Neutron (Cased Hole)	NEU
Neutron (Open Hole)	NEU
Neutron Lifetime	NLL
Neutron Lifetime Gamma Ray	NLL/GR
Neutron Lifetime Neutron	NLL/NEU
Neutron Lifetime G/R-Neutron	NLL/GR/ NEU
Nuclear Flolog	NFL
Nuclear Flolog-Gamma Ray	NFL/GR

Nuclear Flolog-Gamma Ray Neutron	NFL/GR/ NEU
Nuclear Flolog-Neutron	NFL/NEU
Perforating Control	PFC
PFC Gamma Ray	PFC GR
PFC Neutron	PRC NEU
Photon	PHT
Proximity Minilog	PROX/ML
Sidewall Neutron	SWN
Sidewall Neutron-Gamma Ray	SWN/GR
Temperature Differential	TEMP
Temperature-Gamma Ray-Neutron	TEMP/ GR/NEU
Temperature Log	TL TEMP
Temperature Log-Gamma Ray	TEMP/GR
Temperature-Neutron	TEMP/NEU
Total Time Integrator	TT
Tracer Log	TRL
Tracer Log-Neutron	TRL/NEU

SCHLUMBERGER WELL SERVICES

Amplitude Logging	A-BHC
Bore Hole Compensated	BHC
BHC Sonic Logging	BHC
BHC Sonic-Gamma Ray Logging	BHC-GR
BHC Sonic-Variable Density	BHC-VD
Bridge Plug Service	BP
Borehole Televiewer	TVT
Caliper Logging	CAL
Casing Cutter Service	SCE-CC
Cement Bond Logging	CBL
Cement Bond-Gamma Ray Logging	CBL-GR
Cement Bond-Gamma Ray Neutron	CBL-GRN
Cement Bond-Neutron	CBL-N
Cement Bond-Variable Density Logging	CBL-VD
Cement Dump Bailer Service	DB
Computer Processed Interpretation	MCT
Continuous Directional Survey	CDR
Continuous Flowmeter	CFM, PFM
Customer Instrument Service	ICS
Data Transmission	TRD
Density Log	DENL
Depth Determinations	DD
Diamond Core Slicer	SS
Dipmeter	DIPM
Dipmeter-Digital	HDT-D
Directional Survey	DS

Dual Induction-Laterologging	DIL
Electric Logging	ES
Formation Density Logging	FDC
Formation Density-Gamma Ray Logging	FDC-GR
Formation Testing	FT
Gamma Ray Logging	GR
Gamma Ray-Neutron Logging	GRN
Gamma Ray-Sonic Logging	GRS
Gradiomanometer	GM
High-Resolution Thermometer	HRT
Induction-Electron Logging	I-ES
Induction-Gamma Ray Logging	I-GR
Junk Catcher	JB
Log Overlays	OL
Magnetic Taping	TPG
Microlog	ML
Neutron Logging	NL
Orienting Perforating Service	OPR
Perforating-Ceramic DPC	SCE
Perforating-Depth Control	PDC
Perforating-Expendable Shaped Charge	SCE
Perforating-Hyper Jet	SCH
Perforating-Hyper Scallop	SPH
Pressure Control	PC
Production Combination Tool Logging	PCT
Production Packer Service	PPS
Proximity-Microlog	ML
Radioactive Tracer Logging	RTP
Rwa Logging	FAL
Salt Dome Profiling	ES-ULS
Schlumberger	Schl.
Seismic Reference Service	SRS
Sidewall Coring	CST
SNP Neutron Logging	SNP
SNP Neutron-Gamma Ray Logging	SNP-GR
Synergetic Log Systems	MCT
Temperature Logging	T
Temperature-Gamma Ray Logging	T-GR
Thermal Decay Logging	TDT
Thru-Tubing Caliper	C-C
Tubing, Cutter Service	SEC-CC
Variable Density Logging	BHC-VD
Variable Density-Gamma Ray Logging	VD-GR

HALLIBURTON LOG SERVICE (HAL)

Analog Computer Service	**An Cpt. Ser**
Caliper	**Cal**
Compensated Acoustic Velocity Log	**Com AVL**
Compensated Acoustic Velocity Log-Gamma Ray	**Com AVL-G**
Compensated Acoustic Velocity Log-Neutron	**Com AVL-N**
Compensated Density	**Com Den**
Compensated Density Gamma Ray	**Com Den-GR**
Computer Analyzed Log	**CAL**
Contact Caliper	**Cont**
Continuous Drift	**Con Dr.**
Density	**Den**
Density Gamma Ray	**Den-G**
Depth Determination	**DeDet**
Digital Tape Recording	**Dgt Tp Rec**
Dip Log Digital Recording	**Dgt Dip Rec**
Drift	**Dr**
Drill Pipe Electric Log	**DPL**
Electric Log	**EL**
Electro-Magnetic Corrosion Detector	**Cor Det**
Fluid Travel Log	**FTrL**
Formation Tester	**FT**
FoRxo Caliper	**FoRxo**
Frac-Finder Micro-Seismogram	**FF-MSG**
Frac-Finder Micro-Seismogram Gamma	**FF-MSG-G**
Frac-Finder Micro-Seismogram Neutron	**FF-MSG-N**
Gamma Guard	**G-Grd**
Gamma Ray	**GR**
Gamma Ray Depth Control	**GRDC**
Guard	**Grd**
High Temperature Equipment	**HTEq**
Induction Electric Gamma Ray	**IEL-G**
Induction Electric Neutron	**IEL-N**
Induction Electric	**IEL**
Induction Gamma Ray	**Ind-G**
Induction Gamma	**Ind G**
Micro-Seismogram Log, Cased Hole	**MSG-CBL**
Micro-Seismogram Gamma Collar Log, Cased	**MSG-CBL-G**
Micro-Seismogram Neutron Collar Log, Cased	**MSG-CBL-N**
Neutron Log	**NL**
Neutron Depth Control	**NDC**
Precision Temperature	**Pr Temp**
Radiation Guard	**R/A Grd**
Radioactive Tracer	**R/A Tra**
Resistivity Dip	**Dip**
Sidewall Coring	**SWC**
Sidewall Neutron	**SWN**
Sidewall Neutron-Gamma Ray	**SWN-G**
Simultaneous Gamma Ray-Neutron	**GRN**
Special Instrument Service	**Sp Inst Ser**
True Vertical Depth	**TVD**

LOG HEADING NOMENCLATURE

Log Name	Abbrev.	Tool Type
Acoustic Caliper Log	**ACE**	
Aluminum Clay Log		**ACT**
Adaptive Electromagnetic Propagation Log	**ADEPT**	**EPT-G**
Array Sonic Log		**SDT**
Automatic Diverter Flowmeter	**ADF**	**ADF**
Auxilary Measurements Log	**AMS**	**AMS**
Borehole Compensated Sonic Log	**BHC**	**SLT**
Borehole Geometry Log	**BGL**	**BGT**
Bridge Plus	**BP**	**RST**
Casing Collar Log	**CCL**	**CAL, CLL**
Cement Bond-Variable Density Log	**CBL-VDL**	**SLT, SDT, CBT**
Cement Evaluation Log	**CET**	**CET**
Compensated Neutron Log	**CNL**	**CNT-A/H**
Corrosion And Protection Evaluation Log	**CPEL, CPET**	**CPET**
Cyberdip	**CYDIP**	
Cyberlock	**CYL**	
Cyberscan	**CYTDT**	
Deep Propagation Log	**DPL**	**DPT**
Depth Determination Log	**DD**	

Downhole Seismic Array	**DSA**	**DSA-A**
Dual Dipmeter	**SHDT**	**SHDT**
Dual-Burst Thermal Neutron Log	**TDT-P**	**TDT-P**
Dual-Energy Neutron Log	**DNL**	**CNT-G**
Dual Induction SFL Log	**DIL**	**DIT**
Dual Laterolog	**DLL**	**DLT**
Dump Bailer	**DB**	**DB**
Electromagnetic Propagation Time Log	**EPT**	**EPT**
Electromagnetic Thickness Log	**ETT**	**ETT-A**
Enerjet Perforating	**EEJ**	**EEJ**
Formation Density	**FDC**	**PGT**
Formation Micro-scanner	**FMS**	**MEST**
Formation Micro-imager	**FMI**	**FMI**
Four-Arm Caliper Log	**CAL**	**BGT, HDT, SHDT**
Fracture Identi-fication Log	**FIL**	**HDT**
Gamma Ray Log	**GR**	**SGT**
Gamma Ray Spectroscopy Log	**GST**	**GST**
Geochemical Log	**GLT**	**GST, NGT, CNT, ACT**
Guidance Contin-uous Log	**GCT**	**GCT**
Hyperdome Perforating	**HD**	**HD**
Hyperjet Perforating	**HJ, HJII, HJIII**	**HJ, HJII, HJIII**
H₂S Cable	**HS CBL**	
Induction-Spherically Focused Log	**ISF**	**IRT, DIT**
Inclinometer Survey, Directional	**IS**	**GPIT, SHDT, FMS, PMI, HDT**
Junk Basket	**JB**	**JB**
Litho-Density Log	**LDL**	**LDT**
Litho-Density Quicklook	**CYLDT**	
Long-Spacing Sonic Log	**LSS**	**SLT, SDT**
Mechanical Sidewalk Coring or Rotary Cores	**MSCT**	**MSCT**
Microlog	**ML**	**MPT, MLT, PCD**
Microlaterolog	**MLL**	**MPT**
Microspherically Focused Log	**MSFL**	**SRT**
Multifinger Caliper Log	**MFC**	**MFCT**
Multifrequency Electromagnetic Thickness Log	**METL**	**ETT-D**
Multiple Isotope Tracer	**MTT**	
Natural Gamma Ray Spectroscopy Log	**NGS**	**NGS**
Oil Base Mud Dipmeter	**OBDT**	**OBDT**
Perforation Depth Control Log	**PDC**	**CAL, CCL**
Phasor Induction Log	**PIL**	**DIT-E**
Pipe Analysis Log	**PAL**	**PAT**
Pivotgun	**PG**	**PG**
Posiset Thru-Tubing Mechanical Plug-Back	**MPBT**	
Pressure Control	**PC**	**PC**
Production Logging Quicklook	**CYPL**	
Repeat Formation Tester Log	**RFT**	**RFT**
Repeat Formation Tester Quicklook	**CYRFT, RFQL**	
Reservoir Saturation Tool	**RST**	
Seismic Quicklook	**CYWST**	
Sidewall Coring	**CST**	**CST**
Simultaneous Production Log	**PLT**	**PLT**
Stratigraphic High-Resolution Dipmeter Log	**SHDT**	**SHDT**
Temperature Log	**HRT**	**HTT**
Thermal Neutron Decay Time Log	**TDT**	**TDT**
Triaxial Seismic Survey	**SAS**	**SAT**
Ultrajet Perforating	**UJ**	**UJ**
Ultrapack Perforating	**UP**	**UP**
Well Head Equipment (Pressure Control)	**WHE-A,B,C, D,H,BM**	
Well Seismic Survey	**WST**	**WST**

MISCELLANEOUS

Acoustic Amplitude	**AAL**	Casing Inspection/Electro-Magnetic-Detector	**CI**
Acoustic Cement Bond G/R Neutron	**CBL GRN**	Casing Potential Profile	**CPP**
Acoustic Cement Bond Neutron	**CBL N**	Cement Bond	**CBND**
Acoustic Cement Bond	**CBL**	Cement Evaluation	**CEL**
Acoustic Fracture Identification	**AFI**	Cement Top Location	**CTL**
Acoustic or Acoustilog	**SL**	Cemotop	**CTL**
Acoustic Parameter	**ACP**	Channel Survey	**CHNL**
Acoustic Parameter-Depth	**AC PAR D**	Channelmaster	**CML**
Acoustic Parameter-Logging	**AC PAR L**	Channelmaster-Neutron	**CML N**
Acoustic Parameter-16mm Scope	**AC PAR 16**	Chloride	**CL**
Acoustic Scope Picture	**ASL**	Chloride Detection	**CLDL**
Acoustic Signature	**AC SIGN**	Chlorinlog-Gamma Ray	**CHL GR**
Acoustic Velocity	**AVL**	Collar/Collar Correlation	**PDS**
Acoustilog	**ALC**	Compensated Acoustic Velocity	**CAVL**
Acoustilog Caliper-Gamma Ray	**ALC-GR**	Compensated Densilog Caliper	**CDCL**
Acoustilog Caliper-Gamma Ray-Neutron	**ALC-GRN**	Compensated Densilog Caliper-G/R Neutron	**CDLC GRN**
Acoustilog Caliper-Neutron	**ALC-N**	Compensated Densilog Caliper-Gamma Ray	**CDCL GR**
After Pay Out	**APO**	Compensated Densilog Caliper-Minilog	**CDLC M**
Amplitude	**AMP**		
Amplitude Sonic	**ASL**	Compensated Densilog Caliper-Neutron	**CDLC N**
Area of Mutual Interest	**AMI**		
Atlantic Chlorinlog	**A CHL**	Compensated Density Caliper	**CDC**
Audio Logging	**AUD**	Compensated Density Log	**CDL**
BHC Acoustilog Caliper	**BHC ALC**	Compensated Formation Density	**CFD**
BHC Acoustilog Caliper (Thru Casing)	**BHC AL TC**	Compensated Formation Density Caliper	**CFDC**
BHC Acoustilog Caliper G/R Neutron	**BHC ALC GRN**	Compensated Gamma	**CG**
BHC Acoustilog Caliper G/R Neutron (Thru Casing)	**BHC AL GRN TC**	Compensated Neutron Density	**CNDL**
		Compensated Neutron Log	**CNL**
BHC Acoustilog Caliper Gamma Ray	**BHC ALC GR**	Compensated Neutron Log Porosity	**CNLP**
BHC Acoustilog Caliper Neutron	**BHC ALC N**	Dual Induction Gamma Log	**DIGL**
Barrel	**BBL**	Dual Induction Lateral/Dual Induction Focus	**DILL**
Before Pay Out	**BPO**		
Borehole Audio Tracer Survey	**BATS**	Dual Induction Log	**DIL**
Borehole Compensated	**BHC**	Dual Induction SFL	**DIL**
Borehole Compensated Sonic	**BHCS**	Dual Induction Spherically Focused	**DISF**
Borehole Geometry Log	**BGT**		
Borehole Televiewer	**BTL**	Dual Laterolog/Microspherically Focused	**DLL/MSFL**
Bottom Hole Contribution	**BHC**		
Bulk Density	**BLKD**	Dual Laterolog	**DLL**
Caliper	**CALP, CL**	Dual Porosity Compensated Neutron	**DNL**
Caliper Analysis	**CALA**		
Caliper Curve	**CALC**	Dual Resistivity Induction Log	**DRI**
Carbon-Oxygen	**C O**	Dual Sand	**DUSD**
Casing Collar	**CCL**		

Dual Spacing Log	**DSL**	Gamma Spectrometry	**GST**
Electrical	**ES**	Gas Detection	**GASD**
Electrolog	**EL**	Gas Purchase Agreement	**GPA**
Electromagnetic Propagation	**EPT**	Geophone	**GEO**
Epithermal Neutron	**ETN**	Gradiomanometer	**GRMR**
Experimental	**E, XPTL**	Gravity	**GRAV**
Farm In	**FI**	Guard	**GRDL**
Farm Out	**FO**	Gyro Survey	**GYRO**
Flowmeter	**FLO**	Held by Production	**HBP**
Fluid Travel	**FLO**	High Resolution Dipmeter	**HRD**
Fluid Travel Log/Fluid Entry Survey	**FLTR**	Hydrocarbon or Gas Detection	**GT**
		Hydrocarbon or Gas Detection	**HCDS**
Focus	**FOCL**	Inclination	**INCL**
Focused Diplog	**F DIP**	Induction	**IL**
Formation Analysis	**FAL**	Induction-Electric Log	**I-EL**
Formation Density	**FD**	Induction-Lateral	**ILL**
Formation Density Caliper	**FDC**	Induction-Laterolog	**I-LL**
Formation Factor	**FMF**	Induction Electrolog	**IEL**
Formation Tester	**FT**	Induction Electrolog-Gamma Ray	**IEL-GR**
Forxo	**MLL**	Induction Electrolog-Gamma Ray Neutron	**IEL-GRN**
4 Arm High Resolution Diplog	**R H DIP**		
Frac Log	**FRAC L**	Induction Electrolog-Neutron	**IEL-N**
Frac Log-Gamma Ray Log	**FRAC-GR**	Induction Spherically Focused	**ISF**
Fracture Finder	**ASL**	Induction-Electric or Induction-Electro	**IES**
Fracture Finder/Failure ID	**FF**		
Fracture Identification Log	**FIL**	Isotron	**I, ISOL**
Full Bore Flowmeter	**FB FM**	Joint Operating Agreement	**JOA**
Gamma Compensated Density	**GCD**	Joint Venture Agreement	**JVA**
Gamma Gamma	**GG**	Joint Venture Drilling	**JVD**
Gamma Gamma Density	**GGD**	Land Owner Royalty	**LOR**
Gamma Gamma Log	**GGL**	Laser Log	**LASR**
Gamma Guard	**GCRD**	Lateral	**LATL**
Gamma Guard EL	**GGRD**	Laterolog-Gamma Ray-Neutron Log	**LL GR-N**
Gamma Ray	**GR**		
Gamma Ray - Multi-Spaced Neutron Log	**CR-MSN**	Lifetime Log	**LL**
		Limestone Log	**LSL**
Gamma Ray - Neutron	**GRNL**	Limestone Device	**LI**
Gamma Ray - Neutron Log	**GR-N**	Liquid Isotope Injector	**LII**
Gamma Ray - Sonic	**GRSL**	Liquified Natural Gas	**LNG**
Gamma Ray - Tracer Survey	**GRTS**	Liquified Petroleum Gas	**LPG**
Gamma Ray Cased Hole	**GR CH**	Litho-Density	**LDT**
Gamma Ray Depth Control	**GRDC**	Lithology	**LITH**
Gamma Ray Neutron	**GRN**	Logger's Total Depth	**LTD**
Gamma Ray Neutron-Cased Hole	**GRN CH**	Long-Spaced Sonic	**LSS**
Gamma Ray Neutron-Open Hole	**GR/N OH**	Lost Circulation	**LS**
Gamma Ray-Open Hole	**G/R OH**	Manometer	**MAN**
Gamma Ray/Dual Caliper	**GR/D CALIPER**	Measurement for Natural Gas	**MCF**
		Microlog	**ML, MICL**

Microlateral	**MLAT**	Right of Way	**ROW**
Microseismogram	**MSMG**	Salinity	**CL**
Microsonic Gamma Ray	**MSG**	Saraband	**"SBND**
Microspherically Focused Log	**MSFL**	Scattered Gamma Ray	**CTL**
Microsurvey	**MS, MICS**	Scope Picture Analysis	**SPA**
Mini	**ML, MINL**	Section Gauge	**CAL**
Minifocus	**MLL, MINF**	Seismic Reference Survey/ Neutron	**SRSN**
Mobile Picture	**MP**	Seismic Velocity Survey	**SVS**
Mono Electric	**MONO**	Shear Amplitude	**SA**
Mud Log/Focus Log	**MUD**	Sidewall Frac Log	**SFL**
Multi-Shot Survey	**MSS**	Sidewall Neutron	**SN**
Natural Gamma Ray Spectroscopy	**NGS**	Sidewall Neutron Porosity	**SNP**
		Sidewall Sampler	**CST**
Net Revenue Interest	**NRI**	Sonic Caliper Log	**SCL**
Neutron	**N, NEUT**	Sonic Log	**SL, SONL**
Neutron Collar Log	**NCL**	Sonic Seismogram	**SSMG**
Neutron Formation Density	**NFD**	Spectral	**SPCT**
Neutron Lifetime	**NLL**	Spherical	**SPH**
Notice of Acquisition	**NOA**	Spontaneous Potential	**SP**
Nuclear	**NUCR**	Strata	**STRTS- tructural Exploration**
Nuclear Flow	**FLO**		
Nuclear Magnetism	**NML, NMAGL**	Structural Exploration	**SE**
Overriding Royalty	**ORR**	Synergetic	**SYGT**
Overriding Royalty Interest	**ORI**	Televiewer	**TV**
Perforated Log	**PERF**	Temperature	**HRT**
Perforating Depth Control	**PDC**	Temperature Difference Log	**TDL**
Perforating Formation Collar	**PFC**	Temperature Survey Log	**TMPL**
Permalog	**PL**	Thermal Decay Time	**TDT**
Permeability Spinner Survey	**PSS, PRMS**	Three Dimensional	**3D**
		Time Log	**TIME**
Photo	**PHOT**	Tracer Survey	**TRCR**
Photoclinometer	**PHCL**	Ultralong Spacing Electric Log	**ULSEL**
Pipe Analysis Log	**PAL**	Uranium	**URAN**
Pipe Recovery Log	**PRL**	Variable Density	**VD**
Porosity	**POR**	Velocity	**VEL**
Proximity	**PROX**	Velocity Seismic Profile	**VSP**
Proximity Log	**PROXL**	Velocity Survey	**VRS**
Proximity-Microlog	**PML**	Velocity Survey Profile	**VSP**
Quitclaim	**QC**	Viscosity	**VISC**
Radioactive Tracer	**RAT, RTRS**	Water Location Survey	**WLS**
		Wave Form Digitizing	**BHC-WFD**
Refracture	**"REFR**	Wave Form Logging	**BHC-WFL**
Repeat Formation Tester	**RFT**	Well Seismic	**WST**
Resistivity	**RES**	Working Interest	**WI**
Resistivity Water Apparent	**RWA**	X-Y Caliper	**XYCL**

FEDERAL ENVIRONMENTAL ACRONYMS
Compliments of BNA PLUS

AHERA	Asbestos Hazard Emergency Response Act of 1986 (Title II of TSCA)
ANSI	American National Standards Institute
ATSDR	Agency for Toxic Substances and Disease Registry
AWT	Advanced Wastewater Treatment
BACT	Best Available Control Technology
BAT	Best Available Technology
BCT	Best Conventional Pollutant Control Technology
BOD	Biologic Oxygen Demand
BPT	Best Practicable Control Technology
BTU	British Thermal Unit
CAA	Clean Air Act
CAIR	Comprehensive Assessment Information Rule (under TSCA)
CAS	Chemical Abstract Service
CEPP	Chemical Emergency Preparedness Program
CERCLA	Comprehensive Environmental Response, Compensation, and Liability Act (The Superfund Law)
CFR	Code of Federal Regulations
CPSC	Consumer Product Safety Commission
CWA	Clean Water Act
CZM	Coastal Zone Management
DOT-E	Designation of Materials Exempt from Department of Transportation regulations
EP	Extraction Procedure
EPA	Environmental Protection Agency
EPCRA	Emergency Planning and Community Right-To-Know Act (Title III of SARA, commonly called Right-To-Know or SARA Title III)

ERC	Emissions Reduction Credit
FIFRA	Federal Insecticide, Fungicide, and Rodenticide Act
HCS	OSHA Hazard Communication Standard (Worker Right-To-Know)
HMTA	Hazardous Materials Transportation Act
HSWA	Hazardous and Solid Waste Amendments (1984 RCRA Amendments)
IARC	International Agency for Research on Cancer
ITC	Interagency Testing Committee
LAER	Lowest Achievable Emission Rate
LUST	Leaking Underground Storage Tanks
MSDS	Material Safety Data Sheet
MTB	Materials Transportation Bureau of the Department of Transportation
NESHAP	National Emissions Standard for Hazardous Air Pollutants
NIOSH	National Institute for Occupational Safety and Health
NOAA	National Oceanic and Atmospheric Administration
NPDES	National Pollutant Discharge Elimination System
NPRM	Notice of Proposed Rulemaking
NRC	Nuclear Regulatory Commission or National Response Center
NSPS	New Source Performance Standards
NTP	National Toxicology Program
ORM	Other Regulated Materials
OSHA	Occupational Safety and Health Administration
OSHRC	Occupational Safety and Health Review Commission
OTA	Office of Technology Assessment
PEL	Permissible Exposure Limit

pH	Potential of Hydrogen: a measure of acidity and alkalinity
PM	10-micron Particulate Matter
PMN	Premanufacture Notification (under TSCA)
POTWs	Publicly Owned Treatment Works
PRPs	Potentially Responsible Parties
PSD	Prevention of Significant Deterioration
psia	Pounds per Square Inch Absolute
RACT	Reasonably Available Control Technology
RCRA	Resource Conservation and Recovery Act
RQ	Reportable Quantity
RSPA	Research and Special Programs Administration of the Department of Transportation
RTECS	Registry of Toxic Effects of Chemical Substances
SARA	Superfund Amendments and Reauthorization Act of 1987
SDWA	Safe Drinking Water Act
SIP	State Implementation Plan
SNUR	Significant New Use Rule
SOCMI	Synthetic Organic Chemical Industry
SPCC Plan	Spill Prevention Control and Countermeasure Plan
SQG	Small Quantity Generator (of hazardous waste)
SRF	State-administered water pollution control Revolving Funds
SWDA	Solid Waste Disposal Act
TSCA	Toxic Substances Control Act
TSS	Total Suspended Solids (non-filterable)
UIC	Undergound Injection Control
UN	United Nations
UPAC	International Union of Pure and Applied Chemistry

USDW	Underground Source of Drinking Water
UST	Underground Storage Tank
VHAP	Volatile Hazardous Air Pollutant
VOC	Volatile Organic Compound
Z list	OSHA list of hazardous chemicals (29 CFR 1910 Subpart Z, Worker Right-To-Know)

Some Frequently Cited Chemical Abbreviations

CFCs	Chlorofluorocarbons
CO	Carbon Monoxide
NOx	A mixture of nitrous oxide and nitrous dioxide
PCBs	Polychlorinated biphenyls
TCDD	Tetrachlorobenzo-p-dioxin

PIPE COATING TERMINOLOGY AND DEFINITIONS

anode	Corrosion prevention device
C.P.	Cathodic protection or Corrision Protection
dope	Pipe coating
dresser	Mechanical coupling used to join joints or lengths or pipe rather than threading or welding
FB	Flat-bottom mill or shoes
granny rag	Type of coating or method of coating a pipeline in the field rather than factory applied coating
holiday	Hole in the protective coating of a steel pipeline in the field
hot spot	Corrosive area located along the length of a pipeline; usually a wet bog, marsh, or bentonite area
I.P.	Intermediate pressure pipeline
IWRC	Independent wire rope center
jeep	Same as *holiday*

jeeper	Electronic device or instrument used to detect holes (holidays) in the steel pipeline protective coating
overbend	High spot in a pipeline usually installed by field-bending a pipeline joint
P/C or P/W	"Painted and coated" or "painted and wrapped" pipelines—steel pipe with protective external coating of one of several different types
pig	Pipeline cleaning and measuring tool
pig catcher	Used to remove pipeline pig
pig launcher	Used to insert pipeline pig
stub	Length of small-diameter distribution pipeline from the main line to the customer's property or meter location
thin film	Type of epoxy coating for pipeline coating rather than threading or welding
tube turn	Prefabricated piece of pipeline (allows change of direction of a pipeline without field bending the pipeline)
sag	Low spot in a pipeline usually installed by field bending a pipeline joint
WB	Wavy bottom mill or shoes
wrap	Same as dope; protective coating on a steel pipeline
XTC	Extra-coat protective pipeline coating made of polyethylene or polypropylene material

MNEMONICS-SERVICE NAMES & MENEMONICS-COMPUTATIONAL PRODUCTS

Service Menemonic	Service
2CAL	2-Arm Caliper Log
2CEX	2-Arm Caliper Log (Extended Reach)
3CAL	3-Arm Caliper Log
4CAL	4-Arm Caliper Log
4CEX	4-Arm Caliper Log (Extended Reach)
AC	BHC Acoustilog
ACL	Long-Spaced BHC Acoustilog
BAL	Bond Attenuation Log
BHJ	Bottom-Hole Junk Shot
BO	String Shot-Back Off
BO1	Tubing
BO2	Drill Pipe, Wash Pipe, Casing, Liners
BO3	Drill Collars
BP	Bridge Plug
CAC	Circumferential Acoustilog
CBIL	Circumferential Borehole Imaging Log
CBL	Acoustic Cement Bond Log
CC	Chemical Cutters
CC1	1–2-1/16-in. (25-52 mm) Pipe
CC2	2-3/8–4-in. (60-102 mm) Pipe
CC3	4-1/4–5-9/16 in. (108-141 mm) Pipe
CCL	Casing Collar Locator
CDL	Compensated Densilog
CHFT	Cased Hole Formation Tester
CIS	Customer Instrument Service
CN	Compensated Neutron Log
CPP	Casing Potential Profile
CRET	Cement Retainer
CST	Customer Steering Tool
DAC	Digital Array Acousticlog^{SM}
DB	Dump Bailer
DD	Depth Determination
DEL2	Dielectric Log-200 MHz
DEL4	Dielectric Log-47 MHz
DGR	Digital Gamma Ray Log
DIEL	Dielectric Log
DIFL	Dual Induction-Focused Log
DIP	High Resolution (4-Arm) Diplog®
DLL	Dual Laterolog
DMAG	Digital Magnelog
DMGL	Digital Magneline
DPIL	Dual Phase Induction Log
DRB	Dual Receiver Bond Log
DSL	Digital Spectralog
DVL1	Digital Vertiline 4-1/2 in. (114 mm)
DVL2	Digital Vertiline 5-1/2 in. (140 mm)

DVL3	Digital Vertiline 7 in. (178 mm)
DVL4	Digital· Vertiline 8-5/8 in. (219 mm)
DVRT	Digital Vertilog
DVT1	Digital Vertilog 4-1/2 in. (114 mm)
DVT2	Digital Vertilog 5-1/2 in. (140 mm)
DVT3	Digital Vertilog 7 in. (178 mm)
DVT4	Digital Vertilog 8-5/8 in. (219 mm)
DVTL	Digital Vertiline
DWP	Downhole Wireline Packoff
EMO	Electromagnetic Orientation Tool
EMT	Electromagnetic Fishing Tool
FCON	Fluid Conductivity Log
FDBM	Buoyancy Measurement Fluid Density Log
FDDP	Differential Pressure Fluid Density Log
FDN	Nuclear Fluid Density Log
FG	Feeler Gauge
FMBK	Basket Flowmeter
FMCS	Continuous Spinner Flowmeter
FMFI	Folding Impeller Flowmeter
FMT	Formation Multi-Tester
FPM	Magnetic Freepoint Indicator
FPMT	Magna-Tector® Freepoint Indicator
FPST	Spring-Tector™ Freepoint Indicator
FPTM	Tri-Mag Freepoint Indicator
FQPG	Fast Response Quartz Pressure Gauge
GR	Gamma Ray
GUN	Guns
HCS	Hydraulic Cleanout Service
HDIP	Hexdip
HYDL	Hydrolog℠
IEL	Induction Electrolog
JCGR	Junk Catcher Gauge Log
JCS	Jet Cutters
JCS1	Jet Cutters 1–2-7/8-in. (25–73 mm) Pipe
JCS2	Jet Cutters 3–5-1/2-in. (76–140 mm) Pipe
JCS3	Jet Cutters 6 - 7-5/8-in. (152-194 mm) Pipe
LL3	Laterolog
MAG	Magnelog
MFC	Multi-Finger Caliper
MGLN	Magneline
ML	Minilog®
MLL	Micro Laterolog
MSI	MSI Carbon/Oxygen Log
MST	Metal Severing Tool
MVL1	Multi-Channel Vertiline 4-1/2 in. (114 mm)
MVL2	Multi-Channel Vertiline 5-1/2 in. (140 mm)
MVL3	Multi-Channel Vertiline 7 in. (178 mm)
MVL4	Multi-Channel Vertiline 8-5/8 in. (219 mm)
MVRT	Multi-Channel Vertilog
MVT1	Multi-Channel Vertilog 4-1/2 in. (114 mm)
MVT2	Multi-Channel Vertilog 5-1/2 in. (140 mm)
MVT3	Multi-Channel Vertilog 7 in. (178 mm)
MVT4	Multi-Channel Vertilog 8-5/8 in. (219 mm)
MVTL	Multi-Channel Vertiline
NEU	Neutron Log
NFL	Nuclear Flolog
ORIT	Instrument Orientation Log
PDB	Positive Displacement Dump Bailer
PDK	PDK-100® Log
PFC	PFC Gamma Ray Log
PFN	PFC Neutron Log
PHT	Photon Log
POL	Perforating Orientation Log
PPKR	Production Packer
PPL	Polymer Pathfinder Log
PRL	Pipe Recovery Log
PROX	Proximity Log
PRSM	PRISM® Log
RCOR	Rotary Coring Tool
QPG	Quartz Pressure Gauge
SB	Sinker Bar
SBT	Segmented Bond Tool
SL	Spectralog®
SON	Sonan Log

SPG	Strain Pressure Gauge
SRB	Single Receiver Bond Log
SRPL	Surface Recorded Pressure Log
SSH	Surface Shot
ST	FMT Break Off (PVT) Sample Tanks
STHS	FMT Break Off (PVT) Sample Tanks (H_2S)
SWAT	Swing-Arm Tracer Log
SWC	Sidewall Corgun
SWN	Sidewall Epithermal Neutron Log
TBFS	Through-Tubing Borehole Fluid Sampler
TBRT	Thin-Bed Resistivity
TCAL	Through-Tubing (X-Y) Caliper Log
ΔTEMP	Differential Temperature Log
TFLR	TTRM Sub Fluid Resistivity
TRL	Tracerlog
TTBP	Through-Tubing Bridge Plug
TTEM	TTRM Sub Temperature
TTEN	TTRM Sub Tension
TTRM	TTRM Sub
UDIP	Ultrasonic Diplog
VIBR	Vibrator
VL1	Vertiline 3-1/2 in. (89 mm, 4-1/2 in. (114 mm)
VL2	Vertiline 5-1/2 in. (140 mm)
VL3	Vertiline 7 in. (178 mm)
VL4	Vertiline 8-5/8 in. (219 mm)
VPC	Formation Multi-Tester
VRT	Vertilog®
VRT1	Vertilog 3-1/2 in. (89 mm), 4-1/2 in. (114 mm)
VRT2	Vertilog 5-1/2 in. (140 mm)
VRT3	Vertilog 7 in. (178 mm)
VRT4	Vertilog 8-5/8 in. (219 mm)
VTLN	Vertiline
WHI	Water Holdup Indicator
ZDL	Compensated Z-Densilog℠

MNEMONICS COMPUTATIONAL PRODUCTS

Product Mnemonic	**Computational Service**
	Open Hole Analysis
	HORIZON
HCE	Category Estimation
HME	Magnitude Estimation
HLS	Layer/Square
HDI	Category Data Import
HCI	Core Data Import
OPTM	OPTIMA®
LIN	LINEAR
TBA	Thin-Bed Analysis
CLAS	Clay Analysis and Shaly Sand Evaulation (CLASS®)
CLAY	Clay Description (CLAYS)
CRA	Complex Reservoir Analysis
DIE	Dielectric Analysis
RMA	Radioactive Mineral Analysis
FI	Fracture Index
SA	Sandstone Analysis
CS	Coal Seam Analysis
STRA	STRATA LOGIK®
	Acoustic Waveform Processing
DDM	DDBHC Multi-Shot Processing
DDS	DDBHC Slowness Waveform Analysis
MSP	Multi-Shot Processing
SWA	Slowness Waveform Analysis
EWA	Energy Waveform Analysis
WPT	Waveform processing DDBHCΔT
	Rock Properties
MP	Mechanical Properties
SS	Sand Strength
FP	Fracture Migration
DARC	Permeability Estimation
	Image Processing
ACCR	Accelerometer Corrections
	Preprocessing (CIBL℠ PREP)
RSA	Resample
NSE	Noise Reduction
ORI	Orientation
	Borehole Corrections
MUD	Mud Attenuation
ECC	Tool Eccentricity
	Image Enhancement
EDG	Edge Detection
SIG	Sigma Smoothing
THLD	Threshholding

HGM	Histogram Equalization

Interactive Image Processing

SDC	Synthetic Dip Curves (1–4)
SDCA	Additional Dip Curves
OC	Other Curves
SNP	Snap Shot
SNPA	Additional Copies
XSEC	Cross Section
BOUT	Breakout
VIEW	View
CLRP	Color Plot Continuous
RAY	Optimized Grey Scales
RAYP	Additional Playback
IREP	Integrated Report

Diplog Analysis
STRATAGON℠

SOP	Optical
SAT	Automatic
SDIP	STRATA DIP®
DIPC	Basic Diplog Computation
AZD	Azimuth Dipfrac
DS	Directional Survey
TBT	True Bed Thickness
TVT	True Vertical Thickness
TVD	True Vertical Depth

Cased Hole Analysis
PDK-100® Log Analysis (SEARCH)

STL	Time Lapse
LIL	Log-Inject-Log
SOH	With Openhole Data Included
SCH	Basic

MSI C/O Log Analysis (CHES II®)

CTL	Time Lapse
COH	With Openhole Data Included
CHES	Basic
PFLO	PROFLOW
PRM	PRISM®
GAS	Gas Storage

Pipe Evaulation

VTRC	Vertilog/Vertline
CPPC	CPP

Pressure Transient Analysis

LAYR	Multi-Layer Testing
PAN	Pan System Multi-Rate Test
PTA	Pressure Transient Analysis (REALITY)
PDR	Pressure Data Report
NODE	Well Performance Simulation (NODES)
FMTC	FMT Analysis

COMPANIES AND ASSOCIATIONS

U.S. and Canada

AADE	American Association of Drilling Engineers
AAODC	See IADC
AAPG	American Assocation of Petroleum Geologists
AAPL	American Association of Professional Landmen
AAR	Associaton of American Railroads
ABSORB	Alaska Beaufort Sear Oil Spill Response Body
ACMP	Alaska Coastal Management Program
ACS	American Chemical Society
ADDC	Association of Desk and Derrick Clubs
AEC	Atomic Energy Commission
AECRB	Alberta Energy Conservation Resources Board
AGA	American Gas Association
AGI	American Geological Institute
AGTL	Alberta Gas Trunkline Co., Ltd.
AGU	American Geophysical Union
AIChE	American Institute of Chemical Engineers
AIME	American Institute of Mining, Metallurgical and Petroleum Engineers
AISI	American Iron and Steel Institute
ALCOA	Aluminum Company of American

ANSI	American National Standards Institute
AOAC	Association of Official Agricultural Chemists
AOCS	American Oil Chemists Society
AOGA	Alaskan Oil & Gas Association
AOPL	Association of Oil Pipe Lines
AOSC	Association of Oilwell Servicing Contractors
AP&VMA	American Paint & Varnish Manufacturers Assn.
APHA	American Public Health Association
API	American Petroleum Institute
APRA	American Petroleum Refiners Association
APW	Association of Petroleum Writers
ARCO	Atlantic Richfield Co.
ARKLA	Arkansas Louisiana Gas Co.
ASA	American Standards Assoc.
ASCE	American Society of Civil Engineers
ASHRAE	American Society of Heating, Refrigerating, and Air-Conditioning Engineers, Inc.
ASLE	American Society of Lubricating Engineers
ASME	American Society of Mechanical Engineers
ASPG	American Society of Professional Geologists
ASSE	American Society of Safety Engineers
ASTM	American Society for Testing Materials
AWS	American Welding Society
AWWA	American Water Works Association
BLM	Bureau of Land Management
BLS	Bureau of Labor Statistics
BP	British Petroleum
BuMines	Bureau of Mines, U.S. Department of the Interior

CAGC	A combine: Continental Oil Co., Atlantic Richfield Co., Getty Oil Co., and Cities Service Oil Co.
CAODS	Canadian Association of Oilwell Drilling Contractors
CCCOP	Conservation Committee of California Oil Producers
CDS	Canadian Development Corp.
CFR	Coordinating Fuel Research Committee
CFRC	Coordinating Fuel Research Committee
CGA	Clean Gulf Associates
CGTC	Columbia Gas Transmission Co.
CIPA	California Independent Petroleum Association
CL&F	Continental Land and Fur Co.
CNG	Consolidated Natural Gas Supply Company
CNGP	CNG Producing Company
COE	Corps of Engineers
CONOCO	Continental Oil Co.
CORCO	Commonwealth Oil Refining Co., Inc.
CORS	Canadian Operational Research Society
CPA	Canadian Petroleum Association
CRC	Coordinating Research Council, Inc.
DOE	Department of Energy
DOT	Department of Transportation
DNR	Department of Natural Resources
DRAPR	Delaware River Area Petroleum Refineries
Drssr., DA	Dresser Atlas
EMR	Department of Energy, Mines, and Resources (Canada)
EPA	Environmental Protection Agency
ERCB	Energy Resource Conservation Board (Alberta, Canada)

FAA	Federal Aviation Agency
FCC	Federal Communications Commission
FPC	Federal Power Commission
FSIWA	Federation of Sewage and Industrial Wastes Assn.
FTC	Federal Trade Commission
GA	Canadian Gas Association
GAMA	Gas Appliance Manufacturers Association
GE	General Electric Company
GM	General Motors
GNEC	General Nuclear Engineering Co.
GR&DC	Gulf Research and Development Company
IADC	International Association of Drilling Contractors (formerly AAODC)
IAE	Institute of Automotive Engineers
ICC	Interstate Commerce Commission
IEEE	Institute of Electrical and Electronics Engineers
IGT	Institute of Gas Technology
INGAA	Independent Natural Gas Association of America
IOCA	Independent Oil Compounders Association
IOCC	Interstate Oil Compact Commission
IOPA	Independent Oil Producers Agency
IOSA	International Oil Scouts Association
IP	Institute of Petroleum
IPAA	Independent Petroleum Association of America
IPAC	Independent Petroleum Association of Canada
IPE	International Petroleum Exposition
IPP/L	Interprovincial Pipe Line Co.
IRAA	Independent Refiners Association of America
ISA	Instrument Society of America
JCUMWE	Joint Committee on Uniformity of Methods of Water Examination
KERMAC	Kerr-McGee Corp.
KIOGA	Kansas Independent Oil and Gas Association
LL&E	Louisiana Land & Exploration Co.
MIOP	Mandatory Oil Import Program
MMS	Minerals Management Service
NACE	National Association of Corrosion
NACOPS	National Advisory Committee on Petroleum Statistics (Canada)
NAS	National Academy of Science
NASA	National Aeronautical and Space Administration
NEB	National Energy Board (Canada)
NEMA	National Electrical Manufacturers Association
NEPA	National Environmental Policy Act of 1969
NGPA	Natural Gas Processor Association
NCPSA	Natural Gas Processors Suppliers Association
NLGI	National Lubricating Grease Institute
NLPGA	National Liquefied Petroleum Gas Association
NLRB	National Labor Relations Board
NMOCC	New Mexico Oil Conservation Commission
NMOGA	New Mexico Oil and Gas Association
NOFI	National Oil Fuel Institute
NOIA	National Ocean Industries Association
NOJC	National Oil Jobbers Council
NOMADS	National Oil-Equipment Manufacturers and Delegates Society

NPC	National Petroleum Council	**RPI**	Research Planning Institute
NPDES	National Pollution Discharge Elimination System	**RTL**	Refinery Technology Laboratory
NPR	Naval Petroleum Reserve	**SACROC**	Scurry Area Canyon Reef Operators Committee
NPRA	Naval Petroleum Reserve, Alaska	**SAE**	Society of Automotive Engineers
NPRA	National Petroleum Refiners Association	**Schl., Sj**	Schlumberger
NRC	Nuclear Regulatory Commission	**SEG**	Society of Exloration Geophysicists
NSF	National Science Foundation	**SEPM**	Society of Economic Paleontologists and Mineralogists
OCR	Office of Coal Research		
OEP	Office of Emergency Preparedness	**SFER**	Santa Fe Energy Resources, Inc.
OIA	Oil Import Administration	**SGA**	Southern Gas Associaton
OIAB	Oil Import Appeals Board	**SLAM**	A combine: Signal Oil and Gas Co., Louisiana Land & Exploration Co., Amerada Hess Corp., and Marathon Oil Co.
OIC	Oil Information Committee		
OIPA	Oklahoma Independent Petroleum Association		
OOC	Offshore Operators Committee		
OPC	Oil Policy Committee	**SOCAL**	Standard Oil Company of California
ORIAC	Oil Refining Industry Action Committee	**SOHIO**	Standard Oil Co. of Ohio
ORSANCO	Ohio River Valley Water Sanitation Commission	**SPE**	Society of Petroleum Engineers of AIME
OXY	Occidental Petroleum Corp.	**SPEE**	Society of Petroleum Evaluation Engineers
PAD	Petroleum Administration for Defense	**SPWLA**	Society of Professional Well Log Analysts
PESA	Petroleum Equipment Suppliers Association	**STATCAN**	Statistics Canada ex Dominion, Bureau of Statistics (DBS)
PETCO	Petroleum Corporation of Texas		
PGCOA	Pennsylvania Grade Crude Oil Association	**TAPS**	Trans-Alaska Pipeline Systems
PIEA	Petroleum Industry Electrical Association	**TCP**	Trans-Canada Pipe Lines Ltd.
Plato	Pennzoil Louisiana and Texas Offshore	**TETCO**	Texas Eastern Transmission Corp.
PLCA	Pipe Line Contractors Association	**TGT**	Tennessee Gas Transmission Co.
POGO	Pennzoil Offshore Gas Operators	**THUMS**	A combine: Texaco, Inc., Humble Oil & Refining Co. Union Oil Co. of California, Mobil Oil Corp., and Shell Oil Co.
PPI	Plastic Pipe Institute		
PPROA	Panhandle Producers and Royalty Owners Association		
RMOGA	Rocky Mountain Oil and Gas Association	**TIPRO**	Texas Independent Producers and Royalty Owners Association
R-PAT	Regional Petroleum Association	**TRANSCO**	Transcontinental Gas Pipe Line Corp.

TT&T	Texaco Trading & Transportation
UNOCAL	Union Oil of California
UOCO	Union Oil Company
UOP	Universal Oil Products Company
USGS	United States Geological Survey
USP	United States Pharmocopoeia
WeCTOGA	West Central Texas Oil and Gas Association
WOGA	Western Oil & Gas Association
WPC	World Petroleum Congress
WSPA	Western States Petroleum Association
Wx	Welex

COMPANIES AND ASSOCIATIONS

Outside the U.S. and Canada

AAOC	American Asiatic Oil Corp. (Philippines)
ABCD	Asfalti Bitumi Cementi Derivati, S.A. (Italy)
ACNA	Aziende Colori Nazionali Affini (Italy)
A.C.P.H.A.	Association Cooperative pour la Recherche et l'Exploration des Hydrocarbures en Algerie (Algeria)
ADCO-HH	African Drilling Co.-H. Hamouda (Libya)
AGIP S.p.A.	Azienda Generale Italiana Petroli S.p.A. (Italy)
A.H.I. BAU	Allegemeine Hoch-und Ingenieurbau AG (Germany)
AIOC	American International Oil Co. (U.S.A.)
AITASA	Aguas Industriales de Tarragona, S.A. (Spain)
AK CHEMI	GmbH & Co. KG-subsidiary of Associated Octel, Ltd., London, Eng (Germany)
AKU	Algemene Kunstzijde Unie, N.V. (Netherlands)
ATAS	Anadolu Tastiyehanesi A.S. (Turkey)

AUXERAP	Societe Auxiliare de la Regie Autonome des Petroles (France)
AZOLACQ	Societe Chimique d'Engrais et de Produits de Synthese (France)
BAPCO	Bahrain Petroleum Co. Ltd. (Bahrain)
BASF	Badische Anilin & Soda-Fabrik AB (Germany)
BASUCOL	Barranquilla Supply & Co. (Colombia)
B.I.P.M.	Bataafse Internationale Petroleum Mij. N.V. (Netherlands)
BOGOC	Bolivian Gulf Oil Co. (Bolivia)
BORCO	Bahamas Oil Refining Co. (Bahamas)
BP	British Petroleum Co., Ltd. (England)
BRGG	Bureau de Recherches Geologiques et Geophysique (France)
BRGM	Bureau de Recherches Geologique et Minieres (France)
BRIGITTA	Gewerkschaft Brigitta (Germany)
BRP	Bureau de Recherche de Petrole (France)
BRPM	Bureau de Recherches et de Participations Mineres (Morocco)
BSI	British Standards Institute
CALSPAIN	California Oil Co. of Spain (Spain)
CALTEX	Various affiliates of Texaco Inc. and Std. of Calif.
CALVO SOTELO	Empresa Nacional Calvo Sotelo (Spain)
CAMEL	Campagnie Algerienne du Methane Liquide (France, Algeria)
CAMPSA	Compania Arrendataria del Monopolio de Petroleos, S.A. (Spain)
CAPAG	Enterprise Moderne de Canalisations Petrolieres, Aquiferes et Gazieres (France)

CARBESA	Carbon Black Espanola, S.A. (Conoco affiliate)
CAREP	Compagnie Algerienne de Recherche et d'Exploitation Petrolieres (Algiers)
CCC	Compania Carbonos Coloidais (Brazil)
C.E.C.A.	Carbonisation et Charbons Actifs S.A. (France)
CEICO	Central Espanol Ingenieria y Control S.A. (Spain)
CEL	Central European Pipeline (Germany)
CEOA	Centre Europe de'Exploitation de I'OTAN (France)
CEP	Compagnie D'Exploration Petroliere (France)
CEPSA	Compania Espanola de Petroleos, S.A. (Spain)
CETRA	Compagnie Europeanne de Canalisations et de Travaux (France)
CFEM	Compagnie Francaise d'Enterprises Metalliques (France)
CFM	Compagnie Francaise du-Methane (France)
CFMK	Compagnie Ferguson Morrison-Knudsen (France)
CFP	Compagnie Francaise du-Petroles (France)
CFPA	Compagnie Francaise des Petroles (Algeria) (France)
CFPS	Compagnie Francaise de Prospection Sismique (France)
CFR	Compagnie Francaise de Raffinage (France)
CGG	Compagnie Generale de Geophysique (France, Australia, Singapore)
CIAGO	N.V. Chemische Industrie aku-Goodrich (Netherlands)
CIEPSA	Compania de Investigacion y Explotaciones Petroliferas, S.A. (Spain)
CIM	Compagnie Industrielle Maritime (France)

CIMI	Compania Italiana Montaggi Industriali S.p.a. (Italy)
CINSA	Compania Insular del Nitrogena, S.A. (Spain)
CIPAO	Compagnie Industrielle des Petroles de I'A.O. (France)
CIPSA	Compania Iberica de Prospecciones, S A. (Spain)
CIRES	Compania Industrial de Resinas Sinteticas (Portugal)
CLASA	Carburanti Librificanti Affini S.p.A. (Italy)
CMF	Construzioni Metalliche Finsider S.p.A. (Italy)
COCHIME	Compagnie Chimique de la Meterranee (France)
CODI	Colombianos Distribuidores de Combustibles S A. (Colombia)
COFIREP	Compagnie Financiere de Recherches Petrolieres (France)
COFOR	Compagnie Generale de Forages (France)
COLCITO	Colombia-Cities Service Petroleum Corp. (Colombia)
COLPET	Colombian Petroleum Co. (Colombia)
COMEX	Compagnie Maritime d'Expertises (France)
CONSPAIN	Continental Oil Co. of Spain (Spain), Conoco Espanola S.A. (Spain)
COPAREX	Compagnie de Participations, de Recherches et d'Exploitations Petrolieres (France)
COPE	Compagnie Orientale des Petroles d'Egypte (Egypt)
COPEBRAS	Compania Petroquimica Brasileira (Brazil)
COPEFA	Compagnie des Petroles France-Afrique (France)
COPETAO	Compagnie des Petroles Total (Afrique Quest) (France)
COPETMA	Compagnie les Petroles Total (Madagascar)
COPISA	Compania Petrolifera Iberica, Sociedad Anonima (Spain)

COPSEP	Compagnie des Petroles du Sud est Parisien (France)
C.O.R.I.	Compania Richerche Idrocarburi S.p.A. (Italy)
COS	Coordinated Oil Services (France)
CPA	Compagnie des Petroles d'Algeria (Algeria)
CPC	Chinese Petroleum Corporation, Taiwan, China
CPTL	Compagnie des Petroles Total (Libya) (France)
CRAN	Compagnie de Raffinage en Afrique du Nord (Algeria)
CREPS	Compagnie de Recherches et d'Exploitation de Petrole au Sahara (Algeria)
CRR	Compagnie Rhenane de Raffinage (France)
CSRPG	Chambre Syndicale de la Recherche et de la Production du Petrole et du Gaz Naturel (France)
CTIP	Compania Tecnica Industrie Petroli S.p.a. (Italy)
CVP	Corporacion Venezolano del Petroleo (Venezuela)
DCEA	Direction Centrale des Essences des Armees (France)
DEA	Deutsche Erdol-Aktiengesellschaft (Germany)
DEMINEX	Deutsch Erdolversorungs-gesellschaft mbH (Germany) (Trinidad)
DIAMEX	Diamond Chemicals de Mexico, S.A. de C.V. (Mexico)
DICA	Direction des Carburants (France)
DICA	Distilleria Italiana Carburanti Affini (Italy)
DITTA	Macchia Averardo (Italy)
DUPETCO	Dubai Petroleum Company (Trucial States)
E.A.O.R.	East African Oil Refineries, Ltd. (Kenya)
ECF	Essences et Carburants de France (France)

ECOPETROL	Empresa Colombiana de Petroleos (Colombia)
EGTA	Enterprises et Grands Travaux de l'Atlantique (France)
ELF-ERAP	Enterprise de Recherches et d'Activites Petrolieres (France)
ELF-U.I.P.	Elf Union Industrielle des Petroles (France)
ELF—SPAFE	Elf des Petroles D'Afrique Equatoriale (France)
ELGI	M. All. Eotvos Lorand Geofizikai Intezet (Hungary)
ENAP	Empresa Nacional del Petroleo (Chile)
ENCAL	Engenheiros Consultores Associados S.A. (Brazil)
ENCASO	Empresa Nacional Calvo Sotelo de Combustibles Liquidos y Lubricantes, S.A. (Spain)
ENGEBRAS	Engenharia Especializada Brasileira, S.A. (Brazil) (Venezuela)
ENI	Ente Nazionale Idrocarburi (Italy)
ENPASA	Empresa Nacional de Petroleos de Aragon, S.A. (Spain)
ENPENSA	Empresa Nacional de Petroleos de Navarra, S.A. (Spain)
ERAP	Enterprise de Recherches et d'Activites Petrolieres (France)
ESSAF	Esso Standard Societe Anonyme Francaise (France)
ESSOPETROL	Esso Petroleos Espanoles, S.A. (Spain)
ESSO REP	Societe Esso de Recherches et Exploitation Petrolieres (France)
E.T.P.M.	Societe Entrepose G.T.M. pour les Travaux Petroliers Maritimes (France)
EURAFREP	Societe de Recherches et d'Exploitation de Petrole (France)

FERTIBERIA	Fertilizantes de Iberia, S.A. (Spain)		**IAP**	Institut Algerien du Petrole (Algeria)
FFC	Federation Francaise des Carburants (France)		**ICI**	Imperial Chemical Industries Ltd. (England)
FINAREP	Societe Financiere des Petroles (France)		**ICIANZ**	Imperial Chemical Industries of Australia & New Zealand Ltd. (Australia, New Zealand)
FOREX	Societe Forex Forages et Exploitation Petrolieres (United Kingdom)		**ICIP**	Industrie Chimiche Italiane del Petrolio (Italy)
FRANCAREP	Compagnie Franco-Africaine de Recherches Petrolieres (France)		**IEOC**	International Egyptian Oil Co., Inc. (Egypt)
FRAP	Societe de Construction de Feeders, Raffineries, Adductions d'Eau et Pipe-Lines (France)		**IFCE**	Institut Francais des Combustibles et de l'Energie (France)
			IFP	Institut Francaise du Petrole (France)
FRISIA	Erdolwerke Frisia A.G. (Germany)		**IGSA**	Investigaciones Geologicas, S.A. (Spain)
GARRONE	Garrone (Dott. Edoardo) Raffineria Petroli S.a.S. (Italy)		**IIAPCO**	Independent Indonesian American Petroleum Co. (Indonesia)
GBAG	Gelsenberg Benzin (Germany)		**I.L.S.E.A.**	Industria Leganti Stradali et Affini (United Kingdom)
GESCO	General Engineering Services (Colombia)		**I.M.E.**	Industrias Matarazzo de Energia (Brazil)
GHAIP	Ghanian Italian Petroleum Co., Ltd. (Ghana)		**IMEG**	International Management & Engineering Group of Britain Ltd. (United Kingdom)
GO INTL	GO International, Inc.			
GPC	The General Petroleum Co. (Egypt)			
G.T.M.	Les Grands Travaux de Marseille (France)		**IMEG**	Iranian Management & Engineering Group Ltd. (Iran)
HELIECUADOR	Helicopteros Nacionales S.A. (Ecuador)		**IMINOCO**	Iranian Marine International Oil Co. (Iran)
HDC	Hoecsht Dyes & Chemicals Ltd. (India)		**IMS**	Industria Metalurgica de Salvador, S/Z (Brazil)
HIDECA	Hidrocarburos y Derivados C.A. (Brazil, Uraguay & Venezuela)		**I.N.C.I.S.A.**	Impresa Nazionale Condotte Industriali Strade Affini (United Kingdom)
HIP	Hemijska Industrija Pancevo (Yugoslavia)		**INDEIN**	Ingenieria y Desarrolio Industrial S.A. (Spain)
H.I.S.A.	Herramientas Interamericanas, S.A. de C.V. (Mexico)		**INI**	Instituto Nacional de Industria (Spain)
HISPANOIL	Hispanica de Petroleos, S A., (Spain)		**INOC**	Iraq National Oil Co. (Iraq)
			INTERCOL	International Petroleum (Colombia) Ltd. (Colombia)
HOC	Hindustan Organic Chemicals Ltd. (India)		**IODRIC**	International Oceanic Development Research Information Center (Japan)
HYLSA	Hojalata y Lamina, S A. (Mexico)			

IOE & PC	Iranian Oil Exploration & Producing Co. (Iran)
IORC	Iranian Oil Refining Co. N.V. (United Kingdom)
IPAC	Iran Pan American Oil Co. (Iran)
I.P.L.O.M.	Industria Piemontese Lavorazione Oil Minerali (United Kingdom)
IPLAS	Industrija Plastike (Yugoslavia)
IPRAS	Istanbul Petrol Rafinerisi A.S. (Turkey)
IRANOP	Iranian Oil Participants Limited (England)
IROM	Industria Raffinazione Oil Minerali (Italy)
ITOPCO	Iranian Offshore Petroleum Company (United Kingdom)
IROS	Iranian Oil Services, Ltd. (England)
IVP	Instituto Venezolano de Petroquimica (Venezuela)
JAPEX	Japan Petroleum Trading Co. Ltd. (Japan)
KIZ	Kemijska Industrijska Zajednica (Yugoslavia)
KNPC	Kuwait National Petroleum Co. (Arabia)
KSEPL	Kon./Shell Exploration and Production Laboratory (Netherlands)
KSPC	Kuwait Spanish Petroleum Co. (Kuwait)
KUOCO	Kuwait Oil Co., Ltd. (England)
LPACO	Lavan Petroleum Co. (Iran)
LEMIGAS	Lembaga Minjak Dan Gas Bumi (Indonesia)
LINOCO	Libyan National Oil Corp. (Libya)
L.M.B.H.	Lemgaga Kebajoran & Gas Bumi (Libya)
MABANAFT	Marquard & Bahls B.m.b.H. (Germany)
MATEP	Materials Tecnicos de Petroleo S.A. (Brazil)
MAWAG	Mineraloel Aktien Gesellschaft ag (Germany)
MEDRECO	Mediterranean Refining Co. (Lebanon)
MEKOG	N.B. Maatschappij Tot Exploitatie van Kooksoven-gassen (Netherlands)
MENEG	Mene Grande Oil Co. (Venezuela)
METG	Mittelrheinische Ergastransport GmbH (Germany)
M.I.T.I.	Ministry of International Trade and Industry (Japan)
MODEC	Mitsui Ocean Development & Engineering Co. Ltd. (Japan)
MPL	Murco Petroleum Limited (England)
NAKI	Nagynyomasu Kiserleti Intezet (Hungary)
NAM	N.V. Nederlandse Aardolie Mij. (Netherlands)
NAPM	N.V. Nederlands Amerikaanse Pijpleiding Maatschappij (Netherlands)
NCM	Nederlandse Constructiebe-drijven en Machinefrabri-ken N.V. (Netherlands)
NDSM	Nederlandse Dok en Scheepsbouw Maatschappij (Netherlands)
NED.	North Sea Diving Services, N.V. (Netherlands)
NEPTUNE	Soc. de Forages en Mer Neptune (France)
NETG	Nordrheinische Erdgastransport Gesselschaft mbH (Germany)
NEVIKI	Nehezvegyipari Kutato Intezet (Hungary)
NIOC	National Iranian Oil Co. (Iran)
NORDIVE	North Sea Diving Services Ltd. (United Kingdom)
NOSODECO	North Sumatra Oil Development Cooperation Co. Ltd (Indonesia)
NPC	Nederlandse Pijpleiding Constructie Combinatie (Netherlands)
NPCI	National Petroleum Co. of Iran (Iran)

657

N.V.A.I.G.B.	N.V. Algemene Internationale Gasleidingen Bouw (Netherlands)
N.V.G.	Nordsee Versorgungsschiffahrt GmbH (Germany)
NWO	Nord-West Oelleitung GmbH (Germany)
OCCR	Office Central de Chauffe Rationnelle (France)
OEA	Operaciones Especiales Argentinas (Argentina)
OKI	Organsko Kenijska Industrija (Yugoslavia)
OMNIREX	Omnium de Recherches et Exploitations Petrolieres (France)
OMV	Oesterreichische Mineraloelverwaltung A.G. (Australia)
OPEC	Organization of Petroleum Exporting Countries
OTP	Omnium Techniques des Transprots par Pipelines (France)
PCRB	Compagnie des Produits Chimiques et Raffineries de Berre (France)
PEMEX	Petroleos Mexicanos (Mexico)
PERMAGO	Perforaciones Marinas del Golfo S.A. (Mexico)
PETRANGOL	Companhia de Petroleos de Angola (Angola)
PETRESA	Petroquimica Espanola S.A. (Spain)
PETROBRAS	Petroleo Brasileiro S A. (Brazil)
PETROLIBER	Compania Iberica Refinadora de Petroleos, S.A. (Spain)
PETROMIN	General Petroleum and Mineral Organization (Saudi Arabia)
PETRONOR	Refineria de Petroleos del Norte, S.A. (Spain)
PETROPAR	Societe de Participations Petrolieres (France)
PETROREP	Societe Petroliere de Recherches Dans La Region Parisienne (France)
POLICOLSA	Poliolefinas Colombianas S.A. (Colombia)
PREPA	Societe de Prospection et Exploitations Petrolieres en Alsace (France)
PRODESA	Productos de Estireno, S.A. de C.V. (Mexico)
PROTEXA	Construcciones Protexa, S.A. de C.V. (Mexico)
PYDESA	Petroleos y Derivados, S.A. (Spain)
QUIMAR	Quimica del Mar, S.A. (Mexico)
RAP	Regie Autonome des Petroles (France)
RASIOM	Raffinerie Siciliane Olii Minerali (Esso Standard Italiana S.p.A.) (Italy)
RDM	De Rotterdamsche Droogdok Mij. N.V. (Netherlands)
RDO	Rhein-Donau-Oelleitung GmbH (Germany)
REDCO	Rehabilitation, Engineering and Development Co. (Indonesia)
REPESA	Refineria de Petroleos de Escombreras, S.A. (Spain)
REPGA	Recherche et Exploitation de Petrole et de Gaz (France)
RIOGULF	Rio Gulf de Petroleos, S.A. (Spain)
SACOR	Sociedade Anonima Concessionaria da Rafinacao de Petroleos em Portugal (Portugal)
SAEL	Sociedad Anonima Espanola de Lubricantes (Spain)
S.A.F.C.O.	Saudi Arabian Refinery Co. (Saudi Arabia)
SAFREP	Societe Anonyme Francaise de Recherches et d'Exploitation de Petrole (France)
SAIC	Sociedad Anonima Industrial y Commercial (Argentina)
SAM	Societe d'Approvis de Material Patrolier (France)
SAP	Societe Africaine des Petroles (France)

SAPPRO	Societe Anonyme de Pipeline a Produits Petroliers sur Territoire Genevois (Switzerland)
SAR	Societe Africaine de Raffinage (Dakar)
SARAS	S.p.a. Raffinerie Sarde (Italy)
SARL	Chimie Development International (Germany)
SAROC	Saudi Arabia Refinery Co. (Saudi Arabia)
SAROM	Societa Azionaria Raffinazione Olli Minerali (Italy)
SARPOM	Societa per Azioni Raffineria Padana Olii Minerali (Italy)
SASOL	South African Coal, Oil and Gas Corp. Ltd. (South Africa)
S.A.V.A.	Societa Alluminio Veneto per Azioni (Italy)
SCC	Societe Chimiques des Charbonnages (France)
SCI	Societe Chimie Industrielle (France)
SPC	Societe Cherifienne des Petroles (Morocco)
SECA	Societe Europreeme des Carburants (Belgium)
SEHR	Societe d'Exploitation des Hydrocarbures d'Hassi R'Mel (France)
SEPE	Sociedad de Exploracion de Petroleos Espanoles, S.A. (Spain)
SER	Societe Equatoriale de Raffinage (Gabon)
SERCOP	Societe Egyptienne pour le Raffinage et le Commerce du Petrole (Egypt)
SEREPT	Societe de Recherches et d'Exploitation des Petroles en Tunisia (Tunisia)
SER VIPETROL	Transportes y Servicios Petroleros (Ecuador)
SETRAPEM	Societe Equatoriale de Travaux Petroliers Maritimes (France, Germany)
SFPLJ	Societe Francaise de Pipe Line du Jura (France)
SHELLREX	Societe Shell de Rechercheset d'Exploitations (France)
S.I.B.P.	Societe Industrielle Belge des Petroles (Belgium)
SIF	Societe Tunisienne de Sondages, Injections, Forages (Tunisia)
SINCAT	Societa Industriale Cantese S.p.a. (Italy)
SIPSA	Sociedad Investigadora Petrolifera S.A. (Spain)
SIR	Societa Italiana Resine (Italy)
SIREP	Societe Independante de Recherches et d'Exploitation du Petrole (France)
SIRIP	Societe Irano-Italienne des Petroles (Iran)
SITEP	Societe Italo-Tunisienne d'Exploitation (Italy, Tunisia)
SMF	Societe de Fabrication de Material de Forage (France)
SMP	Svenska Murco Petroleum Aktiebolag (Sweden)
SMR	Societe Malagache de Raffinage (Malagasy)
SNGSO	Societe Nationale des Gas de Sud-Ouest (France)
SN MAREP	Societe Nationale de Material pour la Recherche et l'Exploitation du Petrole (France)
SNPA	Societe Nationale des Petroles d'Aquitaine (France)
SN REPAL	Societe Nationale de Recherches et d'Exploitation des Petroles en Algerie (France)
SOCABU	Societe du Caoutchouc Butyl (France)
SOCEA	Societe Eau et Assainissement (France)
SOCIR	Societe Congo-Italienne de Raffinage (Congo Republic)
SOFEI	Societe Francaise d'Enterprises Industrielles (France)
SOGARES	Societe Gabonaise de Realisation de Structures (France)

SOMALGAZ	Societe Mixte Algerienne de Gaz (Algeria)	**SSRP**	Societe Saharienne de Recherches Petrolieres (France)
SOMASER	Societe Maritime de Service (France)	**STEG**	Societe Tunisienne d'Electricite et de Gaz (Tunisia)
SONAP	Sociedade Nacional de Petroleos S.A.R.I. (Portugal)	**STIR**	Societe Tuniso-Italienne de Raffinage (Tunisia)
SONAREP	Sociedade Nacional de Refinacao de Petroleos S.A.R.L. (Mozambique)	**TAL**	Deutsche Transalpine Oelleitung GmbH (Germany)
SONATRACH	Societe Nationale de Transport et de Commercialisation des Hydrocarbures (Algeria, France)	**TAMSA**	Tubos de Acero de Mexico, S.A. (Mexico)
		TATSA	Tanques de Acero Trinity, S.A. (Mexico)
SONPETROL	Sondeos Petroliferos S.A. (Spain)	**TECHINT**	Compania Technica Internacional (Brazil)
SOPEFAL	Societe Petroliere Francaise en Algeria (Algeria)	**TECHNIP**	Compagnie Francaise d'Etudes et de Construction Technip (France)
SOPEG	Societe Petroliere de Gerance (France)		
SOREX	Societe de Recherches et d'Exploitations Petrolieres (France)	**TEXSPAIN**	Texaco (Spain) Inc. (Spain)
		TORC	Thai Oil Refinery Co. (Thailand)
SOTEI	Societe Tunisienne de Enterprises Industrielles (Tunisia)	**T.P.A.O.**	Turkiye Petrolleri A.O. (United Kingdom)
SOTHRA	Societe de Transport du Gaz Naturel D'Hassi-er-r'mel a Arzew (Algeria)	**TRAPIL**	Societe des Transports Petroliers Par Pipeline (France)
SPAFE	Societe des Petroles d'Afrique Equatoriale (France)	**TRAPSA**	Compagnie des Transports par Pipe-Line au Sahara (Algeria)
SPANGOC	Spanish Gulf Oil Co. (Spain)	**UCSIP**	Union des Chambres Syndicales de l'Industrie du Petrole (France)
SPEICHIM	Societe Pour l'Equipment des Industries, Chimiques (France)		
SPG	Societe des Petroles de la Garrone (France)	**UPG**	Union Generale des Petroles (France)
S.P.I.	Societa Petrolifera Italiana (Italy)	**UIE**	Union Industrielle et d'Enterprise (France)
SPIC	Southern Petrochemical Industries Corporation Ltd.	**UNIAO**	Refinaria e Exploracao de Petroleo "UNIAO" S A. (Brazil)
SPLSE	Societe du Pipe-Line Sud Europeen (France)	**URAG**	Unteweser Reederei Gmbh (Germany)
SPM	Societe des Petroles de Madagascar (France)	**URG**	Societe pour l'Utilisation Rationnelle des Gaz (France)
SPV	Societe des Petroles de Valence (France)	**WEPCO**	Western Desert Operating Petroleum Company (Egypt)

YPF   Yacimientos Petroliferos
               Fiscales (Argentina)
YPFB   Yacimientos Petroliferos
               Fiscales Bolivianos
               (Bolivia)

METRIC-ENGLISH SYSTEMS
CONVERSION FACTORS
Basic Dimensions

Metric System
Length—   meter (m)
             kilometer (km)
             centimeter (cm)
             millimeter (mm)
Area—   square meters (m^2)
             square centimeters (cm^2)
Volume—   cubic meters (m^3)
             cubic centimeters (cm^3)
             liters (l)
             milliliters (ml)
Mass—   kilograms (kg)
             grams (g)
             gram-moles (gm-moles)
             kilogram-moles (kg-moles)
Density— kg/m^3, g/cm^3

English System
Length—   inch (in.)
             foot (ft)
             yard (yd)
             mile (mile)
Area—   square inches ($in.^2$)
             square feet (ft^2)
             square miles ($miles^2$)
Volume—   cubic inches ($in.^3$)
             cubic feet (ft^3)
             barrels (bbl)
             U.S. gallons (gal)
             Imperial gallons (Imp. gal)
Mass   pounds (lb)
             pound-moles (lb-moles)
Density— pounds per gallon (lb/gal,
             lb/ft^3)

System Equivalents

Metric System
Length—   1 m = 100 cm = 1,000 mm =
             0.001 km
Area—   1 m^2 = 10,000 cm^2

Volume—   1 m^3 = 1,000,000 cm^3 =
             1,000,000 ml = 1,000 l
Mass—   1 kg = 1,000 g
             1 kg-mole = 1,000 gm-moles
Density— 1 kg/m^3 = 0.001 g/cm^3

English System
Length—   1 ft = 12 in. = 0.333 yd =
             0.000189 miles
             1 mile = 5,280 ft = 1,750 yd
Area—   1 ft^2 = 144 $in.^2$
             1 $mile^2$ = 27,878,400.ft^2
Volume—   1 ft^3 = 1728 $in.^3$ = 0.178 bbl =
             .48 U.S. gal = 6.23 Imp. gal
             1 bbl = 5.61 ft^3 = 42 U.S. gal =
             34.97 Imp. gal
Density— 1 lb/gal = 7.48 lb/ft^3 = 42 lb/bbl

BASIC CONVERSION FACTORS

Length—   1 m = 3.281 ft = 39.37 in.
             1 ft = 0.305 m = 30.5 cm =
             3,050 mm
             1 mile = 1.61 km
             1 km = 0.621 mile
Area—   1 m^2 = 10.76 ft^2 = 1,549 $in.^2$
             1 ft^2 = 0.0929 m^2 = 929.4 cm^2
Volume—   1 m^3 = 35.32 ft^2 = 6.29 bbl
             1 l = 0.035 ft^3 = 61 $in.^3$
             1 ft^3 = 0.0283 m^3 = 28.31
             1 bbl = 0.159 m^3 = 1591
Mass—   1 kg = 2.205 lb
             1 lb = 0.454 kg = 454 g
             1 metric ton = 1,000 kg =
             2,205 lb
Density— 1 kg/m^3 = 0.0624 lb/ft^3
             1 lb/ft^3 = 16.02 kg/m^3 =
             0.01602 g/cm^3
             1 g/cm^3 = 62.4 lb/ft^3
Force—   1 kg force = 2.205 lb force
             1 lb force = 0.454 kg force
Work &
Heat—   1 Btu = 0.252 kilocalories
             (kcal)
             1 kcal = 3.97 Btu
Power—   1 kilowatt (kw) = 860 kcal/
             hr = 3,415 Btu/hr =
             1.341 horsepower (hp)
             1 hp = 0.746 kw = 641 kcal/
             hr = 2,545 Btu/hr

Enthalpy—

1 kcal/kg = 1.8 Btu/lb
1 Btu/lb = 0.556 kcal/kg

Pressure—

1 bar = 14.51 lb/in.² (psi) = 0.987 atmospheres (atm) = 1.02 kg/cm²
1 kg/cm² = 14.22 psi = 0.968 atm
1 psi = 0.0703 kg/cm²

Temperature—

°C = 0.556 (°F − 32)
°K = °C + 273
°F = 1.8°C + 32
°R = °F + 460

Common Oilfield Spellings

A

about-face
aboveground
acknowledgment
acre-feet
aeration
aftercooler
aftertreat
afterwash
air flow (n, adj)
airfoil
air line
airtight
alumna - s fem.
alumnae - pl fem.
alumnus - s masc.
alumni - pl masc.
anti-icer

B

backflow
backlight
back off
back pressure
backup (adj)
backwash
backwater
base line
baseplate
behavior

belowground
bench mark
bench scale
blow-by
blowdown
blowout
boil off
borehole
bottom hole
breakaway
breakdown
break-even (adj)
break even (v)
breakout
breakpoint
breakthrough
breakup (n)
break up (v)
briquette
bubble cap
buildup (n)
build-up (adj)
build up (v)
built in
burn-off
burnout (n)
burnup (n)
bypass
byproduct

C

Caribbean
carry-over
casinghead
center line
changeable
change out (v)
changeover
channeling
charge stock
checklist
check-out (adj)
check out (v)
checkpoint
city-wide
classroom
cleanout
cleanup (n)
clean up (v)
cleanup (adj)
clear-cut
close up (v)
close-up (n, adj)
closeout
coastline
commitment
commingle
controlling
controlled
co-op

co-owner
coproduct
counterbalance
counterbattery
countercurrent
counterflow
country-wide
crisscross
criterion - s
criteria - pl
cross-bedding
cross-country
cross flow
crosshead
crossover
cross plot
cross-reference
cross section
cutdown
cutoff
cutout
cut point
cycle oil

D

datum - s
data - pl
deadman
dead time
deadweight

deepwater (adj)
deep water (n)
de-ethanizer
desiccant
desirable
dew point
doghouse
dogleg
double-jointed
downdip
downdrag
downflow
downgrade
downhole
downstream
downthrown
downtime
drawdown
drawoff (n)
drawworks
drill bit
drill collar
drillhead
drillpipe
drillship
drillsite
drillstem
drillstock
drillstring
drumhead

E
edgewise
end point
en route

F
face-lifting
fail-safe
falloff
farmin (n)
farmout (n)
feedback
feed rate
feedstock
feedwater
fiberglass
fieldman
fill-up
fireflood

firebox
fire wall
firewater (n)
flatbed
flier
flow-control
flow line
flowmeter
flywheel
foam glass
follow-up
forklift
formula
formulas
fourfold
freeze-up
freshwater (adj)
frost line
full time

G
gamma ray
gas oil
gearbox
gearshift
grassroots
gray
ground line
groundwater
guesswork
guideline

H
halfway
hammerblow
handwritten
hardback
headlong
hold-down
holdup
homemade
hookup

I
in between (v)
in-between (n, adj)
industry-wide
infill
inflow·(n)
in-line

inrush
iso-octane
 (no other "iso"
 words)

J
jackknife
jackup (n)
jack up (v)
jobsite
judgment

K
kerosene
know-how
knockout
knowledgeable

L
landfill
landmass
lay barge
laydown (n)
lay-down (adj)
lay down (v)
leakoff
leakproof
left-hand
lengthways
letdown
lightweight
lineup (n)
line up (v)
linkup (n)
link up (v)
lowboy
lockout

M
main line
mainstream
makeup
manageable
man-day
man-hour
man-year
mathematics
measurable
Mediterranean
Midcontinent

Mideast
midyear
mile-wide
millsite
multimillion-dollar
minable
modeling
mountainside
mousehole
movable
mud cake
mud line

N
nationwide
nearby
non-Communist
noticeable

O
observable
oceanfront
offgas
off-line
off-loading
off site
offtake
oil field (n)
oilfield (adj)
oilman
oil well (n)
oil-well (adj)
onshore
on site (n)
on stream (n)
open hole
outfall
overall
overhead
overpressure
overnight
override

P
paperback
passthrough
payout
percent
phaseout
pickup

piggyback
pinchout
pipelay
pipelaying
pipelayer
pipeline
pipe rack
pipe still
powerhouse
predominant
printout
proof-test
pullout (n, adj)
pull out (v)
push button

R
rainwater
rathole
readout
realizable
real time
reconnaissance
reentry
removable
right-of-way
rig-up (n)
rig up (v)
ringwall
riprap
roundoff
round trip
runback
rundown (n)
run down (v)
run-down (adj)
run forward
runoff

S
salable
saltwater (adj)
salt water (n)
salvageable
sandblasting
scale-up
seabed
sealift
seawater
second hand (n)

secondhand (adj)
sendout
set point
setup (n)
severalfold
shipshape
shipside
shipyard
shoreline
shortcut
shortsighted
shortwave
shot hole
shot point
shut down (v)
shutdown (n, adj)
shut in (v)
shutin (n, adj)
shutoff (adj)
side boom
side cut
side draw
side stream
sidetrack
sidewall
sizable
slackline
slip joint
slipstream
slow-up
soleplate
spanwise
spectrum - s
spectra - pl
spot-check
standby
standoff
standpipe
standpoint
standpost
standstill
start-up (n, adj)
start up (v)
steamflood
steam line
step-out (n)
step-up (n, adj)
stepwise
straightforward
straight-run

stratum - s
strata - pl
stiffleg
stopgap
subpoena
superheat
switchgear

T
tailor-made
takeoff
takeover
take-up
tank car
tank truck
teardown
tie-in
tie-down
titleholder
toolpusher
toss-up
towline
trademark
trade name
trade off
trade out
transatlantic
traveling
trouble-free
troubleshoot
trunk line
turn down (v)
turndown (n, adj)
turnkey

U
ultraviolet
underwater
underway
updip
upflow
upstream
up-to-date
usable

W
warm-up (adj)
washout
waste water
water-cooled

water cut
waterflood
waterhead
waterline
waterside
watertight
waveform
wave front
wavelength
weathertight
wellhead
wellbore
wellsite
wellstream
wet out
whipstock
windblow
wind-chill (adj)
wireline
workboat
workbox
work load
workover (n, adj)
work over (v)
worldwide
wraparound

X
X-ray

Y
year-end

STANDARD OIL ABBREVIATOR

API STANDARD OIL-MAPPING SYMBOLS

Location . . . .

Abandoned location . . . . erase symbol

Dry hole . . . .

Oil well . . . .

Abandoned oil well . . . .

Gas well . . . .

Abandoned gas well . . . .

Distillate well . . . .

Abandoned distillate well . . . .

Dual completion—oil . . . .

Dual completion—gas . . . .

Drilled water-input well . . . .

Converted water-input well . . . .

Drilled gas-input well . . . .

Converted gas-input well . . . .

Bottom-hole location . . . .
   (x indicates bottom of hole. Changes in well
      status should be indicated as in symbols
      above.)

Salt-water disposal well . . . .

Courtesy American Petroleum Institute, Division of Production.

MATHEMATICAL SYMBOLS AND SIGNS

$+$	plus	$\therefore$	therefore
$-$	minus	$\because$	because
$\pm$	plus or minus	$:$	is to; divided by
$\times$	multiplied by	$::$	as; equals
$\cdot$	multiplied by	$\vdots$	geometrical proportion
$\div$	divided by	$\propto$	varies as
$/$	divided by	$\doteq$	approaches a limit
$=$	equal to	∞	infinity
$\neq$	not equal to	$\int$	integral
$\approx$	nearly equal to	d	differential
$\cong$	congruent to	∂	partial differential
$\equiv$	identical with	$\sum$	summation of
$\not\equiv$	not identical with	$!$	factorial product
$\Leftrightarrow$	equivalent to	π	pi (3.1416)
$>$	greater than	e	epsilon (2.7183)
$\not>$	not greater than	$\circ$	degree
$<$	less than	$'$	minute; prime
$\not<$	not less than	$''$	second
$\geqq$	greater than or equal to	$\angle$	angle
$\leqq$	less than or equal to	$\llcorner$	right angle
$\sim$	difference between	$\perp$	perpendicular
$\simeq$	difference between	$\bigcirc$	circle
$-:$	difference between	$\frown$	arc
$\sqrt{\ }$	square root	$\triangle$	triangle
$\sqrt[3]{\ }$	cube root	$\square$	square
$\sqrt[n]{\ }$	n th root	$\rectangle$	rectangle

GREEK ALPHABET

A	α	Alpha	N	ν	Nu
B	β	Beta	Ξ	ξ	Xi
Γ	γ	Gamma	O	o	Omicron
Δ	δ	Delta	Π	π	Pi
E	ϵ	Epsilon	P	ρ	Rho
Z	ζ	Zeta	Σ	σ	Sigma
H	η	Eta	T	τ	Tau
Θ	θ	Theta	Y	υ	Upsilon
I	ι	Iota	Φ	ϕ	Phi
K	κ	Kappa	X	χ	Chi
Λ	λ	Lambda	Ψ	ψ	Psi
M	μ	Mu	Ω	ω	Omega

Universal Conversion Factors

compiled and edited by
STEVEN GEROLDE

ACRE: =

0.0015625	square miles or sections
0.004046875	square kilometers
0.1	square furlongs
0.4046875	square hektometers
10	square chains
40.46875	square dekameters
160	square rods
4,046.875	square meters
4,840	square yards
5,645.41213	square varas (Texas)
43,560	square feet
77,440	square spans
100,000	square links
400,000	square hands
404,687.5	square decimeters
40,468.750	square centimeters
6,272,640	square inches
4,046,875,000	square millimeters
0.4046875	hectares
40.46875	ares
4,046.875	centares (centiares)

ACRE FOOT: =

1,233.48766	kiloliters
1,233.48766	cubic meters
1,633.3333	cubic yards
43,560	cubic feet
7,758.34	barrels
12,334.876	hektoliters
35,000	bushels—U.S. (dry)
33,933.16195	bushels—Imperial (dry)
123,348.766	dekaliters
140,000	pecks
325,850.28	gallons—U.S. (liquid)
280,092.5925	gallons—U.S. (dry)
271,325.745265	gallons—Imperial
1,303,401.12	quarts (liquid)
1,120,370.370	quarts (dry)
1,233,487.66	liters
1,233,487.660	cubic decimeters
2,606,802.24	pints
10,427,208.96	gills
12,334,876.6	deciliters
75,271,680	cubic inches
123,348,766	centiliters
1,233,487,660	milliliters
1,233,487,660	cubic centimeters
1,233,487,660,000	cubic millimeters

ACRE FOOT PER DAY: =

7,758.34	barrels per day
323.264167	barrels per hour
5.387736	barrels per minute
0.0897956	barrels per second
325,850.28	gallons (U.S.) per day
13,577.09400	gallons (U.S.) per hour
226.284900	gallons (U.S.) per minute
3.771415	gallons (U.S.) per second
271,325.745265	gallons (Imperial) per day
11,305.238400	gallons (Imperial) per hour
188.42066	gallons (Imperial) per minute
3.140344	gallons (Imperial) per second
1,233.48766	cubic meters per day
51.395319	cubic meters per hour
0.856589	cubic meters per minute
0.0142765	cubic meters per second
1,633.33333	cubic yards per day
68.0555552	cubic yards per hour
1.134259	cubic yards per minute
0.0189043	cubic yards per second
43,560	cubic feet per day
1,815.0	cubic feet per hour
30.25000	cubic feet per minute
0.504167	cubic feet per second
75,271,680.00	cubic inches per day
3,136,320.00	cubic inches per hour
52,272.00	cubic inches per minute
871.200	cubic inches per second

ATMOSPHERE: =

0.103327	hektometers of water @ 60°F.
1.03327	dekameters of water @ 60°F.
10.3327	meters of water @ 60°F.
33.9007	feet of water @ 60°F.
406.8084	inches of water @ 60°F.
103.327	decimeters of water @ 60°F.
1,033.27	centimeters of water @ 60°F.
10,332.7	millimeters of water @ 60°F.
0.00760	hektometers of mercury @ 32°F.
0.0760	dekameters of mercury @ 32°F.
0.760	meters of mercury @ 32°F.
2.49343	feet of mercury @ 32°F.
29.9212	inches of mercury @ 32°F.
7.6	decimeters of mercury @ 32°F.
76	centimeters of mercury @ 32°F.
760	millimeters of mercury @ 32°F.
113,893.88	tons per square hektometer
1,138.9388	tons per square dekameter

ATMOSPHERE: (cont'd)

11.389388	tons per square meter
1.0581	tons per square foot
0.00734792	tons per square inch
0.11389388	tons per square decimeter
0.0011389388	tons per square centimeter
0.000011389388	tons per square millimeter
103,327,000	kilograms per square hektometer
1,033,270	kilograms per square dekameter
10,332.7	kilograms per square meter
959.931252	kilograms per square foot
6.666189	kilograms per square inch
103.327	kilograms per square decimeter
1.03327	kilograms per square centimeter
0.0103327	kilograms per square millimeter
227,774,851.2	pounds per square hektometer
2,277,748.512	pounds per square dekameter
22,777.48512	pounds per square meter
2,116.080	pounds per square foot
14.696	pounds per square inch
227.7748512	pounds per square decimeter
2.277748512	pounds per square centimeter
0.02277748512	pounds per square millimeter
1,033,270,000	hektograms per square hektometer
10,332,700	hektograms per square dekameter
103,327	hektograms per square meter
9,599,31252	hektograms per square foot
66.66189	hektograms per square inch
1,033.27	hektograms per square decimeter
10.3327	hektograms per square centimeter
0.103327	hektograms per square millimeter
10,332,700,000	dekagrams per square hektometer
103,327,000	dekagrams per square dekameter
1,033,270	dekagrams per square meter
95,993.1252	dekagrams per square foot
666.6189	dekagrams per square inch
10,332.7	dekagrams per square decimeter
103.327	dekagrams per square centimeter
1.03327	dekagrams per square millimeter
3,644,397,619.2	ounces per square hektometer
36,443,976.192	ounces per square dekameter
364,439.76192	ounces per square meter
33,857.28	ounces per square foot
235.136	ounces per square inch
3,644.39762	ounces per square decimeter
36.44398	ounces per square centimeter
0.36444	ounces per square millimeter
103,327,000,000	grams per square hektometer
1,033,270,000	grams per square dekameter
10,332,700	grams per square meter
959,931.252	grams per square foot
6,666,189	grams per square inch

ATMOSPHERE: (cont'd)

103,327 . . . . grams per square decimeter
1,033.27 . . . . grams per square centimeter
10.3327 . . . . grams per square millimeter
1,033,270,000,000 . . . . decigrams per square hektometer
10,332,700,000 . . . . decigrams per square dekameter
103,327,000 . . . . decigrams per square meter
9,599,312.52 . . . . decigrams per square foot
66,661.89 . . . . decigrams per square inch
1,033,270 . . . . decigrams per square decimeter
10,332.7 . . . . decigrams per square centimeter
103.327 . . . . decigrams per square millimeter
10,332,700,000,000 . . . . centigrams per square hektometer
103,327,000,000 . . . . centigrams per square dekameter
1,033,270,000 . . . . centigrams per square meter
95,933,125.2 . . . . centigrams per square foot
666,618.9 . . . . centigrams per square inch
10,332,700 . . . . centigrams per square decimeter
103,327 . . . . centigrams per square centimeter
1,033.270 . . . . centigrams per square millimeter
103,327,000,000,000 . . . . milligrams per square hektometer
1,033,270,000,000 . . . . milligrams per square dekameter
10,332,700,000 . . . . milligrams per square meter
959,931,252 . . . . milligrams per square foot
6,666,189 . . . . milligrams per square inch
103,327,000 . . . . milligrams per square decimeter
1,033,270 . . . . milligrams per square centimeter
10,332.7 . . . . milligrams per square millimeter
101,325,000,000,000 . . . . dynes per square hektometer
1,013,250,000,000 . . . . dynes per square dekameter
10,132,500,000 . . . . dynes per square meter
941,343,587 . . . . dynes per square foot
6,537,096 . . . . dynes per square inch
101,325,000 . . . . dynes per square decimeter
1,013,250 . . . . dynes per square centimeter
10,132.5 . . . . dynes per square millimeter
1.01325 . . . . bars

BARREL: =

0.158987 . . . . kiloliters
0.158987 . . . . cubic meters
0.20794 . . . . cubic yards
1.58987 . . . . hectoliters
4.511274 . . . . bushels—U.S. (dry)
4.373766 . . . . bushels—Imperial (dry)
5.6146 . . . . cubic feet
15.89871 . . . . dekaliters
18.045097 . . . . pecks
42 . . . . gallons—U.S. (liquid)
36.09798 . . . . gallons—U.S. (dry)
34.99089 . . . . gallons—Imperial

BARREL: (cont'd)

168	quarts (liquid)
144.408516	quarts (dry)
158.987146	liters
158.987146	cubic decimeters
336	pints
1,344	gills
1,589.87146	deciliters
9,702.0288	cubic inches
15,898.71459456	centiliters
158,987.1459456	milliliters
158,987.1459456	cubic centimeters
158,987,145.9456	cubic millimeters
0.174993	tons (short) of water @ 62°F.
0.1562438	tons (long) of water @ 62°F.
0.1587512	tons (metric) of water @ 62°F.
158.7512	kilograms of water @ 62°F.
349.986	pounds of water @ 62°F.
15.87512	hektograms of water @ 62°F.
1.587512	dekagrams of water @ 62°F.
5,599.776	ounces of water @ 62°F.
0.1587512	grams of water @ 62°F.
0.01587512	decigrams of water @ 62°F.
0.001587512	centigrams of water @ 62°F.
0.0001587512	milligrams of water @ 62°F.
5.1042	sacks of cement
2,449,902	grains
404.25	pounds of salt walter @ 60°F. of 1.155 specific gravity

BARREL OF CEMENT: =

0.158987	kiloliters
0.158987	cubic meters
0.20794	cubic yards
1.58987	hectoliters
4.511274	bushels—U.S. (dry)
4.373766	bushels—Imperial (dry)
5.6146	cubic feet
15.89871	dekaliters
18.045097	pecks
42	gallons—U.S. (liquid)
36.10213	gallons—U.S. (dry)
34.99089	gallons—Imperial
168	quarts (liquid)
144.408516	quarts (dry)
158.987146	liters
158.987146	cubic decimeters
336	pints
1,344	gills
1,589.87146	deciliters
9,702.0288	cubic inches

BARREL OF CEMENT: (cont'd)

15,898.71459456	centiliters
158,987.1459456	milliliters
158,987.1459456	cubic centimeters
158,987,145.9456	cubic millimeters
0.188	tons (short)
0.16796	tons (long)
0.170551	tons (metric)
170.55097	kilograms
376	pounds
1705.5097	hektograms
17,055.097	dekagrams
6,016	ounces
170,550.97	grams
1,705,509.7	decigrams
17,055,097	centigrams
170,550,970	milligrams

BARREL PER DAY: =

0.041667	barrels per hour
0.00069444	barrels per minute
0.000011574	barrels per second
0.1589871	kiloliters per day
0.0066245	kiloliters per hour
0.00011041	kiloliters per minute
0.000001840	kiloliters per second
0.1589871	cubic meters per day
0.0066245	cubic meters per hour
0.00011041	cubic meters per minute
0.000001840	cubic meters per second
0.20794	cubic yards per day
0.0086642	cubic yards per hour
0.0001444	cubic yards per minute
0.0000024067	cubic yards per second
1.589871	hektoliters per day
0.066245	hektoliters per hour
0.0011041	hektoliters per minute
0.00001840	hektoliters per second
5.6146	cubic feet per day
0.233942	cubic feet per hour
0.00389903	cubic feet per minute
0.0000649838	cubic feet per second
15.89871	dekaliters per day
0.66245	dekaliters per hour
0.011041	dekaliters per minute
0.0001840	dekaliters per second
42	gallons (U.S.) per day
1.71875	gallons (U.S.) per hour
0.029167	gallons (U.S.) per minute
0.0004861	gallons (U.S.) per second
34.99089	gallons (Imperial) per day

BARREL PER DAY: (cont'd)

1.45795	gallons (Imperial) per hour
0.024299	gallons (Imperial) per minute
0.00040499	gallons (Imperial) per second
168	quarts (U.S.) per day
6.875	quarts (U.S.) per hour
0.11668	quarts (U.S.) per minute
0.0019444	quarts (U.S.) per second
158.98714	liters per day
6.6245	liters per hour
0.11041	liters per minute
0.001840	liters per second
158.98714	cubic decimeters per day
6.6245	cubic decimeters per hour
0.11041	cubic decimeters per minute
0.001840	cubic decimeters per second
336	pints per day
13.75	pints per hour
0.23336	pints per minute
0.0038888	pints per second
1,344	gills per day
55	gills per hour
0.933344	gills per minute
0.0155552	gills per second
1,589.87146	deciliters per day
66.245	deciliters per hour
1.1041	deciliters per minute
0.01840	deciliters per second
9,702.0288	cubic inches per day
404.2	cubic inches per hour
6.7375	cubic inches per minute
0.112292	cubic inches per second
15,898.7146	centiliters per day
662.45	centiliters per hour
11.041	centiliters per minute
0.1840	centiliters per second
158,987.145946	milliliters per day
6,624.5	milliliters per hour
110.41	milliliters per minute
1.84	milliliters per second
158,987.145946	cubic centimeters per day
6,624.5	cubic centimeters per hour
110.41	cubic centimeters per minute
1.840	cubic centimeters per second
158,987,145.946	cubic millimeters per day
6,624,500	cubic millimeters per hour
110,410	cubic millimeters per minute
1,840	cubic millimeters per second
24	barrels per day
0.016667	barrels per minute
0.000277778	barrels per second
3.81567	kiloliters per day

BARREL PER DAY: (cont'd)

0.1589871	kiloliters per hour
0.0026498	kiloliters per minute
0.000044163	kiloliters per second
3.81567	cubic meters per day
0.1589871	cubic meters per hour
0.0026498	cubic meters per minute
0.000044163	cubic meters per second
4.99056	cubic yards per day
0.20794	cubic yards per hour
0.0034657	cubic yards per minute
0.000057761	cubic yards per second
38.1567	hektoliters per day
1.589871	hektoliters per hour
0.026498	hektoliters per minute
0.00044163	hektoliters per second
134.7504	cubic feet per day
5.6146	cubic feet per hour
0.093577	cubic feet per minute
0.15596	cubic feet per second
381.567	dekaliters per day
15.89871	dekaliters per hour
0.26498	dekaliters per minute
0.0044163	dekaliters per second
1,008	gallons (U.S.) per day
42	gallons (U.S.) per hour
0.7	gallons (U.S.) per minute
0.11667	gallons (U.S.) per second
839.78136	gallons (Imperial) per day
34.99089	gallons (Imperial) per hour
0.58318	gallons (Imperial) per minute
0.0097197	gallons (Imperial) per second
4,032	quarts (U.S.) per day
168	quarts (U.S.) per hour
2.80	quarts (U.S.) per minute
0.046667	quarts (U.S.) per second
3,815.6904	liters per day
158.98714	liters per hour
2.64979	liters per minute
0.044163	liters per second
3,815.6904	cubic decimeters per day
158.98714	cubic decimeters per hour
2.64979	cubic decimeters per minute
0.044163	cubic decimeters per second
8,064	pints per day
336	pints per hour
5.60	pints per minute
0.93333	pints per second
32,256	gills per day
1,344	gills per hour
22.40	gills per minute
0.37333	gills per second

UNIVERSAL CONVERSION FACTORS

BARREL PER DAY: (cont'd)

38,156.904	deciliters per day
1,589.87146	deciliters per hour
26.49786	deciliters per minute
0.44163	deciliters per second
232,848	cubic inches per day
9,702.0288	cubic inches per hour
161.7014	cubic inches per minute
2.695	cubic inches per second
381,569.04	centiliters per day
15,898.7146	centiliters per hour
264.97858	centiliters per minute
4.4163	centiliters per second
3,815,690.4	milliliters per day
158,987.145946	milliliters per hour
2,649.78576	milliliters per minute
44.163	milliliters per second
3,815,690.4	cubic centimeters per day
158,987.145946	cubic centimeters per hour
2,649.78576	cubic centimeters per minute
44.163	cubic centimeters per second
3,815,690,400.0	cubic millimeters per day
158,987,145,946	cubic millimeters per hour
2,649,785.76	cubic millimeters per minute
44,163.096	cubic millimeters per second

BARREL PER MINUTE: =

1,440	barrels per day
60	barrels per hour
0.016667	barrels per second
228.94272	kiloliters per day
9.53928	kiloliters per hour
0.158987	kiloliters per minute
0.0026498	kiloliters per second
228.94272	cubic meters per day
9.53928	cubic meters per hour
0.158987	cubic meters per minute
0.0026498	cubic meters per second
299.43648	cubic yards per day
12.47652	cubic yards per hour
0.20794	cubic yards per minute
0.0034657	cubic yards per second
2,289.4272	hektoliters per day
95.3928	hektoliters per hour
1.58987	hektoliters per minute
0.026498	hektoliters per second
8,085.05280	cubic feet per day
336.8772	cubic feet per hour
5.6146	cubic feet per minute
0.093577	cubic feet per second
22,894.272	dekaliters per day

BARREL PER MINUTE: (cont'd)

953.928	dekaliters per hour
15.8987	dekaliters per minute
0.26498	dekaliters per second
60,480	gallons (U.S.) per day
2,520	gallons (U.S.) per hour
42	gallons (U.S.) per minute
0.7	gallons (U.S.) per second
50,386.7520	gallons (Imperial) per day
2,099.4480	gallons (Imperial) per hour
34.99089	gallons (Imperial) per minute
0.58318	gallons (Imperial) per second
241,920	quarts per day (U.S.)
10,080	quarts per hour (U.S.)
168	quarts per minute (U.S.)
2.80	quarts per second (U.S.)
228,941.48966	liters per day
9,539.22874	liters per hour
158.987146	liters per minute
2.64979	liters per second
228,941.48966	cubic decimeters per day
9,539.22874	cubic decimeters per hour
158.987146	cubic decimeters per minute
2.64979	cubic decimeters per second
483,840	pints per day
20,160	pints per hour
336	pints per minute
5.60	pints per second
1,935,360	gills per day
80,640	gills per hour
1,344	gills per minute
22.40	gills per second
2,289,414.89664	deciliters per day
95,392.28736	deciliters per hour
1,589.87146	deciliters per minute
26.49786	deciliters per second
13,970,921.472	cubic inches per day
582,121.7280	cubic inches per hour
9,702.0288	cubic inches per minute
161.70048	cubic inches per second
22,894,148.9664	centiliters per day
953,922.8736	centiliters per hour
15,898.71459	centiliters per minute
264.97858	centiliters per second
228,941,489.664	milliliters per day
9,539,228.736	milliliters per hour
158,987.14595	milliliters per minute
2,649.78576	milliliters per second
228,941,489.664	cubic centimeters per day
9,539,228.736	cubic centimeters per hour
158,987.14595	cubic centimeters per minute
2,649.78576	cubic centimeters per second
228,941,489,664	cubic millimeters per day

BARREL PER MINUTE: (cont'd)

9,539,228,736	cubic millimeters per hour
158,987,145.946	cubic millimeters per minute
2,649,785.76	cubic millimeters per second

BARREL PER SECOND: =

86,400	barrels per day
3,600	barrels per hour
60	barrels per minute
13,736.47680	kiloliters per day
572.35320	kiloliters per hour
9.53922	kiloliters per minute
0.158987	kiloliters per second
13,736.47680	cubic meters per day
572.35320	cubic meters per hour
9.53922	cubic meters per minute
0.158987	cubic meters per second
17,966.0160	cubic yards per day
748.5840	cubic yards per hour
12.47640	cubic yards per minute
0.20794	cubic yards per second
137,364.7680	hektoliters per day
5,723.5320	hektoliters per hour
95.3922	hektoliters per minute
1.58987	hektoliters per second
485,101.44	cubic feet per day
20,212.560	cubic feet per hour
336.8760	cubic feet per minute
5.6146	cubic feet per second
1,373,648.5440	dekaliters per day
57,235,356	dekaliters per hour
953.92260	dekaliters per minute
15.89871	dekaliters per second
3,628,800	gallons (U.S.) per day
151,200	gallons (U.S.) per hour
2,520	gallons (U.S.) per minute
42	gallons (U.S.) per second
3,023,212.8960	gallons (Imperial) per day
125,967.2040	gallons (Imperial) per hour
2,099.45340	gallons (Imperial) per minute
34.99089	gallons (Imperial) per second
14,515,200	quarts (U.S.) per day
604,800	quarts (U.S.) per hour
10,080	quarts (U.S.) per minute
168	quarts (U.S.) per second
13,736,489.4144	liters per day
572,353.7256	liters per hour
9,539.22876	liters per minute
158.987146	liters per second
13,736,489.4144	cubic decimeters per day
572,353.7256	cubic decimeters per hour

BARREL PER SECOND: (cont'd)

9,539.22876	cubic decimeters per minute
158.987146	cubic decimeters per second
29,030,400	pints per day
1,209,600	pints per hour
20,160	pints per minute
336	pints per second
116,121,600	gills per day
4,838,400	gills per hour
80,640	gills per minute
1,344	gills per second
137,364,894.144	deciliters per day
5,723,537.256	deciliters per hour
95,392.28754	deciliters per minute
1,589.87146	deciliters per second
838,255,288.320	cubic inches per day
34,927,303.6800	cubic inches per hour
582,131.7280	cubic inches per minute
9,702.0288	cubic inches per second
1,373,648,941.440	centiliters per day
57,235,372.56	centiliters per hour
953,922.8754	centiliters per minute
15,898.71459	centiliters per second
13,736,489,414.40	milliliters per day
572,353,725.6	milliliters per hour
9,539,228.760	milliliters per minute
158,987.14594	milliliters per second
13,736,489,414.4	cubic centimeters per day
572,353,725.6	cubic centimeters per hour
9,539,228.760	cubic centimeters per minute
158,987.14595	cubic centimeters per second
13,736,489,414,400	cubic millimeters per day
572,353,725,600	cubic millimeters per hour
9,539,228,754	cubic millimeters per minute
158,987,145.9456	cubic millimeters per second

BTU (60°F.): =

25,030	foot poundals
300,360	inch poundals
777.97265	foot pounds
9,335.67120	inch pounds
0.00027776	ton (short) calories
0.25198	kilogram calories
0.55552	pound calories
2.5198	hektogram calories
25.198	dekagram calories
8.88832	ounce calories
251.98	gram calories
2,519.8	decigram calories
25,198	centigram calories
251.980	milligram calories

BTU (60°F.): (cont'd)

0.000012201	kilowatt days
0.00029283	kilowatt hours
0.01757	kilowatt minutes
1.0546	kilowatt seconds
0.012201	watt days
0.29283	watt hours
17.57	watt minutes
1,054.6	watt seconds
0.11856	ton meters
107.56	kilogram meters
237.12678	pound meters
1,075.6	hektogram meters
10,756	dekagram meters
3,794.02848	ounce meters
107,560	gram meters
1,075,600	decigram meters
10,756,000	centigram meters
107,560,000	milligram meters
0.0011856	ton hektometers
1.0756	kilogram hektometers
2.37127	pound hektometers
10.756	hektogram hektometers
107.56	dekagram hektometers
37.94028	ounce hektometers
1,075.6	gram hektometers
10,756	decigram hektometers
107,560	centigram hektometers
1,075,600	milligram hektometers
0.011856	ton dekameters
10.756	kilogram dekameters
23.7127	pound dekameters
107.56	hektogram dekameters
1,075.6	dekagram dekameters
379.40285	ounce dekameters
10,756	gram dekameters
107,560	decigram dekameters
1,075,600	centigram dekameters
10,756,000	milligram dekameters
0.388977	ton feet
352.887473	kilogram feet
777.97265	pound feet
3,528.874731	hektogram feet
35,288.747308	dekagram feet
12,447.611780	ounce feet
352,887.473080	gram feet
3,528.875	decigram feet
35,288,747	centigram feet
352,887,473	milligram feet
4.667724	ton inches
4,234.649677	kilogram inches
9,335.671800	pound inches

BTU (60°F.): (cont'd)

42,346.496772	hektogram inches
423,464.96772	dekagram inches
149,371.34136	ounce inches
4,234,650	gram inches
42,346,497	decigram inches
423,464,968	centigram inches
4,234,649,677	milligram inches
1.1856	ton decimeters
1.0756	kilogram decimeters
2,371.2678	pound decimeters
10,756	hektogram decimeters
107,560	dekagram decimeters
37,940.2848	ounce decimeters
1,075,600	gram decimeters
10,756,000	decigram decimeters
107,560,000	centigram decimeters
1,075,600,000	milligram decimeters
11.856	ton centimeters
10,756	kilogram centimeters
23,712.678	pound centimeters
107,560	hektogram centimeters
1,075,600	dekagram centimeters
379,402.848	ounce centimeters
10,756,000	gram centimeters
107,560,000	decigram centimeters
1,075,600,000	centigram centimeters
10,756,000,000	milligram centimeters
118.56	ton millimeters
107,560	kilogram millimeters
237,126.780	pound millimeters
1,075,600	hektogram millimeters
10,756,000	dekagram millimeters
3,794,028.48	ounce millimeters
107,560,000	gram millimeters
1,075,600,000	decigram millimeters
10,756,000,000	centigram millimeters
107,560,000,000	milligram millimeters
0.0104028	kiloliter-atmospheres
0.104028	hektoliter-atmospheres
1.040277	dekaliter-atmospheres
10.40277	liter-atmospheres
104.0277	deciliter-atmospheres
1,040.277	centiliter-atmospheres
10,402.77	milliliter-atmospheres
0.0000000000104104	cubic kilometer-atmospheres
0.0000000104104	cubic hektometer-atmospheres
0.000010410	cubic dekameter-atmospheres
0.0104104	cubic meter-atmospheres
0.3676637	cubic feet-atmospheres
635.277597	cubic inch-atmospheres
10.410432	cubic decimeter-atmospheres

BTU (60°F.): (cont'd)

10,410.4320	cubic centimeter-atmospheres
10,410,432	cubic millimeter-atmospheres
1,054.198	joules
0.0003982	Cheval-vapeur hours
0.000016372	horsepower days
0.00039292	horsepower hours
0.0235757	horsepower minutes
1.41451	horsepower seconds
0.0000685	pounds of carbon oxidized with perfect efficiency
0.001030	pounds of water evaporated from and at 212°F.

BTU (60°F.) PER DAY: =

25,030	foot poundals per day
1,042.92	foot poundals per hour
17.3820	foot poundals per minute
0.2897	foot poundals per second
777.97265	foot pounds per day
32.41553	foot pounds per hour
0.54026	foot pounds per minute
0.0090043	foot pounds per second
0.25198	kilogram calories per day
0.010499	kilogram calories per hour
0.00017498	kilogram calories per minute
0.0000029164	kilogram calories per second
8.88832	ounce calories per day
0.37035	ounce calories per hour
0.0061724	ounce calories per minute
0.00010287	ounce calories per second
107.56	kilogram meters per day
4.48164	kilogram meters per hour
0.074694	kilogram meters per minute
0.0012449	kilogram meters per second
10.40277	liter-atmospheres per day
0.43344	liter-atmospheres per hour
0.007224	liter-atmospheres per minute
0.00012040	liter-atmospheres per second
0.3676637	cubic foot-atmospheres per day
0.0153166	cubic foot-atmospheres per hour
0.000255276	cubic foot-atmospheres per minute
0.0000042546	cubic foot-atmospheres per second
0.000016148	Cheval-vapeurs
0.000016372	horsepower
0.000012201	kilowatts
0.012201	watts
1,054.198	joules per day
43.9236	joules per hour
0.73206	joules per minute
0.012201	joules per second

BTU (60°F.) PER DAY: (cont'd)

0.0000685	pounds of carbon oxidized with perfect efficiency per day
0.00000285415	pounds of carbon oxidized with perfect efficiency per hour
0.000000047569	pounds of carbon oxidized with perfect efficiency per minute
0.00000000079282	pounds of carbon oxidized with perfect efficiency per second
0.001030	pounds of water evaporated from and at 212°F. per day
0.000042916	pounds of water evaporated from and at 212°F. per hour
0.00000071526	pounds of water evaporated from and at 212°F. per minute
0.000000011921	pounds of water evaporated from and at 212°F. per second
1.0	BTU per day
0.0416667	BTU per hour
0.00069444	BTU per minute
0.000011574	BTU per second

BTU PER HOUR: =

600,720.1920	foot poundals per day
25,030	foot poundals per hour
417.16680	foot poundals per minute
6.95278	foot poundals per second
18,671.34359	foot pounds per day
777.97265	foot pounds per hour
12.96621	foot pounds per minute
0.21610	foot pounds per second
6.04748	kilogram calories per day
0.25198	kilogram calories per hour
0.0041996	kilogram calories per minute
0.000069994	kilogram calories per second
213.31968	ounce calories per day
8.88832	ounce calories per hour
0.14814	ounce calories per minute
0.0024690	ounce calories per second
2,581.4592	kilogram meters per day
107.56	kilogram meters per hour
1.79268	kilogram meters per minute
0.029878	kilogram meters per second
249.67008	liter-atmospheres per day
10.40277	liter-atmospheres per hour
0.173382	liter-atmospheres per minute
0.0028897	liter-atmospheres per second
8.822304	cubic foot-atmospheres per day
0.3676637	cubic foot-atmospheres per hour
0.00612660	cubic foot-atmospheres per minute
0.00010211	cubic foot-atmospheres per second

BTU PER HOUR: (cont'd)

0.00038754	Cheval-vapeurs
0.00039292	horsepower
0.00029283	kilowatts
0.29283	watts
25,300.5120	joules per day
1,054.198	joules per hour
17.569967	joules per minute
0.292833	joules per second
0.00164402	pounds of carbon oxidized with perfect efficiency per day
0.0000685	pounds of carbon oxidized with perfect efficiency per hour
0.0000011417	pounds of carbon oxidized with perfect efficiency per minute
0.000000019028	pounds of carbon oxidized with perfect efficiency per second
0.02472	pounds of water evaporated from and at 212°F. per day
0.001030	pounds of water evaporated from and at 212°F. per hour
0.000017167	pounds of water evaporated from and at 212°F. per minute
0.00000028611	pounds of water evaporated from and at 212°F. per second
24	BTU per day
1.0	BTU per hour
0.016667	BTU per minute
0.00027778	BTU per second

BTU PER MINUTE: =

36,043,200	foot poundals per day
1,501,800	foot poundals per hour
25,030	foot poundals per minute
417.16667	foot poundals per second
1,120,281	foot pounds per day
46,678.35899	foot pounds per hour
777.97265	foot pounds per minute
12.96621	foot pounds per second
362.854	kilogram calories per day
15.11892	kilogram calories per hour
0.25198	kilogram calories per minute
0.0041997	kilogram calories per second
12,799.180803	ounce calories per day
533.299200	ounce calories per hour
8.88832	ounce calories per minute
0.148139	ounce calories per second
154,886.688000	kilogram meters per day
6,453.6120	kilogram meters per hour
107.56	kilogram meters per minute
1.79267	kilogram meters per second

685

BTU PER MINUTE: (cont'd)

14,980.0320	liter-atmospheres per day
624.168	liter-atmospheres per hour
10.40277	liter-atmospheres per minute
0.17338	liter-atmospheres per second
529.435728	cubic foot-atmospheres per day
22.059822	cubic foot-atmospheres per hour
0.3676637	cubic foot-atmospheres per minute
0.00612773	cubic foot-atmospheres per second
0.023252	Cheval-vapeur hours
0.023575	horsepower
0.01757	kilowatts
17.57	watts
1,518,048	joules per day
63,252	joules per hour
1,054.198	joules per minute
17.569967	joules per second
0.098643	pounds of carbon oxidized with perfect efficiency per day
0.0041101	pounds of carbon oxidized with perfect efficiency per hour
0.0000685	pounds of carbon oxidized with perfect efficiency per minute
0.0000011417	pounds of carbon oxidized with perfect efficiency per second
1.48320	pounds of water evaporated from and at 212°F. per day
0.06180	pounds of water evaporated from and at 212°F. per hour
0.001030	pounds of water evaporated from and at 212°F. per minute
0.000017167	pounds of water evaporated from and at 212°F. per second
1,440	BTU per day
60	BTU per hour
1.0	BTU per minute
0.016667	BTU per second

BTU PER SECOND: =

2,162,592,000	foot poundals per day
90,108,000	foot poundals per hour
1,501,800	foot poundals per minute
25,030	foot poundals per second
67,216,836.960	foot pounds per day
2,800,701.540	foot pounds per hour
46,678.3590	foot pounds per minute
777.97265	foot pounds per second
21,771.0720	kilogram calories per day
907.1280	kilogram calories per hour
15.1188	kilogram calories per minute
0.25198	kilogram calories per second

BTU PER SECOND: (cont'd)

767,950.8480	ounce calories per day
31,997.9520	ounce calories per hour
533.29920	ounce calories per minute
8.88832	ounce calories per second
9,293,184.0	kilogram meters per day
387,216.0	kilogram meters per hour
6,453.60	kilogram meters per minute
107.56	kilogram meters per second
898,799.3280	liter-atmospheres per day
37,449.9720	liter-atmospheres per hour
624.16620	liter-atmospheres per minute
10.40277	liter-atmospheres per second
31,766.143680	cubic foot-atmospheres per day
1,323.589320	cubic foot-atmospheres per hour
22.0598220	cubic foot-atmospheres per minute
0.3676637	cubic foot-atmospheres per second
1.39519	Cheval-vapeurs
1.41454	horsepower
1.055	kilowatts
1,055	watts
91,082,702.2	joules per day
3,795,112.8	joules per hour
63,251.880	joules per minute
1,054.198	joules per second
5.9184	pounds of carbon oxidized with perfect efficiency per day
0.2466	pounds of carbon oxidized with perfect efficiency per hour
0.00411	pounds of carbon oxidized with perfect efficiency per minute
0.0000685	pounds of carbon oxidized with perfect efficiency per second
88.992000	pounds of water evaporated from and at 212°F. per day
3.70800	pounds of water evaporated from and at 212°F. per hour
0.061800	pounds of water evaporated from and at 212°F. per minute
0.001030	pounds of water evaporated from and at 212°F. per second
86,400	BTU per day
3,600	BTU per hour
60	BTU per minute
1	BTU per second

BTU PER SQUARE FOOT PER DAY: =

23,505.12	kilowatts per square hektometer
235.0512	kilowatts per square dekameter
2.350512	kilowatts per square meter
25.30051	kilowatts per square foot

0.17569 . . . . kilowatts per square inch
0.023505 . . . . kilowatts per square decimeter
0.00023505 . . . . kilowatts per square centimeter
0.0000023505 . . . . kilowatts per square millimeter
23,505,120 . . . . watts per square hektometer
235,051.2 . . . . watts per square dekameter
2,350.512 . . . . watts per square meter
25,300.512 . . . . watts per square foot
175.6944 . . . . watts per square inch
23.50512 . . . . watts per square decimeter
0.23505 . . . . watts per square centimeter
0.0023502 . . . . watts per square millimeter
31,537.728 . . . . horsepower per square hektometer
315.37728 . . . . horsepower per square dekameter
3.15377 . . . . horsepower per square meter
33.94829 . . . . horsepower per square foot
0.23576 . . . . horsepower per square inch
0.031538 . . . . horsepower per square decimeter
0.00031538 . . . . horsepower per square centimeter
0.0000031538 . . . . horsepower per square millimeter

BTU PER SQUARE FOOT PER HOUR: =

979.380 . . . . kilowatts per square hektometer
9.7938 . . . . kilowatts per square dekameter
0.097938 . . . . kilowatts per square meter
1.05419 . . . . kilowatts per square foot
0.0073206 . . . . kilowatts per square inch
0.000979380 . . . . kilowatts per square decimeter
0.00000979380 . . . . kilowatts per square centimeter
0.0000000979380 . . . . kilowatts per square millimeter
97,938 . . . . watts per square hektometer
9,793.800 . . . . watts per square dekameter
97.93800 . . . . watts per square meter
1,054.188 . . . . watts per square foot
7.3206 . . . . watts per square inch
0.979380 . . . . watts per square decimeter
0.0097938 . . . . watts per square centimeter
0.000097938 . . . . watts per square millimeter
1,314.0720 . . . . horsepower per square hektometer
13.14072 . . . . horsepower per square dekameter
0.13141 . . . . horsepower per square meter
1.41451 . . . . horsepower per square foot
0.00982332 . . . . horsepower per square inch
0.00131407 . . . . horsepower per square decimeter
0.000013141 . . . . horsepower per square centimeter
0.00000013141 . . . . horsepower per square millimeter

BTU PER SQUARE FOOT PER MINUTE: =

16.323 . . . . kilowatts per square hektometer
0.16323 . . . . kilowatts per square dekameter

BTU PER SQUARE FOOT PER MINUTE: (cont'd)

```
0.0016323 ............................. kilowatts per square meter
0.01757 ............................. kilowatts per square foot
0.00012201 ............................. kilowatts per square inch
0.000016323 ............................. kilowatts per square decimeter
0.00000016323 ............................. kilowatts per square centimeter
0.0000000016323 ............................. kilowatts per square millimeter
16,323 ............................. watts per square hektometer
163.23 ............................. watts per square dekameter
1.6323 ............................. watts per square meter
17.57 ............................. watts per square foot
0.12201 ............................. watts per square inch
0.016323 ............................. watts per square decimeter
0.00016323 ............................. watts per square centimeter
0.0000016323 ............................. watts per square millimeter
21.90118 ............................. horsepower per square hektometer
0.21901 ............................. horsepower per square dekameter
0.0021901 ............................. horsepower per square meter
0.023575 ............................. horsepower per square foot
0.00016372 ............................. horsepower per square inch
0.000021901 ............................. horsepower per square decimeter
0.00000021901 ............................. horsepower per square centimeter
0.0000000021901 ............................. horsepower per square millimeter
```

BTU PER SQUARE FOOT PER SECOND: =

```
0.27205 ............................. kilowatts per square hektometer
0.0027205 ............................. kilowatts per square dekameter
0.000027205 ............................. kilowatts per square meter
0.00029283 ............................. kilowatts per square foot
0.0000020335 ............................. kilowatts per square inch
0.00000027205 ............................. kilowatts per square decimeter
0.0000000027205 ............................. kilowatts per square centimeter
0.000000000027205 ............................. kilowatts per square millimeter
272.05 ............................. watts per square hektometer
2.7205 ............................. watts per square dekameter
0.027205 ............................. watts per square meter
0.29283 ............................. watts per square foot
0.0020335 ............................. watts per square inch
0.00027205 ............................. watts per square decimeter
0.0000027205 ............................. watts per square centimeter
0.000000027205 ............................. watts per square millimeter
0.36502 ............................. horsepower per square hektometer
0.0036502 ............................. horsepower per square dekameter
0.000036502 ............................. horsepower per square meter
0.00039292 ............................. horsepower per square foot
0.0000027287 ............................. horsepower per square inch
0.00000036502 ............................. horsepower per square decimeter
0.0000000036502 ............................. horsepower per square centimeter
0.000000000036502 ............................. horsepower per square millimeter
```

BUSHEL—U.S. (DRY): =

0.035238	kiloliters
0.035238	cubic meters
0.04609	cubic yards
0.304785	barrels—U.S.
0.35238	hectoliters
0.96945	bushels—Imp. (dry)
1.24446	cubic feet
3.5238	dekaliters
4	pecks
9.3088	gallons —U.S. (liquid)
8	gallons—U.S. (dry)
7.81457	gallons—Imp.
37.2353	quarts (liquid)
32	quarts (dry)
35.238	liters
35.238	cubic decimeters
64	pints·(dry)
74.8706	pints (liquid)
299.4824	gills (liquid)
352.38	deciliters
2,150.42	cubic inches
3,523.8	centiliters
35,238	millimeters
35,238	cubic centimeters
35,238,000	cubic millimeters
0.053335	tons (short)
0.047621	tons (long)
0.048385	tons (metric)
48.38492	kilograms
106.67048	pounds
483.84924	hektograms
4,838.4924	dekagrams
7,741.58787	ounces
48,384.924	grams
483,849.24	decigrams
4,838,492.4	centigrams
48,384,924	milligrams

BUSHEL—IMPERIAL: =

0.036348	kiloliters
0.036348	cubic meters
0.047542	cubic yards
0.31439	barrels
0.36348	hectoliters
1.03151	bushels—U.S.
1.2843	cubic feet
3.63484	dekaliters
4.12604	pecks

BUSHEL—IMPERIAL: (cont'd)

9.60212	gallons—U.S. (liquid)
8.25208	gallons—U.S. (dry)
8	gallons—Imp.
38.40858	quarts (liquid)
33.00832	quarts (dry)
36.34835	liters
36.34835	cubic decimeters
66.01664	pints (dry)
76.81716	pints (liquid)
307.26856	gills (liquid)
363.4835	deciliters
2,219.3	cubic inches
3,634.835	centiliters
36,348.35	milliliters
36,348.35	cubic centimeters
36,348,350	cubic millimeters
0.055016	tons (short)
0.049122	tons (long)
0.049910	tons (metric)
49.90953	kilograms
110.031667	pounds
499.095330	hektograms
4,990.95330	dekagrams
7,985.52530	ounces
49,909.53296	grams
499,095.32955	decigrams
4,990,953.29552	centigrams
49,909,532.95524	milligrams

CENTARE (CENTIARE): =

0.0000003831	square miles or sections
0.0000001111	square kilometers
0.000024710	square furlongs
0.00024710	acres
0.0001111	square hektometers
0.00247104	square chains
0.01111	square dekameters
0.039537	square rods
1	square meters
1.19598	square yards
1.39498	square varas (Texas)
10.7639	square feet
19.13580	square spans
27.7104	square links
96.8750	square hands
100	square decimeters
10,000	square centimeters
1,550	square inches
1,000,000	square millimeters
0.0001	hectares
0.01	ares

CENTIGRAM: =

0.000000011231	tons (short)
0.00000000984206	tons (long)
0.00000001	tons (metric)
0.00001	kilograms
0.0000267923	pounds (Troy)
0.000022406	pounds (Avoir)
0.0001	hektograms
0.001	dekagrams
0.000321507	ounces (Troy)
0.000352739	ounces (Avoir)
0.01	grams
0.1	decigrams
10	milligrams
0.1543236	grains
0.00257206	drachmas (fluid)
0.00257206	drams (Troy)
0.0056438	drams (Avoir)
0.006430149	pennyweight
0.00771618	scruples
0.05	carats (metric)

CENTILITERS: =

0.000001	kiloliters
0.00001	cubic meters
0.00001308	cubic yards
0.000062897	barrels
0.0001	hectoliters
0.00028377	bushels—U.S. (dry)
0.00027510	bushels—Imp. (dry)
0.00035314	cubic feet
0.001	dekaliters
0.0026417	gallons—U.S. (liquid)
0.0022707	gallons—U.S. (dry)
0.0021997	gallons—Imp.
0.010567	quarts (liquid)
0.0090828	quarts (dry)
0.01	liters
0.01	cubic decimeters
0.018161	pints
0.072663	gills
0.1	deciliters
0.61025	cubic inches
10	milliliters
10	cubic centimeters
10,000	cubic millimeters
0.33815	ounces (fluid)
2.70518	drams (fluid)
0.0000062137	miles

CENTILITERS: (cont'd)

0.00001	kilometers
0.000049709	furlongs
0.0001	hektometers
0.00049709	chains
0.001	dekameters
0.0019884	rods
0.01	meters
0.010936	yards
0.011811	varas (Texas)
0.032808	feet
0.043744	spans
0.049709	links
0.098424	hands
0.1	decimeters
0.3937	inches
1.00	centimeters
10	millimeters
393.70	mils
10,000	microns
10,000,000	milli-microns
10,000,000	micro-millimeters
100,000,000	Angstrom units
15,531.6	wave lengths of red line of cadmium

CENTIMETERS PER DAY: =

0.000006214	miles per day
0.00000025892	miles per hour
0.0000000043153	miles per minute
0.000000000071921	miles per second
0.00001	kilometers per day
0.00000041667	kilometers per hour
0.0000000069444	kilometers per minute
0.00000000011574	kilometers per second
0.000049709	furlongs per day
0.0000020712	furlongs per hour
0.000000034520	furlongs per minute
0.00000000057534	furlongs per second
0.0001	hektometers per day
0.0000041667	hektometers per hour
0.000000069444	hektometers per minute
0.0000000011574	hektometers per second
0.00049709	chains per day
0.000020712	chains per hour
0.00000034520	chains per minute
0.0000000057534	chains per second
0.001	dekameters per day
0.000041667	dekameters per hour
0.00000069444	dekameters per minute
0.000000011574	dekameters per second
0.0019884	rods per day

CENTIMETERS PER DAY: (cont'd)

0.000082850	rods per hour
0.0000013808	rods per minute
0.000000023014	rods per second
0.01	meters per day
0.00041667	meters per hour
0.0000069444	meters per minute
0.00000011574	meters per second
0.010936	yards per day
0.00045667	yards per hour
0.0000075944	yards per minute
0.00000012657	yards per second
0.011811	varas (Texas) per day
0.00049212	varas (Texas) per hour
0.000008202	varas (Texas) per minute
0.00000013670	varas (Texas) per second
0.032808	feet per day
0.0013670	feet per hour
0.000022783	feet per minute
0.00000037972	feet per second
0.043744	spans per day
0.0018227	spans per hour
0.000030378	spans per minute
0.00000050630	spans per second
0.049709	links per day
0.0020712	links per hour
0.000034520	links per minute
0.00000057534	links per second
0.098424	hands per day
0.0041010	hands per hour
0.000068350	hands per minute
0.0000011392	hands per second
0.1	decimeters per day
0.0041667	decimeters per hour
0.000069444	decimeters per minute
0.0000011574	decimeters per second
1	centimeters per day
0.041667	centimeters per hour
0.00069444	centimeters per minute
0.000011574	centimeters per second
0.3937	inches per day
0.016404	inches per hour
0.00027340	inches per minute
0.0000045567	inches per second
10	millimeters per day
0.41667	millimeters per hour
0.0069444	millimeters per minute
0.00011574	millimeters per second
393.70	mils per day
16.40417	mils per hour
0.27340	mils per minute
0.0045567	mils per second

UNIVERSAL CONVERSION FACTORS

CENTIMETERS PER DAY: (cont'd)

10,000	microns per day
416.66667	microns per hour
6.9444	microns per minute
0.11574	microns per second

CENTIMETERS PER HOUR: =

0.00014914	miles per day
0.000006214	miles per hour
0.00000010357	miles per minute
0.0000000017261	miles per second
0.00024000	kilometers per day
0.00001	kilometers per hour
0.00000016667	kilometers per minute
0.0000000027778	kilometers per second
0.0011930	furlongs per day
0.000049709	furlongs per hour
0.00000082848	furlongs per minute
0.000000013808	furlongs per second
0.0024000	hektometers per day
0.0001	hektometers per hour
0.0000016667	hektometers per minute
0.000000027778	hektometers per second
0.011930	chains per day
0.00049709	chains per hour
0.0000082848	chains per minute
0.00000013808	chains per second
0.024000	dekameters per day
0.001	dekameters per hour
0.000016667	dekameters per minute
0.00000027778	dekameters per second
0.047722	rods per day
0.0019884	rods per hour
0.000033140	rods per minute
0.00000055233	rods per second
0.24000	meters per day
0.01	meters per hour
0.00016667	meters per minute
0.0000027778	meters per second
0.262464	yards per day
0.010936	yards per hour
0.00018227	yards per minute
0.0000030378	yards per second
0.28346	varas (Texas) per day
0.011811	varas (Texas) per hour
0.00019685	varas (Texas) per minute
0.0000032808	varas (Texas) per second
0.78739	feet per day
0.032808	feet per hour
0.00054680	feet per minute
0.0000091133	feet per second

CENTIMETERS PER HOUR: (cont'd)

1.049856	spans per day
0.043744	spans per hour
0.00072907	spans per minute
0.000012151	spans per second
1.19302	links per day
0.049709	links per hour
0.00082848	links per minute
0.000013808	links per second
2.36218	hands per day
0.098424	hands per hour
0.0016404	hands per minute
0.000027340	hands per second
2.40000	decimeters per day
0.1	decimeters per hour
0.0016667	decimeters per minute
0.000027778	decimeters per second
24.00000	centimeters per day
1	centimeters per hour
0.016667	centimeters per minute
0.00027778	centimeters per second
9.44880	inches per day
0.3937	inches per hour
0.0065617	inches per minute
0.00010936	inches per second
240.00000	millimeters per day
10	millimeters per hour
0.16667	millimeters per minute
0.0027778	millimeters per second
9,448.8	mils per day
393.70	mils per hour
6.56167	mils per minute
0.109361	mils per second
240,000	microns per day
10,000	microns per hour
166.66667	microns per minute
2.77778	microns per second

CENTIMETER PER MINUTE: =

0.0089482	miles per day
0.00037284	miles per hour
0.000006214	miles per minute
0.00000010357	miles per second
0.014400	kilometers per day
0.00060000	kilometers per hour
0.00001	kilometers per minute
0.00000016667	kilometers per second
0.071581	furlongs per day
0.0029825	furlongs per hour
0.000049709	furlongs per minute
0.00000082848	furlongs per second

0.14400 . . . . hektometers per day
0.0060000 . . . . hektometers per hour
0.0001 . . . . hektometers per minute
0.0000016667 . . . . hektometers per second
0.71581 . . . . chains per day
0.029825 . . . . chains per hour
0.00049709 . . . . chains per minute
0.0000082848 . . . . chains per second
1.44000 . . . . dekameters per day
0.060000 . . . . dekameters per hour
0.001 . . . . dekameters per minute
0.000016667 . . . . dekameters per second
2.86330 . . . . rods per day
0.11930 . . . . rods per hour
0.0019884 . . . . rods per minute
0.000033140 . . . . rods per second
14.40000 . . . . meters per day
0.60000 . . . . meters per hour
0.01 . . . . meters per minute
0.00016667 . . . . meters per second
15.74784 . . . . yards per day
0.656160 . . . . yards per hour
0.010936 . . . . yards per minute
0.00018227 . . . . yards per second
17.00784 . . . . varas (Texas) per day
0.70866 . . . . varas (Texas) per hour
0.011811 . . . . varas (Texas) per minute
0.00019685 . . . . varas (Texas) per second
47.24352 . . . . feet per day
1.96848 . . . . feet per hour
0.032808 . . . . feet per minute
0.00054680 . . . . feet per second
62.99136 . . . . spans per day
2.62464 . . . . spans per hour
0.043744 . . . . spans per minute
0.00072907 . . . . spans per second
71.58096 . . . . links per day
2.98254 . . . . links per hour
0.049709 . . . . links per minute
0.00083848 . . . . links per second
141.73056 . . . . hands per day
5.90544 . . . . hands per hour
0.098424 . . . . hands per minute
0.0016404 . . . . hands per second
144.00 . . . . decimeters per day
6.00 . . . . decimeters per hour
0.1 . . . . decimeters per minute
0.0016667 . . . . decimeters per second
1,440.00 . . . . centimeters per day
60.00 . . . . centimeters per hour
1.0 . . . . centimeters per minute

CENTIMETER PER MINUTE: (cont'd)

0.016667	centimeters per second
566.92800	inches per day
23.62200	inches per hour
0.3937	inches per minute
0.0065617	inches per second
14,440.00	millimeters per day
600.00	millimeters per hour
10	millimeters per minute
0.16667	millimeters per second
566,928	mils per day
23,622	mils per hour
393.70	mils per minute
6.56167	mils per second
14,440,000	microns per day
600,000	microns per hour
10,000	microns per minute
166.66667	microns per second

CENTIMETER PER SECOND: =

0.53689	miles per day
0.022370	miles per hour
0.00037284	miles per minute
0.000006214	miles per second
0.8640	kilometers per day
0.0360	kilometers per hour
0.0006	kilometers per minute
0.00001	kilometers per second
4.29486	furlongs per day
0.17895	furlongs per hour
0.0029825	furlongs per minute
0.000049709	furlongs per second
8.640	hektometers per day
0.360	hektometers per hour
0.006	hektometers per minute
0.0001	hektometers per second
42.94858	chains per day
1.78952	chains per hour
0.029825	chains per minute
0.00049709	chains per second
86.400	dekameters per day
3.600	dekameters per hour
0.060	dekameters per minute
0.001	dekameters per second
171.79776	rods per day
7.15824	rods per hour
0.11930	rods per minute
0.0019884	rods per second
864.0	meters per day
36.0	meters per hour
0.60	meters per minute

CENTIMETER PER SECOND: (cont'd)

0.01	meters per second
944.8704	yards per day
39.36960	yards per hour
0.656160	yards per minute
0.010936	yards per second
1,020.4704	varas (Texas) per day
42.51960	varas (Texas) per hour
0.70866	varas (Texas) per minute
0.01181	varas (Texas) per second
2,834.6112	feet per day
118.1088	feet per hour
1.96848	feet per minute
0.032808	feet per second
3,779.4816	spans per day
157.4784	spans per hour
2.62464	spans per minute
0.043744	spans per second
4,294.8576	links per day
178.9524	links per hour
2.98254	links per minute
0.049709	links per second
8,503.8336	hands per day
354.3264	hands per hour
5.90544	hands per minute
0.098424	hands per second
8,640.0	decimeters per day
360.0	decimeters per hour
6.0	decimeters per minute
0.1	decimeters per second
86,400	centimeters per day
3,600.0	centimeters per hour
60.0	centimeters per minute
1	centimeters per second
34,015.68	inches per day
1,417.32	inches per hour
23.6220	inches per minute
0.3937	inches per second
864,000	millimeters per day
36,000	millimeters per hour
600	millimeters per minute
10	millimeters per second
34,015,680	mils per day
1,417,320	mils per hour
23,622	mils per minute
393.70	mils per second
864,000,000	microns per day
36,000,000	microns per hour
600,000	microns per minute
10,000	microns per second

CENTIMETER OF MERCURY (0°C.): =

0.0013595	hektometers of water@ 60°F.
0.013595	dekameters of water@ 60°F.
0.13595	meters of water@ 60°F.
0.44604	feet of water@ 60°F.
0.44604	ounces of water@ 60°F.
5.35248	inches of water@ 60°F.
1.35952	decimeters of water@ 60°F.
13.595299	centimeters of water@ 60°F.
135.95299	millimeters of water@ 60°F.
0.0001	hektometers of mercury@ 32°F.
0.001	dekameters of mercury@ 32°F.
0.01	meters of mercury@ 32°F.
0.032808	feet of mercury@ 32°F.
0.032808	ounces of mercury@ 32°F.
0.3937	inches of mercury@ 32°F.
0.1	decimeters of mercury@ 32°F.
1.0	centimeters of mercury@ 32°F.
10.0	millimeters of mercury@ 32°F.
1,498.62505	tons per square hektometer
14.98625	tons per square dekameter
0.14986	tons per square meter
0.013923	tons per square foot
0.000096685	tons per square inch
0.0014986	tons per square decimeter
0.000014986	tons per square centimeter
0.00000014986	tons per square millimeter
1,359,529.9	kilograms per square hektometer
13,595.299	kilograms per square dekameter
135.95299	kilograms per square meter
12.63034	kilograms per square foot
0.087711	kilograms per square inch
1.35953	kilograms per square decimeter
0.013595	kilograms per square centimeter
0.00013595	kilograms per square millimeter
2,997,249.93506	pounds per square hektometer
29,972.49935	pounds per square dekameter
299.7250	pounds per square meter
27.845	pounds per square foot
0.19337	pounds per square inch
2.99725	pounds per square decimeter
0.029973	pounds per square centimeter
0.00029973	pounds per square millimeter
13,595,299	hektograms per square hektometer
135,952.99	hektograms per square dekameter
1,359.5299	hektograms per square meter
126.3034	hektograms per square foot
0.87711	hektograms per square inch
13.5953	hektograms per square decimeter
0.13595	hektograms per square centimeter

CENTIMETER OF MERCURY (0°C.): (cont'd)

0.0013595	hektograms per square millimeter
135,952,990	dekagrams per square hektometer
1,59,529.9	dekagrams per square dekameter
13,595.299	dekagrams per square meter
1,263.034	dekagrams per square foot
8.7711	dekagrams per square inch
135.95299	dekagrams per square decimeter
1.35953	dekagrams per square centimeter
0.013595	dekagrams per square millimeter
47,955,999	ounces per square hektometer
479,560	ounces per square dekameter
4,795.6	ounces per square meter
445.520	ounces per square foot
3.09392	ounces per square inch
47.9560	ounces per square decimeter
0.479560	ounces per square centimeter
0.00479560	ounces per square millimeter
1,359,529,900	grams per square hektometer
13,595,299	grams per square dekameter
135,952.99	grams per square meter
12,630.34	grams per square foot
87.71111	grams per square inch
1,359.5299	grams per square decimeter
13.595299	grams per square centimeter
0.13595	grams per square millimeter
13,595,299,000	decigrams per square hektometer
135,952,990	decigrams per square dekameter
1,359,529.9	decigrams per square meter
126,303.4	decigrams per square foot
877.11111	decigrams per square inch
13,595.299	decigrams per square decimeter
135.95299	decigrams per square centimeter
1.35953	decigrams per square millimeter
135,952,990,000	centigrams per square hektometer
1,359,529,900	centigrams per square dekameter
13,595,299	centigrams per square meter
1,263,034	centigrams per square foot
8,771.1111	centigrams per square inch
135,952.99	centigrams per square decimeter
1,359.5299	centigrams per square centimeter
13.595299	centigrams per square millimeter
1,359,529,900,000	milligrams per square hektometer
13,595,299,000	milligrams per square dekameter
135,952,990	milligrams per square meter
12,630,340	milligrams per square foot
87,711.1111	milligrams per square inch
1,359,529.9	milligrams per square decimeter
13,595.299	milligrams per square centimeter
135.95299	milligrams per square millimeter
0.013333	bars
0.013158	atmospheres

CENTIMETER OF MERCURY (0°C.): (cont'd)

1,322,220,000,000	dynes per square hektometer
13,332,200,000	dynes per square dekameter
133,322,000	dynes per square meter
12,385,916.01	dynes per square foot
86,013.3056	dynes per square inch
1,333,220	dynes per square decimeter
13,332.20	dynes per square centimeter
133.3220	dynes per square millimeter

CENTIMETER PER SECOND PER SECOND: =

0.000006214	miles per second per second
0.00001	kilometers per second per second
0.000049709	furlongs per second per second
0.0001	hektometers per second per second
0.00049709	chains per second per second
0.001	dekameters per second per second
0.0019884	rods per second per second
0.01	meters per second per second
0.010936	yards per second per second
0.011811	varas (Texas) per second per second
0.032808	feet per second per second
0.043744	spans per second per second
0.049709	links per second per second
0.098424	hands per second per second
0.10	decimeters per second per second
0.3937	inches per second per second
1.00	centimeters per second per second
10.00	millimeters per second per second
393.70	mils per second per second
10,000	microns per second per second

CHAIN (SURVEYOR'S OR GUNTER'S): =

0.0125	miles
0.020117	kilometers
0.1	furlongs
0.20117	hektometers
2.0117	dekameters
4	rods
20.117	meters
22	yards
23.76	varas (Texas)
66	feet
88	spans
100	links
198	hands
201.17	decimeters
2,011.7	centimeters
792	inches

CHAIN (SURVEYOR'S OR GUNTER'S): (cont'd)

20,117	millimeters
792,000	mils
21,117,000	microns
21,117,000,000	millimicrons
21,117,000,000	micromillimeters
31,244,672	wave lengths of red line of cadmium
$21,117 \times 10^9$	Angstrom Units

CIRCULAR MIL: =

0.0000000000000019564	square miles or sections
0.0000000000000050671	square kilometers
0.000000000000012521	square furlongs
0.000000000000050671	square hektometers
0.0000000000012521	square chains
0.0000000000050671	square dekameters
0.000000000020034	square rods
0.00000000050671	square meters
0.00000000060602	square yards
0.00000000070687	square varas (Texas)
0.0000000054542	square feet
0.0000000096964	square spans
0.000000012521	square links
0.000000049186	square hands
0.000000050671	square decimeters
0.0000050671	square centimeters
0.0000007854	square inches
0.00050671	square millimeters
0.00000000000000050671	hectares
0.0000000000050671	ares
0.00000000050671	centares (centiares)
0.7854	square mils
0.000001	circular inches
0.001	inches
0.000645143	circular millimeters

CHEVAL-VAPEUR (METRIC HORSEPOWER): =

62,832,926.34	foot poundals
753,995,116.08	inch poundals
1,952,910	foot pounds
23,434,920	inch pounds
0.69727	ton (short) calories
632.551	kilogram calories
1,394.53604	pound calories
6,325.51	hektogram calories
63,255.1	dekagram calories
22,312.57664	ounce calories
632,551	gram calories
6,325,510	decigram calories

CHEVAL-VAPEUR (METRIC HORSEPOWER): (cont'd)

63,255,100	centigram calories
632,551,000	milligram calories
17.6448	kilowatt days
0.7352	kilowatt hours
0.012533	kilowatt minutes
0.00020422	kilowatt seconds
30.633333	watt days
735.2	watt hours
44,120	watt minutes
2,646,720	watt seconds
297.62401	ton meters
270,000	kilogram meters
595,248.02	pound meters
2,700,000	hektogram meters
27,000,000	dekagram meters
9,523,968.32	ounce meters
270,000,000	gram meters
2,700,000,000	decigram meters
27,000,000,000	centigram meters
270,000,000,000	milligram meters
2.97624	ton hektometers
2,700	kilogram hektometers
5,952.4802	pound hektometers
27,000	hektogram hektometers
270,000	dekagram hektometers
95,239.6832	ounce hektometers
2,700,000	gram hektometers
27,000,000	decigram hektometers
270,000,000	centigram hektometers
2,700,000,000	milligram hektometers
29.7624	ton dekameters
27,000	kilogram dekameters
59,524.802	pound dekameters
270,000	hektogram dekameters
2,700,000	dekagram dekameters
952,396.832	ounce dekameters
27,000,000	gram dekameters
270,000,000	decigram dekameters
2,700,000,000	centigram dekameters
27,000,000,000	milligram dekameters
9.07158	ton feet
82,296	kilogram feet
181,431.5965	pound feet
822,960	hektogram feet
8,229,600	dekagram feet
2,902,905.54394	ounce feet
82,296,000	gram feet
822,960,000	decigram feet
8,229,600,000	centigram feet
82,296,000,000	milligram feet
0.75596	ton inches

CHEVAL-VAPEUR (METRIC HORSEPOWER): (cont'd)

6,858	kilogram inches
15,119.29971	pound inches
68,580	hektogram inches
685,800	dekagram inches
241,908.79536	ounce inches
6,858,000	gram inches
68,580,000	decigram inches
685,800,000	centigram inches
6,858,000,000	milligram inches
2,976.24	ton decimeters
2,700,000	kilogram decimeters
5,952,480.2	pound decimeters
27,000,000	hektogram decimeters
270,000,000	dekagram decimeters
95,239,683.2	ounce decimeters
2,700,000,000	gram decimeters
27,000,000,000	decigram decimeters
270,000,000,000	centigram decimeters
2,700,000,000,000	milligram decimeters
29,762.4	ton centimeters
27,000,000	kilogram centimeters
59,524,802	pound centimeters
270,000,000	hektogram centimeters
2,700,000,000	dekagram centimeters
952,396,832	ounce centimeters
27,000,000,000	gram centimeters
270,000,000,000	decigram centimeters
2,700,000,000,000	milligram centimeters
297,624	ton millimeters
270,000,000	kilogram millimeters
595,248,020	pound millimeters
2,700,000,000	hektogram millimeters
27,000,000,000	dekagram millimeters
9,523,968,320	ounce millimeters
270,000,000,000	gram millimeters
2,700,000,000,000	decigram millimeters
27,000,000,000,000	milligram millimeters
26.1298	kiloliter atmospheres
261.298	hektoliter atmospheres
2,612.98	dekaliter atmospheres
26,129.8	liter-atmospheres
261,298	deciliter-atmospheres
2,612,980	centiliter-atmospheres
26,129,800	milliliter-atmospheres
0.00000002613	cubic kilometer-atmospheres
0.00002613	cubic hektometer-atmospheres
0.02613	cubic dekameter-atmospheres
26.1298	cubic meter-atmospheres
922.74776	cubic feet-atmospheres
26,129.8	cubic decimeters-atmospheres
26,129,800	cubic centimeter-atmospheres

CHEVAL-VAPEUR (METRIC HORSEPOWER): (cont'd)

26,129,800,000	cubic millimeter-atmospheres
2,648,700	joules
0.0410967	horsepower days
0.98632	horsepower hours
59.17920	horsepower minutes
3,550.752	horsepower seconds
0.171945	pounds of carbon oxidized with 100% efficiency
2.5877	pounds of water evaporated from and at 212°F.
2,510.152	B.T.U.

CUBIC CENTIMETERS: =

0.00000000000000062137	cubic miles
0.000000000000001	cubic kilometers
0.000000000000122833	cubic furlongs
0.000000000001	cubic hektometers
0.000000000122833	cubic chains
0.000000001	cubic dekameters
0.0000000078613	cubic rods
0.000001	cubic meters
0.000001	kiloliters
0.0000013079	cubic yards
0.0000016479	cubic varas (Texas)
0.0000062897	barrels
0.00001	hectoliters
0.000028378	bushels—U.S. (dry)
0.000027496	bushels—Imperial (dry)
0.000035314	cubic feet
0.000083707	cubic spans
0.0001	dekaliter
0.000113512	pecks
0.000122833	cubic links
0.00026417	gallons—U.S. (liquid)
0.00022705	gallons—U.S. (dry)
0.00021997	gallons—Imperial
0.00095635	cubic hands
0.0010567	quarts (liquid)
0.00090808	quarts (dry)
0.001	liters
0.001	cubic decimeters
0.0021134	pints (liquid)
0.0018162	pints (dry)
0.0084536	gills (liquid)
0.01	deciliters
0.061023	cubic inches
0.1	centiliters
1	milliliters
1	cubic centimeters
1000	cubic millimeters

CUBIC CENTIMETERS: (cont'd)

0.033814	ounces (fluid)
0.27051	drams (fluid)

CUBIC FOOT: =

0.000000000017596	cubic miles
0.000000000028317	cubic kilometers
0.0000000034783	cubic furlongs
0.000000028317	cubic hektometers
0.0000034783	cubic chains
0.000028317	cubic dekameters
0.00022261	cubic rods
0.028317	cubic meters
0.028317	kiloliters
0.037036	cubic yards
0.046656	cubic varas (Texas)
0.17811	barrels
0.28317	hectoliters
0.80358	bushels—U.S. (dry)
0.77860	bushels—Imperial (dry)
2.37033	cubic spans
2.8317	dekaliters
3.2143	pecks
3.48327	cubic links
7.48050	gallons—U.S. (liquid)
6.42937	gallons—U.S. (dry)
6.22889	gallons—Imperial
27.08096	cubic hands
29.92257	quarts (liquid)
25.71410	quarts (dry)
28.317	liters
28.317	cubic decimeters
59.84515	pints (liquid)
51.4934	pints (dry)
239.38060	gills (liquid)
283.17	deciliters
1,727.98829	cubic inches
2,831.7	centiliters
28,317	milliliters
28,317	cubic centimeters
28,317,000	cubic millimeters
957.51104	ounces (fluid)
7,660.03167	drams (fluid)
0.9091	sacks of cement (set)
62.35	pounds of water @ 60°F.
64.3	pounds of salt water
72.0	pounds of salt water @ 60°F. at 1.155 specific gravity
489.542	pounds of steel of 7.851 specific gravity

CUBIC FOOT PER DAY: =

0.17811	barrels per day
0.0074214	barrels per hour
0.00012369	barrels per minute
0.0000020615	barrels per second
0.028317	kiloliters per day
0.0011799	kiloliters per hour
0.000019664	kiloliters per minute
0.00000032774	kiloliters per second
0.028317	cubic meters per day
0.0011799	cubic meters per hour
0.000019664	cubic meters per minute
0.00000032774	cubic meters per second
0.037036	cubic yards per day
0.0015432	cubic yards per hour
0.00002572	cubic yards per minute
0.00000042866	cubic yards per second
0.28317	hektoliters per day
0.011799	hektoliters per hour
0.00019664	hektoliters per minute
0.0000032774	hektoliters per second
1	cubic feet per day
0.041666	cubic feet per hour
0.00069444	cubic feet per minute
0.000011574	cubic feet per second
2.8317	dekaliters per day
0.11799	dekaliters per hour
0.0019664	dekaliters per minute
0.000032774	dekaliters per second
7.48050	gallons per day
0.31169	gallons per hour
0.0051948	gallons per minute
0.000086580	gallons per second
6.22889	gallons (Imperial) per day
0.25954	gallons (Imperial) per hour
0.0043256	gallons (Imperial) per minute
0.000072094	gallons (Imperial) per second
29.92257	quarts per day
1.24679	quarts per hour
0.020780	quarts per minute
0.00034633	quarts per second
28.317	liters per day
1.17988	liters per hour
0.019664	liters per minute
0.00032774	liters per second
28.317	cubic decimeters per day
1.17988	cubic decimeters per hour
0.019664	cubic decimeters per minute
0.00032774	cubic decimeters per second
59.84515	pints per day

CUBIC FOOT PER DAY: (cont'd)

2.49354	pints per hour
0.041559	pints per minute
0.00069265	pints per second
239.38060	gills per day
9.97416	gills per hour
0.166236	gills per minute
0.0027706	gills per second
283.17	deciliters per day
11.79875	deciliters per hour
0.19664	deciliters per minute
0.0032774	deciliters per second
1,727.98829	cubic inches per day
72.0	cubic inches per hour
1.20	cubic inches per minute
0.020	cubic inches per second
2,831.7	centiliters per day
117.98750	centiliters per hour
1.9664	centiliters per minute
0.032774	centiliters per second
28,317	milliliters per day
1,179.8750	milliliters per hour
19.664	milliliters per minute
0.32774	milliliters per second
28,317	cubic centimeters per day
1,179.8750	cubic centimeters per hour
19.664	cubic centimeters per minute
0.32774	cubic centimeters per second
28,317,000	cubic millimeters per day
1,179,875	cubic millimeters per hour
19,664	cubic millimeters per minute
327.74	cubic millimeters per second

CUBIC FOOT PER HOUR: =

4.27464	barrels per day
0.17811	barrels per hour
0.0029685	barrels per minute
0.000049475	barrels per second
0.67961	kiloliters per day
0.028317	kiloliters per hour
0.00047195	kiloliters per minute
0.0000078658	kiloliters per second
0.67961	cubic meters per day
0.028317	cubic meters per hour
0.00047195	cubic meters per minute
0.0000078658	cubic meters per second
0.88888	cubic yards per day
0.037036	cubic yards per hour
0.00061728	cubic yards per minute
0.000010288	cubic yards per second
6.79605	hectoliters per day

CUBIC FOOT PER HOUR: (cont'd)

0.28317	hectoliters per hour
0.0047195	hectoliters per minute
0.000078658	hectoliters per second
24	cubic feet per day
1	cubic feet per hour
0.016667	cubic feet per minute
0.00027778	cubic feet per second
67.96051	dekaliters per day
2.8317	dekaliters per hour
0.047195	dekaliters per minute
0.00078658	dekaliters per second
179.53056	gallons per day
7.48050	gallons per hour
0.12467	gallons per minute
0.0020779	gallons per second
149.48928	gallons (Imperial) per day
6.22889	gallons (Imperial) per hour
0.10381	gallons (Imperial) per minute
0.0017302	gallons (Imperial) per second
718.13952	quarts per day
29.92257	quarts per hour
0.49871	quarts per minute
0.0083118	quarts per second
679.60512	liters per day
28.317	liters per hour
0.47195	liters per minute
0.0078658	liters per second
679.60512	cubic decimeters per day
28.317	cubic decimeters per hour
0.47195	cubic decimeters per minute
0.0078658	cubic decimeters per second
1,436.31360	pints per day
59.84515	pints per hour
0.99744	pints per minute
0.016624	pints per second
5,745.25440	gills per day
239.38060	gills per hour
3.98976	gills per minute
0.066496	gills per second
6,796.05120	deciliters per day
283.17	deciliters per hour
4.71948	deciliters per minute
0.078658	deciliters per second
41,472.0	cubic inches per day
1,727.98829	cubic inches per hour
28.80	cubic inches per minute
0.480	cubic inches per second
67,960.512	centiliters per day
2,831.7	centiliters per hour
47.19480	centiliters per minute
0.78658	centiliters per second

CUBIC FOOT PER HOUR: (cont'd)

679,605.120	milliliters per day
28,317	milliliters per hour
471.94980	milliliters per minute
7.86583	milliliters per second
679,605.120	cubic centimeters per day
28,317	cubic centimeters per hour
471.94980	cubic centimeters per minute
7.86583	cubic centimeters per second
679,605,120	cubic millimeters per day
28,317,000	cubic millimeters per hour
471,949.80	cubic millimeters per minute
7,865.83	cubic millimeters per second

CUBIC FOOT PER MINUTE: =

256.47840	barrels per day
10.68660	barrels per hour
0.17811	barrels per minute
0.0029685	barrels per second
40.77648	kiloliters per day
1.69902	kiloliters per hour
0.028317	kiloliters per minute
0.00047195	kiloliters per second
40.77648	cubic meters per day
1.69902	cubic meters per hour
0.028317	cubic meters per minute
0.00047195	cubic meters per second
53.33213	cubic yards per day
2.22217	cubic yards per hour
0.037036	cubic yards per minute
0.00061727	cubic yards per second
407.76480	hectoliters per day
16.99020	hectoliters per hour
0.28317	hectoliters per minute
0.0047195	hectoliters per second
1,440	cubic feet per day
60	cubic feet per hour
1	cubic feet per minute
0.016667	cubic feet per second
4,077.6480	dekaliters per day
169.9020	dekaliters per hour
2.8317	dekaliters per minute
0.047195	dekaliters per second
10,771.920	gallons per day
448.830	gallons per hour
7.48050	gallons per minute
0.12468	gallons per second
8,969.184	gallons (Imperial) per day
373.7160	gallons (Imperial) per hour
6.22889	gallons (Imperial) per minute
0.10381	gallons (Imperial) per second

CUBIC FOOT PER MINUTE: (cont'd)

43,088.5440 . . . . quarts per day
1,795.3560 . . . . quarts per hour
29.92257 . . . . quarts per minute
0.49871 . . . . quarts per second
40,776.480 . . . . liters per day
1,699.020 . . . . liters per hour
28.3170 . . . . liters per minute
0.47195 . . . . liters per second
40,776.480 . . . . cubic decimeters per day
1,699.020 . . . . cubic decimeters per hour
28.3170 . . . . cubic decimeters per minute
0.47195 . . . . cubic decimeters per second
86,177.0880 . . . . pints per day
3,590.712 . . . . pints per hour
59.84515 . . . . pints per minute
0.99742 . . . . pints per second
344,708.3520 . . . . gills per day
14,362.848 . . . . gills per hour
239.38060 . . . . gills per minute
3.98968 . . . . gills per second
407,764.80 . . . . deciliters per day
16,990.20 . . . . deciliters per hour
283.17 . . . . deciliters per minute
4.7195 . . . . deciliters per second
2,488,303.584 . . . . cubic inches per day
103,769.3160 . . . . cubic inches per hour
1,727.98829 . . . . cubic inches per minute
28.79981 . . . . cubic inches per second
4,077,648.0 . . . . centiliters per day
169,902.0 . . . . centiliters per hour
2,831.7 . . . . centiliters per minute
47.195 . . . . centiliters per second
40,776,480.0 . . . . milliliters per day
1,699,020.0 . . . . milliliters per hour
28,317 . . . . milliliters per minute
471.95 . . . . milliliters per second
40,776,480.0 . . . . cubic centimeters per day
1,699,020.0 . . . . cubic centimeters per hour
28,317 . . . . cubic centimeters per minute
471.95 . . . . cubic centimeters per second
40,776,480,000 . . . . cubic millimeters per day
1,699,020,000 . . . . cubic millimeters per hour
28,317,000 . . . . cubic millimeters per minute
471,950 . . . . cubic millimeters per second

CUBIC FOOT PER SECOND: =

15,388.70400 . . . . barrels per day
641.19600 . . . . barrels per hour
10.68660 . . . . barrels per minute
0.17811 . . . . barrels per second

2,446.58880	kiloliters per day
101.94120	kiloliters per hour
1.69902	kiloliters per minute
0.028317	kiloliters per second
2,446.5880	cubic meters per day
101.94120	cubic meters per hour
1.69902	cubic meters per minute
0.028317	cubic meters per second
3,199.91040	cubic yards per day
133.32960	cubic yards per hour
2.22216	cubic yards per minute
0.037036	cubic yards per second
24,465.8880	hectoliters per day
1,019.41200	hectoliters per hour
16.99020	hectoliters per minute
0.28317	hectoliters per second
86,400	cubic feet per day
3,600	cubic feet per hour
60	cubic feet per minute
1	cubic feet per second
244,658.880	dekaliters per day
10,194.120	dekaliters per hour
169.9020	dekaliters per minute
2.8317	dekaliters per second
646,315.20	gallons per day
26,929.80	gallons per hour
448.830	gallons per minute
7.48050	gallons per second
538,176.096	gallons (Imperial) per day
22,424.004	gallons (Imperial) per hour
373.73340	gallons (Imperial) per minute
6.22889	gallons (Imperial) per second
2,585,310.0480	quarts per day
107,721.2520	quarts per hour
1,795.35420	quarts per minute
29.92257	quarts per second
2,446,588.80	liters per day
101,941.20	liters per hour
1,699.020	liters per minute
28.317	liters per second
2,446,588.80	cubic decimeters per day
101,941.20	cubic decimeters per hour
1,699.020	cubic decimeters per minute
28.317	cubic decimeters per second
5,170,620.960	pints per day
215,442.540	pints per hour
3,590.7090	pints per minute
59.84515	pints per second
20,682,483.84	gills per day
861,770.16	gills per hour
14,362.8360	gills per minute

CUBIC FOOT PER SECOND: (cont'd)

239.38060 . . . . gills per second
24,465,880 . . . . deciliters per day
1,019,412.0 . . . . deciliters per hour
16,990.20 . . . . deciliters per minute
283.17 . . . . deciliters per second
149,298,188.2560 . . . . cubic inches per day
6,220,757.8440 . . . . cubic inches per hour
103,679.29740 . . . . cubic inches per minute
1,727.98829 . . . . cubic inches per second
244,658,880.0 . . . . centiliters per day
10,194,120.0 . . . . centiliters per hour
169,902.0 . . . . centiliters per minute
2,831.7 . . . . centiliters per second
2,446,588,800.0 . . . . milliliters per day
101,941,200.0 . . . . milliliters per hour
1,699,020.0 . . . . milliliters per minute
28,317 . . . . milliliters per second
2,446,588,800.0 . . . . cubic centimeters per day
101,941,200.0 . . . . cubic centimeters per hour
1,699,020.0 . . . . cubic centimeters per minute
28,317 . . . . cubic centimeters per second
2.446,588,800,000 . . . . cubic millimeters per day
101,941,200,000 . . . . cubic millimeters per hour
1,699,020,000 . . . . cubic millimeters per minute
28,317,000 . . . . cubic millimeters per second

CUBIC INCH: =

0.000000000000010183 . . . . cubic miles
0.000000000000016387 . . . . cubic kilometers
0.0000000000020129 . . . . cubic furlongs
0.000000000016387 . . . . cubic hektometers
0.0000000020129 . . . . cubic chains
0.000000016387 . . . . cubic dekameters
0.00000012883 . . . . cubic rods
0.000016387 . . . . cubic meters
0.000016387 . . . . kiloliters
0.000021434 . . . . cubic yards
0.000027000 . . . . cubic varas (Texas)
0.00010307 . . . . barrels
0.00016387 . . . . hectoliters
0.00046503 . . . . bushels—U.S. (dry)
0.00045058 . . . . bushels—Imperial (dry)
0.0005787 . . . . cubic feet
0.0013717 . . . . cubic spans
0.0016387 . . . . dekaliters
0.0018601 . . . . pecks
0.0020129 . . . . cubic links
0.0043290 . . . . gallons—U.S. (liquid)
0.003721 . . . . gallons —U.S. (dry)
0.003607 . . . . gallons (Imperial)

CUBIC INCH: (cont'd)

0.015672	cubic hands
0.017316	quarts (liquid)
0.014881	quarts (dry)
0.016387	liters
0.016387	cubic decimeters
0.034632	pints (liquid)
0.029762	pints (dry)
0.13853	gills (liquid)
0.16387	deciliters
1.63871	centiliters
16.38716	milliliters
16.38716	cubic centimeters
16,387.16	cubic millimeters
4.4329	drams (fluid)
0.2833	pounds of steel (specific gravity—7.851)
0.03607	pounds of water @ 60°F.
0.5541	ounces of fluid
0.041667	pounds of salt water @ 60°F. and 1.155 specific gravity
0.0005261	sacks of cement (set)
0.000016387	kiloliters per day
0.00000068278	kiloliters per hour
0.000000011380	kiloliters per minute
0.00000000018966	kiloliters per second
0.000016387	cubic meters per day
0.00000068278	cubic meters per hour
0.000000011380	cubic meters per minute
0.00000000018966	cubic meters per second
0.000021434	cubic yards per day
0.00000089309	cubic yards per hour
0.000000014885	cubic yards per minute
0.00000000024808	cubic yards per second
0.00010307	barrels per day
0.0000042944	barrels per hour
0.000000071574	barrels per minute
0.0000000011929	barrels per second
0.00016387	hectoliters per day
0.0000068278	hectoliters per hour
0.00000011380	hectoliters per minute
0.0000000018966	hectoliters per second
0.0005787	cubic feet per day
0.000024112	cubic feet per hour
0.00000040187	cubic feet per minute
0.0000000066979	cubic feet per second
0.0016387	dekaliters per day
0.000068278	dekaliters per hour
0.0000011380	dekaliters per minute
0.000000018966	dekaliters per second
0.0043290	gallons per day
0.00018037	gallons per hour

CUBIC INCH: (cont'd)

0.0000030062	gallons per minute
0.000000050104	gallons per second
0.003607	gallons (Imperial) per day
0.00015029	gallons (Imperial) per hour
0.0000025049	gallons (Imperial) per minute
0.000000041748	gallons (Imperial) per second
0.016387	liters per day
0.00068278	liters per hour
0.000011380	liters per minute
0.00000018966	liters per second
0.016387	cubic decimeters per day
0.00068278	cubic decimeters per hour
0.000011380	cubic decimeters per minute
0.00000018966	cubic decimeters per second
0.017316	quarts per day
0.00072151	quarts per hour
0.000012025	quarts per minute
0.00000020042	quarts per second
0.034632	pints per day
0.0014430	pints per hour
0.000024050	pints per minute
0.00000040083	pints per second
0.13853	gills per day
0.0057722	gills per hour
0.000096024	gills per minute
0.0000016034	gills per second
0.16387	deciliters per day
0.0068278	deciliters per hour
0.00011380	deciliters per minute
0.0000018966	deciliters per second
1.0	cubic inches per day
0.041666	cubic inches per hour
0.00069444	cubic inches per minute
0.000011574	cubic inches per second
1.63871	centiliters per day
0.068278	centiliters per hour
0.0011380	centiliters per minute
0.000018966	centiliters per second
16.38716	milliliters per day
0.68278	milliliters per hour
0.011380	milliliters per minute
0.00018966	milliliters per second
16.38716	cubic centimeters per day
0.68278	cubic centimeters per hour
0.011380	cubic centimeters per minute
0.00018966	cubic centimeters per second
16,387.16	cubic millimeters per day
682.7760	cubic millimeters per hour
11.3796	cubic millimeters per minute
0.18966	cubic millimeters per second

CUBIC INCH PER HOUR: =

0.00039328	kiloliters per day
0.000016387	kiloliters per hour
0.00000027311	kiloliters per minute
0.0000000045519	kiloliters per second
0.00039328	cubic meters per day
0.000016387	cubic meters per hour
0.00000027311	cubic meters per minute
0.0000000045519	cubic meters per second
0.00051442	cubic yards per day
0.000021434	cubic yards per hour
0.00000035723	cubic yards per minute
0.0000000059539	cubic yards per second
0.0024737	barrels per day
0.00010307	barrels per hour
0.0000017179	barrels per minute
0.000000028631	barrels per second
0.0039328	hectoliters per day
0.00016387	hectoliters per hour
0.0000027311	hectoliters per minute
0.000000045519	hectoliters per second
0.013889	cubic feet per day
0.0005787	cubic feet per hour
0.000009645	cubic feet per minute
0.00000016075	cubic feet per second
0.039328	dekaliters per day
0.0016387	dekaliters per hour
0.000027311	dekaliters per minute
0.00000045519	dekaliters per second
0.10390	gallons per day
0.0043290	gallons per hour
0.000072150	gallons per minute
0.0000012025	gallons per second
0.086564	gallons (Imperial) per day
0.003607	gallons (Imperial) per hour
0.000060114	gallons (Imperial) per minute
0.0000010019	gallons (Imperial) per second
0.39328	liters per day
0.016387	liters per hour
0.00027311	liters per minute
0.0000045519	liters per second
0.39328	cubic decimeters per day
0.016387	cubic decimeters per hour
0.00027311	cubic decimeters per minute
0.0000045519	cubic decimeters per second
0.41558	quarts per day
0.017316	quarts per hour
0.0002886	quarts per minute
0.000004810	quarts per second
0.83117	pints per day

CUBIC INCH PER HOUR: (cont'd)

0.034632 . . . . pints per hour
0.0005772 . . . . pints per minute
0.000009620 . . . . pints per second
3.32476 . . . . gills per day
0.13853 . . . . gills per hour
0.0023089 . . . . gills per minute
0.000038481 . . . . gills per second
3.93284 . . . . deciliters per day
0.16387 . . . . deciliters per hour
0.0027311 . . . . deciliters per minute
0.000045519 . . . . deciliters per second
24 . . . . cubic inches per day
1.0 . . . . cubic inches per hour
0.016667 . . . . cubic inches per minute
0.00027778 . . . . cubic inches per second
39.32842 . . . . centiliters per day
1.63871 . . . . centiliters per hour
0.027311 . . . . centiliters per minute
0.00045519 . . . . centiliters per second
393.28416 . . . . milliliters per day
16.38716 . . . . milliliters per hour
0.27311 . . . . milliliters per minute
0.0045519 . . . . milliliters per second
393.28416 . . . . cubic centimeters per day
16.38716 . . . . cubic centimeters per hour
0.27311 . . . . cubic centimeters per minute
0.0045519 . . . . cubic centimeters per second
393,284.16 . . . . cubic millimeters per day
16,387.16 . . . . cubic millimeters per hour
273.114 . . . . cubic millimeters per minute
4.5519 . . . . cubic millimeters per second

CUBIC INCH PER MINUTE: =

0.023598 . . . . kiloliters per day
0.00098323 . . . . kiloliters per hour
0.000016387 . . . . kiloliters per minute
0.00000027312 . . . . kiloliters per second
0.023598 . . . . cubic meters per day
0.00098323 . . . . cubic meters per hour
0.000016387 . . . . cubic meters per minute
0.00000027312 . . . . cubic meters per second
0.030865 . . . . cubic yards per day
0.0012860 . . . . cubic yards per hour
0.000021434 . . . . cubic yards per minute
0.00000035723 . . . . cubic yards per second
0.14842 . . . . barrels per day
0.0061841 . . . . barrels per hour
0.00010307 . . . . barrels per minute
0.0000017178 . . . . barrels per second
0.23598 . . . . hectoliters per day

CUBIC INCH PER MINUTE: (cont'd)

0.0098323	hectoliters per hour
0.00016387	hectoliters per minute
0.0000027312	hectoliters per second
0.83333	cubic feet per day
0.034722	cubic feet per hour
0.0005787	cubic feet per minute
0.0000096450	cubic feet per second
2.35976	dekaliters per day
0.098323	dekaliters per hour
0.0016387	dekaliters per minute
0.000027312	dekaliters per second
6.23376	gallons per day
0.25974	gallons per hour
0.0043290	gallons per minute
0.000072150	gallons per second
5.19411	gallons (Imperial) per day
0.21642	gallons (Imperial) per hour
0.003607	gallons (Imperial) per minute
0.000060117	gallons (Imperial) per second
23.59757	liters per day
0.98323	liters per hour
0.016387	liters per minute
0.00027312	liters per second
23.59757	cubic decimeters per day
0.98323	cubic decimeters per hour
0.016387	cubic decimeters per minute
0.00027312	cubic decimeters per second
24.93504	quarts per day
1.038960	quarts per hour
0.017316	quarts per minute
0.0002886	quarts per second
49.87008	pints per day
2.07792	pints per hour
0.034632	pints per minute
0.0005772	pints per second
199.48032	gills per day
8.31168	gills per hour
0.13853	gills per minute
0.0023088	gills per second
235.97568	deciliters per day
9.83232	deciliters per hour
0.16387	deciliters per minute
0.0027312	deciliters per second
1,440	cubic inches per day
60	cubic inches per hour
1	cubic inches per minute
0.016667	cubic inches per second
2,359.7568	centiliters per day
98.3232	centiliters per hour
1.63871	centiliters per minute
0.027312	centiliters per second

CUBIC INCH PER MINUTE: (cont'd)

23,597.568	milliliters per day
983.232	milliliters per hour
16.38716	milliliters per minute
0.27312	milliliters per second
23,597.568	cubic centimeters per day
983.232	cubic centimeters per hour
16.38716	cubic centimeters per minute
0.27312	cubic centimeters per second
23,597,568	cubic millimeters per day
983,232	cubic millimeters per hour
16,387.16	cubic millimeters per minute
273.11933	cubic millimeters per second

CUBIC INCH PER SECOND: =

1.41584	kiloliters per day
0.058993	kiloliters per hour
0.00098322	kiloliters per minute
0.000016387	kiloliters per second
1.41584	cubic meters per day
0.058993	cubic meters per hour
0.00098322	cubic meters per minute
0.000016387	cubic meters per second
1.85190	cubic yards per day
0.077162	cubic yards per hour
0.0012860	cubic yards per minute
0.000021434	cubic yards per second
8.90525	barrels per day
0.37105	barrels per hour
0.0061842	barrels per minute
0.00010307	barrels per second
14.15837	hectoliters per day
0.58993	hectoliters per hour
0.0098322	hectoliters per minute
0.00016387	hectoliters per second
49.99968	cubic feet per day
2.08332	cubic feet per hour
0.034722	cubic feet per minute
0.0005787	cubic feet per second
141.58368	dekaliters per day
5.89932	dekaliters per hour
0.098332	dekaliters per minute
0.0016387	dekaliters per second
374.02560	gallons per day
15.58440	gallons per hour
0.25974	gallons per minute
0.0043290	gallons per second
311.64480	gallons (Imperial) per day
12.98520	gallons (Imperial) per hour
0.21642	gallons (Imperial) per minute
0.003607	gallons (Imperial) per second

CUBIC INCH PER SECOND: (cont'd)

1,415.83680	liters per day
58.99320	liters per hour
0.98322	liters per minute
0.016387	liters per second
1,415.83680	cubic decimeters per day
58.99320	cubic decimeters per hour
0.98322	cubic decimeters per minute
0.016387	cubic decimeters per second
1,496.10240	quarts per day
62.33760	quarts per hour
1.038960	quarts per minute
0.017316	quarts per second
2,992.20480	pints per day
124.6752	pints per hour
2.077920	pints per minute
0.034632	pints per second
11,968.9920	gills per day
498.70800	gills per hour
8.31180	gills per minute
0.13853	gills per second
14,158.36800	deciliters per day
589.93200	deciliters per hour
9.83326	deciliters per minute
0.16387	deciliters per second
86,400	cubic inches per day
3,600	cubic inches per hour
60	cubic inches per minute
1.0	cubic inches per second
141,584.5440	centiliters per day
5,899.3560	centiliters per hour
98.32260	centiliters per minute
1.63871	centiliters per second
1,415,850.6240	milliliters per day
58,993.7760	milliliters per hour
983.22960	milliliters per minute
16.38716	milliliters per second
1,415,850.6240	cubic centimeters per day
58,993.7760	cubic centimeters per hour
983.22960	cubic centimeters per minute
16.38716	cubic centimeters per second
1,415,850,624	cubic millimeters per day
58,993,776	cubic millimeters per hour
983,229.6	cubic millimeters per minute
16,387.16	cubic millimeters per second

CUBIC METER: =

0.00000000062139	cubic miles
0.000000001	cubic kilometers
0.00000012283	cubic furlongs
0.000001	cubic hektometers

CUBIC METER: (cont'd)

0.00012283	cubic chains
0.001	cubic dekameters
0.0078613	cubic rods
1	kiloliters
1.307943	cubic yards
1.64763	cubic varas (Texas)
6.28994	barrels
10	hectoliters
28.37798	bushels (U.S.) dry
27.49582	bushels (Imperial) dry
35.314445	cubic feet
83.70688	cubic spans
100	dekaliters
113.51120	pecks
122.83316	cubic links
264.17762	gallons (U.S.) liquid
227.026407	gallons (U.S.) dry
219.97542	gallons (Imperial)
956.34894	cubic hands
1,000	liters
1,000	cubic decimeters
1,056.71088	quarts (liquid)
908.10299	quarts (dry)
2,113.42176	pints (liquid)
1,816.19834	pints (dry)
8,453.68704	gills (liquid)
10,000	deciliters
61,022.93879	cubic inches
100,000	centiliters
1,000,000	milliliters
1,000,000	cubic centimeters
1,000,000,000	cubic millimeters
2,204.62	pounds of water @ 39°F.
2,201.82790	pounds of water @ 60°F.
2,542.608	pounds of salt water @ 60°F. and 1.155 specific gravity
32.10396	sacks of cement (set)
33,813.54487	ounces (fluid)
270,506.35839	drams (fluid)

CUBIC METERS PER DAY: =

1	kiloliters per day
0.041667	kiloliters per hour
0.00069444	kiloliters per minute
0.000011574	kiloliters per second
1	cubic meters per day
0.041667	cubic meters per hour
0.00069444	cubic meters per minute
0.000011574	cubic meters per second
1.30794	cubic yards per day

CUBIC METERS PER DAY: (cont'd)

0.054497	cubic yards per hour
0.00090828	cubic yards per minute
0.000015138	cubic yards per second
6.289943	barrels per day
0.26208	barrels per hour
0.0043680	barrels per minute
0.000072800	barrels per second
10	hectoliters per day
0.41667	hectoliters per hour
0.0069444	hectoliters per minute
0.00011574	hectoliters per second
35.31444	cubic feet per day
1.47143	cubic feet per hour
0.024524	cubic feet per minute
0.00040873	cubic feet per second
100	dekaliters per day
4.1666	dekaliters per hour
0.069444	dekaliters per minute
0.0011574	dekaliters per second
264.17762	gallons per day
11.0074008	gallons per hour
0.18346	gallons per minute
0.0030576	gallons per second
219.97542	gallons (Imperial) per day
9.16564	gallons (Imperial) per hour
0.15276	gallons (Imperial) per minute
0.0025460	gallons (Imperial) per second
1,000	liters per day
41.66640	liters per hour
0.69444	liters per minute
0.011574	liters per second
1,000	cubic decimeters per day
41.66640	cubic decimeters per hour
0.69444	cubic decimeters per minute
0.011574	cubic decimeters per second
1,056.71088	quarts per day
44.02962	quarts per hour
0.73383	quarts per minute
0.012230	quarts per second
2,113.42176	pints per day
88.059240	pints per hour
1.46765	pints per minute
0.024461	pints per second
8,453.68704	gills per day
352.23696	gills per hour
5.87062	gills per minute
0.097844	gills per second
10,000	deciliters per day
416.66400	deciliters per hour
6.94440	deciliters per minute
0.11574	deciliters per second

CUBIC METERS PER DAY: (cont'd)

61,022.93879	cubic inches per day
2,542.6080	cubic inches per hour
42.3768	cubic inches per minute
0.70628	cubic inches per second
100,000	centiliters per day
4,166.6400	centiliters per hour
69.4440	centiliters per minute
1.15740	centiliters per second
1,000,000	milliliters per day
41,666.40	milliliters per hour
694.440	milliliters per minute
11.57407	milliliters per second
1,000,000	cubic centimeters per day
4,166.64000	cubic centimeters per hour
69.4440	cubic centimeters per minute
1.15740	cubic centimeters per second
1,000,000,000	cubic millimeters per day
4,166,640	cubic millimeters per hour
69,444	cubic millimeters per minute
1,157.4	cubic millimeters per second

CUBIC METERS PER HOUR: =

24	kiloliters per day
1	kiloliters per hour
0.016667	kiloliters per minute
0.00027778	kiloliters per second
24	cubic meters per day
1	cubic meters per hour
0.01667	cubic meters per minute
0.00027778	cubic meters per second
31.39085	cubic yards per day
1.30794	cubic yards per hour
0.021799	cubic yards per minute
0.00036332	cubic yards per second
150.95863	barrels per day
6.289943	barrels per hour
0.10483	barrels per minute
0.0017472	barrels per second
240	hectoliters per day
10	hectoliters per hour
0.16667	hectoliters per minute
0.0027778	hectoliters per second
847.54944	cubic feet per day
35.31444	cubic feet per hour
0.58858	cubic feet per minute
0.0098096	cubic feet per second
2,400	dekaliters per day
100	dekaliters per hour
1.66668	dekaliters per minute
0.027778	dekaliters per second

CUBIC METERS PER HOUR: (cont'd)

6,340.26288	gallons per day
264.17762	gallons per hour
4.40296	gallons per minute
0.073383	gallons per second
5,279.41008	gallons (Imperial) per day
219.97542	gallons (Imperial) per hour
3.66626	gallons (Imperial) per minute
0.06114	gallons (Imperial) per second
24,000	liters per day
1,000	liters per hour
16.66680	liters per minute
0.27778	liters per second
24,000	cubic decimeters per day
1,000	cubic decimeters per hour
16.66680	cubic decimeters per minute
0.27778	cubic decimeters per second
25,360.9920	quarts per day
1,056.71088	quarts per hour
17.61180	quarts per minute
0.29353	quarts per second
50,722.12224	pints per day
2,113.42176	pints per hour
35.22370	pints per minute
0.58706	pints per second
202,888.48896	gills per day
8,453.68704	gills per hour
140.89478	gills per minute
2.34825	gills per second
240,000	deciliters per day
10,000	deciliters per hour
166.6680	deciliters per minute
2.77778	deciliters per second
1,464,550.8480	cubic inches per day
61,022.93879	cubic inches per hour
1,017.04920	cubic inches per minute
16.95082	cubic inches per second
2,400,000	centiliters per day
100,000	centiliters per hour
1,666.680	centiliters per minute
27.77778	centiliters per second
24,000,000	milliliters per day
1,000,000	milliliters per hour
16,666.80	milliliters per minute
277.77778	milliliters per second
24,000,000	cubic centimeters per day
1,000,000	cubic centimeters per hour
16,666.666680	cubic centimeters per minute
277.777778	cubic centimeters per second
24,000,000,000	cubic millimeters per day
1,000,000,000	cubic millimeters per hour

CUBIC METERS PER HOUR: (cont'd)

16,666,800	cubic millimeters per minute
277,777.777778	cubic millimeters per second

CUBIC METERS PER MINUTE: =

1,440	kiloliters per day
60	kiloliters per hour
1	kiloliters per minute
0.016667	kiloliters per second
1,440	cubic meters per day
60	cubic meters per hour
1	cubic meters per minute
0.016667	cubic meters per second
1,883.43360	cubic yards per day
78.47640	cubic yards per hour
1.30794	cubic yards per minute
0.021799	cubic yards per second
9,057.51792	barrels per day
377.39658	barrels per hour
6.289943	barrels per minute
0.10483	barrels per second
14,400	hectoliters per day
600	hectoliters per hour
10	hectoliters per minute
0.16667	hectoliters per second
50,852.4480	cubic feet per day
2,118.8520	cubic feet per hour
35.31444	cubic feet per minute
0.58857	cubic feet per second
144,000	dekaliters per day
6,000	dekaliters per hour
100	dekaliters per minute
1.66667	dekaliters per second
380,415.7728	gallons per day
15,850.65720	gallons per hour
264.17762	gallons per minute
4.40296	gallons per second
316,764.60480	gallons (Imperial) per day
13,198.5252	gallons (Imperial) per hour
219.97542	gallons (Imperial) per minute
3.66626	gallons (Imperial) per second
1,440,000	liters per day
60,000	liters per hour
1,000	liters per minute
16.66667	liters per second
1,440,000	cubic decimeters per day
60,000	cubic decimeters per hour
1,000	cubic decimeters per minute
16.66667	cubic decimeters per second
1,521,663.6672	quarts per day
63,402.65280	quarts per hour

CUBIC METERS PER MINUTE: (cont'd)

1,056.71088	quarts per minute
17.61185	quarts per second
3,043,327.3344	pints per day
126,805.30560	pints per hour
2,113.42176	pints per minute
35.22370	pints per second
12,173,309.8560	gills per day
507,221.2440	gills per hour
8,453.68704	gills per minute
140.89479	gills per second
14,400,000	deciliters per day
600,000	deciliters per hour
10,000	deciliters per minute
166.66667	deciliters per second
87,873,031.8720	cubic inches per day
3,661,376.3280	cubic inches per hour
61,022.93879	cubic inches per minute
1,017.04898	cubic inches per second
144,000,000	centiliters per day
6,000,000	centiliters per hour
100,000	centiliters per minute
1,666.66667	centiliters per second
1,440,000,000	milliliters per day
60,000,000	milliliters per hour
1,000,000	milliliters per minute
16,666.66667	milliliters per second
1,440,000,000	cubic centimeters per day
60,000,000	cubic centimeters per hour
1,000,000	cubic centimeters per minute
16,666.66667	cubic centimeters per second
1,440,000,000,000	cubic millimeters per day
60,000,000,000	cubic millimeters per hour
1,000.000.000	cubic millimeters per minute
16,666,666.66667	cubic millimeters per second

CUBIC METERS PER SECOND: =

86,400	kiloliters per day
3,600	kiloliters per hour
60	kiloliters per minute
1	kiloliters per second
86,400	cubic meters per day
3,600	cubic meters per hour
60	cubic meters per minute
1	cubic meters per second
113,006.0160	cubic yards per day
4,708.5840	cubic yards per hour
78.47640	cubic yards per minute
1.30794	cubic yards per second
543,451.07520	barrels per day
22,643.7948	barrels per hour

CUBIC METERS PER SECOND: (cont'd)

377.39658	barrels per minute
6.28994	barrels per second
864,000	hektoliters per day
36,000	hektoliters per hour
600	hektoliters per minute
10	hektoliters per second
3,051,167.616	cubic feet per day
127,131.9840	cubic feet per hour
2,118.86640	cubic feet per minute
35.31444	cubic feet per second
8,640,000	dekaliters per day
360,000	dekaliters per hour
6,000	dekaliters per minute
100	dekaliters per second
22,824,946.3680	gallons per day
951,039.4320	gallons per hour
15,850.65720	gallons per minute
264.17762	gallons per second
19,005,798.5280	gallons (Imperial) per day
791,908.2720	gallons (Imperial) per hour
13,198.47120	gallons (Imperial) per minute
219.97542	gallons (Imperial) per second
86,400,000	liters per day
3,600,000	liters per hour
60,000	liters per minute
1,000	liters per second
86,400,000	cubic decimeters per day
3,600,000	cubic decimeters per hour
60,000	cubic decimeters per minute
1,000	cubic decimeters per second
91,299,820.0320	quarts per day
3,804,159.16800	quarts per hour
63,402.65280	quarts per minute
1,056.71088	quarts per second
182,599,640.0640	pints per day
7,608,318.3360	pints per hour
126,805.30560	pints per minute
2,113.42176	pints per second
730,398,560.2560	gills per day
30,433,273.3340	gills per hour
507,221.22240	gills per minute
8,453.68704	gills per second
864,000,000	deciliters per day
36,000,000	deciliters per hour
600,000	deciliters per minute
10,000	deciliters per second
5,272,381,911.4560	cubic inches per day
219,682,579.6440	cubic inches per hour
3,661,376.32740	cubic inches per minute
61,022.93879	cubic inches per second
8,640,000,000	centiliters per day

```
360,000,000 .................................... centiliters per hour
6,000,000 .................................... centiliters per minute
100,000 .................................... centiliters per second
86,400,000,000 .................................... milliliters per day
3,600,000,000 .................................... milliliters per hour
60,000,000 .................................... milliliters per minute
1,000,000 .................................... milliliters per second
86,400,000,000 .......................... cubic centimeters per day
3,600,000,000 .......................... cubic centimeters per hour
60,000,000 .......................... cubic centimeters per minute
1,000,000 .......................... cubic centimeters per second
86,400,000,000,000 .......................... cubic millimeters per day
3,600,000,000,000 .......................... cubic millimeters per hour
60,000,000,000 .......................... cubic millimeters per minute
1,000,000,000 .......................... cubic millimeters per second
```

CUBIC POISE CENTIMETER PER GRAM: =

```
0.0000000001 .......................... square kilometers per second
0.00000001 .......................... square hektometers per second
0.000001 .......................... square dekameters per second
0.0001 .......................... square meters per second
0.0010764 .......................... square feet per second
0.1550 .......................... square inches per second
0.01 .......................... square decimeters per second
1.0 .......................... square centimeters per second
100 .......................... square millimeters per second
```

CUBIC POISE FOOT PER POUND: =

```
0.000000006243 ....................... square kilometers per second
0.0000006243 ....................... square hektometers per second
0.00006243 ....................... square dekameters per second
0.006243 ....................... square meters per second
0.067200 ....................... square feet per second
9.67680 ....................... square inches per second
0.6243 ....................... square decimeters per second
62.43 ....................... square centimeters per second
6,243 ....................... square millimeters per second
```

CUBIC POISE INCH PER GRAM: =

```
0.0000000016387 ....................... square kilometers per second
0.00000016387 ....................... square hektometers per second
0.000016387 ....................... square dekameters per second
0.0016387 ....................... square meters per second
0.017639 ....................... square feet per second
2.540 ....................... square inches per second
0.16387 ....................... square decimeters per second
16.387 ....................... square centimeters per second
1,638.7 ....................... square millimeters per second
```

CUBIC YARD: =

0.00000000047509	cubic miles
0.00000000076456	cubic kilometers
0.000000093914	cubic furlongs
0.00000076456	cubic hektometers
0.000093914	cubic chains
0.00076456	cubic dekameters
0.0060105	cubic rods
0.76456	kiloliters
0.76456	cubic meters
1.25971	cubic varas (Texas)
4.80897	barrels
7.64559	hektoliters
21.69666	bushels (U.S.) dry
21.0220	bushels (Imperial) dry
27	cubic feet
63.99891	cubic spans
76.45595	dekaliters
86.78610	pecks
93.91329	cubic links
201.97350	gallons (U.S.) liquid
173.59299	gallons (U.S.) dry
168.18003	gallons (Imperial)
731.18592	cubic hands
764.55945	liters
764.55945	cubic decimeters
807.89400	quarts (liquid)
694.28070	quarts (dry)
1,615.78800	pints (liquid)
1,388.59218	pints (dry)
6,463.15200	gills (liquid)
7,645.5945	deciliters
46,656	cubic inches
76,455.945	centiliters
764,559.45	milliliters
764,559.45	cubic centimeters
764,559,450	cubic millimeters
25,852.79808	ounces (fluid)
206,820.85509	drams (fluid)
24.5457	sacks of cement (set)
1,683.45	pounds of water @ 60°F.
1,736.10	pounds of salt water
1,944	pounds of salt water @ 60°F. and 1.155 specific gravity
13,217.634	pounds of steel of 7.851 specific gravity

CUBIC YARD PER DAY: =

0.76456	kiloliters per day
0.031857	kiloliters per hour

0.00053095	kiloliters per minute
0.0000088491	kiloliters per second
0.76456	cubic meters per day
0.031857	cubic meters per hour
0.00053095	cubic meters per minute
0.0000088491	cubic meters per second
1.0	cubic yards per day
0.04166	cubic yards per hour
0.00069444	cubic yards per minute
0.000011574	cubic yards per second
4.80897	barrels per day
0.20037	barrels per hour
0.0033395	barrels per minute
0.000055659	barrels per second
7.64559	hektoliters per day
0.31857	hektoliters per hour
0.0053095	hektoliters per minute
0.00008849	hektoliters per second
27	cubic feet per day
1.1250	cubic feet per hour
0.018750	cubic feet per minute
0.00031250	cubic feet per second
76.45595	dekaliters per day
3.18568	dekaliters per hour
0.053095	dekaliters per minute
0.00088491	dekaliters per second
201.97350	gallons per day
8.41572	gallons per hour
0.14026	gallons per minute
0.0023377	gallons per second
168.18003	gallons (Imperial) per day
7.00740	gallons (Imperial) per hour
0.11679	gallons (Imperial) per minute
0.0019465	gallons (Imperial) per second
764.55945	liters per day
31.85676	liters per hour
0.53095	liters per minute
0.0088491	liters per second
764.55945	cubic decimeters per day
31.85676	cubic decimeters per hour
0.53095	cubic decimeters per minute
0.0088491	cubic decimeters per second
807.8940	quarts per day
33.66216	quarts per hour
0.56104	quarts per minute
0.0093506	quarts per second
1,615.7880	pints per day
67.32360	pints per hour
1.12206	pints per minute
0.018701	pints per second
6,463.152	gills per day

CUBIC YARD PER DAY: (cont'd)

269.2980 . . . . gills per hour
4.48830 . . . . gills per minute
0.074805 . . . . gills per second
7,645.5945 . . . . deciliters per day
318.56760 . . . . deciliters per hour
5.30946 . . . . deciliters per minute
0.088491 . . . . deciliters per second
46,656 . . . . cubic inches per day
1,944.0 . . . . cubic inches per hour
32.40 . . . . cubic inches per minute
0.5400 . . . . cubic inches per second
76,455.945 . . . . centiliters per day
3,185.6760 . . . . centiliters per hour
53.09460 . . . . centiliters per minute
0.88491 . . . . centiliters per second
764,559.45 . . . . milliliters per day
31,856.760 . . . . milliliters per hour
530.946 . . . . milliliters per minute
8.84907 . . . . milliliters per second
764,559.45 . . . . cubic centimeters per day
31,856.760 . . . . cubic centimeters per hour
530.946 . . . . cubic centimeters per minute
8.84907 . . . . cubic centimeters per second
764,559,450 . . . . cubic millimeters per day
31,856,760 . . . . cubic millimeters per hour
530,946 . . . . cubic millimeters per minute
8,849.07 . . . . cubic millimeters per second

CUBIC YARD PER HOUR: =

18.34963 . . . . kiloliters per day
0.76456 . . . . kiloliters per hour
0.012743 . . . . kiloliters per minute
0.00021238 . . . . kiloliters per second
18.34963 . . . . cubic meters per day
0.76456 . . . . cubic meters per hour
0.012743 . . . . cubic meters per minute
0.00021238 . . . . cubic meters per second
24.0 . . . . cubic yards per day
1.0 . . . . cubic yards per hour
0.016667 . . . . cubic yards per minute
0.00027778 . . . . cubic yards per second
115.41312 . . . . barrels per day
4.80897 . . . . barrels per hour
0.080148 . . . . barrels per minute
0.0013358 . . . . barrels per second
183.49632 . . . . hektoliters per day
7.64559 . . . . hektoliters per hour
0.12743 . . . . hektoliters per minute
0.0021238 . . . . hektoliters per second
648.0 . . . . cubic feet per day

27.0	cubic feet per hour
0.450	cubic feet per minute
0.00750	cubic feet per second
1,834.9632	dekaliters per day
76.45595	dekaliters per hour
1.27428	dekaliters per minute
0.021238	dekaliters per second
4,847.3856	gallons per day
201.97350	gallons per hour
3.36624	gallons per minute
0.056104	gallons per second
4,036.34880	gallons (Imperial) per day
168.18003	gallons (Imperial) per hour
2.80302	gallons (Imperial) per minute
0.046717	gallons (Imperial) per second
18,349.632	liters per day
764.55945	liters per hour
12.7428	liters per minute
0.21238	liters per second
18,349.632	cubic decimeters per day
764.55945	cubic decimeters per hour
12.7428	cubic decimeters per minute
0.21238	cubic decimeters per second
19,389.8880	quarts per day
807.8940	quarts per hour
13.4652	quarts per minute
0.22442	quarts per second
38,778.9120	pints per day
1,615.7780	pints per hour
26.92980	pints per minute
0.44883	pints per second
155,115.648	gills per day
6,463.152	gills per hour
107.7192	gills per minute
1.79532	gills per second
183,496.32	deciliters per day
7,645.5945	deciliters per hour
127.428	deciliters per minute
2.12378	deciliters per second
1,119,744	cubic inches per day
46,656	cubic inches per hour
777.6	cubic inches per minute
12.960	cubic inches per second
1,834,963.2	centiliters per day
76,455.945	centiliters per hour
1,274.28	centiliters per minute
21.23776	centiliters per second
18,349.632	milliliters per day
764,559.45	milliliters per hour
12,742.8	milliliters per minute
212.37763	milliliters per second

CUBIC YARD PER HOUR: (cont'd)

18,349,632	cubic centimeters per day
764,559.45	cubic centimeters per hour
12,742.8	cubic centimeters per minute
212.37763	cubic centimeters per second
18,349,632,000	cubic millimeters per day
764,559,450	cubic millimeters per hour
12,742,800	cubic millimeters per minute
212,377.63	cubic millimeters per second

CUBIC YARD PER MINUTE: =

1,100.96640	kiloliters per day
45.87360	kiloliters per hour
0.76456	kiloliters per minute
0.012743	kiloliters per second
1,100.96640	cubic meters per day
45.87360	cubic meters per hour
0.76456	cubic meters per minute
0.012743	cubic meters per second
1,440	cubic yards per day
60	cubic yards per hour
1.0	cubic yards per minute
0.016667	cubic yards per second
6,924.960	barrels per day
288.540	barrels per hour
4.80897	barrels per minute
0.080150	barrels per second
11,009.66399	hektoliters per day
458.73560	hektoliters per hour
7.64559	hektoliters per minute
0.12743	hektoliters per second
38,880	cubic feet per day
1,620	cubic feet per hour
27.0	cubic feet per minute
0.450	cubic feet per second
110,096.633994	dekaliters per day
4,587.3600	dekaliters per hour
76.45595	dekaliters per minute
1.27427	dekaliters per second
290,842.2720	gallons per day
12,118.4280	gallons per hour
201.97350	gallons per minute
3.36623	gallons per second
242,179.20	gallons (Imperial) per day
10,090.80	gallons (Imperial) per hour
168.18003	gallons (Imperial) per minute
2.80300	gallons (Imperial) per second
1,100,966.39942	liters per day
45,873.59998	liters per hour
764.55945	liters per minute
12.74267	liters per second

CUBIC YARD PER MINUTE: (cont'd)

1,100,966.39942	cubic decimeters per day
45,873.59998	cubic decimeters per hour
764.55945	cubic decimeters per minute
12.74267	cubic decimeters per second
1,163,367.36	quarts per day
48,473.64	quarts per hour
807.8940	quarts per minute
13.4649	quarts per second
2,326,734.72	pints per day
96,947.28	pints per hour
1,615.7880	pints per minute
26.9298	pints per second
9,306,938.88	gills per day
387,789.12	gills per hour
6,463.152	gills per minute
107.71920	gills per second
11,009,663.99424	deciliters per day
458,735.99976	deciliters per hour
7,645.5945	deciliters per minute
127.42667	deciliters per second
67,184,640	cubic inches per day
2,799,360.0	cubic inches per hour
46,656.0	cubic inches per minute
777.6	cubic inches per second
110,096,639.9424	centiliters per day
4,587,359.9976	centiliters per hour
76,455.945	centiliters per minute
1,274.26667	centiliters per second
1,100,966,399.424	milliliters per day
45,873,599.976	milliliters per hour
764,559.45	milliliters per minute
12,742.66667	milliliters per second
1,100,966,399.424	cubic centimeters per day
45,873,599.976	cubic centimeters per hour
764,559.45	cubic centimeters per minute
12,742.66667	cubic centimeters per second
1,100,966,399,424	cubic millimeters per day
45,873,499,976	cubic millimeters per hour
764,559,450	cubic millimeters per minute
12,742,666.67	cubic millimeters per second

CUBIC YARD PER SECOND: =

66,057.93648	kiloliters per day
2,752.41402	kiloliters per hour
45.87357	kiloliters per minute
0.76456	kiloliters per second
66,057.93648	cubic meters per day
2,752.41402	cubic meters per hour
45.87357	cubic meters per minute
0.76456	cubic meters per second

CUBIC YARD PER SECOND: (cont'd)

86,400	cubic yards per day
3,600	cubic yards per hour
60.0	cubic yards per minute
1.0	cubic yards per second
415,495.008	barrels per day
17,312.2920	barrels per hour
288.53820	barrels per minute
4.80897	barrels per second
660,579.3648	hectoliters per day
27,524.1402	hectoliters per hour
458.73567	hectoliters per minute
7.64559	hectoliters per second
2,332,800	cubic feet per day
97,200	cubic feet per hour
1,620	cubic feet per minute
27	cubic feet per second
6,605,793.648	dekaliters per day
275,241.402	dekaliters per hour
4,587.3567	dekaliters per minute
76.45595	dekaliters per second
17,450,510.40	gallons per day
727,104.60	gallons per hour
12,118.410	gallons per minute
201.97350	gallons per second
14,530,754.5920	gallons (Imperial) per day
605,448.1080	gallons (Imperial) per hour
10,090.80180	gallons (Imperial) per minute
168.18003	gallons (Imperial) per second
66,057,936.48	liters per day
2,752,414.02	liters per hour
45,873.567	liters per minute
764.55945	liters per second
66,057,936.48	cubic decimeters per day
2,752,414.02	cubic decimeters per hour
45,873.567	cubic decimeters per minute
764.55945	cubic decimeters per second
69,802,041.6	quarts per day
2,908,418.40	quarts per hour
48,473.640	quarts per minute
807.8940	quarts per second
139,604,083.2	pints per day
5,816,836.8	pints per hour
96,947.280	pints per minute
1,615.7880	pints per second
558,416,332.8	gills per day
23,267,347.20	gills per hour
387,789.120	gills per minute
6,463.152	gills per second
660,579,364.8	deciliters per day
27,524,140.2	deciliters per hour
458,735.67	deciliters per minute

CUBIC YARD PER SECOND: (cont'd)

7,645.5945	deciliters per second
4,031,078,400	cubic inches per day
167,961,600	cubic inches per hour
2,799,360	cubic inches per minute
46,656	cubic inches per second
6,605,793,648	centiliters per day
275,241,402	centiliters per hour
4,587,356.7	centiliters per minute
76,445.945	centiliters per second
66,057,936.480	milliliters per day
2,752,414,020	milliliters per hour
45,873,567	milliliters per minute
764,559.45	milliliters per second
66,057,936,480	cubic centimeters per day
2,752,414,020	cubic centimeters per hour
45,873,565	cubic centimeters per minute
764,559.45	cubic centimeters per second
66,057,936,480,000	cubic millimeters per day
2,752,414,020,000	cubic millimeters per hour
45,873,565,000	cubic millimeters per minute
764,559,450	cubic millimeters per second

DECIGRAM: =

0.0000001102311150	tons (short)
0.0000000984206383	tons (long)
0.0000001	tons (metric)
0.0001	kilograms
0.000267922895	pounds (Troy)
0.00022046223	pounds (Avoir.)
0.001	hektograms
0.01	dekagrams
0.00321507	ounces (Troy)
0.00352739568	ounces (Avoir.)
0.1	grams
10	centigrams
100	milligrams
1.543236	grains
0.0257205	drams (Troy)
0.05643833088	drams (Avoir.)
0.5	carats (metric)

DECILITER: =

0.0001	kiloliters
0.0001	cubic meters
0.0001308	cubic yards
0.00062900	barrels
0.001	hektoliters
0.0028380	bushels (U.S.)

DECILITER: (cont'd)

0.0027513	bushels (Imperial) dry
0.0035316	cubic feet
0.01	dekaliters
0.011352	pecks
0.026418	gallons (U.S.) liquid
0.022706	gallons (U.S.) dry
0.021997	gallons (Imperial)
0.10567	quarts (liquid)
0.090816	quarts (dry)
0.1	liters
0.1	cubic decimeters
0.18163	pints (dry)
0.21134	pints (liquid)
0.84536	gills (liquid)
6.1025	cubic inches
1.0	centiliters
100	milliliters
100	cubic centimeters
100,000	cubic millimeters
3.38147	ounces (fluid)
27.05179	drams (fluid)

DECIMETER: =

0.000062137	miles
0.0001	kilometers
0.00049710	furlongs
0.001	hektometers
0.0049710	chains
0.01	dekameters
0.019884	rods
0.1	meters
0.10935	yards
0.11811	varas (Texas)
0.3280833	feet
0.43744	spans
0.49710	links
0.98425	hands
1	decimeter
10	centimeter
3.93700	inches
100	millimeters
3,937	mils
100,000	microns
100,000,000	millimicrons
100,000,000	micromillimeters
155,316	wave lengths of red line of cadmium
100,000,000,000	Angstrom units

DEGREE (ANGLE): =

```
0.0027778  ....................................  circumferences
0.0027778  ......................................  revolutions
0.01111  ..........................................  quadrants
0.017453  ...........................................  radians
1  ...................................................  hours
60  ...............................................  minutes
3,600  ............................................  seconds
```

DEGREE PER DAY: =

```
0.0027778  ...................................  revolutions per day
0.00011574  ..................................  revolutions per hour
0.0000019290  ...............................  revolutions per minute
0.000000032150  ............................  revolutions per second
0.01111  ......................................  quadrants per day
0.00046292  ...................................  quadrants per hour
0.0000077154  ..............................  quadrants per minute
0.00000012859  .............................  quadrants per second
0.017453  .....................................  radians per day
0.00072720  ..................................  radians per hour
0.000012120  ..................................  radians per minute
0.00000020200  .............................  radians per second
1.0  ..........................................  degrees per day
0.041666  ....................................  degrees per hour
0.00069444  ..................................  degrees per minute
0.000011574  .................................  degrees per second
1.0  ..........................................  hours per day
0.041666  ....................................  hours per hour
0.00069444  ..................................  hours per minute
0.000011574  .................................  hours per second
60  .........................................  minutes per day
2.5  ..........................................  minutes per hour
0.041666  ....................................  minutes per minute
0.00069444  ..................................  minutes per second
3,600  .......................................  seconds per day
150  ..........................................  seconds per hour
2.5  ..........................................  seconds per minute
0.041667  ....................................  seconds per second
```

DEGREE PER HOUR: =

```
0.066667  .....................................  revolutions per day
0.0027778  ....................................  revolutions per hour
0.000046297  .................................  revolutions per minute
0.00000077161  ..............................  revolutions per second
0.26664  ......................................  quadrants per day
0.01111  ......................................  quadrants per hour
0.00018517  ..................................  quadrants per minute
0.0000030861  ...............................  quadrants per second
```

DEGREE PER HOUR: (cont'd)

0.41888	radians per day
0.017453	radians per hour
0.00029088	radians per minute
0.0000048481	radians per second
24	degrees per day
1.0	degrees per hour
0.016667	degrees per minute
0.00027778	degrees per second
24	hours per day
1.0	hours per hour
0.016667	hours per minute
0.00027778	hours per second
1,440	minutes per day
60	minutes per hour
1	minutes per minute
0.016667	minutes per second
86,400	seconds per day
3,600	seconds per hour
60	seconds per minute
1	seconds per second

DEGREE PER MINUTE: =

4.0	revolutions per day
0.16667	revolutions per hour
0.0027778	revolutions per minute
0.000046297	revolutions per second
15.99869	quadrants per day
0.66661	quadrants per hour
0.01111	quadrants per minute
0.00018517	quadrants per second
25.13203	radians per day
1.04717	radians per hour
0.017453	radians per minute
0.00029088	radians per second
1,440	degrees per day
60	degrees per hour
1.0	degrees per minute
0.016667	degrees per second
1,440	hours per day
60	hours per hour
1.0	hours per minute
0.016667	hours per second
86,400	minutes per day
3,600	minutes per hour
60	minutes per minute
1	minutes per second
5,184,000	seconds per day
216,000	seconds per hour
3,600	seconds per minute
60	seconds per second

DEGREE PER MINUTE: (cont'd)

240	revolutions per day
10	revolutions per hour
0.16667	revolutions per minute
0.0027778	revolutions per second
959.904	quadrants per day
39.9960	quadrants per hour
0.6666	quadrants per minute
0.01111	quadrants per second
1,507.9392	radians per day
62.83080	radians per hour
1.047180	radians per minute
0.017453	radians per second
86,400	degrees per day
3,600	degrees per hour
60	degrees per minute
1.0	degrees per second
86,400	hours per day
3,600	hours per hour
60	hours per minute
1.0	hours per second
5,184,000	minutes per day
216,000	minutes per hour
3,600	minutes per minute
60	minutes per second
311,040,000	seconds per day
12,960,000	seconds per hour
216,000	seconds per minute
3,600	seconds per second

DEKAGRAM: =

0.00000984206383	tons (long)
0.00001	tons (metric)
0.0000110231115	tons (short)
0.01	kilograms
0.022046223	pounds (Avoir.)
0.0267922895	pounds (Troy)
0.1	hektograms
0.321507	ounces (Troy)
0.352739568	ounces (Avoir.)
2.572053	drams (Troy)
5.6438330880	drams (Avoir.)
6.430149	pennyweights
7.71618	scruples
10	grams
100	decigrams
154.3234765625	grains
1,000	centigrams
10,000	milligrams
50	carats (metric)

DEKALITER: =

0.01	kiloliters
0.01	cubic meters
0.01308	cubic yards
0.06290	barrels
0.1	hektoliters
0.28380	bushels (U.S.)
0.27513	bushels (Imperial) dry
0.35316	cubic feet
1.1352	pecks
2.6418	gallons (U.S.) liquid
2.2706	gallons (U.S.) dry
2.1997	gallons (Imperial)
10.567	quarts (liquid)
9.0816	quarts (dry)
10	liters
10	cubic decimeters
18.163	pints (dry)
21.134	pints (liquid)
84.536	gills (liquid)
100	deciliters
610.25	cubic inches
1,000	centiliters
10,000	milliliters
10,000	cubic centimeters
100,000,000	cubic millimeters
338.147	ounces (fluid)
2,705.179	drams (fluid)

DEKAMETER: =

0.0062137	miles
0.01	kilometers
0.049710	furlongs
0.1	hektometers
0.49710	chains
1.0	dekameters
1.98838	rods
10	meters
10.935	yards
11.811	varas (Texas)
32.80830	feet
43.744	spans
49.710	links
98.425	hands
100	decimeters
1000	centimeters
393.70	inches
10,000	millimeters
393,700	mils

UNIVERSAL CONVERSION FACTORS

DEKAMETER: (cont'd)

10,000,000	microns
10,000,000,000	millimicrons
10,000,000,000	micromillimeters
15,531,595	wave lengths of red line of cadmium
10,000,000,000 x 10³	Angstrom Units

DRAM (AVOIRDUPOIS): =

0.0000017439	tons (long)
0.0000017718	tons (metric)
0.0000019531	tons (short)
0.001771845	kilograms
0.00390625	pounds (Avoir.)
0.0047471788	pounds (Troy)
0.01771845	hektograms
0.1771845	dekagrams
0.056966146	ounces (Troy)
0.0625	ounces (Avoir.)
0.4557292	drams (Troy)
1.139323	pennyweights
1.3671875	scruples
1.771845	grams
17.71845	decigrams
27.34375	grains
177.1845	centigrams
1,771.845	milligrams
8.85923	carats (metric)

DRAM (FLUID): =

0.0000036966	kiloliters
0.0000036966	cubic meters
0.0000048352	cubic yards
0.000023252	barrels
0.000036966	hektoliters
0.00013055	cubic feet
0.00036966	dekaliters
0.00097658	gallons (U.S.) liquid
0.00081318	gallons (Imperial) liquid
0.00390625	quarts (liquid)
0.0036966	liters
0.0036966	cubic decimeters
0.0078125	pints (liquid)
0.03125	gills (liquid)
0.036966	deciliters
0.225586	cubic inches
0.36966	centiliters
3.69661	milliliters
3.69661	cubic centimeters

DRAM (FLUID): (cont'd)

3,696.61	cubic millimeters
0.125	ounces (fluid)
60	minims

DRAM (TROY OR APOTHECARY): =

0.0000038265308	tons (long)
0.0000038879351	tons (metric)
0.0000042857145	tons (short)
0.0038879351	kilograms
0.008571429	pounds (Avoir.)
0.010416667	pounds (Troy)
0.038879351	hektograms
0.38879351	dekagrams
0.12500	ounces (Troy)
0.1371429	ounces (Avoir.)
2.194286	drams (Avoir.)
2.50	pennyweight
3.0	scruples
3.8879351	grams
38.879351	decigrams
60.0	grains
388.79351	centigrams
3,887.9351	milligrams
19.43968	carats (metric)

FOOT (OR ENGINEER'S LINK): =

0.0001893939	miles
0.0003048006	kilometers
0.00151515	furlongs
0.003048006	hektometers
0.0151515	chains
0.03048006	dekameters
0.0606061	rods
0.3048006	meters
0.33333	yards
0.3600	varas (Texas)
1.33333	spans
1.515152	links
3.00	hands
3.048006	decimeters
30.48006	centimeters
12	inches
304.8006	millimeters
12,000	mils
304,801	microns
304,801,200	millimicrons
304,801,200	micromillimeters
473,404	wave lengths of red line of cadmium
3,048,012,000	Angstrom Units

744

FOOT PER DAY: =

0.0001893939	miles per day
0.0000078914	miles per hour
0.00000013152	miles per minute
0.0000000021921	miles per second
0.0003048006	kilometers per day
0.0000127	kilometers per hour
0.00000021167	kilometers per minute
0.0000000035278	kilometers per second
0.00151515	furlongs per day
0.000063131	furlongs per hour
0.0000010522	furlongs per minute
0.000000017536	furlongs per second
0.003048006	hektometers per day
0.000127	hektometers per hour
0.0000021167	hektometers per minute
0.000000035278	hektometers per second
0.0151515	chains per day
0.00063131	chains per hour
0.000010522	chains per minute
0.00000017536	chains per second
0.03048006	dekameters per day
0.00127	dekameters per hour
0.000021167	dekameters per minute
0.00000035278	dekameters per second
0.0606061	rods per day
0.0025253	rods per hour
0.000042088	rods per minute
0.00000070146	rods per second
0.3048006	meters per day
0.0127	meters per hour
0.00021167	meters per minute
0.0000035278	meters per second
0.33333	yards per day
0.013889	yards per hour
0.00023148	yards per minute
0.0000038580	yards per second
0.3600	varas (Texas) per day
0.015	varas (Texas) per hour
0.00025	varas (Texas) per minute
0.0000041667	varas (Texas) per second
1.0	feet per day
0.041667	feet per hour
0.00069444	feet per minute
0.000011574	feet per second
1.33333	spans per day
0.055556	spans per hour
0.00092593	spans per minute
0.000015432	spans per second
1.515152	links per day
0.0163131	links per hour

FOOT PER DAY: (cont'd)

0.0010522	links per minute
0.000017536	links per second
3.00	hands per day
0.125	hands per hour
0.0020833	hands per minute
0.000034722	hands per second
3.048006	decimeters per day
0.127	decimeters per hour
0.0021167	decimeters per minute
0.000035278	decimeters per second
30.48006	centimeters per day
1.270	centimeters per hour
0.021167	centimeters per minute
0.00035278	centimeters per second
12.0	inches per day
0.50	inches per hour
0.0083333	inches per minute
0.00013889	inches per second
304.8006	millimeters per day
12.70004	millimeters per hour
0.21167	millimeters per minute
0.0035278	millimeters per second

FOOT PER HOUR: =

0.0045455	miles per day
0.0001893939	miles per hour
0.0000031566	miles per minute
0.000000052609	miles per second
0.0073152	kilometers per day
0.0003048006	kilometers per hour
0.00000508	kilometers per minute
0.000000084667	kilometers per second
0.036364	furlongs per day
0.00151515	furlongs per hour
0.000025253	furlongs per minute
0.00000042088	furlongs per second
0.073152	hektometers per day
0.003048006	hektometers per hour
0.0000508	hektometers per minute
0.00000084667	hektometers per second
0.36364	chains per day
0.0151515	chains per hour
0.00025253	chains per minute
0.0000042088	chains per second
0.73152	dekameters per day
0.03048006	dekameters per hour
0.000508	dekameters per minute
0.0000084667	dekameters per second
1.45456	rods per day
0.0606061	rods per hour

UNIVERSAL CONVERSION FACTORS

FOOT PER HOUR: (cont'd)

0.00101010	rods per minute
0.000016835	rods per second
7.31521	meters per day
0.3048006	meters per hour
0.00508	meters per minute
0.000084667	meters per second
8.0000	yards per day
0.33333	yards per hour
0.0055556	yards per minute
0.000092593	yards per second
8.640	varas (Texas) per day
0.3600	varas (Texas) per hour
0.006	varas (Texas) per minute
0.0001	varas (Texas) per second
24.0	feet per day
1.0	feet per hour
0.016667	feet per minute
0.00027778	feet per second
32.000	spans per day
1.33333	spans per hour
0.022222	spans per minute
0.00037037	spans per second
36.36364	links per day
1.51515	links per hour
0.025253	links per minute
0.00042088	links per second
72.0	hands per day
3.0	hands per hour
0.05000	hands per minute
0.00083333	hands per second
73.15214	decimeters per day
3.048006	decimeters per hour
0.0508	decimeters per minute
0.00084667	decimeters per second
731.52144	centimeters per day
30.48006	centimeters per hour
0.5080	centimeters per minute
0.0084667	centimeters per second
288.0	inches per day
12.0	inches per hour
0.200	inches per minute
0.0033333	inches per second
7,315.21440	millimeters per day
304.8006	millimeters per hour
5.080	millimeters per minute
0.084667	millimeters per second

FOOT PER MINUTE: =

0.27273	miles per day
0.011364	miles per hour

747

FOOT PER MINUTE: (cont'd)

0.0001893939	miles per minute
0.0000031566	miles per second
0.43891	kilometers per day
0.018288	kilometers per hour
0.0003048006	kilometers per minute
0.00000508	kilometers per second
2.18182	furlongs per day
0.090909	furlongs per hour
0.0015151	furlongs per minute
0.000025253	furlongs per second
4.38913	hektometers per day
0.18288	hektometers per hour
0.003048006	hektometers per minute
0.0000508	hektometers per second
21.81818	chains per day
0.90909	chains per hour
0.015151	chains per minute
0.00025253	chains per second
43.89129	dekameters per day
1.82880	dekameters per hour
0.03048006	dekameters per minute
0.000508	dekameters per second
87.27273	rods per day
3.63636	rods per hour
0.0606061	rods per minute
0.0010101	rods per second
438.91286	meters per day
18.28804	meters per hour
0.3048006	meters per minute
0.00508	meters per second
480.0	yards per day
20.0	yards per hour
0.33333	yards per minute
0.0055556	yards per second
518.400	varas (Texas) per day
21.60	varas (Texas) per hour
0.3600	varas (Texas) per minute
0.0060000	varas (Texas) per second
1,440.0	feet per day
60.0	feet per hour
1.0	feet per minute
0.016667	feet per second
1,920	spans per day
80.0	spans per hour
1.33333	spans per minute
0.022222	spans per second
2,181.81818	links per day
90.90909	links per hour
1.51515	links per minute
0.025253	links per second
4,320	hands per day

FOOT PER MINUTE: (cont'd)

180.0	hands per hour
3.0	hands per minute
0.050000	hands per second
4,389.12864	decimeters per day
182.88036	decimeters per hour
3.048006	decimeters per minute
0.050800	decimeters per second
43,891.2864	centimeters per day
1,828.80360	centimeters per hour
30.48006	centimeters per minute
0.50800	centimeters per second
17,280	inches per day
720	inches per hour
12.0	inches per minute
0.20000	inches per second
438,912.864	millimeters per day
18,288.0360	millimeters per hour
304.8006	millimeters per minute
5.08001	millimeters per second

FOOT PER SECOND: =

16.36363	miles per day
0.68182	miles per hour
0.011364	miles per minute
0.0001893939	miles per second
26.33477	kilometers per day
1.097282	kilometers per hour
0.018288	kilometers per minute
0.0003048006	kilometers per second
130.90909	furlongs per day
5.45455	furlongs per hour
0.090909	furlongs per minute
0.0015151	furlongs per second
263.34772	hektometers per day
10.97282	hektometers per hour
0.18288	hektometers per minute
0.003048006	hektometers per second
1,309.09090	chains per day
54.54545	chains per hour
0.90909	chains per minute
0.015151	chains per second
2,633.47718	dekameters per day
109.72823	dekameters per hour
1.82880	dekameters per minute
0.03048006	dekameters per second
5,236.36364	rods per day
218.18182	rods per hour
3.63636	rods per minute
0.0606061	rods per second
26,334.77184	meters per day

FOOT PER SECOND: (cont'd)

1,097.28216	meters per hour
18.28804	meters per minute
0.3048006	meters per second
28,800	yards per day
1,200	yards per hour
20.0	yards per minute
0.33333	yards per second
31,104	varas (Texas) per day
1,296.0	varas (Texas) per hour
21.60	varas (Texas) per minute
0.3600	varas (Texas) per second
86,400	feet per day
3,600	feet per hour
60	feet per minute
1	feet per second
115,200	spans per day
4,800	spans per hour
80.0	spans per minute
1.33333	spans per second
130,909.090909	links per day
5,454.54545	links per hour
90.90909	links per minute
1.51515	links per second
259,200	hands per day
10,800	hands per hour
180.0	hands per minute
3.0	hands per second
263,347.7184	decimeters per day
10,972.8216	decimeters per hour
182.88036	decimeters per minute
3.048006	decimeters per second
2,633,477.1840	centimeters per day
109,728.2160	centimeters per hour
1,828.8036	centimeters per minute
30.48006	centimeters per second
1,036,800	inches per day
43,200	inches per hour
720	inches per minute
12	inches per second
26,334,771.8400	millimeters per day
1,097,282.1600	millimeters per hour
18,288.0360	millimeters per minute
304.8006	millimeters per second

FOOT POUNDS: =

32.174	foot poundals
386.088	inch poundals
1.0	foot pounds
12	inch pounds
0.00000035703	ton (short) calories

FOOT POUNDS: (cont'd)

0.00032389	kilogram calories
0.00071406	pound calories
0.0032389	hektogram calories
0.032389	dekagram calories
0.011425	ounce calories
0.32389	gram calories
3.2389	decigram calories
32.389	centigram calories
323.89	milligram calories
0.0000000156925	kilowatt days
0.00000037662	kilowatt hours
0.0000225972	kilowatt minutes
0.00135583	kilowatt seconds
0.0000156925	watt days
0.00037662	watt hours
0.0225972	watt minutes
1.35583	watt seconds
0.0015240	ton meters (short)
0.138255	kilogram meters
0.30480	pound meters
1.38255	hektogram meters
13.8255	dekagram meters
4.8768	ounce meters
138.255	gram meters
1,382.55	decigram meters
13,825.5	centigram meters
138,255	milligram meters
0.000015240	ton hektometers
0.00138255	kilogram hektometers
0.0030480	pound hektometers
0.0138255	hektogram hektometers
0.138255	dekagram hektometers
0.048768	ounce hektometers
1.38255	gram hektometers
13.8255	decigram hektometers
138.255	centigram hektometers
1,382.55	milligram hektometers
0.00015240	ton dekameters
0.0138255	kilogram dekameters
0.030480	pound dekameters
0.138255	hektogram dekameters
1.38255	dekagram dekameters
0.48768	ounce dekameters
13.8255	gram dekameters
138.255	decigram dekameters
1,382.55	centigram dekameters
13,825.5	milligram dekameters
0.0050000	ton feet (short)
0.45359	kilogram feet
1.0	pound feet
4.53592	hektogram feet

FOOT POUNDS: (cont'd)

45.35916	dekagram feet
16	ounce feet
453.59157	gram feet
4,535.91566	decigram feet
45,359.15664	centigram feet
453,591.56642	milligram feet
0.060000	ton inches
5.44308	kilogram inches
12.0	pound inches
54.43104	hektogram inches
544.30992	dekagram inches
192	ounch inches
5,443.09984	gram inches
54,430.98792	decigram inches
544,309.87968	centigram inches
5,443,098.7968	milligram inches
0.015240	ton decimeters
1.38255	kilogram decimeters
3.04800	pound decimeters
13.8255	hektogram decimeters
138.255	dekagram decimeters
48.760	ounce decimeters
1,382.55	gram decimeters
13,825.5	decigram decimeters
138,255	centigram decimeters
1,382,550	milligram decimeters
0.15240	ton centimeters
13.8255	kilogram centimeters
30.480	pound centimeters
138.255	hektogram centimeters
1,382.55	dekagram centimeters
487.60	ounce centimeters
13,825.5	gram centimeters
138,255	decigram centimeters
1,382,550	centigram centimeters
13,825,500	milligram centimeters
1.5240	ton millimeters
138.255	kilogram millimeters
304.80	pound millimeters
1,382.55	hektogram millimeters
13,825.5	dekagram millimeters
4,876.0	ounce millimeters
138,255	gram millimeters
1,382,550	decigram millimeters
13,825,500	centigram millimeters
138,255,000	milligram millimeters
0.000013381	kiloliter-atmospheres
0.00013381	hektoliter-atmospheres
0.0013381	dekaliter-atmospheres
0.013381	liter-atmospheres
0.13381	deciliter-atmospheres

FOOT POUNDS: (cont'd)

1.3381	centiliter-atmospheres
13.381	milliliter-atmospheres
0.000000000000013381	cubic kilometer-atmospheres
0.000000000013381	cubic hektometer-atmospheres
0.000000013381	cubic dekameter-atmospheres
0.000013381	cubic meter-atmospheres
0.00047253	cubic feet-atmospheres
0.013381	cubic decimeter-atmospheres
13.381	cubic centimeter-atmospheres
13,381	cubic millimeter-atmospheres
1.35582	joules (absolute)
0.00000049814	Cheval-vapeur hours
0.0000000210438	horsepower day
0.00000050505	horsepower hours
0.0000303030	horsepower minutes
0.00181818	horsepower seconds
0.00000008808	pounds of carbon oxidized with 100% efficiency
0.0000013256	pounds of water evaporated from and at 212°F.
0.0012854	B.T.U.
13,558,200	ergs
13,558,200	centimeters dynes

FOOT POUND PER DAY: =

32.174	foot poundals per day
1.34058	foot poundals per hour
0.022343	foot poundals per minute
0.00037238	foot poundals per second
1.0	foot pounds per day
0.041667	foot pounds per hour
0.00069444	foot pounds per minute
0.000011574	foot pounds per second
0.00032389	kilogram calories per day
0.000013495	kilogram calories per hour
0.00000022492	kilogram calories per minute
0.0000000037487	kilogram calories per second
0.011425	ounce calories per day
0.00047604	ounce calories per hour
0.0000079340	ounce calories per minute
0.00000013223	ounce calories per second
0.138255	kilogram meters per day
0.0057606	kilogram meters per hour
0.00009601	kilogram meters per minute
0.0000016002	kilogram meters per second
0.013381	liter-atmospheres per day
0.00055754	liter-atmospheres per hour
0.0000092924	liter-atmospheres per minute
0.00000015487	liter-atmospheres per second
0.00047253	cubic foot atmospheres per day

FOOT POUND PER DAY: (cont'd)

0.000019689	cubic foot atmospheres per hour
0.00000032815	cubic foot atmospheres per minute
0.0000000054691	cubic foot atmospheres per second
0.000000021333	Cheval-vapeurs
0.000000021044	horsepower
0.000000015692	kilowatts
0.000015692	watts
1.35582	joules per day
0.056492	joules per hour
0.00094154	joules per minute
0.000015692	joules per second
0.00000008808	pounds of carbon oxidized with perfect efficiency per day
0.0000000036700	pounds of carbon oxidized with perfect efficiency per hour
0.0000000000611667	pounds of carbon oxidized with perfect efficiency per minute
0.0000000000010194	pounds of carbon oxidized with perfect efficiency per second
0.0000013256	pounds of water evaporated from and at 212°F. per day
0.000000055233	pounds of water evaporated from and at 212°F. per hour
0.00000000092056	pounds of water evaporated from and at 212°F. per minute
0.000000000015343	pounds of water evaporated from and at 212°F. per second
0.0012854	BTU per day
0.000053558	BTU per hour
0.00000089264	BTU per minute
0.000000014877	BTU per second

FOOT POUND PER HOUR: =

722.1760	foot poundals per day
32.174	foot poundals per hour
0.53623	foot poundals per minute
0.0089372	foot poundals per second
24.0	foot pounds per day
1.0	foot pounds per hour
0.016667	foot pounds per minute
0.00027778	foot pounds per second
0.0077734	kilogram calories per day
0.00032389	kilogram calories per hour
0.0000053982	kilogram calories per minute
0.000000089969	kilogram calories per second
0.2742	ounce calories per day
0.011425	ounce calories per hour
0.00019042	ounce calories per minute
0.0000031736	ounce calories per second
3.31812	kilogram meters per day

FOOT POUND PER HOUR: (cont'd)

0.138255	kilogram meters per hour
0.0023042	kilogram meters per minute
0.000038404	kilogram meters per second
0.32114	liter-atmospheres per day
0.013381	liter-atmospheres per hour
0.00022302	liter-atmospheres per minute
0.0000037169	liter-atmospheres per second
0.011341	cubic foot atmospheres per day
0.00047253	cubic foot atmospheres per hour
0.0000078755	cubic foot atmospheres per minute
0.00000013126	cubic foot atmospheres per second
0.0000005120	Cheval-vapeurs
0.00000050505	horsepower
0.00000037661	kilowatts
0.00037661	watts
32.53968	joules per day
1.35582	joules per hour
0.022597	joules per minute
0.00037662	joules per second
0.0000021139	pounds of carbon oxidized with perfect efficiency per day
0.00000008808	pounds of carbon oxidized with perfect efficiency per hour
0.0000000014680	pounds of carbon oxidized with perfect efficiency per minute
0.000000000024467	pounds of carbon oxidized with perfect efficiency per second
0.000031814	pounds of water evaporated from and at 212°F. per day
0.0000013256	pounds of water evaporated from and at 212°F. per hour
0.000000022093	pounds of water evaporated from and at 212°F. per minute
0.00000000036822	pounds of water evaporated from and at 212°F. per second
0.030850	BTU per day
0.0012854	BTU per hour
0.000021423	BTU per minute
0.0000035706	BTU per second

FOOT POUND PER MINUTE: =

46,330.56	foot poundals per day
1,930.44	foot poundals per hour
32.174	foot poundals per minute
0.53623	foot poundals per second
1,440	foot pounds per day
60	foot pounds per hour
1.0	foot pounds per minute
0.016667	foot pounds per second
0.46640	kilogram calories per day

FOOT POUND PER MINUTE: (cont'd)

0.019433 . . . . kilogram calories per hour
0.00032389 . . . . kilogram calories per minute
0.0000053982 . . . . kilogram calories per second
16.452 . . . . ounce calories per day
0.6855 . . . . ounce calories per hour
0.011425 . . . . ounce calories per minute
0.00019042 . . . . ounce calories per second
199.0872 . . . . kilogram meters per day
8.29530 . . . . kilogram meters per hour
0.138255 . . . . kilogram meters per minute
0.0023043 . . . . kilogram meters per second
19.26864 . . . . liter-atmospheres per day
0.80286 . . . . liter-atmospheres per hour
0.013381 . . . . liter-atmospheres per minute
0.00022302 . . . . liter-atmospheres per second
0.68044 . . . . cubic foot atmospheres per day
0.028352 . . . . cubic foot atmospheres per hour
0.00047253 . . . . cubic foot atmospheres per minute
0.0000078755 . . . . cubic foot atmospheres per second
0.000030719 . . . . Cheval-vapeurs
0.000030303 . . . . horsepower
0.000022597 . . . . kilowatts
0.022597 . . . . watts
1,952.3808 . . . . joules per day
81.34920 . . . . joules per hour
1.35582 . . . . joules per minute
0.022597 . . . . joules per second
0.00012684 . . . . pounds of carbon oxidized with
perfect efficiency per day
0.0000052848 . . . . pounds of carbon oxidized with
perfect efficiency per hour
0.00000008808 . . . . pounds of carbon oxidized with
perfect efficiency per minute
0.000000001468 . . . . pounds of carbon oxidized with
perfect efficiency per second
0.0019089 . . . . pounds of water evaporated from
and at 212°F. per day
0.000079536 . . . . pounds of water evaporated from
and at 212°F. per hour
0.0000013256 . . . . pounds of water evaporated from
and at 212°F. per minute
0.000000022093 . . . . pounds of water evaporated from
and at 212°F. per second
1.85098 . . . . BTU per day
0.077124 . . . . BTU per hour
0.0012854 . . . . BTU per minute
0.000021423 . . . . BTU per second

FOOT POUND PER SECOND: =

2,779,833.6 . . . . foot poundals per day
115,826.40 . . . . foot poundals per hour

FOOT POUND PER SECOND: (cont'd)

1,930.440	foot poundals per minute
32.174	foot poundals per second
86,400	foot pounds per day
3,600	foot pounds per hour
60	foot pounds per minute
1.0	foot pounds per second
27.9841	kilogram calories per day
1.1660	kilogram calories per hour
0.019433	kilogram calories per minute
0.00032389	kilogram calories per second
987.120	ounce calories per day
41.130	ounce calories per hour
0.68550	ounce calories per minute
0.011425	ounce calories per second
11,945.2320	kilogram meters per day
497.7180	kilogram meters per hour
8.29530	kilogram meters per minute
0.138255	kilogram meters per second
1,156.11840	liter-atmospheres per day
48.17160	liter-atmospheres per hour
0.80286	liter-atmospheres per minute
0.013381	liter-atmospheres per second
40.82659	cubic foot atmospheres per day
1.70111	cubic foot atmospheres per hour
0.028352	cubic foot atmospheres per minute
0.00047253	cubic foot atmospheres per second
0.0018432	Cheval-vapeurs
0.0018182	horsepower
0.0013558	kilowatts
1.35582	watts
117,142.8480	joules per day
4,880.9520	joules per hour
81.34920	joules per minute
1.35582	joules per second
0.0076101	pounds of carbon oxidized with perfect efficiency per day
0.00031709	pounds of carbon oxidized with perfect efficiency per hour
0.0000052848	pounds of carbon oxidized with perfect efficiency per minute
0.00000008808	pounds of carbon oxidized with perfect efficiency per second
0.11453	pounds of water evaporated from and at 212°F. per day
0.0047722	pounds of water evaporated from and at 212°F. per hour
0.000079536	pounds of water evaporated from and at 212°F. per minute
0.0000013256	pounds of water evaporated from and at 212°F. per second
111.05856	BTU per day

FOOT POUND PER SECOND: (cont'd)

4.62744	BTU per hour
0.077124	BTU per minute
0.0012854	BTU per second

FOOT OF WATER AT 60°F: =

0.00304801	hektometers of water at 60°F.
0.0304801	dekameters of water at 60°F.
0.304801	meters of water at 60°F.
1.0	feet of water at 60°F.
12	inches of water at 60°F.
3.04801	decimeters of water at 60°F.
30.4801	centimeters of water at 60°F.
304.801	millimetrs of water at 60°F.
0.002240	hektometers of mercury at 32°F.
0.02240	dekameters of mercury at 32°F.
0.2240	meters of mercury at 32°F.
0.73491	feet of mercury at 32°F.
8.81888	ounces of mercury at 32°F.
0.882612	inches of mercury at 32°F.
2.240	decimeters of mercury at 32°F.
22.40	centimeters of mercury at 32°F.
224.0	millimeters of mercury at 32°F.
3,359.85556	tons per square hektometer
33.59856	tons per square dekameter
0.33599	tons per square meter
0.031214	tons per square foot
0.00021677	tons per square inch
0.0033599	tons per square decimeter
0.000033599	tons per square centimeter
0.00000033599	tons per square millimeter
3,048,010	kilograms per square hektometer
30,480.1	kilograms per square dekameter
304.801	kilograms per square meter
28.31705	kilograms per square foot
0.19665	kilograms per square inch
3.04801	kilograms per square decimeter
0.0304801	kilograms per square centimeter
0.000304801	kilograms per square millimeter
6,719,721.9	pounds per square hektometer
67,197.2	pounds per square dekameter
671.97219	pounds per square meter
62.42832	pounds per square foot
0.433530	pounds per square inch
6.71972	pounds per square decimeter
0.067197	pounds per square centimeter
0.00067197	pounds per square millimeter
30,480,100	hektograms per square hektometer
304,801	hektograms per square dekameter
3,048.01	hektograms per square meter
283.1705	hektograms per square foot

FOOT OF WATER AT 60°F: (cont'd)

1.9665 . . . . hektograms per square inch
30.4801 . . . . hektograms per square decimeter
0.304801 . . . . hektograms per square centimeter
0.00304801 . . . . hektograms per square millimeter
304,801,000 . . . . dekagrams per square hektometer
3,048,010 . . . . dekagrams per square dekameter
30,480.1 . . . . dekagrams per square meter
2,831.705 . . . . dekagrams per square foot
19.655 . . . . dekagrams per square inch
304.801 . . . . dekagrams per square decimeter
3.04801 . . . . dekagrams per square centimeter
0.0304801 . . . . dekagrams per square millimeter
107,515,550.4 . . . . ounces per square hektometer
1,075,155.504 . . . . ounces per square dekameter
10,751.55504 . . . . ounces per square meter
998.85312 . . . . ounces per square foot
6.93648 . . . . ounces per square inch
107.51556 . . . . ounces per square decimeter
1.075156 . . . . ounces per square centimeter
0.0107516 . . . . ounces per square millimeter
3,048,010,000 . . . . grams per square hektometer
30,480,100 . . . . grams per square dekameter
304,801 . . . . grams per square meter
28,317.05 . . . . grams per square foot
196.55 . . . . grams per square inch
3,048.01 . . . . grams per square decimeter
30.4801 . . . . grams per square centimeter
0.304801 . . . . grams per square millimeter
30,480,100,000 . . . . decigrams per square hektometer
304,801,000 . . . . decigrams per square dekameter
3,048,010 . . . . decigrams per square meter
283,170.5 . . . . decigrams per square foot
1,966.5 . . . . decigrams per square inch
30,480.1 . . . . decigrams per square decimeter
304.801 . . . . decigrams per square centimeter
3.04801 . . . . decigrams per square millimeter
304,801,000,000 . . . . centigrams per square hektometer
3,048,010,000 . . . . centigrams per square dekameter
30,480,100 . . . . centigrams per square meter
2,831,705 . . . . centigrams per square foot
19,665 . . . . centigrams per square inch
304,801 . . . . centigrams per square decimeter
3,048.01 . . . . centigrams per square centimeter
30.4801 . . . . centigrams per square millimeter
3,048,010,000,000 . . . . milligrams per square hektometer
30,480,100,000 . . . . milligrams per square dekameter
304,801,000 . . . . milligrams per square meter
28,317,050 . . . . milligrams per square foot
196,650 . . . . milligrams per square inch
3,048,010 . . . . milligrams per square decimeter
30,480.1 . . . . milligrams per square centimeter

759

FOOT OF WATER AT 60°F: (cont'd)

304.801	milligrams per square millimeter
0.029889	bars
0.0294979	atmospheres
821.2	feet of air @ 62°F. and 29.92 barometer pressure
2,988,874,717,500	dynes per square hektometer
29,888,747,175	dynes per square dekameter
298,887,471.75	dynes per square meter
27,767,658.99497	dynes per square foot
192,830.60410	dynes per square inch
2,988,874.7175	dynes per square decimeter
29,888.74718	dynes per square centimeter
298.88747	dynes per square millimeter

GALLON (DRY): =

0.00044040	kiloliters
0.00044040	cubic meters
0.0057601	cubic yards
0.027701	barrels
0.0044040	hektoliters
0.12497	bushels—U.S. (dry)
0.12116	bushels—Imperial (dry)
0.15553	cubic feet
0.44040	dekaliters
1.16342	gallons—U.S. (liquid)
1	gallons—U.S. (dry)
0.96874	gallons—Imperial
4.65368	quarts (liquid)
4	quarts (dry)
4.4040	liters
4.4040	cubic decimeters
9.30736	pints (fluid)
37.22943	gills (fluid)
44.04010	deciliters
268.75	cubic inches
440.40097	centiliters
4,404.00974	milliliters
4,404.00974	cubic centimeters
4,404,009.74	cubic millimeters
0.14139	sacks of cement
71,481	minims
148.91775	ounces (fluid)
1,191.34199	drams (fluid)
0.0043301	tons (long) of water @ 62°F.
0.0043996	tons (metric) of water @ 62°F.
0.0048498	tons (short) of water @ 62°F.
4.40005	kilograms of water @ 62°F.
9.69943	pounds (Avoir) of water @ 62°F.
11.78750	pounds (Troy) of water @ 62°F.
44.00054111	hektograms

GALLON (DRY): (cont'd)

440.0054111	dekagrams
141.45004	ounces (Troy) of water @ 62°F.
155.19091	ounces (Avoir.) of water @ 62°F.
1,131.60029	drams (Troy) of water @ 62°F.
2,483.05454	drams (Avoir.) of water @ 62°F.
2,829.00071	pennyweights of water @ 62°F.
3,394.80086	scruples (Avoir.) of water @ 62°F.
4,400.054111	grams of water @ 62°F.
44,000.54111	decigrams of water @ 62°F.
67,896.022703	grains of water @ 62°F.
440,005.41110	centigrams of water @ 62°F.
4,400,054.1197	milligrams of water @ 62°F.

GALLON (IMPERIAL): =

0.00045460	kiloliters
0.00045460	cubic meters
0.0059459	cubic yards
0.028594	barrels
0.045460	hektoliters
0.12900	bushels—U.S. (dry)
0.125066	bushels—Imperial (dry)
0.16054	cubic feet
0.45460	dekaliters
1.20094	gallons—U.S. (liquid)
1.032184	gallons—U.S. (dry)
1	gallon—Imperial
4.80376	quarts (liquid)
4.12820	quarts (dry)
4.54596	liters
4.54596	cubic decimeters
9.60752	pints (liquid)
38.43008	gills (liquid)
45.4596	deciliters
277.41714	cubic inches
454.596	centiliters
4,545.96	milliliters
4,545.96	cubic centimeters
4,545,960	cubic millimeters
0.14595	sacks of cement
73,785.7536	minims
153.72032	ounces (fluid)
1,299.76256	drams (fluid)
0.0044698	tons (long) of water @ 62°F.
0.0045415	tons (metric) of water @ 62°F.
0.0050061	tons (short) of water @ 62°F.
4.54196	kilograms of water @ 62°F.
10.012237	pounds (Avoir.) of water @ 62°F.
12.16765	pounds (Troy) of water @ 62°F.
45.41955	hektograms of water @ 62°F.
454.19551	dekagrams of water @ 62°F.

GALLON (IMPERIAL): (cont'd)

146.011774 . . . . ounces (Troy) of water @ 62°F.
160.19579 . . . . ounces (Avoir.) of water @ 62°F.
1,168.094195 . . . . drams (Troy) of water @ 62°F.
2,563.13262 . . . . drams (Avoir.) of water @ 62°F.
2,902.23549 . . . . pennyweights of water @ 62°F.
3,504.28258 . . . . scruples of water @ 62°F.
4,541.95508 . . . . grams of water @ 62°F.
45,419.5508 . . . . decigrams of water @ 62°F.
70,085.65746 . . . . grains of water @ 62°F.
454,195.508 . . . . centigrams of water @ 62°F.
4,541,955.08 . . . . milligrams of water @ 62°F.

GALLON (LIQUID): =

0.0037854 . . . . kiloliters
0.0037854 . . . . cubic meters
0.004951 . . . . cubic yards
0.0238095 . . . . barrels
0.037854 . . . . hektoliters
0.10742 . . . . bushels—U.S. (dry)
0.10414 . . . . bushels—Imperial (dry)
0.133681 . . . . cubic feet
0.37854 . . . . dekaliters
1 . . . . gallons—U.S. (liquid)
0.85948 . . . . gallons—U.S. (Dry)
0.83268 . . . . gallons—Imperial
4 . . . . quarts (liquid)
3.43747 . . . . quarts (dry)
3.78544 . . . . liters
3.78544 . . . . cubic decimeters
8 . . . . pints (liquid)
32 . . . . gills (liquid)
37.8544 . . . . deciliters
231 . . . . cubic inches
378.544 . . . . centiliters
3,785.44 . . . . milliliters
3,785.44 . . . . cubic centimeters
3,785,440 . . . . cubic millimeters
0.12153 . . . . sacks of cement
61,440 . . . . minims
128 . . . . ounces (fluid)
1,024 . . . . drams (fluid)
0.0037219 . . . . tons (long) of water @ 62°F.
0.0037816 . . . . tons (metric) of water @ 62°F.
0.0041685 . . . . tons (short) of water @ 62°F.
3.7820 . . . . kilograms of water @ 62°F.
8.337 . . . . pounds (Avoir.) of water @ 62°F.
10.13177 . . . . pounds (Troy) of water @ 62°F.
37.820 . . . . hektograms of water @ 62°F.
378.20 . . . . dekagrams of water @ 62°F.
121.58124 . . . . ounces (Troy) of water @ 62°F.

GALLON (LIQUID): (cont'd)

133.392	ounces (Avoir.) of water @ 62°F.
972.64992	drams (Troy) of water @ 62°F.
2,134.272	drams (Avoir.) of water @ 62°F.
2,431.6284	pennyweights of water @ 62°F.
2,917.94976	scruples (Avoir.) of water @ 62°F.
3,782	grams of water @ 62°F.
37,820	decigrams of water @ 62°F.
58,359	grains of water @ 62°F.
378,200	centigrams of water @ 62°F.
3,782,000	milligrams of water @ 62°F.

GALLON (U.S.) PER DAY: =

0.00037854	kiloliters per day
0.000015773	kiloliters per hour
0.00000026288	kiloliters per minute
0.0000000043813	kiloliters per second
0.00037854	cubic meters per day
0.000015773	cubic meters per hour
0.00000026288	cubic meters per minute
0.0000000043813	cubic meters per second
0.004951	cubic yards per day
0.00020629	cubic yards per hour
0.0000034382	cubic yards per minute
0.000000057303	cubic yards per second
0.02381	barrels per day
0.00099208	barrels per hour
0.000016535	barrels per minute
0.00000027558	barrels per second
0.0037854	hektoliters per day
0.00015773	hektoliters per hour
0.0000026288	hektoliters per minute
0.000000043813	hektoliters per second
0.10742	bushels—U.S. (dry) per day
0.0044758	bushels—U.S. (dry) per hour
0.000074597	bushels—U.S. (dry) per minute
0.0000012433	bushels—U.S. (dry) per second
0.10414	bushels (Imperial) per day
0.00433917	bushels (Imperial) per hour
0.0000723194	bushels (Imperial) per minute
0.00000120532	bushels (Imperial) per second
0.133681	cubic feet per day
0.00557002	cubic feet per hour
0.0000928337	cubic feet per minute
0.00000154723	cubic feet per second
0.378544	dekaliters per day
0.0157725	dekaliters per hour
0.000262875	dekaliters per minute
0.00000438128	dekaliters per second
1.0	gallons—U.S. (liquid) per day
0.0416667	gallons—U.S. (liquid) per hour

GALLON (U.S.) PER DAY: (cont'd)

0.000694444	gallons—U.S. (liquid) per minute
0.0000115741	gallons—U.S. (liquid) per second
0.83268	gallons (Imperial) per day
0.0346950	gallons (Imperial) per hour
0.000578250	gallons (Imperial) per minute
0.00000963750	gallons (Imperial) per second
3.78544	liters per day
0.157726	liters per hour
0.00262877	liters per minute
0.0000438128	liters per second
3.78543	cubic decimeters per day
0.157726	cubic decimeters per hour
0.00262877	cubic decimeters per minute
0.0000438128	cubic decimeters per second
4.0	quarts per day
0.166667	quarts per hour
0.00277778	quarts per minute
0.0000462963	quarts per second
8.0	pints per day
0.333333	pints per hour
0.00555556	pints per minute
0.0000925925	pints per second
32.0	gills per day
1.333333	gills per hour
0.0222222	gills per minute
0.000370370	gills per second
37.8544	deciliters per day
1.57726	deciliters per hour
0.0262877	deciliters per minute
0.000438128	deciliters per second
231.0	cubic inches per day
9.625	cubic inches per hour
0.160417	cubic inches per minute
0.00267361	cubic inches per second
378.544	centiliters per day
15.772608	centiliters per hour
0.262877	centiliters per minute
0.00438128	centiliters per second
3,785.44	milliliters per day
157.726080	milliliters per hour
2.628768	milliliters per minute
0.0438128	milliliters per second
3,785.44	cubic centimeters per day
157.726080	cubic centimeters per hour
2.628768	cubic centimeters per minute
0.0438128	cubic centimeters per second
3,785,440	cubic millimeters per day
157,726.249920	cubic millimeters per hour
2,628.770832	cubic millimeters per minute
43.8128	cubic millimeters per second

GALLON PER HOUR: =

0.0908506	kiloliters per day
0.00378544	kiloliters per hour
0.0000630907	kiloliters per minute
0.00000105151	kiloliters per second
0.0908506	cubic meters per day
0.00378544	cubic meters per hour
0.0000630907	cubic meters per minute
0.00000105151	cubic meters per second
0.11882	cubic yards per day
0.004951	cubic yards per hour
0.000082517	cubic yards per minute
0.0000013753	cubic yards per second
0.57144	barrels per day
0.0238095	barrels per hour
0.00039683	barrels per minute
0.0000066139	barrels per second
0.908506	hektoliters per day
0.0378544	hektoliters per hour
0.000630907	hektoliters per minute
0.0000105151	hektoliters per second
2.57808	bushels—U.S. (dry) per day
0.10742	bushels—U.S. (dry) per hour
0.0017903	bushels—U.S. (dry) per minute
0.000029839	bushels—U.S. (dry) per second
2.49936	bushels (Imperial) per day
0.10414	bushels (Imperial) per hour
0.0017357	bushels (Imperial) per minute
0.000028928	bushels (Imperial) per second
3.20834	cubic feet per day
0.133681	cubic feet per hour
0.0022280	cubic feet per minute
0.000037134	cubic feet per second
9.0850560	dekaliters per day
0.378544	dekaliters per hour
0.00630907	dekaliters per minute
0.000105151	dekaliters per second
24	gallons—U.S. (liquid) per day
1.0	gallons—U.S. (liquid) per hour
0.016667	gallons—U.S. (liquid) per minute
0.00027778	gallons—U.S. (liquid) per second
19.98432	gallons (Imperial) per day
0.83268	gallons (Imperial) per hour
0.013878	gallons (Imperial) per minute
0.0002313	gallons (Imperial) per second
90.850560	liters per day
3.78544	liters per hour
0.0630907	liters per minute
0.00105151	liters per second
90.850560	cubic decimeters per day

GALLON PER HOUR: (cont'd)

3.785440	cubic decimeters per hour
0.0630907	cubic decimeters per minute
0.00105151	cubic decimeters per second
96	quarts per day
4.0	quarts per hour
0.066667	quarts per minute
0.0011111	quarts per second
192	pints per day
8.0	pints per hour
0.13333	pints per minute
0.0022222	pints per second
768	gills per day
32.0	gills per hour
0.53333	gills per minute
0.0088889	gills per second
908.505600	deciliters per day
37.854400	deciliters per hour
0.630907	deciliters per minute
0.0105151	deciliters per second
5,544	cubic inches per day
231	cubic inches per hour
3.85	cubic inches per minute
0.064167	cubic inches per second
9,085.055999	centiliters per day
378.544000	centiliters per hour
6.309067	centiliters per minute
0.105151	centiliters per second
90,850.559990	milliliters per day
3,785.440	milliliters per hour
63.090667	milliliters per minute
1.051511	milliliters per second
90,850.559990	cubic centimeters per day
3,785.440	cubic centimeters per hour
63.090667	cubic centimeters per minute
1.051511	cubic centimeters per second
90,850,560	cubic millimeters per day
3,785,440	cubic millimeters per hour
63,090.666660	cubic millimeters per minute
1,051.511111	cubic millimeters per second

GALLON PER MINUTE: =

5.451034	kiloliters per day
0.227126	kiloliters per hour
0.00378544	kiloliters per minute
0.0000630907	kiloliters per second
5.451034	cubic meters per day
0.227126	cubic meters per hour
0.00378544	cubic meters per minute
0.0000630907	cubic meters per second
7.129440	cubic yards per day

GALLON PER MINUTE: (cont'd)

0.297060	cubic yards per hour
0.004951	cubic yards per minute
0.0000825167	cubic yards per second
34.285680	barrels per day
1.428570	barrels per hour
0.0238095	barrels per minute
0.000396825	barrels per second
54.510336	hektoliters per day
2.271264	hektoliters per hour
0.0378544	hektoliters per minute
0.000630907	hektoliters per second
154.6848	bushels—U.S. (dry) per day
6.4452	bushels—U.S. (dry) per hour
0.10742	bushels—U.S. (dry) per minute
0.00179033	bushels—U.S. (dry) per second
149.9616	bushels (Imperial) per day
6.2484	bushels (Imperial) per hour
0.10414	bushels (Imperial) per minute
0.00173567	bushels (Imperial) per second
192.500640	cubic feet per day
8.020860	cubic feet per hour
0.133681	cubic feet per minute
0.00222802	cubic feet per second
545.103359	dekaliters per day
22.712640	dekaliters per hour
0.378544	dekaliters per minute
0.00630907	dekaliters per second
1,440	gallons—U.S. (liquid) per day
60	gallons—U.S. (liquid) per hour
1.0	gallons—U.S. (liquid) per minute
0.0166667	gallons—U.S. (liquid) per second
1,199.05920	gallons (Imperial) per day
49.960800	gallons (Imperial) per hour
0.83268	gallons (Imperial) per minute
0.0138780	gallons (Imperial) per second
5,451.033594	liters per day
227.126400	liters per hour
3.78544	liters per minute
0.0630907	liters per second
5,451.033594	cubic decimeters per day
227.126400	cubic decimeters per hour
3.78544	cubic decimeters per minute
0.0630907	cubic decimeters per second
5,760	quarts per day
240	quarts per hour
4.0	quarts per minute
0.0666667	quarts per second
11,520	pints per day
480	pints per hour
8.0	pints per minute
0.133333	pints per second

GALLON PER MINUTE: (cont'd)

46,080	gills per day
1,920	gills per hour
32.0	gills per minute
0.533333	gills per second
54,510.335942	deciliters per day
2,271.263998	deciliters per hour
37.854400	deciliters per minute
0.630907	deciliters per second
332,640	cubic inches per day
13,860	cubic inches per hour
231.0	cubic inches per minute
3.850	cubic inches per second
545,103.359424	centiliters per day
22,712.639976	centiliters per hour
378.544000	centiliters per minute
6.309067	centiliters per second
5,451,034	milliliters per day
227,126.399760	milliliters per hour
3,785.440	milliliters per minute
63.090667	milliliters per second
5,451,034	cubic centimeters per day
227,126.399760	cubic centimeters per hour
3,785.440	cubic centimeters per minute
63.090667	cubic centimeters per second
5,451,033,594	cubic millimeters per day
227,126,400	cubic millimeters per hour
3,785,440	cubic millimeters per minute
63,090.6666	cubic millimeters per second

GALLON PER SECOND: =

327.062016	kiloliters per day
13.627584	kiloliters per hour
0.227126	kiloliters per minute
0.00378544	kiloliters per second
327.062016	cubic meters per day
13.627584	cubic meters per hour
0.227126	cubic meters per minute
0.00378544	cubic meters per second
427.766400	cubic yards per day
17.823600	cubic yards per hour
0.297060	cubic yards per minute
0.004951	cubic yards per second
2,057.140800	barrels per day
85.714200	barrels per hour
1.428570	barrels per minute
0.0238095	barrels per second
3,270.620160	hektoliters per day
136.275840	hektoliters per hour
2.271264	hektoliters per minute
0.0378544	hektoliters per second

GALLON PER SECOND: (cont'd)

9,281.08800	bushels—U.S. (dry) per day
386.712000	bushels—U.S. (dry) per hour
6.44520	bushels—U.S. (dry) per minute
0.10742	bushels—U.S. (dry) per second
8,997.696000	bushels (Imperial) per day
374.90400	bushels (Imperial) per hour
6.248400	bushels (Imperial) per minute
0.10414	bushels (Imperial) per second
11,550.038400	cubic feet per day
481.251600	cubic feet per hour
8.020860	cubic feet per minute
0.133681	cubic feet per second
32,706.201600	dekaliters per day
1,362.758400	dekaliters per hour
22.712640	dekaliters per minute
0.378544	dekaliters per second
86,400	gallons—U.S. (liquid) per day
3,600	gallons—U.S. (liquid) per hour
60	gallons—U.S. (liquid) per minute
1.0	gallons—U.S. (liquid) per second
71,943.55200	gallons (Imperial) per day
2,997.64800	gallons (Imperial) per hour
49.96080	gallons (Imperial) per minute
0.83268	gallons (Imperial) per second
327,062.01600	liters per day
13,627.58400	liters per hour
227.12640	liters per minute
3.78544	liters per second
327,062.01600	cubic decimeters per day
13,627.58400	cubic decimeters per hour
227.12640	cubic decimeters per minute
3.78544	cubic decimeters per second
345,600	quarts per day
14,400	quarts per hour
240	quarts per minute
4.0	quarts per second
691,200	pints per day
28,800	pints per hour
480	pints per minute
8.0	pints per second
2,764,800	gills per day
115,200	gills per hour
1,920	gills per minute
32.0	gills per second
3,270,620	deciliters per day
136,275.8400	deciliters per hour
2,271.2640	deciliters per minute
37.8544	deciliters per second
19,958,400	cubic inches per day
831,600	cubic inches per hour
13,860	cubic inches per minute

GALLON PER SECOND: (cont'd)

231	cubic inches per second
32,706,202	centiliters per day
1,362,758	centiliters per hour
22,712.640	centiliters per minute
278.544	centiliters per second
327,062,016	milliliters per day
13,627,584	milliliters per hour
227,126.40	milliliters per minute
3,785.44	milliliters per second
327,062,016	cubic centimeters per day
13,627,584	cubic centimeters per hour
227,126.40	cubic centimeters per minute
3,785.44	cubic centimeters per second
327,061,016,000	cubic millimeters per day
13,627,584,000	cubic millimeters per hour
227,126,400	cubic millimeters per minute
3,785,440	cubic millimeters per second

GRAIN (AVOIRDUPOIS): =

0.000000637755089	tons (long)
0.000000647989857	tons (metric)
0.0000007142857	tons (short)
0.000064798918	kilograms
0.00014285714	pounds (Avoir.)
0.00017361111	pounds (Troy)
0.00064798918	hektograms
0.0064798918	dekagrams
0.00208333	ounces (Troy)
0.0022857	ounces (Avoir.)
0.03657143	drams (Avoir.)
0.0166667	drams (Troy)
0.0416667	pennyweights (Troy)
0.05000	scruples (Troy)
0.064798918	grams
0.64798918	decigrams
6.4798918	centigrams
64.798918	milligrams
0.3240	carats (metric)

GRAIN (AVOIR.) PER BARREL: =

718.956	parts per million
0.000793548	tons (net) per cubic meter
0.71988	kilograms per cubic meter
1.587054	pounds (Avoir.) per cubic meter
25.39320	ounces (Avoir.) per cubic meter
406.29120	drams (Avoir.) per cubic meter
719.88	grams per cubic meter
0.000126161	tons (net) per barrel

GRAIN (AVOIR.) PER BARREL: (cont'd)

0.114449 . . . . kilograms per barrel
0.252316 . . . . pounds (Avoir.) per barrel
4.037124 . . . . ounces (Avoir.) per barrel
64.59390 . . . . drams (Avoir.) per barrel
114.44916 . . . . grams per barrel
0.00000300384 . . . . tons (net) per gallon—U.S.
0.00272498 . . . . kilograms per gallon—U.S.
0.00600753 . . . . pounds (Avoir.) per gallon—U.S.
0.096122 . . . . ounces (Avoir.) per gallon—U.S.
1.53795 . . . . drams (Avoir.) per gallon—U.S.
2.724982 . . . . grams per gallon—U.S.
0.000000793548 . . . . tons (net) per liter
0.00071988 . . . . kilograms per liter
0.00158705 . . . . pounds (Avoir.) per liter
0.0253932 . . . . ounces (Avoir.) per liter
0.406291 . . . . drams (Avoir.) per liter
0.71988 . . . . grams per liter
0.000000000793548 . . . . tons (net) per cubic centimeter
0.00000071980 . . . . kilograms per cubic centimeter
0.00000158705 . . . . pounds (Avoir.) per cubic centimeter
0.0000253932 . . . . ounces (Avoir.) per cubic centimeter
0.000406291 . . . . drams (Avoir.) per cubic centimeter
0.00071988 . . . . grams per cubic centimeter

GRAIN (AVOIR.) PER CUBIC CENTIMETER: =

0.00452219 . . . . parts per million
0.00000000499137 . . . . tons (net) per cubic meter
0.00000452800 . . . . kilograms per cubic meter
0.00000998248 . . . . pounds (Avoir.) per cubic meter
0.000159722 . . . . ounces (Avoir.) per cubic meter
0.00255555 . . . . drams (Avoir.) per cubic meter
0.00452800 . . . . grams
0.000000000793545 . . . . tons (net) per barrel
0.000000719878 . . . . kilograms per barrel
0.00000158705 . . . . pounds (Avoir.) per barrel
0.0000253933 . . . . ounces (Avoir.) per barrel
0.000406292 . . . . drams (Avoir.) per barrel
0.000719879 . . . . grams per barrel
0.0000000000188940 . . . . tons (net) per gallon—U.S.
0.00000001714 . . . . kilograms per gallon—U.S.
0.0000000377873 . . . . pounds (Avoir.) per gallon—U.S.
0.000000604602 . . . . ounces (Avoir.) per gallon—U.S.
0.00000967362 . . . . drams (Avoir.) per gallon—U.S.
0.00001714 . . . . grams per gallon—U.S.
0.00000000000499137 . . . . tons (net) per liter
0.00000000452800 . . . . kilograms per liter
0.00000000998248 . . . . pounds (Avoir.) per liter
0.000000159722 . . . . ounces (Avoir.) per liter
0.00000255555 . . . . drams (Avoir.) per liter
0.00000452800 . . . . grams per liter

GRAIN (AVOIR.) PER CUBIC CENTIMETER: (cont'd)

0.0000000000000499137 . . . . tons (net) per cubic centimeter
0.00000000000452800 . . . . kilograms per cubic centimeter
0.00000000000998248 . . . . pounds (Avoir.) per cubic centimeter
0.000000000159722 . . . . ounces (Avoir.) per cubic centimeter
0.00000000255555 . . . . drams (Avoir.) per cubic centimeter
0.00000000452800 . . . . grams per cubic centimeter

GRAIN (AVOIR.) PER CUBIC METER: =

4,522.192260 . . . . parts per million
0.00499137 . . . . tons (net) per cubic meter
4.528004 . . . . kilograms per cubic meter
9.982479 . . . . pounds (Avoir.) per cubic meter
159.721781 . . . . ounces (Avoir.) per cubic meter
2,555.548489 . . . . drams (Avoir.) per cubic meter
4,528.004167 . . . . grams per cubic meter
0.000793545 . . . . tons (net) per barrel
0.71978 . . . . kilograms per barrel
1.587053 . . . . pounds (Avoir.) per barrel
25.393280 . . . . ounces (Avoir.) per barrel
406.291949 . . . . drams (Avoir.) per barrel
719.878693 . . . . grams per barrel
0.0000188940 . . . . tons (net) per gallon—U.S.
0.01714 . . . . kilograms per gallon—U.S.
0.0377873 . . . . pounds (Avoir.) per gallon—U.S.
0.604602 . . . . ounces (Avoir.) per gallon—U.S.
9.673618 . . . . drams (Avoir.) per gallon—U.S.
17.14 . . . . grams per gallon—U.S.
0.00000499137 . . . . tons (net) per liter
0.00452800 . . . . kilograms per liter
0.00998248 . . . . pounds (Avoir.) per liter
0.159722 . . . . ounces (Avoir.) per liter
2.555548 . . . . drams (Avoir.) per liter
4.528004 . . . . grams per liter
0.00000000499137 . . . . tons (net) per cubic centimeter
0.00000452800 . . . . kilograms per cubic centimeter
0.00000998248 . . . . pounds (Avoir.) per cubic centimeter
0.000159722 . . . . ounces (Avoir.) per cubic centimeter
0.00255555 . . . . drams (Avoir.) per cubic centimeter
0.00452800 . . . . grams

GRAIN (AVOIR.) PER GALLON—U.S.: =

17.118 . . . . parts per million
0.000018894 . . . . tons (net) per cubic meter
0.01714 . . . . kilograms per cubic meter
0.037787 . . . . pounds (Avoir.) per cubic meter
0.60460 . . . . ounces (Avoir.) per cubic meter
9.67360 . . . . drams (Avoir.) per cubic meter
17.14 . . . . grams per cubic meter

GRAIN (AVOIR.) PER GALLON—U.S.: (cont'd)

0.00000300384 .... tons (net) per barrel
0.00272498 .... kilograms per barrel
0.00600753 .... pounds (Avoir.) per barrel
0.096122 .... ounces (Avoir.) per barrel
1.53795 .... drams (Avoir.) per barrel
2.72498 .... grams per barrel
0.000000071520 .... tons (net) per gallon—U.S.
0.000064881 .... kilograms per gallon—U.S.
0.00014304 .... pounds (Avoir.) per gallon—U.S.
0.0022886 .... ounces (Avoir.) per gallon—U.S.
0.036617 .... drams (Avoir.) per gallon—U.S.
0.064881 .... grams per gallon—U.S.
0.000000018894 .... tons (net) per liter
0.00001714 .... kilograms per liter
0.000037787 .... pounds (Avoir.) per liter
0.00060460 .... ounces (Avoir.) per liter
0.0096735 .... drams (Avoir.) per liter
0.01714 .... grams per liter
0.000000000018894 .... tons (net) per cubic centimeter
0.00000001714 .... kilograms per cubic centimeter
0.000000037787 .... pounds (Avoir.) per cubic centimeter
0.00000060460 .... ounces (Avoir.) per cubic centimeter
0.0000096735 .... drams (Avoir.) per cubic centimeter
0.00001714 .... grams per cubic centimeter

GRAIN (AVOIR.) PER LITER: =

4.522192 .... parts per million
0.00000499137 .... tons (net) per cubic meter
0.00452800 .... kilograms per cubic meter
0.00998248 .... pounds (Avoir.) per cubic meter
0.159722 .... ounces (Avoir.) per cubic meter
2.555548 .... drams (Avoir.) per cubic meter
4.52804 .... grams per cubic meter
0.00000793545 .... tons (net) per barrel
0.000719878 .... kilograms per barrel
0.00158705 .... pounds (Avoir.) per barrle
0.0253933 .... ounces (Avoir.) per barrel
0.406292 .... drams (Avoir.) per barrel
0.719879 .... grams per barrel
0.0000000188940 .... tons (net) per gallon—U.S.
0.00001714 .... kilograms per gallon—U.S.
0.0000377873 .... pounds (Avoir.) per gallon—U.S.
0.00604602 .... ounces (Avoir.) per gallon—U.S.
0.00967362 .... drams (Avoir.) per gallon—U.S.
0.01714 .... grams per gallon—U.S.
0.0000000049937 .... tons (net) per liter
0.00000452800 .... kilograms per liter
0.00000998248 .... pounds (Avoir.) per liter
0.000159722 .... ounces (Avoir.) per liter
0.00255555 .... drams (Avoir.) per liter

GRAIN (AVOIR.) PER LITER: (cont'd)

```
0.00452800 . . . . . . . . . . . . . . . . . . . . . . . . . . . . . . . . . . . grams per liter
0.00000000000499137 . . . . . . . . . . . . . . . . tons (net) per cubic centimeter
0.00000000452800 . . . . . . . . . . . . . . . . . . . . kilograms per cubic centimeter
0.00000000998248 . . . . . . . . . . . . . . . pounds (Avoir.) per cubic centimeter
0.000000159722 . . . . . . . . . . . . . . . . . ounces (Avoir.) per cubic centimeter
0.00000255555 . . . . . . . . . . . . . . . . . . drams (Avoir.) per cubic centimeter
0.00000452800 . . . . . . . . . . . . . . . . . . . . . . . . grams per cubic centimeter
```

GRAM: =

```
0.00000098426 . . . . . . . . . . . . . . . . . . . . . . . . . . . . . . . . . . . tons (long)
0.000001 . . . . . . . . . . . . . . . . . . . . . . . . . . . . . . . . . . . . . . tons (metric)
0.00000110231 . . . . . . . . . . . . . . . . . . . . . . . . . . . . . . . . . . . tons (short)
0.001 . . . . . . . . . . . . . . . . . . . . . . . . . . . . . . . . . . . . . . . . . kilograms
0.00220462 . . . . . . . . . . . . . . . . . . . . . . . . . . . . . . . . . pounds (Avoir.)
0.00267923 . . . . . . . . . . . . . . . . . . . . . . . . . . . . . . . . . . pounds (Troy)
0.01 . . . . . . . . . . . . . . . . . . . . . . . . . . . . . . . . . . . . . . . . . hektograms
0.1 . . . . . . . . . . . . . . . . . . . . . . . . . . . . . . . . . . . . . . . . . . dekagrams
0.0321507 . . . . . . . . . . . . . . . . . . . . . . . . . . . . . . . . . . . . ounces (Troy)
0.03527392 . . . . . . . . . . . . . . . . . . . . . . . . . . . . . . . . . . ounces (Avoir.)
0.257206 . . . . . . . . . . . . . . . . . . . . . . . . . . . . . . . . . . . . . drams (Troy)
0.564383 . . . . . . . . . . . . . . . . . . . . . . . . . . . . . . . . . . . . . drams (Avoir.)
0.6430149 . . . . . . . . . . . . . . . . . . . . . . . . . . . . . . . . . . . . pennyweights
0.771618 . . . . . . . . . . . . . . . . . . . . . . . . . . . . . . . . . . . . . . . . scruples
5.0 . . . . . . . . . . . . . . . . . . . . . . . . . . . . . . . . . . . . . . . carats (metric)
10.0 . . . . . . . . . . . . . . . . . . . . . . . . . . . . . . . . . . . . . . . . . decigrams
15.4324 . . . . . . . . . . . . . . . . . . . . . . . . . . . . . . . . . . . . . . . . . . grains
100 . . . . . . . . . . . . . . . . . . . . . . . . . . . . . . . . . . . . . . . . . . centigrams
1,000 . . . . . . . . . . . . . . . . . . . . . . . . . . . . . . . . . . . . . . . . . milligrams
```

GRAM CALORIE (MEAN): =

```
99.334 . . . . . . . . . . . . . . . . . . . . . . . . . . . . . . . . . . . . . . foot poundals
1,192.008 . . . . . . . . . . . . . . . . . . . . . . . . . . . . . . . . . . . . inch poundals
3.0874 . . . . . . . . . . . . . . . . . . . . . . . . . . . . . . . . . . . . . . . . foot pounds
37.0488 . . . . . . . . . . . . . . . . . . . . . . . . . . . . . . . . . . . . . . . inch pounds
0.0000001 . . . . . . . . . . . . . . . . . . . . . . . . . . . . . . . ton (short) calories
0.001 . . . . . . . . . . . . . . . . . . . . . . . . . . . . . . . . . . . . kilogram calories
0.00204622 . . . . . . . . . . . . . . . . . . . . . . . . . . . . . . . . . pound calories
0.01 . . . . . . . . . . . . . . . . . . . . . . . . . . . . . . . . . . . . hektogram calories
0.1 . . . . . . . . . . . . . . . . . . . . . . . . . . . . . . . . . . . . . dekagram calories
0.0327395 . . . . . . . . . . . . . . . . . . . . . . . . . . . . . . . . . . . ounce calories
1.0 . . . . . . . . . . . . . . . . . . . . . . . . . . . . . . . . . . . . . . . . . gram calorie
10.0 . . . . . . . . . . . . . . . . . . . . . . . . . . . . . . . . . . . . . decigram calories
100.0 . . . . . . . . . . . . . . . . . . . . . . . . . . . . . . . . . . . centigram calories
1,000 . . . . . . . . . . . . . . . . . . . . . . . . . . . . . . . . . . . . milligram calories
0.00000004845 . . . . . . . . . . . . . . . . . . . . . . . . . . . . . . . kilowatt days
0.0000011628 . . . . . . . . . . . . . . . . . . . . . . . . . . . . . . . . kilowatt hours
0.0000697680 . . . . . . . . . . . . . . . . . . . . . . . . . . . . . . kilowatt minutes
0.004186 . . . . . . . . . . . . . . . . . . . . . . . . . . . . . . . . . kilowatt seconds
```

GRAM CALORIE (MEAN): (cont'd)

0.00004845	watt days
0.0011628	watt hours
0.0697680	watt minutes
4.186	watt seconds
0.00042685	ton meters
0.42685	kilogram meters
0.941043	pound meters
4.2685	hektogram meters
42.685	dekagram meters
15.056688	ounce meters
426.85	gram meters
4,268.5	decigram meters
42,685	centigram meters
426,850	milligram meters
0.0000042685	ton hektometers
0.0042685	kilogram hektometers
0.00941043	pound hektometers
0.042685	hektogram hektometers
0.42685	dekagram hektometers
0.150567	ounce hektometers
4.2685	gram hektometers
42.685	decigram hektometers
426.85	centigram hektometers
4,268.5	milligram hektometers
0.000042685	ton dekameters
0.042685	kilogram dekameters
0.0941043	pound dekameters
0.42685	hektogram dekameters
4.2685	dekagram dekameters
1.50567	ounce dekameters
42.685	gram dekameters
426.85	decigram dekameters
4,268.5	centigram dekameters
42,685	milligram dekameters
0.00140042	ton feet
1.400424	kilogram feet
0.0308704	pound feet
14.004236	hektogram feet
140.042357	dekagram feet
0.493985	ounce feet
1,400.423566	gram feet
14,004.235661	decigram feet
140,042.356605	centigram feet
1,400,423.0	milligram feet
0.0168050	ton inches
16.805083	kilogram inches
37.0488557	pound inches
168.050828	hektogram inches
1,680.0508284	dekagram inches
592.78169	ounce inches
16,805.082792	gram inches

GRAM CALORIE (MEAN): (cont'd)

168,050.82792 .... decigram inches
1,680,508 .... centigram inches
16,805,082 .... milligram inches
0.0042685 .... ton decimeters
4.2685 .... kilogram decimeters
9.410430 .... pound decimeters
42.685 .... hektogram decimeters
426.85 .... dekagram decimeters
150.566885 .... ounce decimeters
4,268.5 .... gram decimeters
42,685 .... decigram decimeters
426,850 .... centigram decimeters
4,268,500 .... milligram decimeters
0.042685 .... ton centimeters
42.685 .... kilogram centimeters
94.104303 .... pound centimeters
426.85 .... hektogram centimeters
4,268.5 .... dekagram centimeters
1,505.668846 .... ounce centimeters
42,685.0 .... gram centimeters
426,850 .... decigram centimeters
4,268,500 .... centigram centimeters
42,685,000 .... milligram centimeters
0.42685 .... ton millimeters
426.85 .... kilogram millimeters
941.043029 .... pound millimeters
4,268.5 .... hektogram millimeters
42,685 .... dekagram millimeters
15,056.688464 .... ounce millimeters
426,850 .... gram millimeters
4,268,500 .... decigram millimeters
42,685,000 .... centigram millimeters
426,850,000 .... milligram millimeters
0.000041311 .... kiloliter-atmospheres
0.00041311 .... hektoliter-atmospheres
0.0041311 .... dekaliter-atmospheres
0.041311 .... liter-atmospheres
0.41311 .... deciliter-atmospheres
4.1311 .... centiliter-atmospheres
41.311 .... milliliter-atmospheres
0.0000000000000415977 .... cubic kilometer-atmospheres
0.0000000000415977 .... cubic hektometer-atmospheres
0.0000000415977 .... cubic dekameter-atmospheres
0.0000415977 .... cubic meter-atmospheres
0.001469 .... cubic feet-atmospheres
2.538415 .... cubic inch-atmospheres
0.0415977 .... cubic decimeter-atmospheres
41.597673 .... cubic centimeter-atmospheres
41,597.673 .... cubic millimeter-atmospheres
4.185829 .... joules per gram
0.00000158097 .... Cheval-vapeur hours

GRAM CALORIE (MEAN): (cont'd)

0.0000000649708	horsepower days
0.0000015593	horsepower hours
0.0000935980	horsepower minutes
0.00561588	horsepower seconds
0.000000271842	pounds of carbon oxidized with 100% efficiency
0.00000408756	pounds of water evaporated from and at 212°F.
1.8	BTU (mean) per pound
0.0039685	BTU
41,858,291	ergs

GRAM CALORIE: =
(15°C. per square centimeter per second for temperature gradient of 1°C. per centimeter): =

4.185829	joules (absolute) per square centimeter per second for temperature gradient of 1°C per centimeter
0.80620	BTU (mean) per square foot per second for a temperature gradient of 1°F. per inch

GRAM WEIGHT SECOND PER SQUARE CENTIMETER: =

980.665	poises
98,066.5	centipoises

GRAM PER CUBIC CENTIMETER: =

100,000	kilograms per cubic meter
2,831.7	kilograms per cubic foot
16.38776	kilograms per cubic inch
0.001	kilograms per cubic centimeter
2,204.62	pounds per cubic meter
62.42822	pounds per cubic foot
0.036127	pounds per cubic inch
0.0022046	pounds per cubic centimeter
35,273.92	ounces per cubic meter
998.85159	ounces per cubic foot
0.57804	ounces per cubic inch
0.035274	ounces per cubic centimeter
1,000,000	grams per cubic meter
28,317.0	grams per cubic foot
16.38705	grams per cubic inch
1.0	grams per cubic centimeter
15,432,400	grains per cubic meter
436,999.27080	grains per cubic inch
252.89148	grains per cubic inch
15.4324	grains per cubic centimeter

GRAM PER CUBIC FOOT: =

0.035314	kilograms per cubic meter
0.001	kilograms per cubic foot
0.00000057870	kilograms per cubic inch
0.000000035314	kilograms per cubic centimeter
0.077854	pounds per cubic meter
0.0022046	pounds per cubic foot
0.0000012758	pounds per cubic inch
0.000000077854	pounds per cubic centimeter
1.24568	ounces per cubic meter
0.035274	ounces per cubic foot
0.000020413	ounces per cubic inch
0.00000124578	ounces per cubic centimeter
35.31444	grams per cubic meter
1.0	grams per cubic foot
0.00057870	grams per cubic inch
0.0000353144	grams per cubic centimeter
544.98657	grains per cubic meter
15.4324	grains per cubic foot
0.0089307	grains per cubic inch
0.00054499	grains per cubic centimeter

GRAM PER CUBIC INCH: =

61.203	kilograms per cubic meter
1.72799	kilograms per cubic foot
0.001	kilograms per cubic inch
0.000061203	kilograms per cubic centimeter
134.53131	pounds per cubic meter
3.80952	pounds per cubic foot
0.0022046	pounds per cubic inch
0.00013453	pounds per cubic centimeter
2,152.52530	ounces per cubic meter
60.95306	ounces per cubic foot
0.035274	ounces per cubic inch
0.0021525	ounces per cubic centimeter
610,230	grams per cubic meter
17,279.88291	grams per cubic foot
1.0	grams per cubic inch
0.61023	grams per cubic centimeter
941,731.3452	grains per cubic meter
26,667.0065002	grains per cubic foot
15.4324	grains per cubic inch
0.94173	grains per cubic centimeter

GRAM PER CUBIC METER: =

0.001	kilograms per cubic meter
0.00028317	kilograms per cubic foot

GRAM PER CUBIC METER: (cont'd)

0.00000016387 . . . . kilograms per cubic inch
0.000000001 . . . . kilograms per cubic centimeter
0.00220462 . . . . pounds per cubic meter
0.000062428 . . . . pounds per cubic foot
0.000000036127 . . . . pounds per cubic inch
0.00000000220462 . . . . pounds per cubic centimeter
0.035274 . . . . ounces per cubic meter
0.00099885 . . . . ounces per cubic foot
0.00000057803 . . . . ounces per cubic inch
0.000000035274 . . . . ounces per cubic centimeter
1.0 . . . . grams per cubic meter
0.028317 . . . . grams per cubic foot
0.000016387 . . . . grams per cubic inch
0.000001 . . . . grams per cubic centimeter
15.4324 . . . . grains per cubic meter
0.437 . . . . grains per cubic foot
0.00025289 . . . . grains per cubic inch
0.000015434 . . . . grains per cubic centimeter

GRAM PER CENTIMETER: =

0.100 . . . . kilograms per meter
0.03048006 . . . . kilograms per foot
0.00254005 . . . . kilograms per inch
0.001 . . . . kilograms per centimeter
0.22046 . . . . pounds per meter
0.067197 . . . . pounds per foot
0.0054014 . . . . pounds per inch
0.0022046 . . . . pounds per centimeter
3.52736 . . . . ounces per meter
1.075152 . . . . ounces per foot
0.086422 . . . . ounces per inch
0.035274 . . . . ounces per centimeter
100 . . . . grams per meter
30.48006 . . . . grams per foot
2.54005 . . . . grams per inch
1.0 . . . . grams per centimeter
1,543.24 . . . . grains per meter
470.38048 . . . . grains per foot
39.19822 . . . . grains per inch
15.4324 . . . . grains per centimeter

GRAM PER FOOT: =

0.0032808 . . . . kilograms per meter
0.001 . . . . kilograms per foot
0.000083333 . . . . kilograms per inch
0.000032808 . . . . kilograms per centimeter
0.0072329 . . . . pounds per meter
0.0022046 . . . . pounds per foot

GRAM PER FOOT: (cont'd)

0.00018372 .... pounds per inch
0.000072329 .... pounds per centimeter
0.11573 .... ounces per meter
0.035274 .... ounces per foot
0.0029395 .... ounces per inch
0.0011573 .... ounces per centimeter
3.28083 .... grams per centimeter
1.0 .... grams per foot
0.083333 .... grams per inch
0.0328083 .... grams per centimeter
50.63113 .... grains per meter
15.4324 .... grains per foot
1.28603 .... grains per inch
0.50631 .... grains per centimeter

GRAM PER INCH: =

0.03937 .... kilograms per meter
0.012 .... kilograms per foot
0.001 .... kilograms per inch
0.00039370 .... kilograms per centimeter
0.086796 .... pounds per meter
0.026455 .... pounds per foot
0.0022046 .... pounds per inch
0.00086796 .... pounds per centimeter
1.38874 .... ounces per meter
0.42329 .... ounces per foot
0.035274 .... ounces per inch
0.013887 .... ounces per centimeter
39.37000 .... grams per meter
12.0 .... grams per foot
1.0 .... grams per inch
0.39370 .... grams per centimeter
607.57359 .... grains per meter
185.1888 .... grains per foot
15.4324 .... grains per inch
6.075736 .... grains per centimeter

GRAM PER LITER: =

8.34543 .... pounds per 1000 gallons
1,000 .... parts per million
0.00110230 .... tons (net) per cubic meter
1.0 .... kilograms per cubic meter
2.2046099 .... pounds (Avoir.) per cubic meter
35.273758 .... ounces (Avoir.) per cubic meter
564.379806 .... drams (Avoir.) per cubic meter
1,000 .... grams per cubic meter
15,432.258039 .... grains per cubic meter
0.0000312140 .... tons (net) per cubic foot
0.0283169 .... kilograms per cubic foot

GRAM PER LITER: (cont'd)

0.0624280 . . . . pounds (Avoir.) per cubic foot
0.998848 . . . . ounces (Avoir.) per cubic foot
15.981559 . . . . drams (Avoir.) per cubic foot
28.316846 . . . . grams per cubic foot
436.995689 . . . . grains per cubic foot
0.000175254 . . . . tons (net) per barrel
0.158988 . . . . kilograms per barrel
0.350508 . . . . pounds (Avoir.) per barrel
5.608134 . . . . ounces (Avoir.) per barrel
89.730060 . . . . drams (Avoir.) per barrel
158.987766 . . . . grams per barrel
2,543.556 . . . . grains per barrel
0.00000417271 . . . . tons (net) per gallon—U.S.
0.00378542 . . . . kilograms per gallon—U.S.
0.00834543 . . . . pounds (Avoir.) per gallon—U.S.
0.133527 . . . . ounces (Avoir.) per gallon—U.S.
2.136430 . . . . drams (Avoir.) per gallon—U.S.
3.785423 . . . . grams per gallon—U.S.
58.418 . . . . grains per gallon—U.S.
0.00000110231 . . . . tons (net) per liter
0.001 . . . . kilograms per liter
0.00220462 . . . . pounds (Avoir.) per liter
0.0352739 . . . . ounces (Avoir.) per liter
0.564383 . . . . drams (Avoir.) per liter
15.4324 . . . . grains per liter
0.00000000110231 . . . . tons (net) per cubic centimeter
0.000001 . . . . kilograms per cubic centimeter
0.00000220462 . . . . pounds (Avoir.) per cubic centimeter
0.0000352739 . . . . ounces (Avoir.) per cubic centimeter
0.000564383 . . . . drams (Avoir.) per cubic centimeter
0.001 . . . . grams per cubic centimeter
0.0154324 . . . . grains per cubic centimeter

GRAM PER METER: =

0.001 . . . . kilograms per meter
0.00030480 . . . . kilograms per foot
0.000025401 . . . . kilograms per inch
0.00001 . . . . kilograms per centimeter
0.0022046 . . . . pounds per meter
0.00067197 . . . . pounds per foot
0.0000560 . . . . pounds per inch
0.000022046 . . . . pounds per centimeter
0.035274 . . . . ounces per meter
0.010752 . . . . ounces per foot
0.0008960 . . . . ounces per inch
0.00035274 . . . . ounces per centimeter
1.0 . . . . grams per meter
0.30480 . . . . grams per foot
0.025401 . . . . grams per inch
0.01 . . . . grams per centimeter

GRAM PER METER: (cont'd)

15.4324	grains per meter
4.70380	grains per foot
0.39198	grains per inch
0.15432	grains per centimeter

GRAM PER SQUARE CENTIMETER: =

10	kilograms per square meter
0.92903	kilograms per square foot
0.0064516	kilograms per square inch
0.001	kilograms per square centimeter
22.046	pounds per square meter
2.048140	pounds per square foot
0.014223	pounds per square inch
0.0022046	pounds per square centimeter
352.7392	ounces per square meter
32.77053	ounces per square foot
0.22756	ounces per square inch
0.035274	ounces per square centimeter
10,000	grams per square meter
929.03	grams per square foot
6.45156	grams per square inch
1.0	grams per square centimeter
154,324	grains per square meter
14,337.16257	grains per square foot
99.56299	grains per square inch
15.4324	grains per square centimeter
0.73556	millimeters of mercury @ 0°C.
0.00096784	atmospheres
980.665	dynes

GRAM PER SQUARE FOOT: =

0.010764	kilograms per square meter
0.001	kilograms per square foot
0.0000069444	kilograms per square inch
0.0000010764	kilograms per square centimeter
0.023730	pounds per square meter
0.00220462	pounds per square foot
0.000015310	pounds per square inch
0.0000023730	pounds per square centimeter
0.37968	ounces per square meter
0.035274	ounces per square foot
0.00024496	ounces per square inch
0.000037968	ounces per square centimeter
10.76387	grams per square meter
1.0	grams per square foot
0.0069444	grams per square inch
0.00107639	grams per square centimeter
166.11235	grains per square meter

GRAM PER SQUARE FOOT: (cont'd)

15.4324 .... grains per square foot
0.10717 .... grains per square inch
0.016611 .... grains per square centimeter

GRAM PER SQUARE INCH: =

1.55 .... kilograms per square meter
0.1440 .... kilograms per square foot
0.001 .... kilograms per square inch
0.000155 .... kilograms per square centimeter
3.41713 .... pounds per square meter
0.31746 .... pounds per square foot
0.0022046 .... pounds per square inch
0.00034171 .... pounds per square centimeter
54.67470 .... ounces per square meter
5.0794444 .... ounces per square foot
0.035274 .... ounces per square inch
0.0054675 .... ounces per square centimeter
1,550 .... grams per square meter
143.99965 .... grams per square foot
1.0 .... grams per square inch
0.155 .... grams per square centimeter
23,920.099878 .... grains per square meter
2,222.25834 .... grains per square foot
15.4324 .... grains per square inch
2.39201 .... grains per square centimeter

GRAM PER SQUARE METER: =

0.001 .... kilograms per square meter
0.000092903 .... kilograms per square foot
0.00000064516 .... kilograms per square inch
0.0000001 .... kilograms per square centimeter
0.0022046 .... pounds per square meter
0.00020481 .... pounds per square foot
0.0000014222 .... pounds per square inch
0.00000022046 .... pounds per square centimeter
0.035274 .... ounces per square meter
0.0032771 .... ounces per square foot
0.000022757 .... ounces per square inch
0.0000035274 .... ounces per square centimeter
1.0 .... grams per square meter
0.092903 .... grams per square foot
0.00064521 .... grams per square inch
0.0001 .... grams per square centimeter
15.4324 .... grains per square meter
1.43372 .... grains per square foot
0.0099563 .... grains per square inch
0.00154324 .... grains per square centimeter

HECTARE: =

0.003861	square miles or sections
0.010	square kilometers
0.247104	square furlongs
1.0	square hektometers
24.71044	square chains
100	square dekameters
395.367	square rods
10,000	square meters
11,959.888	square yards
13,949.8	square varas (Texas)
107,639	square feet
191,358	square spans
277,104.4	square links
968,750	square hands
1,000,000	square decimeters
100,000,000	square centimeters
15,500,016	square inches
10,000,000,000	square millimeters
100	ares
1,000	centares (centiares)
2.471044	acres

HEKTOLITER: =

0.000000000053961	cubic miles
0.0000000001	cubic kilometers
0.000000000431688	cubic furlongs
0.0000001	cubic hektometers
0.0000122835	cubic chains
0.0001	cubic dekameters
0.000786142	cubic rods
0.1	kiloliters
0.1	cubic meters
0.13080	cubic yards
0.164759	cubic varas (Texas)
0.628976	barrels
1.0	hektoliters
2.8378	bushels—U.S. (dry)
2.7497	bushel—Imperial (dry)
3.53145	cubic feet
8.370844	cubic spans
10.0	dekaliters
11.3513	pecks
12.283475	cubic links
26.417762	gallons—U.S. (liquid)
22.702	gallons—U.S. (dry)
21.998	gallons—Imperial
95.635866	cubic hands
100	liters

HEKTOLITER: (cont'd)

100	cubic decimeters
105.6710	quarts—U.S. (liquid)
90.8102	quarts—U.S. (dry)
211.34	pints—U.S. (liquid)
181.62	pints—U.S. (dry)
845.38	gills (liquid)
1,000	deciliters
61,025	cubic inches
10,000	centiliters
100,000	milliliters
100,000	cubic centimeters
100,000,000	cubic millimeters
220.46223	pounds of water @ 39°F.
3,381.47	ounces (fluid)
27,051.79	drams (fluid)

HEKTOGRAM: =

0.000098426	tons (long)
0.0001	tons (metric)
0.000110231	tons (short)
0.1	kilograms
0.220462	pounds (Troy)
0.267923	pounds (Avoir.)
1.0	hektograms
10	dekagrams
3.21507	ounces (Troy)
3.527392	ounces (Avoir.)
25.7206	drams (Troy)
56.4383	drams (Avoir.)
64.30149	pennyweights (Troy)
77.1618	scruples (Troy)
500	carats (metric)
1,000	decigrams
1,543.24	grains
10,000	centigrams
100,000	milligrams

HORSEPOWER: =

47,520,000	foot pounds per day
1,980,000	foot pounds per hour
33,000	foot pounds per minute
550	foot pounds per second
570,240,000	inch pounds per day
23,760,000	inch pounds per hour
396,000	inch pounds per minute
6,600	inch pounds per second
15,390.720	kilogram calories (mean) per day
641.280	kilogram calories (mean) per hour

HORSEPOWER: (cont'd)

10.688	kilogram calories (mean) per minute
0.178133	kilogram calories (mean) per second
33,930.724525	pounds calories (mean) per day
1,413.780189	pound calories (mean) per hour
23.563003	pound calories (mean) per minute
0.392716	pound calories (mean) per second
542,891.59248	ounce calories (mean) per day
22,620.483024	ounce calories (mean) per hour
377.008048	ounce calories (mean) per minute
6.283456	ounce calories (mean) per second
15,390,720	gram calories (mean) per day
641,280	gram calories (mean) per hour
10,688	gram calories (mean) per minute
178.133	gram calories (mean) per second
61,081.344	BTU (mean) per day
2,545.5600	BTU (mean) per hour
42.41760	BTU (mean) per minute
0.70696	BTU (mean) per second
0.7452	kilowatts (g=980)
0.74570	kilowatts (g=980.665)
745.2	watts (g=980)
745.70	watts (g=980.665
1.0139	horsepower (metric)
1.0139	Cheval-vapeur hours
0.174	pounds carbon oxidized with 100% efficiency
2.62	pounds water evaporated from and @ 212°F.
635.769600	kiloliter-atmospheres per day
24.490400	kiloliter-atmospheres per hour
0.441507	kiloliter-atmospheres per minute
0.00735844	kiloliter-atmospheres per second
635,769.599962	liter-atmospheres per day
26,490.399998	liter-atmospheres per hour
441.506667	liter-atmospheres per minute
7.358844	liter-atmospheres per second
635,769,599.962	milliliter-atmospheres per day
26,490,399.998	milliliter-atmospheres per hour
441,506.666667	milliliter-atmospheres per minute
7,358.444444	milliliter-atmospheres per second

INCH: =

0.00001578	miles
0.00002540	kilometers
0.000126263	furlongs
0.0002560	hektometers
0.00126263	chains
0.002540	dekameters
0.00505051	rods
0.02540	meters
0.027777	yards
0.030000	varas (Texas)

INCH: (cont'd)

0.083333 . . . . feet
0.111111 . . . . spans
0.126263 . . . . links
0.25000 . . . . hands
0.2540 . . . . decimeters
2.5400 . . . . centimeters
1 . . . . inches
25.40 . . . . millimeters
1000 . . . . mils
25,400 . . . . microns
39,450.33 . . . . wave lengths of red line of cadmium
25,400,000 . . . . millimicrons
25,400,000 . . . . micromillimeters
254,000,000 . . . . Angstrom Units

INCH OF MERCURY @ 32°F. =

0.00345349 . . . . hektometers of water @ 60°F.
0.0345349 . . . . dekameters of water @ 60°F.
0.345349 . . . . meters of water @ 60°F.
1.132944 . . . . feet of water @ 60°F.
13.595326 . . . . inches of water @ 60°F.
3.45349 . . . . decimeters of water @ 60°F.
34.5349 . . . . centimeters of water @ 60°F.
345.349 . . . . millimeters of water @ 60°F.
0.000254 . . . . hektometers of mercury @ 32°F.
0.00254 . . . . dekameters of mercury @ 32°F.
0.0254 . . . . meters of mercury @ 32°F.
0.0833325 . . . . feet of mercury @ 32°F.
1 . . . . inches of mercury @ 32°F.
0.254 . . . . decimeters of mercury @ 32°F.
2.54 . . . . centimeters of mercury @ 32°F.
25.4 . . . . millimeters of mercury @ 32°F.
3806.515240 . . . . tons per square hektometer
38.065152 . . . . tons per square dekameter
0.380652 . . . . tons per square meter
0.0353645 . . . . tons per square foot
0.000245581 . . . . tons per square inch
0.00380652 . . . . tons per square decimeter
0.0000380652 . . . . tons per square centimeter
0.000000380652 . . . . tons per square millimeter
3,453,490 . . . . kilograms per square hektometer
34,534.9 . . . . kilograms per square dekameter
345.349 . . . . kilograms per square meter
32.0811278 . . . . kilograms per square foot
0.222786 . . . . kilograms per square inch
3.45349 . . . . kilograms per square decimeter
0.0345349 . . . . kilograms per square centimeter
0.000345349 . . . . kilograms per square millimeter
7,613,030 . . . . pounds per square hektometer
76,130.300609 . . . . pounds per square dekameter

INCH OF MERCURY @ 32°F.: (cont'd)

761.303006	pounds per square meter
70.726441	pounds per square foot
0.491161	pounds per square inch
7.613030	pounds per square decimeter
0.0761303	pounds per square centimeter
0.000761303	pounds per square millimeter
34,534,900	hektograms per square hektometer
345,349	hektograms per square dekameter
3,453.49	hektograms per square meter
320.811278	hektograms per square foot
2.227864	hektograms per square inch
34.5349	hektograms per square decimeter
0.345349	hektograms per square centimeter
0.00345349	hektograms per square millimeter
345,349,000	dekagrams per square hektometer
3,453,490	dekagrams per square dekameter
34,534.90	dekagrams per square meter
3208.112776	dekagrams per square foot
22.278639	dekagrams per square inch
345.3490	dekagrams per square decimeter
3.453490	dekagrams per square centimeter
0.03453490	dekagrams per square millimeter
121,808.481	ounces per square hektometer
1,218,085	ounces per square dekameter
12,180.85	ounces per square meter
1,131.623063	ounces per square foot
7.858573	ounces per square inch
121.8085	ounces per square decimeter
1.218085	ounces per square centimeter
0.0121809	ounces per square millimeter
3,453,490,000	grams per square hektometer
34,534,900	grams per square dekameter
345,349	grams per square meter
32,081.127762	grams per square foot
222.786665	grams per square inch
3,453.49	grams per square decimeter
34.5349	grams per square centimeter
0.345349	grams per square millimeter
34,534,900,000	decigrams per square hektometer
345,349,000	decigrams per square dekameter
3,453,490	decigrams per square meter
320.811278	decigrams per square foot
2,227.866675	decigrams per square inch
34,534.90	decigrams per square decimeter
345.3490	decigrams per square centimeter
3.453490	decigrams per square millimeter
345,349,000,000	centigrams per square hektometer
3,453,490,000	centigrams per square dekameter
34,534,900	centigrams per square meter
3,208,113	centigrams per square foot
22,278.666751	centigrams per square inch

INCH OF MERCURY @ 32°F.: (cont'd)

```
345,349 . . . . . . . . . . . . . . . . . . . . . . . . . centigrams per square decimeter
3,453.49 . . . . . . . . . . . . . . . . . . . . . . . centigrams per square centimeter
34.5349 . . . . . . . . . . . . . . . . . . . . . . . centigrams per square millimeter
3,453,490,000,000 . . . . . . . . . . . . . . . . milligrams per square hektometer
34,534,900,000 . . . . . . . . . . . . . . . . . . milligrams per square dekameter
345,349,000 . . . . . . . . . . . . . . . . . . . . . . milligrams per square meter
32,081,128 . . . . . . . . . . . . . . . . . . . . . . . . milligrams per square foot
222,786.667776 . . . . . . . . . . . . . . . . . . . . milligrams per square inch
3,453,490 . . . . . . . . . . . . . . . . . . . . . . milligrams per square decimeter
34,534.90 . . . . . . . . . . . . . . . . . . . . . . milligrams per square centimeter
345.3490 . . . . . . . . . . . . . . . . . . . . . . milligrams per square millimeter
0.0338659 . . . . . . . . . . . . . . . . . . . . . . . . . . . . . . . . . . . . . . bars
0.0334214 . . . . . . . . . . . . . . . . . . . . . . . . . . . . . . . . . atmospheres
3,383,845,567,678 . . . . . . . . . . . . . . . . . . . dynes per square hektometer
33,838,455,677 . . . . . . . . . . . . . . . . . . . . . dynes per square dekameter
338,384,557 . . . . . . . . . . . . . . . . . . . . . . . . . dynes per square meter
31,460,290 . . . . . . . . . . . . . . . . . . . . . . . . . . dynes per square foot
218,471.233172 . . . . . . . . . . . . . . . . . . . . . . . dynes per square inch
3,383,846 . . . . . . . . . . . . . . . . . . . . . . . . dynes per square decimeter
33,838.455677 . . . . . . . . . . . . . . . . . . . . . dynes per square centimeter
338.384557 . . . . . . . . . . . . . . . . . . . . . . dynes per square millimeter
930.464111 . . . . . . . . . . . . . feet of water @ 62°F. and 29.92 Barom. Press
```

INCH OF WATER @ 60°F.: =

```
0.000254 . . . . . . . . . . . . . . . . . . . . . . . . . hektometers of water @ 60°F.
0.00254 . . . . . . . . . . . . . . . . . . . . . . . . . . dekameters of water @ 60°F.
0.0254 . . . . . . . . . . . . . . . . . . . . . . . . . . . . . meters of water @ 60°F.
0.0833332 . . . . . . . . . . . . . . . . . . . . . . . . . . . feet of water @ 60°F.
1 . . . . . . . . . . . . . . . . . . . . . . . . . . . . . . . . . inches of water @ 60°F.
0.254 . . . . . . . . . . . . . . . . . . . . . . . . . . . . decimeters of water @ 60°F.
2.54 . . . . . . . . . . . . . . . . . . . . . . . . . . . . centimeters of water @ 60°F.
25.4 . . . . . . . . . . . . . . . . . . . . . . . . . . . . millimeters of water @ 60°F.
0.0000186820 . . . . . . . . . . . . . . . . . . . . hektometers of mercury @ 32°F.
0.000186820 . . . . . . . . . . . . . . . . . . . . . dekameters of mercury @ 32°F.
0.00186820 . . . . . . . . . . . . . . . . . . . . . . . meters of mercury @ 32°F.
0.00612925 . . . . . . . . . . . . . . . . . . . . . . . . feet of mercury @ 32°F.
0.0735510 . . . . . . . . . . . . . . . . . . . . . . . . inches of mercury @ 32°F.
0.0186820 . . . . . . . . . . . . . . . . . . . . . . . decimeters of mercury @ 32°F.
0.186820 . . . . . . . . . . . . . . . . . . . . . . . centimeters of mercury @ 32°F.
1.868197 . . . . . . . . . . . . . . . . . . . . . . . millimeters of mercury @ 32°F.
279.973171 . . . . . . . . . . . . . . . . . . . . . . . tons per square hektometer
2.799732 . . . . . . . . . . . . . . . . . . . . . . . . . tons per square dekameter
0.0279973 . . . . . . . . . . . . . . . . . . . . . . . . . tons per square meter
0.00260110 . . . . . . . . . . . . . . . . . . . . . . . . . tons per square foot
0.0000180627 . . . . . . . . . . . . . . . . . . . . . . . . tons per square inch
0.000279973 . . . . . . . . . . . . . . . . . . . . . . . tons per square decimeter
0.00000279973 . . . . . . . . . . . . . . . . . . . . . tons per square centimeter
0.0000000279973 . . . . . . . . . . . . . . . . . . . . tons per square millimeter
254,000 . . . . . . . . . . . . . . . . . . . . . . . . kilograms per square hektometer
2,540 . . . . . . . . . . . . . . . . . . . . . . . . . . kilograms per square dekameter
```

INCH OF WATER @ 60°F.: (cont'd)

25.4	kilograms per square meter
2.359600	kilograms per square foot
0.0163861	kilograms per square inch
0.254	kilograms per square decimeter
0.00254	kilograms per square centimeter
0.0000254	kilograms per square millimeter
559,946	pounds per square hectometer
5,599.463120	pounds per square dekameter
55.994631	pounds per square meter
5.202004	pounds per square foot
0.0361250	pounds per square inch
0.559946	pounds per square decimeter
0.00559946	pounds per square centimeter
0.0000559946	pounds per square millimeter
2,540,000	hektograms per square hektometer
25,400	hektograms per square dekameter
254	hektograms per square meter
23.596005	hektograms per square foot
0.163862	hektograms per square inch
2.54	hektograms per square decimeter
0.0254	hektograms per square centimeter
0.000254	hektograms per square millimeter
25,400,000	dekagrams per square hektometer
254,000	dekagrams per square dekameter
2,540	dekagrams per square meter
235.960045	dekagrams per square foot
1.638617	dekagrams per square inch
25.40	dekagrams per square decimeter
0.2540	dekagrams per square centimeter
0.002540	dekagrams per square millimeter
8,959,141	ounces per square hektometer
89,591.41	ounces per square dekameter
895.9141	ounces per square meter
83.232058	ounces per square foot
0.5780	ounces per square inch
8.959141	ounces per square decimeter
0.0895914	ounces per square centimeter
0.000895914	ounces per square millimeter
254,000,000	grams per square hektometer
2,540,000	grams per square dekameter
25,400	grams per square meter
2,359.600439	grams per square foot
16.386192	grams per square inch
254	grams per square decimeter
2.54	grams per square centimeter
0.0254	grams per square millimeter
2,540,000,000	decigrams per square hektometer
25,400,000	decigrams per square dekameter
254,000	decigrams per square meter
23,596,004	decigrams per square foot
163.861920	decigrams per square inch

INCH OF WATER @ 60°F.: (cont'd)

2,540	decigrams per square decimeter
25.40	decigrams per square centimeter
0.2540	decigrams per square millimeter
25,400,000,000	centigrams per square hektometer
254,000,000	centigrams per square dekameter
2,540,000	centigrams per square meter
235,960	centigrams per square foot
1,638.619198	centigrams per square inch
25,400	centigrams per square decimeter
254	centigrams per square centimeter
2.54	centigrams per square millimeter
254,000,000,000	milligrams per square hektometer
2,540,000,000	milligrams per square dekameter
25,400,000	milligrams per square meter
2,359,600	milligrams per square foot
16,396.192	milligrams per square inch
254,000	milligrams per square decimeter
2,540	milligrams per square centimeter
25.40	milligrams per square millimeter
0.00245562	bars
0.00245818	atmospheres
248,885,374,188	dynes per square hektometer
2,488,853,742	dynes per square dekameter
24,888,537	dynes per square meter
2,313,937	dynes per square foot
16,069.0079	dynes per square inch
248,885	dynes per square decimeters
2,488.853742	dynes per square centimeter
24.888537	dynes per square millimeter
68.44	feet of water @ 62°F. and 29.92 Barom. Press.

JOULE (ABSOLUTE): =

23.730	foot poundals
284.760	inch pounds
0.73756	foot pounds
8.85072	inch pounds
0.000000263331	ton (net) calories
0.00023889	kilogram calories (mean)
0.00526661	pound calories
0.00842658	ounce calories
0.23889	gram calories (mean)
238.89	milligram calories
0.0000000115740	kilowatt days
0.0000002778	kilowatt hours
0.0000166667	kilowatt minutes
0.001	kilowatt seconds
0.0000115740	watt days
0.0002778	watt hours
0.0166667	watt minutes
1	watt seconds

JOULE (ABSOLUTE): (cont'd)

0.000112366	ton (net) meters
0.101937	kilogram meters
0.224733	pounds meters
3.595721	ounce meters
101.937	gram meters
101,937	milligram meters
0.000368654	ton (net) feet
0.334438	kilogram feet
0.737311	pound feet
11.796960	ounce feet
334.438274	gram feet
334,438.273531	milligram feet
0.00442385	ton (net) inches
4.013259	kilogram inches
8.847732	pound inches
141.563520	ounce inches
4,013.259288	gram inches
4,013,259	milligram inches
0.0112366	ton (net) centimeters
10.1937	kilogram centimeters
22.4733	pound centimeters
359.5721	ounce centimeters
10,193.7	gram centimeters
10,193,700	milligram centimeters
0.112366	ton (net) millimeters
101.937	kilogram millimeters
224.733	pound millimeters
3,595.721	ounce millimeters
101,937	gram millimeters
101,937,000	milligram millimeters
0.0000098705	kiloliter-atmosphere
0.000098705	hektoliter-atmosphere
0.0003485	cubic foot-atmosphere
0.00098705	dekaliter-atmosphere
0.0098705	liter-atmosphere
0.098705	deciliter-atmosphere
0.98705	centiliter-atmosphere
9.8705	millimeter-atmosphere
1	joules
0.0000003775	Cheval-vapeur hours
0.0000000155208	horsepower days
0.0000003725	horsepower hours
0.0000223500	horsepower minutes
0.00134100	horsepower seconds
0.0000000642	pounds of carbon oxidized with perfect efficiency
0.0000009662	pounds of water evaporated from and at 212°F.
0.0009480	BTU (mean)
100,000,000	ergs

KILOGRAMS: =

0.000984206	tons (long)
0.001	tons (metric)

UNIVERSAL CONVERSION FACTORS
KILOGRAMS: (cont'd)

0.00110231	tons (short)
1	kilograms
2.679229	pounds (Troy)
2.204622	pounds (Avoir.)
10	hektograms
100	dekagrams
32.150742	ounces (Troy)
35.273957	ounces (Avoir.)
1,000	grams
257.21	drams (Troy)
564.38	drams (Avoir)
643.01	pennyweights (Troy)
771.62	scruples (Troy)
5,000	carats (metric)
10,000	decigrams
15,432.4	grains
100,000	centigrams
1,000,000	milligrams

KILOGRAM CALORIE (MEAN): =

99,334	foor poundals
1,192,008	inch poundals
3,087.4	foot pounds
37,048.8	inch pounds
0.0001	ton (short) calories
1	kilogram calories
2.04622	pound calories
10	hektogram calories
100	dekagram calories
32.7395	ounce calories
1,000	gram calories
10,000	decigram calories
100,000	centigram calories
1,000,000	milligram calories
0.00004845	kilowatt days
0.0011628	kilowatt hours
0.0697680	kilowatt minutes
4.186	kilowatt seconds
0.04845	watt days
1.1628	watt hours
69.7680	watt minutes
4,186	watt seconds
0.42685	ton meters
426.85	kilogram meters
941.043	pound meters
4,268.5	hektogram meters
42,685	dekagram meters
15,056.688	ounce meters
426,850	gram meters
4,268,500	decigram meters
42,685,000	centigram meters

KILOGRAM CALORIE (MEAN): (cont'd)

426,850,000	milligram meters
0.0042685	ton hektometers
4.2685	kilogram hektometers
9.41043	pound hektometers
42.685	hektogram hektometers
426.85	dekagram hektometers
150.567	ounce hektometers
4,268.5	gram hektometers
42,685	decigram hektometers
426,850	centigram hektometers
4,268,500	milligram hektometers
0.042685	ton dekameters
42.685	kilogram dekameters
94.1043	pound dekameters
426.85	hektogram dekameters
4,268.5	dekagram dekameters
1,505.67	ounce dekameters
42,685	gram dekameters
426,850	decigram dekameters
4,268,500	centigram dekameters
42,685,000	milligram dekameters
1.40042	ton feet
1,400.424	kilogram feet
30.8704	pound feet
14,004.236	hektogram feet
140,042.357	dekagram feet
493,985	ounce feet
1,400,424	gram feet
14,004,236	decigram feet
140,042,357	centigram feet
1,400,423,566	milligram feet
16.8050	ton inches
16,805.083	kilogram inches
307.445	pound inches
168,050.828	hektogram inches
1,680,051	dekagram inches
5,927.117	ounce inches
16,805,083	gram inches
168,050,828	decigram inches
1,680,508,279	centigram inches
16,805,082,792	milligram inches
4.2685	ton decimeters
4,268.5	kilogram decimeters
9,410.430	pound decimeters
42,685	hektogram decimeters
426,850	dekogram decimeters
150,566.885	ounce decimeters
4,268,500	gram decimeters
42,685,000	decigram decimeters
426,850,000	centigram decimeters
4,268,500,000	milligram decimeters

KILOGRAM CALORIE (MEAN): (cont'd)

42.685	ton centimeters
42,685	kilogram centimeters
94,104.303	pound centimeters
426,850	hektogram centimeters
4,268,500	dekagram centimeters
1,505,669	ounce centimeters
42,685,000	gram centimeters
426,850,000	decigram centimeters
4,268,500,000	centigram centimeters
42,685,000,000	milligram centimeters
426.85	ton millimeters
426,850	kilogram millimeters
941,043.029	pound millimeters
4,268,500	hektogram millimeters
42,685,000	dekagram millimeters
15,056,688	ounce millimeters
426,850,000	gram millimeters
4,268,500,000	decigram millimeters
42,685,000,000	centigram millimeters
426,850,000,000	milligram millimeters
0.041311	kiloliter-atmosphere
0.41311	hektoliter-atmosphere
4.1311	dekaliter-atmosphere
41.311	liter-atmosphere
413.11	deciliter-atmosphere
4,131.1	centiliter-atmosphere
41,311	milliliter-atmosphere
0.0000000000415977	cubic kilometer-atmosphere
0.0000000415977	cubic hektometer-atmosphere
0.0000415977	cubic dekameter-atmosphere
0.0415977	cubic meter-atmosphere
1.469	cubic feet-atmosphere
2,538.415	cubic feet-atmosphere
41.5977	cubic decimeter-atmosphere
41,597.673	cubic centimeter-atmosphere
41,597,673	cubic millimeter-atmosphere
4,185.8291	joules
0.00158097	Cheval-vapeur hours
0.0000649708	horsepower days
0.0015593	horsepower hours
0.0935980	horsepower minutes
5.61588	horsepower seconds
0.00029909	pounds of carbon oxidized with 100% efficiency
0.004501	pounds of water evaporated from at 212°F.
1.800	BTU (mean) per pound
3.9685	BTU (mean)
41,858,291,000	ergs

KILOGRAM CALORIE (MEAN) PER MINUTE: =

4,443,725	foot pound per day
185,155.2	foot pound per hour

KILOGRAM CALORIE (MEAN) PER MINUTE: (cont'd)

3,085.920	foot pound per minute
51.432	foot pound per second
0.0935980	horsepower
0.069680	kilowatts
69.7680	watts
232.71	foot poundals
2,792.52	inch poundals
7.2330	foot pounds
86.7960	inch pounds
0.0000023427	ton (net) calories
0.0023427	hektogram calories
0.00516477	pound calories
0.023427	hektogram calories
0.23427	dekagram calories
0.0826363	ounce calories
2.3427	gram calories (mean)
23.427	decigram calories
234.27	centigram calories
2,343.7	kilowatt days
0.000000113479	kilowatt days
0.0000027235	kilowatt hours
0.000163410	kilowatt minutes
0.00980460	kilowatt seconds
0.000113479	watt days
0.0027235	watt hours
0.163410	watt minutes
9.80460	watt seconds
0.001	ton meters
1	kilogram meters
2.204622	pound meters
10	hektogram meters
100	dekagram meters
35.273957	ounce meters
1,000	gram meters
10,000	decigram meters
100,000	centigram meters
1,000,000	milligram meters
0.00001	ton hektometers
0.01	kilogram hektometers
0.02204622	pound hektometers
0.1	hektogram hektometers
1	dekagram hektometers
0.352740	ounce hektometers
10	gram hektometers
100	decigram hektometers
1,000	centigram hektometers
10,000	milligram hektometers
0.0001	ton dekameters
0.1	kilogram dekameters
0.2204622	pound dekameters
1	hektogram dekameters

KILOGRAM CALORIE (MEAN) PER MINUTE: (cont'd)

10 . . . . dekagram dekameters
3.527396 . . . . ounce dekameters
100 . . . . gram dekameters
1,000 . . . . decigram dekameters
10,000 . . . . centigram dekameters
100,000 . . . . milligram dekameters
0.00328084 . . . . ton feet
3.280843 . . . . kilogram feet
7.233020 . . . . pound feet
32.80843 . . . . hektogram feet
328.0843 . . . . dekagram feet
115.728320 . . . . ounce feet
3,280.843 . . . . gram feet
32,808.43 . . . . decigram feet
328,084.3 . . . . centigram feet
3,280,843 . . . . milligram feet
0.0393701 . . . . ton inches
39.370116 . . . . kilogram inches
86.796236 . . . . pound inches
393.70116 . . . . hektogram inches
3,937.0116 . . . . dekagram inches
1,388.739776 . . . . ounce inches
39,370.116 . . . . gram inches
393,701.16 . . . . decigram inches
3,937,012 . . . . centigram inches
39,370,116 . . . . milligram inches
0.01 . . . . ton decimeters
10 . . . . kilogram decimeters
22.046223 . . . . pound decimeters
100 . . . . hektogram decimeters
1,000 . . . . dekagram decimeters
352.739568 . . . . ounce decimeters
10,000 . . . . gram decimeters
100,000 . . . . decigram decimeters
1,000,000 . . . . centigram decimeters
10,000,000 . . . . milligram decimeters
0.1 . . . . ton centimeters
100 . . . . kilogram centimeters
220.46223 . . . . pound centimeters
1,000 . . . . hektogram centimeters
10,000 . . . . dekagram centimeters
3,527.39568 . . . . ounce centimeters
100,000 . . . . gram centimeters
1,000,000 . . . . decigram centimeters
10,000,000 . . . . centigram centimeters
100,000,000 . . . . milligram centimeters
1 . . . . ton millimeters
1,000 . . . . kilogram millimeters
2,204.6223 . . . . pound millimeters
10,000 . . . . hektogram millimeters
100,000 . . . . dekagram millimeters

KILOGRAM CALORIE (MEAN) PER MINUTE: (cont'd)

35,273.9568	ounce millimeters
1,000,000	gram millimeters
10,000,000	decigram millimeters
100,000,000	centigram millimeters
1,000,000,000	milligram millimeters
0.000096782	kiloliter-atmospheres
0.00096782	hektoliter-atmospheres
0.0096782	dekaliter-atmospheres
0.096782	liter-atmospheres
0.96782	deciliter-atmospheres
9.6782	centiliter-atmospheres
96.782	milliliter-atmospheres
0.0000000000000967790	cubic kilometer-atmospheres
0.0000000000967790	cubic hektometer-atmospheres
0.0000000967790	cubic dekameter-atmospheres
0.0000967790	cubic meter-atmospheres
0.0034177	cubic foot-atmospheres
5.905746	cubic inch-atmospheres
0.0967790	cubic decimeter-atmospheres
96.779011	cubic centimeter-atmospheres
96,779.011	cubic millimeter-atmospheres
9.80665	joules
0.000000154324	Cheval-vapeur hours
0.000000152208	horsepower days
0.0000036530	horsepower hours
0.000219180	horsepower minutes
0.0131508	horsepower seconds
0.00000063718	pounds of carbon oxidized with 100% efficiency
0.0000095895	pounds of water evaporated form and at 212°F.
4.216948	BTU (mean) per pounds
0.0092972	BTU (mean)
98,066,500	ergs

KILOGRAM METER PER SECOND: =

0.0098046	kilowatts
9.8046	watts
0.0131508	horsepower

KILOGRAM METER PER MINUTE: =

0.000163410	kilowatts
0.163410	watts
0.00021980	horsepowers

KILOGRAM PER METER: =

1.774004	tons (net) per mile
1.102311	tons (net) per kilometer
0.00110231	tons (net) per meter

KILOGRAM PER METER: (cont'd)

0.001007956	tons (net) per yard
0.000335985	tons (net) per foot
0.0000279987	tons (net) per inch
0.0000110231	tons (net) per centimeter
0.00000110231	tons (net) per millimeter
1,609.349954	kilograms per mile
1,000	kilograms per kilometer
1	kilograms per meter
0.914403	kilograms per yard
0.304801	kilograms per foot
0.0254001	kilograms per inch
0.01	kilograms per centimeter
0.001	kilograms per millimeter
3,548.00896449	pounds (Avoir.) per mile
2,204.622341	pounds (Avoir.) per kilometer
2.204622	pounds (Avoir.) per meter
2.0159127	pounds (Avoir.) per yard
0.671971	pounds (Avoir.) per foot
0.0559976	pounds (Avoir.) per inch
0.0220462	pounds (Avoir.) per centimeter
0.00220462	pounds (Avoir.) per millimeter
56,768.143440	ounces (Avoir.) per mile
35,273.957456	ounces (Avoir.) per kilometer
35.273957	ounces (Avoir.) per meter
32.254604	ounces (Avoir.) per yard
10.751535	ounces (Avoir.) per foot
0.895961	ounces (Avoir.) per inch
0.352740	ounces (Avoir.) per centimeter
0.0352740	ounces (Avoir.) per millimeter
1,609,349.954	grams per mile
1,000,000	grams per kilometer
1,000	grams per meter
914.403	grams per yard
304.801127	grams per foot
25.4001	grams per inch
10	grams per centimeter
1	grams per millimeter
24,836,063	grains per mile
15,432,356	grains per kilometer
15,432.356387	grains per meter
14,111.388900	grains per yard
4,703.79630	grains per foot
391.983025	grains per inch
154.323564	grains per centimeter
15.432356	grains per millimeter

KILOGRAM PER SQUARE CENTIMETER: =

10	meters of water @ 60°F.
32.80833	feet of water @ 60°F.
393.69996	inches of water @ 60°F.

KILOGRAM PER SQUARE CENTIMETER: (cont'd)

1,000	centimeters of water @ 60°F.
10,000	millimeters of water @ 60°F.
0.735499	meters of mercury @ 32°F.
2.413053	feet of mercury @ 32°F.
28.956632	inches of mercury @ 32°F.
73.54985	centimeters of mercury @ 32°F.
735.49845	millimeters of mercury @ 32°F.
10,000	kilograms per square meter
929.034238	kilograms per square foot
6.451626	kilograms per square inch
1.0	kilograms per square centimeter
0.01	kilograms per square millimeter
22,046.233	pounds per square meter
2,048.1696	pounds per square foot
14.2234	pounds per square inch
2.2046223	pounds per square centimeter
0.0220462	pounds per square millimeter
352,739.568	ounces per square meter
32,770.7136	ounces per square foot
227.5744	ounces per square inch
35.273957	ounces per square centimeter
0.352740	ounces per square millimeter
10,000,000	grams per square meter
929,034.230	grams per square foot
6,451.626597	grams per square inch
1,000	grams per square centimeter
10	grams per square millimeter
154,324,000	grains per square meter
14,337,228	grains per square foot
99,564.0822955	grains per square inch
15,432.5	grains per square centimeter
154.324	grains per square millimeter
0.967778	atmosphere

KILOGRAM PER SQUARE FOOT: =

0.0107638	meters of water @ 60°F.
0.0353518	feet of water @ 60°F.
0.423774	inches of water @ 60°F.
1.076387	centimeters of water @ 60°F.
10.76387	millimeters of water @ 60°F.
0.000791682	meters of mercury @ 32°F.
0.00259738	feet of mercury @ 32°F.
0.0311689	inches of mercury @ 32°F.
0.0791682	centimeters of mercury @ 32°F.
0.791682	millimeters of mercury @ 32°F.
10.76387	kilograms per square meter
1	kilograms per square foot
0.00694445	kilograms per square inch
0.00107639	kilograms per square centimeter
0.0000107639	kilograms per square millimeter

UNIVERSAL CONVERSION FACTORS
KILOGRAM PER SQUARE FOOT: (cont'd)

23.730265 .... pounds per square meter
2.204622 .... pounds per square foot
0.0153099 .... pounds per square inch
0.00237302 .... pounds per square centimeter
0.0000237302 .... pounds per square millimeter
379.684288 .... ounces per square meter
35.273957 .... ounces per square foot
0.244958 .... ounces per square inch
0.0379685 .... ounces per square centimeter
0.000379685 .... ounces per square millimeter
10,763.87 .... grams per square meter
1,000 .... grams per square foot
6.94445 .... grams per square inch
1.076387 .... grams per square centimeter
0.0107639 .... grams per square millimeter
166,112.347388 .... grains per square meter
15,432.4 .... grains per square foot
107.169481 .... grains per square inch
16.611235 .... grains per square centimeter
0.166112 .... grains per square millimeter
0.00104170 .... atmospheres

KILOGRAM PER SQUARE INCH: =

1.549987 .... meters of water @ 60°F.
5.090659 .... feet of water @ 60°F.
61.023456 .... inches of water @ 60°F.
154.999728 .... centimeters of water @ 60°F.
1,549.99728 .... millimeters of water @ 60°F.
0.114002 .... meters of mercury @ 32°F.
0.374022 .... feet of mercury @ 32°F.
4.488322 .... inches of mercury @ 32°F.
11.40002 .... centimeters of mercury @ 32°F.
114.00022 .... millimeters of mercury @ 32°F.
1,549.99728 .... kilograms per square meter
144 .... kilograms per square foot
1 .... kilograms per square inch
0.1549997 .... kilograms per square centimeter
0.001549997 .... kilograms per square millimeter
3,417.158160 .... pounds per square meter
317.4655568 .... pounds per square foot
2.2046223 .... pounds per square inch
0.341715 .... pounds per square centimeter
0.00341715 .... pounds per square millimeter
54,674.537472 .... ounces per square meter
5,079.449808 .... ounces per square foot
35.273957 .... ounces per square inch
5.467464 .... ounces per square centimeter
0.0546746 .... ounces per square millimeter
1,549,997 .... grams per square meter
144,000 .... grams per square foot

KILOGRAM PER SQUARE INCH: (cont'd)

1,000	grams per square inch
154.999728	grams per square centimeter
1.549997	grams per square millimeter
23,920,178	grains per square meter
2,222,266	grains per square foot
15,432.4	grains per square inch
2,392.017840	grains per square centimeter
23.920178	grains per square millimeter
0.150005	atmospheres

KILOGRAM PER SQUARE KILOMETER: =

0.000000001	meters of water @ 60°F.
0.00000032843	feet of water @ 60°F.
0.00000003937	inches of water @ 60°F.
0.0000001	centimeters of water @ 60°F.
0.000001	millimeters of water @ 60°F.
0.0000000000735499	meters of mercury @ 32°F.
0.000000000241305	feet of mercury @ 32°F.
0.00000000289570	inches of mercury @ 32°F.
0.00000000735499	centimeters of mercury @ 32°F.
0.0000000735499	millimeters of mercury @ 32°F.
0.000001	kilograms per square meter
0.0000000929034	kilograms per square foot
0.000000000645163	kilograms per square inch
0.0000000001	kilograms per square centimeter
0.000000000001	kilograms per square millimeter
0.00000220462	pounds per square meter
0.000000204817	pounds per square foot
0.00000000142234	pounds per square inch
0.000000000220462	pounds per square centimeter
0.00000000000220462	pounds per square millimeter
0.0000352740	ounces per square meter
0.00000327707	ounces per square foot
0.0000000227574	ounces per square inch
0.00000000352740	ounces per square centimeter
0.0000000000352740	ounces per square millimeter
0.001	grams per square meter
0.0000929034	grams per square foot
0.000000645163	grams per square inch
0.0000001	grams per square centimeter
0.000000001	grams per square millimeter
0.0154324	grains per square meter
0.00143372	grains per square foot
0.00000995640	grains per square inch
0.00000154324	grains per square centimeter
0.0000000154324	grains per square millimeter
0.0000000000967778	atmospheres

KILOGRAM PER SQUARE METER: =

0.001	meters of water @ 60°F.
0.0032843	feet of water @ 60°F.

KILOGRAM PER SQUARE METER: (cont'd)

0.03937 . . . . inches of water @ 60°F.
0.1 . . . . centimeters of water @ 60°F.
1 . . . . millimeters of water @ 60°F.
0.0000735499 . . . . meters of mercury @ 32°F.
0.000241300 . . . . feet of mercury @ 32°F.
0.00289570 . . . . inches of mercury @ 32°F.
0.0073549 . . . . centimeters of mercury @ 32°F.
0.0735499 . . . . millimeters of mercury @ 32°F.
1 . . . . kilograms per square meter
0.0929034 . . . . kilograms per square foot
0.000645163 . . . . kilograms per square inch
0.0001 . . . . kilograms per square centimeter
0.000001 . . . . kilograms per square millimeter
2.204622 . . . . pounds per square meter
0.204817 . . . . pounds per square foot
0.00142234 . . . . pounds per square inch
0.000220462 . . . . pounds per square centimeter
0.00000220462 . . . . pounds per square millimeter
35.273957 . . . . ounces per square meter
3.277071 . . . . ounces per square foot
0.0227574 . . . . ounces per square inch
0.00352740 . . . . ounces per square centimeter
0.0000352740 . . . . ounces per square millimeter
1,000 . . . . grams per square meter
92.903423 . . . . grams per square foot
0.645163 . . . . grams per square inch
0.1 . . . . grams per square centimeter
0.001 . . . . grams per square millimeter
15,432.4 . . . . grains per square meter
1,433.7228 . . . . grains per square foot
9.956408 . . . . grains per square inch
1.543240 . . . . grains per square centimeter
0.0154324 . . . . grains per square millimeter
0.0000967778 . . . . atmospheres

KILOGRAM PER SQUARE MILLIMETER: =

1,000 . . . . meters of water @ 60°F.
3,280.833 . . . . feet of water @ 60°F.
39,369.996 . . . . inches of water @ 60°F.
100,000 . . . . centimeters of water @ 60°F.
1,000,000 . . . . millimeters of water @ 60°F.
73,549845 . . . . meters of mercury @ 32°F.
241.3053 . . . . feet of mercury @ 32°F.
2,895.6632 . . . . inches of mercury @ 32°F.
7,354.9845 . . . . centimeters of mercury @ 32°F.
73,549.845 . . . . millimeters of mercury @ 32°F.
1,000,000 . . . . kilograms per square meter
92,903.4238 . . . . kilograms per square foot
645.1626 . . . . kilograms per square inch
100 . . . . kilograms per square centimeter

KILOGRAM PER SQUARE MILLIMETER: (cont'd)

1 . . . . kilograms per square millimeter
2,204,622 . . . . pounds per square meter
204,816.6 . . . . pounds per square foot
1,422.34 . . . . pounds per square inch
220,46223 . . . . pounds per square centimeter
2,204622 . . . . pounds per square millimeter
35,273,957 . . . . ounces per square meter
3,277,071 . . . . ounces per square foot
22,757.44 . . . . ounces per square inch
3,527.3957 . . . . ounces per square centimeter
35.273957 . . . . ounces per square millimeter
1,000,000,000 . . . . grams per square meter
92,903,423 . . . . grams per square foot
645,162.6597 . . . . grams per square inch
100,000 . . . . grams per square centimeter
1,000 . . . . grams per square millimeter
15,432,400,000 . . . . grains per square meter
1,433,722,800 . . . . grains per square foot
9,956,408 . . . . grains per square inch
1,543,240 . . . . grains per square centimeter
15,432.4 . . . . grains per square millimeter
96.7778 . . . . atmospheres

KILOMETER: =

0.53961 . . . . miles (nautical)
0.62137 . . . . miles (statute)
1 . . . . kilometers
4.970974 . . . . furlongs
10 . . . . hektometers
49.709741 . . . . chains
100 . . . . dekameters
198.838579 . . . . rods
1,000 . . . . meters
1,093.6 . . . . yards
1,181.1 . . . . varas (Texas)
3,280.8 . . . . feet
4,374.440070 . . . . spans
4,970.974310 . . . . links
9,842.50 . . . . hands
10,000 . . . . decimeters
100,000 . . . . centimeters
39,370 . . . . inches
1,000,000 . . . . millimeters
39,370,000 . . . . mils
1,000,000,000 . . . . microns
1,000,000,000,000 . . . . millimicrons
1,000,000,000,000 . . . . micromillimeters
546.81 . . . . fathoms

KILOMETER PER DAY: =

0.62137	miles per day
0.0258904	miles per hour
0.000431507	miles per minute
0.00000719718	miles per second
1.0	kilometers per day
0.0416667	kilometers per hour
0.000694444	kilometers per minute
0.0000115741	kilometers per second
4.970974	furlongs per day
0.207124	furlongs per hour
0.00345207	furlongs per minute
0.0000575344	furlongs per second
10	hektometers per day
0.416667	hektometers per hour
0.00694444	hektometers per minute
0.00115741	hektometers per second
49.709741	chains per day
2.071239	chains per hour
0.0345207	chains per minute
0.00057534	chains per second
100	dekameters per day
4.166667	dekameters per hour
0.0694444	dekameters per minute
0.00115741	dekameters per second
198.838579	rods per day
8.284941	rods per hour
0.138082	rods per minute
0.00230137	rods per second
1,000	meters per day
41.666667	meters per hour
0.694444	meters per minute
0.0115741	meters per second
1,093.6	yards per day
45.566667	yards per hour
0.759444	yards per minute
0.0126574	yards per second
1,181.1	varas (Texas) per day
49.21250	varas (Texas) per hour
0.820208	varas (Texas) per minute
0.0136701	varas (Texas) per second
3,280.8	feet per day
136.7	feet per hour
2.278333	feet per minute
0.0379722	feet per second
4,374.440070	spans per day
182.268336	spans per hour
3.0378056	spans per minute
0.0506301	spans per second
4,970.974310	links per day

KILOMETER PER DAY: (cont'd)

207.123916	links per hour
3.452065	links per minute
0.0575344	links per second
9,842.50	hands per day
410.104166	hands per hour
6.835069	hands per minute
0.113918	hands per second
10,000	decimeters per day
416.666664	decimeters per hour
6.944444	decimeters per minute
0.115741	decimeters per second
100,000	centimeters per day
4,166.666640	centimeters per hour
69.444444	centimeters per minute
1.157407	centimeters per second
39,370	inches per day
1,640.416666	inches per hour
27.340278	inches per minute
0.455671	inches per second
1,000,000	millimeters per day
41,666.6664	millimeters per hour
694.444444	millimeters per minute
11.574074	millimeters per second

KILOMETER PER HOUR: =

14.912880	miles per day
0.62137	miles per hour
0.0103562	miles per minute
0.000172603	miles per second
24	kilometers per day
1	kilometers per hour
0.0166667	kilometers per minute
0.000277778	kilometers per second
119.303376	furlongs per day
4.970974	furlongs per hour
0.0828496	furlongs per minute
0.00138083	furlongs per second
240	hektometers per day
10	hektometers per hour
0.166667	hektometers per minute
0.00277778	hektometers per second
1,193.0337599	chains per day
49.709741	chains per hour
0.828496	chains per minute
0.0138083	chains per second
2,400	dekameters per day
100	dekameters per hour
1.666667	dekameters per minute
0.0277778	dekameters per second
4,772.125895	rods per day

KILOMETER PER HOUR: (cont'd)

198.838579	rods per hour
3.313976	rods per minute
0.0552329	rods per second
24,000	meters per day
1,000	meters per hour
16.666667	meters per minute
0.277778	meters per second
26,246.4	yards per day
1,093.6	yards per hour
18.226667	yards per minute
0.303777	yards per second
28,346.399997	varas (Texas) per day
1,181.1	varas (Texas) per hour
19.685	varas (Texas) per minute
0.328083	varas (Texas) per second
78,739.199997	feet per day
3,280.8	feet per hour
54.68	feet per minute
0.911333	feet per second
104,986.561622	spans per day
4,374.440070	spans per hour
72.907334	spans per minute
1.215122	spans per second
119,303.375990	links per day
4,970.974310	links per hour
82.849567	links per minute
1.380826	links per second
236,219.999933	hands per day
9,842.5	hands per hour
164.0416666	hands per minute
2.734028	hands per second
240,000	decimeters per day
10,000	decimeters per hour
166,666667	decimeters per minute
2.777778	decimeters per second
2,400,000	centimeters per day
100,000	centimeters per hour
1,666.666667	centimeters per minute
27.777778	centimeters per second
944,879.999904	inches per day
39,370	inches per hour
656.166667	inches per minute
10.936111	inches per second
24,000,000	millimeters per day
1,000,000	millimeters per hour
16,666.666667	millimeters per minute
277.777778	millimeters per second
0.5396	knots

KILOMETER PER MINUTE: =

894.772800	miles per day
37.282200	miles per hour

KILOMETER PER MINUTE: (cont'd)

0.62137	miles per minute
0.0103562	miles per second
1,440	kilometers per day
60	kilometers per hour
1	kilometers per minute
0.0166667	kilometers per second
7,158.202560	furlongs per day
298.258440	furlongs per hour
4.970974	furlongs per minute
0.0828496	furlongs per second
14,400	hektometers per day
600	hektometers per hour
10	hektometers per minute
0.166667	hektometers per second
71,582.0256029	chains per day
2,982.584400	chains per hour
49.709741	chains per minute
0.828496	chains per second
144,000	dekameters per day
6,000	dekameters per hour
100	dekameters per minute
1.666667	dekameters per second
286,327.553184	rods per day
11,930.314716	rods per hour
198.838579	rods per minute
3.313976	rods per second
1,440,000	meters per day
60,000	meters per hour
1,000	meters per minute
16.666667	meters per second
1,574,784	yards per day
65,616.000012	yards per hour
1,093.6	yards per minute
18.226667	yards per second
1,700,784	varas (Texas) per day
70,866.0	varas (Texas) per hour
1,181.1	varas (Texas) per minute
19.685	varas (Texas) per second
4,724,352	feet per day
196,848	feet per hour
3,280.0	feet per minute
54.68	feet per second
6,299,194	spans per day
262,466.404200	spans per hour
4,374.440070	spans per minute
72.907335	spans per second
7,158,203	links per day
298,258.440012	links per hour
4,970.974310	links per minute
82.849567	links per second
14,173,200	hands per day

KILOMETER PER MINUTE: (cont'd)

590,550 . . . . hands per hour
9,842.5 . . . . hands per minute
164.0416667 . . . . hands per second
14,400,000 . . . . decimeters per day
600,000 . . . . decimeters per hour
10,000 . . . . decimeters per minute
166.666667 . . . . decimeters per second
144,000,000 . . . . centimeters per day
6,000,000 . . . . centimeters per hour
100,000 . . . . centimeters per minute
1,666.666667 . . . . centimeters per second
56,692,800 . . . . inches per day
2,362,200 . . . . inches per hour
39,370 . . . . inches per minute
656.166667 . . . . inches per second
1,440,000,000 . . . . millimeters per day
60,000,000 . . . . millimeters per hour
1,000,000 . . . . millimeters per minute
16,666.666667 . . . . millimeters per second

KILOMETER PER SECOND: =

53,686.3680 . . . . miles per day
2,236.93200 . . . . miles per hour
37.28220 . . . . miles per minute
0.62137 . . . . miles per second
86,400 . . . . kilometers per day
3,600 . . . . kilometers per hour
60 . . . . kilometers per minute
1 . . . . kilometers per second
429,492.180384 . . . . furlongs per day
17,895.506400 . . . . furlongs per hour
298.258440 . . . . furlongs per minute
4.970974 . . . . furlongs per second
864,000 . . . . hektometers per day
36,000 . . . . hektometers per hour
600 . . . . hektometers per minute
10 . . . . hektometers per second
4,294,922 . . . . chains per day
178,955.064 . . . . chains per hour
2,982.58440 . . . . chains per minute
49.709741 . . . . chains per second
8,640,000 . . . . dekameters per day
360,000 . . . . dekameters per hour
6,000 . . . . dekameters per minute
100 . . . . dekameters per second
17,179,653 . . . . rods per day
715,818.88440 . . . . rods per hour
11,930.314740 . . . . rods per minute
198.838579 . . . . rods per second
86,400,000 . . . . meters per day

3,600,000 . . . . meters per hour
60,000 . . . . meters per minute
1,000 . . . . meters per second
94,487,040 . . . . yards per day
3,936,960 . . . . yards per hour
65,616 . . . . yards per minute
1,093.6 . . . . yards per second
102,047,040 . . . . varas (Texas) per day
4,251,960 . . . . varas (Texas) per hour
70,866 . . . . varas (Texas) per minute
1,181.1 . . . . varas (Texas) per second
283,461,120 . . . . feet per day
11,810,880 . . . . feet per hour
196,848 . . . . feet per minute
3,280.8 . . . . feet per second
377,951,622 . . . . spans per day
15,747,984 . . . . spans per hour
262,466.404200 . . . . spans per minute
4,374.440070 . . . . spans per second
429,492,180 . . . . links per day
17,895,508 . . . . links per hour
298,258.440 . . . . links per minute
4,970.974310 . . . . links per second
850,392,000 . . . . hands per day
35,433,000 . . . . hands per hour
590,550 . . . . hands per minute
9,842.5 . . . . hands per second
864,000,000 . . . . decimeters per day
36,000,000 . . . . decimeters per hour
600,000 . . . . decimeters per minute
10,000 . . . . decimeters per second
8,640,000,000 . . . . centimeters per day
360,000,000 . . . . centimeters per hour
6,000,000 . . . . centimeters per minute
100,000 . . . . centimeters per second
3,401,568,000 . . . . inches per day
141,732,000 . . . . inches per hour
2,362,200 . . . . inches per minute
39,370 . . . . inches per second
86,400,000,000 . . . . millimeters per day
3,600,000,000 . . . . millimeters per hour
60,000,000 . . . . millimeters per minute
1,000,000 . . . . millimeters per second

KILOMETER PER HOUR PER SECOND: =

0.000172594 . . . . miles per second per second
0.0002778 . . . . kilometers per second per second
0.002778 . . . . hektometers per second per second
0.02778 . . . . dekameters per second per second
0.2778 . . . . meters per second per second

KILOMETER PER HOUR PER SECOND: (cont'd)

0.303767	yards per second per second
0.9113	feet per second per second
2.778	decimeters per second per second
27.78	centimeters per second per second
10.9356	inches per second per second
277.8	millimeters per second per second

KILOLITER: =

1	kiloliters
1	cubic meters
1.3080	cubic yards
10	hektoliters
28.378	bushels (U.S.) dry
27.497	bushels (Imperial) dry
35.316	cubic feet
100	dekaliters
11.3513	pecks
264.18	gallons (U.S.) liquid
227.0574264	gallons (U.S.) dry
219.977402	gallons (Imperial)
1,056.72	quarts (liquid)
908.110825	quarts (dry)
1,000	liters
1,000	cubic decimeters
2,113.44	pints (liquid)
8,453.76	gills (liquid)
10,000	deciliters
34,607.58	cubic inches
100,000	centiliters
1,000,000	milliliters
1,000,000	cubic centimeters
1,000,000,000	cubic millimeters
1.101234	tons (short) of water @ 62°F.
0.983252	tons (long) of water @ 62°F.
1.0	tons (metric) of water @ 62°F.
1,000	kilograms of water @ 62°F.
2,202.46866	pounds (Avoir.) of water @ 62°F.
2,676.611	pounds (Troy) of water @ 62°F.
10,000	hektograms of water @ 62°F.
100,000	dekagrams of water @ 62°F.
32,119.331983	ounces (Troy) of water @ 62°F.
35,239.49856	ounces (Avoir.) of water @ 62°F.
256,954.655866	drams (Troy) of water @ 62°F.
563,831.97696	drams (Avoir.) of water @ 62°F.
642,386.639664	pennyweights of water @ 62°F.
770,863.967597	scruples (Avoir.)
1,000,000	grams of water @ 62°F.
10,000,000	decigrams of water @ 62°F.
15,417,281	grains of water @ 62°F.
100,000,000	centigrams of water @ 62°F.

KILOLITER: (cont'd)

1,000,000,000 . . . . milligrams of water @ 62°F.
33,815.04 . . . . ounces (fluid)
270,520.32 . . . . drams (fluid)
32.105795 . . . . sacks of cement

KILOWATT: =

63,725,184 . . . . foot pounds per day
2,655,216 . . . . foot pounds per hour
44,253.60 . . . . foot pounds per minute
737.56 . . . . foot pounds per second
764,702,208 . . . . inch pounds per day
31,862,592 . . . . inch pounds per hour
531,043.20 . . . . inch pounds per minute
8,850.72 . . . . inch pounds per second
20,640.09600 . . . . kilogram calories (mean) per day
860.004 . . . . kilogram calories (mean) per hour
14.33340 . . . . kilogram calories (mean) per minute
0.23889 . . . . kilogram calories (mean) per second
45,503.615916 . . . . pound calories (mean) per day
1,895.983996 . . . . pound calories (mean) per hour
31.599733 . . . . pound calories (mean) per minute
0.526662 . . . . pound calories (mean) per second
728,057.85472 . . . . ounce calories (mean) per day
30,335.743936 . . . . ounce calories (mean) per hour
505.595728 . . . . ounce calories (mean) per minute
8.426592 . . . . ounce calories (mean) per second
20,640,096 . . . . gram calories (mean) per day
860,004 . . . . gram calories (mean) per hour
14,333.40 . . . . gram calories (mean) per minute
238.89 . . . . gram calories (mean) per second
81,930.52800 . . . . BTU (mean) per day
3,413.77200 . . . . BTU (mean) per hour
56.89620 . . . . BTU (mean) per minute
0.94827 . . . . BTU (mean) per second
1 . . . . kilowatts
1,000 . . . . watts
3,600,000 . . . . joules
1.341 . . . . horsepower
1.3597 . . . . horsepower (metric)
1.3597 . . . . Cheval-vapeur hours
0.234 . . . . pounds carbon oxidized with 100% efficiency
3.52 . . . . pounds water evaporated from and at 212°F.
852.647040 . . . . kiloliter-atmospheres per day
35.52695 . . . . kiloliter-atmospheres per hour
0.592116 . . . . kiloliter-atmospheres per minute
0.0098686 . . . . kiloliter-atmospheres per second
852,647 . . . . liter-atmospheres per day
35,526.95 . . . . liter-atmospheres per hour
592.116 . . . . liter-atmospheres per minute
9,8686 . . . . liter-atmospheres per second

UNIVERSAL CONVERSION FACTORS
KILOWATT: (cont'd)

8,808,000	kilogram meters per day
367,000	kilogram meters per hour
6,116.666667	kilogram meters per minute
101.944444	kilogram meters per second

LINK (SURVEYORS): =

0.0001250	miles
0.000201168	kilometers
0.001	furlongs
0.00201168	hektometers
0.01	chains
0.0201168	dekameters
0.04	rods
0.201168	meters
0.22	yards
0.23760	varas (Texas)
0.66	feet
0.879998	spans
1.98	hands
2.011684	decimeters
20.11684	centimeters
7.92	inches
201.1684	millimeters
7,920	mils
201,168	microns
201,168,400	millimicrons
201,168,400	micromillimeters
312,447	wave lengths of red line of cadmium
201,168,400,000	Angstrom Units

LITER: =

0.001	kiloliters
0.001	cubic meters
0.0013080	cubic yards
0.00628995	barrels
0.01	hektoliters
0.028378	bushels (U.S.) dry
0.027497	bushels (Imperial) dry
0.0353144	cubic feet
0.1	dekaliters
0.113512	pecks (U.S.) dry
0.264178	gallons (U.S.) liquid
0.22702	gallons (U.S.) dry
0.21998	gallons (Imperial)
1.056710	quarts (liquid)
0.908102	quarts (dry)
1	cubic decimeters
1.8162	pints (U.S.) dry

LITER: (cont'd)

2.1134	pints (U.S.) liquid
7.0392	gills (Imperial)
8.4538	gills (U.S.)
10	deciliters
61.025	cubic inches
100	centiliters
1,000	milliliters
1,000	cubic centimeters
1,000,000	cubic millimeters
33.8147	ounces (U.S.) fluid
35.196	ounces (Imperial) fluid
270.5179	drams (fluid)
16,231.0740	minims
2.20462	pounds of water at maximum density

LITER ATMOSPHERE: =

2,404.59243	foot poundals
28,855.10916	inch poundals
74.738589	foot pounds
896.863068	inch pounds
0.0000242701	ton calories
0.0242071	kilogram calories
0.0533676	pound calories
0.242071	hektogram calories
2.42071	dekagram calories
0.853881	ounce calories
24.2071	gram calories (mean)
242.071	decigram calories
2,420.71	centigram calories
24,207.1	milligram calories
0.00000117258	kilowatt days
0.0000281419	kilowatt hours
0.00168852	kilowatt minutes
0.101311	kilowatt seconds
0.00117258	watt days
0.0281419	watt hours
1.688516	watt minutes
101.310932	watt seconds
0.010333	ton meters
10.333	kilogram meters
22.780362	pounds meters
103.33	hektogram meters
1,033.3	dekagram meters
364.485798	ounce meters
10,333	gram meters
103,330	decigram meters
1,033,300	centigram meters
10,333,000	milligram meters
0.00010333	ton hektometers
0.10333	kilogram hektometers

0.227804	pound hektometers
1.0333	hektogram hektometers
10.333	dekagram hektometers
3.644858	ounce hektometers
103.33	gram hektometers
1,033.3	decigram hektometers
10,333	centigram hektometers
103,330	milligram hektometers
0.0010330	ton dekameters
1.0333	kilogram dekameters
2.278036	pound dekameters
10.333	hektogram dekameters
103.33	dekagram dekameters
36.448583	ounce dekameters
1033.3	gram dekameters
10,333	decigram dekameters
103,330	centigram dekameters
1,033,300	milligram dekameters
0.0339001	ton feet
33.900951	kilogram feet
74.738796	pound feet
339.009507	hektogram feet
3,390.095072	dekagram feet
1,195.820731	ounce feet
33,900.950719	gram feet
339,009.50719	decigram feet
3,390,095	centigram feet
33,900,951	milligram feet
0.406811	ton inches
406.811409	kilogram inches
896.865507	pound inches
4,068.114086	hektogram inches
40,681.140863	dekagram inches
14,349.848105	ounce inches
406,811.408628	gram inches
4,068,114	decigram inches
40,681,141	centigram inches
406,811,409	milligram inches
0.10333	ton decimeters
103.33	kilogram decimeters
227.803622	pound decimeters
1,033.3	hektogram decimeters
10,333	dekagram decimeters
3,644.857956	ounce decimeters
103,330	gram decimeters
1,033,300	decigrams decimeters
10,333,000	centigram decimeters
103,330,000	milligram decimeters
1.0333	ton centimeters
1,033.3	kilogram centimeters
2,278.036223	pound centimeters

LITER ATMOSPHERE: (cont'd)

10,333	hektogram centimeters
103,330	dekagram centimeters
36,448.579561	ounce centimeters
1,033,300	gram centimeters
10,333,000	decigram centimeters
103,330,000	centigram centimeters
1,033,300,000	milligram centimeters
10.333	ton millimeters
10,333	kilogram millimeters
22,780.362226	pound millimeters
103,330	hektogram millimeters
1,033,300	dekagram millimeters
364,485.795614	ounce millimeters
10,333,000	gram millimeters
103,330,000	decigram millimeters
1,033,300,000	centigram millimeters
10,333,000,000	milligram millimeters
0.001	kiloliter-atmospheres
0.01	hektoliter-atmospheres
0.1	dekaliter-atmospheres
1	liter-atmospheres
10	deciliter-atmospheres
100	centiliter-atmospheres
1,000	millimeter-atmospheres
0.000000000001	cubic kilometer-atmospheres
0.000000001	cubic hektometer-atmospheres
0.000001	cubic dekameter-atmospheres
0.001	cubic meter-atmospheres
0.035319	cubic foot-atmospheres
61.025	cubic inch-atmospheres
1	cubic decimeter-atmospheres
1,000	cubic centimeter-atmospheres
1,000,000	cubic millimeter-atmospheres
101.328	joules (absolute)
0.00003827	Cheval-vapeur hours
0.00000157277	horsepower days
0.000037745	horsepower hours
0.00226479	horsepower minutes
0.135887	horsepower seconds
0.00000658398	pounds of carbon oxidized with 100% efficiency
0.00009907	pounds of water evaporated from and at 212°F.
43.573724	BTU (mean) per pound
0.09607	BTU (mean)
1,013,321,145	ergs

LITER PER DAY: =

0.001	kiloliters per day
0.0000416667	kiloliters per hour
0.000000694444	kiloliters per minute
0.0000000115741	kiloliters per second
0.001	cubic meters per day

LITER PER DAY: (cont'd)

0.0000416667	cubic meters per hour
0.000000694444	cubic meters per minute
0.0000000115741	cubic meters per second
0.0013080	cubic yards per day
0.0000545000	cubic yards per hour
0.000000908333	cubic yards per minute
0.0000000151389	cubic yards per second
0.00628996	barrels per day
0.000262082	barrels per hour
0.00000436803	barrels per minute
0.0000000728005	barrels per second
0.01	hektoliters per day
0.000416667	hektoliters per hour
0.00000694444	hektoliters per minute
0.000000115741	hektoliters per second
0.028378	bushels (U.S.—dry) per day
0.00118242	bushels (U.S.—dry) per hour
0.0000197069	bushels (U.S.—dry) per minute
0.000000328449	bushels (U.S.—dry) per second
0.027497	bushels (Imperial—dry) per day
0.00114571	bushels (Imperial—dry) per hour
0.0000190951	bushels (Imperial—dry) per minute
0.000000318252	bushels (Imperial—dry) per second
0.0353144	cubic feet per day
0.00147143	cubic feet per hour
0.000024539	cubic feet per minute
0.000000408731	cubic feet per second
0.1	dekaliters per day
0.00416667	dekaliters per hour
0.0000694444	dekaliters per minute
0.00000115741	dekaliters per second
0.113512	pecks (U.S.—dry) per day
0.00472967	pecks (U.S.—dry) per hour
0.0000788278	pecks (U.S.—dry) per minute
0.000000131380	pecks (U.S.—dry) per second
0.264178	gallons (U.S.—liquid) per day
0.0110074	gallons (U.S.—liquid) per hour
0.000183457	gallons (U.S.—liquid) per minute
0.00000305762	gallons (U.S.—liquid) per second
0.22702	gallons (U.S.—dry) per day
0.00945917	gallons (U.S.—dry) per hour
0.000157653	gallons (U.S.—dry) per minute
0.00000262755	gallons (U.S.—dry) per second
0.21998	gallons (Imperial) per day
0.00916583	gallons (Imperial) per hour
0.000152764	gallons (Imperial) per minute
0.00000254606	gallons (Imperial) per second
1.056710	quarts (liquid) per day
0.00440296	quarts (liquid) per hour
0.000733826	quarts (liquid) per minute
0.0000122304	quarts (liquid) per second

LITER PER DAY: (cont'd)

0.908102	quarts (dry) per day
0.0378376	quarts (dry) per hour
0.000630626	quarts (dry) per minute
0.0000105104	quarts (dry) per second
1	liters per day
0.0416667	liters per hour
0.000694444	liters per minute
0.0000115741	liters per second
1	cubic decimeters per day
0.0416667	cubic decimeters per hour
0.00069444	cubic decimeters per minute
0.0000115741	cubic decimeters per second
1.8162	pints (U.S.—dry) per day
0.0756750	pints (U.S.—dry) per hour
0.00126125	pints (U.S.—dry) per minute
0.0000210208	pints (U.S.—dry) per second
2.1134	pints (U.S.—liquid) per day
0.0880583	pints (U.S.—liquid) per hour
0.00146764	pints (U.S.—liquid) per minute
0.0000244606	pints (U.S.—liquid) per second
7.0392	gills (Imperial) per day
0.293300	gills (Imperial) per hour
0.00488833	gills (Imperial) per minute
0.0000814722	gills (Imperial) per second
8.4538	gills (U.S.) per day
0.352242	gills (U.S.) per hour
0.00587069	gills (U.S.) per minute
0.0000978449	gills (U.S.) per second
10	deciliters per day
0.416667	deciliters per hour
0.00694444	deciliters per minute
0.000115741	deciliters per second
61.025	cubic inches per day
2.542708	cubic inches per hour
0.0423785	cubic inches per minute
0.000706308	cubic inches per second
100	centiliters per day
4.166667	centiliters per hour
0.0694444	centiliters per minute
0.00115741	centiliters per second
1,000	milliliters per day
41.666667	milliliters per hour
0.694444	milliliters per minute
0.0115741	milliliters per second
1,000	cubic centimeters per day
41.666667	cubic centimeters per hour
0.694444	cubic centimeters per minute
0.0115741	cubic centimeters per second
1,000,000	cubic millimeters per day
41,666.666400	cubic millimeters per hour
694.444444	cubic millimeters per minute

LITER PER DAY: (cont'd)

11.574074	cubic millimeters per second
33.8147	ounces (U.S.) fluid per day
1.408946	ounces (U.S.) fluid per hour
0.0234824	ounces (U.S.) fluid per minute
0.000391374	ounces (U.S.) fluid per second
35.196	ounces (Imperial—fluid) per day
1.466500	ounces (Imperial—fluid) per hour
0.0244417	ounces (Imperial—fluid) per minute
0.000407361	ounces (Imperial—fluid) per second
270.5179	drams (fluid) per day
11.271579	drams (fluid) per hour
0.187860	drams (fluid) per minute
0.00313099	drams (fluid) per second
16,231.0740	minims per day
676.294747	minims per hour
11.271579	minims per minute
0.187860	minims per second

LITER PER HOUR: =

0.0240	kiloliters per day
0.001	kiloliters per hour
0.0000166667	kiloliters per minute
0.000000277778	kiloliters per second
0.0240	cubic meters per day
0.001	cubic meters per hour
0.0000166667	cubic meters per minute
0.000000277778	cubic meters per second
0.031392	cubic yards per day
0.0013080	cubic yards per hour
0.0000218	cubic yards per minute
0.000000363333	cubic yards per second
0.150959	barrels per day
0.00628995	barrels per hour
0.000104833	barrels per minute
0.00000174721	barrels per second
0.24	hektoliters per day
0.01	hektoliters per hour
0.000166667	hektoliters per minute
0.00000277778	hektoliters per second
0.681072	bushels (U.S.—dry) per day
0.028378	bushels (U.S.—dry) per hour
0.000472967	bushels (U.S.—dry) per minute
0.00000788278	bushels (U.S.—dry) per second
0.659928	bushels (Imperial—dry) per day
0.027497	bushels (Imperial—dry) per hour
0.000458283	bushels (Imperial—dry) per minute
0.00000763806	bushels (Imperial—dry) per second
0.84746	cubic feet per day
0.0353144	cubic feet per hour
0.000588573	cubic feet per minute

LITER PER HOUR: (cont'd)

0.00000980956	cubic feet per second
2.4	dekaliters per day
0.1	dekaliters per hour
0.00166667	dekaliters per minute
0.0000277778	dekaliters per second
2.724288	pecks (U.S.—dry) per day
0.113512	pecks (U.S.—dry) per hour
0.00189187	pecks (U.S.—dry) per minute
0.0000315311	pecks (U.S.—dry) per second
6.340272	gallons (U.S.—liquid) per day
0.264178	gallons (U.S.—liquid) per hour
0.00440297	gallons (U.S.—liquid) per minute
0.0000733828	gallons (U.S.—liquid) per second
5.448480	gallons (U.S.—dry) per day
0.22702	gallons (U.S.—dry) per hour
0.00378367	gallons (U.S.—dry) per minute
0.0000630611	gallons (U.S.—dry) per second
5.279520	gallons (Imperial) per day
0.21998	gallons (Imperial) per hour
0.00366633	gallons (Imperial) per minute
0.0000611056	gallons (Imperial) per second
25.361040	quarts (liquid) per day
1.056710	quarts (liquid) per hour
0.0176118	quarts (liquid) per minute
0.000293531	quarts (liquid) per second
21.794448	quarts (dry) per day
0.908102	quarts (dry) per hour
0.0151350	quarts (dry) per minute
0.000252251	quarts (dry) per second
24	liters per day
1	liters per hour
0.0166667	liters per minute
0.000277778	liters per second
24	cubic decimeters per day
1	cubic decimeters per hour
0.0166667	cubic decimeters per minute
0.000277778	cubic decimeters per second
43.58880	pints (U.S.—dry) per day
1.8162	pints (U.S.—dry) per hour
0.0302700	pints (U.S.—dry) per minute
0.0005045	pints (U.S.—dry) per second
50.721600	pints (U.S.—liquid) per day
2.1134	pints (U.S.—liquid) per hour
0.0352233	pints (U.S.—liquid) per minute
0.000587056	pints (U.S.—liquid) per second
168.940800	gills (Imperial) per day
7.0392	gills (Imperial) per hour
0.117320	gills (Imperial) per minute
0.00195533	gills (Imperial) per second
202.891200	gills (U.S.) per day
8.4538	gills (U.S.) per hour

LITER PER HOUR: (cont'd)

0.140897 . . . . gills (U.S.) per minute
0.00234828 . . . . gills (U.S.) per second
240 . . . . deciliters per day
10 . . . . deciliters per hour
0.166667 . . . . deciliters per minute
0.00277778 . . . . deciliters per second
1,464.6 . . . . cubic inches per day
61.025 . . . . cubic inches per hour
1.0170833 . . . . cubic inches per minute
0.0169514 . . . . cubic inches per second
2,400 . . . . centiliters per day
100 . . . . centiliters per hour
1.666667 . . . . centiliters per minute
0.0277778 . . . . centiliters per second
24,000 . . . . milliliters per day
1,000 . . . . milliliters per hour
16.66667 . . . . milliliters per minute
0.277778 . . . . milliliters per second
24,000 . . . . cubic centimeters per day
1,000 . . . . cubic centimeters per hour
16.666667 . . . . cubic centimeters per minute
0.277778 . . . . cubic centimeters per second
24,000,000 . . . . cubic millimeters per day
1,000,000 . . . . cubic millimeters per hour
16,666.666667 . . . . cubic millimeters per minute
277.777778 . . . . cubic millimeters per second
811.552800 . . . . ounces (U.S.) fluid per day
33.8147 . . . . ounces (U.S.) fluid per hour
0.563578 . . . . ounces (U.S.) fluid per minute
0.00939297 . . . . ounces (U.S.) fluid per second
844.704 . . . . ounces (Imperial—fluid) per day
35.196 . . . . ounces (Imperial—fluid) per hour
0.586600 . . . . ounces (Imperial—fluid) per minute
0.00977667 . . . . ounces (Imperial—fluid) per second
6,492.429599 . . . . drams (fluid) per day
270.5179 . . . . drams (fluid) per hour
4.508632 . . . . drams (fluid) per minute
0.0751439 . . . . drams (fluid) per second
389,545.775424 . . . . minims per day
16,231.0740 . . . . minims per hour
270.5179 . . . . minims per minute
4.508632 . . . . minims per second

LITER PER MINUTE: =

1.440 . . . . kiloliters per day
0.0600 . . . . kiloliters per hour
0.001 . . . . kiloliters per minute
0.0000166667 . . . . kiloliters per second
1.440 . . . . cubic meters per day
0.0600 . . . . cubic meters per hour

LITER PER MINUTE: (cont'd)

0.001	cubic meters per minute
0.0000166667	cubic meters per second
1.883520	cubic yards per day
0.0784800	cubic yards per hour
0.0013080	cubic yards per minute
0.0000218	cubic yards per second
9.0575383	barrels per day
0.377397	barrels per hour
0.00628996	barrels per minute
0.000104833	barrels per second
14.40	hektoliters per day
0.600	hektoliters per hour
0.01	hektoliters per minute
0.000166667	hektoliters per second
40.864320	bushels (U.S.—dry) per day
1.702680	bushels (U.S.—dry) per hour
0.028378	bushels (U.S.—dry) per minute
0.000472967	bushels (U.S.—dry) per second
39.595680	bushels (Imperial—dry) per day
1.649820	bushels (Imperial—dry) per hour
0.027497	bushels (Imperial—dry) per minute
0.000458283	bushels (Imperial—dry) per second
50.852736	cubic feet per day
2.118864	cubic feet per hour
0.0353144	cubic feet per minute
0.000588573	cubic feet per second
144.0	dekaliters per day
6.0	dekaliters per hour
0.1	dekaliters per minute
0.00166667	dekaliters per second
163.457280	pecks (U.S.—dry) per day
6.810720	pecks (U.S.—dry) per hour
0.113512	pecks (U.S.—dry) per minute
0.00189187	pecks (U.S.—dry) per second
380.416320	gallons (U.S.—liquid) per day
15.850680	gallons (U.S.—liquid) per hour
0.264178	gallons (U.S.—liquid) per minute
0.00440297	gallons (U.S.—liquid) per second
326.908800	gallons (U.S.—dry) per day
13.621200	gallons (U.S.—dry) per hour
0.22702	gallons (U.S.—dry) per minute
0.00378367	gallons (U.S.—dry) per second
316.771200	gallons (Imperial) per day
13.198800	gallons (Imperial) per hour
0.21998	gallons (Imperial) per minute
0.00366633	gallons (Imperial) per second
1,521.662400	quarts (liquid) per day
63.402600	quarts (liquid) per hour
1.056710	quarts (liquid) per minute
0.0176118	quarts (liquid) per second
1,307.666880	quarts (dry) per day

LITER PER MINUTE: (cont'd)

54.486120 . . . . quarts (dry) per hour
0.908102 . . . . quarts (dry) per minute
0.0151350 . . . . quarts (dry) per second
1,440 . . . . liters per day
60 . . . . liters per hour
1 . . . . liters per minute
0.0166667 . . . . liters per second
1,440 . . . . cubic decimeters per day
60 . . . . cubic decimeters per hour
1 . . . . cubic decimeters per minute
0.0166667 . . . . cubic decimeters per second
2,615.328 . . . . pints (U.S.—dry) per day
108.972 . . . . pints (U.S.—dry) per hour
1.8162 . . . . pints (U.S.—dry) per minute
0.0302700 . . . . pints (U.S.—dry) per second
3,043.296 . . . . pints (U.S.—liquid) per day
126.804 . . . . pints (U.S.—liquid) per hour
2.1134 . . . . pints (U.S.—liquid) per minute
0.0352233 . . . . pints (U.S.—liquid) per second
10,136.448 . . . . gills (Imperial) per day
422.352 . . . . gills (Imperial) per hour
7.0392 . . . . gills (Imperial) per minute
0.117320 . . . . gills (Imperial) per second
12,173.472003 . . . . gills (U.S.) per day
507.228 . . . . gills (U.S.) per hour
8.4538 . . . . gills (U.S.) per minute
0.140897 . . . . gills (U.S.) per second
14,400 . . . . deciliters per day
600 . . . . deciliters per hour
10 . . . . deciliters per minute
0.166667 . . . . deciliters per second
87,875.999971 . . . . cubic inches per day
3,661.500 . . . . cubic inches per hour
61.025 . . . . cubic inches per minute
1.0170833 . . . . cubic inches per second
144,000 . . . . centiliters per day
6,000 . . . . centiliters per hour
100 . . . . centiliters per minute
1.666667 . . . . centiliters per second
1,440,000 . . . . milliliters per day
60,000 . . . . milliliters per hour
1,000 . . . . milliliters per minute
16.666667 . . . . milliliters per second
1,440,000 . . . . cubic centimeters per day
60,000 . . . . cubic centimeters per hour
1,000 . . . . cubic centimeters per minute
16.666667 . . . . cubic centimeters per second
1,440,000,000 . . . . cubic millimeters per day
60,000,000 . . . . cubic millimeters per hour
1,000,000 . . . . cubic millimeters per minute
16,666.666667 . . . . cubic millimeters per second

LITER PER MINUTE: (cont'd)

48,693.167997	ounces (U.S.) fluid per day
2,028.882	ounces (U.S.) fluid per hour
33.8147	ounces (U.S.) fluid per minute
0.563578	ounces (U.S.) fluid per second
50,682.240	ounces (Imperial—fluid) per day
2,111.760	ounces (Imperial—fluid) per hour
35.196	ounces (Imperial—fluid) per minute
0.586600	ounces (Imperial—fluid) per second
389,545.776029	drams (fluid) per day
16,231.0740	drams (fluid) per hour
270.5179	drams (fluid) per minute
4.508632	drams (fluid) per second
23,372,747	minims per day
973,864.440	minims per hour
16,231.0740	minims per minute
270.5179	minims per second

LITER PER SECOND: =

86.4	kiloliters per day
3.60	kiloliters per hour
0.060	kiloliters per minute
0.001	kiloliters per second
86.4	cubic meters per day
3.60	cubic meters per hour
0.060	cubic meters per minute
0.001	cubic meters per second
113.0112	cubic yards per day
4.708800	cubic yards per hour
0.0784800	cubic yards per minute
0.0013080	cubic yards per second
543.451886	barrels per day
22.643829	barrels per hour
0.377397	barrels per minute
0.00628995	barrels per second
864	hektoliters per day
36	hektoliters per hour
0.60	hektoliters per minute
0.01	hektoliters per second
2,451.859200	bushels (U.S.—dry) per day
102.160800	bushels (U.S.—dry) per hour
1.702680	bushels (U.S.—dry) per minute
0.028378	bushels (U.S.—dry) per second
2,375.740800	bushels (Imperial—dry) per day
98.989200	bushels (Imperial—dry) per hour
1.649820	bushels (Imperial—dry) per minute
0.27497	bushels (Imperial—dry) per second
3,051.164160	cubic feet per day
127.131840	cubic feet per hour
2.118864	cubic feet per minute
0.0353144	cubic feet per second

8,640	dekaliters per day
360	dekaliters per hour
6.0	dekaliters per minute
0.1	dekaliters per second
9,807.436800	pecks (U.S.—dry) per day
408.643200	pecks (U.S.—dry) per hour
6.810720	pecks (U.S.—dry) per minute
0.113512	pecks (U.S.—dry) per second
22,824.979200	gallons (U.S.—liquid) per day
951.040800	gallons (U.S.—liquid) per hour
15.850680	gallons (U.S.—liquid) per minute
0.264178	gallons (U.S.—liquid) per second
19,614.52800	gallons (U.S.—dry) per day
817.27200	gallons (U.S.—dry) per hour
13.62120	gallons (U.S.—dry) per minute
0.22702	gallons (U.S.—dry) per second
19,006.27200	gallons (Imperial) per day
791.92800	gallons (Imperial) per hour
13.19880	gallons (Imperial) per minute
0.21998	gallons (Imperial) per second
91,299.744	quarts (liquid) per day
3,804.156	quarts (liquid) per hour
63.402600	quarts (liquid) per minute
1.056710	quarts (liquid) per second
78,460.01280	quarts (dry) per day
3,269.167200	quarts (dry) per hour
54.486120	quarts (dry) per minute
0.908102	quarts (dry) per second
86,400	liters per day
3,600	liters per hour
60	liters per minute
1	liters per second
86,400	cubic decimeters per day
3,600	cubic decimeters per hour
60	cubic decimeters per minute
1	cubic decimeters per second
156,920	pints (U.S.—dry) per day
6,538.320	pints (U.S.—dry) per hour
108.9720	pints (U.S.—dry) per minute
1.8162	pints (U.S.—dry) per second
182,598	pints (U.S.—liquid) per day
7,608.2400	pints (U.S.—liquid) per hour
126.8040	pints (U.S.—liquid) per minute
2.1134	pints (U.S.—liquid) per second
608,187	gills (Imperial) per day
25,341.12	gills (Imperial) per hour
422.3520	gills (Imperial) per minute
7.0392	gills (Imperial) per second
730,408	gills (U.S.) per day
30,433.68	gills (U.S.) per hour
507.2280	gills (U.S.) per minute

LITER PER SECOND: (cont'd)

8.4538	gills (U.S.) per second
864,000	deciliters per day
36,000	deciliters per hour
600	deciliters per minute
10	deciliters per second
5,272,560	cubic inches per day
219,690	cubic inches per hour
3,661.500	cubic inches per minute
61.025	cubic inches per second
8,640,000	centiliters per day
360,000	centiliters per hour
6,000	centiliters per minute
100	centiliters per second
86,400,000	milliliters per day
3,600,000	milliliters per hour
60,000	milliliters per minute
1,000	milliliters per second
86,400,000	cubic centimeters per day
3,600,000	cubic centimeters per hour
60,000	cubic centimeters per minute
1,000	cubic centimeters per second
86,400,000,000	cubic millimeters per day
3,600,000,000	cubic millimeters per hour
60,000,000	cubic millimeters per minute
1,000,000	cubic millimeters per second
2,921,590	ounces (U.S.) fluid per day
121,733	ounces (U.S.) fluid per hour
2,028.8820	ounces (U.S.) fluid per minute
33.8147	ounces (U.S.) fluid per second
3,040,934	ounces (Imperial—fluid) per day
126,706	ounces (Imperial—fluid) per hour
2,111.76	ounces (Imperial—fluid) per minute
35.196	ounces (Imperial—fluid) per second
23,372,747	drams (fluid) per day
973,864	drams (fluid) per hour
16,231.0740	drams (fluid) per minute
270.5179	drams (fluid) per second
1,402,364,794	minims per day
58,431,866	minims per hour
973,864	minims per minute
16,231.0740	minims per second

METER: =

0.00053961	miles (nautical)
0.00062137	miles (statute)
0.001	kilometers
0.00497097	furlongs
0.01	hektometers
0.0497097	chains
0.1	dekameters

METER: (cont'd)

0.198839	rods
1	meters
1.093611	yards
1.811	varas (Texas)
3.280833	feet
4.374440	spans
4.970974	links
9.84250	hands
10	decimeters
100	centimeters
39.370	inches
1,000	millimeters
39,370	mils
1,000,000	microns
1,000,000,000	millimicrons
1,000,000,000	micromillimeters
1,553,164	wave lengths of red line cadmium
1,000,000	Angstrom Units
0.54681	fathoms

METER PER DAY: =

0.00053961	miles (nautical) per day
0.0000224837	miles (nautical) per hour
0.000000374729	miles (nautical) per minute
0.00000000624549	miles (nautical) per second
0.00062137	miles (statute) per day
0.0000258904	miles (statute) per hour
0.000000431507	miles (statute) per minute
0.00000000719178	miles (statute) per second
0.001	kilometers per day
0.0000416667	kilometers per hour
0.000000694444	kilometers per minute
0.0000000115741	kilometers per second
0.00497097	furlongs per day
0.000207124	furlongs per hour
0.00000345206	furlongs per minute
0.0000000575344	furlongs per second
0.01	hektometers per day
0.000416667	hektometers per hour
0.00000694444	hektometers per minute
0.000000115741	hektometers per second
0.0497097	chains per day
0.00207124	chains per hour
0.0000345206	chains per minute
0.000000575344	chains per second
0.1	dekameters per day
0.00416667	dekameters per hour
0.0000694444	dekameters per minute
0.00000115741	dekameters per second

METER PER DAY: (cont'd)

0.198839	rods per day
0.00828496	rods per hour
0.000138083	rods per minute
0.00000230138	rods per second
1	meters per day
0.0416667	meters per hour
0.000694444	meters per minute
0.0000115741	meters per second
1.093611	yards per day
0.0455671	yards per hour
0.000759452	yards per minute
0.0000126575	yards per second
1.1811	varas (Texas) per day
0.0492125	varas (Texas) per hour
0.000820208	varas (Texas) per minute
0.0000136701	varas (Texas) per second
3.280833	feet per day
0.133680	feet per hour
0.00222801	feet per minute
0.0000371334	feet per second
4.374440	spans per day
0.182268	spans per hour
0.00303781	spans per minute
0.0000506301	spans per second
4.970974	links per day
0.207124	links per hour
0.00345206	links per minute
0.0000575344	links per second
10	decimeters per day
0.416667	decimeters per hour
0.00694444	decimeters per minute
0.000115741	decimeters per second
100	centimeters per day
4.166667	centimeters per hour
0.0694444	centimeters per minute
0.00115741	centimeters per second
39.370	inches per day
1.640417	inches per hour
0.0273403	inches per minute
0.000455671	inches per second
1,000	millimeters per day
41.666667	millimeters per hour
0.694444	millimeters per minute
0.0115741	millimeters per second
39,370	mils per day
1,640.416667	mils per hour
27.340278	mils per minute
0.455671	mils per second
1,000,000	microns per day
41,666.666400	microns per hour
694.444444	microns per minute

METER PER DAY: (cont'd)

11.574074	microns per second
1,000,000	Angstrom Units per day
41,666.666400	Angstrom Units per hour
694.444444	Angstrom Units per minute
11.574074	Angstrom Units per second

METER PER HOUR: =

0.0129506	miles (nautical) per day
0.00053961	miles (nautical) per hour
0.00000899350	miles (nautical) per minute
0.000000149892	miles (nautical) per second
0.0149129	miles (statute) per day
0.00062137	miles (statute) per hour
0.0000103562	miles (statute) per minute
0.000000172603	miles (statute) per second
0.0240	kilometers per day
0.001	kilometers per hour
0.0000166667	kilometers per minute
0.000000277778	kilometers per second
0.119303	furlongs per day
0.00497097	furlongs per hour
0.0000828495	furlongs per minute
0.00000138083	furlongs per second
0.240	hektometers per day
0.01	hektometers per hour
0.000166667	hektometers per minute
0.00000277778	hektometers per second
1.193033	chains per day
0.0497097	chains per hour
0.000828495	chains per minute
0.0000138083	chains per second
2.4	dekameters per day
0.1	dekameters per hour
0.00166667	dekameters per minute
0.0000277778	dekameters per second
4.772136	rods per day
0.198839	rods per hour
0.00331398	rods per minute
0.0000552331	rods per second
24	meters per day
1	meters per hour
0.0166667	meters per minute
0.000277778	meters per second
26.246664	yards per day
1.093611	yards per hour
0.0182268	yards per minute
0.000303781	yards per second
28.346400	varas (Texas) per day
1.1811	varas (Texas) per hour
0.0196850	varas (Texas) per minute

METER PER HOUR: (cont'd)

0.000328083	varas (Texas) per second
78.74	feet per day
3.280833	feet per hour
0.0546806	feet per minute
0.000911343	feet per second
104.986560	spans per day
4.374440	spans per hour
0.0729073	spans per minute
0.00121512	spans per second
119.303280	links per day
4.970974	links per hour
0.0828495	links per minute
0.00138083	links per second
240	decimeters per day
10	decimeters per hour
0.166667	decimeters per minute
0.00277778	decimeters per second
2,400	centimeters per day
100	centimeters per hour
1.666667	centimeters per minute
0.0277778	centimeters per second
944.880	inches per day
39.370	inches per hour
0.656167	inches per minute
0.0109361	inches per second
24,000	millimeters per day
1,000	millimeters per hour
16.666667	millimeters per minute
0.277778	millimeters per second
944,879.999004	mils per day
39,370	mils per hour
656.166667	mils per minute
10.936111	mils per second
24,000,000	microns per day
1,000,000	microns per hour
16,666.666667	microns per minute
277.777778	microns per second
24,000,000	Angstrom Units per day
1,000.000	Angstrom Units per hour
16,666.666667	Angstrom Units per minute
277.777778	Angstrom Units per second

METER PER MINUTE: =

0.777038	miles (nautical) per day
0.0323766	miles (nautical) per hour
0.00053961	miles (nautical) per minute
0.00000899350	miles (nautical) per second
0.894773	miles (statute) per day
0.0372822	miles (statute) per hour
0.00062137	miles (statute) per minute

METER PER MINUTE: (cont'd)

0.0000103562	miles (statute) per second
1.440	kilometers per day
0.0600	kilometers per hour
0.001	kilometers per minute
0.0000166667	kilometers per second
7.158197	furlongs per day
0.298258	furlongs per hour
0.00497097	furlongs per minute
0.0000828495	furlongs per second
14.40	hektometers per day
0.600	hektometers per hour
0.01	hektometers per minute
0.000166667	hektometers per second
71.581968	chains per day
2.982582	chains per hour
0.0497097	chains per minute
0.000828495	chains per second
144.0	dekameters per day
6.0	dekameters per hour
0.1	dekameters per minute
0.00166667	dekameters per second
286.328160	rods per day
11.930340	rods per hour
0.198839	rods per minute
0.00331398	rods per second
1,440	meters per day
60	meters per hour
1	meters per minute
0.0166667	meters per second
1,574.799840	yards per day
65.616660	yards per hour
1.093611	yards per minute
0.0182269	yards per second
1,700.784	varas (Texas) per day
70.866	varas (Texas) per hour
1.1811	varas (Texas) per minute
0.0196850	varas (Texas) per second
4,724.400	feet per day
196.85	feet per hour
3.280833	feet per minute
0.0546806	feet per second
6,299.193600	spans per day
262.466400	spans per hour
4.374440	spans per minute
0.0729073	spans per second
7,158.196800	links per day
298.258200	links per hour
4.970974	links per minute
0.0828495	links per second
14,400	decimeters per day
600	decimeters per hour

METER PER MINUTE: (cont'd)

10	decimeters per minute
0.166667	decimeters per second
144,000	centimeters per day
6,000	centimeters per hour
100	centimeters per minute
1.666667	centimeters per second
56,692.800	inches per day
2,362.200	inches per hour
39.370	inches per minute
0.656167	inches per second
1,440,000	millimeters per day
60,000	millimeters per hour
1,000	millimeters per minute
16.666667	millimeters per second
56,692,800	mils per day
2,362,200	mils per hour
39,370	mils per minute
656.166667	mils per second
1,440,000,000	microns per day
60,000,000	microns per hour
1,000,000	microns per minute
16,666.666667	microns per second
1,440,000,000	Angstrom Units per day
60,000,000	Angstrom Units per hour
1,000,000	Angstrom Units per minute
16,666.666667	Angstrom Units per second

METER PER SECOND: =

46.622304	miles (nautical) per day
1.942596	miles (nautical) per hour
0.0323766	miles (nautical) per minute
0.00053961	miles (nautical) per second
53.686368	miles (statute) per day
2.236932	miles (statute) per hour
0.0372822	miles (statute) per minute
0.00062137	miles (statute) per second
86.4	kilometers per day
3.60	kilometers per hour
0.060	kilometers per minute
0.001	kilometers per second
429.491808	furlongs per day
17.895492	furlongs per hour
0.298258	furlongs per minute
0.00497097	furlongs per second
864	hektometers per day
36	hektometers per hour
0.60	hektometers per minute
0.01	hektometers per second
4,294.918080	chains per day
178.954920	chains per hour

METER PER SECOND: (cont'd)

2.982582	chains per minute
0.0497097	chains per second
8,640	dekameters per day
360	dekameters per hour
6.0	dekameters per minute
0.1	dekameters per second
17,179.689600	rods per day
715.820400	rods per hour
11.930340	rods per minute
0.198839	rods per second
86,400	meters per day
3,600	meters per hour
60	meters per minute
1	meters per second
94,487.990400	yards per day
3,936.999600	yards per hour
65.616660	yards per minute
1.093611	yards per second
102,047.0400	varas (Texas) per day
4,251.9600	varas (Texas) per hour
70.8660	varas (Texas) per minute
1.1811	varas (Texas) per second
283,463.971200	feet per day
11,810.998800	feet per hour
196.849980	feet per minute
3.280833	feet per second
377,951.616000	spans per day
15,747.984000	spans per hour
262.466400	spans per minute
4.374440	spans per second
429,492.153600	links per day
17,895.506400	links per hour
298.258440	links per minute
4.970974	links per second
864,000	decimeters per day
36,000	decimeters per hour
600	decimeters per minute
10	decimeters per second
8,640,000	centimeters per day
360,000	centimeters per hour
6,000	centimeters per minute
100	centimeters per second
3,401,568	inches per day
141,732.0	inches per hour
2,362.200	inches per minute
39.370	inches per second
86,400,000	millimeters per day
3,600,000	millimeters per hour
60,000	millimeters per minute
1,000	millimeters per second
3,401,568,000	mils per day

METER PER SECOND: (cont'd)

141,732,000	mils per hour
2,362,200	mils per minute
39,370	mils per second
86,400,000,000	microns per day
3,600,000,000	microns per hour
60,000,000	microns per minute
1,000,000	microns per second
86,400,000,000	Angstrom Units per day
3,600,000,000	Angstrom Units per hour
60,000,000	Angstrom Units per minute
1,000,000	Angstrom Units per second

METER OF MERCURY @ 32°F.:=

0.135964	hektometers of water @ 60°F.
1.359639	dekameters of water @ 60°F.
13.596390	meters of water @ 60°F.
44.604005	feet of water @ 60°F.
535.247985	inches of water @ 60°F.
135.963901	decimeters of water @ 60°F.
1,359.639013	centimeters of water @ 60°F.
13,596.39013	millimeters of water @ 60°F.
0.01	hektometers of mercury @ 32°F.
0.1	dekameters of mercury @ 32°F.
1	meters of mercury @ 32°F.
3.280833	feet of mercury @ 32°F.
39.37	inches of mercury @ 32°F.
10	decimeters of mercury @ 32°F.
100	centimeters of mercury @ 32°F.
1,000	millimeters of mercury @ 32°F.
149,862.505	tons per square hektometer
1,498.62505	tons per square dekameter
14.986251	tons per square meter
1.392300	tons per square foot
0.00966852	tons per square inch
0.149863	tons per square decimeter
0.00149863	tons per square centimeter
0.0000149863	tons per square millimeter
135,963,901	kilograms per square hektometer
1,359,639	kilograms per square dekameter
13,596.39	kilograms per square meter
1,263.034001	kilograms per square foot
8.771085	kilograms per square inch
135.9639	kilograms per square decimeter
1.359639	kilograms per square centimeter
0.0135964	kilograms per square millimeter
299,724,991	pounds per square hektometer
2,997,249	pounds per square dekameter
29,972.49	pounds per square meter
2,784.499982	pounds per square foot
19.337009	pounds per square inch

METER OF MERCURY @ 32°F.: (cont'd)

299.7249 . . . . pounds per square decimeter
2.997249 . . . . pounds per square centimeter
0.0299725 . . . . pounds per square millimeter
13,596,390 . . . . hektograms per square hektometer
135,963.90 . . . . hektograms per square dekameter
1,359.6390 . . . . hektograms per square meter
12,630.340015 . . . . hektograms per square foot
87.711006 . . . . hektograms per square inch
13.596390 . . . . hektograms per square decimeter
0.135964 . . . . hektograms per square centimeter
0.00135964 . . . . hektograms per square millimeter
13,596,390,130 . . . . dekagrams per square hektometer
135,963,901 . . . . dekagrams per square dekameter
1,359,639 . . . . dekagrams per square meter
126,303.4 . . . . dekagrams per square foot
877.110017 . . . . dekagrams per square inch
13,596.39 . . . . dekagrams per square decimeter
135.9639 . . . . dekagrams per square centimeter
1.359639 . . . . dekagrams per square millimeter
4,795,599,897 . . . . ounces per square hektometer
47,955,998 . . . . ounces per square dekameter
479,560 . . . . ounces per square meter
44,552.0 . . . . ounces per square foot
309.392019 . . . . ounces per square inch
4,795.5998 . . . . ounces per square decimeter
47.955998 . . . . ounces per square centimeter
0.479560 . . . . ounces per square millimeter
135,963,901,300 . . . . grams per square hektometer
1,359,639,013 . . . . grams per square dekameter
13,596,390 . . . . grams per square meter
1,263,034 . . . . grams per square foot
8,771.111 . . . . grams per square inch
135,963.90 . . . . grams per square decimeter
1,359.6390 . . . . grams per square centimeter
13.596390 . . . . grams per square millimeter
1,359,639,013 x 10³ . . . . decigrams per square hektometer
13,596,390,130 . . . . decigrams per square dekameter
135,963,900 . . . . decigrams per square meter
12,630,340 . . . . decigrams per square foot
87,711.11 . . . . decigrams per square inch
1,359,639 . . . . decigrams per square decimeter
13,596.390 . . . . decigrams per square centimeter
135.96390 . . . . decigrams per square millimeter
1,359,639,013 x 10⁴ . . . . centigrams per square hektometer
135,963,901,300 . . . . centigrams per square dekameter
1,359,639,013 . . . . centigrams per square meter
126,303.40 . . . . centigrams per square foot
877,111.1 . . . . centigrams per square inch
13,596,390 . . . . centigrams per square decimeter
135,963.9 . . . . centigrams per square centimeter
1,359.6390 . . . . centigrams per square millimeter

METER OF MERCURY @ 32°F.: (cont'd)

1,359,639,013 x 10^5	milligrams per square hektometer
1,359,639,013 x 10^3	milligrams per square dekameter
13,596,390,130	milligrams per square meter
1,263,034	milligrams per square foot
8,771,111	milligrams per square inch
135,963,901	milligrams per square decimeter
1,359,639	milligrams per square centimeter
13,596.390	milligrams per square millimeter
133,222 x 10^9	dynes per square hektometer
133,222 x 10^7	dynes per square dekameter
133,222 x 10^5	dynes per square meter
1,238,591,617	dynes per square foot
8,601,331	dynes per square inch
133,222,000	dynes per square decimeter
1,332,220	dynes per square centimeter
13,322.20	dynes per square millimeter
1.333300	bars
1.315801	atmospheres

METER OF WATER @ 60°F.:=

0.01	hektometers of water @ 60°F.
0.1	dekameters of water @ 60°F.
1.0	meters of water @ 60°F.
3.280833	feet of water @ 60°F.
39.37	inches of water @ 60°F.
10	decimeters of water @ 60°F.
100	centimeters of water @ 60°F.
1,000	millimeters of water @ 60°F.
0.000735510	hektometers of mercury @ 32°F.
0.00735510	dekameters of mercury @ 32°F.
0.0735510	meters of mercury @ 32°F.
2.413086	feet of mercury @ 32°F.
28.957029	inches of mercury @ 32°F.
0.735510	decimeters of mercury @ 32°F.
7.355103	centimeters of mercury @ 32°F.
73.55103	millimeters of mercury @ 32°F.
11,022.543742	tons per square hektometer
110.225437	tons per square dekameter
1.102254	tons per square meter
0.102405	tons per square foot
0.000711128	tons per square inch
0.0110225	tons per square decimeter
0.000110225	tons per square centimeter
0.00000110225	tons per square millimeter
10,000,000	kilograms per square hektometer
100,000	kilograms per square dekameter
1,000	kilograms per square meter
92.897452	kilograms per square foot
0.645121	kilograms per square inch
10	kilograms per square decimeter

METER OF WATER @ 60°F.: (cont'd)

0.10	kilograms per square centimeter
0.001	kilograms per square millimeter
22,045,074	pounds per square hektometer
220,450	pounds per square dekameter
2,204.5074	pounds per square meter
204.802897	pounds per square foot
1.422241	pounds per square inch
22.045074	pounds per square decimeter
0.220451	pounds per square centimeter
0.00220451	pounds per square millimeter
100,000,000	hektograms per square hektometer
1,000,000	hektograms per square dekameter
10,000	hektograms per square meter
928.974717	hektograms per square foot
6.45147	hektograms per square inch
100	hektograms per square decimeter
1	hektograms per square centimeter
0.01	hektograms per square millimeter
1,000,000,000	dekagrams per square hektometer
10,000,000	dekagrams per square dekameter
100,000	dekagrams per square meter
9,289.746972	dekagrams per square foot
64.512351	dekagrams per square inch
1,000	dekagrams per square decimeter
10	dekagrams per square centimeter
0.1	dekagrams per square millimeter
352,721,381	ounces per square hektometer
3,527,213	ounces per square dekameter
35,272.13	ounces per square meter
3,276.846123	ounces per square foot
22.755860	ounces per square inch
352.7213	ounces per square decimeter
3.527213	ounces per square centimeter
0.0352721	ounces per square millimeter
10,000,000,000	grams per square hektometer
100,000,000	grams per square dekameter
1,000,000	grams per square meter
92,897.469283	grams per square foot
645.124379	grams per square inch
10,000	grams per square decimeter
100	grams per square centimeter
1	grams per square millimeter
100,000,000,000	decigrams per square hektometer
1,000,000,000	decigrams per square dekameter
10,000,000	decigrams per square meter
928,975	decigrams per square foot
6,451.243790	decigrams per square inch
100,000	decigrams per square decimeter
1,000	decigrams per square centimeter
10	decigrams per square millimeter
10×10^{11}	centigrams per square hektometer

METER OF WATER @ 60°F.: (cont'd)

10,000,000,000	centigrams per square dekameter
100,000,000	centigrams per square meter
9,289,747	centigrams per square foot
64,512.437904	centigrams per square inch
1,000,000	centigrams per square decimeter
10,000	centigrams per square centimeter
100	centigrams per square millimeter
10×10^{12}	milligrams per square hektometer
10×10^{10}	milligrams per square dekameter
1,000,000,000	milligrams per square meter
92,897,468	milligrams per square foot
645,124	milligrams per square inch
10,000,000	milligrams per square decimeter
100,000	milligrams per square centimeter
1,000	milligrams per square millimeter
9,798,617,182,254	dynes per square hektometer
97,986,171,823	dynes per square dekameter
979,861,718	dynes per square meter
91,099,700	dynes per square foot
632,637	dynes per square inch
9,798,617	dynes per square decimeter
97,986.171823	dynes per square centimeter
979.861718	dynes per square millimeter
0.0966778	bars
0.0967785	atmospheres

MICRON: =

0.000000000621259	miles (statute)
0.000000001	kilometers
0.00000000497097	furlongs
0.00000001	hektometers
0.0000000497097	chains
0.0000001	dekameters
0.000000198839	rods
0.000001	meters
0.00000109358	yards
0.0000011811	varas (Texas)
0.0000032808	feet
0.00000437440	spans
0.00000497097	links
0.0000098425	hands
0.00001	decimeters
0.0001	centimeters
0.00003937	inches
0.001	millimeters
0.039370	mils
1	microns
1,000	millimicrons
1,000	micromillimeters
1.553159	wave lengths of red line of cadmium
10,000	Angstrom Units

MIL: =

0.00000001578	miles (statute)
0.0000000254	kilometers
0.000000126263	furlongs
0.000000254	hektometers
0.00000126263	chains
0.00000254	dekameters
0.00000505051	rods
0.0000254	meters
0.000027777	yards
0.00003	varas (Texas)
0.000083333	feet
0.000111111	spans
0.000126263	links
0.00025	hands
0.000254	decimeters
0.00254	centimeters
0.001	inches
0.0254	millimeters
1	mils
25.4	microns
25,400	millimicrons
25,400	micromillimeters
39.450445	wave lengths of red line of cadmium
25.4	Angstrom Units

MILE (STATUTE): =

0.86836	miles (nautical)
1	miles (statute)
1.60935	kilometers
8	furlongs
16.0935	hektometers
80	chains
160,935	dekameters
320	rods
1,609.35	meters
1,760	yards
1,900.8	varas (Texas)
5,280	feet
7,040	spans
8,000	links
15,840	hands
16,093.5	decimeters
160,935	centimeters
63,360	inches
1,609,350	millimeters
63,360,000	mils
1,609,344,000	microns
$1,609,344 \times 10^6$	millimicrons

MILE (STATUTE): (cont'd)

$1,609,344 \times 10^6$	micromillimeters
2,499,572,909	wave lengths of red line of cadmium
$16,093,440 \times 10^6$	Angstrom Units

MILE (STATUTE) PER DAY: =

0.86836	miles (nautical) per day
0.0361817	miles (nautical) per hour
0.000603028	miles (nautical) per minute
0.0000100505	miles (nautical) per second
1	miles (statute) per day
0.0416667	miles (statute) per hour
0.000694444	miles (statute) per minute
0.0000115741	miles (statute) per second
1.60935	kilometers per day
0.0670562	kilometers per hour
0.00111760	kilometers per minute
0.0000186267	kilometers per second
8	furlongs per day
0.333333	furlongs per hour
0.00555556	furlongs per minute
0.0000925926	furlongs per second
16.0935	hektometers per day
0.670562	hektometers per hour
0.0111760	hektometers per minute
0.000186267	hektometers per second
80	chains per day
3.333333	chains per hour
0.0555556	chains per minute
0.000925926	chains per second
160.935	dekameters per day
6.705625	dekameters per hour
0.111760	dekameters per minute
0.00186267	dekameters per second
320	rods per day
13.333333	rods per hour
0.222222	rods per minute
0.00370370	rods per second
1,609.35	meters per day
67.056250	meters per hour
1.117604	meters per minute
0.0186267	meters per second
1,760	yards per day
73.333333	yards per hour
1.222222	yards per minute
0.0203703	yards per second
1,900.8	varas (Texas) per day
79.2	varas (Texas) per hour
1.320	varas (Texas) per minute
0.0220	varas (Texas) per second
5,280	feet per day

MILE (STATUTE) PER DAY: (cont'd)

220	feet per hour
3.666667	feet per minute
0.06111111	feet per second
7,040	spans per day
293.333333	spans per hour
4.888889	spans per minute
0.0814814	spans per second
8,000	links per day
333.333333	links per hour
5.555556	links per minute
0.0925925	links per second
15,840	hands per day
660	hands per hour
11	hands per minute
0.183333	hands per second
16,093.5	decimeters per day
670.56250	decimeters per hour
11.176042	decimeters per minute
0.186267	decimeters per second
160,935	centimeters per day
6,705.624996	centimeters per hour
111.760417	centimeters per minute
1.862674	centimeters per second
63,360	inches per day
2,640	inches per hour
44	inches per minute
0.733333	inches per second
1,609,350	millimeters per day
67,056.249960	millimeters per hour
1,117.604166	millimeters per minute
18.626736	millimeters per second

MILE (STATUTE) PER HOUR: =

20.84064	miles (nautical) per day
0.86836	miles (nautical) per hour
0.0144727	miles (nautical) per minute
0.000241211	miles (nautical) per second
24	miles (statute) per day
1	miles (statute) per hour
0.0166667	miles (statute) per minute
0.000277778	miles (statute) per second
38.6244	kilometers per day
1.60935	kilometers per hour
0.0268225	kilometers per minute
0.000447042	kilometers per second
192	furlongs per day
8	furlongs per hour
0.133333	furlongs per minute
0.00222222	furlongs per second
386.244	hektometers per day

MILE (STATUTE) PER HOUR: (cont'd)

16.0935	hektometers per hour
0.268225	hektometers per minute
0.00447042	hektometers per second
1,920	chains per day
80	chains per hour
1.333333	chains per minute
0.0222222	chains per second
3,862.44	dekameters per day
160.935	dekameters per hour
2.682250	dekameters per minute
0.0447042	dekameters per second
7,680	rods per day
320	rods per hour
5.333333	rods per minute
0.0888889	rods per second
38,624.4	meters per day
1,609.35	meters per hour
26.8225	meters per minute
0.447042	meters per second
42,240	yards per day
1,760	yards per hour
29.333333	yards per minute
0.488889	yards per second
45,619.2	varas (Texas) per day
1,900.8	varas (Texas) per hour
31.68	varas (Texas) per minute
0.528	varas (Texas) per second
126,720	feet per day
5,280	feet per hour
88	feet per minute
1.466667	feet per second
168,960	spans per day
7,040	spans per hour
117.333333	spans per minute
1.955556	spans per second
192,000	links per day
8,000	links per hour
133.333333	links per minute
2.222222	links per second
380,160	hands per day
15,840	hands per hour
264	hands per minute
4.40	hands per second
386,244	decimeters per day
16,093.5	decimeters per hour
268.225	decimeters per minute
4.470417	decimeters per second
3,862,440	centimeters per day
160,935	centimeters per hour
2,682.25	centimeters per minute
44.704167	centimeters per second

MILE (STATUTE) PER HOUR: (cont'd)

1,520,640	inches per day
63,360	inches per hour
1,056	inches per minute
17.6	inches per second
38,624,400	millimeters per day
1,609,350	millimeters per hour
26,822.5	millimeters per minute
447.0416667	millimeters per second

MILE (STATUTE) PER MINUTE: =

1,250.438400	miles (nautical) per day
52.101600	miles (nautical) per hour
0.86836	miles (nautical) per minute
0.0144727	miles (nautical) per second
1,440	miles (statute) per day
60	miles (statute) per hour
1	miles (statute) per minute
0.01666667	miles (statute) per second
2,317.464	kilometers per day
96.561	kilometers per hour
1.60935	kilometers per minute
0.0268225	kilometers per second
11,520	furlongs per day
480	furlongs per hour
8	furlongs per minute
0.133333	furlongs per second
23,174.64	hektometers per day
965.61	hektometers per hour
16.0935	hektometers per minute
0.268225	hektometers per second
115,200	chains per day
4,800	chains per hour
80	chains per minute
1.333333	chains per second
231,746	dekameters per day
9,656.1	dekameters per hour
160.935	dekameters per minute
2.682250	dekameters per second
460,800	rods per day
19,200	rods per hour
320	rods per minute
5.333333	rods per second
2,317,464	meters per day
96,561	meters per hour
1,609.35	meters per minute
26.82250	meters per second
2,534,400	yards per day
105,600	yards per hour
1,760	yards per minute
29.333333	yards per second

MILE (STATUTE) PER MINUTE: (cont'd)

2,737,152	varas (Texas) per day
114,048	varas (Texas) per hour
1,900.8	varas (Texas) per minute
31.680	varas (Texas) per second
7,603,200	feet per day
316,800	feet per hour
5,280	feet per minute
88	feet per second
10,137,600	spans per day
422,400	spans per hour
7,040	spans per minute
117.333333	spans per second
11,520,000	links per day
480,000	links per hour
8,000	links per minute
133.333333	links per second
22,809,600	hands per day
950,400	hands per hour
15,840	hands per minute
264.0	hands per second
23,174,640	decimeters per day
965,610	decimeters per hour
16,093.5	decimeters per minute
268.225	decimeters per second
231,746,400	centimeters per day
9,656,100	centimeters per hour
160,935	centimeters per minute
2,682.25	centimeters per second
91,238,400	inches per day
3,801,600	inches per hour
63,360	inches per minute
1,056.0	inches per second
2,317,464,000	millimeters per day
96,561,000	millimeters per hour
1,609,350	millimeters per minute
26,822.5	millimeters per second

MILE (STATUTE) PER SECOND: =

75,026.304	miles (nautical) per day
3,126.096	miles (nautical) per hour
52.10260	miles (nautical) per minute
0.86836	miles (nautical) per second
86,400	miles (statute) per day
3,600	miles (statute) per hour
60	miles (statute) per minute
1	miles (statute) per second
139,048	kilometers per day
5,793.660	kilometers per hour
96.561	kilometers per minute
1.60935	kilometers per second

MILE (STATUTE) PER SECOND: (cont'd)

691,200 . . . . furlongs per day
28,800 . . . . furlongs per hour
480 . . . . furlongs per minute
8 . . . . furlongs per second
1,390,480 . . . . hektometers per day
57,936.6 . . . . hektometers per hour
965.61 . . . . hektometers per minute
16.0935 . . . . hektometers per second
6,912,000 . . . . chains per day
288,000 . . . . chains per hour
4,800 . . . . chains per minute
80 . . . . chains per second
13,904,800 . . . . dekameters per day
579,366 . . . . dekameters per hour
9,656.1 . . . . dekameters per minute
160.935 . . . . dekameters per second
27,648,000 . . . . rods per day
1,152,000 . . . . rods per hour
19,200 . . . . rods per minute
320 . . . . rods per second
139,048,000 . . . . meters per day
5,793,660 . . . . meters per hour
96,561.0 . . . . meters per minute
1,609.35 . . . . meters per second
152,064,000 . . . . yards per day
6,336,000 . . . . yards per hour
105,600 . . . . yards per minute
1,760 . . . . yards per second
164,229,120 . . . . varas (Texas) per day
6,842,880 . . . . varas (Texas) per hour
114,048 . . . . varas (Texas) per minute
1,900.8 . . . . varas (Texas) per second
456,192,000 . . . . feet per day
19,008,000 . . . . feet per hour
316,800 . . . . feet per minute
5,280 . . . . feet per second
608,256,000 . . . . spans per day
25,344,000 . . . . spans per hour
422,400 . . . . spans per minute
7,040 . . . . spans per second
691,200,000 . . . . links per day
28,800,000 . . . . links per hour
480,000 . . . . links per minute
8,000 . . . . links per second
1,390,478,000 . . . . decimeters per day
57,936,600 . . . . decimeters per hour
965,610 . . . . decimeters per minute
16,093.5 . . . . decimeters per second
13,904,780,000 . . . . centimeters per day
579,366,000 . . . . centimeters per hour
9,656,100 . . . . centimeters per minute

MILE (STATUTE) PER SECOND: (cont'd)

160,935	centimeters per second
5,474,304,000	inches per day
228,096,000	inches per hour
3,801,600	inches per minute
63,360	inches per second
139,047,800,000	millimeters per day
5,793,660,000	millimeters per hour
96,561,000	millimeters per minute
1,609,350	millimeters per second

MIL: =

0.0000000137061	miles (nautical)
0.00000001578	miles (statute)
0.0000000254	kilometers
0.000000126263	furlongs
0.000000254	hektometers
0.00000126263	chains
0.00000254	dekameters
0.00000505052	rods
0.0000254	meters
0.0000277778	yards
0.00003	varas (Texas)
0.0000833333	feet
0.000111111	spans
0.000126263	links
0.000250	hands
0.000254	decimeters
0.00254	centimeters
0.001	inches
0.0254	millimeters
1	mils
25.4	microns
25,400	millimicrons
25,400	micromillimeters
39.450445	wave lengths of red line of cadmium
25.4	Angstrom Units

MILLIMETER: =

0.00000053961	miles (nautical)
0.00000062137	miles (statute)
0.000001	kilometers
0.00000497097	furlongs
0.00001	hektometers
0.0000497097	chains
0.0001	dekameters
0.000198839	rods
0.001	meters
0.00109361	yards
0.0011811	varas (Texas)

UNIVERSAL CONVERSION FACTORS

MILLIMETER: (cont'd)

0.00328083	feet
0.00437444	spans
0.00497097	links
0.00984250	hands
0.01	decimeters
0.1	centimeters
0.039370	inches
1	millimeters
39.370	mils
1,000	microns
1,000,000	millimicrons
1,000,000	micromillimeters
1,553.164	wave lengths of red line of cadmium
1,000	Angstrom Units

MINUTE (ANGLE): =

0.00018519	quadrants
0.000290888	radians
0.0166667	degrees
60	seconds
0.0000462963	circumference or revolutions

MILLIGRAM: =

0.00000000098426	tons (long)
0.000000001	tons (metric)
0.00000000110231	tons (short)
0.000001	kilograms
0.00000267923	pounds (Troy)
0.00000220462	pounds (Avoir.)
0.00001	hektograms
0.0001	dekagrams
0.0000321507	ounces (Troy)
0.0000352739	ounces (Avoir.)
0.001	grams
0.01	decigrams
0.1	centigrams
1	milligrams
0.0154324	grains
0.000257206	drams (Troy)
0.000564383	drams (Avoir.)
0.000643015	pennyweights
0.000771618	scruples
0.005	carats (metric)

MILLIGRAM PER LITER: =

1	parts per million

MILLILITER: =

0.000001	kiloliters
0.000001	cubic meters
0.0000013080	cubic yards
0.00000628995	barrels
0.00001	hektoliters
0.000028378	bushels (U.S.—dry)
0.000027497	bushels (Imperial—dry)
0.000353144	cubic feet
0.0001	dekaliters
0.000113512	pecks (U.S.—dry)
0.000264178	gallons U.S.—liquid)
0.00022702	gallons (U.S.—dry)
0.00021998	gallons (Imperial)
0.00105671	quarts (liquid)
0.000908102	quarts (dry)
0.001	liters
0.001	cubic decimeters
0.0018162	pints (U.S.—dry)
0.0021134	pints (U.S.—liquid)
0.0073092	gills (Imperial)
0.0084538	gills (U.S.)
0.01	deciliters
0.0610234	cubic inches
0.1	centiliters
1	milliliters
1	cubic centimeters
1,000	cubic millimeters
0.0338147	ounces (U.S.—fluid)
0.035196	ounces (Imperial—fluid)
0.270518	drams (fluid)
16.231074	minims
0.00220462	pounds of water @ maximum density

OHM (ABSOLUTE): =

0.00000000000111263	electrostatic cgs unit or statohm
0.000001	megohm (absolute)
0.99948	International ohm
1,000,000	microhms (absolute)
1,000,000,000	electromagnetic cgs or abohms

ohm per kilometer = 0.3048 ohms per 1,000 feet
ohm per 1,000 feet = 3.280833 ohms per kilometer
ohm per 1,000 yards = 1.0936 ohms per kilometer

OUNCE (AVOIRDUPOIS): =

0.0000279018	tons (long)
0.0000283495	tons (metric)

OUNCE (AVOIRDUPOIS): (cont'd)

0.00003125	tons (short)
0.0282495	kilograms
0.0759549	pounds (Troy)
0.0625	pounds (Avoir.)
0.283495	hektograms
2.834953	dekagrams
0.9114583	ounces (Troy)
1	ounces (Avoir.)
28.349527	grams
283.49527	decigrams
2,834.9527	centigrams
28,349.527	milligrams
437.5	grains
7.29166	drams (Troy)
16	drams (Avoir.)
18.22917	pennyweights
21.875	scruples
141.75	carats (metric)

OUNCE (FLUID): =

0.0000295729	kiloliters
0.0000295729	cubic meters
0.0000386814	cubic yards
0.000186012	barrels
0.000295729	hektoliters
0.000839221	bushels (U.S.—dry)
0.000813167	bushels (Imperial—dry)
0.00104435	cubic feet
0.00295729	dekaliters
0.00335688	pecks (U.S.—dry)
0.00781252	gallons U.S.—liquid)
0.00671365	gallons (U.S.—dry)
0.00650545	gallons (Imperial)
0.03125	quarts (liquid)
0.0268552	quarts (dry)
0.0295729	liters
0.0295729	cubic decimeters
0.0537104	pints (U.S.—dry)
0.0625	pints (U.S.—liquid)
0.208170	gills (Imperial)
0.25	gills (U.S.)
0.295729	deciliters
1.80469	cubic inches
2.957294	centiliters
29.572937	milliliters
29.572937	cubic centimeters
29,572.9372	cubic millimeters
1	ounces (U.S.—fluid)
1.0408491	ounces (Imperial—fluid)

OUNCE (FLUID): (cont'd)

8	drams (fluid)
480	minims
0.0651972	pounds of water @ maximum density

OUNCE (TROY): =

0.0000306122	tons (long)
0.0000311034	tons (metric)
0.000034285	tons (short)
0.0311035	kilograms
0.0833333	pounds (Troy)
0.0685714	pounds (Avoir.)
0.311035	hektograms
3.110348	dekagrams
1	ounces (Troy)
1.09714	ounces (Avoir.)
31.103481	grams
311.03481	decigrams
3,110,3481	centigrams
31,103.481	milligrams
480	grains
8	drams (Troy)
17.55428	drams (Avoir.)
20	pennyweights
24	scruples
155.52	carats (metric)

OUNCE (WEIGHT) PER SQUARE INCH: =

0.000439419	hektometers of water @ 60°F.
0.00439419	dekameters of water @ 60°F.
0.0439419	meters of water @ 60°F.
0.144174	feet of water @ 60°F.
1.730092	inches of water @ 60°F.
0.439419	decimeters of water @ 60°F.
4.394188	centimeters of water @ 60°F.
43.941875	millimeters of water @ 60°F.
0.0000323219	hektometers of mercury @ 32°F.
0.000323219	dekameters of mercury @ 32°F.
0.00323219	meters of mercury @ 32°F.
0.0106042	feet of mercury @ 32°F.
0.127250	inches of mercury @ 32°F.
0.0323219	decimeters of mercury @ 32°F.
0.323219	centimeters of mercury @ 32°F.
3.232188	millimeters of mercury @ 32°F.
484.379356	tons per square hektometer
4.843794	tons per square dekameter
0.0484379	tons per square meter
0.00450014	tons per square foot
0.0000312500	tons per square inch
0.000484379	tons per square decimeter

OUNCE (WEIGHT) PER SQUARE INCH: (cont'd)

0.00000484379 . . . . tons per square centimeter
0.0000000484379 . . . . tons per square millimeter
439,419 . . . . kilograms per square hektometer
4,394.1875 . . . . kilograms per square dekameter
43.941875 . . . . kilograms per square meter
4.0823252 . . . . kilograms per square foot
0.0283494 . . . . kilograms per square inch
0.439419 . . . . kilograms per square decimeter
0.00439419 . . . . kilograms per square centimeter
0.0000439419 . . . . kilograms per square millimeter
968,758 . . . . pounds per square hektometer
9,687.58 . . . . pounds per square dekameter
96.8758 . . . . pounds per square meter
90 . . . . pounds per square foot
0.0625 . . . . pounds per square inch
0.968758 . . . . pounds per square decimeter
0.00968758 . . . . pounds per square centimeter
0.0000968758 . . . . pounds per square millimeter
4,394,190 . . . . hektograms per square hektometer
43,941.875 . . . . hektograms per square dekameter
439.41875 . . . . hektograms per square meter
40.823252 . . . . hektograms per square foot
0.283494 . . . . hektograms per square inch
4.394188 . . . . hektograms per square decimeter
0.043919 . . . . hektograms per square centimeter
0.000439419 . . . . hektograms per square millimeter
43,941,900 . . . . dekagrams per square hektometer
439,419 . . . . dekagrams per square dekameter
4,394.1875 . . . . dekagrams per square meter
408.232519 . . . . dekagrams per square foot
2.834944 . . . . dekagrams per square inch
43.941875 . . . . dekagrams per square decimeter
0.439419 . . . . dekagrams per square centimeter
0.00439419 . . . . dekagrams per square millimeter
15,500,139 . . . . ounces per square hektometer
155,001 . . . . ounces per square dekameter
1,550.0139 . . . . ounces per square meter
144 . . . . ounces per square foot
1 . . . . ounces per square inch
15.500139 . . . . ounces per square decimeter
0.155001 . . . . ounces per square centimeter
0.00155001 . . . . ounces per square millimeter
439,419,000 . . . . grams per square hektometer
4,394,190 . . . . grams per square dekameter
43,941.875 . . . . grams per square meter
4,082.325187 . . . . grams per square foot
28.349438 . . . . grams per square inch
439.4187 . . . . grams per square decimeter
4.394187 . . . . grams per square centimeter
0.0439419 . . . . grams per square millimeter
4,394,190,000 . . . . decigrams per square hektometer

OUNCE (WEIGHT) PER SQUARE INCH: (cont'd)

43,941,900 . . . . decigrams per square dekameter
439,419 . . . . decigrams per square meter
40,823.25187 . . . . decigrams per square foot
283.494375 . . . . decigrams per square inch
4,394.1875 . . . . decigrams per square decimeter
43.94187 . . . . decigrams per square centimeter
0.439419 . . . . decigrams per square millimeter
43,941,900,000 . . . . centigrams per square hektometer
439,419,000 . . . . centigrams per square dekameter
4,394,190 . . . . centigrams per square meter
408,233 . . . . centigrams per square foot
2,834.943750 . . . . centigrams per square inch
43,941.875 . . . . centigrams per square decimeter
439.41875 . . . . centigrams per square centimeter
4.394188 . . . . centigrams per square millimeter
439,419,000,000 . . . . milligrams per square hektometer
4,394,190,000 . . . . milligrams per square dekameter
43,941,900 . . . . milligrams per square meter
4,082,325 . . . . milligrams per square foot
28,349.43750 . . . . milligrams per square inch
439,419 . . . . milligrams per square decimeter
4,394.1875 . . . . milligrams per square centimeter
43.941875 . . . . milligrams per square millimeter
430,920,000,000 . . . . dynes per square hektometer
4,309,200,000 . . . . dynes per square dekameter
43,092,000 . . . . dynes per square meter
4,003,324 . . . . dynes per square foot
28,050.875 . . . . dynes per square inch
430,920 . . . . dynes per square decimeter
4,309.2 . . . . dynes per square centimeter
43.092 . . . . dynes per square millimeter
0.00430919 . . . . bars
0.00425288 . . . . atmosphere

PARTS PER MILLION: =

0.00000110231 . . . . tons (net) per cubic meter
0.001 . . . . kilograms per cubic meter
0.00220462 . . . . pounds (Avoir.) per cubic meter
0.0352739 . . . . ounces (Avoir.) per cubic meter
1.0 . . . . grams per cubic meter
15.4324 . . . . grains per cubic meter
0.000000175250 . . . . tons (net) per barrel
0.000158984 . . . . kilograms per barrel
0.000350499 . . . . pounds (Avoir.) per barrel
0.00560799 . . . . ounces (Avoir.) per barrel
0.158984 . . . . grams per barrel
2.453505 . . . . grains per barrel
0.0000000312133 . . . . tons (net) per cubic foot
0.0000283162 . . . . kilograms per cubic foot
0.0000624264 . . . . pounds (Avoir.) per cubic foot

PARTS PER MILLION: (cont'd)

0.000998823	ounces (Avoir.) per cubic foot
0.0283162	grams per cubic foot
0.436987	grains per cubic foot
0.00000000417262	tons (net) per gallon (U.S.—liquid)
0.00000378524	kilograms per gallon (U.S.—liquid)
0.00000834522	pounds (Avoir.) per gallon (U.S.—liquid)
0.000133524	ounces (Avoir.) per gallon (U.S.—liquid)
0.00378534	grams per gallon (U.S.—liquid)
0.0584168	grains per gallons (U.S.—liquid)
0.00000000501107	grains per gallon (Imperial—liquid)
0.00000454585	kilograms per gallon (Imperial—liquid)
0.0000100221	pounds (Avoir.) per gallon (Imp.—liquid)
0.000160355	ounces (Avoir.) per gallon (Imp.—liquid)
0.00454597	grams per gallon (Imperial—liquid)
0.0701552	grains per gallon (Imperial—liquid)
0.00000000110231	tons (net) per liter
0.000001	kilograms per liter
0.00000220462	pounds (Avoir.) per liter
0.0000352739	ounces (Avoir.) per liter
0.001	grams per liter
0.0154324	grains per liter
0.0000000000180663	tons (net) per cubic inch
0.0000000163867	kilograms per cubic inch
0.0000000361264	pounds (Avoir.) per cubic inch
0.000000578023	ounces (Avoir.) per cubic inch
0.0000163867	grams per cubic inch
0.000252886	grains per cubic inch
8.345	pounds per million gallons

POISE: =

1	gram per centimeter per second

PENNYWEIGHT: =

0.00000153061	tons (long)
0.00000155517	tons (metric)
0.00000171429	tons (net
0.00155517	kilograms
0.0041667	pounds (Troy)
0.00342857	pounds (Avoir.)
0.0155517	hektograms
0.155517	dekagrams
0.05	ounces (Troy)
0.0548571	ounces (Avoir.)
1.55517	grams
15.5517	decigrams
155.517	centigrams
1,555.17	milligrams
24	grains

PENNYWEIGHT: (cont'd)

0.4	drams (Troy)
0.877714	drams (Avoir.)
1	pennyweights
1.2	scruples
7.776	carats (metric)

POUND (TROY): =

0.000367347	tons (long)
0.000373242	tons (metric)
0.000411429	tons (net
0.373242	kilograms
1	pounds (Troy)
0.822857	pounds (Avoir.)
3.732418	hektograms
37.324176	dekagrams
12	ounces (Troy)
13.165714	ounces (Avoir.)
373.241762	grams
3,732.417621	decigrams
37,324.176213	centigrams
373,242	milligrams
5,760	grains
96	drams (Troy)
210.651425	drams (Avoir.)
240	pennyweights
288	scruples
1,866.239964	carats (metric)

POUND (AVOIRDUPOIS): =

0.000446429	tons (long)
0.000453593	tons (metric)
0.0005	tons (net
0.453592	kilograms
1.215278	pounds (Troy)
1	pounds (Avoir.)
4.535924	hektograms
45.359243	dekagrams
14.5833	ounces (Troy)
16	ounces (Avoir.)
453.592428	grams
4,535.92428	decigrams
45,359.2428	centigrams
453,592	milligrams
7,000	grains
116.666675	drams (Troy)
256	drams (Avoir.)
291.6667	pennyweights
350.1	scruples
2,268	carats (metric)

POUND OF CARBON OXIDIZED WITH 100% EFFICIENCY: =

365,245,567	foot poundals
4,382,946,806	inch poundals
11,352,418	foot pounds
136,229,016	inch pounds
3.676937	ton calories
3,676.937455	kilogram calories
8,106.271602	pound calories
36,769.374545	hektogram calories
367,694	dekagram calories
129,700	ounce calories
3,676,937	gram calories (mean)
36,769,375	decigram calories
367,693,745	centigram calories
3,676,937,455	milligram calories
0.178109	kilowatt days
4.274614	kilowatt hours
256.477745	kilowatt minutes
15,388.634345	kilowatt seconds
178.109039	watt days
4,274.613901	watt hours
256,477	watt minutes
15,388,624	watt seconds
1,569.531035	ton meters
1,569,513	kilogram meters
3,460,223	pounds meters
15,695,310	hektogram meters
156,953,104	dekagram meters
55,363,570	ounce meters
1,569,531,035	gram meters
15,695,310,350	decigram meters
156,953,103,500	centigram meters
1,569,531,035 x 10³	milligram meters
15.695310	ton hektometers
15,695.31035	kilogram hektometers
34,602.288580	pound hektometers
156,953	hektogram hektometers
1,569,531	dekagram hektometers
553,636	ounce hektometers
15,695,310	gram hektometers
156,953,104	decigram hektometers
1,569,531,035	centigram hektometers
15,695,310,350	milligram hektometers
156.907535	ton dekameters
156,907	kilgram dekameters
346,022	pound dekameters
1,569,075	hektogram dekameters
15,690.754	dekagram dekameters
5,536,358	ounce dekameters
156,907,535	gram dekameters

POUND OF CARBON OXIDIZED WITH 100% EFFICIENCY: (cont'd)

1,569,075,350 . . . . decigram dekameters
15,690,753,500 . . . . centigram dekameters
156,907,535 x 10^3 . . . . milligram dekameters
5,149.255690 . . . . ton feet
5,149,256 . . . . kilogram feet
11,352,449 . . . . pound feet
51,492,557 . . . . hektogram feet
514,925,569 . . . . dekagram feet
181,639,190 . . . . ounce feet
5,149,255,690 . . . . gram feet
51,492,556,895 . . . . decigram feet
514,925,568,950 . . . . centigram feet
51,492,556.895 x 10^2 . . . . milligram feet
61,792,556845 . . . . ton inches
61,792,557 . . . . kilogram inches
136,229,386 . . . . pound inches
617,925,568 . . . . hektogram inches
6,179,255,685 . . . . dekagram inches
2,179,670,177 . . . . ounce inches
61,792,556,845 . . . . gram inches
617,925,568,450 . . . . decigram inches
61,792,556,845 x 10^2 . . . . centigram inches
61,792,556,845 x 10^3 . . . . milligram inches
15,695.31035 . . . . ton decimeters
15,695,310 . . . . kilogram decimeters
34,602,231 . . . . pound decimeters
156,953,104 . . . . hektogram decimeters
1,569,531,035 . . . . dekagram decimeters
553,635,699 . . . . ounce decimeters
15,695,310,350 . . . . gram decimeters
156,953,103,500 . . . . decigram decimeters
1,569,531,035 x 10^3 . . . . centigram decimeters
1,569,531,035 x 10^4 . . . . milligram decimeters
156,953 . . . . ton centimeters
156,953,104 . . . . kilogram centimeters
346,022,312 . . . . pound centimeters
1,569,531,035 . . . . hektogram centimeters
15,695,310,350 . . . . dekagram centimeters
5,536,356,992 . . . . ounce centimeters
156,953,103,500 . . . . gram centimeters
1,569,531,035 x 10^3 . . . . decigram centimeters
1,569,531,035 x 10^4 . . . . centigram centimeters
1,569,531,035 x 10^5 . . . . milligram centimeters
1,569,531 . . . . ton millimeters
1,569,531,035 . . . . kilogram millimeters
3,460,223,121 . . . . pound millimeters
15,695,310,350 . . . . hektogram millimeters
156,953,103,500 . . . . dekagram millimeters
55,363,569,923 . . . . ounces millimeters
1,569,531,035 x 10^3 . . . . gram millimeters
1,569,531,035 x 10^4 . . . . decigram millimeters

POUND OF CARBON OXIDIZED WITH 100% EFFICIENCY: (cont'd)

$1,569,531,035 \times 10^5$	centigram millimeters
$1,569,531,035 \times 10^6$	milligram millimeters
151.895	kiloliter-atmospheres
1,518.95	hektoliter-atmospheres
15,189.5	dekaliter-atmospheres
151,895	liter-atmospheres
1,518,950	deciliter-atmospheres
15,189,500	centiliter-atmospheres
151,895,000	milliliter-atmospheres
0.000000151895	cubic kilometer-atmospheres
0.000151895	cubic hektometer atmospheres
0.151895	cubic dekameter-atmospheres
151.895	cubic meter-atmospheres
5,364.779505	cubic foot-atmospheres
9,269,392	cubic inch-atmospheres
151,895	cubic decimeter-atmospheres
151,895,000	cubic centimeter-atmospheres
$151,895 \times 10^6$	cubic millimeter-atmospheres
15,391,217	joules (absolute)
5.813022	Cheval-vapeur hours
0.238896	horespower days
5,733277	horsepower hours
344.0102771	horsepower minutes
20,640.555865	horsepower seconds
1	pounds of carbon oxidized with 100% efficiency
15.0482377	pounds of water evaporated from and at 212° F.
6,618,631	BTU (mean) per pound
14,592.55265	BTU (mean)
$153,918,415,320 \times 10^3$	ergs

POUND (PRESSURE) PER SQUARE INCH: =

0.0070307	hektometers of water @ 60°F.
0.070307	dekameters of water @ 60°F.
0.70307	meters of water @ 60°F.
2.306787	feet of water @ 60°F.
27.681473	inches of water @ 60°F.
7.0307	decimeters of water @ 60°F.
70,307	centimeters of water @ 60°F.
703.07	millimeters of water @ 60°F.
0.00051715	hektometers of mercury @ 32°F.
0.0051715	dekameters of mercury @ 32°F.
0.051715	meters of mercury @ 32°F.
0.169667	feet of mercury @ 32°F.
2.0360	inches of mercury @ 32°F.
0.51715	decimeters of mercury @ 32°F.
5.1715	centimeters of mercury @ 32°F.
51.715	millimeters of mercury @ 32°F.
7,750.0696898	tons per square hektometer
77.500697	tons per square dekameter
0.775007	tons per square meter
0.0720023	tons per quare foot

POUND (PRESSURE) PER SQUARE INCH: (cont'd)

0.0005 . . . . tons per square inch
0.00775007 . . . . tons per square decimeter
0.0000775007 . . . . tons per square centimeter
0.000000775007 . . . . tons per square millimeter
7,030,700 . . . . kilograms per square hektometer
70,307 . . . . kilograms per square hektometer
703.07 . . . . kilograms per square meter
65.317203 . . . . kilograms per square foot
0.453592 . . . . kilograms per square inch
7.0307 . . . . kilograms per square decimeter
0.070307 . . . . kilograms per square centimeter
0.00070307 . . . . kilograms per square millimeter
15,500,130 . . . . pounds per square hektometer
155,001 . . . . pounds per square dekameter
1,550.0130 . . . . pounds per square meter
144 . . . . pounds per square foot
1 . . . . pounds per square inch
15.500130 . . . . pounds per square decimeter
0.155001 . . . . pounds per square centimeter
0.00155001 . . . . pounds per square millimeter
70,307,000 . . . . hektograms per square hektometer
703,070 . . . . hektograms per square dekameter
7,030.70 . . . . hektograms per square meter
653.172168 . . . . hektograms per square foot
4.535933 . . . . hektograms per square inch
70.3070 . . . . hektograms per square decimeter
0.70307 . . . . hektograms per square centimeter
0.0070307 . . . . hektograms per square millimeter
703,070,000 . . . . dekagrams per square hektometer
7,030,700 . . . . dekagrams per square dekameter
70,307 . . . . dekagrams per square meter
6,531.721544 . . . . dekagrams per square foot
45.359332 . . . . dekagrams per square inch
7,030.7 . . . . dekagrams per square decimeter
70.307 . . . . dekagrams per square centimeter
0.70307 . . . . dekagrams per square millimeter
248,002,217 . . . . ounces per square hektometer
2,480,022 . . . . ounces per square dekameter
24,800.22 . . . . ounces per square meter
2,303.985941 . . . . ounces per square foot
16 . . . . ounces per square inch
248.0022 . . . . ounces per square decimeter
2.480022 . . . . ounces per square centimeter
0.0248002 . . . . ounces per square millimeter
7,030,700,000 . . . . grams per square hektometer
70,307,000 . . . . grams per square dekameter
703,070 . . . . grams per square meter
65,317.215135 . . . . grams per square foot
453.593927 . . . . grams per square inch
7,030.7 . . . . grams per square decimeter
70.307 . . . . grams per square centimeter

POUND (PRESSURE) PER SQUARE INCH: (cont'd)

0.70307	grams per square millimeter
70,307,000,000	decigrams per square hektometer
703,070,000	decigrams per square dekameter
7,030,700	decigrams per square meter
653,172	decigrams per square foot
4,535.939265	decigrams per square inch
70,307	decigrams per square decimeter
703.07	decigrams per square centimeter
7.0307	decigrams per square millimeter
703,070,000,000	centigrams per square hektometer
7,030,700,000	centigrams per square dekameter
70,307,000	centigrams per square meter
6,531,720	centigrams per square foot
45,359.392595	centigrams per square inch
703,070	centigrams per square decimeter
7,030.70	centigrams per square centimeter
70.3070	centigrams per square millimeter
$70,307 \times 10^8$	milligrams per square hektometer
$70,307 \times 10^6$	milligrams per square dekameter
$70,307 \times 10^4$	milligrams per square meter
65,317,200	milligrams per square foot
453,594	milligrams per square inch
70,307,000	milligrams per square decimeter
703,070	milligrams per square centimeter
7,030.70	milligrams per square millimeter
$68,947 \times 10^8$	dynes per square hektometer
68,947,000,000	dynes per square dekameters
689,470,000	dynes per square meter
64,053,184	dynes per square foot
448,814	dynes per square inch
6,894,700	dynes per square decimeter
68,947	dynes per square centimeter
689.47	dynes per square millimeter
0.068947	bars
0.068046	atmospheres

POUND (WEIGHT) PER SQUARE FOOT: =

0.0000488243	hektometers of water @ 60°F.
0.000488243	dekameters of water @ 60°F.
0.00488243	meters of water @ 60°F.
0.0160194	feet of water @ 60°F.
0.192232	inches of water @ 60°F.
0.0488243	decimeters of water @ 60°F.
0.488243	centimeters of water @ 60°F.
4.882431	millimeters of water @ 60°F.
0.00000359132	hektometers of mercury @ 32°F.
0.0000359132	dekameters of mercury @ 32°F.
0.000359132	meters of mercury @ 32°F.
0.00117824	feet of mercury @ 32°F.
0.0141389	inches of mercury @ 32°F.

POUND (WEIGHT) PER SQUARE FOOT: (cont'd)

0.00359132	decimeters of mercury @ 32°F.
0.0359132	centimeters of mercury @ 32°F.
0.359132	millimeters of mercury @ 32°F.
53.819930	tons per square hektometer
0.538199	tons per square dekameter
0.00538199	tons per square meter
0.0005000016	tons per square foot
0.00000347222	tons per square inch
0.0000538199	tons per square decimeter
0.000000538199	tons per square centimeter
0.00000000538199	tons per square millimeter
48,824.305552	kilograms per square hektometer
488.243056	kilograms per square dekameter
4.882431	kilograms per square meter
0.453592	kilograms per square foot
0.00314994	kilograms per square inch
0.0488243	kilograms per square decimeter
0.000488243	kilograms per square centimeter
0.00000488243	kilograms per square millimeter
107,640	pounds per square hektometer
1,076.397917	pounds per square dekameter
10.763979	pounds per square meter
1	pounds per square foot
0.00694444	pounds per square inch
0.107640	pounds per square decimeter
0.00107640	pounds per square centimeter
0.0000107640	pounds per square millimeter
488,243	hektograms per square hektometer
4,882.430555	hektograms per square dekameter
48.824306	hektograms per square meter
4.535918	hektograms per square foot
0.0314994	hektograms per square inch
0.488243	hektograms per square decimeter
0.00488243	hektograms per square centimeter
0.0000488243	hektograms per square millimeter
4,882,431	dekagrams per square hektometer
48,824.305552	dekagrams per square dekameter
488.243056	dekagrams per square meter
45.359178	dekagrams per square foot
0.314994	dekagrams per square inch
4.882431	dekagrams per square decimeter
0.0488243	dekagrams per square centimeter
0.000488243	dekagrams per square millimeter
1,722,238	ounces of per square hektometer
17,222.376179	ounces of per square dekameter
172.223762	ounces per square meters
16	ounces per square foot
0.111111	ounces per square inch
1.722238	ounces per square decimeter
0.0172224	ounces per square centimeter
0.000172224	ounces per square millimeter

POUND (WEIGHT) PER SQUARE FOOT: (cont'd)

48,824,306	grams per square hektometer
488,243	grams per square dekameter
4,882.430555	grams per square meter
453.591783	grams per square foot
3.149953	grams per square inch
48.824306	grams per square decimeter
0.488243	grams per square centimeter
0.00488243	grams per square millimeter
488,243,056	decigrams per square hektometer
4,882,431	decigrams per square dekameter
48,824.305552	decigrams per square meter
4,535.917833	decigrams per square foot
3.149953	decigrams per square inch
488.243056	decigrams per square decimeter
4.882431	decigrams per square centimeter
0.0488243	decigrams per square millimeter
4,882,430,555	centigrams per square hektometer
48,824,306	centigrams per square dekameter
488,243	centigrams per square meter
45,359.178330	centigrams per square foot
31.499535	centigrams per square inch
4,882.430555	centigrams per square decimeter
48.824306	centigrams per square centimeter
0.488243	centigrams per square millimeter
48,824,305,552	milligrams per square hektometer
488,243,056	milligrams per square dekameter
4,882.431	milligrams per square meter
453,592	milligrams per square foot
314.995347	milligrams per square inch
48,824.305552	milligrams per square decimeter
488.243056	milligrams per square centimeter
4.882431	milligrams per square millimeter
47,879,860,000	dynes per square hektometer
478,798,600	dynes per square dekameter
4,787,986	dynes per square meter
444,814	dynes per square foot
3,116.763889	dynes per square inch
47,879.861108	dynes per square decimeter
478.798611	dynes per square centimeter
4.787986	dynes per square millimeter
0.00047880	bars
0.00047254	atmospheres

POUND PER INCH: =

19.685057	tons (net) per kilometer
31.680146	tons (net) per mile
0.0196851	tons (net) per meter
0.018	tons (net) per yard
0.006	tons (net) per foot
0.0005	tons (net) per inch

POUND PER INCH: (cont'd)

0.000196851	tons (net) per centimeter
0.0000196851	tons (net) per millimeter
17,857.985915	kilograms per kilometer
28,739.749194	kilograms per mile
17.857985	kilograms per meter
16.329328	kilograms per yard
5.442109	kilograms per foot
0.453592	kilograms per inch
0.178579	kilograms per centimeter
0.0178579	kilograms per millimeter
39,370.11300	pounds (Avoir.) per kilometer
63,360.291357	pounds (Avoir.) per mile
39.370113	pounds (Avoir.) per meter
36	pounds (Avoir.) per yard
12	pounds (Avoir.) per foot
1	pounds (Avoir.) per inch
0.393701	pounds (Avoir.) per centimeter
0.0393701	pounds (Avoir.) per millimeter
629,921.80800	ounces (Avoir.) per kilometer
1,013,764	ounces (Avoir.) per mile
629.9211808	ounces (Avoir.) per meter
576	ounces (Avoir.) per yard
192	ounces (Avoir.) per foot
16	ounces (Avoir.) per inch
6.299218	ounces (Avoir.) per centimeter
0.629921	ounces (Avoir.) per millimeter
17,857,985	grams per kilometer
28,739,794	grams per mile
17,857.9851	grams per meter
16,329.327396	grams per yard
5,443.109132	grams per foot
453.5924277	grams per inch
178.579851	grams per centimeter
17.857985	grams per millimeter
275,590,791	grains per kilometer
443,522,039	grains per mile
275,590.791	grains per meter
252,000	grains per yard
84,000	grains per foot
7,000	grains per inch
2,755.90791	grains per centimeter
275.590791	grains per millimeter

POUND (AVOIR.) PER FOOT: =

1.640239	tons (net) per kilometer
2.639719	tons (net) per mile
0.00164024	tons (net) per meter
0.00149984	tons (net) per yard
0.000499946	tons (net) per foot
0.0000416622	tons (net) per inch

POUND (AVOIR.) PER FOOT: (cont'd)

0.0000164024	tons (net) per centimeter
0.00000164024	tons (net) per millimeter
1,488.161203	kilograms per kilometer
2,394.71280	kilograms per mile
1.488161	kilograms per meter
1.360777	kilograms per yard
0.453592	kilograms per foot
0.0377994	kilograms per inch
0.00148816	kilograms per centimeter
0.00148816	kilograms per millimeter
3,280.8	pounds (Avoir.) per kilometer
5,280	pounds (Avoir.) per mile
3.280833	pounds (Avoir.) per meter
3	pounds (Avoir.) per yard
1	pounds (Avoir.) per foot
0.0833333	pounds (Avoir.) per inch
0.0328083	pounds (Avoir.) per centimeter
0.00328083	pounds (Avoir.) per millimeter
52,493.328	ounces (Avoir.) per kilometer
84,480	ounces (Avoir.) per mile
52.493328	ounces (Avoir.) per meter
48	ounces (Avoir.) per yard
16	ounces (Avoir.) per foot
1.333333	ounces (Avoir.) per inch
0.524933	ounces (Avoir.) per centimeter
0.0524933	ounces (Avoir.) per millimeter
1,488,161	grams per kilometer
2,394,713	grams per mile
1,488.161	grams per meter
1,360.777283	grams per yard
453.5924277	grams per foot
37.799369	grams per inch
14.881612	grams per centimeter
1.488161	grams per millimeter
22,965,899	grains per kilometer
36,960,000	grains per mile
22,965.899250	grains per meter
21,000	grains per yard
7,000	grains per foot
583.333333	grains per inch
229.658993	grains per centimeter
22.965899	grains per millimeter

POUND (WEIGHT) PER CUBIC FOOT: =

17,657.261726	tons (net) per cubic hektometer
17.657262	tons (net) per cubic dekameter
0.01765731	tons (net) per cubic meter
0.0005	tons (net) per cubic foot
0.000000289352	tons (net) per cubic inch
0.0000176573	tons (net) per cubic decimeter

POUND (WEIGHT) PER CUBIC FOOT: (cont'd)

0.0000000176573	tons (net) per cubic centimeter
0.0000000000176573	tons (net) per cubic millimeter
16,018,400	kilograms per cubic hektometer
16,018.4	kilograms per cubic dekameter
16,0184	kilograms per cubic meter
0.453593	kilograms per cubic foot
0.000262496	kilograms per cubic inch
0.0160184	kilograms per cubic decimeter
0.0000160184	kilograms per cubic centimeter
0.0000000160184	kilograms per cubic millimeter
35,314,445	pounds per cubic hektometer
35,314.445	pounds per cubic dekameter
35.314445	pounds per cubic meter
1	pounds per cubic foot
0.000578704	pounds per cubic inch
0.0353144	pounds per cubic decimeter
0.0000353144	pounds per cubic centimeter
0.0000000353144	pounds per cubic millimeter
160,184,000	hektograms per cubic hektometer
160,184	hektograms per cubic dekameter
160.184	hektograms per cubic meter
4.535934	hektograms per cubic foot
0.00262496	hektograms per cubic inch
0.160184	hektograms per cubic decimeter
0.000160184	hektograms per cubic centimeter
0.000000160184	hektograms per cubic millimeter
1,601,840,000	dekagrams per cubic hektometer
1,601,840	dekagrams per cubic dekameter
1,601.840	dekagrams per cubic meter
45.359342	dekagrams per cubic foot
0.0262496	dekagrams per cubic inch
1.60184	dekagrams per cubic decimeter
0.00160184	dekagrams per cubic centimeter
0.00000160184	dekagrams per cubic millimeter
565,031,120	ounces per cubic hektometer
565,031	ounces per cubic dekameter
565.031120	ounces per cubic meter
16	ounces per cubic foot
0.00925926	ounces per cubic inch
0.565031	ounces per cubic decimeter
0.000565031	ounces per cubic centimeter
0.000000565031	ounces per cubic millimeter
16,018,400,000	grams per cubic hektometer
16,018,400	grams per cubic dekameter
16,018.4	grams per cubic meter
453.593422	grams per cubic foot
0.262496	grams per cubic inch
16.0184	grams per cubic decimeter
0.0160184	grams per cubic centimeter
0.0000160184	grams per cubic millimeter
160,184,000,000	decigrams per cubic hektometer

POUND (WEIGHT) PER CUBIC FOOT: (cont'd)

160,184,000	decigrams per cubic dekameter
160,184	decigrams per cubic meter
4,535.934220	decigrams per cubic foot
2.624962	decigrams per cubic inch
160.184	decigrams per cubic decimeter
0.160184	decigrams per cubic centimeter
0.000160184	decigrams per cubic millimeter
160,184 x 10^7	centigrams per cubic hektometer
1,601,840,000	centigrams per cubic dekameter
1,601,840	centigrams per cubic meter
45,359.342205	centigrams per cubic foot
26.249619	centigrams per cubic inch
1,601.840	centigrams per cubic decimeter
1.60184	centigrams per cubic centimeter
0.00160184	centigrams per cubic millimeter
160,184 x 10^8	milligrams per cubic hektometer
16,018,400,000	milligrams per cubic dekameter
16,018,400	milligrams per cubic meter
453,593	milligrams per cubic foot
262.496193	milligrams per cubic inch
16,018.4	milligrams per cubic decimeter
16.0184	milligrams per cubic centimeter
0.0160184	milligrams per cubic millimeter
157,085,886 x 10^9	dynes per cubic hektometer
157,086 x 10^6	dynes per cubic dekameter
157,086,000	dynes per cubic meter
444,820	dynes per cubic foot
257.418981	dynes per cubic inch
157,086	dynes per cubic decimeter
157.0858859	dynes per cubic centimeter
0.157086	dynes per cubic millimeter

POUND PER CUBIC INCH: =

30,511,748	tons (net) per cubic hektometer
30,511.748042	tons (net) per cubic dekameter
30.511748	tons (net) per cubic meter
0.864001	tons (net) per cubic foot
0.0005	tons (net) per cubic inch
0.0305117	tons (net) per cubic decimeter
0.0000305117	tons (net) per cubic centimeter
0.0000000305117	tons (net) per cubic millimeter
27,679,795,200	kilograms per cubic hektometer
27,679,795	kilograms per cubic dekameter
27,679.7952	kilograms per cubic meter
783.809037	kilograms per cubic foot
0.453593	kilograms per cubic inch
27.679795	kilograms per cubic decimeter
0.0276798	kilograms per cubic centimeter
0.0000276798	kilograms per cubic millimeter
61,023,360,960	pounds per cubic hektometer
61,023,361	pounds per cubic dekameter

POUND PER CUBIC INCH: (cont'd)

61,023.360960 . . . . pounds per cubic meter
1,728 . . . . pounds per cubic foot
1 . . . . pounds per cubic inch
61.0233610 . . . . pounds per cubic decimeter
0.0610234 . . . . pounds per cubic centimeter
0.0000610234 . . . . pounds per cubic millimeter
276,797,952,000 . . . . hektograms per cubic hektometer
276,797,952 . . . . hektograms per cubic dekameter
276,798 . . . . hektograms per cubic meter
7,838.0903748 . . . . hektograms per cubic foot
4.535934 . . . . hektograms per cubic inch
276.797952 . . . . hektograms per cubic decimeter
0.276798 . . . . hektograms per cubic centimeter
0.000276798 . . . . hektograms per cubic millimeter
$276,797,952 \times 10^4$ . . . . dekagrams per cubic hektometer
2,767,979,520 . . . . dekagrams per cubic dekameter
2,767,980 . . . . dekagrams per cubic meter
78,380.903748 . . . . dekagrams per cubic foot
45.359342 . . . . dekagrams per cubic inch
2,767.97952 . . . . dekagrams per cubic decimeter
2.767980 . . . . dekagrams per cubic centimeter
0.00276798 . . . . dekagrams per cubic millimeter
976,373,775,360 . . . . ounces per cubic hektometer
975,373,775 . . . . ounces per cubic dekameter
976,374 . . . . ounces per cubic meter
27,648 . . . . ounces per cubic foot
16 . . . . ounces per cubic inch
976.373775 . . . . ounces per cubic decimeter
0.976374 . . . . ounces per cubic centimeter
0.000976374 . . . . ounces per cubic millimeter
$276,797,952 \times 10^5$ . . . . grams per cubic hektometer
27,679,795,200 . . . . grams per cubic dekameter
27,679,795 . . . . grams per cubic meter
783,809 . . . . grams per cubic foot
453.593425 . . . . grams per cubic inch
27,679.7952 . . . . grams per cubic decimeter
27.679795 . . . . grams per cubic centimeter
0.0276798 . . . . grams per cubic millimeter
$276,797,952 \times 10^6$ . . . . decigrams per cubic hektometer
$276,797,952 \times 10^3$ . . . . decigrams per cubic dekameter
276,797,952 . . . . decigrams per cubic meter
7,838,090 . . . . decigrams per cubic foot
4,535.934249 . . . . decigrams per cubic inch
276,798 . . . . decigrams per cubic decimeter
276.797952 . . . . decigrams per cubic centimeter
0.276798 . . . . decigrams per cubic millimeter
$276,797,952 \times 10^7$ . . . . centigrams per cubic hektometer
$276,797,952 \times 10^4$ . . . . centigrams per cubic dekameter
2,767,979,520 . . . . centigrams per cubic meter
78,380,904 . . . . centigrams per cubic foot
45,359.342495 . . . . centigrams per cubic inch

POUND PER CUBIC INCH: (cont'd)

2,767,980	centigrams per cubic decimeter
2,767.979520	centigrams per cubic centimeter
2.767980	centigrams per cubic millimeter
276,797,952 x 10^8	milligrams per cubic hektometer
276,797,952 x 10^5	milligrams per cubic dekameter
276,797,952 x 10^2	milligrams per cubic meter
783,809,040	milligrams per cubic foot
453,593	milligrams per cubic inch
27,679.795	milligrams per cubic decimeter
27,679.7952	milligrams per cubic centimeter
27.679795	milligrams per cubic millimeter
27,144,411 x 10^9	dynes per cubic hektometer
27,144,411 x 10^6	dynes per cubic dekameter
27,144,411 x 10^3	dynes per cubic meter
768,648,960	dynes per cubic foot
444,820	dynes per cubic inch
271,444,110	dynes per cubic decimeter
271,444	dynes per cubic centimeter
271.444411	dynes per cubic millimeter

POUND OF WATER EVAPORATED FROM AND AT 212°F.: =

24,271,651	foot poundals
291,259,807	inch poundals
754,402	foot pounds
9,052,822	inch pounds
0.244343	ton calories
244.343393	kilogram calories
538.685776	pound calories
2,443.433927	hektogram calories
24,434.339270	dekagram calories
8,618.949444	ounce calories
244,343	gram calories (mean)
2,443,434	decigram calories
24,434,339	centigram calories
244,343,393	milligram calories
0.0118359	kilowatt days
0.284061	kilowatt hours
17.0437064	kilowatt minutes
1,022.620366	kilowatt seconds
11.835871	watt days
284.0607707	watt hours
17,043.706381	watt minutes
1,022,620	watt seconds
104.32	tons meters
104,320	kilogram meters
229,942	pound meters
1,043,200	hektogram meters
10,432,000	dekagram meters
3,679,073	ounce meters
104,320,000	gram meters

POUND OF WATER EVAPORATED FROM AND AT 212°F.: (cont'd)

1,043,200,000	decigram meters
10,432,000,000	centigram meters
104,320,000,000	milligram meters
1.0432	ton hektometers
1,043.2	kilogram hektometers
2,299.424641	pounds hektometers
10,432	hektogram hektometers
104,320	dekagram hektometers
36,790.753232	ounce hektometers
1,043,200	gram hektometers
10,432,000	decigram hektometers
104,320,000	centigram hektometers
1,043,200,000	milligram hektometers
10.432	ton dekameters
10,432	kilogram dekameters
22,994.24641	pound dekameters
104,320	hektogram dekameters
1,043,200	dekagram dekameters
367,907	ounce dekameters
10,432,000	gram dekameters
104,320,000	decigram dekameters
1,043,200,000	centigram dekameters
10,432,000,000	milligram dekameters
342.183304	ton feet
342,183	kilogram feet
754,271	pound feet
3,421,830	hektogram feet
34,218,300	dekagram feet
12,070,463	ounce feet
342,183,000	gram feet
3,421,830,000	decigram feet
34,218,300,000	centigram feet
342,183,000,000	milligram feet
4,106.298562	ton inches
4,106,299	kilogram inches
9,052,847	pound inches
41,062,986	hektogram inches
410,629,856	dekagram inches
144,845,540	ounce inches
4,106,298,562	gram inches
41,062,985,620	decigram inches
410,629,856,200	centigram inches
$4,106,298,562 \times 10^3$	milligram inches
1,043.2	ton decimeters
1,043,200	kilogram decimeters
2,299,425	pound decimeters
10,432,000	hektogram decimeters
104,320,000	dekagram decimeters
36,790,700	ounce decimeters
1,043,200,000	gram decimeters
10,432,000,000	decigram decimeters

POUND OF WATER EVAPORATED FROM AND AT 212°F.: (cont'd)

104,320,000,000 . . . . centigram decimeters
$10,432 \times 10^8$ . . . . milligram decimeters
10,432 . . . . ton centimeters
10,432,000 . . . . kilogram centimeters
22,994250 . . . . pound centimeters
104,320,000 . . . . hektogram centimeters
1,043,200,000 . . . . dekagram centimeters
367,907,000 . . . . ounce centimeters
10,432,000,000 . . . . gram centimeters
104,320,000,000 . . . . decigram centimeters
$10,432 \times 10^8$ . . . . centigram centimeters
$10,432 \times 10^9$ . . . . milligram centimeters
104,320 . . . . ton millimeters
104,320,000 . . . . kilogram millimeters
229,942,500 . . . . pound millimeters
1,043,200,000 . . . . hektogram millimeters
10,432,000,000 . . . . dekagram millimeters
3,679,070,000 . . . . ounce millimeters
104,320,000,000 . . . . gram millimeters
$10,432 \times 10^8$ . . . . decigram millimeters
$10,432 \times 10^9$ . . . . centigram millimeters
$10,432 \times 10^{10}$ . . . . milligram millimeters
10.0938730 . . . . kiloliter-atmospheres
100.938730 . . . . hektoliter-atmospheres
1,009.387298 . . . . dekaliter-atmospheres
10,093.872982 . . . . liter-atmospheres
100,939 . . . . deciliter-atmospheres
1,009,387 . . . . centiliter-atmospheres
10,093,873 . . . . milliliter-atmospheres
0.0000000100939 . . . . cubic kilometer-atmospheres
0.0000100939 . . . . cubic hektometer-atmospheres
0.0100939 . . . . cubic dekameter-atmospheres
10.0938730 . . . . cubic meter-atmospheres
356.505500 . . . . cubic foot-atmospheres
615,0979 . . . . cubic inch-atmospheres
10,093.872982 . . . . cubic decimeter-atmospheres
10,093,873 . . . . cubic centimeter-atmospheres
10,093,872,982 . . . . cubic millimeter-atmospheres
1,022,792 . . . . joules (absolute)
0.386293 . . . . Cheval-vapeur hours
0.0158753 . . . . horsepower days
0.380993 . . . . horsepower hours
22.860503 . . . . horsepower minutes
1,371.626098 . . . . horsepower seconds
0.0664530 . . . . pounds of carbon oxidized with 100% efficiency
pounds of water evaporated from and at 212°F.
439,828 . . . . BTU (mean) per pound
969.718377 . . . . BTU (mean)

POUND OF WATER AT 60°F.: =

0.0000454248	kiloliters
0.0000454248	cubic meters
0.00059412	cubic yards
0.0028572	barrels
0.00454248	hektoliters
0.0128904	bushels (U.S.—dry)
0.0124968	bushels (Imperial—dry)
0.0160417	cubic feet
0.0454248	dekaliters
0.12	gallons (U.S.—liquid)
0.103138	gallons (U.S.—dry)
0.0999216	gallons (Imperial)
0.48	quarts (liquid)
0.412496	quarts (dry)
0.454248	liters
0.454248	cubic decimeters
0.96	pints
3.84	gills
4.54248	deciliters
27.72	cubic inches
45.4248	centiliters
454.248	milliliters
454.248	cubic centimeters
454,248	cubic millimeters
7,372.80	minims
15.36	ounces (fluid)
122.88	drams (fluid)
0.000446628	tons (long)
0.000453792	tons (metric)
0.000500220	tons (short)
0.453792	kilograms
1	pounds (Avoir.)
1.215812	pounds (Troy)
4.53792	hektograms
45.3792	dekagrams
14.589749	ounces (Troy)
16	ounces (Avoir.)
116.717990	drams (Troy)
256	drams (Avoir.)
35.0153971	scruples
453,792	grams
4,537.92	decigrams
7,000	grains
45,379.2	centigrams
453,792	milligrams
27.6798	inches of water

POUNDS PER MILLION GALLONS: =

0.000000132090	tons (net) per cubic meter
0.00011983	kilograms per cubic meter

POUNDS PER MILLION GALLONS: (cont'd)

0.000264180	pounds (Avoir.) per cubic meter
0.00422687	ounces (Avoir.) per cubic meter
0.11983	grams per cubic meter
1.849264	grains per cubic meter
0.0000000210002	tons (net) per barrel
0.0000190510	kilograms per barrel
0.0000420003	pounds (Avoir.) per barrel
0.000672005	ounces (Avoir.) per barrel
0.0190510	grams per barrel
0.294004	grains per barrel
0.00000000370029	tons (net) per cubic foot
0.00000339313	kilograms per cubic foot
0.00000748056	pounds (Avoir.) per cubic foot
0.00119689	ounces (Avoir.) per cubic foot
0.00339313	grams per cubic foot
0.0523642	grains per cubic foot
0.0000000005	tons (net) per gallon (U.S.—liquid)
0.000000453585	kilograms per gallon (U.S.—liquid)
0.000001	pounds (Avoir.) per gallon (U.S.—liquid)
0.000016	ounces (Avoir.) per gallon (U.S.—liquid)
0.000453585	grams per gallon (U.S.—liquid)
0.007	grains per gallon (U.S.—liquid)
0.000000000600477	tons (net) per gallon (Imperial—liquid)
0.000000544729	kilograms per gallon (Imperial—liquid)
0.00000120095	pounds (Avoir.) per gallon (Imp.—liquid)
0.0000192153	ounces (Avoir.) per gallon (Imp.—liquid)
0.000544729	grams per gallon (Imperial—liquid)
0.00840670	grains per gallon (Imperial—liquid)
0.000000000132090	tons (net) per liter
0.00000011983	kilograms per liter
0.000000264180	pounds (Avoir.) per liter
0.00000422687	ounces (Avoir.) per liter
0.00011983	grams per liter
0.00184926	grains per liter
0.00000000000216453	tons (net) per cubic inch
0.00000000196362	kilograms per cubic inch
0.00000000432903	pounds (Avoir.) per cubic inch
0.0000000692645	ounces (Avoir.) per cubic inch
0.00000196362	grams per cubic inch
0.0000303033	grains per cubic inch

POUND WEIGHT SECOND PER SQUARE FOOT: =

478.8	poises

POUND WEIGHT PER SECOND PER SQUARE INCH: =

68,950	poises

QUADRANT (ANGLE): =

324,000	seconds
5,400	minutes
90	degrees
1.57080	radians
0.25	circumference of revolution
0.7854	pi (π)

QUART (U.S.—DRY): =

0.000110089	kiloliters
0.000110089	cubic meters
0.00143986	cubic yards
0.00692448	barrels
0.0110089	hektoliters
0.0312402	bushels (U.S.—dry)
0.0302863	bushels (Imperial—dry)
0.0388775	cubic feet
0.110089	dekaliters
0.290823	gallons (U.S.—liquid)
0.249956	gallons (U.S.—dry)
0.242162	gallons (Imperial)
1.163290	quarts (liquid)
1	quarts (dry)
1.100889	liters
1.100889	cubic decimeters
2	pints (U.S.—dry)
2.326580	pints (U.S.—liquid)
7.749187	gills (Imperial)
9.306320	gills (U.S.)
11.00888839	deciliters
67.18	cubic inches
110.0888839	centiliters
1,100.888839	millimeters
1,100.888839	cubic centimeters
1,100,889	cubic millimeters (fluid)
17,868.135060	minims (fluid)
37.225281	ounces (fluid)
297.802251	drams (fluid)
0.00108241	tons (long)
0.00109977	tons (metric)
0.00121230	tons (short)
1.0997744	kilograms
2.424587	pounds (Avoir.)
2.946547	pounds (Troy)
10.997744	hektograms
109.977441	dekagrams
35.358561	ounces (Troy)
38.793396	ounces (Avoir.)
282.868492	drams (Troy)

QUART (U.S.—DRY): (cont'd)

620.694342	drams (Avoir.)
707.171230	pennyweights
848.605475	scruples
1,099.774407	grams
10,997.744067	decigrams
16,972.110905	grains
109,977	centigrams
1,099,774	milligrams

QUART (U.S.—LIQUID): =

0.0000946358	kiloliters
0.0000946358	cubic meters
0.00123775	cubic yards
0.0059525	barrels
0.00946358	hektoliters
0.026855	bushels (U.S.—dry)
0.026035	bushels (Imperial—dry)
0.0334203	cubic feet
0.0946358	dekaliter
0.25	gallons (U.S.—liquid)
0.21487	gallons (U.S.—dry)
0.20817	gallons (Imperial)
1	quarts (liquid)
0.859368	quarts (dry)
0.946358	liters
0.946358	cubic decimeters
1.718733	pints (U.S.—dry)
2	pints (U.S.—liquid)
6.66144	gills (Imperial)
8	gills (U.S.)
9,46358	deciliters
57.75	cubic inches
94.6358	centiliters
946.358	milliliters
946.358	cubic centimeters
946,358	cubic millimeters
15,360	minims
32	ounces (fluid)
256	drams (fluid)
0.000930475	tons (long) water @ 62°F.
0.0009454	tons (metric) water @ 62°F.
0.00104213	tons (short) water @ 62°F.
0.9454	kilgrams water @ 62°F.
2.08425	pounds (Avoir.) water @ 62°F.
2.532943	pounds (Troy) water @ 62°F.
9.455	hektograms water @ 62°F.
94.55	dekagrams water @ 62°F.
30.39531	ounces (Troy) water @ 62°F.
33.348	ounces (Avoir.) water @ 62°F.
243.16248	drams (Troy) water @ 62°F.

QUART (U.S.—LIQUID): (cont'd)

533.568 .... drams (Avoir.) water @ 62°F.
607.9062 .... pennyweights water @ 62°F.
729.48744 .... scruples water @ 62°F.
945.5 .... grams water @ 62°F.
9,455 .... decigrams water @ 62°F.
14,589.75 .... grains water @ 62°F.
94,550 .... centigrams water @ 62°F.
945,500 .... milligrams water @ 62°F.

QUIRE: =

25 sheets

RADIAN: =

206,265 .... seconds or inches
3,437.75 .... minutes
57.29578 .... degrees
0.637 .... quadrants
0.159155 .... circumference or revolutions
0.5 .... pi (π)
57° 17′ 44.8″ .... (In degrees, minutes, and seconds)

RADIANS PER SECOND: =

57.29578 .... degrees per second
3,437.7468 .... degrees per minute
206,265 .... degrees per hour
4,950.355 .... degrees per day
0.637 .... quadrants per second
38.22 .... quadrants per minute
2,293.2 .... quadrants per hour
55,036.8 .... quadrants per day
0.159155 .... revolutions per second
9.5493 .... revolutions per minute
572.958 .... revolutions per hour
13,750.992 .... revolutions per day

RADIAN PER SECOND PER SECOND: =

57.29578 .... degrees per second per second
3,437.7468 .... degrees per minute per second
206,265 .... degrees per minute per minute
0.637 .... quadrants per second per second
38.22 .... quadrants per minute per second
2.293.2 .... quandrants per minute per minute
0.159155 .... revolutions per second per second
9.5493 .... revolutions per minute per second
572.958 .... revolutions per minute per minute

REAM: =

500 . . . . sheets
20 . . . . quires

REVOLUTION: =

1,296,000 . . . . seconds or inches
21,600 . . . . minutes
360 . . . . degrees
6.2832 . . . . radians
4 . . . . quadrants
2 . . . . Pi (π)
1 . . . . circumference

REVOLUTIONS PER SECOND PER SECOND: =

360 . . . . degrees per second per second
21,600 . . . . degrees per minute per second
1,296,000 . . . . degrees per minute per minute
6.2832 . . . . radians per second per second
376.9920 . . . . radians per minute per second
22,619.52 . . . . radians per minute per minute
4 . . . . quadrants per second per second
240 . . . . quadrants per minute per second
14,400 . . . . quadrants per minute per minute
1 . . . . revolutions per second per second
60 . . . . revolutions per minute per second
3,600 . . . . revolutions per minute per minute

REVOLUTIONS PER MINUTE PER MINUTE: =

0.1 . . . . degrees per second per second
6 . . . . degrees per minute per second
360 . . . . degrees per minute per minute
0.0017453 . . . . radians per second per second
0.104718 . . . . radians per minute per second
6.2832 . . . . radians per minute per minute
0.00111111 . . . . quadrants per second per second
0.06666667 . . . . quadrants per second per second
4 . . . . quadrants per minute per minute
0.000277778 . . . . revolutions per second per second
0.0166667 . . . . revolutions per minute per second
1 . . . . revolutions per minute per minute

REVOLUTIONS PER SECOND: =

111,974,400,000 . . . . seconds or inches per day
4,665,600,000 . . . . seconds or inches per hour

REVOLUTIONS PER SECOND: (cont'd)

77,760,000	seconds or inches per minute
1,296,000	seconds or inches per second
1,866,240,000	minutes per day
77,760,000	minutes per hour
1,296,000	minutes per minute
21,600	minutes per second
31,104,000	degrees per day
1,296,000	degrees per hour
21,600	degrees per minute
360	degrees per second
542,868	radians per day
22,619.52	radians per hour
376.9920	radians per minute
6.2832	radians per second
345,600	quadrants per day
14,400	quadrants per hour
240	quadrants per minute
4	quadrants per second
172,800	pi (π) per day
7,200	pi (π) per hour
120	pi (π) per minute
2	pi (π) per second
86,400	revolutions or circumferences per day
3,600	revolutions or circumferences per hour
60	revolutions or circumferences per minute
1	revolutions or circumferences per second

REVOLUTIONS PER MINUTE: =

1,866,240,000	seconds or inches per day
77,760,000	seconds or inches per hour
1,296,000	seconds or inches per minute
21,600	seconds or inches per second
31,104,00	minutes per day
1,296,000	minutes per hour
21,600	minutes per minute
360	minutes per second
518,400	degrees per day
21,600	degrees per hour
360	degrees per minute
6	degrees per second
9,047.808	radians per day
376.992	radians per hour
6.2832	radians per minute
0.10472	radians per second
5,760	quadrants per day
240	quadrants per hour
4	quadrants per minute
0.0666667	quadrants per second
2,880	pi (π) per day

REVOLUTIONS PER MINUTE: (cont'd)

120 . . . . pi (π) per hour
2 . . . . pi (π) per minute
0.0333333 . . . . pi (π) per second
1,440 . . . . revolutions or circumferences per day
60 . . . . revolutions or circumferences per hour
1 . . . . revolutions or circumferences per minute
0.016667 . . . . revolutions or circumferences per second

REVOLUTION PER HOUR: =

31,104,000 . . . . seconds or inches per day
1,296,000 . . . . seconds or inches per hour
21,600 . . . . seconds or inches per minute
360 . . . . seconds or inches per second
518,400 . . . . minutes per day
21,600 . . . . minutes per hour
360 . . . . minutes per minute
6 . . . . minutes per second
8,640 . . . . degrees per day
360 . . . . degrees per hour
6 . . . . degrees per minute
0.1 . . . . degrees per second
150.796512 . . . . radians per day
6.2832 . . . . radians per hour
0.104720 . . . . radians per minute
0.00174533 . . . . radians per second
96 . . . . quadrants per day
4 . . . . quadrants per hour
0.0666667 . . . . quadrants per minute
0.00111111 . . . . quadrants per second
48 . . . . pi (π) per day
2 . . . . pi (π) per hour
0.0333333 . . . . pi (π) per minute
0.000555556 . . . . pi (π) per second
24 . . . . revolutions or circumferences per day
1 . . . . revolutions or circumferences per hour
0.0166667 . . . . revolutions or circumferences per minute
0.000277778 . . . . revolutions or circumferences per second

REVOLUTIONS PER DAY: =

1,296,000 . . . . seconds or inches per day
54,000 . . . . seconds or inches per hour
900 . . . . seconds or inches per minute
15 . . . . seconds or inches per second
21,600 . . . . minutes per day
900 . . . . minutes per hour
15 . . . . minutes per minute
0.25 . . . . minutes per second
360 . . . . degrees per day

REVOLUTIONS PER DAY: (cont'd)

15	degrees per hour
0.25	degrees per minute
0.00416667	degrees per second
6.2618	radians per day
0.2618	radians per hour
0.00436333	radians per minute
0.0000727222	radians per second
4	quadrants per day
0.166667	quadrants per hour
0.00277778	quadrants per minute
0.0000462963	quadrants per second
2	pi (π) per day
0.0833333	pi (π) per hour
0.00138889	pi (π) per minute
0.0000231481	pi (π) per second
1	revolutions or circumferences per day
0.0416667	revolutions or circumferences per hour
0.000694446	revolutions or circumferences per minute
0.0000115741	revolutions or circumferences per second

ROD: =

0.00271363	miles (nautical)
0.003125	miles (statute)
0.00502922	kilometers
0.025	furlongs
0.0502922	hektometers
0.25	chains
0.502922	dekameters
1	rods
5.0292188	meters
5.5	yards
5.94	varas (Texas)
16.5	feet
22	spans
25	links
49.5	hands
50.292188	decimeters
502.921875	centimeters
198	inches
6,029.21875	millimeters
198,000	mils
5,029,219	microns
50,029,218,750	millimicrons
5,029,218,750	micromillimeters
7,811,165	wave lengths of red line of cadmium
50,292,187,500	Angstrom Units

SACK CEMENT: =

0.19592	barrels
94	pounds (Avoir.)

SACK CEMENT: (cont'd)

8.22857	gallons (U.S.—liquid)
1.1	cubic feet (set)
1,900.8	cubic inches
3.15	specific gravity
0.484	cubic feet (absolute volume)

SEAWATER GRAVITY: =

1.02 to 1.03

SECOND, FOOT (WATER): =

2,446.594024	kiloliters per day
101.941418	kiloliters per hour
1.699024	kiloliters per minute
0.0283171	kiloliters per second
2,446.594024	cubic meters per day
101.941418	cubic meters per hour
1.699024	cubic meters per minute
0.0283171	cubic meters per second
3,200.144983	cubic yards per day
133.339374	cubic yards per hour
2.222323	cubic yards per minute
0.0370378	cubic yards per second
15,388.959079	barrels per day
641.206628	barrels per hour
10.686777	barrels per minute
0.178113	barrels per second
24,465.940237	hektoliters per day
1,019.414177	hektoliters per hour
16.990236	hektoliters per minute
0.283171	hektoliters per second
69,429.445206	bushels (U.S.—dry) per day
2,892.893550	bushels (U.S.—dry) per hour
48.214893	bushels (U.S.—dry) per minute
0.803582	bushels (U.S.—dry) per second
67,273.995871	bushels (Imperial—dry) per day
2,803.0831613	bushels (Imperial—dry) per hour
46.718053	bushels (Imperial—dry) per minute
0.778634	bushels (Imperial—dry) per second
86,400	cubic feet per day
3,600	cubic feet per hour
60	cubic feet per minute
1	cubic feet per second
244,659	dekaliters per day
10,194.141766	dekaliters per hour
169.902363	dekaliters per minute
2.831706	dekaliters per second
277,718	pecks (U.S.—dry) per day
11,571.574201	pecks (U.S.—dry) per hour

SECOND, FOOT (WATER): (cont'd)

192.859570 . . . . pecks (U.S.—dry) per minute
3.214326 . . . . pecks (U.S.—dry) per second
646,336 . . . . gallons (U.S.—liquid) per day
26,930.679834 . . . . gallons (U.S.—liquid) per hour
448.844664 . . . . gallons (U.S.—liquid) per minute
7,480744 . . . . gallons (U.S.—liquid) per second
555,426 . . . . gallons (U.S.—dry) per day
23,142.740636 . . . . gallons (U.S.—dry) per hour
385.712344 . . . . gallons (U.S.—dry) per minute
6.428539 . . . . gallons (U.S.—dry) per second
538,202 . . . . gallons (Imperial) per day
22,425.0730560 . . . . gallons (Imperial) per hour
373.751218 . . . . gallons (Imperial) per minute
6.229187 . . . . gallons (Imperial) per second
2,585,340 . . . . quarts (liquid) per day
107,723 . . . . quarts (liquid) per hour
1,795.375258 . . . . quarts (liquid) per minute
29.922921 . . . . quarts (liquid) per second
2,221,757 . . . . quarts (dry) per day
92,573.205256 . . . . quarts (dry) per hour
1,542.886754 . . . . quarts (dry) per minute
25.713779 . . . . quarts (dry) per second
2,446,594 . . . . liters per day
101,941 . . . . liters per hour
1,699.0236276 . . . . liters per minute
28.317060 . . . . liters per second
2,446,594 . . . . cubic decimeters per day
101,941 . . . . cubic decimeters per hour
1,699.0236276 . . . . cubic decimeters per minute
28.317060 . . . . cubic decimeters per second
4,443,513 . . . . pints (U.S.—dry) per day
185,146 . . . . pints (U.S.—dry) per hour
3,085.766712 . . . . pints (U.S.—dry) per minute
51.429445 . . . . pints (U.S.—dry) per second
5,170,639 . . . . pints(U.S.—liquid) per day
215,443 . . . . pints (U.S.—liquid) per hour
3,590.716535 . . . . pints (U.S.—liquid) per minute
59.845276 . . . . pints (U.S.—liquid) per second
17,222,068 . . . . gills (Imperial) per day
717,586 . . . . gills (Imperial) per hour
11,959.767119 . . . . gills (Imperial) per minute
199.329452 . . . . gills (Imperial) per second
20,683,007 . . . . gills (U.S.) per day
861,792 . . . . gills (U.S.) per hour
14,363.205943 . . . . gills (U.S.) per minute
239.386766 . . . . gills (U.S.) per second
24,465,940 . . . . deciliters per day
1,019,414 . . . . deciliters per hour
16,990.236276 . . . . deciliters per minute
283.170605 . . . . deciliters per second
149,303,400 . . . . cubic inches per day

SECOND, FOOT (WATER): (cont'd)

6,220,975	cubic inches per hour
103,683	cubic inches per minute
1,728.048146	cubic inches per second
244,659,402	centiliters per day
10,194,142	centiliters per hour
169,902	centiliters per minute
2,831.706046	centiliters per second
2,446,594,024	milliliters per day
101,941,418	milliliters per hour
1,699,023	milliliters per minute
28,317.06046	milliliters per second
2,446,594,024	cubic centimeters per day
101,941,418	cubic centimeters per hour
1,699,023	cubic centimeters per minute
28,317.06046	cubic centimeters per second
2,446,594,024 x 10³	cubic millimeters per day
101,941,417,656	cubic millimeters per hour
1,699,023,627	cubic millimeters per minute
28,317,060	cubic millimeters per second
82,730,841	ounces (U.S.) fluid per day
3,447,121	ounces (U.S.) fluid per hour
57,451.974260	ounces (U.S.) fluid per minute
957.532904	ounces (U.S.) fluid per second
86,110,312	ounces (Imperial—fluid) per day
3,587,941	ounces (Imperial—fluid) per hour
59,798.835597	ounces (Imperial—fluid) per minute
996.647260	ounces (Imperial—fluid) per second
661,847,490	drams (fluid) per day
27,576,966	drams (fluid) per hour
459,616	drams (fluid) per minute
7,660,271730	drams (fluid) per second
39,710,848,658	minims per day
1,654,618,682	minims per hour
27,576,966	minims per minute
459,616	minims per second
1.98347	acre feet
1	acre, inch per hour

SECOND, (ANGLE): =

1	seconds
0.0166667	minutes
0.000277778	degrees
0.0000048414	radians
0.00000308651	quadrants
0.00000154321	pi (π)
0.000000771607	circumference or revolutions

SQUARE CENTIMETER: =

0.00000000003831	square miles or sections
0.0000000001	square kilometers

SQUARE CENTIMETER: (cont'd)

0.00000000247104	square furlongs
0.0000000247104	acres
0.00000001	square hektometers or hectares
0.000000247104	square chains
0.000001	square dekameters or acres
0.00000395367	square rods
0.0001	square meters or centares
0.00011960	square yards
0.000139498	square varas (Texas)
0.00107639	square feet
0.00247104	square links
0.01	square decimeters
1	square centimeters
0.1550	square inches
100	square millimeters
155,000	square mils
197,350	circular mils
127.32	circular millimeters

SQUARE CHAIN: =

0.00015625	square miles or sections
0.000404687	square kilometers
0.01	square furlongs
0.1	acres
0.0404687	square hektometers or hectares
1	square chains
4.046873	square dekameters or acres
16	square rods
404.6873	square meters or centares
484	square yards
564.530690	square varas (Texas)
4,356	square feet
10,000	square links
40,468.73	square decimeters
4,046,873	square centimeters
627,265	square inches
404,687,300	square millimeters
627,265,000,000	square mils
798,650,386,550	circular mils
515,247,870	circular millimeters

SQUARE DECIMETER: =

0.000000003831	square miles or sections
0.00000001	square kilometers
0.000000247104	square furlongs
0.00000247104	acres
0.000001	square hektometers or hectares
0.0000247104	square chains

SQUARE DECIMETER: (cont'd)

0.0001	square dekameters or acres
0.000395367	square rods
0.01	square meters or centares
0.011960	square yards
0.0139498	square varas (Texas)
0.107639	square feet
0.247104	square links
1	square decimeters
100	square centimeters
15.5	square inches
10,000	square millimeters
15,500,000	square mils
19,735,000	circular mils
12,732	circular millimeters

SQUARE DEKAMETER: =

0.00003831	square miles or sections
0.0001	square kilometers
0.00247104	square furlongs
0.0247104	acres
0.01	square hektometers or hectares
0.247104	square chains
1	square dekameters or acres
3.95367	square rods
100	square meters or centares
119.60	square yards
139.498	square varas (Texas)
1,076.39	square feet
2,471.04	square links
10,000	square decimeters
1,000,000	square centimeters
155,000	square inches
100,000,000	square millimeters
155,000,000,000	square mils
197,350,000,000	circular mils
127,320,000	circular millimeters

SQUARE FOOT: =

0.0000000355913	square miles or sections
0.0000000929034	square kilometers
0.00000229568	square furlongs
0.0000229568	acres
0.00000929030	square hektometers or hectares
0.000229568	square chains
0.000929034	square dekameters or acres
0.00367309	square rods
0.0929034	square meters or centares
0.111111	square yards

SQUARE FOOT: (cont'd)

0.129598	square varas (Texas)
1	square feet
2.29568	square links
9.290341	square decimeters
929.0341	square centimeters
144	square inches
92,903.41152	square millimeters
144,000,000	square mils
183,346,560	circular mils
118,285	circular millimeters

SQUARE FURLONG: =

0.015625	square miles or sections
0.040687	square kilometers
1	square furlongs
10	acres
4.04687	square hektometers or hectares
100	square chains
404.6873	square dekameters or acres
1,600	square rods
40,468.73	square meters or centares
48,400	square yards
56,453.069	square varas (Texas)
435,600	square feet
1,000,000	square links
4,046,873	square decimeters
404,687,300	square centimeters
62,726,500	square inches
40,468,730,000	square millimeters
$627,265 \times 10^8$	square mils
$79,865,038,655 \times 10^3$	circular mils
51,524,787,000	circular millimeters

SQUARE HEKTOMETER: =

0.003831	square miles or sections
0.01	square kilometers
0.247104	square square furlongs
2.471044	acres
1	square hektometers or hectares
24.71044	square chains
100	square dekameters or acres
395.367	square rods
10,000	square meters or centares
11,959.8	square yards
13,949.8	square varas (Texas)
107,639	square feet
247,104	square links
1,000,000	square decimeters

SQUARE HEKTOMETER: (cont'd)

100,000,000	square centimeters
15,500,000	square inches
1×10^{10}	square millimeters
155×10^{11}	square mils
$19,735 \times 10^{8}$	circular mils
$12,732 \times 10^{6}$	circular millimeters

SQUARE INCH: =

0.00000000024908	square miles or sections
0.00000000064516̶3	square kilometers
0.0000000159422	furlongs
0.000000159422	acres
0.0000000645163	square hektometers or hectares
0.00000159423	square chains
0.00000645163	square dekameters or acres
0.0000255076	square rods
0.000645163	square meters or centares
0.000771605	square yards
0.000899986	square varas (Texas)
0.00694444	square feet
0.0159423	square links
0.0645163	square decimeters
6.451626	square centimeters
1	square inches
645.16258	square millimeters
1,000,000	square mils
1,273,240	circular mils
821.423611	circular millimeters

SQUARE KILOMETER: =

0.383101	square miles or sections
1	square kilometers
24.71044	square furlongs
247.1044	acres
100	square hektometers or hectares
2,471.044	square chains
10,000	square dekameters or acres
39,536.7	square rods
1,000,000	square meters or centares
1,195,980	square yards
1,394,980	square varas (Texas)
10,764,000	square feet
24,710,440	square links
100,000,000	square decimeters
1×10^{10}	square centimeters
1,550,000,000	square inches
1×10^{12}	square millimeters
155×10^{13}	square mils

SQUARE KILOMETER: (cont'd)

19,735 10^{10}	circular mils
12,732 x 10^8	circular millimeters

SQUARE LINK: =

0.000000015625	square miles or sections
0.0000000404687	square kilometers
0.000001	square furlongs
0.00001	acres
0.00000404687	square hektometers or hectares
0.0001	square chains
0.000404687	square dekameters or acres
0.0016	square rods
0.040469	square meters or centares
0.0484	square yards
0.0564531	square varas (Texas)
0.4356	square feet
1	square links
4.046873	square decimeters
404.6873	square centimeters
62.7265	square inches
40,468.73	square millimeters
62,726,500	square mils
79,863,380	circular mils
51,524.787	circular millimeters

SQUARE METER: =

0.0000003831	square miles or sections
0.000001	square kilometers
0.0000247104	square furlongs
0.000247104	acres
0.0001	square hektometers or hectares
0.00247104	square chains
0.01	square dekameters or acres
0.0395367	square rods
1	square meters or centares
1.19598	square yards
1.39498	square varas (Texas)
10.7639	square feet
19.13580	square spans
24.71044	square links
96.8750	square hands
100	square decimeters
10,000	square centimeters
1,550	square inches
1,000,000	square millimeters
1,550,000,000	square mils
197,350,000	circular mils
1,273,200	circular millimeters

SQUARE MIL: =

0.0000000000000000024908	square miles or sections
0.0000000000000000064516	square kilometers
0.0000000000000159422	square furlongs
0.000000000000159422	acres
0.00000000000000064516	square hektometers or hectares
0.00000000000159423	square chains
0.0000000000645163	square dekameters or acres
0.0000000000255076	square rods
0.00000000000645163	square meters or centares
0.000000000771605	square yards
0.000000000899986	square varas (Texas)
0.00000000694444	square feet
0.0000000159423	square links
0.0000000645163	square decimeters
0.00000645163	square centimeters
0.000001	square inches
0.000645163	square millimeters
1	square mils
1.273224	circular mils
0.000821424	circular millimeters

SQUARE MILE OR SECTION: =

1	square miles or sections
2.589998	square kilometers
64	square furlongs
640	acres
258.9998	square hektometers or hectares
6,400	square chains
25,899.98	square dekameters or acres
102,400	square rods
2,589,998	square meters or centares
3,097,600	square yards
3,612,995	square varas (Texas)
27,878,400	square feet
64,000,000	square links
258,999,800	square decimeters
25,899,980,000	square centimeters
4,014,489,600	square inches
2,589,998,000,000	square millimeters
40,144,969 x 10^8	square mils
51,114 x 10^{10}	circular mils
32,976 x 10^8	circular millimeters

SQUARE MILLIMETER: =

0.0000000000003831	square miles or sections
0.000000000001	square kilometers

SQUARE MILLIMETER: (cont'd)

0.0000000000247104	square furlongs
0.000000000247104	acres
0.0000000001	square hektometers or hectares
0.0000000024104	square chains
0.00000001	square dekameters or acres
0.0000000395367	square rods
0.000001	square meters or centares
0.0000011960	square yards
0.00000139498	square varas (Texas)
0.0000107639	square feet
0.0000247104	square links
0.0001	square decimeters
0.01	square centimeters
0.00155	square inches
1	square millimeters
1,550	square mils
1,973.5	circular mils
1.2732	circular millimeters

SQUARE ROD: =

0.00000976563	square miles or sections
0.0000252929	square kilometers
0.000625	square furlongs
0.00625	acres
0.00252929	square hektometers or hectares
0.0625	square chains
0.252929	square dekameters or acres
1	square rods
25.2929	square meters or centares
30.25	square yards
35.283168	square varas (Texas)
272.25	square feet
625	square links
2,529.29	square decimeters
252,929	square centimeters
39,204	square inches
25,292,900	square millimeters
39,202,600,000	square mils
49,913,691,182	circular mils
32,202,992	circular millimeters

SQUARE VARA (TEXAS): =

0.000000274622	square miles or sections
0.000000716843	square kilometers
0.0000177135	square furlongs
0.000177135	acres
0.0000716843	square hektometers or hectares
0.00177135	square chains

SQUARE VARA (TEXAS): (cont'd)

0.00716843	square dekameters or acres
0.0283416	square rods
0.716843	square meters or centares
0.857332	square yards
1	square varas (Texas)
7.716	square feet
17.713467	square links
71.684263	square decimeters
7,168.426344	square centimeters
1,111.104	square inches
716,843	square millimeters
1,111,104,000	square mils
1,414,702,057	circular mils
912,687	circular millimeters

SQUARE YARD: =

0.000000322831	square miles or sections
0.000000836131	square kilometers
0.0000206612	square furlongs
0.000206612	acres
0.0000836131	square hektometers or hectares
0.00206612	square chains
0.00836131	square dekameters or acres
0.0330579	square rods
0.836131	square meters or centares
1	square yards
1.166382	square varas (Texas)
9	square feet
20.66112	square links
83.61306	square decimeters
8,361.306	square centimeters
1,296	square inches
836,131	square millimeters
1,296,000,000	square mils
1,650,119,040	circular mils
1.064,565	circular millimeters

TEMPERATURE, ABSOLUTE IN CENTIGRADE OR KELVIN: =

temperature in C° + 273.18°

TEMPERATURE, ABSOLUTE IN FAHRENHEIT OR RANKIN: =

temperature in F° + 459.59°

TEMPERATURE, DEGREES CENTIGRADE: =

5/9 (Temp. F° - 32°)
5/4 (Temp. Reaumur)

889

TEMPERATURE, DEGREES FAHRENHEIT: =

> 9/5 (Temp. C° + 32°)
> 9/4 (Temp. Reaumur + 32°)

TEMPERATURE, DEGREES REAUMUR: =

> 4/9 (Temp. F° - 32°)
> 4/5 (Temp. C°)

Degree Centigrade: =

> 0.8 or 4/5 degree Reaumur
> 1.00 degrees absolute, Kelvin
> 1.8 or 9/5 degrees Fahrenheit

Degree Fahrenheit: =

> 0.44444 or 4/9 degree Reaumur
> 0.55556 or 5/9 degree Centigrade

Degree Reaumur: =

> 1.25 or 5/4 degrees Centigrade
> 2.25 or 9/4 degrees Fahrenheit

TONS (LONG): =

1	tons (long)
1.0160470	tons (metric)
1.12	tons (net)
1,016.0470	kilograms
2,722.22	pounds (Troy)
2,240	pounds (Avoir.)
10,160.470	hektograms
101,605	dekagrams
32,667	ounces (Troy)
35,840	ounces (Avoir.)
1,016,047	grams
10,160,470	decigrams
101,604,700	centigrams
1,016,047,000	milligrams
15,680,000	grains
261,333	drams (Troy)
573,440	drams (Avoir.)
653,333	pennyweights
784,022	scruples
5,080,430	carats (metric)
6.19755	barrels of water @ 60°F.

TONS (LONG): (cont'd)

7.33627	barrels of oil @ 36° API
28.607	cubic feet
260.02971	gallons (U.S.—liquid)

TONS (METRIC): =

0.984206	tons (long)
1	tons (metric)
1.10231	tons (net)
1,000	kilograms
2,679.23	pounds (Troy)
2,204.622341	pounds (Avoir.)
10,000	hektograms
100,000	dekagrams
32,150.76	ounces (Troy)
35,273.96	ounces (Avoir.)
1,000,000	grams
10,000,000	decigrams
100,000,000	centigrams
1,000,000,000	milligrams
15,432,365	grains
257,206	drams (Troy)
564,384	drams (Avoir.)
643,015	pennyweights
771,618	scruples
5,000,086	carats (metric)
6.297	barrels of water @ 60°F.
7.454	barrels of oil @ 36° API
29.0662	cubic feet
264.474	gallons (U.S.—liquid)

TONS (NET): =

0.892858	tons (long)
0.907185	tons (metric)
1	tons (net)
907.184872	kilograms
2,430.56	pounds (Troy)
2,000	pounds (Avoir.)
9,071.84872	hektograms
90,718.4872	dekagrams
29,166.66	ounces (Troy)
32,000	ounces (Avoir.)
907,185	grams
9,701,849	decigrams
90,718,487	centigrams
907,184,872	milligrams
14,000,000	grains
233,333	drams (Troy)
512,000	drams (Avoir.)

TONS (NET): (cont'd)

583,333 . . . . pennyweights
700,020 . . . . scruples
4,536,000 . . . . carats (metric)
5.71255 . . . . barrels of water @ 60°F.
6.76216 . . . . barrels of oil @ 36° API
32.04 . . . . cubic feet
239.9271 . . . . gallons (U.S.—liquid)

TON OF REFRIGERATION: =

7,208,640,000 . . . . foot poundals per day
300,360,000 . . . . foot poundals per hour
5,006,000 . . . . foot poundals per minute
83,433.334 . . . . foot poundals per second
224,056,200 . . . . foot pounds per day
9,335,672 . . . . foot pounds per hour
15,594.53 . . . . foot pounds per minute
2,593.242 . . . . foot pounds per second
72,570.8 . . . . kilogram calories per day
3,023,784 . . . . kilogram calories per hour
50.396 . . . . kilogram calories per minute
0.83994 . . . . kilogram calories per second
2,559,836 . . . . ounce calories per day
107,060 . . . . ounce calories per hour
1,777.664 . . . . ounce calories per minute
29.62780 . . . . ounce calories per second
30,977,400 . . . . kilogram meters per day
1,290,722 . . . . kilogram meters per hour
21,512 . . . . kilogram meters per minute
358,534 . . . . kilogram meters per second
2,996,006 . . . . liter-atmospheres per day
124,834 . . . . liter-atmospheres per hour
2,080.554 . . . . liter-atmospheres per minute
34.676 . . . . liter-atmospheres per second
1,058,693,800 . . . . cubic foot atmospheres per day
44,112,200 . . . . cubic foot atmospheres per hour
735,200 . . . . cubic foot atmospheres per minute
12,253.4 . . . . cubic foot atmospheres per second
4.655 . . . . Cheval-vapeur hours
4.715 . . . . horsepowers
3.514 . . . . kilowatts
3.514 . . . . watts
303,609,600 . . . . joules per day
12,650,400 . . . . joules per hour
210,840 . . . . joules per minute
3,514 . . . . joules per second
19.7286 . . . . pounds of carbon oxidized with
100% efficiency per day
0.82202 . . . . pounds of carbon oxidized with
100% efficiency per hour

TONS OF REFRIGERATION: (cont'd)

0.0137	pounds of carbon oxidized with 100% efficiency per minute
0.00022834	pounds of carbon oxidized with 100% efficiency per second
296.64	pounds of water evaporated from and at 212°F. per day
12.36	pounds of water evaporated from and at 212°F. per hour
0.206	pounds of water evaporated from and at 212°F. per minute
0.0034334	pounds of water evaporated from and at 212°F. per second
288,000	BTU per day
12,000	BTU per hour
200	BTU per minute
3.3334	BTU per second

TONS (NET) OF WATER PER DAY: =

0.0907226	kiloliters per day
0.00378023	kiloliters per hour
0.0000630030	kiloliters per minute
0.00000105004	kiloliters per second
0.0907226	cubic meters per day
0.00378023	cubic meters per hour
0.0000630030	cubic meters per minute
0.00000105004	cubic meters per second
1.186579	cubic yards per day
0.0494404	cubic yards per hour
0.000824014	cubic yards per minute
0.0000137335	cubic yards per second
5.706411	barrels per day
0.237766	barrels per hour
0.00396285	barrels per minute
0.0000660467	barrels per second
0.907226	hektoliters per day
0.0378023	hektoliters per hour
0.00630030	hektoliters per minute
0.0000105004	hektoliters per second
32.0385859	cubic feet per day
1.334931	cubic feet per hour
0.0222490	cubic feet per minute
0.000370809	cubic feet per second
9.0722588	dekaliters per day
0.378023	dekaliters per hour
0.00630030	dekaliters per minute
0.000105004	dekaliters per second
239.664469	gallons (U.S.) per day
9.986099	gallons (U.S.) per hour
0.166433	gallons (U.S.) per minute

TONS (NET) OF WATER PER DAY: (cont'd)

0.00277388	gallons (U.S.) per second
199.563810	gallons (Imperial) per day
8.315159	gallons (Imperial) per hour
0.138586	gallons (Imperial) per minute
0.00230977	gallons (Imperial) per second
90.722588	liters per day
3.780228	liters per hour
0.0630030	liters per minute
0.00105004	liters per second
90.722588	cubic decimeters per day
3.780228	cubic decimeters per hour
0.0630030	cubic decimeters per minute
0.00105004	cubic decimeters per second
958.657876	quarts per day
39.944877	quarts per hour
0.665740	quarts per minute
0.011095	quarts per second
1,917.315752	pints per day
79.888076	pints per hour
1.331470	pints per minute
0.0221913	pints per second
7,669.263008	gills per day
319.551826	gills per hour
5.325824	gills per minute
0.0887645	gills per second
907.225881	deciliters per day
37.802277	deciliters per hour
0.630030	deciliters per minute
0.0105004	deciliters per second
55,362.492339	cubic inches per day
2,306.770514	cubic inches per hour
38.446974	cubic inches per minute
0.640767	cubic inches per second
9,072.258810	centiliters per day
378.0227670	centiliters per hour
6.300300	centiliters per minute
0.105004	centiliters per second
90,722.588095	milliliters per day
3,780.227670	milliliters per hour
63.00299561	milliliters per minute
1.0500419	milliliters per second
90,722.588095	cubic centimeters per day
3,780.227670	cubic centimeters per hour
63.00299561	cubic centimeters per minute
1.0500419	cubic centimeters per second
90,722,588	cubic millimeters per day
3,780,227	cubic millimeters per hour
63,002.995611	cubic millimeters per minute
1,050.0419380	cubic millimeters per second
2,000	pounds (Avoir.) per day
83.333333	pounds (Avoir.) per hour

TONS (NET) OF WATER PER DAY: (cont'd)

1.388888	pounds (Avoir.) per minute
0.0231481	pounds (Avoir.) per second
25.744757	bushels (U.S.—dry) per day
1.0726902	bushels (U.S.—dry) per hour
0.0178783	bushels (U.S.—dry) per minute
0.000297975	bushels (U.S.—dry) per second
24.958658	bushels (Imperial—dry) per day
1.0399521	bushels (Imperial—dry) per hour
0.0173323	bushels (Imperial—dry) per minute
0.000288868	bushels (Imperial—dry) per second

TONS (NET) OF WATER PER HOUR: =

2.177342	kiloliters per day
0.0907226	kiloliters per hour
0.00151204	kiloliters per minute
0.0000252007	kiloliters per second
2.177342	cubic meters per day
0.0907226	cubic meters per hour
0.00151204	cubic meters per minute
0.0000252007	cubic meters per second
28.476932	cubic yards per day
1.186579	cubic yards per hour
0.0197763	cubic yards per minute
0.000329611	cubic yards per second
136.953864	barrels per day
5.706411	barrels per hour
0.0951061	barrels per minute
0.00158512	barrels per second
21.773421	hektoliters per day
0.907226	hektoliters per hour
0.0151204	hektoliters per minute
0.000252007	hektoliters per second
768.925102	cubic feet per day
32.0385859	cubic feet per hour
0.533972	cubic feet per minute
0.00889970	cubic feet per second
217.734211	dekaliters per day
9.0722588	dekaliters per hour
0.151204	dekaliters per minute
0.00252007	dekaliters per second
5,751.947256	gallons (U.S.) per day
239.664469	gallons (U.S.) per hour
3.994488	gallons (U.S.) per minute
0.0665740	gallons (U.S.) per second
4,789.531441	gallons (Imperial) per day
199.56381	gallons (Imperial) per hour
3.326054	gallons (Imperial) per minute
0.0554344	gallons (Imperial) per second
2,177.342114	liters per day
90.722588	liters per hour

TONS (NET) OF WATER PER HOUR: (cont'd)

1.512043	liters per minute
0.0252007	liters per second
2,177.342114	cubic decimeters per day
907.225881	cubic decimeters per hour
1.512043	cubic decimeters per minute
0.0252007	cubic decimeters per second
23,007.789024	quarts (liquid) per day
958.657876	quarts (liquid) per hour
15.977711	quarts (liquid) per minute
0.266291	quarts (liquid) per second
46,015.578048	pints (liquid) per day
1,917.315752	pints (liquid) per hour
31.954464	pints (liquid) per minute
0.532282	pints (liquid) per second
184,062	gills per day
7669.263008	gills per hour
127.820251	gills per minute
2.130353	gills per second
21,773.421143	deciliters per day
9,072.258810	deciliters per hour
15.120431	deciliters per minute
0.252007	deciliters per second
1,328,700	cubic inches per day
55,362.492339	cubic inches per hour
922.708206	cubic inches per minute
15.378550	cubic inches per second
217,734	centiliters per day
90,722.588095	centiliters per hour
151.204313	centiliters per minute
2.520072	centiliters per second
2,177,342	milliliters per day
907,226	milliliters per hour
1,512.0431349	milliliters per minute
25.200719	milliliters per second
2,177,343	cubic centimeters per day
907,226	cubic centimeters per hour
1,512.0431349	cubic centimeters per minute
25.200719	cubic centimeters per second
2,177,342,114	cubic millimeters per day
907,225,881	cubic millimeters per hour
15,120.431349	cubic millimeters per minute
25,200.718915	cubic millimeters per second
48,000	pounds (Avoir.) per day
2,000	pounds (Avoir.) per hour
33.333333	pounds (Avoir.) per minute
0.555556	pounds (Avoir.) per second
617.874174	bushels (U.S.—dry) per day
25.744757	bushels (U.S.—dry) per hour
0.429071	bushels (U.S.—dry) per minute
0.00715135	bushels (U.S.—dry) per second
599.00778724	bushels (Imperial) per day

TONS (NET) OF WATER PER HOUR: (cont'd)

24.958658	bushels (Imperial) per hour
0.415986	bushels (Imperial) per minute
0.00693301	bushels (Imperial) per second

TONS (NET) OF WATER MINUTE: =

1,305.415814	kiloliters per day
54.434991	kiloliters per hour
0.907226	kiloliters per minute
0.0151207	kiloliters per second
1,306.415814	cubic meters per day
54.434991	cubic meters per hour
0.907226	cubic meters per minute
0.0151207	cubic meters per second
1,708.673452	cubic yards per day
71.194727	cubic yards per hour
1.186579	cubic yards per minute
0.0197764	cubic yards per second
8,217.231850	barrels per day
342.384660	barrels per hour
57.0641101	barrels per minute
0.0951061	barrels per second
13,064.158138	hektoliters per day
544.349908	hektoliters per hour
9.0722588	hektoliters per minute
0.151207	hektoliters per second
46,135.563668	cubic feet per day
1,922.315153	cubic feet per hour
32.0385859	cubic feet per minute
0.533972	cubic feet per second
130,642	dekaliters per day
5,443.499084	dekaliters per hour
90.722588	dekaliters per minute
1.512067	dekaliters per second
345,117	gallons (U.S.) per day
14,379.86814	gallons (U.S.) per hour
239.664469	gallons (U.S.) per minute
3.994488	gallons (U.S.) per second
287,372	gallons (Imperial) per day
11,973.828603	gallons (Imperial) per hour
199.563810	gallons (Imperial) per minute
3.326064	gallons (Imperial) per second
1,306,416	liters per day
54,434.990844	liters per hour
907.225881	liters per minute
15.120671	liters per second
1,306,416	cubic decimeters per day
54,434.990844	cubic decimeters per hour
907.225881	cubic decimeters per minute
151.206710	cubic decimeters per second
1,380,467	quarts per day

TONS (NET) OF WATER MINUTE: (cont'd)

57,519.472560	quarts per hour
958.657876	quarts per minute
15.977711	quarts per second
2,760,935	pints per day
115.039	pints per hour
1,917.315752	pints per minute
31.954464	pints per second
11,043,739	gills per day
460,156	gills per hour
7,669.263008	gills per minute
127.820970	gills per second
13,064,158	deciliters per day
544,350	deciliters per hour
9,072.258810	deciliters per minute
1,512.067104	deciliters per second
79,721,989	cubic inches per day
3,321,750	cubic inches per hour
55,362.492339	cubic inches per minute
922.708206	cubic inches per second
130,641,581	centiliters per day
5,443,499	centiliters per hour
90,722.588095	centiliters per minute
15,120.671014	centiliters per second
1,306,415,814	milliliters per day
54,434,991	milliliters per hour
907,226	milliliters per minute
151,207	milliliters per second
1,306,415,814	cubic centimeters per day
54,434,991	cubic centimeters per hour
907,226	cubic centimeters per minute
151,207	cubic centimeters per second
1,306,415,814 x 10³	cubic millimeters per day
54,434,991,000	cubic millimeters per hour
907,226,000	cubic millimeters per minute
151,207,000	cubic millimeters per second
2,880,000	pounds (Avoir.) per day
120,000	pounds (Avoir.) per hour
2,000	pounds (Avoir.) per minute
33.333333	pounds (Avoir.) per second
37,072.450454	bushels (U.S.) per day
1,544.685436	bushels (U.S.) per hour
25.744757	bushels (U.S.) per minute
0.429071	bushels (U.S.) per second
35,940.467234	bushels (Imperial) per day
1,497.519468	bushels (Imperial) per hour
24.958658	bushels (Imperial) per minute
0.415986	bushels (Imperial) per second

TONS (NET) OF WATER PER SECOND: =

783,849	kiloliters per day
3,266.0395345	kiloliters per hour

TONS (NET) OF WATER PER SECOND: (cont'd)

54.434991	kiloliters per minute
0.907226	kiloliters per second
783,849	cubic meters per day
3,266.0395345	cubic meters per hour
54.434991	cubic meters per minute
0.907226	cubic meters per second
102,520	cubic yards per day
4,271.683630	cubic yards per hour
71.194727	cubic yards per minute
1.186579	cubic yards per second
493,034	barrels per day
20,543.0796248	barrels per hour
342.384660	barrels per minute
5.706411	barrels per second
783,849	hektoliters per day
32,660.390552	hektoliters per hour
544,340322	hektoliters per minute
9.0722588	hektoliters per second
2,768,134	cubic feet per day
115,339	cubic feet per hour
1,922.315153	cubic feet per minute
32.0385859	cubic feet per second
7,838,494	dekaliters per day
326,604	dekaliters per hour
5,443.398425	dekaliters per minute
90.722588	dekaliters per second
20,707,010	gallons (U.S.) per day
862,792	gallons (U.S.) per hour
14.379868	gallons (U.S.) per minute
239.664469	gallons (U.S.) per second
17,242,313	gallons (Imperial) per day
718,430	gallons (Imperial) per hour
11,973.828603	gallons (Imperial) per minute
199.563810	gallons (Imperial) per second
78,384,937	liters per day
3,266,039	liters per hour
54,433.984253	liters per minute
907.225881	liters per second
78,384,937	cubic decimeters per day
3,266,039	cubic decimeters per hour
54,433.984253	cubic decimeters per minute
907.225881	cubic decimeters per second
82,828,040	quarts per day
3,451,168	quarts per hour
57,519.47256	quarts per minute
958.657876	quarts per second
165,656,081	pints per day
6,902,337	pints per hour
115,039	pints per minute
1,917.315752	pints per second
662,624,324	gills per day

TONS (NET) OF WATER PER SECOND: (cont'd)

27,609.347	gills per hour
460,156	gills per minute
7,669.263008	gills per second
783,849,373	deciliters per day
32,660,391	deciliters per hour
544,340	deciliters per minute
9,072.258810	deciliters per second
4,783,319,338	cubic inches per day
199,304,972	cubic inches per hour
3,321,750	cubic inches per minute
55,362.492339	cubic inches per second
7,838,493,732	centiliters per day
326,603,906	centiliters per hour
5,443,398	centiliters per minute
90,722.588095	centiliters per second
78,384,937,325	milliliters per day
3,266,039,055	milliliters per hour
54,433,984	milliliters per minute
907,226	milliliters per second
78,384,937,325	cubic centimeters per day
3,266,039,055	cubic centimeters per hour
54,433,984	cubic centimeters per minute
907,226	cubic centimeters per second
78,384,937,325 x 10^3	cubic millimeters per day
3,266,039,055 x 10^3	cubic millimeters per hour
54,433,984,000	cubic millimeters per minute
907,226,000	cubic millimeters per second
172,800,000	pounds (Avoir.) per day
7,200,000	pounds (Avoir.) per hour
120,000	pounds (Avoir.) per minute
2,000	pounds (Avoir.) per second
2,224,347	bushels (U.S.—dry) per day
92,681.126136	bushels (U.S.—dry) per hour
1,544.685436	bushels (U.S.—dry) per minute
25.744757	bushels (U.S.—dry) per second
2,156,428	bushels (Imperial) per day
89,851.168086	bushels (Imperial) per hour
150.605153	bushels (Imperial) per minute
24.958658	bushels (Imperial) per second
1,544.685436	bushels (U.S.—dry) per minute
25.744757	bushels (U.S.—dry) per second
2,156,428	bushels (Imperial) per day
89,851.168086	bushels (Imperial) per hour
150.605153	bushels (Imperial) per minute
24.958658	bushels (Imperial) per second

VARA (TEXAS): =

0.000456840	miles (nautical)
0.000526094	miles (statute)
0.000846670	kilometers

VARA (TEXAS): (cont'd)

0.00420875	furlongs
0.00846670	hektometers
0.0420875	chains
0.0846670	dekameters
0.168350	rods
0.846670	meters
0.925926	yards
1	varas (Texas)
2.777778	feet
3.703704	spans
4.208754	links
8.333333	hands
8.466700	decimeters
84.667003	centimeters
33.333333	inches
846.670032	millimeters
33,333.333333	mils
846,670	microns
846,670,032	millimicrons
846,670,032	micromillimeters
1,315,011	wavelengths of red line of cadmium
8,466,700,319	Angstrom units

WATT: =

63,725.184	foot pounds per day
2,655.22	foot pounds per hour
44.2536	foot pounds per minute
0.73756	foot pounds per second
764.702	inch pounds per day
31,862.5920	inch pounds per hour
531.04320	inch pounds per minute
8.85072	inch pounds per second
20.640096	kilogram calories (mean) per day
0.860004	kilogram calories (mean) per hour
0.0143334	kilogram calories (mean) per minute
0.00023889	kilogram calories (mean) per second
45.503616	pound calories (mean) per day
1.895984	pound calories (mean) per hour
0.0315997	pound calories (mean) per minute
0.000526662	pound calories (mean) per second
728.0578547	ounce calories (mean) per day
30.335744	ounce calories (mean) per hour
0.505596	ounce calories (mean) per minute
0.00842659	ounce calories (mean) per second
20,640.096	gram calories (mean) per day
860.004	gram calories (mean) per hour
14.3334	gram calories (mean) per minute
0.23889	gram calories (mean) per second
81.930528	BTU (mean) per day
3.413772	BTU (mean) per hour

WATT: (cont'd)

0.056896	BTU (mean) per minute
0.00094827	BTU (mean) per second
0.001	kilowatts
1	watts
3,600	joules per hour
0.001341	horsepowers
0.0013597	horsepowers (metric)
0.0013597	Cheval-vapeur hours
0.000234	pounds carbon oxidized with 100% efficiency
0.00352	pounds water evaporated from and at 212°F.
0.852647	kiloliter-atmospheres per day
0.0355270	kiloliter-atmospheres per hour
0.000592116	kiloliter-atmospheres per minute
0.0000098686	kiloliter-atmospheres per second
852.647	liter-atmospheres per day
35.52695	liter-atmospheres per hour
0.592116	liter-atmospheres per minute
0.0098686	liter-atmospheres per second
8,808	kilogram meters per day
367.1	kilogram meters per hour
6.116667	kilogram meters per minute
0.101944	kilogram meters per second

WATT HOUR: =

2,655.22	foot pounds
31,982,64	inch pounds
0.860004	kilogram calories (mean)
1.895984	pound calories (mean)
30.385744	ounce calories (mean)
860.004	gram calories (mean)
3.413772	BTU (mean)
0.001	kilowatt hours
3,600	joules
0.001341	horsepower hours
0.0013597	horsepower hours (metric)
0.0355270	kiloliter-atmospheres
35.53695	liter-atmospheres
367.1	kilogram meters

WATT PER SQUARE INCH: =

8.1913	BTU per square foot per minute
6,372.6	foot pounds per square foot per minute
0.19310	horsepowers per square foot

YARD: =

0.000483387	miles (nautical)
0.000568182	miles (statute)
0.000914404	kilometers
0.00454545	furlongs
0.00914404	hektometers
0.0454545	chains
0.0914404	dekameters
0.181818	rods
0.914404	meters
1	yards
1.08	varas (Texas)
3	feet
4	spans
4.54545	links
9	hands
9.144036	centimeters
36	inches
914.40360	millimeters
36,000	mils
914,404	microns
914,403,600	millimicrons
914,403,600	micromillimeters
1,420,212	wavelengths of red line of cadmium
9,144.036345	Angstrom Units

PERSONAL CONVERSION FACTORS